Earth Materials

INTRODUCTION TO MINERALOGY AND PETROLOGY

The fundamental concepts of mineralogy and petrology are explained in this highly illustrated, full-color textbook, to create a concise overview for students studying Earth materials. The relationship between minerals and rocks and how they relate to the broader Earth, materials, and environmental sciences is interwoven throughout. Beautiful photos of specimens and CrystalViewer's three-dimensional illustrations allow students to easily visualize minerals, rocks, and crystal structures. Review questions at the end of chapters allow students to check their understanding. The importance of Earth materials to human cultural development and the hazards they pose to humans are discussed in later chapters. This ambitious, wide-ranging book is written by two world-renowned textbook authors, each with more than 40 years of teaching experience, who bring that experience here to clearly convey the important topics.

Cornelis ("Kase") Klein is Emeritus Professor in the Department of Earth and Planetary Sciences at the University of New Mexico, Albuquerque. He received his B.Sc. and M.Sc. in geology from McGill University, Canada, and his Ph.D. from Harvard University, also in geology. He has been a member of the geology faculty at Harvard University; Indiana University, Bloomington; and the University of New Mexico; he has taught courses in mineralogy at all these universities. His published books include *Manual of Mineralogy* (19th–21st eds.); *Manual of Mineral Science* (22nd–23rd eds.); and *Minerals and Rocks: Exercises in Crystal and Mineral Chemistry, Crystallography, X-Ray Powder Diffraction, Mineral and Rock Identification, and Ore Mineralogy* (3rd ed., 2008). He has received two awards for excellence in teaching from the University of New Mexico.

Anthony R. Philpotts is Emeritus Professor of Geology and Geophysics at the University of Connecticut, a Visiting Fellow in the Department of Geology and Geophysics at Yale University, and Adjunct Professor in the Department of Geosciences at the University of Massachusetts. He received his B.Sc. and M.Sc. in geology from McGill University, Canada, and his Ph.D. from the University of Cambridge. He has taught igneous and metamorphic petrology courses at McGill University and the University of Connecticut for more than 40 years. His published books include *Principles of Igneous and Metamorphic Petrology* (2nd ed., 2009, Cambridge University Press) and *Petrography of Igneous and Metamorphic Rocks* (1989, 2003).

Cueva de Los Cristales (Cave of the Crystals) in Naica, Chihuahua, Mexico. The main chamber of the cave contains enormous selenite crystals, some of the largest natural crystals ever found. Photograph © Carsten Peter.

Earth Materials

INTRODUCTION TO MINERALOGY AND PETROLOGY

Cornelis Klein
University of New Mexico, Emeritus

Anthony R. Philpotts
University of Connecticut, Emeritus

CAMBRIDGE
UNIVERSITY PRESS

CAMBRIDGE UNIVERSITY PRESS
Cambridge, New York, Melbourne, Madrid, Cape Town,
Singapore, São Paulo, Delhi, Mexico City

Cambridge University Press
32 Avenue of the Americas, New York, NY 10013-2473, USA

www.cambridge.org
Information on this title: www.cambridge.org/9780521145213

© Cornelis Klein and A. R. Philpotts 2013

First published 2013

Printed in the United States of America

A catalog record for this publication is available from the British Library.

Library of Congress Cataloging in Publication Data

Klein, Cornelis, 1937–
Earth materials : introduction to mineralogy and petrology / Cornelis Klein, Anthony Philpotts.
   p.   cm.
Includes bibliographical references and index.
ISBN 978-0-521-76115-4 (hardback) – ISBN 978-0-521-14521-3 (paperback)
1. Mineralogy – Textbooks.   2. Petrology – Textbooks.   I. Philpotts, Anthony R. (Anthony Robert), 1938–   II. Title.
QE363.2.K529   2012
553–dc23   2011044843

ISBN 978-0-521-76115-4 Hardback
ISBN 978-0-521-14521-3 Paperback

Additional resources for this publication at www.cambridge.org/earthmaterials

Cover: Photograph of a polished surface of a rock type known as garbenschiefer, from the German words *Garbe*, meaning sheaf, and *Schiefer*, meaning schist. It is a metamorphosed igneous rock and consists of coarse black hornblende sheaves and reddish-brown garnets in a fine-grained matrix of plagioclase, quartz, chlorite, and muscovite. This rock is quarried in Ashfield, Massachusetts, as "dimension stone," and is commercially known as "Crowsfoot" Ashfield Stone. It is part of the Ordovician Hawley Formation, which has a minimum age of 462 million years. Field of view: ~15 cm by 20 cm. Photograph courtesy of Marc Klein.

Cornelis Klein dedicates this book to his two children and their immediate families. His son and daughter-in-law, Marc and Laura Klein, and their two children, Alaxandra and Hugh. And to his daughter and son-in-law, Stephanie and Jack Stahl, and Stephanie's three sons, Max, Miles, and Bo Peponis.

Anthony R. Philpotts dedicates this book to his three daughters, Liane, Marlaine, and Alison.

Contents

Preface

Over the past two decades, many curriculum changes have occurred in geology, Earth science, and environmental science programs in universities. Many of these have involved the compression of separate one-semester courses in mineralogy, optical mineralogy, and petrology into a single-semester offering that combines mineralogy and petrology, commonly called Earth Materials. Such a course is a challenge to the instructor (or a team of instructors) and the students. This is especially so when few, if any, textbooks for such a one-semester course have been available.

This text, *Earth Materials*, is an introduction to mineralogy and petrology in which both subjects are covered with a roughly even balance. To keep this textbook reasonably short and applicable to a one-semester course, we decided against providing a shallow survey of everything and instead concentrated on what we consider the most fundamental aspects of the various subjects.

In the writing of this text, we assumed that the students who enroll in an Earth materials course would have previously taken an introductory physical geology course, as well as a course in college-level chemistry.

Coverage

Basic aspects of mineralogy must precede the coverage of petrology. This sequence is obvious from the chapter headings. After a brief, general introduction in Chapter 1, minerals and rocks are broadly defined in Chapter 2. That is followed by three chapters that relate to various mineralogical aspects and concepts. Chapter 3 covers the identification techniques that students must become familiar with to recognize unknown minerals in the laboratory and in the field. It also includes discussion of two common instrumental techniques: X-ray powder diffraction and electron beam methods. Chapter 4 covers the most fundamental aspects of crystal chemistry, and Chapter 5 is a short introduction to basic aspects of crystallography. Chapter 6 covers optical mineralogy. This subject is included so that instructors who plan to introduce thin sections of rocks in their course can give their students quick access to the fundamentals of optical mineralogy and the optical properties of rock-forming minerals.

The sequencing of subsequent systematic mineralogy chapters is completely different from that most commonly used in mineralogy textbooks. In these chapters, minerals are discussed in groups based first on chemistry (native elements, oxides, silicates, and so on) and, subsequently, for the silicates, on structural features (layer, chain, and framework silicates, and so on). Here, the decision was made to group systematic mineralogy descriptions as part of the three major rock types: igneous, sedimentary, and metamorphic. This allows for the closest possible integration of mineralogy and petrology.

Chapter 7 gives systematic mineralogical data on 29 of the most common igneous minerals, including, in order of decreasing abundance, silicates, oxides, a few sulfides, and a phosphate. This is followed by Chapter 8, which presents the most fundamental aspects of the formation of igneous rocks. Chapter 9 addresses the occurrence of igneous rock types, their classification, and plate tectonic settings.

This approach is repeated with respect to sedimentary and metamorphic minerals and rocks. Chapter 10 gives systematic mineralogical descriptions of 14 common sedimentary minerals as well as phosphorite and soil. (The siliciclastic components of sedimentary rocks are discussed in Chapter 7, which deals with igneous minerals). Chapter 11 deals with the formation, transport, and lithification of sediment, and Chapter 12 discusses sedimentary rock classification, as well as the occurrence and plate tectonic setting of sedimentary rocks.

Chapter 13 gives the systematic mineralogy of 26 of the most common metamorphic minerals, all of which are silicates, except for one, an oxide. Chapter 14 addresses the causes of metamorphism, gives rock classifications, and relates their occurrence to plate tectonic settings.

Chapter 15 gives systematic mineralogical descriptions of selected minerals that are of economic importance. Chapter 16 gives a brief overview of some selected resources of Earth materials, and Chapter 17 discusses the health effects of several minerals and chemical elements, and the hazards presented by certain rock-forming processes.

In the chapters that deal mainly with systematic mineralogy (Chapters 7, 10, 13, and 15), the main emphasis is on geologic occurrence (paragenesis), chemistry and atomic structure, physical properties that are pertinent to hand specimen identification (in laboratory sessions associated with an Earth materials course), and uses in industry and manufacturing. Hand specimen photographs and atomic structure illustrations are given for each mineral discussed.

This text is meant to be not only a supplement to lectures but also a reference source in the applied laboratory sessions of the course. Basic concepts in crystal chemistry, crystallography, and the origin of various rock types are best presented by the instructor in lectures in the classroom. Mineral and rock identification and classification schemes, however, are best learned in the laboratory with hand specimens and thin sections, using those parts of the book that specifically address the applied aspects.

All chapters begin with a boxed overview of what follows and end with a summary and set of review questions. When a new term is first encountered in the text, it is printed in bold type to signify that its definition is included in the glossary at the end of the text.

CRYSTALVIEWER

The atomic structure illustrations, which are static images in this text, can also be viewed as interactive visualizations in CrystalViewer, a crystal structures visualization program for Mac and Windows. CrystalViewer is designed to provide the missing "third dimension" for crystal structure illustrations in the book. Each structure can be rotated and scaled with the computer mouse, and it is hoped that such interactive exploration will lead to an improved visual understanding of the complex three-dimensional atomic arrangements of minerals. The program contains 105 structure illustrations, which are distributed over two files. The first file, with the title "Learning," contains 24 structures that are referenced with figure numbers from Chapters 2, 4, and 5. These 24 structures illustrate basic aspects of crystal chemistry. The other file, entitled "Reference" with 81 crystal structures, is arranged in alphabetical order, by mineral name. This file contains the structures of the rock-forming minerals discussed in Chapters 5, 7, 10, 13, and 15. These structures complement the structure illustrations in the text that show unit cell outlines, space group

PYROXENE

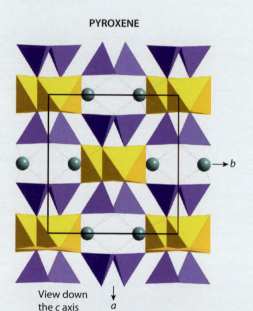

View down
the *c* axis

QUARTZ

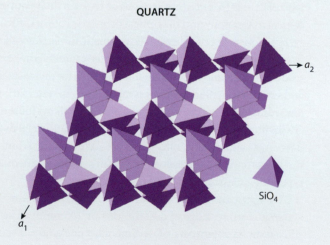

notation, and legends with atomic site occupancies. The files and the CrystalViewer download are at www.cambridge.org/earthmaterials/crystalviewer.

Our overall goal was the production of an accessible, highly illustrated and visually attractive, condensed and well-integrated mineralogy-petrology textbook suitable for one-semester Earth materials courses. It is our hope that we have succeeded.

Acknowledgments

Cornelis Klein thanks Charles Langmuir, Professor in the Department of Earth and Planetary Sciences at Harvard University, for granting him permission (together with a professional photographer, David Nufer, of David Nufer Photography in Albuquerque, New Mexico) to access and photograph specimens from the Harvard Mineralogy Collections. David and I spent three full days there and with the full-time and very attentive help of Carl Francis (curator of the Harvard Mineralogy Museum and Collections) – whose enormous knowledge of the collections allowed us to locate the most appropriate specimens quickly – we completed all of the necessary hand specimen photography of the minerals for this text. Overnight lodging for our four nights in Cambridge, Massachusetts, was generously provided by Leverett House, one of the college houses of which I had been Allston Burr Senior Tutor between 1966 and 1970. We are most grateful to JoAnn DiSalvo Haas and Lauren Brandt for having provided us with some great student rooms.

Throughout the two-year period devoted to the writing of my sections of this text, many colleagues, be it at the University of New Mexico or elsewhere, have been helpful and generous with their time in reviewing sections of text while still in progress. They appear here in alphabetical order: Adrian Brearley, Jonathan Callender, Brian Davis, Amy Ellwein, Maya Elrick, Dave Gutzler, Rhian Jones, Bruce Loeffler, Matt Nyman, Frans Rietmeijer, Malcolm Ross, Jane Selverstone, and Mary Simmons.

I am grateful to David Palmer of CrystalMaker Software Limited, Yarnton, Oxfordshire, England, for providing expertise and guidance in the design of the crystal structure visualization program that accompanies this textbook.

This book would not have been possible without the support and patient understanding of my wife, Shirley Morrison. The word processing of my part of this text was most efficiently and enthusiastically accomplished by Mabel Chavez of Santo Domingo Pueblo, New Mexico.

Anthony R. Philpotts would like to thank the many reviewers who have painstakingly struggled through what we have written and suggested improvements. We have tried to incorporate as many of these as possible within the limits set by the length of the book. I would particularly like to thank Grant Cawthorn for one of the most thorough reviews I have ever received. His knowledge of igneous rocks and the photographs he provided have greatly benefited the book. Dan Kontak, Tony Morse, Brian Robins, and Jane Selverstone also offered valuable advice, as did numerous anonymous reviewers. I am grateful to all of them.

While writing this book, I have greatly appreciated interactions with many colleagues. Jay Ague, Brian Skinner, and Leo Hickey at Yale University, and Sheila Seaman, Mike Rhodes, and Tony Morse at the University of Massachusetts have all provided me with geological insights. I have also learned a considerable amount about sedimentary rocks from Randy Steinen, formerly of the University of Connecticut, and Paul Olsen, of the Lamont-Dougherty Earth Observatory of Columbia University.

I am grateful to the many petrology students I have had over the years. Their many questions and interests played a big role in how I taught the courses and in no small way have determined what, and how, petrology is presented in this book.

Last, none of my part of this book would have been possible without the support of my wife, who allowed me to disappear into my study for fully two years. She is owed an enormous debt of gratitude, especially in view of the fact that when I finished revising my previous book (*Principles of Igneous and Metamorphic Petrology*), I promised her that it was definitely the last one!

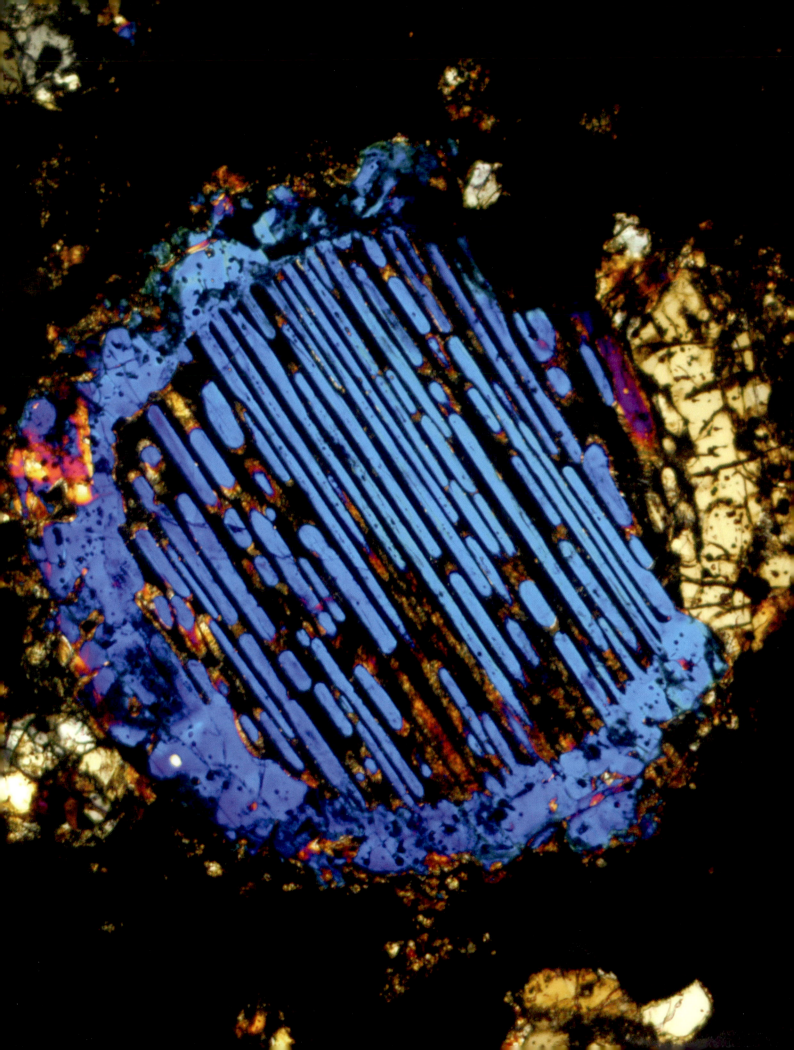

1 Introduction

The Earth has had a long history – 4.56 billion years, give or take a few million – but most of its chemical elements were created at a much earlier time. Most of this book is devoted to how atoms (or ions) fit together to form minerals, the basic building blocks of the Earth, and how such minerals became part of what we refer to as rocks. In this introductory chapter, however, we briefly cover the evidence for where Earth's chemical elements came from. Their creation occurred earlier in the history of the universe and was associated with processes taking place in stars and, in particular, those accompanying the death of massive stars. Material from these earlier stars was dispersed into space, and only much later did it come together to form the solar system and planet Earth.

We also review the basic large-scale internal structure of the Earth. Although we never see material from deep in the planet, its composition and movement play important roles in determining processes that create the Earth materials that we see in the crust. Ever since its formation, the planet has been cooling, and a direct consequence of this process has been plate tectonics. We learn in later chapters that almost all new Earth materials are formed in specific plate tectonic settings, and it is, therefore, important to review this material before delving into the details of minerals and rocks.

This is a cross-section of a spherical chondrule composed mainly of the mineral olivine, Mg_2SiO_4, with interstitial glass and pyroxene. It is a piece of the Allende carbonaceous chondrite meteorite, which exploded over northern Mexico in 1969. The blue and yellow of the olivine are interference colors created by the interaction of transmitted polarized light with the crystal structure. Material such as this is believed to have accreted in the disk surrounding the Sun to form the planet Earth 4.6 billion years ago. Width of field is 2.6 mm. (Photograph courtesy of Rhian Jones, University of New Mexico.)

This book provides an introduction to the study of the solids that make up planet Earth. These materials consist of naturally occurring chemical compounds, known as **minerals**, and their aggregates, **rocks**. Only through the study of minerals and rocks can we learn about the history of the Earth, and this knowledge is also important because of the extensive use made of Earth materials in everyday life, such as the fabrication of tools; the manufacture of vehicles; and their use as construction materials, sources of energy, and soils for agriculture. This knowledge is clearly important in the search for mineral resources, but the general public needs to know the finite nature of many of our natural resources to make informed decisions.

Many different processes are involved in forming a rock from a group of minerals. These processes are normally divided into three general categories: ones involving molten material, which we call **igneous**; ones involving the weathering of rock and transport of sediment, which we call **sedimentary**; and those that modify rocks through changes in temperature, pressure, and fluids inside the Earth, which we call **metamorphic**. Throughout the book, we first introduce how to identify the minerals that are common in each of these main types of rock, and then we discuss the processes that lead to the formation of those rocks. These processes, many of which are intimately related to plate tectonics, have played important roles in the evolution of the planet.

In the following chapters, we deal with these main types of Earth materials, but in this first chapter we look at where the materials that constitute the Earth came from, and we then review the Earth's major structural units. The wide compositional range of the many minerals and rocks found on Earth must in some way reflect the composition of the Earth as a whole. It is natural, then, to wonder where the chemical elements that constitute the Earth came from and what determined their abundances.

The Earth and solar system were formed 4.56 billion years ago, but none of the original planet has been preserved. The oldest rocks found to date are about 4 billion years old, although individual minerals have been found that are 4.4 billion years old. We are, therefore, missing about half a billion years of history. Fortunately, the study of distant stars provides glimpses into earlier times, and meteorites provide actual samples of material from which our planet is believed to have formed. Therefore, we start our study of Earth materials by briefly examining what astronomical and meteoritic studies tell us about the early history of the Earth.

1.1 Formation of Earth's chemical elements in supernovae

Earth materials are formed from chemical elements that have had a long history and whose origins we can explain through studies of distant stars and meteorites. Stars are born from the condensation mainly of hydrogen, and they spend most of their life fusing the hydrogen to form helium. Their lives can end in various ways depending on the mass of the star, with the more massive ones ending in cataclysmic explosions known as **supernovae**, during which elements heavier than iron are created. These explosions disperse material throughout space and form the raw material from which new stars and solar systems are formed. The Earth and other terrestrial planets in the solar system are formed from the chemical elements that were left over from these early supernovae and other evolved stars that collected together to form the Sun and solar system 4.6 billion years ago.

When the universe began with the **Big Bang** ~14 billion years ago, only light elements such as hydrogen and helium and trace amounts of lithium, beryllium, and boron were formed. Subsequently, processes in stars formed heavier elements through nuclear fusion. Small stars, such as the Sun, fuse hydrogen atoms together to form helium, and late in their life, the helium atoms may fuse together to form carbon, but none of the heavier elements is formed. Stars that are more than eight times the mass of the Sun have greater gravitational attraction and can generate higher pressures and temperatures in their cores, which lead to additional nuclear reactions that create elements as heavy as iron. Once a star reaches the iron stage, it implodes under its own gravitational attraction and then explodes to form a supernova (Fig. 1.1). In these cataclysmic explosions, all the elements heavier than iron are formed.

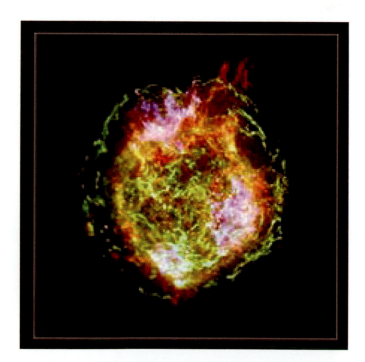

Figure 1.1 Cassiopeia A is a supernova remnant. This image, taken with NASA's Chandra X-ray Observatory, allows us to identify the elements being created in the exploding star by the characteristic wavelengths of X-rays emitted by each element, which have been converted in this image to false colors. (Photograph courtesy of NASA.)

Although supernovae occur in our part of the galaxy only once every few hundred years, they are relatively common in the center of the galaxy and in other galaxies. They have consequently been well documented. For example, NASA's **Hubble space telescope** has provided spectacular photographs of these exploding stars, and the orbiting Chandra X-ray observatory allows us to identify the actual elements that are produced in a supernova (Fig. 1.1).

The debris from a supernova initially forms clouds and jets of gas that are hurled out at enormous speeds from the exploding star (Fig. 1.1). These expanding clouds can remain visible for thousands of years. For example, the Crab Nebula is a supernova remnant from an explosion witnessed by Chinese astronomers in 1054 CE (common era). Today, this cloud is still expanding at the incredible velocity of 1800 km/s. Eventually, the material ejected from a supernova is dispersed throughout space, and it is from such material that our solar system was formed.

1.2 Birth of the solar system and Earth

Most of the dispersed matter in the universe consists of hydrogen, with the heavier elements formed in stars constituting only a very small fraction. If the dispersed matter becomes clustered, it develops a gravitational field, which causes more material to be attracted to it. These gas and dust clouds contract under gravitational forces and form what is known as a **nebula**, which eventually collapses into a flattened rotating disk (Fig. 1.2). If

the nebula is sufficiently large, pressures and temperatures in its core are raised by gravitational collapse to a point at which nuclear fusion begins (10 000 000 Kelvin-degrees absolute), and a star is born. The critical mass required for fusion is ~80 times the mass of the planet Jupiter.

The nebula that would become our solar system was formed slightly more than 4.56 billion years ago. Most of the material in that nebula collapsed inward to form the Sun, but some remained in the solar disc to form planets, moons, asteroids, meteorites, and comets. In the inner part of the disc, where temperatures were higher, elements like carbon, nitrogen, and hydrogen were present as gases, and solid material was composed of rock-forming elements such as silicon, magnesium, iron, and oxygen. Because silicon, magnesium, and iron are much less abundant than carbon, nitrogen, and hydrogen, the **terrestrial planets** (Mercury, Venus, Earth, Mars) and the asteroids, which formed in the inner solar system, are small. Farther out in the solar disc where temperatures were lower, ices of water, carbon dioxide, ammonia, and methane could also form, and because these involved the more abundant elements in the solar nebula, they formed the much larger outer **gas giant planets** (Jupiter, Saturn, Uranus, and Neptune).

Because the Sun constitutes 99.9% of the mass of the solar system, its composition must be essentially the same as that of the nebula from which it formed. We can determine the composition of the outer part of the Sun from the strength of absorption lines in the electromagnetic spectrum that are characteristic of the elements. The Sun vigorously convects, so analyses of the **photosphere**, the light-emitting part of the Sun, are believed to represent a large part of the Sun. However, heavier elements are concentrated toward its core, so estimates of the solar system's bulk composition from analyses of the photosphere take into account this distribution (Fig. 1.3(A)).

The Sun is composed largely of hydrogen (74%) and helium (24%), with oxygen and carbon being the next most abundant elements, and all other elements being extremely minor. The planets, asteroids, meteorites, and comets were formed from particles in the solar nebula that accreted to form larger bodies. In the inner part of the solar disc, only the less volatile solids were available to form the terrestrial planets. Their compositions will differ from solar abundances in that they are depleted in the more volatile elements (Fig. 1.3(B)).

Meteorites are natural objects from space that impact the Earth's surface. During the early part of Earth's history, these impacts led to the accretion of the planet. With time, the frequency of meteorite impacts decreased. Meteorites still impact the Earth and provide us with samples of the primordial material from which the Earth was most likely formed. Most come from the asteroid belt that lies between the orbits of Mars and Jupiter, but a very small number are composed of material that was blasted from the surface of the Moon and Mars by large meteorite impacts. Some meteorites are clearly fragments of planetary bodies

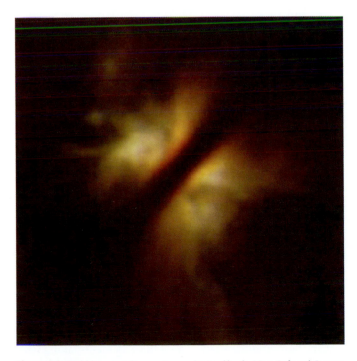

Figure 1.2 Hubble space telescope image obtained by the Near-Infrared Camera and Multi-Object Spectrometer (NICMOS) shows a star (IRAS 04302+2247) that is hidden by a nebular disc of material (diagonal dark region) similar in mass and size to the one that formed our solar system. Light from the star illuminates gas and dust that is still being pulled into the nebular disk. (NASA HST image.)

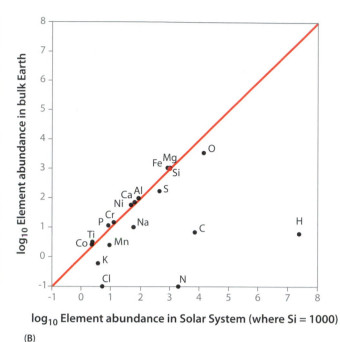

Element	Bulk Solar System	Bulk Earth
H	2.431×10^7	6
He	2.343×10^6	-
O	14130	3494
C	7079	7
Ne	2148	-
N	1950	0.1
Mg	1020	1061
Si	1000	1000
Fe	838	1066
S	445	169
Ar	103	-
Al	84	97
Ca	63	71
Na	58	10
Ni	48	58
Cr	13	15
Mn	9.2	2.5
P	8.4	11.5
Cl	5.2	0.1
K	3.7	0.6
Ti	2.4	3.2
Co	2.3	2.6

(A)

(B)

Figure 1.3 **(A)** Estimated abundances of elements in the bulk solar system (after Lodders, 2003) and the bulk Earth (after Newsom, 1995) based on the abundance of silicon in the solar system being 1000. **(B)** Logarithmic plot of element abundances in bulk solar system versus those in the bulk Earth. The less volatile elements have the same abundances in both bodies and hence plot near the red line (slope = 1), whereas the more volatile elements are depleted in the Earth.

that grew large enough to partially melt and undergo differentiation to produce iron-rich cores and silicate mantles, whereas others never grew large enough to differentiate.

The most common type of meteorite is called a **chondrite**, because it contains small (millimeter-size) spheres known as **chondrules**, which consist of minerals composed mainly of silicon, oxygen, magnesium, and iron (Fig. 1.4 and chapter-opening photograph). Chondrules are never found in terrestrial rocks. They are thought to have formed by flash heating and

Figure 1.4 A piece of the Axtell carbonaceous chondrite, which was found in Texas in 1943. The chondrite contains many small (millimeter-size), light-gray chondrules and a dark-gray matrix. The large white object in the center is an example of an inclusion, one of the oldest pieces of rock that formed in the solar system, 4.567 billion years ago. (Photograph courtesy of the Institute of Meteoritics, University of New Mexico.)

melting of primordial dust particles in the solar disc at temperatures near 2000°C. What caused the heating is uncertain, but we know that the melting and subsequent cooling must have been rapid, over a period of hours, because of the form of the crystals, as, for example, the barred texture of the olivine in the chapter-opening photograph.

Chondrules are among the first pieces of rock that formed in the solar system. The oldest objects in chondrites are so-called refractory inclusions (Fig. 1.4). These are made of exotic minerals rich in low volatility elements such as calcium, aluminum, and titanium. They are about 2 million years older than chondrules, and their formation ages are taken to be the age of the solar system itself, 4.567 billion years. Chondrites also contain rare, tiny mineral grains that are the debris from supernovae that took place before the solar system formed. The Earth is believed to have formed from accretion of material similar to that found in chondrites. Their composition is, therefore, used along with constraints set by the mass of the Earth, its moment of inertia, and known seismic discontinuities (see the following section) to estimate the Earth's bulk composition (Fig. 1.3(A)). This estimate shows the Earth to be depleted in volatile constituents (e.g., H, C, N, O) compared with the bulk solar system, but the relative abundances of the less volatile elements are similar (Fig. 1.3(B)). For example, the abundances of magnesium, silicon, and iron are all about the same in the solar system (Sun) and the Earth. These three elements, along with oxygen, make up most of the Earth, with other elements being minor constituents. Therefore, it should not come as a surprise that many rock-forming minerals are compounds that include these four elements.

1.3 Accretion and early history of the Earth

Planet Earth is believed to have formed by the accretion of primordial solar material similar in composition to chondrites. As the planet grew larger, the kinetic energy of accreting material was converted to heat in the planet. Some of these early bombardments were so large that they actually knocked material off the planet. One such collision with a Mars-size body before 4.45 billion years ago removed material from Earth to form the Moon.

During this early accretionary stage, the Earth was hot. Not only did accretion generate enormous amounts of heat, but also radioactive decay provided additional heat. Another important source of heat was the formation of the molten iron-nickel core. The oldest rocks on Earth indicate that a strong magnetic field already existed 4 billion years ago. The magnetic field is generated by convection in the molten metallic outer core, which must, therefore, have been present at that time. The energy released by sinking iron and nickel to form the core generated sufficient heat to melt a large fraction of the Earth. Arguments based on the abundance of radioisotopes of hafnium and tungsten indicate that the Earth accreted in about 10 million years and that **core formation** was completed by about 30 million years after accretion of the planet. The combined effects of accretion, radioactive decay, and core formation guaranteed that the Earth had an extremely hot birth, and early in its history the surface would have been completely molten. Since that time, the Earth has been cooling, and the dissipation of heat has been the most important planetary process, which has made the Earth a dynamic planet.

The planet began to cool and solidify, and because of chemical variations and changes due to increased pressure with depth, a zoned planet was produced (Fig. 1.5). We have already seen that iron and nickel sank to form the core in the first 30 million years. As they sank, an equivalent volume of lower-density hot material would have risen toward the surface, which would have allowed its heat to be radiated into space and helped cool the planet. This cooling by transfer of hot material to cooler regions is known as convective cooling, and even though the outer part of the planet is now solid, convective cooling still remains the most effective way the planet has of getting rid of heat. The solidification of the planet has been a long, slow process, and it continues today, with the outer part of the core still molten. During this convective cooling of the planet, igneous processes redistributed elements, and the result is a compositionally layered planet.

1.4 Internal structure of the Earth

The main evidence for layering in the planet comes from the study of the paths and velocities of seismic compressional (P) and shear (S) waves passing through the Earth. This evidence is discussed in all introductory geology texts and is not repeated here. Instead, we simply review the main findings of these studies as they relate to the Earth's internal structure (Fig. 1.5).

The Earth's radius is 6371 km, almost half of which (3483 km) is occupied by the metallic core, which is composed predominantly of iron and nickel but must also contain small amounts of light elements, such as silicon, oxygen, sulfur, and hydrogen. The core is slowly crystallizing from the bottom up, with the solid **inner core** having a radius of 1220 km. The temperature in the inner core is estimated to be above 5000 K, which is considerably hotter than the outer part of the core, which is about 4000 K. The inner core is solid not because of temperature but because of the extremely high pressure at the center of the Earth (364 GPa [billion pascal]; see Sec. 8.4). As the liquid in the **outer core** crystallizes onto the inner core, it liberates the **latent heat of crystallization** of iron and nickel, which helps drive the convection cells in the outer core, where the Earth's magnetic field is generated.

Above the core lies the largest unit in the earth, the **mantle**. Although the mantle is solid, it behaves as a plastic material that slowly convects. What is not clear is whether convection currents pass all the way through the mantle or convect in two separate layers, the lower and upper mantle. The division between these two parts of the mantle is the prominent seismic discontinuity at a depth of ~660 km. The subduction of lithospheric plates into the mantle generates earthquakes that can be traced to a depth of 660 km but no deeper. Does this mean that material from the upper mantle cannot penetrate into the lower mantle, or does it simply mean that rocks below this depth are not sufficiently brittle to generate earthquakes? These two possible explanations have led to the two-layer mantle convection model and the whole-mantle convection model, which are illustrated in the left and right halves of Figure 1.5(C), respectively.

Recent studies of seismic velocities in the mantle favor at least some subducted slabs penetrating to the depth of the core-mantle boundary, where they may correlate with depressions on that boundary (Fig. 1.5(C)). Immediately above the core-mantle boundary is the 100 km to 300 km-thick **D″ (D double prime) layer**, which may be the graveyard of subducted slabs. This is a complex layer, but toward its base is a 5 km to 40 km-thick zone with ultralow seismic velocities, which undoubtedly indicate the presence of partially melted rock. It is possible that magmas that rise at hot spots such as Hawaii have their source in this zone.

The upper mantle is bounded on its lower side by the 660 km seismic discontinuity. Between this depth and another prominent discontinuity at 410 km is called the **transition zone** (Fig. 1.5(B)). Above this is the uppermost mantle, which terminates at the base of the **crust** at the prominent **Mohorovičić discontinuity** (or **Moho**). An extremely important zone marked by low seismic velocities occurs in the uppermost

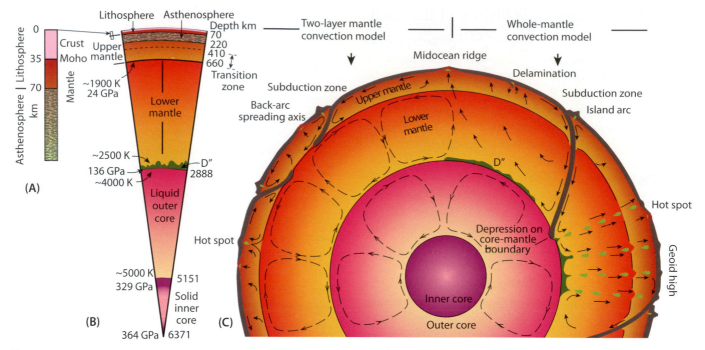

Figure 1.5 Major structural units of the Earth. (**A**) Cross-section through the lithosphere. (**B**) Section through the Earth showing the depth, pressure, and temperature at major seismic discontinuities. (**C**) Cross-section through the Earth showing the two-layer and whole-mantle convection models (left and right, respectively). Bright green indicates parts of the mantle undergoing partial melting as it rises toward Earth's surface.

mantle between depths of 20 to 50 km beneath oceans and 70 to 220 km beneath continents. This zone is known as the **asthenosphere**, from the Greek word *asthenēs*, meaning "weak." Its low velocities are attributed to the presence of very small amounts of melt, which weakens the rock. Above the asthenosphere, the uppermost mantle and overlying crust form the relatively strong **lithosphere** (Fig. 1.5(A)). The asthenosphere is of importance because it is on this weak layer that the lithospheric plates move around the surface of the Earth to give us **plate tectonics**.

Finally, the crust is the outermost layer of the Earth. It is from 25 to 70 km thick beneath continents and 7 to 10 km thick beneath oceans. The rocks in the continental crust are less dense than those in the oceanic crust, and as a result of **isostasy** (buoyancy), continents stand higher than ocean floors.

1.5 Cooling of the planet and plate tectonics

We know that the Earth's interior is hot and that the planet is still cooling. Deep drill holes indicate that the temperature in the Earth increases by about 25°C per kilometer but can range from 10 to 60°C/km. This is known as the **geothermal gradient**. We also know that heat flows from high to low temperature and must, therefore, be escaping from the Earth.

Knowing the **thermal conductivity** of rocks (0.005 cal/cm s °C), we can calculate that a geothermal gradient of 25°C would result in 1.25×10^{-6} calories escaping from 1 square centimeter of the Earth's surface every second. By expressing this value as 394 kilocalories per square meter per year, we can better appreciate how small a quantity of heat this is. Recall that a **calorie** is the quantity of heat required to raise 1 gram of water 1 degree centigrade. We are perhaps more familiar with the calorie when used for the energy content of food, but the food calorie is actually a kilocalorie (kcal = 1000 cal). For example, when we see that a McDonald's Quarter Pounder hamburger contains 410 calories, this is actually 410 kcal, which is almost the same as the amount of heat flowing from a square meter of the Earth's surface in an entire year. Despite its low value, this heat flow is sufficient to make the Earth a dynamic planet. It drives convection in the mantle and the movement of lithospheric plates, which results in plate tectonics and the processes that create and destroy rocks.

Calculations of the heat flow from the Earth have been performed on more than 24 000 drill holes over all of the continents and the ocean floor. These data have been synthesized by the **International Heat Flow Commission** to create the map shown in Figure 1.6 (see "Online Resources" at end of chapter). It uses SI rather than cgs units of heat flow; that is, milliwatts per square meter (mW/m²). If we convert the heat flow discussed in the previous paragraph (1.25×10^{-6} cal/cm² s) to SI units, it becomes 50.2 mW/m². On the map, this value can be seen to be at the high end of the pale blue regions and is typical of most continental areas.

The heat-flow map shows that heat is not lost evenly from the planet. Indeed, most escapes from new ocean floor formed along

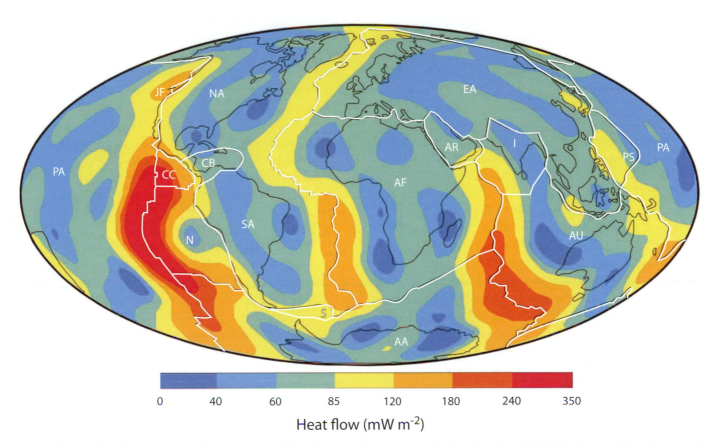

Figure 1.6 Global heat flow as synthesized by the International Heat Flow Commission (www.geophysik.rwth-aachen.de/IHFC/heatflow.html). For discussion of the synthesis, see Pollack et al. (1993). Tectonic plate boundaries are shown by white lines, and the names of the plates are shown by letters, the full names of which are given in the text.

divergent tectonic plate boundaries. These regions, which constitute only 30% of the Earth's surface, account for 50% of all heat lost from the planet. This shows that plate tectonics is intimately related to the planet's cooling. Near the Earth's surface, heat is transferred through rocks by **conduction**; that is, thermal vibrations of atoms are transferred to adjoining atoms down the temperature gradient. This is a slow process because rocks have such low thermal conductivity. In the Earth's mantle, however, solid rocks are plastic and capable of moving, albeit very slowly. In this region, heat can be transferred by moving hot low-density rock from depth toward the surface to replace cooler higher-density rock that sinks. This sets up **convection**, which more efficiently transfers heat in the mantle than does conduction. The thermally driven mantle convection creates stresses in the lithosphere that result in plate tectonics. The motion of tectonic plates, in turn, controls processes that ultimately are responsible for the formation of igneous, metamorphic, and sedimentary rocks.

1.6 Plate tectonics and the formation of rocks

The lithosphere, which consists of the crust and upper mantle, is about 100 km thick. It is broken into eight major plates and numerous smaller ones. The major plates, which are indicated

in Figure 1.6 by letters, include, in alphabetical order, the African (AF), Antarctic (AA), Australian (AU), Eurasian (EA), Indian (I), North American (NA), South American (SA), and Pacific (PA) plates. Some of the more important smaller plates include the Arabian (AR), Caribbean (CB), Cocos (CC), Juan de Fuca (JF), Nazca (N), Philippine Sea (PS), and Scotia (S) plates.

Plate boundaries are of three types. At **divergent plate boundaries**, plates move apart and new crust is created by molten material rising from the mantle and solidifying to form new ocean floor, as occurs along the Mid-Atlantic Ridge and East Pacific Rise. At convergent plate boundaries, plates converge and crust is destroyed as it is subducted into the mantle, as happens to the Pacific plate, where it is subducted beneath the Aleutian Islands and Japan. Along **transform plate boundaries**, plates grind past each other along major faults, such as the San Andreas. Useful animations of these plate motions and reconstructions of past plate positions are given on the U.S. Geological Survey's Web site referred to in "Online Resources" at the end of the chapter.

The rates at which plates diverge from each other can be measured from magnetic anomalies on the ocean floor (see USGS animations). New crust formed at midocean ridges develops magnetic anomalies due to periodic reversals in the Earth's magnetic field. As plates diverge, these anomalies are split apart, and by correlating similar age anomalies on either side of the spreading axis, the rate of divergence can be determined. These **relative plate velocities** can be as much as 160 mm/year but are typically about 40 mm/year, which is of the same order of magnitude as the rate at which your fingernails grow.

Absolute plate velocities are not as easily determined but can be determined if we assume that **hot spots**, such as Hawaii or Yellowstone, have sources deep in the mantle that do not move relative to each other, which appears to be a good first approximation. As a lithospheric plate moves across a stationary hot spot, a string of volcanoes results that produce a **hot-spot track**. For example, the Hawaiian Island chain was created by the Pacific plate moving across the Hawaiian hot spot (Fig. 9.41). By dating the volcanoes along this track, the Pacific plate is shown to be moving at 95 mm/year in a direction N 59° W.

Most rocks are formed by processes related to plate tectonics, and many are formed near plate boundaries. We can think of the location in which a rock is formed as a rock factory, and as we will see in later chapters, most of these factories are intimately related to specific plate tectonic settings. By way of introduction, we briefly describe these main plate tectonic settings.

1.6.1 Divergent plate boundaries

The world's most productive rock factories occur at divergent plate boundaries in oceanic regions, where molten material rises from the mantle and cools and solidifies to form new oceanic crust (Fig. 1.7(A)). The cooling gives these regions high heat flow (Fig. 1.6). While still hot, the new crust has lower density than older, colder crust, so it isostatically stands high and forms a **midocean ridge**. These ridges typically have a median valley, which slowly widens as the plates diverge and is the locus of much of the lava that erupts along these ridges. Much of the cooling of the new crust results from circulation of seawater through the rocks, which changes their minerals and results in metamorphism. As the plates diverge and cool, they become denser, and the ocean floor deepens. The new ocean floor receives a constant flux of sediment, which consists mainly of the bodies of **pelagic organisms** that sink to the bottom on dying. Because the ocean floor gets older away from a spreading axis, the thickness of this layer of sediment gradually thickens.

If divergence occurs in a continental plate, the stretched lithosphere fails in the crust along **normal faults** to form a **rift valley**, whereas the asthenosphere rises from below (Fig. 1.7(B)). Sediment eroded from the highlands is washed into these basins, and magma formed in the rising asthenosphere forms igneous rocks. The sedimentary rocks are usually slightly metamorphosed.

1.6.2 Convergent plate boundaries

The world's second most productive rock factory occurs at convergent plate boundaries, where one plate is subducted beneath

another. If both plates involve oceanic lithosphere, the older, and hence denser, ocean crust is subducted beneath the younger ocean crust. If one plate is continental and the other oceanic, the oceanic plate, being denser, is subducted beneath the continent, as illustrated in Figure 1.7(C). When continental plates collide, the less dense crustal rocks resist sinking into the mantle and are crumpled to form major orogenic belts. Convergent plate boundaries have the greatest relief on the planet, ranging from the highest mountains to the deepest oceanic trenches.

When oceanic rocks are subducted, the increased pressure causes metamorphism and release of water, which rises into the overriding plate, where it results in partial melting. The melt, being less dense than the solid rocks, rises toward the surface and forms the volcanoes that characterize most convergent plate boundaries. Radioactive decay in the thickened crust and the passage of magma through the crust raises its temperature and, along with the stresses caused by convergence, causes metamorphism in the orogenic belt. Erosion of mountains produces large amounts of sediment, much of which is rapidly transported to **forearc basins**. Behind many convergent plate boundaries, the lithosphere may be stretched, with formation of **back-arc basins**, which become sites of sediment deposition. If back-arc spreading continues, the lithosphere may rift apart and form new ocean floor. In oceanic trenches, deep ocean sediment may accumulate and, along with sediment spilling over from forearc basins, become intensely deformed and metamorphosed at the convergent plate boundary.

1.6.3 Transform boundaries

Very few new rocks are formed at transform plate boundaries, but existing rocks become highly deformed by the shearing action of one plate grinding past the other. Irregularities along transform boundaries can result in formation of pull-apart basins, which become sites of sediment deposition (Fig. 1.7(D)).

1.6.4 Mantle plumes and hot spots

Regions of long-lasting volcanism at relatively stationary positions on the planet are known as hot spots. As lithospheric plates pass over them, they leave a hot-spot track. The Hawaiian hot spot is one of the most prominent, but many others occur around the world. It has been postulated that hot spots form above **mantle plumes** that rise from deep in the mantle, possibly from as deep as the D″ layer near the core-mantle boundary. On approaching the base of the lithosphere, these plumes are supposed to flatten out to form a large mushroom-shaped head (Fig. 1.7(E)).

Considerable controversy surrounds the existence of mantle plumes. Early attempts to identify them seismically failed but did show that, if they exist, they must have small diameters at depth. Recent seismic studies, however, have been able to image the plume beneath Hawaii. The enormous amounts of heat

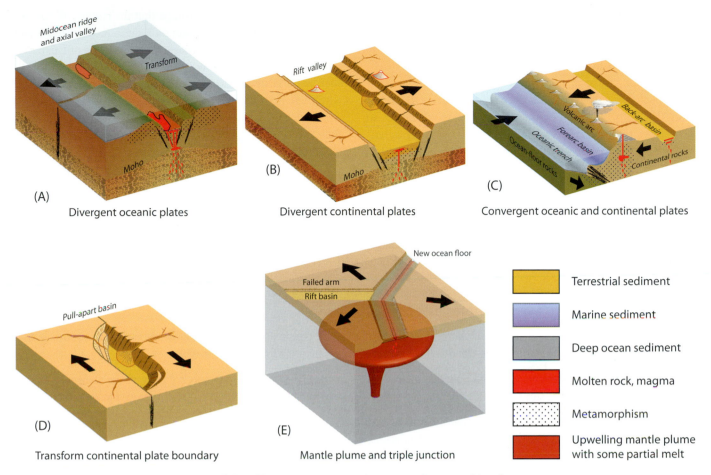

Figure 1.7 Types of tectonic plate boundaries and sites of formation of igneous, sedimentary, and metamorphic rocks.

released at hot spots, such as Hawaii and Yellowstone, necessitate a deep source in the planet.

Large mantle plumes in the geologic past are believed to have caused periods of extensive volcanism, during which huge volumes of new volcanic rocks were created in what are known as **large igneous provinces** (LIPs). The lithosphere above these plumes would have been heated and become less dense and hence risen to form large domes. As a plume head spreads, the lithosphere would have been rifted apart to form a **triple junction** (Fig. 1.7(E)). Commonly at these triple junctions, two of the rifts remain active and create new ocean floor, whereas the other rift becomes inactive or widens only slowly and is known as a failed rift. A mantle plume forming such a triple junction may be present beneath the Afar region of Ethiopia, with the Red Sea and Gulf of Aden being the active rifts and the East African rift being the failed arm (Box 8.1).

1.7 Outline of subsequent chapters

In the following chapters, we discuss the mineralogical makeup of all the major rock types and discuss how and where they are formed and what uses we make of them. We start by learning about the physical properties of minerals, which are used to identify them in hand specimens. Next we learn about the chemical makeup of minerals and the way atoms fit together to form crystalline structures, as well as their external crystal form and internal atomic arrangement. Another chapter provides instruction on how to use the polarizing petrographic microscope, one of the most useful tools for studying minerals and rocks.

Armed with these tools on how to identify minerals, we proceed through the following chapters by first studying the common rock-forming minerals that occur in a particular rock type, which is followed by chapters dealing with the formation and classification of that rock type. We begin by discussing igneous minerals and rocks, because these would have been the first Earth materials formed and constitute the most abundant material in the crust. This is followed by sedimentary minerals and rocks, and finally we deal with metamorphic minerals and rocks.

Throughout the book, we make reference to uses that are made of minerals and rocks, but in the penultimate chapter, we focus on some of the most important Earth materials that we make use of on a daily basis. This includes construction materials, clays for ceramics, metals from ore minerals, and energy sources.

In the final chapter, we discuss the effects of Earth materials on human health, which can be positive or negative. This is a huge topic that we can only briefly touch on. Through food, we obtain nutrients from the Earth that are essential to our well-being. Feeding the ever-growing world population is an agricultural challenge that requires the use of fertilizers that come from the Earth. Some minerals are hazardous to our health. In some cases, these hazardous materials are part of the natural environment, but in others they are the result of human activity.

After reading this book, we hope that the reader will have an appreciation of how Earth materials came into existence and how we make use of them. A basic understanding of Earth materials is essential not only to the professional working with such materials but also to every human, because we interact with these materials every day, through tools we use, materials with which we construct, energy we use, and food we eat. Hopefully, this book will provide a basic understanding of Earth materials that will be useful to a wide spectrum of readers.

Summary

This introductory chapter briefly summarized where the chemical elements that form the Earth came from, the planet's main structural units, and the plate tectonic settings in which new rocks are generated.

- Most of the elements in the Earth, especially those denser than helium, were formed by processes occurring in stars and during explosion of massive stars (supernovae) early in the history of the universe.

- A solar system forms from the dispersed matter in space when a cloud of gas and dust collapses gravitationally into a rotating disc, with a star forming at the center of the cloud and the planets forming from the material in the disc.

- The Earth and solar system were formed from the debris from a supernova 4.56 billion years ago, but none of the earliest-formed material in the planet has survived, so we turn to meteorites to learn about the material that probably accreted to form the planet.

- The terrestrial planets are small because they are formed from the least abundant elements in the solar system, whereas the outer gaseous giant planets also include the more abundant elements.

- Chondritic meteorites, which contain small spheres of minerals composed of oxygen, silicon, magnesium, and iron, are thought to be composed of the same material that accreted to form the planet Earth. These four elements are the most abundant elements in the Earth.

- Heat from the accretion of the planet, radioactive decay, and core formation would have been sufficient to melt the outer part of the Earth early in its history. Since then, cooling has been the most important planetary process.

- The major divisions of the Earth are its core, mantle, and crust. The core, which is composed predominantly of iron and nickel, consists of an inner solid and an outer liquid core. The mantle is divided into upper and lower by a prominent seismic discontinuity at a depth of 660 km. The Mohorovičić discontinuity marks the boundary between the mantle and crust, at a depth of 7 to 10 km beneath oceans and 25 to 70 km beneath continents.

- The outer ~100 km of the Earth forms the relatively strong lithosphere, which moves in a number of plates over the much weaker asthenosphere creating plate tectonics.

- The Earth is slowly ridding itself of internal heat at a rate of about 394 kcal/m^2 year. As a result of plate tectonics, most of the internal heat is released from new ocean floor formed at divergent plate boundaries.

- Eight major and numerous smaller lithospheric plates have boundaries that are either divergent, convergent, or transform.

- Relative velocities of plate divergence are on the order of 40 mm/year.

- The largest volumes of new crust are formed at divergent plate boundaries, most along midocean ridges.

- At convergent boundaries, oceanic crust is subducted into the mantle, but new crust is formed by the eruption of many volcanoes.

- Sediments eroded from continents are deposited in basins that are formed at convergent plate boundaries, rift valleys at divergent plate boundaries and back-arc spreading axes, and pull-apart basins along transform boundaries.

- Mantle plumes are hypothesized to bring hot rocks up from deep in the mantle. Hot spots are believed to form from melts that rise from these plumes toward the surface.

- The spreading head of a mantle plume may be responsible for rifting of the lithosphere at triple junctions.

Review questions

1. Where were most of Earth's chemical elements formed?

2. Why are the terrestrial inner planets small compared with the outer gas giant planets?

3. What is a chondrule, and why do we use the composition of chondritic meteorites to determine the composition of the Earth?

4. What caused the Earth to be so hot early in its history that its outer part would have been molten?

5. What are the major divisions of the Earth?

6. What two important layers in the Earth are responsible for plate tectonics?

7. What are the main types of plate boundaries, and what types of rock might be formed there?

ONLINE RESOURCES

www.geophysik.rwth-aachen.de/IHFC/heatflow.html – This is the Web site of the International Heat Flow Commission, where you can download the heat-flow map of the world and see other information relating to the Earth's internal heat.

http://www.nature.nps.gov/geology/usgsnps/animate/pltecan. html – This U.S. Geological Survey Web site provides useful animations of many aspects of plate tectonics.

FURTHER READING

Cosmochemistry. *Elements* 7, no. 1 (2011). This issue comprises articles on the origin of the minerals and elements that make up the planet Earth and the primordial solar nebula.

Lodders, K. (2003). Solar system abundances and condensation temperatures of the elements. *The Astrophysical Journal*, 591, 1220–1247.

Newsom, H. E. (1995). Composition of the solar system, planets, meteorites, and major terrestrial reservoirs. In *Global Earth Physics: A Handbook of Physical Constants*, ed. T. J. Ahrens, American Geophysical Union Reference Shelf 1, 159–189.

Pollack, H. N., Hurter, S. J., and Johnson, J. R. (1993). Heat flow from the Earth's interior: Analysis of the global data set. *Reviews of Geophysics*, 31, 267–280.

2 Materials of the solid Earth

This chapter introduces minerals and rocks, the solid building materials of planet Earth. We define minerals and rocks and give examples of each from common daily uses. The main purpose of this chapter is to explain what minerals and rocks are with minor reference to specific names. Although we use several rock and mineral names in giving examples, these need not be memorized, because we encounter them in later chapters. In this chapter, it is important to understand what minerals and rocks are and to appreciate their differences. The examples of minerals and rocks that we encounter on a daily basis are given simply to emphasize the importance of Earth materials to our daily lives. Indeed, human cultural evolution is normally classified on the basis of the Earth materials used for making tools (e.g., Stone Age, Bronze Age, Iron Age). Although we have benefited from the use of Earth materials, some materials pose potential health hazards (e.g., asbestos). Rocks have provided the main source of construction material for large buildings throughout history, and even though modern buildings are mostly made of concrete and steel, the concrete is made with limestone. We end the chapter with a brief discussion of where rocks are formed. Rocks are the direct product of plate tectonic processes, and characteristic sets of igneous, sedimentary, and metamorphic rocks form in specific plate tectonic settings.

Ocean waves pounding the rocky coast of the Caribbean island of Aruba. Wave action is one of several physical processes that, together with chemical reactions, reduce rocks to fragments and mineral particles that are transported into deeper water, where they accumulate to form sedimentary layers on the ocean floor. The gently sloping terrace, part of which has recently collapsed, is an ancient wave-cut terrace formed during the former high stand of sea level in the last interglacial period 120 000 years ago. The terrace is cut into limestone that was formed by a coral reef that surrounded the igneous core of the island. A dark basaltic lava forms the coastal promontory just above the crest of the surging wave.

This text has the title *Earth Materials*. An inclusive definition of Earth materials encompasses the following: **minerals** and **rocks**, which are the solid parts; **soils**, the unconsolidated materials above bedrock; **fossil fuels**, which include all the hydrocarbons used for fuel and energy – petroleum, natural gas, and coal; the various forms of H_2O, in salt water and freshwater, in glaciers and ice caps; and the atmosphere, the mixture of gases that surrounds the Earth.

This book is an introduction to the solid materials that compose the Earth, but short discussions of soils (Chapter 10), coal (Chapter 12), and H_2O and fossil fuels (Chapter 16) are included.

Solid materials are found abundantly underfoot on hiking trails in mountainous regions, but they are also present on beaches all around the globe. The solid materials in mountain chains, most commonly seen as outcrops in highway cuts, are rocks. Some common rock types are granite, basalt, limestone, and sandstone. The loose, solid materials on beaches consist mostly of the mineral quartz (with some minor components such as feldspar and mica grains mixed in) or calcite. Almost all beach sands are quartz rich, but some in the Caribbean consist mainly of granular calcite.

This book deals extensively with various aspects of minerals and rocks, and our main objective is to understand their origin and identification. Therefore, we are concerned with the questions of why and how chemical matter (**elements** and **ions**) is organized into particular minerals and rocks, and how minerals and rocks react to physical changes (e.g., in temperature and/or pressure) in the Earth's interior and to chemical changes in those exposed to the atmosphere. Parts of the text are also devoted to mineral and rock descriptions, because they are essential as basic knowledge for further discussion and for their identification in the laboratory or in the field.

2.1 Definition of a mineral

A mineral is a naturally occurring solid, with an ordered atomic arrangement and a definite (but commonly not fixed) chemical composition. Almost all minerals are formed by inorganic chemical processes.

This definition is most easily understood if we assess each of its clauses in sequence.

Naturally occurring means that minerals are formed by processes that occur in nature and are not produced by laboratory processes. Many gemstones that occur naturally are today routinely manufactured in the laboratory. These are referred to as **synthetic** gems. Almost all reputable jewelers exhibit and sell naturally occurring gems such as emerald, sapphire, and so on. Specialty stores, however, might sell their synthetic counterparts. Such laboratory-produced materials should be identified with the prefix *synthetic*, as in synthetic emerald.

The word *solid* excludes liquids and gases. In solids, chemical elements (generally as ions, that is, electrically charged atoms) occur in fixed and regular patterns, which is not the case for liquids and gases. One very rare mineral that does occur as a liquid is mercury, Hg, in its elemental form. Mercury occurs in liquid spheres and globules and is a unique exception to the definition of a mineral as a solid.

An *ordered atomic arrangement* requires that the internal structure of minerals consists of regularly repeated three-dimensional patterns of atoms, ions, or ionic groups that are held together by various chemical bonds. Solids that lack such a regular internal repeated pattern are referred to as **amorphous**.

A definite (but commonly not fixed) chemical composition. The word *definite* means that all minerals have fixed ratios of cations (positive charge) to anions (negative charge). For example, in the simple mineral composition of quartz, SiO_2, the cation (Si^{4+}) to anion (O^{2-}) ratio is 1:2. This is referred to as fixed because all quartz has this specific cation-anion ratio. A slightly more complex composition is shown by the mineral olivine, which is normally represented by the formula $(Mg, Fe)_2SiO_4$. The cation-anion ratios in this formula are fixed as follows: $(Mg^{2+}, Fe^{2+}) = 2$, $Si^{4+} = 1$, and $O^{2-} = 4$, which results in the ratio 2:1:4. Given that in the atomic structure of olivine the Si^{4+} occurs as $(SiO_4)^{4-}$ anionic complexes with a tetrahedral outline, these ratios can be rewritten for $(Mg, Fe)_2SiO_4$ as 2:1. In olivine, the Mg^{2+} and Fe^{2+} ions (which have the same electrical charge and similar ionic sizes and can, therefore, be housed in the same atomic site in the olivine structure) can substitute for each other freely, which leads to the clause "but commonly not fixed." Most mineral compositions have such chemical substitutions in their formulas; therefore, the majority of mineral compositions show considerable chemical variability, although the overall cation-anion ratios remain fixed.

"Formed by inorganic processes" reflects the fact that all common rock-forming minerals are the result of inorganic chemical processes. This does not mean that there are no exceptions to this statement. The mineral calcite, $CaCO_3$, which is the constituent of mollusk shells, has been deposited by the organic soft parts inside the shell, and as such it is of organic origin. Bones and teeth in the human body consist of a form of the mineral apatite, $Ca_5(PO_4, CO_3)_3(OH, O, F)$, which in this instance is of organic origin. Those two examples are exceptions to the statement that minerals are formed by inorganic processes. These organically formed species are normally included in any listing, or catalog, of minerals.

An example of an organic material that is definitely excluded, although commonly sold as a gem in jewelry stores, is amber. This is because amber does not fit most of the criteria outlined in the preceding mineral definition. It not only is always of organic origin but also is a hard, brittle, fossil resin that lacks an ordered atomic internal structure and has a highly variable (not fixed) chemical composition. Because it is a very attractive solid material, especially when polished, with an appealing yellow to brown

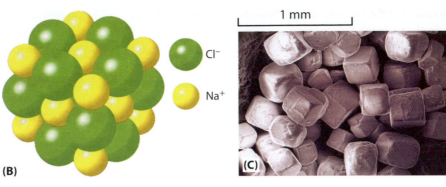

Figure 2.1 (**A**) A standard container of table salt, NaCl, with some salt in a Petri dish. (**B**) A perspective view of the internal atomic structure of NaCl with an overall cubic outline. (**C**) Salt grains from the Petri dish in (A) photographed by scanning electron microscopy. This shows perfect cubic cleavage fragments with somewhat beveled corners and edges as a result of abrasion. The abrasion is due to the juggling of salt grains in the large container during transport. (Photograph courtesy of Adrian Brearley, University of New Mexico.)

color, it is sold among other gems, most of which are inorganic in origin and fit the definition of a mineral, as given already.

2.1.1 Examples of some familiar minerals

At this stage, let us review a few minerals that you may have some familiarity with: halite, quartz, talc, chrysotile (one of several "asbestos" minerals), and garnet. This brief overview of the five minerals will aid in your understanding of the definition of a mineral given in the prior section.

All five minerals are naturally occurring inorganic solids. They also have regular atomic internal structures, and their composition (or range of composition) is a function of the

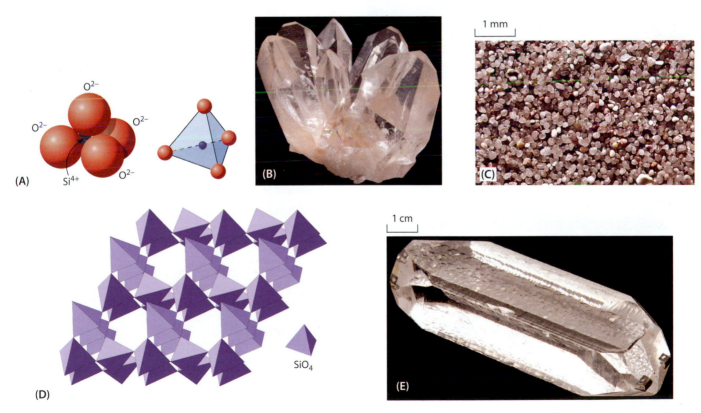

Figure 2.2 (**A**) On the left, a drawing of a central Si^{4+} ion surrounded by four O^{2-} ions; on the right, a schematic representation of the same, showing the outline of a tetrahedron (from the Greek *tetra*, meaning "four," and *hedron*, meaning "plane"), a four-sided geometric shape. (**B**) Euhedral (meaning "well formed") crystals of quartz, SiO_2. (**C**) Beach sand consisting mainly of well-rounded translucent and milky quartz grains. Several of the colored grains are quartz with colored inclusions and/or feldspar. (**D**) A perspective view of the internal, atomic structure of quartz, consisting of a network of linked $(SiO_4)^{4+}$ tetrahedra. (**E**) A synthetic crystal of quartz, which, due to the conditions of growth in high-temperature vessels in the laboratory, has a very different external shape from crystals that occur in nature (see Fig. 2.2(B)). Long dimension of crystal is 8 cm.

internal structure as well as the chemical composition of the source from which each mineral crystallized.

Halite, also known as table salt, has the composition NaCl (Fig. 2.1(A)). Halite is a member of a group of minerals known as **halides** in which the anion is Cl^-, Br^-, F^-, or I^-. Halite has a simple, highly symmetric structure (see Fig. 2.1(B)) that leads to a cubic external shape (the grains in saltshakers, when observed with magnification, are all little cubic fragments, known as cleavage fragments, not little cubic crystals; Fig. 2.1(C)). Naturally occurring halite has a composition very close to NaCl, because K^+, which occurs in the mineral sylvite, KCl, with the same structure as NaCl, has an ionic radius much larger than that of Na^+. The two cations, therefore, have difficulty substituting for each other in the same structural space. Halite is extracted from sedimentary rocks known as evaporite deposits. It is also extracted

from large evaporative ponds (created by the diking of shallow seas) along the coast of countries with arid, desertlike climates, through evaporation of seawater. This type of halite is produced by natural processes in a manmade system. However, it is still referred to by its mineral name even though humans extract it.

The other four minerals in our list are all silicates. In these the $(SiO_4)^{4+}$ ionic group is the basic building block, in the form of a tetrahedron (a four-sided geometrical body), which occurs in the structure of almost all silicates (Fig. 2.2(A)). The manner in which these tetrahedra are linked to one another and to cations in the silicate structure is what creates the large variety of silicates.

Quartz, SiO_2, is familiar in many ways (Fig. 2.2(B)). It forms the rounded sand grains on beaches (Fig. 2.2(C)); it is the loose granular material used in sandboxes; and as crushed,

(A)

(B)

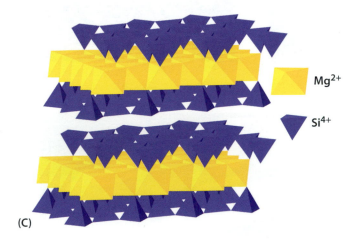

- Mg^{2+}
- Si^{4+}

(C)

100 μm

(D)

Figure 2.3 (**A**) Hand specimens of talc, $Mg_3Si_4O_{10}(OH)_2$. (**B**) Commercial containers of baby (talcum) powder and cosmetic facial powder. Two dishes, with talc and facial powder, respectively. (**C**) A perspective view of the layered internal atomic structure of talc. The Mg^{2+} ions are sandwiched between two opposing sheets of $(SiO_4)^{4+}$ tetrahedra. (**D**) Talcum powder from the Petri dish in (B) photographed by scanning electron microscope. This shows the somewhat variable grain size of the tiny (very soft) flakes at about 400× magnification. (Photograph courtesy of Adrian Brearley, University of New Mexico.)

angular grains, it is the abrasive component of many standard sandpapers. Its structure consists of a close linking of $(SiO_4)^{4-}$ tetrahedra (Fig. 2.2(D)) that is devoid of any other chemical elements. It is so consistently of constant SiO_2 composition that it is commonly referred to as a pure substance. This is in contrast to the compositional variability of many other silicates.

Quartz sand is an important component of cement in the construction industry, and it is a source of material for making silica glass. The manufacture of synthetic quartz crystals is also an important industry. Because perfect quartz crystals (*perfect* meaning "no defects" of any sort in clear quartz; i.e., no cracks, no bubbles, no inclusions of fluids or of other minerals) are difficult to obtain consistently from natural occurrences, synthetic quartz is routinely manufactured (Fig. 2.2(E)). This quartz is used for highly specialized electronic applications, such as the timing mechanism in quartz watches.

Talc, $Mg_3Si_4O_{10}(OH)_2$, is a chemically more complex mineral than quartz (Fig. 2.3(A)). This mineral, packaged and sold as talcum powder, is a material that has likely been part of your life since birth (Fig. 2.3(B)). Talc is a versatile mineral used in cosmetic products, in ceramics, in the paper industry, in vulcanizing rubber, and in paint production. The atomic structure of talc consists of sheets of $(SiO_4)^{4-}$ tetrahedra (cross-linked by Mg^{2+} ions in between), producing a flat sandwichlike arrangement (Fig. 2.3(C)). These structural layers are responsible for the thin, somewhat flexible folia that pure talc exhibits in nature. In the manufacture of cosmetic products, the original talc is ground very finely in roller mills to preserve the original platy structure of the mineral during milling (Fig. 2.3(D)). In most cosmetic products made mainly of talc, some minor other materials are added, such as coloring agents and scent. Talc has very little variation in its chemical composition; as such, the formula $Mg_3Si_4O_{10}(OH)_2$ applies to most of its occurrences.

Chrysotile, $Mg_3Si_2O_5(OH)_4$, is a mineral that you may not have heard of. Instead, you may have read articles, or seen programs on television, on the health issues associated with asbestos. The word *asbestos* is a commercial term that includes six naturally occurring minerals, of which chrysotile is by far the

(A)

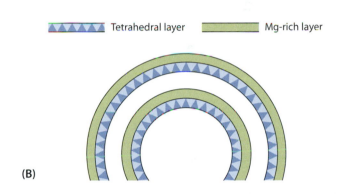

Tetrahedral layer   Mg-rich layer

(B)

100 nm

(C)

5 nm

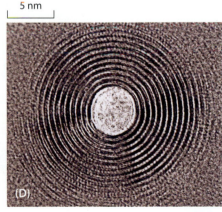

(D)

(E)

Figure 2.4 (A) A hand specimen of chrysotile, $Mg_3Si_2O_5(OH)_4$. Note the very fibrous and flexible nature of this naturally occurring substance. (B) A schematic representation of the coiling of the atomic layers that produce a tubelike habit (the fibrous habit) of chrysotile. The layers differ from that of talc (Fig. 2.3(C)) in that an opposing tetrahedral sheet is missing. (C) Transmission electron microscope image of synthetic chrysotile fibers showing that their internal structure is comparable to that of drinking straws. (D) High-resolution transmission electron microscope image of a cross-section of a drinking straw, showing the coiled nature of the structure. ((C) and (D) courtesy of Alain Baronnet, CINaM-CNRS, Centre Interdisciplinaire de Nanosciences de Marseilles, France.) (E) Laboratory gloves, used in high-temperature experiments, woven from very flexible and strongly insulating, naturally occurring chrysotile fibers.

most common. About 95% of asbestos that was used commercially was chrysotile (Fig. 2.4(A)). The other 5% consisted of a very different structure type, known as an amphibole (with the mineral crocidolite being the most common constituent). The chrysotile structure consists of atomic layers that are coiled to form hollow tubes like a drinking straw (Fig. 2.4(B)–(D)). These flexible fibers can occur naturally as bundles, similar to yarn (Fig. 2.4(A)) and can, therefore, be woven into cloth, which was used in suits and gloves for firefighters (Fig. 2.4(E)) before it was banned in the United States. The main use of chrysotile was as insulation of the internal steel structures of skyscrapers (to protect against meltdown in case of fires) and in many building materials. It has long been shown that chrysotile is rather harmless, and it is clear that banning this material from many products has been unwise. The health controversy about asbestos is discussed in Chapter 17. The chemical composition of chrysotile is rather constant and is well represented by the formula $Mg_3Si_2O_5(OH)_4$.

Garnet is the birthstone for January and has been commonly used as a red gemstone, especially in jewelry of the Victorian Era (1837–1901). Garnet often shows up in antique and estate jewelry (Fig. 2.5(A)). This common association of the red color in garnet jewelry gives the impression that garnet as a gem is generally of a reddish color. In reality, garnet includes a closely related group of rock-forming minerals of a wide range of colors, from pink to brown, yellow, green, orange, and black. Garnet commonly occurs in good crystal forms, one of which is known as a dodecahedron, consisting of twelve parallelogram, diamond-shaped faces (*dodeca*, from the Greek meaning "twelve"; Fig. 2.5(B)). Garnet is also familiar because it is used in a variety of abrasive applications (because of its high hardness and sharp broken edges) as in garnet paper (used in specialized applications in the finishing of leather, hard rubber, and plastics).

The internal, atomic structure of garnet is a compact and complex arrangement of independent $(SiO_4)^{4-}$ tetrahedra (Fig. 2.5(C)) bonded to cations such as Ca^{2+}, Fe^{2+}, Mn^{2+}, Mg^{2+}, as well as Al^{3+}, Fe^{3+}, or Cr^{3+}. The compactness of its structure is reflected in its density (or specific gravity), which is considerably higher than that of other silicates. As seen from the foregoing listing of ions that may occur in its structure, the chemical composition of garnet is highly variable. Six specific garnet compositions include the following:

$$Mg_3Al_2(SiO_4)_3 - pyrope$$

$$Fe_3Al_2(SiO_4)_3 - almandine$$

$$Mn_3Al_2(SiO_4)_3 - spessartine$$

$$Ca_3Al_2(SiO_4)_3 - grossular$$

$$Ca_3Fe_2^{3+}(SiO_4)_3 - andradite$$

$$Ca_3Cr_2(SiO_4)_3 - uvarovite$$

In other words, garnets are a good example of a mineral group in which the cation-to-anion proportions are fixed, at 3:2:3, but

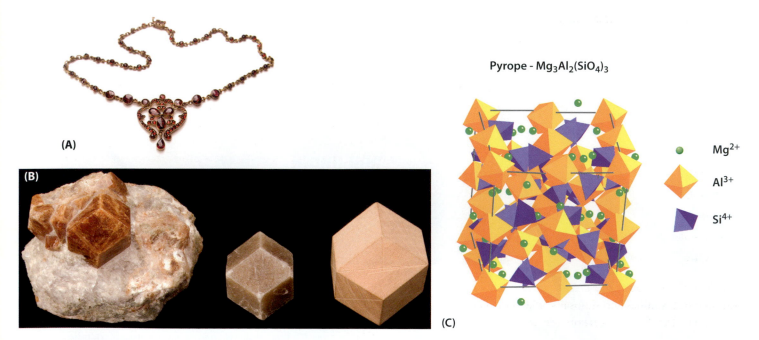

Pyrope - Mg₃Al₂(SiO₄)₃

Mg^{2+}

Al^{3+}

Si^{4+}

(A)

(B)

(C)

Figure 2.5 (A) A garnet necklace of Victorian design. This type of red garnet is referred to as Bohemian garnet in the trade. It is a red garnet of pyrope composition. (Courtesy of Ron Beauchamp, Beauchamp Jewelers, Albuquerque, NM.) (B) Examples of the relatively common occurrence of garnet in well-formed (euhedral) dodecahedrons. Two almost perfect crystals, as well as a small wooden model of the same dodecahedron, used in laboratory teaching of crystal forms. (C) A perspective view of the complex and densely packed atomic structure of a garnet of composition $Mg_3Al_2(SiO_4)_3$, known as pyrope. The overall outline of the structure is that of a cube.

the amounts of a specific cation, in a specific atomic (structural) site, are highly variable.

Garnets have also been synthesized in the laboratory. The reason for such synthetic production was that some garnets with very unusual compositions (that do not occur in nature) were found to be fairly good imitations of diamond. These synthetic garnets (with an internal structure identical to that of natural garnet) have extremely unusual compositions that include elements such as yttrium (to produce YAG, an abbreviation for yttrium aluminum garnet) and gadolinium and gallium (GGG, a gadolinium gallium garnet). These synthetic simulants of diamond have been surpassed in popularity since the synthesis of cubic zirconia, ZrO_2, in 1976, as an almost perfect inexpensive simulant for natural diamond, with the composition of carbon, C.

The name garnet is very dear to one of the authors, Anthony R. Philpotts, because the maiden name of his wife, Doreen, is Garnet, and they now live in Connecticut, whose state mineral is garnet.

2.2 How are minerals classified?

Since the mid-nineteenth century, minerals have been systematically grouped into different chemical classes on the basis of their chemical composition. Such groupings are based on the composition of the dominant anion or anion complex. For example, minerals with sulfur, S^{2-}, as an anion are grouped under sulfides; those with $(SiO_4)^{4-}$, as an anionic group, are silicates. The broadest classification subdivisions used in all mineralogical literature are based on the anionic unit in the chemical formula as indicated in the following:

- Native elements   Au, Cu, Pt
- Sulfides   ZnS, FeS_2, $CuFeS_2$
- Sulfosalts Cu_3AsS_4, $Cu_{12}Sb_4S_{13}$
- Oxides Al_2O_3, Fe_3O_4, TiO_2
- Hydroxides $Mg(OH)_2$, $FeO(OH)$
- Halides   NaCl, KCl, CaF_2
- Carbonates $CaCO_3$, $CaMg(CO_3)_2$
- Phosphates $Ca_5(PO_4)_3(F, Cl, OH)$
- Sulfates $CaSO_4 \cdot 2H_2O$
- Tungstates $CaWO_4$
- Silicates Mg_2SiO_4, Al_2SiO_5

Approximately 4150 minerals are known, of which 1140 are silicates; 624, sulfides and sulfosalts; 458, phosphates; 411, oxides and hydroxides; 234, carbonates; and 90, native elements.[1]

[1] These data were provided by J. A. Mandarino, Toronto, Ontario, Canada in March 2006. A complete listing of all chemical classes is given in Klein and Dutrow, 2008, p. 333.

In this text, we systematically review only about 85 minerals, of which the majority are silicates. The remaining minerals are members of the carbonate, oxide, hydroxide, phosphate, sulfide, and native element classes. The reason for this small number is that these 85 represent the most common minerals that enter into the composition of the most abundant rock types. This small number of minerals is, therefore, referred to as the **rock-forming minerals**. Combinations of these minerals reflect the composition of the most common rock types, and they are the basis for rock classifications.

2.3 How are minerals named?

The naming of minerals is an interesting but not a scientific process. Many common mineral names date from times when there were essentially no scientific tools to classify, define, or describe a mineral quantitatively.

The name *quartz* is of obscure origin but dates back to the Middle Ages, when it was first applied to describe waste material produced during mining (such waste is now referred to as gangue). The name *feldspar* is derived from the word *feldtspat*, given in 1740 to the presence of spar (spath) in tilled fields overlying granite bedrock in Sweden. The name *garnet* is very likely derived from the Latin *granatum*, for "pomegranate," because of its similarity in color to that of the fruit pulp. Since these early days, when quartz, feldspar, garnet, and other common mineral names were chosen, the process by which a new mineral species must be uniquely defined has become a very demanding, quantitative scientific process, but the naming of a new mineral is still left to the investigator who discovered the new mineral species.

In the twentieth century, there has always been a clear set of regulations that mineralogists have had to follow in the characterization of a newly discovered mineral. Since 1959, the Commission on New Minerals and New Mineral Names of the International Mineralogical Association (IMA) has been internationally recognized for the purpose of controlling the introduction of new minerals and new mineral names and of rationalizing mineral nomenclature. This commission maintains a Web site (http://pubsites.uws.edu.au/ima-cnmnc/). The abbreviation CNMNC stands for its new name: Commission on New Minerals, Nomenclature, and Classification. If an investigator thinks that he or she has discovered a new mineral, a scientifically complete report on all aspects of the new mineral's chemical and physical properties must be submitted for review to the commission. An international committee of mineralogists reviews the report.

The naming of the mineral by the original investigator is by far the easiest and most fun part. Minerals can be named after the original locality of the find (e.g., aragonite, $CaCO_3$, – with an internal structure that is distinctly different from that of calcite, also $CaCO_3$ – after a locality in the Aragon region of Spain); after some major chemical constituent (as in molybdenite, MoS_2, because molybdenum is the only cation in the structure); after all of the cations, K^+, Al^{3+}, and Si^{4+}, in a mineral (as in kalsilite, $KAlSiO_4$); or after a famous personality (as in sillimanite, Al_2SiO_5, after Benjamin Silliman, 1779–1864, professor of chemistry and mineralogy at Yale University). If you are so lucky as to discover a new mineral, the choice of the name is pretty much up to you.

2.4 What is a crystal, and what is the crystalline state?

The majority of minerals occur as relatively small, irregular grains in an interlocking pattern with other minerals as the constituents of various rock types. These grains are generally less than 1 cm in diameter and are irregular in shape. So, where do the beautifully shaped large crystals come from that are commonly exhibited in natural history museums and high-quality rock shops (Fig. 2.6)? These have most commonly been carefully extracted (mined) from unique geologic occurrences known as **veins** (most desirable with open spaces known as **vugs**) and as **pegmatites**.

Veins are lenticular or tabular bodies, commonly with some coarse-grained minerals, which may cut across the structure seen in the bedrock. Veins are mostly the result of mineral deposition in free spaces in a fracture (fissure) that either was opened or was gradually opening during the period of mineralization. Open spaces (or vugs) are commonly the sites from which prospectors and mineral collectors retrieve beautifully crystallized mineral specimens, with large crystals that grew in these open spaces (Fig. 2.7). Pegmatites are exceptionally coarse-grained igneous rocks, usually found as irregular dikes, lenses, or veins. In general, their mineral grain size is well more than 1 cm in diameter (see Fig. 8.25). All three – veins, vugs, and pegmatites – are the main source of well-formed large crystals.

What is a crystal? A **crystal** can be defined as a mineral, or other crystalline chemical compound (a solid with a specific chemical composition), with an external shape bounded by smooth plane surfaces (Fig. 2.8). Such a crystal shape is the outward expression of its ordered internal, atomic arrangement. Crystals can be described as perfect, good, or malformed,

Figure 2.6 A hand specimen of beautifully crystallized green microcline feldspar (known as amazonite) and smoky quartz crystals. The crystals in this museum-quality specimen grew freely in an open space (a vug) without interference from adjoining minerals.

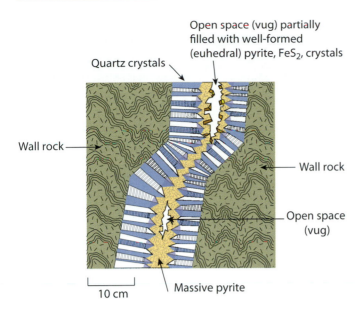

Figure 2.7 A schematic cross-section of a vein that is the result of open space filling. Well-formed quartz crystals grew perpendicular to the walls of the host rock. Some open spaces (vugs) still remain. One is empty, and the other is partially filled with well-formed cubic crystals of pyrite.

depending on how well the external shape reflects the overall symmetry inherent in the internal atomic arrangement. Perfect crystals or crystal groupings are what museum curators collect for mineral exhibits open to the public.

Three other terms that are commonly applied to crystals to describe the perfection (or lack thereof) of their external shape are **euhedral**, **subhedral**, and **anhedral**.

Euhedral, interchangeable with *perfect*, describes a crystal with excellent crystal shape (*eu*, from the Greek, meaning "good," and *hedron*, meaning "plane").

Subhedral describes mineral grains (and crystals) that are partly bounded by good crystal faces, as well as faces that may have been crowded by adjacent mineral grains (*sub*, from the Latin, meaning "less than").

Anhedral is applied to mineral grains that lack well-formed crystal faces as a result of having grown against adjacent minerals during their crystallization (*an*, from the Greek, meaning "without"). The majority of mineral grains that compose various rock types are anhedral to subhedral because such grains were generally in contact with neighboring grains (without open spaces) during their growth.

The term **crystalline** is applicable to all solid materials that exhibit an internal, ordered arrangement of the constituent atoms

Figure 2.8 Examples of some very well-formed (euhedral) crystals. From left to right: a milky, translucent crystal of gypsum; a red crystal of tourmaline; a transparent, light blue crystal of aquamarine (a gem variety of the mineral beryl); and a red crystal of ruby (a gem variety of the mineral corundum).

(or ions) forming a repetitive three-dimensional pattern. This crystalline state may be expressed in esthetically pleasing large, euhedral crystals (Fig. 2.8). Most rock-forming minerals that are part of various rock types occur as anhedral to subhedral grains, lacking any outwardly observable symmetry that might reflect their internal structure. Even these anhedral grains, however, have internal ordered atomic structures with specific chemical compositions that reflect their temperature or pressure of formation. In other words, all minerals (not just well-formed crystals) are part of the crystalline state. The crystallinity of small-grained, anhedral minerals can be proved by optical techniques, by X-ray diffraction methods, electron beam methods, and others (see Chapter 3). Minerals (and other materials) that lack the ordered internal arrangement are referred to as amorphous.

2.5 What is a rock?

A rock is simply a naturally occurring consolidated mixture of minerals. It is the solid material that makes up the Earth. We saw in Chapter 1 that the Earth's composition reflects the abundance of elements formed in a supernova and later modified by accretionary process during formation of the solar system. We have also seen that a mineral has a restricted compositional range. It is unlikely, therefore, that in any given part of the Earth, this bulk composition can be incorporated into a single mineral. Instead, a number of minerals must form to accommodate all of the elements present, and this mixture is a rock.

What do we mean when we say that rock is consolidated? In general, rocks are harder, and in many cases, much harder, than the soils that blanket the Earth's surface. Where highways cut through rock, drilling and blasting are often required to remove the material. Some rocks are quite soft but are still more competent than loose surficial materials. It is the strength of rocks that allow high mountain ranges to stand tall or quarry walls or road cuts to stand at steep angles (Fig. 2.9).

A rock's strength is determined by the way the individual mineral grains are intergrown with one another and the strength of the individual minerals. For example, the grains of quartz in Figure 6.14(A) have sutured boundaries, so they fit together like the pieces in a jigsaw puzzle. The rock shown in Figure 6.5, from the Giant's Causeway, in Northern Ireland, consists of an interlocking network of feldspar and pyroxene crystals. Because of the even distribution and random orientation of the crystals, this rock has no planes of weakness and, as a result, is one of the toughest rocks known. Quarrying of this rock type, which is popularly known as **trap rock** (from the Dutch word *trap*, meaning "stairs," which often describes the topography on these bodies of rock) is a major worldwide industry because, when crushed, the rock forms the most durable aggregate to mix with asphalt for paving roads (Fig. 2.9).

The way in which individual mineral grains are intergrown in a rock is referred to as **texture** (Fig. 2.10(A)). Larger-scale

Figure 2.9 The active north face of Tilcon's North Branford quarry, Connecticut, where a thick basaltic lava flow is quarried for its extremely durable rock that is used as aggregate and mixed with asphalt to pave roads. This quarry produces 20 000 tons of crushed trap-rock aggregate per day.

features, which commonly reflect uneven distributions of minerals such as compositional layering, are referred to as **structure** (Fig. 2.10(B)). Textures and structures develop when rocks are formed, and hence they reveal much about their origin. The texture and structure of rocks, along with their composition and absolute age, preserve the record from which Earth's history is revealed. Being able to correctly read the textures and structures in rocks is, therefore, of paramount importance. We discuss here only a few of the most important points about rocks and how they are formed.

2.6 How do rocks form? Classification into igneous, sedimentary, and metamorphic

Although the solid Earth is made of rocks and rocks are found everywhere, the formation of a rock is a rare event. In all likelihood, if you look at a geological map of the region where you live, you will find rocks that formed a long time ago, and there may have been only a couple of times in all of Earth's history that rocks were formed in your region. Rock formation is a rare occurrence because the conditions required for their formation occur only in a few places on Earth at any given time.

Many rocks form in response to the Earth's continuous attempt to cool down from its fiery beginning. We saw in Chapter 1 that early in Earth's history the planet was almost totally molten. Its surface would have been covered by molten rock, a **magma ocean**. Since then, enormous amounts of heat have been lost into space, and almost all of the Earth's crust and mantle have solidified. The cooling still causes the solid mantle to convect, which in turn moves lithospheric plates across the surface of the planet. This plate motion sets up conditions necessary for the formation of rocks.

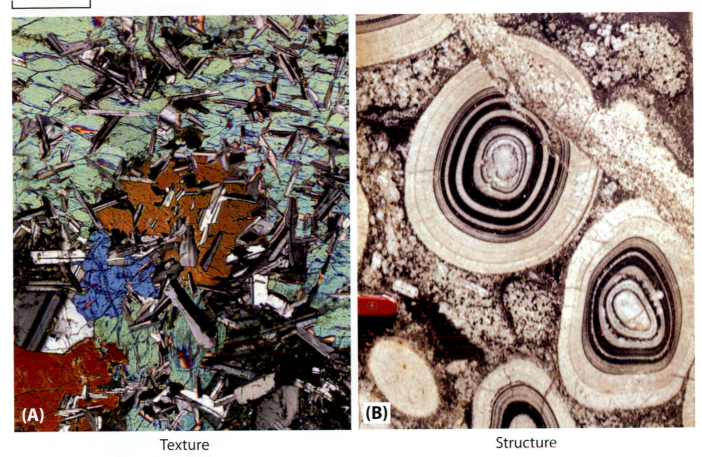

Texture   Structure

Figure 2.10 **(A)** *Texture* refers to the way individual mineral grains are intergrown. It is seen best in thin sections of the rock under a polarizing microscope, as seen here in this photomicrograph of a basaltic rock from the Bird River Sill, South Africa. This texture is described as ophitic, meaning that small plagioclase feldspar crystals (gray) are completely embedded in larger crystals of pyroxene (brightly colored grains). Photograph taken under crossed polarized light (see Chapter 6 for explanation of polarized light). **(B)** *Structure* refers to larger-scale features that are commonly evident at the hand specimen or outcrop scale. This sample of granite from Kangasala, Finland, exhibits an orbicular structure, where concentrations of light and dark minerals form concentric rhythmic layers around a central inclusion. Swiss army knife for scale.

Throughout Earth's history, the largest production of rocks has involved the partial melting of the mantle, followed by the buoyant rise of the melt into the crust, and the eventual solidification of the melt. Molten rock is known as **magma**, and rocks that form from the solidification of magma are known as **igneous rocks**. Most of the crust is made of rocks that were initially igneous. The formation of igneous rocks transfers heat from within the Earth to near the surface, where it can be more easily released into space.

Although igneous rocks are the most abundant rocks in the crust as a whole, most shallow crustal rocks are formed from the consolidation of sediment and are known as **sedimentary rocks**. Energy from the sun drives the hydrologic cycle, which causes weathering and transport of sediment. A sedimentary rock forms if this sediment finds a basin in which to accumulate. Most sediment accumulates in the sea, but some can be trapped in continental basins that develop in response to plate tectonics.

Tectonic plates continuously move across the surface of the Earth at rates of centimeters per year. When rocks, whether of igneous or sedimentary origin, find themselves in new environments where pressures and temperatures are different from where they formed, minerals change to equilibrate with their new environment, and a new rock is formed. Rocks that are formed in response to changes in environmental factors, such as pressure, temperature, or fluid composition, are described as **metamorphic rocks**. Most of the reactions involved in metamorphism consume heat and, therefore, help cool the Earth. Although we began by saying that most rocks in the crust are of igneous origin, many of these have subsequently been metamorphosed and, along with metamorphosed sedimentary rocks, constitute ~60% of the crust.

Distinguishing between igneous, sedimentary, and metamorphic rocks is usually simple at both the outcrop and microscopic scales. Most igneous rocks form relatively homogeneous massive bodies, which if crystallized in the crust may be relatively coarse grained. If the magma erupts onto the surface, the grain size is much finer, but the rock forms characteristic volcanic structures. Most sedimentary rocks are layered, with the layering generally paralleling the surface on which deposition occurs; this is usually close to horizontal, especially when the sediment is fine grained. Metamorphic rocks have textures that indicate recrystallization of minerals. If only heating is involved, the recrystallized grains form equidimensional polyhedra. If directed pressures are involved, as would occur, for example, at convergent plate boundaries, the metamorphic rocks commonly produce a prominent **foliation**, such as that found in **slate**, which is used for roofing tiles.

Under the microscope, the three types of rock are usually easily distinguished. Igneous rocks typically consist of minerals that show a sequence of crystallization. The earliest crystallizing minerals tend to be well formed (euhedral), and because they start growing first, they tend to be larger than other crystals in the rock. These early formed crystals are known as **phenocrysts**. They are surrounded by other later crystallizing minerals that form a **groundmass** (Fig. 2.11(A)). In contrast, the grains of many sedimentary rocks show evidence of transport and rounding through abrasion (compare Figs. 2.2(C) and 2.11(B)). These grains are the most resistant minerals to abrasion, with quartz being by far the most common. The sedimentary grains are commonly cemented together by quartz or a carbonate mineral. Transport also tends to separate grains on the basis of size and density. As a result, most sedimentary rocks are layered, with material in individual layers sharing similar transport properties. The largest volumes of metamorphic rock are formed near convergent plate boundaries, where subducting slabs are raised to such high pressures that they release fluids that rise into the overriding slab, where they promote metamorphism. Because these rocks develop on a regional scale, they are described as regional metamorphic rocks. This is in contrast to metamorphic rocks developed on a local scale near igneous intrusions where heat and fluids escaping from the cooling body produce contact metamorphic rocks. Regional metamorphic rocks formed near convergent plate boundaries undergo recrystallization in a directed stress field, which causes elongate and platy minerals to align perpendicular

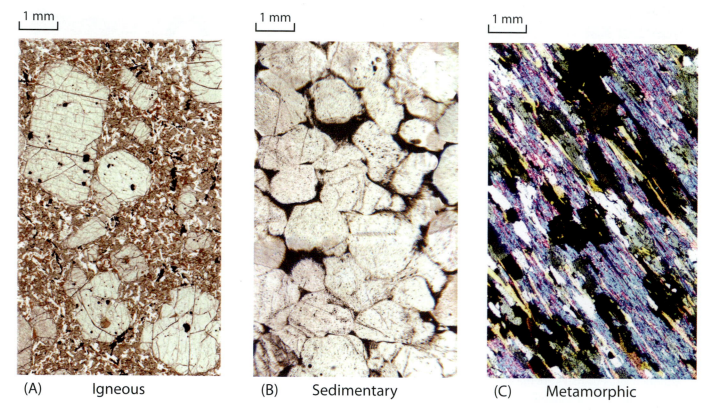

| 1 mm | 1 mm | 1 mm |

(A)   Igneous   (B)   Sedimentary   (C)   Metamorphic

Figure 2.11 Typical igneous, sedimentary, and metamorphic rocks as seen in photomicrographs of thin sections under low magnification. **(A)** Basalt from the 1959, Kilauea Iki, Hawaii, lava lake, containing phenocrysts of olivine (clear) in a groundmass of brown pyroxene and opaque ilmenite. The olivine phenocrysts have inclusions of a still earlier opaque chromium-rich spinel. Plane polarized light. **(B)** Photomicrograph of a thin section of sandstone showing subrounded grains of quartz surrounded by dark clay particles and quartz cement. Plane polarized light. **(C)** When Avalon and North America collided to form the Appalachian Mountains in southern New England, metamorphism of muddy sedimentary rocks resulted in the growth of mica crystals (bright blue and red) that were preferentially oriented perpendicular to the maximum compressive stress. The result was a strongly foliated rock that we call schist. Crossed polarized light.

1 cm

Figure 2.12 Sedimentary, igneous, and metamorphic rocks from Montreal, Quebec. The gray rock is Ordovician limestone, formed from the accumulation of calcium carbonate mud on the ancient continental shelf of North America. In the Cretaceous Period, this rock was intruded by hot mantle-derived basaltic magma, which solidified to form black igneous rock. Heat from the intrusion drove hydrocarbons out of the limestone at the contacts with the **igneous** rock to produce a white metamorphic rock, marble.

to the maximum compressive stress. Regionally metamorphosed rocks are, therefore, typically foliated (Fig. 2.11(C)), in contrast to contact metamorphic rocks, which typically exhibit no foliation whatsoever (Fig. 2.12).

All three major rock types are illustrated in the sample shown in Figure 2.12 from Montreal, Quebec. The main rock type in this sample is a sedimentary rock formed by the accumulation of calcium carbonate ($CaCO_3$) mud near the shore of North America during the Ordovician Period (450 Ma). Because it is composed of calcium carbonate, it is called a **limestone** (lime being CaO). It is dark gray because it contains hydrocarbons, which were formed from the remains of animal and plant matter that accumulated with the mud. During the Cretaceous Period (110 Ma), magma from the Earth's mantle intruded this gray limestone along numerous fractures to form small black **dikes**. As the magma cooled and crystallized to form the black basaltic igneous rock, the liberated heat expelled the hydrocar-

bons from the limestone, converting it to a white contact metamorphic rock called **marble**. The relation between the three main rock types in this sample leads to a clear relative chronology; that is, the limestone is the oldest rock because it was intruded by the basalt, and the metamorphism is clearly a consequence of the intrusion of the basaltic magma. Although not evident in the photograph, the limestone contains fossils that are indicative of the Ordovician Period, and the igneous rock has been dated as Cretaceous by absolute methods based on the radioactive decay of uranium to lead. In this way, we piece together Earth's history from the record preserved in rocks.

2.7 Examples of some familiar rocks

Most people are familiar with a number of common rocks and their uses. The most familiar rocks are those used for building materials, but rocks have many other important uses in

Figure 2.13 Rapakivi granite is one of the most common decorative building stones. It is shipped from quarries in southern Finland all over the world. Its characteristic texture is created by large rounded pink crystals of alkali feldspar rimmed by grayish-green plagioclase feldspar.

Figure 2.14 Basaltic lava erupting from Pu'u 'O'o in the east rift zone of Kilauea on the Big Island of Hawaii in March 2004. (**A**) Aerial view of the summit crater with steam rising from several small active vents. The crater rim was breached in the foreground and lava poured down the slope to produce the patchy light and dark colored rock. Once a crust had formed the lava continued to flow beneath the surface and extrude on the lower slopes of the volcano. (**B**) Molten lava that has traveled from the summit of Pu'u 'O'o, which is seen in the background, extrudes from beneath the bulbous crust of the flow. By traveling beneath a surface crust, lava retains its heat and is able to flow greater distances.

everyday life, such as supplying us with a variety of raw materials, including fossil fuels, or acting as reservoirs for water and oil. In this section, we consider just a few familiar rock types.

Perhaps the most familiar rock of all is **granite**. It has been used as a building stone since Egyptian times, and, because of its great durability, it has been popular for tombstones and other memorial structures. Even today when large buildings are made of steel and reinforced concrete, granite is still used as a decorative facing stone. One of the most common of these stones is **rapakivi granite**, which is shipped from southern Finland all over the world. It has a characteristic appearance with large round phenocrysts of pink alkali feldspar surrounded by narrow rims of greenish-gray plagioclase feldspar in a groundmass of pink alkali feldspar and quartz (Fig. 2.13). In the polished stone business, the word *granite* has come to mean any hard homogeneous igneous rock and includes rocks that are very different in composition from granite. In geological usage, granite refers to a coarse-grained igneous rock composed of approximately two-thirds alkali feldspar and one-third quartz. The continental crust has a composition very similar to granite. Most large bodies of granite have formed by melting of the lower crust.

Although granite may be the most familiar igneous rock, by far the most abundant rock on the surface of the planet is the volcanic rock known as **basalt**. Most of the ocean floors and large parts of continents are underlain by basaltic lava flows. Because the lava cools quickly under water and on land, the rock is fine grained, and because it contains ~50% iron-bearing minerals, it is a dark color (see Sec. 3.3), in contrast to the light color of granite, which contains only small amounts of iron-bearing minerals. We see basalt erupted from volcanoes such as those in Hawaii (Fig. 2.14), but the largest volumes of basalt, especially that on the ocean floor, have erupted from long

fissures rather than central vents. Although basalt may not be as familiar a rock as granite, the chances are that you come in contact with it far more than any other rock type. Worldwide, basalt is the preferred rock for aggregate to be mixed with asphalt for paving roads (see trap-rock quarry, Fig. 2.9). It is by far the most durable rock for this purpose and is always used as long as a source is not too distant to make transportation costs prohibitive. It is also a dense rock, and hence the aggregate is commonly used for railway beds and to stabilize freshly excavated slopes along, for example, highway cuts through surficial material that may be prone to slump.

A number of sedimentary rocks play important roles in modern life. Limestone is perhaps the most familiar sedimentary rock. Massive varieties have long been a popular building stone, because it is soft and easily quarried, yet is remarkably strong.

Figure 2.15 (A) Rainwater percolating through fractures in limestone causes solution and the development of karst topography. In this scene from the Cirque de Mourèze, Languedoc-Roussillon, France, pillars of dolomitic (CaMg(CO₃)₂) limestone have been left standing between the joints along which solution occurred. (B) Beneath the surface, solution of limestone forms caves. At times when the water becomes supersaturated with calcium carbonate, calcite is precipitated to form stalactites, where water drips from the roof, and stalagmites, where drips land on the floor of the cave. Large open channels, such as the one seen here in Harrison's Cave, Barbados, make many limestones important aquifers.

Unlike most other rocks, limestone slowly dissolves in rainwater – only rock salt is more soluble. When limestone is exposed to weathering, it rapidly develops solution features, which characterize what is known as **karst** topography (Fig. 2.15(A)). Beneath the surface, solution leads to the development of caves (Fig. 2.15(B)). Solution plays an important role in converting beds of limestone into important aquifers and reservoirs for oil. For example, the Cretaceous Edwards limestone in Texas was exposed to a period of weathering shortly after its deposition and developed extensive solution channels. The limestone was later covered by clay, which was formed from the weathering of volcanic ash. The clay was subsequently converted into an impervious rock known as **shale**. The shale now caps and confines the water in the Edwards limestone, making it one of the world's largest aquifers, supplying water to more than 2 million people in south-central Texas. Caves, with their stalactites and stalagmites (Fig. 2.15(B)), attest to how easily $CaCO_3$ can be moved around by solutions. In some regions where there is a source of heat, large convecting groundwater systems can be set up, and when the hot water comes out on the surface, it forms a geyser. If the circulating water passes through carbonate-bearing rocks, it is typically supersaturated in $CaCO_3$ when it pours out on the surface and cools. The result is precipitation of layers of calcite to form a rock known as **travertine**. Travertine is commonly

Figure 2.16 (A) Statue of a lion in the Piazza della Signoria, Florence, carved from travertine limestone, which is formed by deposition of $CaCO_3$ from hot springs. The layering seen in the lion is the result of the continuous buildup of successive layers of calcium carbonate. (B) Travertine tends to be highly porous and may precipitate around the roots of plants that later decay and disappear, leaving tubular holes as seen in this close-up of the pedestal beneath the lion's tail. Photographs in (A) and (B), courtesy of Laura Crossey, University of New Mexico. (C) Michelangelo's unfinished *Blockhead Slave* shows the partially completed sculpted body in a 9-foot-high block of Carrara marble. The statue is in the Galleria dell'Accademia, Florence, Italy.

Figure 2.17 Spider rock, in Canyon de Chelly (pronounced "Shay"), Arizona, consists of hard cross-bedded sandstone that was formed from sand dunes that drifted across a desert on the western side of the Pangaea supercontinent during the Permian.

of sedimentary rocks that were formed when the vegetation in coastal swamps and deltas was buried beneath layers of sand and pebbles to form a rock sequence consisting of coal, sandstone, and **conglomerate**, respectively (Fig. 2.18).

One of the most common metamorphic rocks is **slate**, which is a fine-grained rock that can be split into extremely thin but large sheets. It is composed largely of small crystals of mica, whose parallel alignment (see Fig. 2.11(C)) gives the rock its ability to cleave. It is also chemically resistant to weathering but is soft and can be easily cut into desired shapes. Before the relatively recent introduction of asphalt roofing tiles, slate was one of the most widely used materials for covering roofs. Huge quarries, such as those in Wales (Fig. 2.19), attest to the amount of material removed for this purpose. Perhaps the largest sheets of slate are used in the manufacture of high-quality pool (billiard) tables, where it is important to have an absolutely flat surface that will not warp. Blackboards in classrooms were also originally made from slate, but these are now usually made of green frosted glass. As the temperature of metamorphism increases, mica crystals grow larger and become visible in the hand specimen to form a rock known as **schist**. With further increases in temperature, mica reacts with other minerals and contributes to the formation of garnet (Sec. 2.1.1), which is one of the most common high-temperature metamorphic minerals. The appearance of garnet and other granular minerals at the expense of mica converts foliated schist into a hard layered rock known as **gneiss**, and if the temperature is high enough, the rock can actually begin to melt and form a mixed metamorphic-igneous

rather porous and, being made of the soft mineral calcite, is easily cut into blocks for building purposes or flooring tiles. It has also been used for statues, as seen in the example of the lion in Figure 2.16(A). Limestone is also the raw material from which cement is made. Cement, in turn, is mixed with sand and gravel in roughly 30:70 proportions to form concrete. Today, the construction industry uses, worldwide, a staggering 8 billion tons of concrete per year; that is, more than 1 ton of concrete for every person on the planet each year. This enormous production is accompanied by the release of carbon dioxide into the atmosphere, which is of concern because of its greenhouse effect on the climate. **Sandstone** is another familiar sedimentary rock. Most sandstone is composed of quartz (Fig. 2.11(B)), a mineral that is almost chemically inert. Sandstone is, therefore, a very resistant rock, commonly forming prominent cliffs (Fig. 2.17) and providing a source of building stone. When the grains of quartz in sandstone are not completely cemented, the rock can have a high porosity, which makes it a good aquifer or reservoir rock for oil. Another important sedimentary rock is **coal**, one of the main sources of fossil fuel. Coal beds occur in sequences

Figure 2.18 Coal bed in a syncline exposed in a road cut near, McAdoo, Pennsylvania. At the far end of the road cut, alternating green and red sandstone and shale are overlain by a thin bed of coal, which in turn is overlain by lighter colored sandstone that coarsens upward and, at the top of the road cut, contains small pebbles of quartz (a rock known as conglomerate). Nearby, thicker beds of coal have been mined for the much-prized slightly metamorphosed coal known as anthracite. The syncline is cut by at least two thrust faults.

Figure 2.19 One of the many quarries in northern Wales from which slate has been extracted for roofing tiles.

rock known as **migmatite**. Migmatites commonly exhibit interesting structures that make them popular in the stone industry for use as decorative kitchen countertops. For example, the countertop illustrated in Figure 2.20 is of a garnet-rich migmatite from Brazil. The wispy white patches are composed of granite that segregated as a liquid from the solid garnet-rich part of the rock. The patches pinch and swell as a result of deformation that occurred at the time of melting. In contrast, a later dike of granite that cuts the rock can be seen to be un-deformed. The quarrying of high-grade metamorphic and igneous rocks for the manufacture of countertops has become a big business, with Brazil, India, and China being some of the major producers. Another common metamorphic rock is marble, which is

Figure 2.20 Kitchen countertop made from high-grade metamorphic rock from Espirito Santo, Brazil. The rock contains abundant red garnet (see Fig. 2.5). It also contains irregular light-colored streaks of granite, which were formed when this rock reached a high enough temperature to begin melting. Much of the deformation that can be seen in this rock probably took place while it was partly molten. A straight, un-deformed granite dike cutting the metamorphic rock must have intruded at a later date. (Photograph courtesy of Tim Warzyniec and Amy Ellwein, University of New Mexico.)

formed from the metamorphism of limestone (Fig. 2.12). It is used both as a construction material and as a decorative facing stone. It also is the favorite material of sculptors, because it is soft enough to carve easily, yet has considerable strength. One of the most famous sources of marble is from Carrara, ~100 km west-northwest of Florence, Italy. This is where Michelangelo spent considerable time looking for just the right piece of marble for his next statue (Fig. 2.16(C)).

2.8 Plate tectonics and the generation of rocks

Plate tectonics, which is driven by the cooling of the planet, sets up the conditions necessary for the formation of igneous, sedimentary, and metamorphic rocks. Most rocks are formed at specific plate tectonic locations, which we can think of as **rock factories** – the places where rocks are made in response to plate tectonic-related processes. At each factory, suites of rocks are formed that are characteristic of the particular plate tectonic setting. In later chapters, we see specifically what these rock associations are, but in this section we outline where these rock factories are located with respect to plate tectonics and mention only briefly the rocks they produce.

Figure 2.21(A) shows a schematic cross-section through the outer part of the Earth at a **convergent plate boundary** where an oceanic lithospheric plate is being subducted beneath a continental plate. At the left side of the diagram, the **midocean ridge** is a **divergent plate boundary**. Although the figure is static, it is important to keep in mind that the plates are moving (yellow arrows) at rates of centimeters per year. This motion develops relief on the surface of the planet and disturbs the temperature and pressure distributions in the rocks beneath. These disturbances are the primary causes for rock formation; rocks form in an attempt to eliminate the perturbations caused by plate motion. The main plate tectonic settings where rocks are formed are listed across the top of the cross-section in larger print, and the actual rock types formed are given below in smaller print. Photographs of particular rock types associated with particular settings are keyed to the cross-section by bold letters (B–G).

2.8.1 Midocean-ridge rock factory

By far the largest rock factories on Earth occur along midocean ridges, where tectonic plates move apart (Fig. 2.21(B)). The divergence makes room for magma to rise along fractures to form steeply dipping sheetlike bodies of igneous rock known as dikes and to erupt on the ocean floor along the rift valleys located on the crest of the oceanic ridges. As much as 20 km³ of new rock are formed per year along the 65 000-km-long midocean ridge system. This is an order of magnitude more than is erupted subaerially. The igneous rock formed here is always of

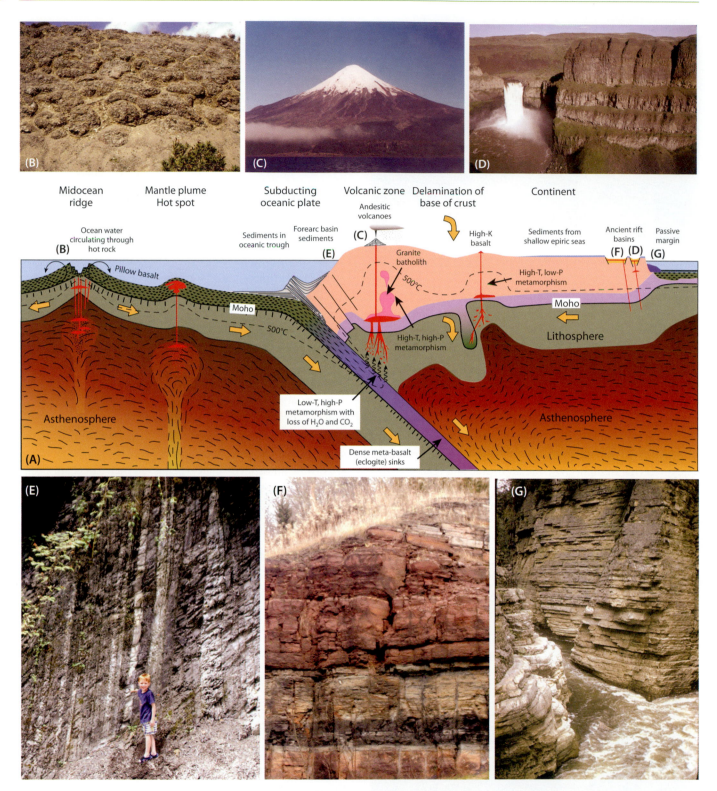

Figure 2.21 (**A**) Cross-section of the lithosphere extending from a midocean ridge (left) through a convergent plate boundary (center) to a passive continental margin (right). Letters in the cross-section (B–G) refer to the photographs of rocks formed at specific plate tectonic settings. (**B**) Pillows of basaltic lava erupted on the ocean floor in an outcrop on the island of Cyprus, where part of the Mediterranean ocean floor is exposed. (**C**) When a subducting plate reaches a depth of about 100 km, water is released that rises into the overlying wedge and causes melting. This magma rises to form conical andesitic volcanoes, such as Osorno in Chile. (**D**) Volcanic rocks associated with rift basins are commonly erupted from long fissures to form thick basaltic lava flows, known as flood basalts, as seen here at Palouse Falls, in eastern Washington. (**E**) Steeply dipping beds of Ordovician Martinsburg shale and sandstone near Shippensburg, Pennsylvania. These sedimentary rocks were tipped on end during the Appalachian orogeny, which also began to convert the shale into a metamorphic rock with a slaty cleavage dipping to the left in the photograph. (**F**) Red arkosic sandstone and black shale in the Mesozoic Hartford Rift Basin in Connecticut. (**G**) Potsdam Sandstone in Ausable Chasm, New York, was deposited as a quartz-rich sand on the Cambrian shore of North America.

basaltic composition, and because it is formed at a midocean ridge, it is commonly called a **midocean ridge basalt** (or MORB for short). This is the most abundant rock on the surface of the planet, covering most of the ocean floor. Because it erupts beneath water, its surface cools rapidly and commonly develops meter-sized blobs of lava known as **pillows** (Fig. 2.21(B)). Almost no sedimentary rocks are formed in this environment. Midocean ridges are typically far from any continental source of sediment. The sediment that does eventually accumulate on this newly formed ocean floor consists of silica-rich mud formed from the skeletons of **radiolaria**, minute pelagic organisms that build their bodies from silica. This radiolarian ooze accumulates slowly, so only as the ocean floor gets older away from the divergent plate boundary does it attain significant thickness.

Another important rock-forming process takes place in the vicinity of midocean ridges. The ridges stand high relative to older ocean floor because they are still hot and, thus, float higher than does older denser oceanic crust. Seawater circulates through the newly formed crust and helps cool it. As much as 34% of the total amount of heat lost through the ocean floor is estimated to be removed by this circulation, and the entire volume of water in the world's oceans circulates through oceanic ridges every 8 million years. This enormous circulation of hot water through the rock leads to **hydrothermal alteration**, with many of the original igneous minerals being converted to hydrous and carbonate minerals. These newly formed minerals play an all-important role in triggering rock-forming processes when the oceanic plate is subducted at a convergent plate boundary.

2.8.2 Convergent-plate-boundary rock factory

Although midocean ridges are the most productive rock factories on Earth, most rocks they produce are subducted back into the mantle at convergent plate boundaries (Fig. 2.21(A)). As slabs subduct, the increased pressure causes the hydrothermal alteration minerals that were formed in the basalt as it cooled at the midocean ridge to react and liberate water and carbon dioxide, which rise into the overlying mantle wedge. These volatile constituents, especially water, lower the melting points of rocks, and magma is formed. The magma rises toward the surface and forms a chain of conical-shaped volcanoes, which are built of lava and ash ejected during volcanic explosions. The main rock type is known as **andesite**, which contains a little more silica than does basalt but otherwise looks very similar (Fig. 2.21(C)). The higher silica content makes the magma more viscous than basaltic magma, which accounts for the steep slope on the volcanoes, but it also prevents gas escaping easily from the magma. These magmas typically contain a considerable amount of water, which was derived from the subducting oceanic plate, and when they approach the surface, exsolving gas can cause violent volcanic explosions, which

deposit layers of ash over large areas. Not all magma rises to the surface to form volcanoes. Some may pond near the base of the continental crust, where it causes melting of crustal rocks to produce granitic magma, which may solidify at depth to form large bodies known as **batholiths** (Fig. 2.21(A)). Granitic batholiths form the core of many mountain chains developed at convergent plate boundaries.

The relief created in the overriding plate at convergent boundaries, both through folding and thrusting of existing rocks and through the growth of new volcanoes, creates ideal conditions for the generation of sediment. Rivers transport the sediment rapidly to the ocean, with the finest sediment making it out into the deep waters of the **forearc basin** formed above the subducting slab. The rapid accumulation of coarser sediment near the shore leads to unstable conditions, with the sediment slumping and moving to deeper water in the form of **mudflows** and **turbidity currents**. These dense suspensions are able to transport coarser sediment great distances out into the forearc basin. Deposition from turbidity currents results in **graded layers** that vary from coarse at the base to fine at the top. These sediments are converted to shales and sandstones that become highly deformed as they are scraped off the subducting plate and accreted to the orogenic wedge. Figure 2.21(E) shows steeply dipping shales and sandstones of the Ordovician Martinsburg formation in Pennsylvania, which were deposited in a forearc basin that developed during the closing of the Iapetus Ocean.

2.8.3 Continental divergent-plate-boundary rock factory (rift valley)

Divergent plate boundaries can occur within continental plates, with formation of **rift valleys**. The East African Rift is a modern example of such a boundary. Figure 2.21(A) shows paleo-rift basins, such as those that developed in eastern North America and Morocco when Pangaea rifted apart to form the Atlantic Ocean. The relief along rift valleys results in rapid transport of sediment into the rift, and as these depositional basins are on continents, the oxygen in the atmosphere keeps iron in the sediment in its ferric state, which stains many of these rocks red. Some sediment containing organic matter, however, is black because it accumulates under anaerobic (no oxygen) conditions in deep lakes that develop along rift valleys (Fig. 2.21(F)). Because the transport distance from highland to basin is commonly short, the harder minerals, such as quartz, have insufficient time to abrade and eliminate the softer minerals such as feldspar. Consequently, the sandstones contain considerable amounts of feldspar and are known as feldspathic sandstone. As sedimentary basins form on the Earth's surface in response to rifting, the mantle beneath rises, decompresses, and partially melts to form basaltic magma, which erupts from long fissures paralleling the rifts. This develops outpourings of basaltic magma that form large flat lava flows known as **flood basalts** (Fig. 2.21(D)).

2.8.4 Mantle plume hot-spot rock factory

Other important igneous rock factories occur at **hot spots**, which form above what are believed to be **mantle plumes** that rise from possibly as deep as the core-mantle boundary (Fig. 2.21(A)). Hot spots remain relatively stationary on the planet while tectonic plates move independently above them. This independence might suggest that hot spots are not part of the plate tectonic cycle, but they undoubtedly play a major role in the convective transfer of heat through the mantle and crust. The best-known hot spot is located beneath the island of Hawaii (Fig. 2.14), with its trail of islands and seamounts bearing witness to the motion of the Pacific plate across the hot spot (Fig. 9.52). Basaltic rocks are the primary igneous rock formed above hot spots, but if the hot spot rises beneath a continental crust, as it does beneath Yellowstone National Park, for example, a wide range of igneous rocks can form, including granite (Fig. 9.24).

2.8.5 Passive-margin rock factories

When continents rift apart to form an ocean, **passive margins** form; that is, the edge of the continent simply abuts the new ocean floor that was formed along the rift. The relief along such margins is typically low, and the supply of sediment to the newly formed continental margin is slow. Wave action along the shoreline has ample time to abrade soft minerals and to sort the sediment on the basis of grain size. The result is a sequence of flat-lying or gently dipping sedimentary rocks that range from quartz-rich sandstone to shale (Fig. 2.21(G)). Limestone may also be part of the sequence if the sediments accumulate where the climate is warm enough to promote abundant growth of organisms with carbonate skeletons or shells. The Bahamas is a modern-day example of such an environment. As passive margins migrate away from spreading axes, the oceanic crust cools and subsides, making room for deposition of considerable thicknesses of sediment.

2.8.6 Epeiric-sea rock factories

The rate of tectonic plate motion is not constant and at times is significantly greater than at others, as indicated by variations in the width of magnetic anomalies on the ocean floor. For example, the Atlantic Ocean is opening at a couple of centimeters per year. In contrast, India split away from Madagascar at the end of the Cretaceous at ~10 cm/year toward Asia. During periods of rapid plate motion, midocean ridges become more active and more new hot ocean crust is created. As a result, ocean ridges become larger and displace ocean water onto the continents to form shallow **epeiric seas**. Deposition in these shallow seas can result in extensive beds of flat-lying sedimentary rocks that do not owe their present location to a specific plate tectonic setting but nonetheless are formed in response to plate tectonic processes. Much of the central part of North America is covered with sedimentary rocks formed in these shallow seas. Minor flooding of the continent can also occur as a result of the melt-back of continental ice sheets and to local tectonic conditions.

2.8.7 Metamorphic rock factories

In a static Earth, the temperature everywhere would increase at the same rate with depth, so that a given temperature would always be found at the same depth. The Earth, however, is dynamic and the temperature variation with depth changes dramatically with plate tectonic setting. For example, if we were to drill a deep hole in a stable continental region, we would find that a temperature of 500°C would be reached at a depth of ~20 km (based on a common geothermal gradient of 25°C/km). The dashed line in Figure 2.21(A), which represents the location of this temperature, is known as an **isotherm** (equal temperature). If we trace the isotherm across the figure, we see that its depth changes considerably. Beneath the midocean ridge, it is much shallower because of the presence of all the new hot igneous rock. Where the oceanic plate is subducted beneath the continental plate, the isotherm is transported to considerable depth. In the overlying plate, the production and rise of magma into the crust causes the isotherm to move to shallower depths. The isotherm can also be perturbed if parts of the lower continental crust delaminate and sink into the mantle. The hot mantle that rises to replace the delaminated slab causes isotherms to be raised in that region. The result is that the 500°C isotherm occurs at very different depths, and consequently pressures, depending on the plate tectonic setting. Changes in temperature, pressure, and fluid composition produce metamorphic rocks.

The main plate tectonic settings for the development of regional metamorphic rocks are shown in Figure 2.21(A). Ocean-floor rocks, on being subducted, experience a rapid increase in pressure with little increase in temperature. This produces metamorphic rocks that are characterized by minerals that are stable at low temperature and high pressure. One such mineral is the blue amphibole glaucophane, which makes many of these rocks blue. Water and carbon dioxide released from these metamorphic rocks rise into the overriding mantle wedge, where they help promote metamorphic reactions in the overlying rocks. The water is also essential in causing melting, and as magma rises into the overlying crust, it transports heat with it. This, in turn, brings about metamorphism at high temperatures and pressures. Above zones of delamination, the temperatures of rocks at relatively shallow depths (low pressure) can be raised to elevated temperatures to bring about metamorphism at high temperatures and low pressures. The rise of the mantle beneath rift valleys produces a slightly elevated temperature gradient, which can cause low-temperature metamorphism.

In later chapters, we see what specific rock types are developed in these various plate tectonic settings. In the meantime, it is important to keep in mind the role that plate tectonic processes play in the formation of rocks.

Summary

This chapter introduced the solid materials that constitute the Earth, that is, minerals and rocks. We distinguished between minerals and rocks and gave some common examples of both that are encountered on a daily basis. We discussed how minerals and rocks are classified, and how their origin is closely related to plate tectonic processes. The following are the main points discussed in this chapter:

- Minerals are the basic building blocks of the Earth. A mineral is a naturally occurring solid, with a definite (but commonly not fixed) chemical composition and an ordered atomic arrangement. All minerals have a definite crystalline structure; that is, they have an ordered three-dimensional arrangement of the constituent atoms, which imposes the chemical restrictions on their composition.

- Minerals are classified into a number of chemical groups, with the silicates being by far the largest. Despite the large number of minerals on Earth, only 85 occur as essential constituents of rocks. These are known as the rock-forming minerals.

- Because of the restricted chemical composition of minerals, several minerals must normally be present to account for the composition of the Earth at any given locality. This assemblage of minerals is what we refer to as a rock. The assemblage is normally restricted to consolidated masses to distinguish it from loose surficial materials. The solid Earth is made of rock.

- Rocks can form by crystallization of molten rock known as magma, in which case they are known as igneous. They can form through the cementing and consolidation of sediment, in which case they are known as sedimentary. Rocks that undergo mineralogical and textural changes in response to changes in environmental factors such as temperature, pressure, and fluid composition create new rocks known as metamorphic.

- Rocks are formed in response to plate tectonic processes. As a result, common associations of igneous, sedimentary, and metamorphic rocks develop that are characteristic of specific plate tectonic settings.

- The largest quantities of new rock are formed at midocean ridges, where divergent plates cause basaltic magma to rise from the mantle.

- The second-largest production of new rocks occurs at convergent plate boundaries, where andesitic volcanoes are the surface expression of intense igneous activity at depth, with granite batholiths being emplaced into the core of orogenic belts. Sediment eroded from these belts and transported into forearc basins is accreted onto continents as it is scraped off the subducting slab.

- Hot spots, rift basins, passive continental margins, and flooded parts of continents are other important settings for rock formation.

- All rocks, when exposed to changes in temperature, pressure, or fluid composition, can undergo mineralogical changes to produce metamorphic rocks. Changes in environmental factors are caused mainly by plate tectonic motion, but local changes can occur near igneous intrusions.

Review questions

1. What is a mineral, and why is it important to specify naturally occurring?

2. Why does table salt (NaCl) break into cubelike cleavage fragments?

3. What is the composition, shape, and electrical charge on the basic building unit of all silicate minerals?

4. What are some common uses for the mineral quartz (SiO_2)?

5. How does the arrangement of silica tetrahedra in the crystal structures of quartz and talc differ?

6. Although the cation-to-anion ratio in garnet is fixed, what allows it to have a wide compositional range, and why are such widely different compositions still classified as the same mineral?

7. Can minerals be synthesized in the laboratory?

8. On what basis are minerals classified into different chemical groups, and of these, which forms the largest group?

9. What is a crystal, and does the crystalline state require that crystal faces be present?

10. What is a rock, and how does it differ from a mineral?

11. What is the difference between a rock's texture and structure?

12. From a geological map of the area where you live, determine how many rock-forming periods your home area has experienced and approximately what fraction of geologic time these periods represent.

13. What single process has been operating continuously since the beginning of the Earth and has been responsible for the formation of most rocks and drives plate tectonics?

14. Give brief definitions of igneous, sedimentary, and metamorphic rocks, and describe how you might distinguish between them.

15. What general distinction can you make between the igneous rocks of the ocean floor and those of the continental crust?

16. What sedimentary rock do we make the greatest use of in the modern world?

17. What feature characterizes most regionally metamorphosed rocks?

18. Where does the largest production of new rock take place on Earth?

19. Water plays what important roles when circulating through fractures in newly formed oceanic crust?

20. What common volcanic rock forms above subducting tectonic plates, and what role does water play in its origin?

21. Why are granite batholiths common above subducting tectonic plates?

22. What is the source of sediment in forearc basins and what is a common characteristic of these sedimentary beds?

23. Why are the sedimentary rocks of rift valleys commonly red colored and still contain a considerable amount of the relatively soft mineral feldspar?

24. From where do mantle plumes derive their heat?

25. How are shallow epeiric seas that flood continents related to plate tectonics?

26. What plate tectonic processes deflect isotherms up or down in the Earth?

27. Trace the role played by water and carbon dioxide in metamorphic rocks in going from their formation near midocean ridges to the production of granite by partial melting of continental crust above subducting plates.

ONLINE RESOURCES

Useful tips on identifying rocks and minerals can be found on the Mineralogical Society of America's Web site at http://www.minsocam.org.

FURTHER READING

In this book, we give only an introduction to the most important topics in mineralogy and petrology. Numerous books give in-depth coverage of these subjects. The following books provide the reader with extensive coverage of these subjects and lead to the relevant literature.

Mineralogy

Dyar, M. D., Gunter, M. E., and Tasa, D. (2008). *Mineralogy and Optical Mineralogy*, Mineralogical Society of America, Chantilly, VA.

Klein, C., and Dutrow, B. (2008). *Manual of Mineral Science*, 23rd ed., John Wiley and Sons, New York.

Perkins, D. (2010). *Mineralogy*, 3rd ed., Prentice Hall, Upper Saddle River, NY.

Wenk, H. R., and Bulakh, A. (2004). *Minerals: Their Constitution and Origin*, Cambridge University Press, Cambridge.

Petrology

Blatt, H., Tracy, R., and Owens, B. (2006). *Petrology: Igneous, Sedimentary, Metamorphic*, 3rd ed., W. H. Freeman and Company, New York. This book gives a general coverage of igneous, sedimentary, and metamorphic rocks.

Boggs, S., Jr. (2009). *Petrology of Sedimentary Rocks*, 2nd ed., Cambridge University Press, Cambridge. This book gives an in-depth coverage of sedimentary rocks.

Davis, E., and Elderfield, H., eds. (2004). *Hydrogeology of the Oceanic Lithosphere*, Cambridge University Press,

Cambridge. This book provides the reader with a selection of topics dealing with water in the oceanic crust.

Neuendorf, K. K. E., Mehl, J. P., Jr., and Jackson, J. A., eds. (2005). *Glossary of Geology*, 5th ed., American Geological Institute, Alexandria, VA. A valuable resource for the definition of geological terms.

Philpotts, A. R., and Ague, J. J. 2009. *Principles of Igneous and Metamorphic Petrology*, 2nd ed., Cambridge University Press, Cambridge. A detailed and quantitative treatment of igneous and metamorphic rocks.

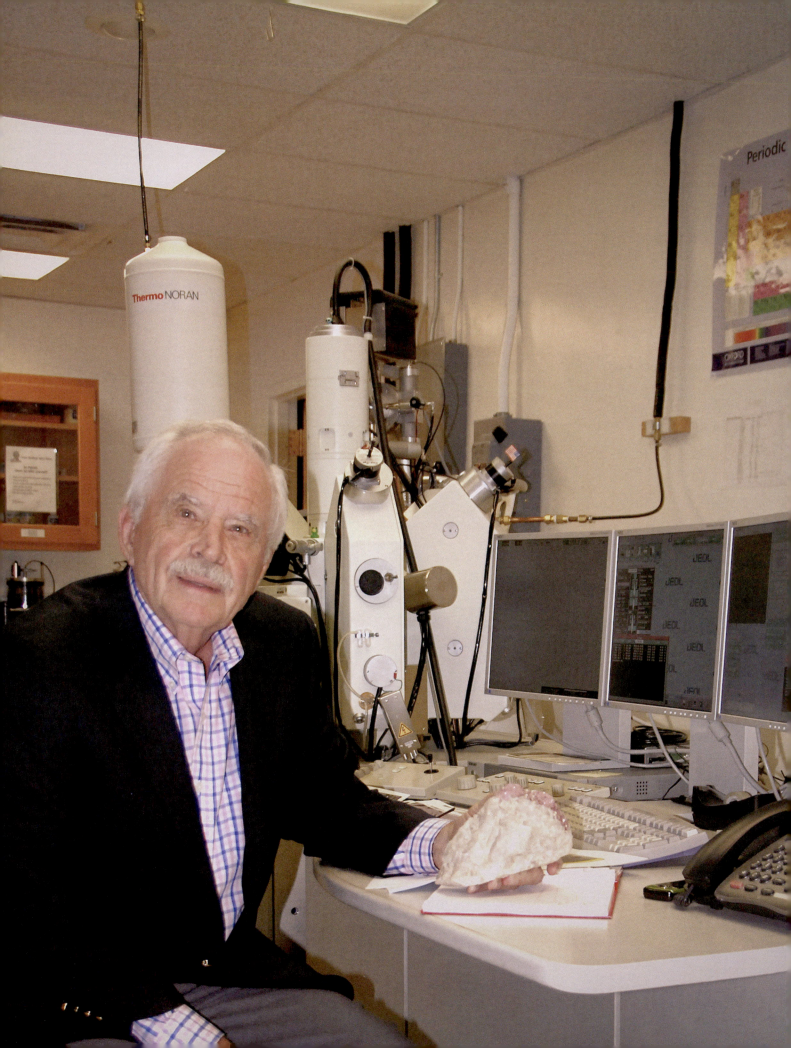

3

How are minerals identified?

The ability to recognize common minerals in hand specimens is basic to much that follows in this text. The tests that you will become familiar with are exactly the same as those used worldwide by field and research geologists, be it in the field, in the research laboratory, or in the home office. Your course instructor may discuss the various techniques for the unique identification of a mineral (or several minerals as in a rock) in lectures, but personal expertise can be gained best by hands-on work that you will likely do in the laboratory that accompanies your course. There you will have small test samples that you will be allowed to scratch (to test hardness), to hold in your hand ("to heft" the specimen so as to assess its average specific gravity), and to evaluate their reaction to dilute HCl, and so on. Even better is that you will probably also see much larger, better mineral hand specimens in which you can observe other properties such as crystal form, habit, cleavage, range of color, and state of aggregation. You will learn to combine these observations and develop the skills to identify unknown minerals correctly.

Once you know what mineral you are dealing with, or which several minerals, as most common rocks contain a mix of minerals, you will be well prepared for understanding how rocks are classified and the conditions under which different rock types (igneous, sedimentary, and metamorphic) are formed. For example, once you have identified all four major minerals in a specific, relatively coarse-grained, light-colored rock as (1) quartz, (2) two types of feldspar, and (3) a mica, you can conclude that you are dealing with a granite. Such knowledge of common rock-forming minerals is basic to much of what is presented in this text. The four later chapters that deal with the systematic mineralogy of igneous rocks (Chapter 7); sedimentary rocks (Chapter 10); metamorphic rocks (Chapter 13); and ore deposits, coarse-grained pegmatites, and quartz veins (Chapter 15) give detailed information on the diagnostic properties that allow us to identify minerals.

We introduce you to various observations and tests that can be made of minerals in hand specimens. In many instances a combination of the results of several of these leads to the

Cornelis Klein with a hand specimen of white quartz edged by pink lepidolite (a lithium-containing mica) from the Harding Mine, near Taos, New Mexico. He is seated in front of an electron microprobe analysis instrument (see Box 3.1 and Sec. 3.8) manufactured by JEOL Ltd., Tokyo, that is located in the Department of Earth and Planetary Sciences and Institute of Meteoritics, University of New Mexico, Albuquerque, New Mexico.

identification of the mineral at hand. If not, the use of additional instrumental techniques may result in a unique identification. The subjects discussed in this chapter are the following:

Habit
State of aggregation ⎬ visual observations
Color and luster

Play of color
Chatoyancy, labradorescence, asterism ⎬ visual observations
Fluorescence   using an ultraviolet light source
Streak   using a streak plate
Cleavage   breaking along crystallographic directions; visual observation
Hardness
Specific gravity ⎬ using some test equipment
Magnetism, solubility in HCl, radioactivity

Instrumental techniques

X - ray powder diffraction
Electron beam techniques : SEM, EMPA, and TEM

Identification of minerals in hand specimens is the main emphasis of this chapter. The methods used are careful visual evaluation and basic tests with easily available tools. Only at the end of this chapter do we briefly introduce some sophisticated instrumental methods that are used in the quantitative characterization of minerals and other crystalline solids. Chapter 6 is devoted to the study of minerals with a polarizing optical microscope.

The identification of an unknown mineral in a hand specimen begins with making observations that allow us to assess a specimen's overall form (or crystal habit if it is well crystallized), state of aggregation, and color. Those properties that allow us to identify a mineral or at least narrow down the possibilities are said to be **diagnostic**. Color is probably the first property the observer sees, followed by the overall shape of the mineral. But, though instantly noted, color is not a reliable diagnostic property in most minerals, because many (chemically variable) mineral groups exhibit a range of colors.

Important **physical properties** that characterize a mineral and allow us to separate one from another in hand specimens are the following:

- **Habit**
- **State of aggregation**
- **Color**
- **Luster**
- **Cleavage**
- **Hardness**
- **Specific gravity (or relative density)**

Each of these properties is discussed in this chapter, but it must be recognized that the actual process of mineral identification is best learned in the laboratory part of the course you are enrolled in. There you will tune your observational skills through the study of labeled mineral specimens as well as unknowns. Here, we first introduce those properties that can be evaluated by observation only – habit, state of aggregation, color, and cleavage – and subsequently discuss properties such as hardness, specific gravity, magnetism, radioactivity, and solubility in hydrochloric acid, all of which require testing tools.

3.1 Habit

If a mineral specimen is well crystallized – meaning that it shows well-developed crystal faces – then the crystal form can be used to help identify the mineral, and we can say that the crystal form is diagnostic in our mineral identification. The external crystal form is an outward expression of the ordered internal atomic arrangement. We look at crystal forms and their inherent symmetry in detail in Chapter 5, but here we concentrate on properties that can be assessed without specific knowledge of crystallography.

Various habits are described by different adjectives, and several of these are described here and illustrated with photographs in Figure 3.1.

Prismatic means that the mineral has an elongate habit with the bounding faces forming a prismlike shape, as is common in members of the pyroxene and amphibole groups of silicates.

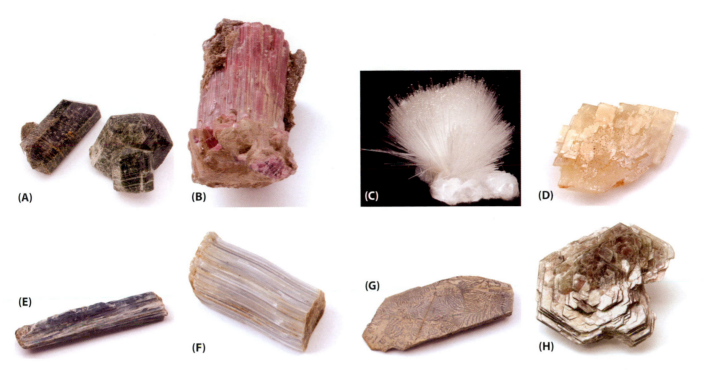

Figure 3.1 Photographic illustrations of many of the mineral habits listed in Section 3.1. (**A**) Prismatic, as in diopside crystals. (**B**) Columnar, as in a pink tourmaline crystal. (**C**) Acicular (or needlelike), as in this tufted aggregate of mesolite fibers (mesolite is a member of the zeolite group of minerals). (**D**) A tabular crystal of barite. (**E**) A bladed crystal of kyanite. (**F**) Fibrous celestite. (**G**) A dendritic pattern of manganese oxide minerals on a flat surface of siltstone. (**H**) Foliated muscovite as a rosette of crystals.

Columnar exhibits rounded columns, as is common in tourmaline.

Acicular means "needlelike," as is common for natrolite, a member of the zeolite group. The word is derived from the Latin word *acicula*, meaning "needle."

Tabular describes crystal masses that are flat like a board, as commonly seen in barite.

Bladed refers to crystal shapes that are elongate and flat, as in a knife blade. Kyanite shows this commonly.

Capillary applies to minerals that form hairlike, or threadlike, thin crystals, as shown by millerite.

Fibrous refers to threadlike masses, as exhibited by chrysotile, the most common mineral included in the commercial term *asbestos* (see Fig. 2. 4A).

Dendritic describes minerals that show a treelike branching pattern, as is common in manganese oxide minerals. The term is derived from the Greek word *dendron*, meaning "tree."

Foliated refers to a stack of thin leaves or plates that can be separated from each other, as in mica and graphite.

Massive describes a mineral specimen that is totally devoid of crystal faces.

3.2 State of aggregation

Most mineral specimens, unless unusually well crystallized, appear as aggregates of smaller grains. Such occurrences are best described by some additional adjectives (photographs of the appearance of several of these are given in Fig. 3.2).

Granular applies to rock and mineral specimens that consist of mineral grains of approximately equal dimensions, as in the rock **dunite**, composed essentially of granular olivine grains.

Compact describes a specimen that is so fine-grained that the state of aggregation is not obvious, as in specimens of clay minerals, or chert (flint).

Banded is said of a mineral specimen that shows bands of different color or texture but that may or may not differ in mineral composition. A banded **agate** may show various differently colored bands, but each of the bands is composed of the same silica, SiO_2, known as chalcedony. In contrast, banded iron-formations commonly show banding, on a millimeter scale, of two or three different minerals, chert (light colored) and hematite (red) and/or magnetite (black).

Mammillary is from the Latin word *mamma*, meaning "breast," and describes minerals that occur as smoothly rounded masses resembling breasts, or portions of spheres. Examples are goethite and hematite.

Botryoidal is from the Greek word *botrys*, meaning "band or cluster of grapes" or "having a surface of spherical shapes." In a botryoidal appearance, the rounded prominences are generally of a smaller scale than those described as mammillary. Common as the outer surface of chalcedony, a microcrystalline variety of quartz, SiO_2.

Figure 3.2 Photographic illustrations of many of the states of aggregation listed in Section 3.2. (**A**) Granular grains of yellow-green olivine in a volcanic bomb of dunite (a rock type made up essentially of olivine). (**B**) Finely banded as in this polished slab of agate. (**C**) Botryoidal, as shown by the outer, almost-black surface of chalcedony. (**D**) Botryoidal goethite. (**E**) Reniform hematite. (**F**) A geode with an outer lining of slightly blue chalcedony and a cavity lined with milky white, translucent quartz crystals. (**G**) Oolitic and pisolitic limestones. (**H**) Pisolitic bauxite.

Reniform describes the surface of a mineral aggregate that resembles that of a kidney. Seen in hematite. Derived from the Latin wood *renis*, meaning "kidney."

Stalactitic is a term used for a mineral that is made up of forms like small stalactites. Some limonite occurs this way, as does rhodochrosite.

Geode is a rock cavity partially filled with minerals. Common mineral fill is well-crystallized quartz or the purple variety of quartz, amethyst. Agates may be partly or completely filled rock cavities, and they may show attractive color banding in the chalcedony, microcrystalline SiO_2, that lines the outer cavity.

Oolitic describes the occurrence of mineral grains in rounded masses the size of fish roe. Derived from the Greek word *oön*, meaning "egg." This structure may occur in iron ore, made up mainly of hematite, and known as oolitic iron ore, or in sedimentary rocks known as oolitic limestone, which consists of millimeter-size spheres of rhythmically precipitated calcite.

Pisolitic applies to rounded mineral grains the size of a pea. From the Greek word *pisos*, meaning "pea." Bauxite, a major ore of aluminum, is commonly pisolitic. Pisolitic grains are larger than those described as oolitic.

3.3 Color and luster

Of all the physical properties that a mineral possesses, color is the most easily observed. For some minerals, such as those shown in Figure 3.3, color is highly diagnostic. However, in most minerals color can be variable and as such proves an unreliable diagnostic property.

Color is the response of the eye to the visible light range of the electromagnetic spectrum (Box 3.1). Visible light ranges in wavelength from about 400 to 750 nanometers (where 1 nm = 10 Ångstroms, Å). **White light** is a mixture of all these wavelengths. The following equation relates the energy of radiation (E) to its wavelength (λ):

$$E = \frac{hc}{\lambda}, \tag{3.1}$$

where c is the speed of light and h is the Planck constant, which has a value of 4.135×10^{-15} electron volts × seconds (in SI units: 1.054×10^{-34} joule × seconds). What this equation shows is that when the wavelength becomes longer, the energy is reduced, and when the wavelength becomes shorter, the energy increases.

When white light strikes a mineral, it may be scattered, or reflected; it may be refracted; and it may be transmitted and absorbed (Fig. 3.4). If almost all of the light is reflected and/or scattered, the mineral will have a luster described as **metallic luster**. This is exhibited by the surface of steel, copper, silver, and gold. Such materials are described as **opaque** because little light passes into the mineral. Metallic luster is also shown by many oxides and sulfides (if not tarnished). For example, pyrite (FeS_2), galena (PbS), and hematite (Fe_2O_3) are all opaque to light and have a metallic luster (Fig. 3.5).

Figure 3.3 Examples of a few minerals that can be uniquely identified by their color. From left in a clockwise direction: turquoise, turquoise in color; malachite, green; rhodochrosite, pink; lazurite, blue (known in the gem trade as lapis lazuli); and sulfur, yellow.

When most of the light that impinges on a mineral is transmitted, as in most light-colored **translucent** minerals (meaning that it is capable of transmitting light but is not transparent), the luster is described as **nonmetallic luster**. There is no sharp division between metallic and nonmetallic, but after some experience in the laboratory with metallic and nonmetallic specimens, the eye evaluates these distinctions quickly. Nonmetallic luster is a distinguishing property of silicates and carbonates. The color of most minerals with a metallic luster varies little, but those with a nonmetallic luster may show a wide range.

The luster of nonmetallic minerals is commonly described by one of the following two terms: **vitreous**, from the Latin word *vitrum*, for "glass," meaning with the luster of glass, and **resinous**, with the luster of resin.

If the incoming light to a transparent, nonmetallic mineral suffers no absorption, the mineral will be colorless. If, however,

Figure 3.4 The various processes that occur when light passes through a piece of a transparent substance (adapted from Nassau, 1980).

Figure 3.5 Metallic luster of hematite, Fe_2O_3. This type of hematite is commonly referred to as specularite.

BOX 3.1 | THE ELECTROMAGNETIC SPECTRUM

Visible light is only a small part of this spectrum (see illustration) and ranges from short wavelengths of 400 nm at the violet end to long wavelengths of 750 nm at the red end. This short spectral range has produced an enormous amount of scientific information about Earth materials, and the Earth itself, through the unaided eye, as well as with hand lenses, various types of optical microscopes, and many other optical instruments.

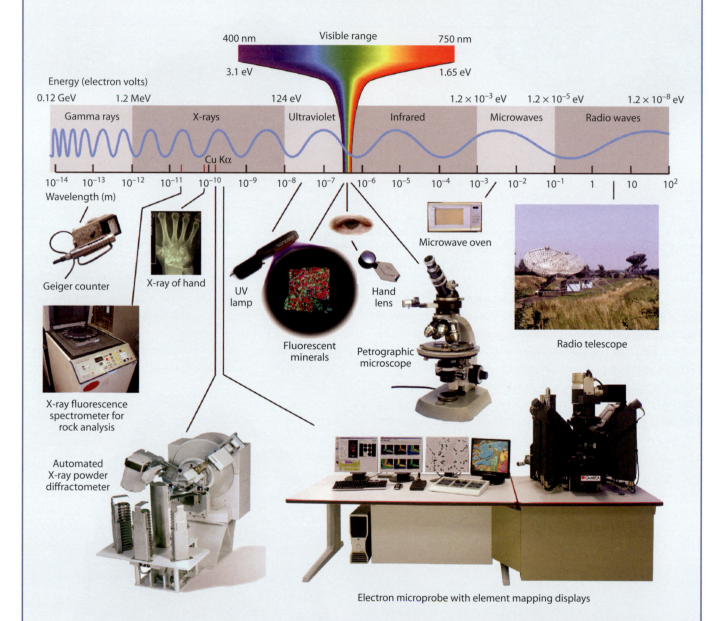

The overall spectrum covers a wide range of energies and wavelengths, stretching from gamma rays that are given off during radioactive decay of minerals (with energies of millions of electron volts (MeV) and wavelengths of 10^{-13} m), to radio waves detected by radio telescopes (with energies of only 10^{-7} eV and wavelengths of meters).

X-rays occur at wavelengths much shorter than the visible spectrum. These are used in medical applications as well as in X-ray diffraction studies of the crystalline state. An example of this is an X-ray powder diffractometer (Sec. 3.8). Other sources of still-higher-energy X-rays are used to bombard samples of unknown composition so as to cause their atoms to give off X-rays that are characteristic of the constituent atoms. This is the technique used in X-ray fluorescence spectrometers for the chemical analysis of powdered rock samples.

Atoms can also be caused to give off characteristic X-radiation by being bombarded with high-energy electrons. The electron microprobe analyzer (EMPA, Sec. 3.8) uses a focused beam of electrons to analyze very small regions (on the order of a few microns in size) in mineral grains, or very fine-grained minerals, under high magnification.

Electromagnetic radiation that is slightly shorter than the visible range falls in the ultraviolet range. This is the radiation that causes your skin to burn, but it can also be used to make some minerals give of fluorescent radiation in the visible range. Radiation with wavelengths longer than red light passes into the infrared, which we commonly refer to as heat waves. At still longer wavelengths are the microwaves, which we make use of in microwave ovens and in telecommunications. Once wavelengths go beyond 10 cm, they enter the radio range, which is used primarily for communication. (The photograph of the X-ray powder diffractometer is courtesy of PANalytical, Chester, NY, and that of the electron microprobe is courtesy of Cameca SAS, Gennevilliers, France).

parts of the visible spectrum are absorbed, the mineral will be colored. The color of a mineral is the eye's perception of the wavelengths of light that are either transmitted or reflected (by the mineral) before reaching the eye. When part of the visible light spectrum is absorbed, during its path through a mineral, the part that is not absorbed is responsible for the color perceived by the eye. The absorption of specific wavelengths of visible light inside a transparent, or translucent, mineral results from the interaction of the incoming electromagnetic radiation with specific chemical elements housed in the crystal structure. Most of the reactions that cause light absorption are the result of wavelengths of visible light inside a transparent, or translucent, mineral and the interaction of that incoming electromagnetic radiation with specific chemical elements housed in the crystal structure.[1] For example, the mineral corundum, Al_2O_3, is normally colorless, but when it contains trace amounts of Cr^{3+}, it becomes red, known as ruby, and when it contains trace amounts of Fe^{2+} and Ti^{4+}, it is blue and is called sapphire. The blue color of fluorite, CaF_2, is caused by the interaction of electromagnetic radiation with defects in its structure.

Several major rock-forming mineral groups show specific colors because they have in their structures large amounts of certain chemical elements known as the **transition elements**. Transition elements have atomic numbers between 21 and 30 (see the periodic table on the inside cover of this book) and include Sc, Ti, V, Cr, Mn, Fe, Co, Ni, Cu, and Zn. These are also referred to as the **chromophore elements**, with *chromophore* derived from two Greek words meaning "color causing." Some of these transition elements (especially Cr, Mn, and Fe) are major components (not trace elements) in, for example, garnet, olivine, pyroxenes, and amphiboles.

[1] A trace element is a chemical element that is not essential in a mineral but is found in small quantities in its structure. Although not quantitatively defined, it generally constitutes <0.1 weight % of the mineral's chemical composition.

Examples include the following:

Almandine garnet – $Fe_3Al_2(SiO_4)_3$: red
Uvarovite garnet – $Ca_3Cr_2(SiO_4)_3$: bright green
Spessartine garnet – $Mn_3Al_2(SiO_4)_3$: orange, and
Olivine – $(Mg, Fe)_2SiO_4$: yellow-green to brownish green, as a function of increasing Fe^{2+} content.

3.3.1 Reasons for color

The color of minerals results mainly from the interaction of visible light (electromagnetic radiation) with specific atoms or atomic sites in a crystal structure. The most important of these interactions include the following:

- **Crystal field transitions**
- **Molecular orbital transitions**
- **Color centers**

Crystal field transitions are interactions between the energy of white light and *d* orbitals of certain elements (chromophores) that are only partially filled with electrons. When these empty, *d* orbitals are surrounded by other atoms in a crystal; the empty orbitals have a higher or lower energy than the filled ones, depending on their coordination (the geometric shape defined by the surrounding atoms; such as octahedral, see Sec. 4.3). These energy differences correspond to wavelengths of photons in visible light. If the radiant energy of the visible light shining on the mineral is high enough, it will raise an electron from a lower-energy *d* state to a higher-energy *d* state, and some of the light will be absorbed in this process.

Let us briefly illustrate the absorption of light in olivine $(Mg,Fe)_2SiO_4$. If white light shines on a transparent member of the olivine series (a green gem, peridot, is part of this series, in which about 10% of the total (Mg + Fe) is Fe), most of the absorbance takes place in the infrared region, with some occurring in the visible range as well. The absorbance of the red

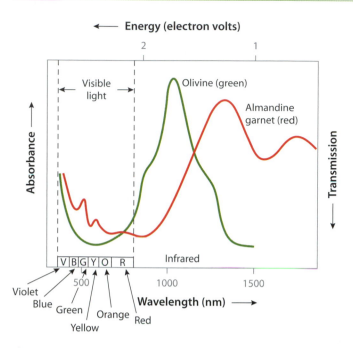

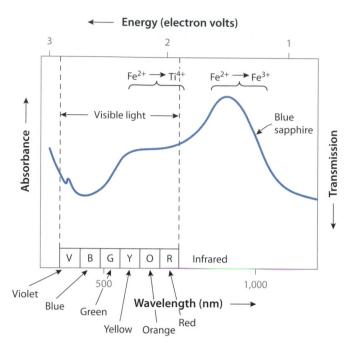

Figure 3.6 The visible and infrared absorption spectra of two Fe^{2+}-containing minerals, red almandine garnet and green olivine. Almandine contains as much as 35 weight % FeO. Olivine has a variable range of FeO, but most common olivines contain at least 10 weight % FeO. Major absorption peaks for both minerals appear in the infrared region. In the region of the visible spectrum, almandine has the least absorption in the orange to red region, which results in its red color. Olivine has the least absorption in the yellow to green region, resulting in the yellow-green color of the gem mineral peridot. (From Loeffler and Burns, 1976; reproduced with permission of *American Scientist*.)

Figure 3.7 The visible and infrared absorption spectrum of sapphire, Al_2O_3, due to the molecular orbital transitions between the trace elements Fe and Ti housed in its structure. The least absorption in this spectrum is in the blue range of the visible region causing the blue color of gem sapphire. (From Loeffler and Burns, 1976; reproduced with permission of *American Scientist*.)

component of visible light results in the mainly green color of peridot, as well as other olivines of similar composition (Fig. 3.6). In garnet of almandine composition, $Fe_3Al_2(SiO_4)_3$, the light absorption that is also caused by Fe^{2+} in a specific site in the structure occurs still farther into the infrared (Fig. 3.6), but it also occurs toward the blue end of the visible range, which results in the mineral's deep red color.

Molecular orbital transitions result from the transfer (hopping back and forth) of electrons between adjacent cations (that have variable charges) in a crystal structure. Such electrons are not in orbitals that center on specific constituent atoms; instead, they are delocalized. The blue color of sapphire, which has the composition of corundum, Al_2O_3, with only trace amounts of Fe^{2+} and Ti^{4+}, is caused by electrons that pass back and forth between the Fe^{2+} and Ti^{4+} that are located in adjacent sites normally occupied by Al^{3+}. In one instant, the electron charge is distributed as Fe^{2+} and Ti^{4+}, and in the next instant as Fe^{3+} and Ti^{3+}. The part of the visible light spectrum that is absorbed as a result of this charge transfer results in the blue color in sapphire (Fig. 3.7). The blue colors of kyanite, glaucophane (a member of the amphibole group), and crocidolite (a highly fibrous member of the amphibole group that is one of the six minerals grouped as "asbestos") are all due to these molecular orbital transitions.

Color centers refer to defects in mineral structures, such as ionic omissions (known as vacancies) that may become filled with an excess electron to balance the charge of the missing ion. For example, in fluorite, CaF_2, there may be some vacancies due to the absence of F^- anions. If these vacancies are filled with excess electrons, the presence of such electrons may cause the absorption of part of the spectrum, which in the case of fluorite results in a blue color. For a complete discussion of these topics, see "Further Reading" at the end of this chapter.

The foregoing explains the main reasons for a range of color in many mineral groups. The following three segments evaluate color as seen in tests (streak test and fluorescence as a result of ultraviolet irradiation) and in the unique play of color in gem opal.

Play of color

Play of color is best shown by precious opal (Fig. 3.8). Opal has the composition $SiO_2 \cdot nH_2O$. In electron microscope studies published in 1976 (Darragh, Gaskin, and Sanders, 1976), it was finally discovered what causes the play of color for which precious opal is so highly prized. Precious opal consists of amorphous silica, SiO_2, in spheres of about 3000 Å diameter (Fig. 3.9), with small amounts of water. These spheres, which are cemented by amorphous silica with a slightly different water content, are stacked together in a regularly packed array. The interaction of visible light with such a regular array of spheres causes diffraction effects that result in a play of color (Fig. 3.10).

Figure 3.8 A close-up of a polished surface of precious opal.

10,000 Å

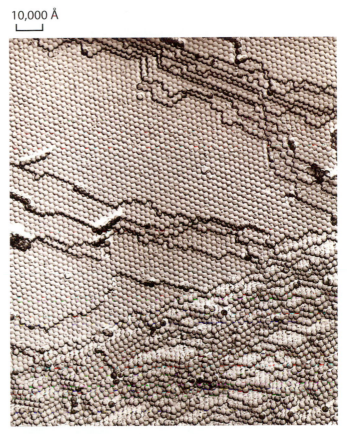

Figure 3.9 A scanning electron microscope image of opal with a chalky appearance. The amorphous spheres of $SiO_2 \cdot nH_2O$ composition are ~2500 Å in diameter. (Photograph courtesy of A. J. Gaskin, CSIRO, Perth, Australia; see Darragh et al., 1976.)

Chatoyancy, labradorescence, and asterism

The three optical effects of chatoyancy, labradorescence, and asterism are commonly quite subtle, but mineral specimens that show these effects to great perfection are much sought after in the gem trade. So as to exhibit these effects to their fullest, such gems are cut en cabochon. This means that they are polished with a convex top and a flat base but without facets.

These optical effects are the result of some very fine-grained physical discontinuities or symmetrically oriented inclusions inside the mineral specimen. Some minerals, in reflected light, show a silky or milky sheen that is the result of closely spaced fibers, inclusions, or cavities. When the specimen (or polished stone) is tilted, a narrow beam of light moves from side to side as a result of the tilting. This is known as **chatoyancy** and is best seen in the satin spar variety of gypsum, $CaSO_4 \cdot 2H_2O$.

A common mineral series known as the plagioclase feldspars, ranging in composition from $NaAlSi_3O_8$ to $CaAl_2Si_2O_8$, shows an internal iridescence in part of its compositional range. It is most pronounced in that part of the chemical series known as labradorite (Fig. 3.11(A)). Here the iridescence is the result

of the scattering of light from a microstructure that consists of a regular pattern of closely spaced, parallel planar lamellae (known as **exsolution lamellae**), which have slightly different chemical compositions in adjoining lamellae. The scattered light "diffracts" from these microstructures, producing colors

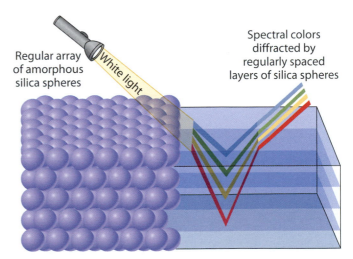

Regular array of amorphous silica spheres

White light

Spectral colors diffracted by regularly spaced layers of silica spheres

Figure 3.10 The diffraction of visible light, into various colors, due to regularly spaced planes, caused by the regular stacking of amorphous spheres in precious opal.

ranging from blue to green, yellow, and red. This is known as **labradorescence** and is a highly diagnostic property in the identification of feldspar specimens that are part of a specific compositional range of the plagioclase series.

Asterism is a six-rayed optical phenomenon that is the result of light reflected from minute inclusions arranged in a star-like (six-rayed) pattern. It is best seen in star rubies and star sapphires that have been cut en cabochon. The inclusions in these gems consist of very fine needles of rutile, TiO_2, that are arranged parallel to the sixfold symmetry inherent in the crystal structure of corundum, Al_2O_3, of which ruby and sapphire are the colored gem varieties. Rose quartz, SiO_2, may show this as well because of tiny included needles (consisting of dumortierite) arranged at 120° to each other, as a result of the hexagonal symmetry of the quartz structure (Fig. 3.11(B)).

Fluorescence

Fluorescence is the emission of visible light in some minerals, as a result of exposure to ultraviolet (UV) light. Electrons in the mineral structure that become excited by the UV radiation are raised to higher energy levels. When they fall back to a lower (intermediate) energy level, they emit visible light with a specific energy and hence a narrow range of wavelengths, which can produce vivid colors, such as the bright green of willemite, Zn_2SiO_4, from Franklin Furnace, New Jersey. Because fluorescence is commonly caused by trace constituents in minerals, it is not a diagnostic property by which we can identify the host mineral.

Streak

For some minerals with metallic luster, the color of the streak (the powdered mineral) may be helpful in its identification. The fine powder of the streak of a metallic mineral may be quite different from that of its color in a hand specimen. For example, iron-black hematite, Fe_2O_3, crystals with a bright metallic luster show a red streak (Fig. 3.12). Chromite, $FeCr_2O_4$, with a metallic luster and iron-black to brownish-black color, has a dark brown streak. The small scratch that is used in a hardness test may show the color of the powdered mineral, but it is best seen when the mineral in question is rubbed against an unglazed white porcelain plate. Metallic minerals that show a colored streak are not completely opaque, and when streaked or scratched, the particles of powder are small enough that some light passes through them and creates the color. These same minerals, when observed on extremely thin edges of grains

(A)

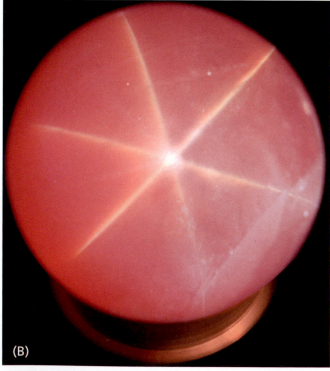

(B)

Figure 3.11 (A) Iridescence, also known as labradorescence, as shown by plagioclase feldspar. (B) A rose quartz sphere showing asterism. Sphere diameter is 6 cm.

Figure 3.12 A black, metallic crystal of hematite, Fe_2O_3. The streak of all hematite is red as shown on an unglazed porcelain plate.

under the microscope, exhibit these same colors. Opaque grains of hematite appear red on thin edges and chromite appears brown. Minerals with a nonmetallic luster usually show a streak that is close to white or colorless.

3.4 Cleavage

As noted previously, crystal form is the outward expression of the regular atomic, internal structure of a mineral. Cleavage is also a property that is directly related to crystal structure and represents the breaking of a mineral along crystallographic planes. The types of cleavage are expressed by various terms, such as *cubic cleavage*, but there is also a scientific notation that relates the cleavage planes to crystallographic directions in the crystal structure. This notation is discussed in Section 5.3.3.

Cleavage is the result of weak bonding or large interplanar spacing across atomic planes in a crystal structure. It is evident in smooth plane surfaces, the directions and number of which are the expression of the underlying symmetry (in the structure) of the mineral in question. Cleavage can result from the impact of a hammer blow or a knife blade on a mineral. Many minerals cleave with diagnostic patterns that, when recognized, are helpful in their identification.

Some of the common terms that describe the type of cleavage are the following:

- **Planar**
- **Prismatic**
- **Cubic**
- **Rhombohedral**
- **Octahedral**

Planar cleavage, as the term implies, describes cleavage along a single planar direction. This is best shown by minerals belonging to the mica group. Figure 3.13 shows a large sheet of muscovite mica. When such a large specimen is crushed (with a

hammer or in a commercial milling process), it produces large numbers of much smaller, flaky grains as shown in front of the large mica sheet in Figure 3.13. This planar cleavage property is a key feature of the mica group and is commonly diagnostic.

Prismatic cleavage consists of two different cleavage directions whose lines of intersection are commonly parallel to a specific crystallographic direction. This coincides with the axis of elongation of the cleavage fragments. Figure 3.14 shows two different crystals that are vertically elongate with two sets of cleavage traces on the top faces of the crystals. The two crystals also show vertical, straight cleavage lines extending along the elongate vertical faces of the crystals. The two cleavage directions (producing four planes of cleavage, in two sets of parallel planes) result in prismatic fragments with various internal angles. The internal angle between the two cleavage directions in the pyroxene crystal illustrated in Figure 3.14(A) is ~90° (actually the angles range from 87° to 88°, but appear as 90° to the unaided eye). In Figure 3.14(B), an amphibole crystal is shown with two vertical cleavage directions that have internal angles of 124° and 56°. The angular differences between the prismatic cleavage directions in these two major rock-forming mineral groups, pyroxenes and amphiboles, are highly

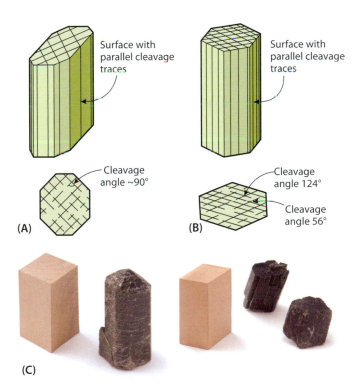

(A)   (B)

(C)

Figure 3.14 Perspective sketches of pyroxene (**A**) and amphibole crystals (**B**) that show pronounced prismatic cleavage traces on the top faces of the crystals. With the top faces redrawn (parallel to the page) the actual angles between the two cleavage directions can be shown. These are at about 90° to each other in the pyroxene crystal, and at 124° and 56° to each other in the amphibole. (**C**) Photograph (on the left) of a pyroxene crystal and (on the right) two amphibole crystals. The external prismatic faces are parallel to their internal cleavage directions. The two small wooden models illustrate (on the left) the pyroxene cleavage angles and (on the right) amphibole cleavage angles.

Figure 3.13 A large sheet of muscovite mica, which, when crushed, produces many smaller flaky fragments as a result of excellent basal cleavage.

Figure 3.15 Excellent cubic cleavage (three cleavage directions at 90° to one another) in halite, NaCl. Blue and white translucent cleavage fragments and a small wooden model of a cube.

Figure 3.16 Excellent rhombohedral cleavage (three cleavage directions, but not at 90° to one another) as shown by calcite, $CaCO_3$. This optically clear, colorless variety of calcite is known as Iceland spar. A small wooden rhombohedral model shows the planar orientations that deviate from those seen in a cube.

diagnostic in their identification. Pyroxene and amphibole grains, in rocks with an average grain size of a few millimeters, occur commonly as dark-colored, anhedral grains (ranging in color from dark green to black), and the only diagnostic characteristic such grains may show is cleavage. With the aid of a hand lens (or jeweler's loupe), these angular differences may be observed and otherwise similarly appearing grains can be classified as either a pyroxene or amphibole.

Feldspars, another major rock-forming mineral group, also exhibit pronounced prismatic cleavage directions at about 90° to one another. These cleavages result in fragments with a blocky outline. The name *orthoclase*, $KAlSi_3O_8$, is derived from two Greek words, *orthos*, meaning "upright," and *klasis*, meaning "fracture," in allusion to its pronounced prismatic and blocky cleavage pattern.

Cubic cleavage results in mineral fragments that have cubic outlines on account of three cleavage directions at 90° to one another. The overall shape of such fragments ranges from almost-perfect cubes to those with blocky outlines. Halite, NaCl (Figs. 2.1(C) and 3.15), and galena, PbS, both of which crystallize in the isometric crystal system (with crystals of overall cubic symmetry) are two of the best examples of excellent cubic cleavage in common minerals.

Rhombohedral cleavage results in fragments with an external shape with six sides. A cube has six sides as well, but in a rhombohedron, the six sides (representing three different cleavage directions) are not at 90° to one another, as in a cube. Instead, a rhombohedral shape can be viewed as a cube that has been extended (or compressed) along one of the four diagonal directions in a cube (that run from corner to corner). Many carbonates show rhombohedral cleavage, and some calcite, $CaCO_3$, shows it to perfection (Fig. 3.16).

Octahedral cleavage is the result of breakage along four different directions, caused by four sets of parallel planes, forming the shape of an octahedron. The term *octahedron* is derived from two Greek words, *octo*, meaning "eight," and *hedra*, meaning "plane." As such, it describes a highly symmetric shape (or form) with eight bounding faces. Fluorite, CaF_2, provides the best example of this type of cleavage (Fig. 3.17).

Many minerals show no cleavage, and only a relatively small number show it to perfection. Quartz, SiO_2, is a good example of a mineral that shows no planar surfaces upon breaking. Instead, it may show a fracture pattern that is described as **conchoidal fracture** (Fig. 3.18), which resembles the curvature of the interior surface of a shell (conchoidal from the Latin word *concha*, meaning "shell"). This is the same fracture pattern that is seen in glass. The reason for this curved and irregular fracture pattern is the absence of clearly defined planes of weakness in the crystal structure of quartz or in the amorphous internal atomic arrangement of glass.

Having discussed physical properties of minerals that can be assessed without any special tools, we now describe some other highly diagnostic properties that require some commonly

Figure 3.17 Excellent octahedral cleavage in fluorite, CaF_2. The wooden octahedron illustrates the same four cleavage directions.

Figure 3.18 Conchoidal fracture as seen in a specimen of clear quartz.

available basic test equipment. These properties are hardness, specific gravity, magnetism, solubility in hydrochloric acid, and radioactivity.

3.5 Hardness

Hardness is the resistance to abrasion, or indentation. In a mineral, it is the resistance of a smooth surface to the scratching by a sharp point or edge, or by another mineral. In mineralogy, hardness is designated as **H**. A hardness test is an assessment of the reaction of a crystal structure to stress without rupture (cleavage and fracture represent rupture). The scratching of a number of metallic minerals may result in a groove, because of plastic flow of the crystal structure under pressure. As a result,

native metals such as gold, silver, and copper are commonly described by adjectives such as **malleable** and **ductile**. In contrast, nonmetallic minerals react by microfracturing on a fine scale, and they are described as **brittle**.

In 1824, an Austrian mineralogist, F. Mohs, selected ten common minerals as a scale to determine relative hardness. These are numbered from 1 to 10 in order of increasing relative hardness (see Table 3.1), known as the **Mohs hardness scale**. If a mineral is said to have a hardness of 6, it means that can be scratched by quartz (at 7) and the other three minerals above it in the hardness scale. If the hardness is given as 4½, the mineral is harder than fluorite but softer than apatite.

It is not uncommon to refer to a specific mineral as a **hard mineral**, meaning that it is as hard as or harder than quartz, with $H = 7$ on the relative hardness scale (Table 3.1). Hardness is a property that has much influence on how minerals react to physical weathering processes. Those that are hard (with $H > 7$) generally survive well during the various chemical and physical processes that are part of such cycles. Quartz is abundant as a detrital constituent of sandstones and quartz conglomerates. Similarly, ruby and sapphire, gem varieties of the mineral corundum, with $H = 9$, are very resistant during weathering and are commonly mined from gravel deposits. Similarly, diamond, with $H = 10$, can be found in river gravels and sandy beaches. Gem minerals not only must have beauty, luster, and brilliance but also must be durable, which is a function of hardness, toughness, and stability. Ideally, therefore, gems should be harder than quartz, a qualification that is met by only 10 or 12 major gemstones. Gems of lesser hardness are subject to abrasion. This is particularly true when a gemstone is set in a ring.

In practice, the hardness of an unknown mineral (one that you are trying to identify) can be approximated by the use of a few common objects: a fingernail, a penny, a pocketknife, and

Table 3.1 Mohs hardness scale minerals.

Hardness number (**H**)	Mineral name	Chemical formula	Remarks
1	Talc	$Mg_3Si_4O_{10}(OH)_2$	Soft, greasy feel; flakes are left on the fingers
2	Gypsum	$CaSO_4 \cdot 2H_2O$	Can be easily scratched by the fingernail *fingernail hardness ~2.2*
3	Calcite	$CaCO_3$	Can be easily scratched with a knife and just scratched by a copper penny *copper penny hardness ~3.2*
4	Fluorite	CaF_2	Less easily scratched by a knife than calcite
5	Apatite	$Ca_5(PO_4)_3(F, Cl, OH)$	Is scratched by a knife with difficulty *pocket knife hardness ~5.1* *glass plate hardness ~5.5*
6	Orthoclase	$KAlSi_3O_8$	Not scratched by a knife and will scratch ordinary glass
7	Quartz	SiO_2	Scratches glass easily *porcelain streak plate hardness ~7*
8	Topaz	$Al_2SiO_4(F, OH)_2$	Scratches glass very easily[a]
9	Corundum	Al_2O_3	Cuts glass[a]
10	Diamond	C	Used as a glass cutter[a]

[a] There are few minerals that are as hard as, or harder than, quartz, and these include several of the highly prized gems.

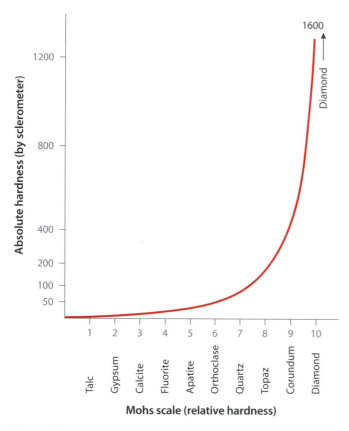

Figure 3.19 Comparison of the Mohs relative hardness scale to absolute measurements of hardness.

a glass plate. These objects all have different relative hardnesses and are useful in identifying minerals that have a hardness lower than that of quartz (7), as shown in Table 3.1.

The Mohs hardness scale is nonlinear when compared to an absolute scale (Fig. 3.19) obtained by an instrument known as a **sclerometer**. This test consists of the microscopic measurement of the width of a scratch made by a diamond under a fixed load and drawn across the face of the mineral specimen. The hardness number is the weight in grams required to produce a standard scratch. Figure 3.19 shows that the absolute hardness of diamond is 15 times greater than that of quartz, and that of diamond is about 4 times harder than the next-hardest gemstones (ruby or sapphire, both gem varieties of corundum).

Hardness plays an important role in the selection of natural and synthetic materials used as abrasives. An abrasive is a substance that is used to grind, polish, abrade, scour, and clean. Minerals commonly used as abrasives are industrial diamond, corundum, garnet, and quartz.

3.6 Specific gravity (relative density)

Specific gravity (designated as **G**) is the comparative weight of a mineral and is an important diagnostic property. It is defined as the ratio of the weight of a volume of a substance to the weight

of the same volume of some other material taken as a standard. For solids, the standard reference material is water. Thus, a mineral with a specific gravity (**G**) of 4 weighs four times as much as the same volume of water. For accurate density determinations, the temperature of the water must be 4°C (39.2°F), the temperature at which water has its maximum density. Density and specific gravity are commonly used interchangeably, but **density**, defined as mass per unit volume (g/cm³), requires scientific units, whereas specific gravity is a number (without units) that expresses a unitless ratio. Although the **G** of galena would be given as 2.65, its density is 2.65 g/cm³.

Accurate values for specific gravity of minerals that are insoluble in water can be obtained by a specially designed balance known as the **Jolly balance** (invented in 1874 by Phillip Gustav von Jolly, a professor at the University of Munich, Germany). Because specific gravity is a ratio, this balance allows for the direct measurement of this ratio, that of the mineral specimen in air versus that in water. These ratios, obtained by such a balance, are listed to two decimal places. For example, the **G** of quartz = 2.65, and the **G** of olivine (Mg_2SiO_4 to Fe_2SiO_4) ranges from 3.27 to 4.27. You may have access to such a Jolly balance in your laboratory work that accompanies your course, but here we introduce only the more qualitative assessment of specific gravity.

3.6.1 Specific gravity and atomic structure

The specific gravity of a mineral is a function of the atomic weights of the chemical elements in its crystal structure and the way in which those elements (atoms or ions) are packed together, called atomic packing. The influence of the increasing weight of a cation in a group of **isostructural** minerals (*isostructural* meaning that the structures being compared are the same, from the Greek *isos*, meaning "same") on their resulting specific gravity is shown by three oxides:

Mineral	**G**	Cation atomic weight
Rutile, TiO_2	4.2	47.9
Pyrolusite, MnO_2	4.8	54.9
Cassiterite, SnO_2	6.9	118.7

The influence of the atomic packing on the specific gravity of a mineral is clearly shown by the two different structural arrangements of carbon in graphite and diamond. The atoms in the soft mineral graphite are held together by two different bond types. One is strong within layers of the structure and one is very weak, which results in relatively large atomic spacings between closely bonded layers, thus giving rise to **G** = 2.2. In contrast, for diamond, which is tightly bonded throughout, **G** = 3.5.

The specific gravity of a mineral can be helpful in its identification, and with practice one can develop a relative sense

Table 3.2 Some rock-forming minerals arranged according to light, average, and heavy values of specific gravity (**G**).

Light	Average	Heavy
Bauxite (an ore of aluminum) 2.0–2.5	Microcline 2.5	Barite 4.5
	Albite 2.6	Pyrolusite 4.7
Halite 2.1	Quartz 2.6 Calcite 2.7	Pyrite 5.0
Serpentine 2.2–2.6	Plagioclase 2.6–2.8	Magnetite 5.2
Sodalite 2.3	Dolomite 2.85 Tremolite 3.0–3.3 Apatite 3.1–3.2	Galena 7.4–7.6 Silver 10.5 Gold 15.0–19.3

of the difference in weight of homogeneous specimens (consisting of a single mineral, not an intergrowth). For example, a specimen of anhydrite, $CaSO_4$, with **G** = 2.9, will seem light when compared with a specimen of barite, $BaSO_4$, with **G** = 4.5. Because of its high specific gravity, barite is extensively used as a weighting agent for muds circulating in rotary drilling of oil and gas wells. These drilling muds are a mixture of water, clay, and barite.

Minerals with a metallic luster have an average specific gravity of about 5. For example, pyrite, FeS_2, has **G** = 5; chalcopyrite, $CuFeS_2$, has **G** = 4.2; and native copper has **G** = 8.9. Nonmetallic minerals, which include most of the common rock-forming minerals such as quartz, feldspar, pyroxenes, and amphiboles, typically have specific gravities ranging from 2.6 to 3.3. The **G** of quartz = 2.6; for feldspar, it ranges from 2.6 to 2.7; and for diopside (a calcic pyroxene), it is 3.2. In assessing the relative **G** of a mineral specimen, one must keep in mind that the mass of the specimen is important because a large chunk of quartz may appear heavy even though its specific gravity is not high, just because it is a large piece. So, if assessed by hefting a specimen, experience will allow for distinctions among light, medium, and heavy, but specific gravity can be highly diagnostic only if it measured with an instrument such as a Jolly balance. Table 3.2 gives a listing of some rock-forming minerals arranged according to qualitative evaluations.

3.7 Magnetism, solubility in acid, and radioactivity

The following are some additional quick and simple tests that allow us to identify, uniquely, some minerals and mineral groups.

Magnetism

Two common minerals are distinctly magnetic: magnetite, Fe_3O_4, and pyrrhotite, $Fe_{1-x}S$. Magnetite is strongly attracted to a magnet; pyrrhotite much less so. Both are opaque, but magnetite is black and pyrrhotite is bronze. A small hand magnet reveals their presence in a specimen that contains a considerable amount of either mineral, but if they are present in only small quantities, then the removal of a few grains from the specimen (with a pocketknife or needle) for individual testing is often diagnostic. A full description of the various causes of magnetic behavior in minerals is given in various mineralogy textbooks, referenced at the end of this chapter. Especially informative and detailed discussions of magnetism are given in the journal *Elements* (*Mineral Magnetism* listed under "Further Reading" at the end of this chapter).

Solubility in HCl

Several carbonate minerals show **effervescence**, also known as fizz or bubbling, when a drop of dilute HCl is placed on them. This is due to the following chemical reaction:

$$2H^+ + CO_3^{2-} \rightarrow H_2O + CO_2 \uparrow$$
$$\text{acid} \quad \text{in calcite} \quad \text{water} \quad \text{gas} \tag{3.2}$$

Calcite and the other structural form of $CaCO_3$, aragonite, both react strongly. Several other carbonates, such as dolomite, magnesite, siderite, and rhodochrosite, effervesce only in hot dilute HCl. This test is especially useful when some fine- to medium-grained, light-colored mineral, with no other discernible diagnostic properties, is distributed among other grains in a rock. If the fizz test works, the mineral is one of several carbonates.

Radioactivity

Uraninite, UO_2, and the massive variety known as pitchblende, may both contain U and Th as major constituents. These elements are radioactive; that is, they change into daughter elements by emitting alpha particles, beta particles, and gamma rays from the nucleus of the atom. These rays are easily detected with a Geiger or scintillation counter (Box 3.1). If the test is positive, one is clearly dealing with a radioactive mineral.

3.8 Instrumental methods for the quantitative characterization of minerals

All the tests described in the previous sections may be helpful in the identification of some minerals, but many times the results are not specific and lead only to the broad conclusion that the mineral at hand is a member of some large mineral group. The test results may suggests that your unknown might be an amphibole, or a pyroxene, or some kind of feldspar, and commonly there is still the question of which of the amphiboles, which of the pyroxenes, or which member of the feldspars are present in the rock.

Two instrumental tests that may quickly provide a unique answer to those questions are optical microscopy, which we discuss in Chapter 6, and X-ray powder diffraction. Equipment for both procedures is commonly present in many geology departments. Optical microscopy, using a polarizing microscope, is commonly a quick technique that helps in the unique identification of a mineral that is part of a mineral group, but only if you are familiar with both its theoretical background and use of the microscope and its manipulations, which you can learn in Chapter 6. Here it suffices to illustrate just two examples of the unique identification of two minerals that you have decided (as a result of applicable tests described earlier) belong to the feldspar group. One sample is nonmetallic; has a blocky cleavage; is light beige in color; has an **H** of about 6; and might be potassium feldspar (K-feldspar), $KAlSi_3O_8$. There are three types of K-feldspar, known as sanidine, orthoclase, and microcline. So which one might it be? If this feldspar is part of a rock that is under study, a thin section of that rock (a slice of a rock, about 0.03 mm thick, mounted between two glass slides) is probably available. Put the thin section under the microscope and a very diagnostic pattern (of twinning, see Chapter 6) is seen in doubly polarized light (Fig. 3.20(A)). This pattern uniquely identifies the K-feldspar as the microcline type. The other sample has a light grayish color with all the appropriate physical properties of a feldspar. If you put a thin section of it under the polarizing microscope, and if, with crossed polarizers, you see a fine pattern of repetitive thin lamellae, then it must be a member of the plagioclase series (Fig. 3.20(B)). If your course includes an introduction to optical microscopy, then you will learn aspects of this versatile identification technique.

Optical microscopic techniques are used extensively in the research laboratories of companies involved in oil and gas exploration and in mining. Similarly, centers of materials research employ optical techniques in addition to many of the other instrumental methods.

3.8.1 X-ray powder diffraction

As noted earlier, X-ray powder diffraction equipment is available in most geology departments, and some instructors present students in an introductory course to this powerful mineral identification technique. So here, we give a short introduction to the underlying principles, the instrumentation, and the interpretation of the results of X-ray powder diffraction.

In the brief introduction to mineral identification by optical microscopy in this chapter, we have seen how distinctive twin patterns (Fig. 3.20(A)–(B)) led to the unique identifications of two different feldspars. Here we use a similar approach, showing how two distinctly different X-ray powder diffraction patterns can lead to the unique identification of an unknown mineral. Suppose you have a dark green to black, fine-grained volcanic rock with some white mineral filling vesicles (gas cavities). You have scratched it and it has an **H** ~ 7, but it is too fine-grained to recognize any crystal form or cleavage. You think it might be quartz or one of its two high-temperature forms (known as **polymorphs**, from the Greek words *poly* and *morpho*, meaning "many forms"; the term implies different atomic structures), tridymite and cristobalite. If you had a thin section of this mineral assemblage, optical study would reveal the identity of the mineral in question. But if you do not have a thin

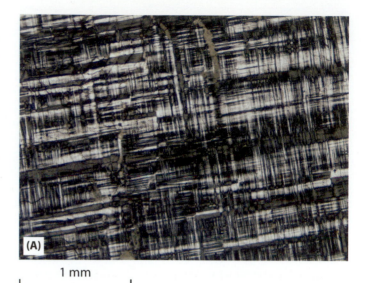

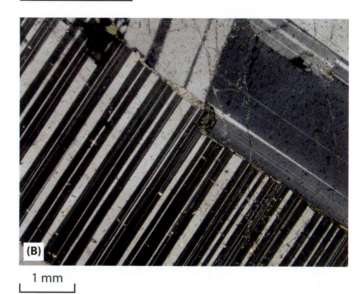

Figure 3.20 **(A)** A photomicrograph of a relatively large, single grain of K-feldspar, $KAlSi_3O_8$ (a photomicrograph is a photograph taken with a camera mounted on the vertical tube of an optical microscope). This photograph, which was taken in doubly polarized light (see Chapter 6), shows a distinctive tartan pattern. This uniquely identifies the K-feldspar as one of three structure types of $KAlSi_3O_8$, namely microcline. The pattern is the result of twinning (discussed in Chapter 5). **(B)** A photomicrograph of several intergrown feldspar grains, one of which displays a regular pattern of light and dark, parallel strips (known as twin lamellae, from the Latin word *lamella*, meaning "plate"). This pattern, known as albite twinning, was photographed in doubly polarized light. It is present in all members of the plagioclase feldspar ($NaAlSi_3O_8$ to $CaAl_2Si_2O_8$) series.

section, you can pry out some of the light-colored grains from the vesicles, grind them to a fine powder, mount the powder on a glass slide (or in a special sample holder) in an X-ray powder diffractometer, and observe the results of the experiment on a graphical display. The last two illustrations in Box 3.2 show the distinctly different X-ray powder diffraction patterns for quartz and cristobalite. There are large numbers of files of

X-ray diffraction patterns of minerals that can be searched on-line. You can compare your own pattern with these reference patterns, such that if your unknown produced the (B) trace, it is cristobalite; if it resulted in the (A) trace, it is quartz; and if it resembles neither, it is likely to be tridymite. A brief introduction to X-ray powder diffraction is given in Box 3.2.

BOX 3.2 | X-RAY POWDER DIFFRACTION TECHNIQUE

X-rays are part of the electromagnetic spectrum (Box 3.1), with wavelengths ranging from about 100 to 0.02 Ångstroms, where 1 Å = 0.1 nm (nanometers). Visible light, by comparison, has wavelengths between 7200 Å and 4000 Å, and hence is much less energetic and penetrating than X-radiation. X-rays are generated in vacuum-sealed X-ray tubes in which electrons, under high accelerating voltage, bombard a target material. When the energy of the electron beam (inside the X-ray tube) is great enough to dislodge electrons from the inner electron shells of the target material, those electrons are expelled, leaving vacancies that are filled by other electrons from surrounding electron shells. When electrons transition from outer to inner electron shells, X-radiation is emitted of specific wavelengths, related to the target material. The radiation spectrum has a very intense K$\overline{\alpha}$ (produced by transitions from L- to K-shells) and a less intense Kβ peak (resulting from transitions from M- to K-shells; see the first illustration). The intense K$\overline{\alpha}$ peak (consisting of a very intense Kα_1 and a much less intense, but close in λ, Kα_2 peak that are averaged out as K$\overline{\alpha}$) is the source of X-radiation used to irradiate crystalline materials and to produce X-ray diffraction effects. The most common target material in X-ray diffractometer instruments, used for general identification purposes, is copper, with a K$\overline{\alpha}$ wavelength of 1.5418 Å. Other available target materials in X-ray tubes are iron and molybdenum.

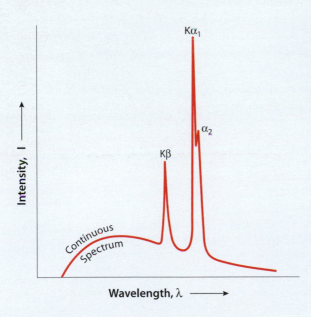

When an intense X-ray beam is aimed at a crystalline material, the X-rays are scattered by its ordered internal atomic arrangement. These effects are known as **diffraction**. The relation of these diffraction effects to a crystal structure is expressed by the **Bragg equation**, so named after Sir W. L. Bragg (1862–1942), an English crystallographer who determined the first crystal structures by X-ray techniques in 1914. He and his son Sir W. H. Bragg (1890–1971) received the Nobel Prize in Physics in 1915. The Bragg equation is

$$n\lambda = 2d\sin\theta, \tag{3.3}$$

which relates the interplanar (atomic) spacing (d) inside a crystalline material to a diffraction angle (θ) as a function of the wavelength (λ) of the X-rays used; n is an integer such as 1, 2, or 3. The letter d is a symbolic representation of all possible distances between differently oriented crystallographic (planar) directions inside a crystalline substance.

The second illustration shows some of the basic mechanical components of an X-ray powder diffractometer in which the powdered sample is rotated through angle θ while the detector moves at an angle of 2θ, beginning at about 5° 2θ in an increasing direction of 2θ. The third diagram is a schematic

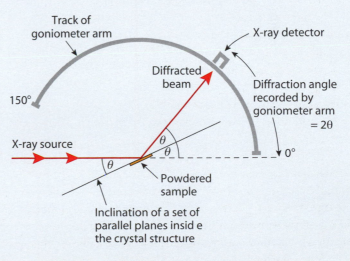

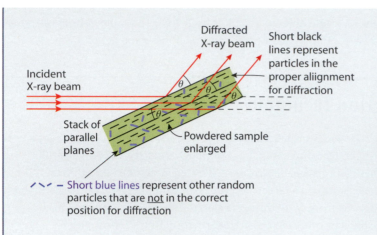

Incident
X-ray beam

Diffracted
X-ray beam

Short black
lines represent
particles in the
proper aliignment
for diffraction

θ θ θ

θ

Stack of
parallel
planes

Powdered sample
enlarged

∕ ∖ ∕ − Short blue lines represent other random
particles that are not in the correct
position for diffraction

illustration of an enlarged powdered sample being irradiated. This finely powdered sample consists of an infinitely large number of tiny crystalline particles that (because of their random orientation inside the sample) have a multitude of internal crystallographic planes in parallel positions. These planes, in combination, produce the diffraction effects. The angle of incidence, θ, is the same as the angle of diffraction, θ. The instrument records the intensity of diffracted peaks as a function of 2θ. The wavelength of the incoming radiation is a known quantity (as for Cu K $\overline{\alpha}$ = 1.5418 Å), so d can be computed using the above Bragg equation (all of this is presently done by an online computer). The end result of all this is a listing of d's against their intensity (I). Such a listing can be regarded as a fingerprint of the mineral that was X-rayed.

The last two figures show the distinct qualitative differences in the diffractometer tracings (intensity of diffraction effect, I, versus 2θ angle) of two types of SiO_2. The first is for low quartz, the common form of quartz. The second is for cristobalite, a high-temperature form of SiO_2 (or quartz) found mainly in some volcanic rocks. Both patterns are obtained from the Powder Diffraction File™ and are used with permission from © JCPDS-International Centre for Diffraction Data, 2009. All rights reserved. A subsequent listing of d versus I provides a quantitative statement derived therefrom. Such listings, of d versus I for an unknown, can be compared with similar listings for all known crystalline materials, which are archived in the Powder Diffraction File (PDF) published by the International Center for Diffraction Data (ICDD). More than 291 000 calculated and experimental patterns for natural and synthetic crystalline materials are on file.

In short, this X-ray technique provides, in most cases, a unique identification of the unknown mineral at hand.

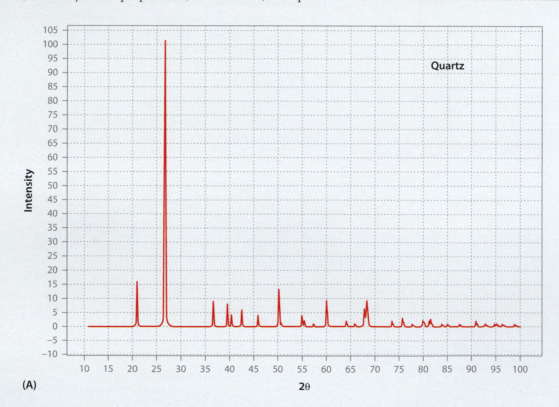

(A)

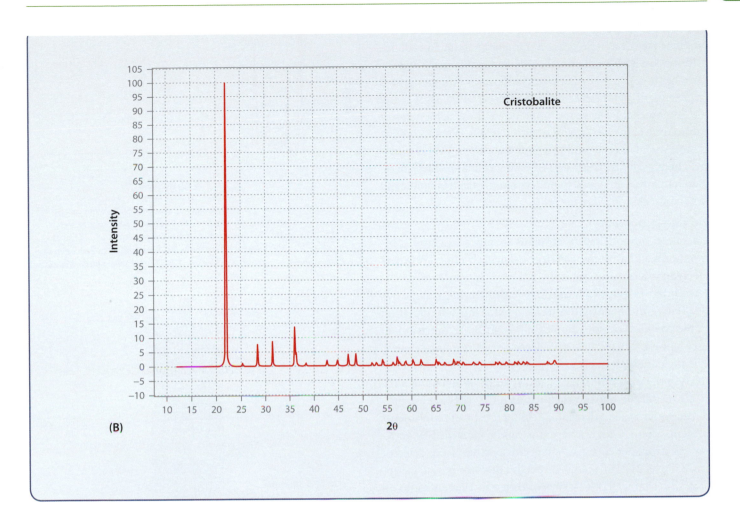

(B)

3.8.2 Electron beam techniques: Scanning electron microscopy, electron microprobe analysis, and transmission electron microscopy

If the mineral under study is very fine- to extremely fine-grained (e.g., talc in talcum powder, clay minerals) and after having obtained test results that still do not completely characterize the unknown under investigation, you may want to obtain a very high magnification image of the unknown. Here is where an electron beam technique, known as **scanning electron microscopy (SEM)** comes in. A very useful aspect of SEM imaging is the large depth of field, resulting in high-definition photographs (Figs. 2.1(C), 2.3(D), and 3.10). Such photographs, especially of very fine-grained materials, can provide much information about the symmetry content (the crystallography) of the unknown.

An SEM instrument has an electron optical column that produces a very fine electron beam that can be focused and scanned (i.e., rastered) across a selected area of the specimen or over small particles (or grains) mounted on a specimen holder.

The maximum magnification that can be attained by a modern SEM is about 1 million times, at a resolution of about 1 nm.

When the electron beam impacts the surface of a solid, it produces various types of radiation that are recorded by detectors inside the instrument. The detectors are capable of measuring the following signals: X-ray radiation, secondary electrons (SE), back-scattered electrons (BSE), and electrons absorbed by the specimen (known as specimen current).

Mineral composition

This leads us to the question of the chemical composition of the mineral that we identified through a combination of techniques. Let us take a look at a possible identification sequence. We have a hand specimen of, for example, a very dark green to black mineral with well-developed cleavage directions of ~90° to one another, giving rise to stubby shapes (Fig. 3.14(A)–(B)). It has an **H** of about 5½, so it may be a pyroxene, but which specific member of this large and complex mineral group? A thin section of this same mineral specimen or some loose grains mounted in an oil show a pronounced green color under the microscope. Measurement of an extinction angle under the microscope (see Chapter 6) gives a number of about 45°. All this information leads to the conclusion that the pyroxene at hand is probably augite. But what is its composition? Augites show a large range in major elements such as Al, Fe^{2+}, Fe^{3+}, Mg, Ca, and Na. So now that we have narrowed down the general name of the unknown, the next major question is still, What technique is most useful in determining its mineral chemistry? Quantitative chemical information on minerals is fundamental to all research in mineralogy and to understanding the formation and origin of rocks.

Electron microprobe

Chemical compositions of minerals are nowadays most commonly obtained by the electron beam technique known as **electron microprobe analysis (EMPA)**. This instrument was introduced in the 1960s and has since become a standard, qualitative and quantitative technique for the determination of the chemical composition of solids. An EMPA instrument uses a finely focused beam of electrons as the energy source. The incident high-energy electrons displace inner-shell electrons in the atoms of the solid under study. When outer-shell electrons fill these inner-shell vacancies, X-ray rays are produced with wavelengths that are characteristic of the elements present. These X-ray wavelengths can be identified by an X-ray spectrometer or by an energy-dispersive X-ray detector (EDS; Fig. 3.21). An energy-dispersive detector is used to separate the characteristic X-rays of different elements into an energy spectrum, and EDS software is used to analyze the energy spectrum to determine the abundance of specific chemical elements. Figure 3.22 is an illustration of the qualitative EDS spectrum of our unknown pyroxene that we suspect is

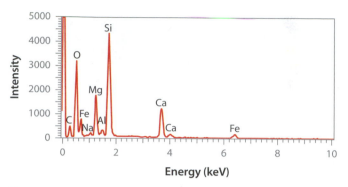

Figure 3.22 An elemental spectrum of augite, obtained through energy-dispersive X-ray spectroscopy (EDS) on an electron microprobe analyzer (EPMA). This qualitative spectrum identifies the presence of Si, O, Ca, Al, Fe, Mg, and Na. The C peak is due to the carbon coating on the specimen to make it conductive to the electron beam. (Courtesy of Mike Spilde, Institute of Meteoritics, University of New Mexico.)

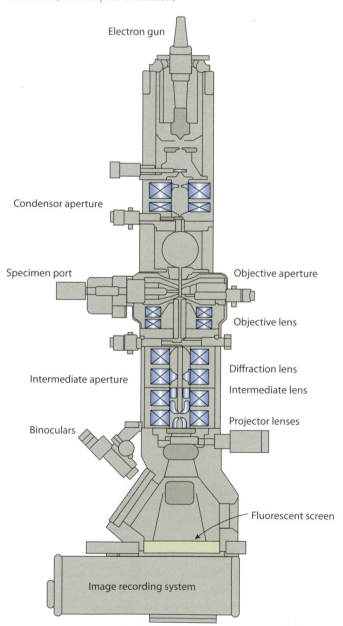

Figure 3.23 A schematic illustration of a transmission electron microscope (TEM). All lenses are electromagnetic lenses.

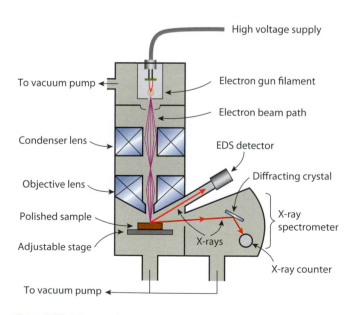

Figure 3.21 Schematic illustration of an electron microprobe analyzer (EMPA). It shows the electron optical column and the two possible read-out systems for X-ray analysis (X-ray spectrometer and EDS).

augite. This spectrum shows the presence of six chemical elements, among them Si, Ca, Na, Al, Mg, and Fe. This suggests strongly that the specimen is indeed augite. A representative augite formula is (Ca, Na) (Fe, Mg, Al) Si_2O_6. Further evaluation of this spectrum (by comparison to that of a standard mineral of known composition) would lead to a quantitative estimate of each of these elements.

Transmission electron microscopy

Transmission electron microscopy (TEM) is a microscopic technique whereby a beam of electrons is transmitted through an ultrathin specimen. The interaction of the electrons with the specimen produces a highly magnified image that is focused on a fluorescent screen or a camera. An illustration of the instrument is given in Figure 3.23. Transmission electron microscopy is capable of imaging objects as small as a single layer of atoms, which is a resolution of tens of thousands of times better than that of a light microscope. Examples of TEM images are in Figures 2.4(C) and 3.24. Figure 3.24 illustrates the application of the TEM to the study of microorganisms and nanoparticles that may be responsible for chemical reactions that have a dramatic effect on environmental safety. In this image the hematite nanoparticles range in size from 11 to 43 nm.

High-resolution transmission electron microscopy (HR-TEM) is an imaging mode of the TEM that allows for the resolution of the crystal structure of a sample on an atomic scale. At present, the highest resolution is at about 0.8 Å, and new developments will improve that to 0.5 Å. Examples of HRTEM images are in Figures 2.4(D) and 3.25. The former image illustrates, with very high resolution, the coiled nature of the structure of chrysotile. Figure 3.25 illustrates the results of a mineral reaction on an almost atomic scale. This is a hydration reaction of enstatite (a member of the pyroxene group) to form the hydrous mineral talc:

$$MgSiO_3 + SiO_2 + H_2O \rightarrow Mg_3Si_4O_{10}(OH)_2$$
enstatite silica vapor talc (3.4)

0.5 μm

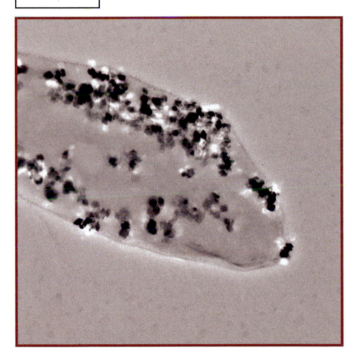

Figure 3.24 A transmission electron micrograph (TEM image) of hematite nanoparticles, Fe_2O_3, to which a specific species of bacterium is attached. This species respirates on oxygen, but in anoxic environments it can respirate on metal oxides. Therefore, in this picture, the bacterium literally breathes on the oxide mineral particles, without which the bacteria would die. In this respiration process, the Fe^{3+} in the hematite is reduced to Fe^{2+} and, as a result, the mineral particles dissolve. (Photograph courtesy of Michael Hochella, Virginia Polytechnic Institute; from Bose et al., 2009.)

10 nm

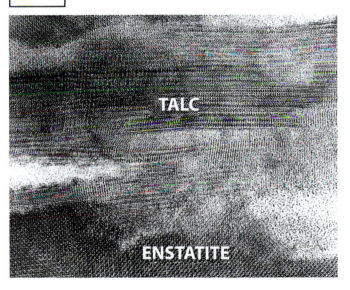

Figure 3.25 A high-resolution transmission electron microscope (HRTEM) image of the hydration reaction of enstatite (a pyroxene) to form talc. The resolution in this image is so great that mineral domains are visible at an almost atomic scale. The blocky structure outlines (lower left corner) are enstatite, and the layered domains are talc. (Photograph courtesy of Adrian Brearley, University of New Mexico.)

Summary

This chapter introduced various aspects of what is involved in the observation and testing of an unknown mineral, in a hand specimen, so as to arrive at a positive identification thereof.

- Mineral descriptions make use of standard adjectives that relate to mineral habit and state of aggregation. These adjectives are worth learning because they are a part of many systematic mineral descriptions.

- Although the color of a mineral is instantly observed, color is not a good diagnostic property.

- Two types of luster, metallic and nonmetallic, are easily assessed and very helpful in distinguishing metallic minerals such as some native elements, oxides, and sulfides from the majority of minerals that are nonmetallic.

- The overall color of a mineral is the result of that part of the visible spectrum that is least absorbed by the mineral as white light travels through the mineral specimen.

- Color in minerals can result from the presence of major elements but commonly results from the presence of trace elements.

- The color of the streak of a mineral is that of its fine powder.

- The play of color in opal results from diffraction of white light into its various colors as it passes through the well-layered arrangement of large amorphous spheres of SiO_2 (with small amounts of water).

- Labradorescence, which occurs in a specific compositional region of the plagioclase feldspar series, is the result of white light that is scattered from microscopic, closely spaced parallel lamellae.

- Cleavage, the breaking of a mineral along crystallographic planes, can be a highly diagnostic property in minerals. The cleavage directions are directly related to crystallographic planes of weakness in the crystal structure.

- A hardness test assesses the reaction of a crystal structure to stress, without rupture. It is very useful in distinguishing between major mineral groups.

- The relative hardness scale that is numbered from 1 to 10 is named after its inventor, the Austrian mineralogist F. Mohs.

- Specific gravity measurements, when done with a specialized balance on small and pure samples, are highly diagnostic. However, in hand specimens, the best you can develop is a sense of relative heft, which allows you to distinguish minerals with large differences in specific gravity.

- Magnetism is the force exerted between a magnet and a mineral specimen. Only two minerals are distinctly magnetic: magnetite and pyrrhotite.

- Solubility in cold or hot HCl is a very diagnostic test for carbonate minerals.

- Radioactivity of hand specimens can be measured using a Geiger or scintillation counter.

- X-ray powder diffraction instrumentation, which is available in most geology departments, can provide, fairly quickly, a unique mineral name for an unknown that had been elusive using other less sophisticated tests.

Review questions

1. What is meant by each of the following adjectives: *prismatic, acicular, dendritic, granular, botryoidal,* and *oolitic*?

2. What distinguishes metallic from nonmetallic luster?

3. What is meant by *vitreous*?

4. What is meant by the term *chromophore elements*?

5. If in white light a specific mineral specimen exhibits a yellow-green color (e.g., the gem olivine called peridot), which part of the optical spectrum is least absorbed?

6. What is meant by the term *labradorescence*?

7. What is asterism?

8. Halite has excellent cubic cleavage – what does that mean?

9. Members of the mineral group mica have perfect planar cleavage – how does that show itself?

10. What is the name of the relative hardness scale? What is the range of numbers on that scale?

11. Metallic and nonmetallic minerals respond very differently to a hardness test. What are these differences?

12. Specific gravity (or density) is a function 1) of an atomic property as well as 2) a crystal structure arrangement? What are these?

13. What mineral groups tend to have high values of specific gravity?

14. A mineral is listed as having **H** = 1 and excellent planar cleavage. Which mineral is this?

15. Another mineral has **H** = 10 and perfect octahedral cleavage. Which mineral is this?

FURTHER READING

The following journal references and mineralogy textbooks are further reading on the origin of color, play of color in opal, and magnetism. They also lead to further reading on instrumental techniques. The last entry is to a laboratory manual with many mineral related exercises.

Bose, S., Hochella, M. F., Gorby, Y. A., Kennedy, D. W., McCready, D. E., Madden, A. S., and Lower, B. H. (2009). Bioreduction of Hematite Nanoparticles by the Dissimilatory Iron Reducing Bacterium *Shewanella Oneidensis MR-1. Geochemica et Cosmochimica Acta,* 73, 962–976.

Darragh, P. J., Gaskin, P. J., and Sanders, J. V. (1976). Opals. *Scientific American,* 234, 84–96.

Dyar, M. D., Gunter, M. E., and Tasa, D. (2008). *Mineralogy and Optical Mineralogy,* Mineralogical Society of America, Chantilly, VA.

Goldstein, J., Newbury, D., Joy, D., Lyman, C., Echlin, P., Lifshin, E., Sawyer, L., and Michael, J. (2003). *Scanning Electron Microscopy and X-Ray Microanalysis,* 3rd ed., Kluwer Academic/Plenum Publishing, New York.

Jenkins, R., and Snyder, R. L. (1996). *Introduction to X-Ray Powder Diffractometry,* John Wiley and Sons, New York.

Klein, C. (2008). *Minerals and Rocks: Exercises in Crystal and Mineral Chemistry, X-Ray Powder Diffraction, Mineral and Rock Identification, and Ore Mineralogy,* John Wiley and Sons, New York.

Klein, C., and Dutrow, B. (2008). *Manual of Mineral Science,* 23rd ed., John Wiley and Sons, New York.

Loeffler, B. J., and Burns, R. G. (1976). Shedding light on the color of gems and minerals. *American Scientist,* 64, 636–647.

Mineral Magnetism: From Microbes to Meteorites. (2009). *Elements* 5, no. 4, 209–246.

Nassau, K. (1978). The Origin of Color in Minerals. *American Mineralogist,* 63, 219–229.

Nassau, K. (1980). The Causes of Color. *Scientific American,* 243, 124–156.

Perkins, D. (2011). *Mineralogy,* 3rd ed., Prentice Hall, Upper Saddle River, NJ.

Wenk, H. R., and Bulakh, A. (2004). *Minerals: Their Constitution and Origin,* Cambridge University Press, Cambridge.

Williams, D. B., and Carter, C. B. (1996). *Transmission Electron Microscopy,* Plenum Press, New York.

Laboratory Manual

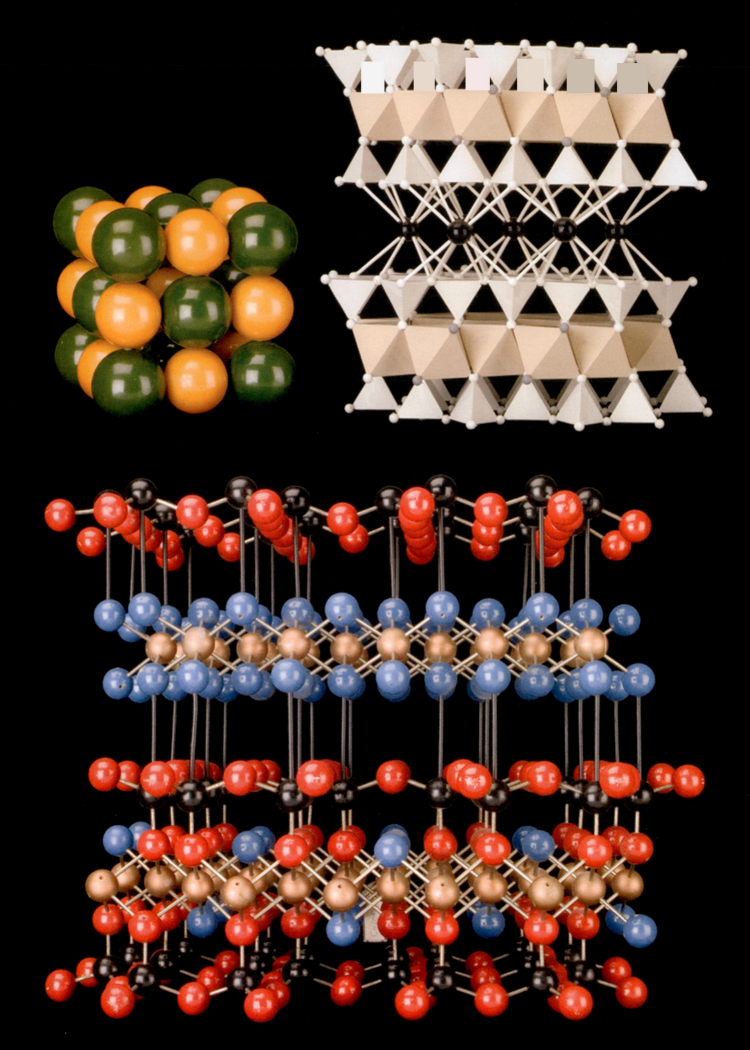

CHAPTER

4

Fundamentals of crystal structures

In this chapter, we introduce various fundamental aspects of the architecture of crystal structures of minerals. We begin with their building blocks, which are the naturally occurring chemical elements, be they atoms or ions. We then discuss the size of the atoms and ions, and we point out that the size of ions is influenced by the number of an ion's closest neighbors. This leads to the concepts of **coordination number**, **C.N.**, and coordination polyhedron. We then discuss the five rules, known as Pauling's rules, that relate to how polyhedra stack together to form a crystal structure. We also address the forces that hold crystal structures together, known as bonding forces. Finally, we address the various factors that influence the variability in the chemical composition of many mineral groups.

Examples of physical models used in the teaching of crystal structures. These models, built of wooden or plastic spheres, plastic polyhedra, and connecting rods are commonly available in the laboratories associated with mineralogy courses. Similar but interactive electronic images are available on CrystalViewer that accompanies this text. On the top left is a close-packed structure model of halite. On the top right is a polyhedral model of muscovite, a mica (a layer silicate). On the bottom is a ball-and-stick model of chlorite, also a layer silicate. (David Nufer Photography, Albuquerque, New Mexico.)

Crystal structures represent the most fundamental aspects of minerals. In classifications, minerals are organized by chemical groups (see Sec. 2.2) as well as structure types (see Sec. 7.4). Because there are about 1140 known silicates, this large chemical group is subdivided by structure type.

Many of the minerals treated in this text are silicates, and gaining an understanding of their various structure types is essential. This text is accompanied by CrystalViewer, which incorporates all the crystal structures of the minerals that are systematically treated in Chapters 7, 10, 13, and 15. Furthermore, each systematic mineral description is accompanied by an illustration of its structure, its mineral name, its chemical formula, and a shorthand notation known as the space group. This chapter provides you with the basic aspects of crystal structures, which will provide you with a good understanding of the crystal structures both in CrystalViewer and in the crystal structure illustrations in this text.

4.1 Naturally occurring chemical elements

In Section 2.1, we introduced the definition of mineral in which one of the clauses states with a definite, but commonly not fixed, chemical composition. That led to the concept of minerals with a fixed composition such as quartz, SiO_2, and minerals with a variable composition, such as members of the olivine series, ranging from Mg_2SiO_4 to Fe_2SiO_4, with the general formula $(Mg, Fe)_2SiO_4$.

If one were to ask, "What are the chemical elements that minerals are composed of?," the first suggestion might be "Consult the periodic table of the elements" (inside front cover of this book). The periodic table lists 118 elements. An element is a substance that cannot be decomposed into simpler materials by chemical reactions. Of these 118 elements, 30 have been synthesized and do not occur naturally. Elements 43, 61, 85, 87, and 93–118 are radioactive, have no stable isotopes, and decay relatively quickly. Consequently, they are present in natural materials in only ultra-trace quantities. Many of the elements in the periodic table are also rare, and Table 4.1 lists 29 of the most common ones. This list, therefore, represents those naturally occurring elements that are most commonly incorporated into mineral structures at the time of their origin. Thirteen of these make up 99% of the Earth's crust, as shown in Table 4.2

4.2 Atomic and ionic radii

In the crystal structures of minerals, **atoms** or **ions** are the basic building blocks. The manner in which these are arranged inside a crystal structure is strongly influenced by their size. The crystal structure of a native element such as gold, Au, in which all the atoms are of the same element, and therefore of the same size, is very different from that of quartz, SiO_2, which consists

Table 4.1 The 29 most common, naturally occurring elements and their common ionic states.

Atomic number	Element	Ion	Atomic number	Element	Ion
3	Lithium	Li^+	24	Chromium	Cr^{3+}
4	Beryllium	Be^{2+}	25	Manganese	Mn^{2+}
5	Boron	B^{3+}			Mn^{4+}
6	Carbon	C^{4+}	26	Iron	Fe^{2+}
8	Oxygen	O^{2-}			Fe^{3+}
9	Fluorine	F^-	27	Cobalt	Co^{2+}
11	Sodium	Na^+	28	Nickel	Ni^{2+}
12	Magnesium	Mg^{2+}	29	Copper	Cu^+
13	Aluminum	Al^{3+}			Cu^{2+}
14	Silicon	Si^{4+}	30	Zinc	Zn^{2+}
15	Phosphorus	P^{5+}	38	Strontium	Sr^{2+}
16	Sulfur	S^{2-}	40	Zirconium	Zr^{4+}
		S^{6+}	47	Silver	Ag^+
17	Chlorine	Cl^-	56	Barium	Ba^{2+}
19	Potassium	K^+	82	Lead	Pb^{2+}
20	Calcium	Ca^{2+}	92	Uranium	U^{4+}
22	Titanium	Ti^{4+}			

of two different ions of very different size. This means that if we wish to gain an understanding of the various ways in which the internal structures of minerals are arranged, we must first look into the sizes of both atoms and ions.

The sizes of both atoms and ions are determined by crystallographers by using specific X-ray diffraction techniques that are different from the X-ray powder diffraction method discussed in Section 3.8 and Box 3.2. In determining atomic and/or ionic radii, a small, perfect single crystal of a mineral is irradiated by X-rays, and the resulting X-ray diffraction pattern becomes the basis for the measurement of distances between atoms and/or ions. Atomic radii are obtained from the X-ray diffraction spacing measured between atoms of pure metals.

Table 4.2 Thirteen of the most common elements (exclusive of hydrogen) that make up 99% of the Earth's crust.

O	Ti
K	Al
Na	Si
Ca	P
Mn	S
Fe	C
Mg	

Note: Elements are listed in order of decreasing size of their most common ionic state. Ionic radii are given in Table 4.4.

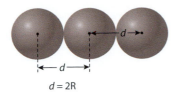

$$d = 2R$$

Figure 4.1 Schematic illustration of the determination of the radius of a metal atom in a close-packed structure. $R_X = d/2$

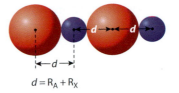

$$d = R_A + R_X$$

Figure 4.2 Schematic illustration of the determination of the radius of a cation (R_X) when the distance (d) between the centers of an anion-cation pair, as well as the radius of the anion (R_A), is known. $R_X = d - R_A$.

Figure 4.1 shows that the radius of an atom is half the length measured between the centers of two identical atoms, where the shape of the atom is assumed to be spherical. Table 4.3 gives atomic radii for some common elements.

The determination of the size of ions in the internal structure of, for example, a silicate, is more complicated than determining the atomic radius of an element in a metal. In silicates, which are built of oxygen **anions**, O^{2-}, and various **cations**, the ionic radius of a cation can be obtained only if the radius of the anion (O^{2-}) is already known, as illustrated in Figure 4.2. The distance between the centers of the cation (X) and the anion (A) is the sum of the two different radii ($R_A + R_X$; where R is the radius in Ångstroms or nanometers; 1 nanometer = 10Å). If the radius of one of the two is known, then the radius of the other can be obtained. Because so many common minerals contain oxygen as an anion (e.g., silicates, oxides, sulfates, phosphates), knowing the size of the anion radius of O^{2-} allows for the determination of the radius of the various cations. In 1927, Linus Pauling established a radius for O^{2-} as 1.40 Å when the oxygen anion is surrounded by six of the same cations (e.g., Mg^{2+}) as neighbors. The number of neighboring ions surrounding a central ion has a considerable effect on its radius size. In other words, an ion inside a crystal structure cannot be assumed to behave as a rigid sphere. It expands somewhat as a function of an increase in the number of closest ion neighbors, and it shrinks as a function of a decrease in the number of ion neighbors. In other words, the size of a central ion is influenced by the space provided by surrounding ions. From extensive review of the radius of oxygen in many oxygen-containing compounds, Shannon (1976) concluded that an oxygen anionic radius for O^{2-} of 1.26 Å is preferable over the 1.40 Å radius originally established by Linus Pauling. Using the 1.26 Å radius allows for the calculation of cation radii (Fig. 4.2) from the equation $R_X = d - R_A$, where d = the interatomic distance (the distance between the centers of the two ions) and R_X and R_A are cation and anion radii, respectively. These calculated cation radii (referred to by Shannon as crystal radii or effective radii) appear to correspond best to the physical size of ions in solids. Table 4.4 is a listing of the 29 most common elements in their most common ionic state and gives the radius for each ion. The top of the table, at the right, notes that the radius is a function of the **coordination number**. A coordination number is the number of closest neighbors that surround a central ion. In other words, if a cation such as Fe^{2+}, or Mg^{2+}, is surrounded by six closest neighbors of O^{2-} (as is the case in the crystal structure of olivine ((Mg^{2+}, Fe^{2+})$_2SiO_4$), it is said to have a coordination number (**C.N.**) of VI (Roman numerals are most commonly used). In the case of C^{4+} in the (CO_3)$^{2-}$ group of carbonates, the central C^{4+} cation is surrounded by three O^{2-} anions; therefore, the coordination number, **C.N.**, is III. The five columns on the right of Table 4.4 record the variability of the radii of ions in crystal structures as a function of **C.N.** As was noted earlier, these radii show that cations show an increase in size as a function of increasing **C.N.**

In conclusion, it is instructive to compare the radius values in Tables 4.3 and 4.4. This shows that the radius of a cation is considerably smaller than its atomic radius. This results from the loss of one or more outer electrons in the formation of the cation, which leads to a reduction in the overall size of the electron cloud. An anion, because it gains electrons, is larger than the corresponding neutral atom.

As we said earlier, elements (as atoms or ions) are the basic building blocks of crystal structures. Now that we have a clear picture of their size range, and the dependence of the size of an ion on the number of its closest neighbors, the question that arises is, "How do these building blocks arrange themselves inside a crystal structure?" That discussion follows.

Table 4.3 Atomic radii in Ångstroms for 12-fold coordination.

Atom	Radius	Atom	Radius
Li	1.57	Cr	1.29
Be	1.12	Mn	1.37
Na	1.91	Fe	1.26
Mg	1.60	Cu	1.28
Al	1.43	Ag	1.44
K	2.35	Sn	1.58
Ca	1.97	Pt	1.39
Ti	1.47	Au	1.44

Source: Wells (1991)

Table 4.4 Radii of common ions (in Ångstroms) as a function of coordination number.

Atomic number	Element	Ion	Radius as a function of coordination number				
			III	IV	VI	VIII	XII
3	Lithium	Li^+		0.73	0.90	1.06	
4	Beryllium	Be^{2+}	0.30	0.41	0.59		
5	Boron	B^{3+}	0.15	0.25	0.41		
6	Carbon	C^{4+}	0.06	0.29	0.30		
8	Oxygen	O^{2-}	1.22	1.24	1.26	1.28	
9	Fluorine	F^-	1.16	1.17	1.19		
11	Sodium	Na^+		1.13	1.16	1.32	1.53
12	Magnesium	Mg^{2+}		0.71	0.86	1.03	
13	Aluminum	Al^{3+}		0.53	0.68		
14	Silicon	Si^{4+}		0.40	0.54		
15	Phosphorus	P^{3+}			0.58		
		P^{5+}		0.31	0.52		
16	Sulfur	S^{2-}			1.70		
		S^{4+}			0.51		
		S^{6+}		0.26	0.43		
17	Chlorine	Cl^-			1.67		
19	Potassium	K^+		1.51	1.52	1.65	1.78
20	Calcium	Ca^{2+}			1.14	1.26	1.48
22	Titanium	Ti^{4+}		0.56	0.65	0.88	
24	Chromium	Cr^{3+}			0.76		
25	Manganese	Mn^{2+}		0.80	0.97	1.10	
		Mn^{4+}		0.53	0.67		
26	Iron	Fe^{2+}		0.77	0.92	1.06	
		Fe^{3+}		0.63	0.78	0.92	
27	Cobalt	Co^{2+}		0.72	0.88	1.04	
28	Nickel	Ni^{2+}		0.69	0.83		
29	Copper	Cu^+		0.74	0.91		
		Cu^{2+}		0.71	0.87		
30	Zinc	Zn^{2+}		0.74	0.88	1.04	
38	Strontium	Sr^{2+}			1.32	1.40	1.58
40	Zirconium	Zr^{4+}		0.73	0.86	0.98	
47	Silver	Ag^+		1.14	1.29	1.42	
56	Barium	Ba^{2+}			1.49	1.56	1.75
82	Lead	Pb^{2+}			1.33	1.43	1.63
92	Uranium	U^{3+}			1.17		
		U^{4+}			1.03	1.14	1.31
		U^{6+}		0.66	0.87	1.00	

Note: These data represent the crystal radii reported by Shannon (1976). In textbooks such as Klein and Dutrow (2008) and Dyar et al. (2008), the traditional radii (based on the radius of oxygen = 1.40 Å) are reported. The difference between crystal radii and traditional radii is a constant factor of 0.14 Å.

4.3 What factors control the packing of ions (and atoms) in mineral structures?

The best approach to this question is to consider atoms (and ions), as a first approximation, as variably sized spheres with specific electronic charges. This allows for an evaluation of how ions (and atoms) surround themselves with neighboring ions (or atoms) without consideration of the details of their electronic configuration about a central nucleus.

In a stable structure, the cation is the center of a geometric shape, and the locations of the adjoining anions describe what is known as a **coordination polyhedron**. As noted in the prior section, the number of anions in the polyhedron is the coordination number (C.N.) of the cation with respect to the anions (expressed in Roman numerals, as in Table 4.4). The actual number (e.g., IV, VI) is a function of the relative sizes of the cations and anions involved. Each coordination number reflects a specific polyhedral shape.

Here we illustrate the change in coordination number about a cation by keeping the size of the larger anion fixed throughout while increasing the size of the central cation from very small to that of the size of the anion. This mirrors the presence of the large O^{2-} anions (assumed to be of constant radius) and a wide range of cation sizes in silicate, oxide, and carbonate structures. This approach is similar to using ping-pong balls, with a constant diameter of 40 mm (20 mm radius) as the anions and evaluating how ball bearings (of continually increasing size) can be packed between the ping-pong balls. The radius of the cation is denoted as R_A; that of the anion as R_X. Relative values of radius ratio (the ratio of the size of the cation divided by that of the anion: R_A/R_X) are commonly good predictors of coordination. Here we approach this in a qualitative, visual fashion. Geometric derivations of the limiting radius ratio values are given in standard mineralogy textbooks.

Figure 4.3(A) shows a **linear** array when the cation is very small with respect to the anion. Here the cation can have only two adjoining anion neighbors. The limiting value of R_A/R_X (for this linear coordination) is 0.155. This means that when the radius of the cation is less than 15.5% of the radius of the anion, a linear arrangement (II) occurs. With a ratio bigger than this, the next-larger coordination results (as in Fig. 4.3(B)). With R_A/R_X between 0.155 and 0.225, **triangular coordination** results.

The cation is then large enough to surround itself with three anions (of fixed radius), resulting in a **C.N. = III**. Further increase in the size of the central cation results in **tetrahedral coordination** (C.N. = IV), as shown in Figure 4.3(C). This occurs between R_A/R_X values of 0.225 and 0.414. Yet further increase in the size of the cation results in 6-fold coordination (**C.N. = IV**) between R_A/R_X values of 0.414 and 0.732. This is known as **octahedral coordination** and is shown in Figure 4.3(D) with an octahedral coordination polyhedron (**C.N. = VI**). With R_A/R_X values between 0.732 and 1.0, **cubic coordination** (C.N. = VIII) results (Fig. 4.3(E)). When the cation size becomes the same as that of the anion ($R_A/R_X = 1.0$, as in Fig. 4.3(F)), structures result that are known as **closest packing**, in which the central ion or atom is surrounded by 12 closest neighbors of the same size (C.N. = XII). This type of structural arrangement is most commonly found in native metals, where all atoms are of the same size. Table 4.5 lists the most common types of coordination polyhedra found in the crystal structures of rock-forming minerals. This table, with ranges of ionic radii and corresponding numbers for C.N., is most helpful in the study of the crystal structures of minerals.

The atomic packing scheme referred to as closest packing needs some further discussion. The two different ways of stacking together spheres of equal size are discussed in Boxes 4.1 and 4.2.

Two oxides, corundum (Al_2O_3) and hematite (Fe_2O_3), have structures in which the oxygen ions are in hexagonal closest packing with the metal ions in octahedrally coordinated interstices.

To conclude this section on coordination polyhedra, let us look at three different renditions of the structure of diopside, $CaMgSi_2O_6$, a member of the pyroxene group. All three of

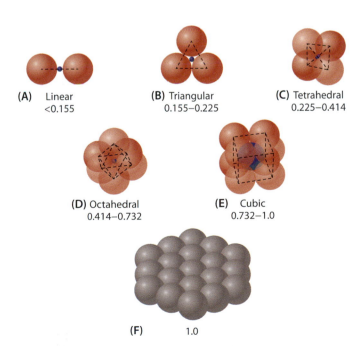

(A) Linear
<0.155

(B) Triangular
0.155–0.225

(C) Tetrahedral
0.225–0.414

(D) Octahedral
0.414–0.732

(E) Cubic
0.732–1.0

(F) 1.0

Figure 4.3 Coordination geometry as a function of limiting R_A(cation)/R_X(anion) ratios.

Table 4.5 Common ions in rock-forming minerals (exclusive of hydrogen) and their **C.N.** as a function of decreasing ionic size.

Ion	C.N. with oxygen	Ionic radius in Å
O^{2-}		1.26[IV]
K^+	VIII – XII	1.65 [VIII] – 1.78[XII]
Na^+	VI – VIII ⎫ octahedral to cubic	1.16[VI] – 1.32[VIII]
Ca^{2+}	VI – VIII ⎭	1.14[VI] – 1.26[VIII]
Mn^{2+}	VI ⎫	0.97
Fe^{2+}	VI	0.92
Mg^{2+}	VI ⎬ octahedral	0.86
Fe^{3+}	VI	0.78
Ti^{4+}	VI	0.65
Al^{3+}	VI ⎭	0.68
Al^{3+}	IV ⎫	0.53
Si^{4+}	IV ⎬ tetrahedral	0.40
P^{5+}	IV	0.31
S^{6+}	IV ⎭	0.26
C^{4+}	III triangular	0.06

Note: Ionic radii taken from Table 4.4.

BOX 4.1 | HEXAGONAL CLOSEST PACKING (HCP)

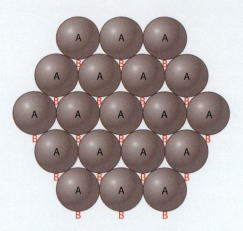

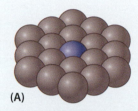

(A)

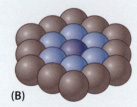

(B)

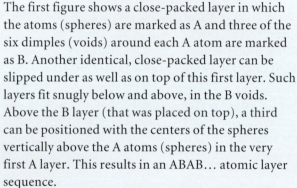

The first figure shows a close-packed layer in which the atoms (spheres) are marked as A and three of the six dimples (voids) around each A atom are marked as B. Another identical, close-packed layer can be slipped under as well as on top of this first layer. Such layers fit snugly below and above, in the B voids. Above the B layer (that was placed on top), a third can be positioned with the centers of the spheres vertically above the A atoms (spheres) in the very first A layer. This results in an ABAB… atomic layer sequence.

It is important to realize that the ABAB… sequence, as well as the one still to be discussed (in Box 4.2), leads to 12-fold coordination of the central atom. The stacking sequence in the second illustration shows this. The central sphere in (A) has six closest neighbors as shown in (B). If a close-packed layer is slipped below the original (filling the B voids) as in (C), the central ion has three additional closest neighbors, making the total number of closest neighbors 9. This is shown in (D). With an identical close-packed layer placed on top, three more close-packed neighbors are added, resulting in **C.N.** = XII (E). The close-packed coordination package that results from this three-layer stacking is shown in (F).

This 12-fold configuration of the coordination in an ABAB…sequence is represented in a close-packed package (based on a three-layer sequence) in (G). When this is redrawn in a more open ball-and-stick representation (H), the array shows as a hexagonal outline. This ABAB… sequence, therefore, is known as **hexagonal closest packing**, or **HCP**.

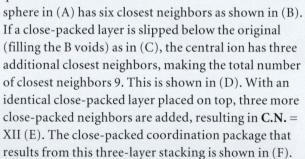

(C)

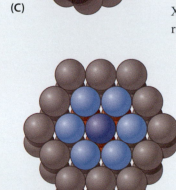

(D)

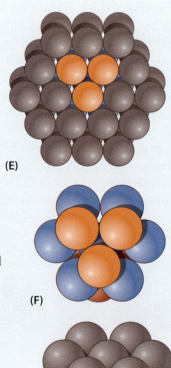

(E)

(F)

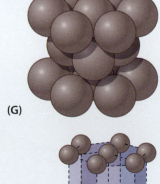

(G)

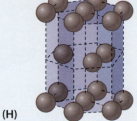

(H)

BOX 4.2 | ## CUBIC CLOSEST PACKING (CCP)

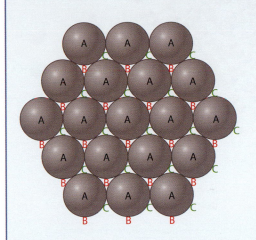

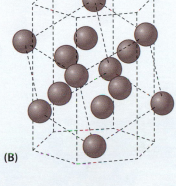

(A)

The first illustration shows the same close-packed layer as was shown in Box 4.1 but with the dimples (or voids) between the atoms (A) identified by B and C. Here, we stack two layers on top of the original A layer, first by adding a layer that fills the B voids, and subsequently a layer that fills the C voids. This results in the ABCABC…stacking sequence, known as **cubic closest packing**, or **CCP**. This ABCABC… stacking sequence leads to a different 12-fold coordination arrangement from that shown in Box 4.1. This is shown in the following three figures (A, B, and C), in which (A) represents a close-packed arrangement. (B) and (C) are expanded ball-and-stick representations of the same showing a tilted cube, in which all six cube faces are centered with an atom. The native metals Au, Cu, Ag, and Pt all have this CCP structure.

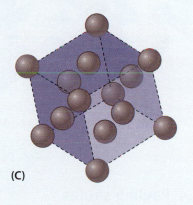

(B)

(C)

these renderings are available in CrystalViewer that accompanies this text. The crystal structures in CrystalViewer can be manipulated on the screen, to allow for views of the structures from all possible angles. Figure 4.4(A) is the most realistic representation showing the close packing of oxygens with Ca^{2+}, Mg^{2+}, and Si^{4+} in interstices (an **interstice** is a space or opening). This representation, however, provides essentially no insight into the packing schemes of the cations. Figure 4.4(B) is a ball-and-stick representation that reveals the positions of the ions and the bonds between them. On careful study of this image, it becomes evident that the small Si^{4+} ions are in tetrahedral coordination. The larger spheres (Mg^{2+}) are in octahedral coordination, and the largest spheres (Ca^{2+}) occur in an irregular coordination polyhedron with eight oxygen neighbors; this is not a regular but a distorted cube. The polyhedral rendering of the diopside structure in Figure 4.4(C) is derived from the information given in Figure 4.4(B), but fundamental aspects of the pyroxene structure become much more visible. The octahedral-tetrahedral chains parallel to the c axis are self-evident. The Ca positions are shown as spheres because representing their distorted 8-fold coordination polyhedra would make the image less understandable. In this text, especially in the discussion of rock-forming silicates, polyhedral representations are used most commonly because they offer the clearest picture of how the tetrahedral (SiO_4) polyhedra are interwoven with other cations.

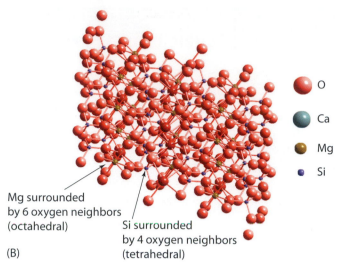

(A)

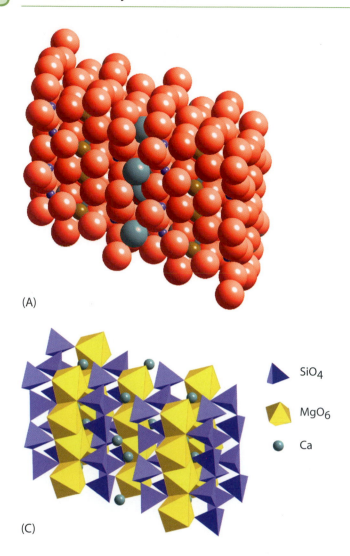

SiO₄

MgO₆

Ca

(C)

Mg surrounded
by 6 oxygen neighbors
(octahedral)

Si surrounded
by 4 oxygen neighbors
(tetrahedral)

(B)

O
Ca
Mg
Si

Figure 4.4 (**A**) A perspective view of a space-filling representation of the structure of diopside, $CaMgSi_2O_6$. The large turquoise colored spheres represent Ca^{2+}, the brown spheres are Mg^{2+}, and the small purple spheres are Si^{4+}. The large red spheres are oxygens. (**B**) Ball-and-stick representation of the same structure in the same orientation as in (A). Same color scheme as well. (**C**) Polyhedral representation of the diopside structure. This reveals the linked octahedral (Mg^{2+}) and tetrahedral (Si^{4+}) chains that run parallel to the *c* axis. The Ca^{2+} ions are represented by the turquoise spheres.

In the subsequent section, we explore some general principles that underlie crystal structures. These have been summarized by Linus Pauling in the form of five rules.

4.4 Pauling's rules

In 1929, Linus Pauling formulated five rules that provide some broad generalizations about the structure of solids, especially for those with ionic bonding:

1. The coordination principle
2. The electrostatic valency principle
3. Sharing of polyhedral elements I
4. Sharing of polyhedral elements II
5. The principle of parsimony.

Pauling was a prolific scientist whose chemical research into inorganic and organic structures using X-ray single crystal techniques has provided enormous insight into the physics and chemistry of the crystalline state. This field of research is commonly known as **crystal chemistry**. In 1954, he received the Nobel Prize in Chemistry for his research into the nature of the chemical bond and its application to the elucidation of complex substances. In

1962, he was awarded a second Nobel, the Nobel Peace Prize, because of his enormous commitment to nuclear disarmament and his very strong views on war. He was also well known for his research into the application and benefits of vitamin C.

Here are the five rules, each of which is discussed in subsequent paragraphs:

Rule 1. The coordination principle. Cation coordination numbers are determined by the ratio of the cation to anion radii, and cation-anion distances are equal to the sum of their effective radii.

Rule 2. The electrostatic valency principle. The strength of an ionic bond is equal to the ionic charge divided by the coordination number.

Rule 3. Sharing of polyhedral elements (I). Sharing of edges or faces between coordinating polyhedra is inherently unstable.

Rule 4. Sharing of polyhedral elements (II). Cations of high valence (charge) and small coordination number tend not to share polyhedral elements with each other.

Rule 5. The principle of parsimony. The number of different types of constituents in a crystal tends to be small.

Rule 1. The coordination principle.

This principle states, as we have discussed in Section 4.2, that the relative sizes of the cation and anion determine how they coordinate, or pack together. It implies that the relative ratios of R_A/R_X (R_A = cation radius and R_X = anion radius) are good predictors of how structures tend to arrange themselves in terms of different coordination polyhedra.

Let us take two examples with radii taken from Table 4.4. The radius ratio of sodium and chlorine in halite, NaCl, is arrived at as follows:

$$R_{Na}^{+[VI]} = 1.16; R_{Cl}^{-[VI]} = 1.67$$
$$R_{Na}^{+}/R_{Cl}^{-} = 1.16/1.67 = 0.69, \text{ which predicts } \mathbf{C.N.} = VI, \text{ as is}$$
seen in the structure of NaCl (Fig. 2.1(B)).

Another example is diopside, $CaMgSi_2O_6$ (see Fig. 4.4). The relevant radii are $R_{Ca}^{2+[VI]} = 1.14; R_{Mg}^{2+[VI]} = 0.86, R_{Si}^{4+[IV]} = 0.40$, and $R_O^{2-[VI]} = 1.26$. The ratio $R_{Ca}^{2+}/R_O^{2-} = 1.14/1.26 = 0.90$; the ratio $R_{Mg}^{2+}/R_O^{2-} = 0.86/1.26 = 0.68$; and the ratio $R_{Si}^{4+}/R_O^{2-} = 0.40/1.26 = 0.32$. When these radius ratios are compared with their limiting values, as given in Figures 4.3 and 4.5, the 0.90 ratio falls between 0.732 and 1.0, predicting that eight oxygen neighbors surround a central Ca^{2+} (**C.N.** = VIII). This is indeed the case in diopside, where eight oxygens surround the Ca^{2+} in the shape of a distorted cube. The 0.68 ratio falls between 0.414 and 0.732, predicting six oxygen neighbors about a central Mg^{2+} (**C.N.** = VI). This results in octahedral coordination polyhedra, as seen in Figure 4.4(C) for magnesium. The ratio of 0.32 falls between 0.225 and 0.414 and predicts that four oxygens surround the central Si^{4+} cation (**C.N.** = IV). That is the case in diopside (Fig. 4.4(C)) as it is in almost all silicate structures.

Although we have, until now, discussed only geometrically regular polyhedra, actual coordination polyhedra, as measured and determined in X-ray structure analysis, are almost always somewhat distorted. This is especially true if a small, highly charged cation is surrounded by considerably larger and polarizable anions (anions that instead of being truly spherical have a less symmetric shape as a result of directional electrostatic forces). Furthermore, although coordinations of III, IV, VI, VIII, and XII are most common, V, VII, and IX coordination may also occur. The mineral andalusite, Al_2SiO_5, has Al^{3+} in octahedral coordination (**C.N.** = VI) but also in 5-fold coordination (**C.N.** = V). These uncommon coordinations occur in complex structures in which the anions are not closely packed, which allows for the distortion of the more normal, regular polyhedron. It must also be remembered that radius ratio considerations apply best to mainly ionically bonded structures. If other bonding mechanism predominate, radius ratio predictions of coordination may not be correct.

The second part of rule 1 states that the cation anion distances are equal to the sum of their effective radii. That was shown in Figure 4.2.

At this stage, it is helpful to look back at Table 4.5, which lists 13 of the most common elements in the Earth's crust, their ionic sizes, and coordination polyhedra. This information is basic to understanding of atomic packing of mineral structures.

Rule 2. **The electrostatic valency principle.**

This rule states that the strength of a bond can be calculated by dividing the ion's charge (or valence) by its coordination number (**C.N.**); that is, bond strength (e.v.) = Z/**C.N.** (where Z is the charge on the ion, and e.v. is the electrostatic valence). This number is a measure of the strength of the bonds that exist between an ion and its closest neighbors (expressed as an absolute value; no + or − sign). We noted already that in the structure of halite, NaCl, Na^+ is in 6-fold coordination with surrounding Cl^- anions, in the shape of an octahedron. That means that each of the six bonds reaching the central Na^+ has an e.v. of 1/6. Six of these, $6 \times 1/6 = 1$, together neutralize the charge on the central Na^+ (Fig. 4.6(A)). We also showed that in the structure of diopside, $CaMgSi_2O_6$, Ca^{2+} is surrounded by eight oxygen neighbors in the shape of a distorted cube. The strength of each of the eight bonds that radiate from Ca^{2+} to the eight oxygen neighbors, turns out to be 2/8 = 1/4. Eight of these, $8 \times 1/4 = 2$, exactly neutralize the charge on the central Ca^{2+} (Fig. 4.6(B)).

The structures of minerals can be grouped on the basis of the types of bond strengths in them. Structures in which all of the bonds are of the same strength throughout are referred to as **isodesmic** (from two Greek words, *isos* and *desmo*, meaning "the same bond"). If the bond strengths inside one type of polyhedron in a structure are different from that of other bonds, the structure is referred to as **anisodesmic**, from the Greek

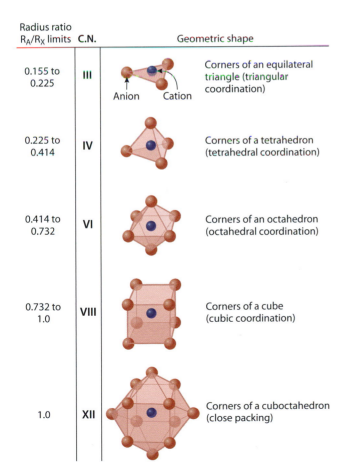

Radius ratio R_A/R_X limits	C.N.	Geometric shape	
0.155 to 0.225	III		Corners of an equilateral triangle (triangular coordination)
0.225 to 0.414	IV		Corners of a tetrahedron (tetrahedral coordination)
0.414 to 0.732	VI		Corners of an octahedron (octahedral coordination)
0.732 to 1.0	VIII		Corners of a cube (cubic coordination)
1.0	XII		Corners of a cuboctahedron (close packing)

Figure 4.5 Coordination numbers and their polyhedral representations.

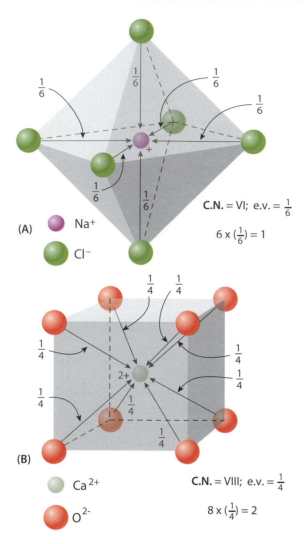

(A)   Na$^+$ C.N. = VI; e.v. = $\frac{1}{6}$

Cl$^-$ $6 \times (\frac{1}{6}) = 1$

(B)   Ca^{2+} C.N. = VIII; e.v. = $\frac{1}{4}$

O^{2-} $8 \times (\frac{1}{4}) = 2$

Figure 4.6 Examples of coordination polyhedra and the distribution of bond strengths about the central cation. (**A**) The halite, NaCl, structure with six Cl$^-$ surrounding a central Na$^+$. The six bonds each with e.v. = ⅙ together exactly neutralize the charge of the Na$^+$. (**B**) Cubic coordination of eight oxygens about a central Ca^{2+}. This occurs in the structure of diopside, CaMgSi$_2$O$_6$, in the shape of a distorted cube. Eight bonds each with e.v. = ¼ ($8 \times ¼ = 2$) together exactly neutralize the Ca^{2+} in the center.

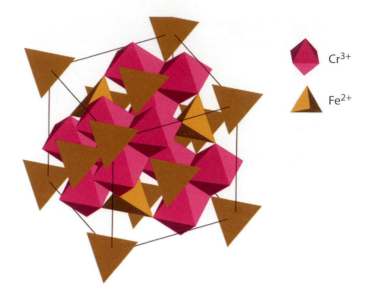

Cr^{3+}

Fe^{2+}

Figure 4.7 Perspective view of the polyhedral representation of the structure of chromite, FeCr$_2$O$_4$. The outline is a cubic unit cell. The bonds inside the tetrahedra (around Fe^{2+}) are of the same strengths as those inside the octahedra (around Cr^{3+}). Such a structure is classified as isodesmic.

electronic forces known as stabilization energies are the cause of this deviation.

Anisodesmic bond strengths occur in the structures of several major mineral groups. Examples are carbonates, sulfates, and phosphates. All members of the group known as carbonates have C^{4+} in triangular coordination with oxygen (Fig. 4.8). In this arrangement the bond strengths (e.v.) between the central C^{4+} and the three closest oxygen neighbors are 4/3 = 1⅓. This value is greater than one-half of the charge on the oxygen, leaving a residual charge of only −2/3 on the oxygens to bond with other cations. This means that in all carbonate structures, the (CO$_3$)$^{2-}$ polygons (triangles) are internally more strongly bonded than they are to other surrounding cations. Figure 4.9, a polyhedral illustration of the structure of calcite, CaCO$_3$, shows the (CO$_3$)$^{2-}$ triangles and the octahedral coordination of Ca^{2+}.

word *aniso*, meaning "unequal." A third category is known as **mesodesmic**, from the Greek root *meso*, meaning "middle." Examples of each of these follow.

Chromite, FeCr$_2$O$_4$, is a member of the spinel group of oxides, in which Fe^{2+} is in tetrahedral coordination and Cr^{3+} in octahedral coordination with oxygen (Fig. 4.7). Therefore, the bond strengths around the Fe^{2+} are 2/4 = 1/2, and the bond strengths around the octahedral Cr^{3+} are 3/6 = 1/2. With all bonds of the same strength, the structure of chromite is referred to as isodesmic. Furthermore, the structures of members of the spinel group are examples of the fact that radius ratios do not always correctly predict coordination polyhedra. Fe^{2+}, which is larger than Cr^{3+}, would be expected to reside in the octahedra, and Cr^{3+}, which is smaller, in the tetrahedra. Other

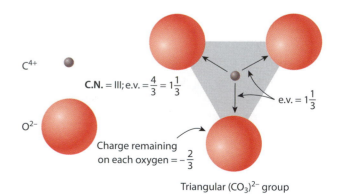

C^{4+}

O^{2-}

C.N. = III; e.v. = $\frac{4}{3} = 1\frac{1}{3}$

e.v. = $1\frac{1}{3}$

Charge remaining on each oxygen = $-\frac{2}{3}$

Triangular (CO$_3$)$^{2-}$ group

Figure 4.8 Illustration of the central C^{4+} cation in triangular (CO$_3$)$^{2-}$ coordination and the e.v. of the three bonds that radiate from the C^{4+}. In this case, the e.v. of the bonds, at 1⅓, is greater than half the charge on the oxygen anions.

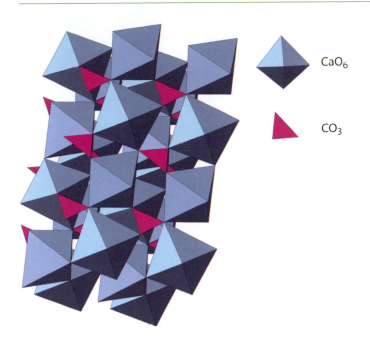

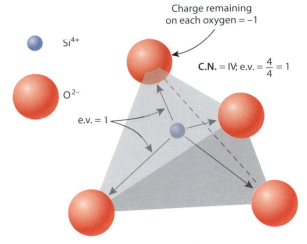

Tetrahedral $(SiO_4)^{4-}$ group

Figure 4.10 Illustration of the central Si^{4+} cation in a tetrahedral $(SiO_4)^{4-}$ group and the e.v. of the four bonds that radiate from the Si^{4+}. The e.v. of each of these bonds is $4/4 = 1$, which is exactly equal to one-half of the total bonding energy available on each oxygen ion ($\frac{1}{2} \times 2 = 1$). This allows for an apical oxygen linking up with another (SiO_4) tetrahedron. Such a linked oxygen is known as a bridging oxygen.

Figure 4.9 Perspective view of the polyhedral representation of the structure of calcite, $CaCO_3$. When looking into the inner part of this structure (not along the edges), it is clear that each corner of the triangular $(CO_3)^{2-}$ group is shared with two adjoining octahedra.

This shows that each oxygen is coordinated to two Ca^{2+} as well as to a carbon ion at the center of a triangle. Such oxygens, with a residual charge of $-2/3$ (at the corners of a triangle), share this charge between two octahedra. Therefore, each octahedral bond to a central Ca^{2+} has an e.v. of 1/3, which with $6 \times 1/3 = 2$ exactly neutralizes the central Ca^{2+}.

In the structure of sulfates, the small S^{6+} ion occurs in 4-fold (tetrahedral) coordination as $(SO_4)^{2-}$ groups. The e.v. of each of the four bonds radiating from the central S^{6+} is $6/4 = 1\frac{1}{2}$, which leaves a residual charge of only 1/2 on each of the four surrounding oxygens. The same holds for phosphates where P^{5+} occurs at the center of tetrahedral $(PO_4)^{3-}$ groups. The four bonds in the tetrahedral group each has an e.v. of $5/4 = 1\frac{1}{4}$, leaving a residual charge of only $-3/4$ on each of the surrounding oxygens. All of the foregoing examples represent anisodesmic bonding.

The linkage of $(SiO_4)^{4-}$ tetrahedra in silicate structures is described as **mesodesmic**. Figure 4.10 illustrates the $(SiO_4)^{4-}$ tetrahedron and shows that because the e.v. of the bonds between the central Si^{4+} and the four surrounding oxygens is $4/4 = 1$, the remaining charge on the four oxygens is exactly **half** of that of oxygen, which is -2. This allows some or all of the four tetrahedral oxygens to be shared between pairs of (SiO_4) tetrahedra. This ability to link (SiO_4) tetrahedra into various patterns is the basis for the structural classification of silicates. Figure 2.2(D) shows the three-dimensional linkage of (SiO_4) tetrahedra in the structure of quartz. This is referred to as a **network structure**. Figure 2.3(C) shows infinitely extending sheets of (SiO_4) tetrahedra in the structure of talc. This is one of many types of silicate

structures known as **sheet silicates**. Figure 4.4(C) illustrates the occurrence of **chains** of tetrahedra in the structure of diopside.

Rule 3. Sharing of polyhedral elements (I).

This rule states that crystal structures become less stable when coordinating polyhedra share edges and become even more unstable when they share faces. The reason for this is that when polyhedra share edges, or faces, the central cations in the polyhedra come closer together and tend to repel each other (at this shorter distance) on account of their positive charges. This is shown in Figure 4.11. This effect is large for cations with a high valence and a low coordination number. This rule relates directly to the previous comments about silicate structures. Their tetrahedra are linked only through oxygens at their respective apices (the plural of the word *apex*, meaning "tip"). Such oxygens that provide a link between two tetrahedra are known as **bridging oxygens**. Tetrahedra are rarely joined across edges and never share faces.

Rule 4. Sharing of polyhedral elements (II).

This rule is an extension of rule 3. In structures containing various cations, those of high charge (valence) and small coordination number tend not to share polyhedral elements. If they do, the shared edge contracts and the centers of the cations are displaced away from the shared edge. This minimizes cation-cation repulsion. This is shown in Figure 4.12.

Rule 5. The principle of parsimony.

This principle states that crystal structures tend to have a limited set of distinctly different cation and anion sites. This means

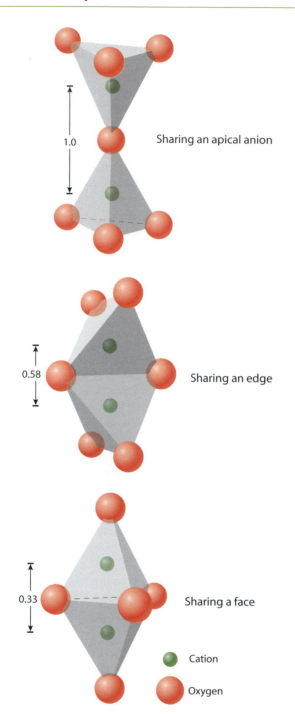

Sharing an apical anion

1.0

Sharing an edge

0.58

Sharing a face

0.33

● Cation

● Oxygen

Figure 4.11 Change in the relative distance between two cations (each centered in a tetrahedron) when the geometry is changed from linking across an apex, an edge, and a face. The distances shown reflect the relative decrease in cation-cation distance due to sharing.

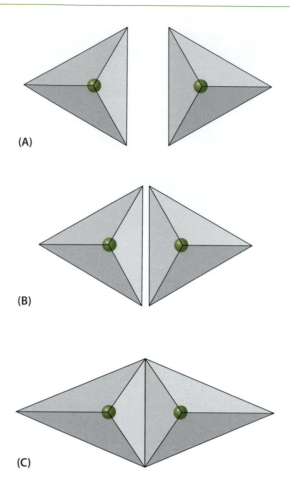

(A)

(B)

(C)

Figure 4.12 (**A**) Two well-shaped tetrahedra with a central ion of high valence. (**B**) The two tetrahedra approach each other but are still very regular in shape. (**C**) One edge is joined. The repulsion caused by the two high-valence cations shortens the shared edge and in the process distorts both tetrahedra so as to allow for larger separation between the two cations.

that in the structure of minerals of complex composition, in which many cations can replace each other (depending on ionic radius and valence), but with a limited number of atomic sites, there is extensive replacement (also known as **substitutions**, or **solid solution**) in such sites. This is best illustrated using the crystal structure of an amphibole (Fig. 4.13). The

amphibole group of minerals is one of the more complex in composition because of the wide range of cations it can house. These cations range from large Na^+ and Ca^{2+} (see Table 4.4) to intermediate Mn^{2+}, Fe^{2+}, Mg^{2+}, Fe^{3+}, Ti^{4+}, and Al^{3+}, to small Al^{3+} and Si^{4+}. The polyhedral structure illustrated in Figure 4.13 has regular tetrahedral sites that accommodate Si^{4+} and/or Al^{3+}. The regular octahedra house Mg^{2+}, Fe^{2+}, Fe^{3+}, Ti^{4+}, and Al^{3+}. The blue spheres (commonly referred to as the M_4 site) are the locations for Ca^{2+} as well as some Na^+ and Mn^{2+}. The green spheres (commonly known as the A site) house the bulk of the Na^+. This means that there are four different cation sites over which nine cations are distributed according to their radius ratios with respect to oxygen. An additional site shown in Figure 4.13, by a small pink sphere, is where $(OH)^-$ is housed. Amphiboles are hydrous minerals with the following general formula: $Na_{0-1}(Ca, Na, Mn)_2(Fe, Mg, Al, Ti)_5(Si, Al)_8O_{22}(OH)_2$. The preceding illustrates that there is a mix of ions in several distinct cation sites, and this chemical mix is the result of the restricted number of atomic sites available in the amphibole structure.

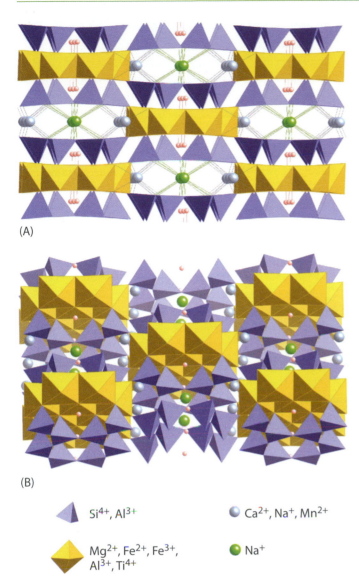

(A)

(B)

▲ Si⁴⁺, Al³⁺   ● Ca²⁺, Na⁺, Mn²⁺

◆ Mg²⁺, Fe²⁺, Fe³⁺,   ● Na⁺
Al³⁺, Ti⁴⁺

Figure 4.13 Polyhedral representation of the structure of amphibole. **(A)** The structure viewed down the vertical, c, axis. The tetrahedra are all regular and essentially identical in size. So are the octahedra. The blue spheres represent a site for irregular VI to VIII coordination, and the green spheres show the largest site with X to XII coordination. The assignment of cations to these various sites is discussed in the text. **(B)** The same structure as in (A) but now tilted slightly upward, toward the viewer, to show that the amphibole structure consists of infinitely extending chains (along the vertical, c, axis) of tetrahedra and octahedra.

4.5 What forces hold crystal structures together?

In prior sections, we discussed naturally occurring elements and their presence in mineral structures as atoms and/or ions. Subsequently, we introduced the coordination number (**C.N.**) and the concept of the coordination polyhedron, the geometric shape that describes the distribution of oppositely charged ions about a central ion. That led to the discussion of Pauling's rules. We now address how atoms and ions are bonded together to produce minerals with hardness values ranging from 1 to 10 on the Mohs hardness scale, as discussed in Section 3.5. Hardness

is a direct function of the strengths of the bonds inside a crystal structure. To this end, we first give a short overview of the electronic configuration of atoms and ions, followed by a discussion of their bonding mechanisms.

4.5.1 Electronic configuration of atoms and ions

Atoms are the smallest particles that retain the characteristics of the elements. (A table of the elements, the periodic table, is given on the inside of the front cover of this text). Small nuclei of high mass that form the center of atoms (consisting of neutrons and protons) are surrounded by a much larger volume populated by electrons.

In the earliest model of the atom, developed by Niels Bohr in 1913, the electrons are visualized as circling the nucleus in orbits (as in a planetary system; Fig. 4.14) or energy levels at fixed distances from the nucleus. The electrons are distributed in shells defined by n (e.g., $n = 1$, $n = 2$, $n = 3$, etc.) which is known as the quantum number (Fig. 4.15). In 1926, Erwin Schrödinger (1887–1961) began a field of study called wave mechanics or quantum mechanics. Schrödinger solved a mathematical equation called the wave equation, which is discussed in appendix 2 of Klein and Dutrow (2008). Here, we avoid this advanced mathematical equation but briefly discuss its results. Different types of waves are described as orbitals to distinguish them from Bohr's orbits. Each orbital has a characteristic energy and is viewed as a region around the nucleus where an electron can be expected to be found. The wave theory, therefore, visualizes the motion of electrons only in terms of the probability of finding a certain electron within a small volume. It does not depict the movement of electrons in well-defined orbits. The wave functions that describe the orbitals are characterized by three quantum numbers:

1. **The principal quantum number, n.** The energy levels in an atom are arranged as shells, as determined by the principal

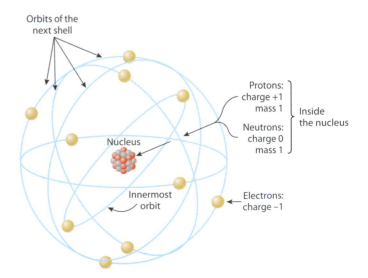

Figure 4.14 The Bohr model of the atom. Neutrons and protons in the nucleus and electrons circling outside.

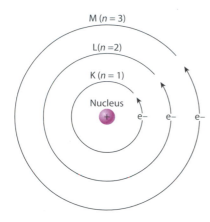

M (n = 3)

L (n = 2)

K (n = 1)

Nucleus
+
e– e– e–

Figure 4.15 In the Bohr model, the electrons are able to travel along certain specific orbits of fixed energy. The orbits are identified as K, L, M, N, … shells with specific quantum numbers, $n = 1, 2, 3, 4 … \infty$.

quantum number, n. Letters are commonly associated with these shells as follows (Fig. 4.15):

Principal quantum number	1	2	3	4
Letter classification	K	L	M	N

2. **The azimuthal quantum number, l.** The wave theory predicts that each main shell is composed of one or more subshells, each of which is specified by the azimuthal quantum number, l. This quantum number determines the shape of the orbital and, to some degree, its energy. It is common practice to give letters to the various values of l.

Value of ℓ	0	1	2	3	4	5	6
Subshell designation	s	p	d	f	g	h	i

3. **The magnetic quantum number, m.** Each subshell is composed of one or more orbitals. The magnetic quantum number restricts the orientation and shape of each type of orbital. It has integer values that range from $-l$ to $+l$. When $l = 0$, only one value of m is allowed, $m = 0$. Therefore, an s subshell consists of only one orbital (an s orbital). A p subshell ($l = 1$) contains three orbitals corresponding to $m = -1, 0,$ and $+1$. A d subshell ($l = 2$) is composed of

five orbitals and an f subshell ($l = 3$) of seven. A summary of these quantum numbers is given in Table 4.6, and a tabulation of the electron configuration of elements 1–54 is given in Table 4.7. More detailed discussion of the electronic structure of atoms is found in standard mineralogy and chemistry textbooks.

Now a few words about **isotopes**. Atoms of the same element (with the same number of protons) but with different numbers of neutrons are called isotopes. The addition of a neutron to the nucleus does not change the electrical charge of the atom but does affect its mass. For example, carbon (atomic number, $Z = 6$) has three isotopes, the most common of which has a nucleus with six protons and six neutrons and is known as ^{12}C. A much rarer isotope of carbon is ^{13}C, with six protons and seven neutrons (Fig. 4.16). Still another isotope of carbon, ^{14}C, with six protons and eight neutrons, is radioactive and decays to ^{14}N. This decay is the basis of the carbon-14 absolute dating technique, which is commonly used in archeological studies.

In the foregoing section, we dealt briefly with some basic aspects of the configuration of electrons about the atomic nucleus. This is meant as a quick review of what you learned in prior chemistry courses. You might ask, "What of this picture is essential to the understanding of bonding mechanisms in mineral structures?" The answer to this is the **valence electrons**. These are the electrons available for chemical bonding, and they represent the "mortar" that holds mineral structures together (see the partially filled outer columns in Table 4.7).

4.5.2 Chemical bonding

Atoms have a strong tendency to achieve stable, completely filled outer shells with a noble-gas configuration (see Table 4.7). This can be attained by gaining or losing electrons or by sharing of electrons. The redistribution of electrons that leads to a more stable configuration between two or more atoms is basic

Table 4.6 Summary of the three quantum numbers.

Principal quantum number, n (shell)	Azimuthal quantum number, l (subshell)	Subshell designation	Magnetic quantum number, m	Number of orbitals in subshell	Maximum number of electrons
1 (K)	0	$1s$	0	1	2
2 (L)	0	$2s$	0	1	2 ⎫ 8
	1	$2p$	$-1, 0, +1$	3	6 ⎭
3 (M)	0	$3s$	0	1	2 ⎫
	1	$3p$	$-1, 0, +1$	3	6 ⎬ 18
	2	$3d$	$-2, -1, 0, +1, +2$	5	10 ⎭
4 (N)	0	$4s$	0	1	2 ⎫
	1	$4p$	$-1, 0, +1$	3	6 ⎬ 32
	2	$4d$	$-2, -1, 0, +1, +2$	5	10 ⎪
	3	$4f$	$-3, -2, -1, 0, +1, +2, +3$	7	14 ⎭

Table 4.7 Electron configurations of atoms with atomic number 1–54.

Shell	K	L		M			N				O					P		Q	
Element	1s	2s	2p	3s	3p	3d	4s	4p	4d	4f	5s	5p	5d	5f	5g	6s	6p	6d	7s
1. H	1																		
2. He	2																		
3. Li	2	1																	
4. Be	2	2																	
5. B	2	2	1																
6. C	2	2	2																
7. N	2	2	3																
8. O	2	2	4																
9. F	2	2	5																
10. Ne	2	2	6																
11. Na	2	2	6	1															
12. Mg	2	2	6	2															
13. Al	2	2	6	2	1														
14. Si	2	2	6	2	2														
15. P	2	2	6	2	3														
16. S	2	2	6	2	4														
17. Cl	2	2	6	2	5														
18. Ar	2	2	6	2	6														
19. K	2	2	6	2	6		1												
20. Ca	2	2	6	2	6		2												
21. Sc	2	2	6	2	6	1	2												
22. Ti	2	2	6	2	6	2	2												
23. V	2	2	6	2	6	3	2												
24. Cr	2	2	6	2	6	5	1												
25. Mn	2	2	6	2	6	5	2												
26. Fe	2	2	6	2	6	6	2												
27. Co	2	2	6	2	6	7	2												
28. Ni	2	2	6	2	6	8	2												
29. Cu	2	2	6	2	6	10	1												
30. Zn	2	2	6	2	6	10	2												
31. Ga	2	2	6	2	6	10	2	1											
32. Ge	2	2	6	2	6	10	2	2											
33. As	2	2	6	2	6	10	2	3											
34. Se	2	2	6	2	6	10	2	4											
35. Br	2	2	6	2	6	10	2	5											
36. Kr	2	2	6	2	6	10	2	6											
37. Rb	2	2	6	2	6	10	2	6			1								
38. Sr	2	2	6	2	6	10	2	6			2								
39. Y	2	2	6	2	6	10	2	6	1		2								
40. Zr	2	2	6	2	6	10	2	6	2		2								
41. Nb	2	2	6	2	6	10	2	6	4		1								
42. Mo	2	2	6	2	6	10	2	6	5		1								
43. Tc	2	2	6	2	6	10	2	6	5		2								
44. Ru	2	2	6	2	6	10	2	6	7		1								
45. Rh	2	2	6	2	6	10	2	6	8		1								
46. Pd	2	2	6	2	6	10	2	6	10										
47. Ag	2	2	6	2	6	10	2	6	10		1								
48. Cd	2	2	6	2	6	10	2	6	10		2								
49. In	2	2	6	2	6	10	2	6	10		2	1							
50. Sn	2	2	6	2	6	10	2	6	10		2	2							
51. Sb	2	2	6	2	6	10	2	6	10		2	3							
52. Te	2	2	6	2	6	10	2	6	10		2	4							
53. I	2	2	6	2	6	10	2	6	10		2	5							
54. Xe	2	2	6	2	6	10	2	6	10		2	6							

Note: This range of atomic numbers is similar to that of Table 4.4.

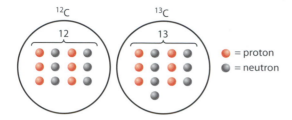

Figure 4.16 Schematic illustration of the content of the nucleus of two different isotopes of carbon, namely ^{12}C and ^{13}C.

to bonding mechanisms. The most important bonding types in mineralogy are classified as follows:

- **Covalent**
- **Ionic**
- **Metallic**
- **Van der Waals bonds**

Covalent bonding

We begin with the covalent bonding mechanism responsible for the toughness and high hardness of carbon in the form of diamond. The geometric coordination polyhedron of carbon atoms in the crystal structure of diamond is a tetrahedron, as shown in Figure 4.17(A). It consists of a central carbon atom surrounded by four closest carbon neighbors. These four carbon atoms, at the apices of a tetrahedron, share outer electrons with the central carbon, which results in a stable noble-gas configuration. This electron sharing results in what is known as a covalent bond.

In Table 4.7, carbon shows two electrons in the $2p$ orbital of the L shell. The noble gas neon (atomic number 10) has six electrons in this same location. To achieve this same noble-gas configuration, the central carbon in Figure 4.17(A) shares one electron with each of its four carbon neighbors as shown in Figure 4.17(B). This results in the central atom having achieved the highly stable noble-gas configuration of neon. This same electron sharing exists throughout the structure of diamond.

Another example of covalent bonding is in the formation of a chlorine gas molecule from two atoms of chlorine. In Table 4.7, a chlorine atom (atomic number 17) has five outer electrons in the $3p$ orbital of the M shell. When two of these atoms share one of their electrons in the outer shell, a molecule of Cl_2 with the noble-gas configuration of argon results (Fig. 4.18).

The covalent bond is the strongest of the chemical bonds and in minerals this results in great stability, insolubility, and high melting points. Such minerals are nonconductors of electricity because all available electrons are captured in the bonding mechanism.

Ionic bonding

The ionic bond is one of electron exchange instead of sharing as in covalent bonding. This bond results when one or more electrons in the valence shell of an atom are transferred to a valence shell of another. This results in electrostatic forces between two oppositely charged ions. A good example of this bonding

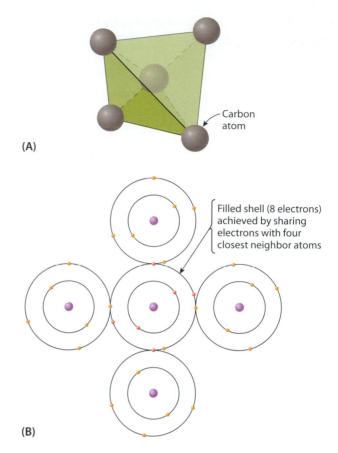

(A)

(B)

Filled shell (8 electrons) achieved by sharing electrons with four closest neighbor atoms

Figure 4.17 (**A**) Tetrahedral coordination of a central carbon with four closest carbon neighbors in the structure of diamond. (**B**) The sharing of one electron with each of four closest carbon neighbors in the structure of diamond.

mechanism is found in the crystal structure of halite, NaCl. Sodium (atomic number 11) has a single valence electron in the $3s$ orbital of the M shell which is easily lost (see Table 4.7). This leaves the atom with a single positive charge and with the noble-gas configuration of neon. Chlorine (atomic number 17) has five electrons in the $3p$ orbital of the M shell, and by gaining one electron, it attains the noble-gas configuration of argon. In this

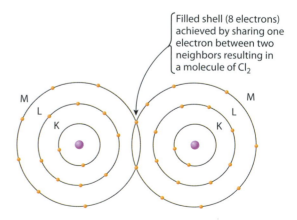

Filled shell (8 electrons) achieved by sharing one electron between two neighbors resulting in a molecule of Cl_2

Figure 4.18 The sharing of one electron between two atoms of chlorine resulting in the formation of a stable Cl_2 molecule.

process, it acquires a negative charge. In other words, if atoms of Na and Cl are in close proximity, an exchange of electrons takes place, as in Figure 4.19. The formation of this ionic bond is the result of the exchange of electrons, which leads to the formation of positive (cation) and negative (anion) charges. The physical properties of NaCl are totally different from those of the elemental constituents, Na a metal, and (Cl_2) a gas. As such, the properties of ionic structures are the properties of the ions, not the elements.

Structures that are bonded by ionic (electrostatic) forces are generally less strong than those in which the covalent bond dominates. Ionic bonding produces structures of moderate hardness, fairly high melting points, and poor electrical conductivity. The lack of electrical conductivity is the result of the stability of the ions that neither lose nor gain electrons. The electrostatic charge of the ions in the structure is evenly spread over their external shell which allows each cation (+ charge) and each anion (– charge) to surround itself with the maximum number of oppositely charged ions that space will permit.

Metallic bonding

The metallic bond is responsible for the unique properties of metals. Examples of this bonding mechanism are found in gold, silver, and copper. These metals have large radii: Au-1.44 Å; Ag-1.44 Å; and Cu-1.28 Å (from Table 4.3). The valence electrons for Au occur in the $6s$ orbital of the P shell; for Ag they occur in $5s$ orbital of the O shell; and for Cu they are found in the $4s$ orbital of the N Shell (for Ag and Cu, see Table 4.7). These outer-valence electrons in the relatively large electron clouds of metal atoms owe no affinity to any particular nucleus. As such, they are free to drift through the metal structure. This type of structure can be visualized as positively charged nuclei (of the metal atoms) with filled electron orbitals and a cloud of free-moving electrons among them. Instead of sharing electrons (as in the covalent bond) or exchanging electrons (as in the ionic bond), the drifting electrons are shared by all of the atoms.

Metals have unique properties that are the direct result of this metallic bonding mechanism. They range from plastic to ductile; they have a generally low hardness and metallic luster. Their high specific gravity is the result of both their atomic weights and their packing in close-packed structures (HCP and CCP; see Sec. 4.3). They have high electrical conductivity and are malleable, which means that they tend to deform when hammered. Ionically and covalently bonded structures are brittle and break on impact.

Van der Waals bonding

Carbon occurs naturally in two different structures (known as polymorphs; see Sec. 5.8) namely, diamond and graphite. Diamond is the prime example of covalent bonding, whereas graphite is made up of parallel sheets of strongly bonded carbon atoms that are separated from each other by a considerable distance (Fig. 4.20). Within the sheets, the C-C distance is 1.42 Å, but the distance between the sheets is 3.35 Å. This large distance is the result of a very weak bonding mechanism between them. This is known as the van der Waals bond. Neutral molecules form cohesive structures held together by such forces. This type of bond does not involve electrons but results from weak electrostatic (residual) charges on units of structures.

The clay mineral group is one in which this type of bond is prevalent. Clay minerals (Fig. 4.21) consist of infinitely extending layers made of tetrahedral and octahedral sheets that are strongly bonded internally, but the bonds that bind one layer to the other are weak van der Waals bonds. This accounts for their excellent cleavage and low hardness (**H** of kaolinite, $Al_2Si_2O_5(OH)_4$, = 2). The van der Waals bond is the weakest of the four bond types.

In the preceding discussion of chemical bonding, we have treated each type separately, but we must point out that actual mineral structures may consist of a mixture of bond types as well as gradations between bond types as illustrated in Figure 4.22. The bonding in diamond is totally covalent, but the bonding mechanism in silicates is about 50% ionic and 50% covalent. In graphite and clay minerals, the bonding within the layers of these structures is a mixture of ionic and

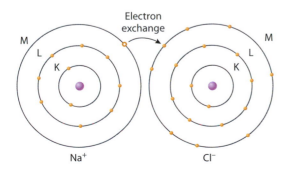

Figure 4.19 The exchange of an electron from an atom of Na to one of Cl, resulting in the stable ionic structure of halite, NaCl.

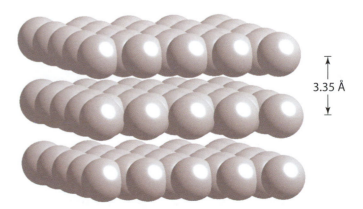

Figure 4.20 Perspective view of a close packed-structure representation of graphite. The spacing between successive layers is 3.35 Å.

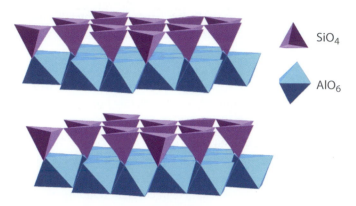

Figure 4.21 Perspective view of the polyhedral representation of the structure of kaolinite, $Al_2Si_2O_5(OH)_4$, one of several clay minerals. The tetrahedral (SiO_4)–octahedral (AlO_6) package is neutral, and the large spacing between the layers represents van der Waals bonding.

covalent, but the weak bonding between the layers is of the van der Waals type. In many metals the bonding is essentially metallic, but in sulfides it is a mixture of covalent and metallic bonding. Table 4.8 is a summary of selected properties as a function of bond type.

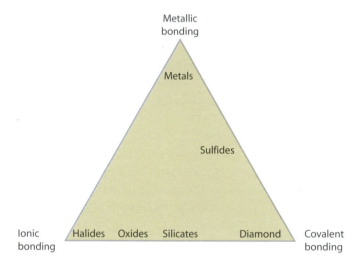

Figure 4.22 Depiction of the gradational and/or mixed nature of bond types in many mineral structures. The corners represent pure bonding of one type.

4.6 Atomic substitutions

At the end of Section 4.4, we discussed the implications of Pauling's rule number 5, known as the principle of parsimony. This addresses the fact that mineral structures have a limited set of atomic (or ionic) sites in which a much larger number of elements may need to be housed. As an example, we used the amphibole structure (Fig. 4.13), which has four distinct structural sites over which nine chemical elements may be distributed. In other words, chemical elements substitute for each other in a specific structural site. This atomic substitution is commonly referred to as solid solution, which means that a mineral may range in chemical composition within some finite limits. This relates directly to a clause in the definition of a mineral (Sec. 2.1) that states that a mineral "has a definite (but commonly not fixed) chemical composition." An example of such is the mineral olivine. Its structure (Fig. 4.23) consists of independent tetrahedra and chains of linked octahedra. The tetrahedral sites are occupied by Si^{4+} and the octahedra accommodate Mg^{2+} and/or Fe^{2+}. This leads to a general formula for olivine as $(Mg, Fe)_2SiO_4$ with definite cation to anion ratios of 2:1:4. However, the statement $(Mg, Fe)_2$ implies that there is a mineral with composition Mg_2SiO_4 as well as one with Fe_2SiO_4, and a whole range of olivines of variable Mg-Fe ratios in between. The fixed compositions, Mg_2SiO_4 and Fe_2SiO_4, are commonly referred to as **end members**, and the series of intermediate compositions, between the end members, is known as the extent of solid solution. This means that in the structure of olivine Mg^{2+} and Fe^{2+} can substitute for each other, in the octahedral sites, in all proportions. Examples of such intermediate compositions are $(Mg_{0.6}Fe_{0.4})_2SiO_4$ and $(Mg_{0.3}Fe_{0.7})_2SiO_4$. Solid solution can be defined as a range of composition in a mineral in which specific atomic sites may be occupied by two or more different chemical elements (or ions) in variable proportions.

4.6.1 Factors responsible for the extent of atomic substitution (solid solution)

The compositional variability of a mineral results from atomic (ionic) substitutions in a specific site of its structure. The four

Table 4.8 Characteristic properties resulting from the principal chemical bond types.

Examples and properties	Covalent	Ionic	Metallic	Van der Waals
Examples	Diamond, C Sphalerite, ZnS Organic molecules	Halite, NaCl Fluorite, CaF_2 Most minerals	Most metals	Weakest bond in minerals; clays and graphite
Hardness	Very hard; brittle	Medium to high	Low to medium; sectile, ductile, malleable	Very low; crystals soft
Electrical conductivity	Insulators	Poor conductors	Good conductors	Insulators
Melting point (m.p.)	High m.p.	Moderate to high m.p.	Variable m.p.	Low m.p.
Structure and symmetry	Generally lower coordination and lower symmetry	High coordination and symmetry	Very high coordination and symmetry	Low symmetry

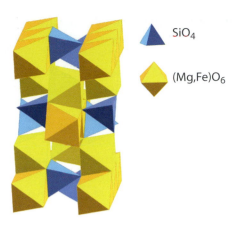

Figure 4.23 Perspective view of the polyhedral representation of the structure of olivine $(Mg, Fe)_2SiO_4$ showing independent (nonlinked) tetrahedra of composition $(SiO_4)^{4-}$ and chains of linked octahedra in which Mg^{2+} and/or Fe^{2+} reside.

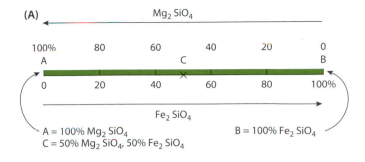

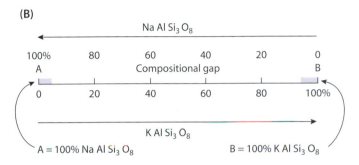

Figure 4.24 (A) Example of a complete solution series as in olivine. The Mg end member is known as forsterite and the Fe end member as fayalite. All compositions intermediate between these two end members are possible. (B) Example of the very limited solid solution between two types of feldspar, at low temperature. $NaAlSi_3O_8$ is known as albite and $KAlSi_3O_8$ as microcline. The large compositional range in which no homogeneous feldspar compositions exist is known as a compositional gap. The scales used in both (A) and (B) are molecular percentage of the formula unit, or atomic percentage of the cation involved. Both are equivalent.

factors that influence the amount of atomic substitution (or solid solution) are the following:

1. The comparative sizes of the atoms, ions, or ionic groups involved
2. Their electric charges (valence)
3. The temperature at which the substitution takes place
4. The availability of the atoms (ions) involved

We now discuss each of these general aspects of solid solution.

The size of the atoms or ions. A wide range of substitution is possible if the size differences between the substituting elements are less than about 15%. If they differ by 15% to 30%, substitution is limited, and if they differ by more than 30%, solid solution is generally not possible. Table 4.9 gives examples of the extent of solid solution as a function of the difference in ionic radii. In graphical illustrations, the extent of solid solution is commonly illustrated by a **bar diagram**. Figure 4.24

Table 4.9 Extent of solid solution as a function of ionic radius.

Radius of $Mg^{2+[VI]} = 0.86Å$ Radius of $Fe^{2+[VI]} = 0.92Å$	difference 6.5% - complete solid solution
Radius of $Fe^{2+[VI]} = 0.92Å$ Radius of $Mn^{2+[VI]} = 0.97Å$	difference 5% - complete solid solution
Radius of $Mg^{2+[VI]} = 0.86Å$ Radius of $Mn^{2+[VI]} = 0.97Å$	difference 11% - generally not complete
Radius of $Fe^{2+[VI]} = 0.92Å$ Radius of $Ca^{2+[VI]} = 1.14Å$	difference 19% - partial
Radius of $Na^{2+[VIII]} = 1.32Å$ Radius of $K^{+[VIII]} = 1.65Å$	difference 20% - partial

Note: Radii taken from Table 4.4. Square brackets contain the coordination numbers. The percentage difference is calculated as follows: $0.92 - 0.86 = 0.06$; $0.06/0.92 \times 100 = 6.5\%$.

shows examples of complete and partial solid solution using such compositional bars. These same composition bars are used as the horizontal axis in temperature-composition illustrations that are known as **variation diagrams**. Such diagrams plot the variation in some physical property along the vertical axis with the horizontal axis being a composition bar. Figure 4.25 shows the variation in specific gravity (**G**) as well as three refractive indices of the olivine series as a function of composition.

The charge on the ions. When the charges of the ions that substitute for each other are the same, the crystal structure remains overall electrically neutral. The first four ion pairs listed in Table 4.9 have identical charges. If, however, ions of different charge (but similar size) substitute for each other, additional substitutions elsewhere in the structure must occur to maintain overall neutrality. This is known as **coupled substitution**. In the crystal structure of the plagioclase feldspar series, ranging in composition from albite, $NaAlSi_3O_8$, to anorthite, $CaAl_2Si_2O_8$, substitution of Na^+ by Ca^{2+} is coupled with simultaneous substitution of Al^{3+} for Si^{4+} as follows:

$$Na^+ + Si^{4+} \rightleftarrows Ca^{2+} + Al^{3+}$$

The overall charge on either side of this equation is 5^+, so the structure remains neutral. This coupled substitutional

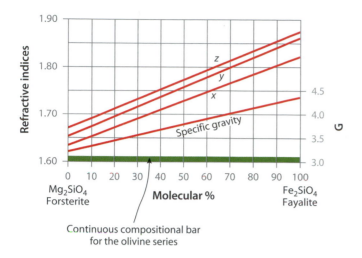

Figure 4.25 Variation diagram for refractive indices and specific gravity (**G**) as a function of the continuous compositional range (complete solid solution) between two end members of the olivine series.

mechanism implies that when one Ca^{2+} replaces an Na^+, a simultaneous substitution of Si^{4+} by Al^{3+} must occur. Figure 4.26 helps illustrate where this happens in the structure of the plagioclase series. The $Si^{4+} \rightleftarrows Al^{3+}$ exchange happens inside the tetrahedral framework and the $Na^+ \rightleftarrows Ca^{2+}$ replacement occurs in a large (relatively open) site of irregular (9-fold) coordination.

Such coupled substitutions are not limited to members of the feldspar group. They occur in many silicate structures. For example, in the structure of diopside, $CaMg_2SiO_6$ (a member of the pyroxene group), there is some substitution of Na^+ for the Ca^{2+}, and the concurrent substitution of Mg^{2+} by Al^{3+}. This coupled substitution is as follows: $Ca^{2+} Mg^{2+} \rightleftarrows Na^+ Al^{3+}$. The overall charge on either side of this equation is the same, namely 4+. Although neutrality is maintained in this substitution, it does not result in an extensive solid solution series between $CaMgSi_2O_6$ and $NaAlSi_2O_6$ (jadeite) because the sizes of the ions involved are quite different.

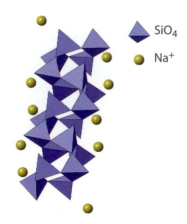

Figure 4.26 Perspective view of a small section of the structure of albite, $NaAlSi_3O_8$. It shows the infinitely extending framework of tetrahedra that house Si^{4+} and/or Al^{3+}. The spheres are the large sites where Na^+ and/or Ca^{2+} occur in irregular 9-fold coordination.

The temperature at which atomic substitution occurs. The crystal structures of minerals, as well as of most other materials, expand at high temperatures. At higher temperatures, when thermal vibrations inside a structure are greater, the size of atomic sites available for substitution increases. As a result, a structure at a higher temperature is more tolerant of divergent ionic (atomic) sizes than one at a lower temperature. This means that if there is only limited solid solution between two end member compositions at low temperatures, it may increase at higher temperatures. At the highest temperature, but below the melting temperature of the chemical system, this may well lead to a complete solid solution series. This is schematically illustrated in Figure 4.27, which is known as a **temperature-composition diagram**. The horizontal axis is the compositional bar (with a compositional gap) between two members of the pyroxene group: enstatite, $MgSiO_3$, and augite, $Ca(Mg, Fe)Si_2O_6$. The vertical axis represents increasing temperature. The central domed region is one in which homogeneous pyroxene compositions are not possible. Above about 1160°C there is a complexity at the top left, but the general picture is that at about 1250°C the dome is closed. This means that above this temperature a complete solid solution series exists between the two end members of the composition bar. The compositional gap is known as a **miscibility gap**, and this feature is responsible for what are known as exsolution textures in similar mineral systems that cool from high temperatures (see chapter opening photo in Chapter 6 and Sec. 7.12). The lack of solid solution at the lowest temperature in this diagram is understood if first we rewrite the enstatite formula as

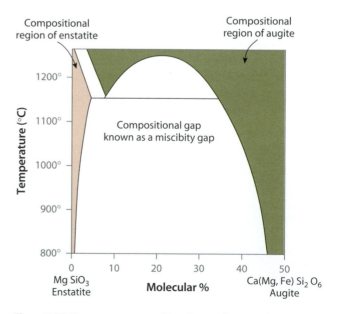

Figure 4.27 Temperature-composition diagram for two end member compositions in the pyroxene group: enstatite, $MgSiO_3$ (also written as $Mg_2Si_2O_6$) and augite, $Ca(Mg, Fe)Si_2O_6$. At low temperatures (~800°C), solid solution between these two compositions is limited. At about 1160°C there is more atomic substitution in enstatite and considerably more in augite. At the very top of the diagram (~1250°C), solid solution between the two pyroxene compositions is essentially complete (except for the complication at the far left).

$Mg_2Si_2O_6$ (instead of $MgSiO_3$), and second if we provide coordination formation for the substituting cations. This results in $Mg^{VI}Mg^{VI}Si_2O_6$ (for enstatite) and $Ca^{VIII}(Mg^{VI}, Fe^{VI})Si_2O_6$ (for augite). At low temperatures, the larger Ca^{2+} cannot be accommodated in the smaller coordination site for Mg^{2+}, and Mg^{2+} does not fit well in the larger Ca^{2+} site. At high temperatures, divergent ionic sizes are more easily accommodated.

The availability of the ions. To make solid solution possible, a ready supply of the substituting ions must be available. In a chemical environment, for example, where Mn is rare but Fe is abundant, Mn^{2+} has little chance substituting for Fe^{2+}.

4.6.2 Types of solid solution

Solid solution occurs by three different mechanisms. These are known as

- **substitutional**
- **interstitial**
- **omission solid solution**

Substitutional solid solution. This includes simple cationic and anionic substitutions as well as coupled substitution. Examples of cationic substitutions (simple and coupled) were discussed in Section 4.6. An example of complete solid solution between two anions occurs in the chemical system KCl-KBr, where the radius of $Cl^{-[VI]} = 1.67$ Å and of $Br^{-[VI]} = 1.82$ Å. The size difference between these two anions is 0.15 Å, which recalculates as a percentage size difference of $0.15/1.67 \times 100\% = 9\%$. This being well below a 15% difference results in a complete anionic solid solution series.

Interstitial solid solution. This occurs in structures in which a specific atomic site is normally empty but that under different chemical circumstances can house additional ions, ionic groups, or molecules. A good example of this is in the structure of beryl, $Be_3Al_2Si_6O_{18}$. This structure (Fig. 4.28) consists of tetrahedral silicate rings (Si_6O_{18}) that are cross-linked by Al in octahedrally and Be in tetrahedrally coordinated polyhedra. The silicate rings, which are stacked directly above each other, along the vertical, c, axis of the structure, provide large tubular openings with a diameter of about 4.4 Å. These can house H_2O as well as CO_2 molecules that are weakly bonded with the oxygens of the surrounding Si_6O_{18} rings. Large ions such as Na^+, K^+, and Cs^+ can also be housed inside these silicate rings, but these ions are more strongly bonded than neutral molecules because they are involved in coupled substitution such as $Si^{4+} \rightleftarrows Al^{3+} +$

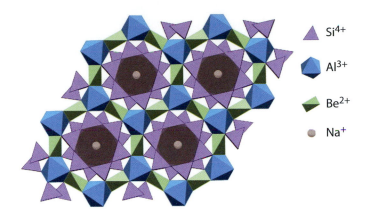

Figure 4.28 Polyhedral representation of the structure of beryl, $Be_3Al_2Si_6O_{18}$. This view is looking down the vertical, c, axis of the structure. Na^+ is housed in the large hexagonally shaped openings. With 0.3 atomic percentage Na in this structure, the formula becomes (with coupled substitution of Al^{3+}, see text) $Na_3Be_3Al_2(Si_{5.7}Al_{0.3})O_{18}$.

Na^+. The charges on both sides of this equation are the same, +4. The Al^{3+} substitutes for Si^{4+} in the tetrahedral positions, and the large Na^+ is accommodated at the center of the large vertical, normally open (or vacant) channels of the beryl structure. The structure in Figure 4.28 shows the Na^+ ions in the centers of the rings. This is referred to as interstitial solid solution because additional molecules or ions (not normally present) are housed in the interstices (this word means "opening") of the structure.

Omission solid solution. This refers, as the word *omission* implies, to unfilled or vacant atomic sites in a crystal structure. Such structures are commonly referred to as **defect** structures. The best mineralogical example of this is found in the structure of pyrrhotite, $Fe_{1-x}S$. If this structure showed no vacant atomic sites, its formula would be FeS. The subscript $(1-x)$ means that there is a variable number of vacancies in the octahedral site normally occupied by Fe^{2+}, with x ranging from 0 to 0.2. The structural sites in which S occurs are always completely filled, and the vacancies occur only in the sites normally occupied by Fe^{2+}. With some Fe^{2+} absent from the structure of pyrrhotite, there must be an additional substitutional mechanism to keep the overall structure neutral. It is likely that some of the remaining Fe is oxidized to Fe^{3+} resulting in a pyrrhotite formula of

$$(Fe^{2+}_{1-3x}\ Fe^{3+}_{2x})\square_x S,$$

where $\square$ represents vacancies in the Fe sites.

Summary

This chapter presented some basic aspects of the crystal structures of minerals. It described how atoms (or ions, or ionic groups) are packed in regular three-dimensional arrays and what bonding mechanisms hold crystal structures together. Finally, we looked into the reasons for the variability of the chemical composition of many minerals:

- In rock-forming minerals the basic building blocks are the 29 elements in Table 4.1, listed in their most common ionic state with associated radii. Although 99% of the Earth's crust is made up of only 13 elements (Table 4.2), the additional elements listed in Table 4.1 represent minor and trace elements in minerals.

- In the "packing" of atoms (or ions) to form regular atomic (ionic) arrangements inside crystal structures, it is important to pay close attention to the distribution of oppositely charged ions (e.g., anions) about a central ion (e.g., cation). That led to our discussion of the coordination number (**C.N.**) and coordination polyhedron.

- Coordination polyhedra range from triangular, to tetrahedral, to octahedral, to cubic and finally to hexagonal and cubic closest packing (HCP and CCP). It is important to recognize that the size of an ion (its radius) is dependent on the number of closest neighbors (its **C.N.**) and that cations tend to increase in size somewhat as a function of the increasing size of the anionic polyhedron that surrounds it.

- Knowing the size (the radius) of a cation, we can predict with considerable certainty the type of polyhedron it will form in oxygen-rich compounds such as silicates, oxides, carbonates, sulfates, and so on. This is conveyed in Table 4.5.

- Next we discussed Pauling's rules. Rule 1 states that the cation coordination number is determined by the ratio of cation to anion radii ($R_X:R_A$) and that the cation-anion distance is the sum of their effective radii. This was also discussed in Section 4.3.

- Rule 2 states that the strength (e.v.) of an ionic bond is equal to the ionic charge divided by the coordination number (e.v. = $Z/C.N.$). For example, the e.v. between Na^+ and Cl^-, in the crystal structure of halite, is 1/6 because each Na^+ is surrounded by six closest Cl^- neighbors in octahedral (VI) coordination.

- Rules 3 and 4 relate to the sharing of polyhedral elements. Sharing of edges and faces of adjoining polyhedra destabilizes a crystal structure. Polyhedra with cations of high charge (valence) at their center tend not to share any polyhedral elements with each other because of the repulsive forces between the central cations.

- Rule 5, the principal of parsimony, addresses the fact that crystal structures have a limited number of cation sites with a specific **C.N.** and coordination polyhedron. This means that if a large number of different cations are to be housed in a crystal structure, they must substitute for each other in appropriate sites.

- The question of what holds crystal structures together is discussed in Section 4.5.1 with a brief overview of the electronic configuration of atoms and ions. This short section is basic chemistry that you have probably learned in prior chemistry courses.

- That section is followed by the various mechanisms that bond ions (or atoms) together. These are covalent, ionic, metallic, and van der Waals bonding mechanisms. Covalent (electron sharing) is the strongest bond. Ionic, the result of electron exchange, is less strong but very common in minerals such as silicates, oxides, and carbonates. Metallic bonding in metals is a unique bond that results in the malleability and high electrical conductivity of metals. Van der Waals bonding, a residual bond, is commonly the weakest bond in a crystal structure.

- The chapter ends with a discussion of the factors that influence the extent of atomic substitution in crystal structures. Such factors are (1) the comparative sizes of atoms (and ions) substituting for each other; (2) their valences (electric charges); and (3) the temperature at which the atomic substitution takes place.

- If ionic (atomic) sizes are close to the same (<15% difference in size) and charges are identical, solid solution will be extensive. Considerable divergence in size leads to partial or no solid solution.

- If the charges on the ions substituting for each other are not the same, additional charge balance must occur inside the crystal structure through coupled substitution.

- The higher the temperature at which substitution occurs in the solid state, the more tolerant the structure is of accommodating ions (or atoms) of divergent size.

- The types of solid solution are known as substitutional, interstitial, and omission solid solution.

Review questions

1. Why is oxygen the first element listed in Table 4.2? Give two reasons.

2. How is the radius of a cation obtained in an oxygen-rich compound if the radius of oxygen is known?

3. What is meant by coordination number (**C.N.**)?

4. What is a coordination polyhedron?

5. Give the range of shapes of coordination polyhedra.

6. If Mg^{2+} occurs in 6-fold coordination in a crystal structure, how is that written out in the scientific notation used in Table 4.4?

7. What is the size of that same Mg^{2+}?

8. What is meant by closest packing?

9. What are two different closest-packing schemes?

10. How is bond strength defined?

11. What is the strength of the bonds that surround a central Fe^{2+} in octahedral coordination with oxygen?

12. Why do tetrahedra consisting of (SiO_4) only share corners, not edges, or faces?

13. What is meant by the term *mesodesmic*?

14. In the structures of what mineral group do mesodesmic bonds occur?

15. What electrons, in the nuclear structure of elements, are of prime importance in chemical bonding?

16. What happens to electrons in a covalent bond?

17. What happens to electrons in an ionic bond?

18. What is the role of electrons in metallic bonding, where there is malleability and high electrical conductivity?

19. What silicate mineral group has weak structures due to van der Waals bonding?

20. What are two basic parameters that allow for ease of atomic substitution of one ion for another?

21. What is meant by *coupled substitution*?

22. Why does an increase in temperature of a solid (a mineral) allow for increased atomic substitution by various elements (or ions)?

23. What is meant by *interstitial solid solution*?

24. What is meant by *omission solid solution*?

FURTHER READING

The references that follow are mainly standard texts that allow for further study of the subjects covered in this chapter. The article by Shannon (1976) is the source of all ionic radii.

Dyar, M. D., Gunter, M. E., and Tasa, D. (2008). *Mineralogy and Optical Mineralogy*, Mineralogical Society of America, Chantilly, VA.

Klein, C. (2008). *Minerals and Rocks: Exercises in Crystal and Mineral Chemistry, Crystallography, X-Ray Powder Diffraction, Mineral and Rock Identification, and Ore Mineralogy*, John Wiley and Sons, New York.

Klein, C., and Dutrow, B. (2008). *Manual of Mineral Science*, 23rd ed., John Wiley and Sons, New York.

Pauling, L. (1960). *The Nature of the Chemical Bond and the Structure of Molecules and Crystals: An Introduction to Modern structural Chemistry*, W. H. Freeman and Company, San Francisco.

Perkins, D. (2011). *Mineralogy*, 3rd ed., Prentice Hall, Upper Saddle River, NJ.

Shannon, R. D. (1976). Revised Effective Ionic Radii and Systematic Studies of Interatomic Distances in Halides and Chalcogenides. *Acta Crystallographica*, A32, 751–767.

Wells, A. F. (1991). *Structural Inorganic Chemistry*, 5th ed., Clarendon Press, Oxford, UK.

Wenk, H. R., and Bulakh, A. (2004). *Minerals: Their Constitution and Origin*. Cambridge University Press, Cambridge.

5 Introduction to crystallography

When minerals occur as well-developed (euhedral) crystals, they reveal symmetry. Figure 2.5(B) shows garnets as perfect dodecahedra (a special form in the isometric system). Figure 2.6 shows very well-formed but less symmetric crystals of green microcline, and Figure 2.8 shows four different crystals with very different outward appearances and symmetry contents. These well-developed crystals are the outward expression of the underlying internal order of their crystal structures.

Symmetry deals with repetition of objects (atoms, ions, ionic groups) through reflection, rotation, inversion, and translation. The study of the external form and the crystal structure of crystalline solids and the principles that govern their growth, shape, and internal atomic arrangement is called **crystallography**. Here we begin with aspects of the external symmetry (**morphology**) of crystals, because this is generally simpler to evaluate than the more complex symmetry content of the underlying crystal structure. We proceed with that in Section 5.7 of this chapter.

Three different types of carbon. A gem quality octahedron of diamond, 9.2 ct in size (1 carat = 0.2 grams), from Kimberley, South Africa. A fibrous specimen of graphite from Roger's Ford, Mt. Burgess, Canada. Diamond and graphite are two polymorphs of carbon (polymorph is derived from two Greek words, *poly* and *morpho*, meaning "many forms"; see Sec. 5.8) with very different crystal structures. Also a round, 116 gram specimen of **bort** from Bahia, Brazil (bort is a microcrystalline brown-black aggregate of diamond used in industry). (David Nufer Photography, Albuquerque, New Mexico.)

The basic aspects of crystallography that are described in this chapter are probably new to most readers. Chemical concepts that were treated in Chapter 4 were mostly not new because of earlier chemistry courses. In crystallography, very basic aspects such as symmetry elements are later incorporated into more complex concepts. In other words, earlier learned materials sequentially build into later, conceptually more difficult ideas.

We begin with symmetry elements that can be seen in well-formed (euhedral) crystals. We subsequently assess combinations of symmetry elements and see how these can be grouped and referred to various sets of reference axes. This then leads to what are called crystal classes (or point groups). And yet further on, we evaluate symmetry elements and translations in crystal structures, instead of just those that are represented by morphological crystals.

This necessary, sequential approach is best expressed by the concept map that follows:

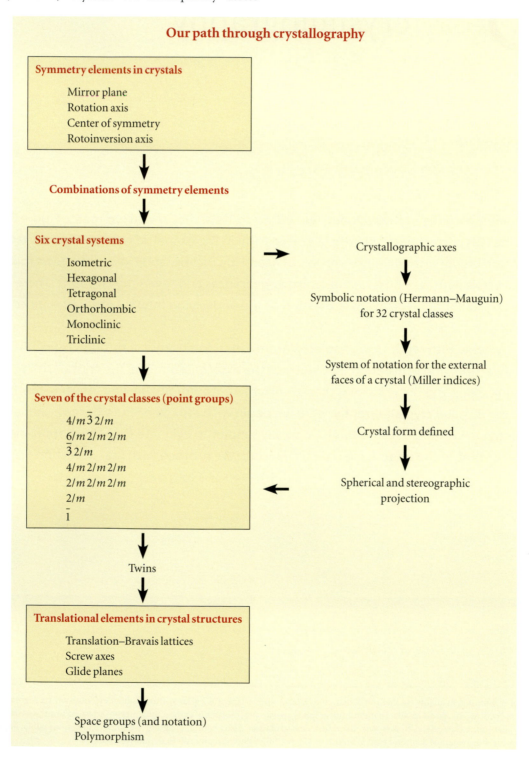

Our path through crystallography

Symmetry elements in crystals
- Mirror plane
- Rotation axis
- Center of symmetry
- Rotoinversion axis

↓

Combinations of symmetry elements

↓

Six crystal systems
- Isometric
- Hexagonal
- Tetragonal
- Orthorhombic
- Monoclinic
- Triclinic

→ Crystallographic axes

↓

Symbolic notation (Hermann–Mauguin) for 32 crystal classes

↓

System of notation for the external faces of a crystal (Miller indices)

↓

Seven of the crystal classes (point groups)
- $4/m\,\bar{3}\,2/m$
- $6/m\,2/m\,2/m$
- $\bar{3}\,2/m$
- $4/m\,2/m\,2/m$
- $2/m\,2/m\,2/m$
- $2/m$
- $\bar{1}$

Crystal form defined

↓

← Spherical and stereographic projection

↓

Twins

↓

Translational elements in crystal structures
- Translation–Bravais lattices
- Screw axes
- Glide planes

↓

Space groups (and notation)
Polymorphism

5.1 Symmetry elements and operations

The geometric feature that expresses the symmetry of an ordered arrangement is known as a **symmetry element**. The process that results in this symmetry element is known as a symmetry operation. Although these terms imply motion, the faces on a crystal or the atoms inside a crystal structure do not move. The operations are conceptual analytical elements that help us better visualize the geometry of crystals and crystal structures. Four of these symmetry elements and operations are as follows:

Element	Operation
Mirror plane (m)	Reflection through a plane
Rotation axis	Rotation about an axis by some fraction $360°/n$, where $n = 1, 2, 3, 4,$ or 6.
Center of symmetry (i)	Inversion through a point
Rotoinversion axis	Rotation with inversion

The first two symmetry elements are illustrated in Figure 5.1 using some common objects as well as a euhedral (well-formed) crystal.

Mirrors.

These reflect an object, or a specific crystal face, or a unit of structure, into its mirror image. Mirrors are identified by the letter m in drawings of crystals and crystal structures. Their presence is also shown by bold solid lines in illustrations.

Rotation axes.

These are imaginary lines about which an object, or a crystal face, or an atomic arrangement (inside the crystal structure), is rotated and repeats itself once or several times during a complete (360°) rotation. Rotation axes are numbered as $n = 1, n = 2, n = 3, n = 4,$ and $n = 6$, where n represents a divisor of 360°. For example, $n = 1$ means that the crystal face, or an atom inside a structure, is repeated every 360°. And $n = 4$ means that the crystal face, or atom, is repeated four times during a 360° rotation, that is, every 90° of rotation about the axis. A list of rotation axes, their angles of rotation, and their standard graphical symbols is given in Figure 5.2.

The operations involved in a center of symmetry and rotoinversion are illustrated in Figure 5.3.

Center of symmetry.

This is also known as inversion (i). This symmetry element is difficult to recognize inside a crystal model or image of a crystal structure, but in a wooden model of an external crystal form (as is commonly used in laboratories dealing with crystallography), it is easily identified. With a wooden crystal block laid down on the table on any face, observe whether an equivalent face of the same size and shape is present in a horizontal position at the top of the crystal model. If so, it contains a center of symmetry. This is equivalent to saying that every face on a crystal with a center of symmetry has an equivalent, parallel face on the opposite side. The cube illustrated in Box 5.1 consists of three pairs of parallel faces. This crystal contains a center of symmetry because every face, when laid down on a table, has an equivalent face parallel to the table top. The operation of inversion is illustrated in Fig. 5.3(A).

Rotoinversion axes.

These are imaginary lines about which a crystal face or an atom (or atom cluster) in a crystal structure is rotated as well as inverted. Rotoinversion axes are numbered as $n = \bar{1}, n = \bar{2}, n = \bar{3}, n = \bar{4},$ and $n = \bar{6}$. The numbers reflect the same angles of rotation as in the rotation statements. For example, $n = \bar{4}$ is the result of 4-fold rotation (four repeats at

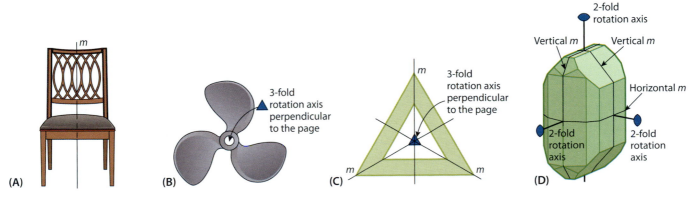

Figure 5.1 Symmetry elements in some common objects and in a euhedral crystal of olivine. (**A**) The chair displays a vertical mirror (m). (**B**) A ship's propeller with a 3-fold rotation axis perpendicular to the page. (**C**) An equilateral triangle (used in drafting) contains a 3-fold rotation axis perpendicular to the page and three mirrors (m) that intersect in the 3-fold axis. (**D**) The olivine crystal displays three mutually perpendicular 2-fold axes as well as three mirror planes, each pair of which intersects along the 2-fold axes.

Type of rotation axis	Angle of rotation	Symbol or letter
1-fold rotation = 1	360°	none
2-fold rotation = 2	180°	◆
3-fold rotation = 3	120°	▲
4-fold rotation = 4	90°	■
6-fold rotation = 6	60°	⬡
1-fold rotoinversion = $\bar{1}$[a]	360°	i[a]
2-fold rotoinversion = $\bar{2}$[b]	180°	m[b]
3-fold rotoinversion = $\bar{3}$[c]	120°	◬
4-fold rotoinversion = $\bar{4}$	90°	◨
6-fold rotoinversion = $\bar{6}$[d]	60°	⬡

[a] i = inversion, which is equivalent to a center of symmetry.

[b] m = mirror; m is used instead of $\bar{2}$ in the description of the symmetry of crystals.

[c] $\bar{3}$ is equivalent to a 3-fold rotation axis in combination with a center of symmetry (i).

[d] $\bar{6}$ is equivalent to a 3-fold rotation axis with a mirror perpendicular to it; expressed as $3/m$.

Figure 5.2 Rotation and rotoinversion axes: their angles of rotation, symbols, and letters used in illustrations. (Adapted from *International Tables for Crystallography*, 1983, vol. A)

90° each) combined with inversions through the center of a morphological crystal, or inside a crystal structure. In practice, rotoinversion axes are more difficult to recognize than rotation axes. An example of a 2-fold rotoinversion operation ($\bar{2}$) is given in Figure 5.3(B).

All of the foregoing symmetry elements (or operations) are best studied and recognized in the laboratory using wooden crystal blocks modeled after real crystals. Well-illustrated exercises to this effect are given in the laboratory manual by Klein (2008), *Minerals and Rocks*. Three-dimensional animations of these same symmetry operations are found in *Mineralogy Tutorials*, a CD-ROM that accompanies Klein and Dutrow (2008).

5.2 Combinations of symmetry elements

The first two objects in Figure 5.1 each contain but one symmetry element. The chair has a vertical mirror and the ship's propeller has a 3-fold rotation axis along the length of its shaft. The equilateral triangle (as used in drafting; Fig. 5.1(C)) and the olivine crystal (Fig. 5(D)) both have higher symmetry as a result of a combination of symmetry elements. The equilateral triangle has a 3-fold rotation axis (perpendicular to the plane on which it lies) and three mirror lines at 60° to each other, intersecting in a point that locates the 3-fold axis. The olivine crystal has three 2-fold rotation axes at 90° to each other, as well as three mirror planes, each of which is perpendicular to one of the rotation axes.

A manufactured object such as a crystal paperweight in the shape of a cube has the same symmetry content as well-formed cube of halite. Both exhibit many mirrors and various rotation and rotoinversion axes, as well as a center of symmetry. These symmetry elements are also known as **point group** elements. The word *point* indicates that the symmetry elements inside a crystal all intersect in the center of that crystal (see Fig. 5.1(D)), a central point.

In the nineteenth century, through the extensive and careful study of naturally occurring crystal shapes, it was proved that there are 32 unique combinations of symmetry elements. These are known as the 32 **crystal classes**, or point groups. Table 5.1 provides a listing of these.

The symmetry content of a natural object is expressed in the left column of Table 5.1 as a listing of how many symmetry elements there are of each of the four types (mirror, rotation axis, center of symmetry, and rotoinversion axis). Rotation axes are abbreviated by R, with a subscript, such as R_4. This describes a 4-fold rotation axis. If it is preceded by a number such as 3, resulting in $3R_4$, this means this specific crystal shape has three 4-fold rotation axes. Similarly, rotoinversion axes are shown by R, but with an overbar: $\bar{R}$. This symbol may also be preceded by a number, as in $1\,\bar{R}_3$. Mirrors are listed as m, and the number that precedes them indicates how many mirrors there are. The presence of a center of symmetry is listed as i, inversion, or as part of a rotoinversion operation, where $\bar{R} = R + i$.

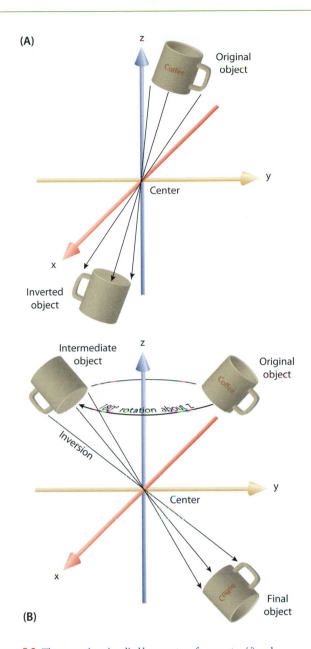

(A)

(B)

Figure 5.3 The operations implied by a center of symmetry (*i*) and rotoinversion. (**A**) The original upper object is replicated in the lower object by inversion through a center (located at the intersection of three axes). (**B**) Objects related by 180° (2-fold) rotation and inversion (resulting in rotoinversion).

The middle column in Table 5.1 is a symbolic representation of the symmetry elements listed in the left-hand column. This is referred to as the **Hermann-Mauguin** (or **international**) notation, to which we return (Box 5.1) after having discussed the meaning of the entries in the third column of Table 5.1.

5.3 The six crystal systems

The 32 symmetry combinations (which include some single symmetry elements as at the bottom of the left column in Table 5.1) can be grouped into six **crystal systems** (in the right

column). The grouping of the 32 crystal classes into six crystal systems is based on the presence of symmetry elements (or combinations thereof) that are unique to each crystal system. The first five listings (in the left column) show the presence of four rotation or rotoinversion axes ($4R_3$ or $4\,\overline{R}_3$) that allow for their grouping into the **isometric system**. All entries on the left with R_6, $\overline{R}_6$, R_3, or $\overline{R}_3$ are grouped in the **hexagonal system**. All those with $1R_4$ or $\overline{R}_4$ are in the **tetragonal system**. The **orthorhombic system** groups three crystal classes in which R_2s or *m*s coincide with three mutually perpendicular directions with binary (R_2 or *m*) symmetry. The **monoclinic system** includes three crystal classes, two of which contain one R_2 and one a single *m*. **Triclinic**, the system with the lowest symmetry, groups two crystal classes, one with only a center of symmetry (*i*), the other with no symmetry.

The grouping of the 32 crystal classes into six crystal systems is visually a little more obvious after you have been introduced to the Hermann-Mauguin notation for crystal classes (Box 5.1), as tabulated in the second column of Table 5.1.

In summary, the 32 crystal classes can be grouped into six crystal systems on the basis of common symmetry characteristics:

isometric
hexagonal
tetragonal
orthorhombic
monoclinic
triclinic

5.3.1. Crystallographic axes

With all the possible symmetry combinations grouped into six crystal systems (Table 5.1), a set of **crystallographic axes** compatible with each of the six systems can be developed. The choices are depicted in Figure 5.4. In five of the crystal systems, a set of three imaginary reference lines are designated as the crystal axes, except in the hexagonal system, where four crystallographic axes, a_1, a_2, a_3, and *c* are used. In systems with high symmetry such that two or three dimensions along the chosen axes are the same, axes of identical lengths are labeled as $a_1 = a_2$ and/or $a_1 = a_2 = a_3$. In the most general case, the three axes are labeled as *a*, *b*, and *c*, indicating that each of the axes is of a different length. The ends of the axes are labeled as plus (+) or minus (−), with the positive end of *a* to the front, the positive end of *b* to the right, and the positive end of *c* to the top. The opposite ends are negative. The angles between the positive ends of the axes are labeled with Greek letters α, β, and γ. The angle between axial directions *b* and *c* is α; between *a* and *c* is β; and between *a* and *b* it is γ. The relative lengths of the crystallographic axes and the angles between them are stated in Figure 5.4 below the appropriate illustration.

Table 5.1 A listing of all possible symmetry contents (32 of these), their representation in Hermann-Mauguin notation, and their grouping into six crystal systems. The 32 symmetry combinations are known as crystal classes as well as point groups. R means rotation axis. The subscript refers to 1, 2, 3, 4, and 6-fold rotation. m represents mirrors, and i = inversion (center of symmetry).

Symmetry content	Crystal class (in Hermann-Mauguin notation; see Box 5.1)	Crystal system
$3R_4, 4\bar{R}_3, 6R_2, 9m$	$4/m\,\bar{3}\,2/m$	
$(1\bar{R}_3 = 1R_3 + i)$		
$3\bar{R}_4, 4R_3, 6m$	$\bar{4}\,3\,m$	
$3R_4, 4R_3, 6R_2$	$4\,3\,2$	Isometric
$3R_2, 3m, 4\bar{R}_3$		
$(1\bar{R}_3 = 1R_3 + i)$	$2/m\,\bar{3}$	
$3R_2, 4R_3$	23	
$1R_6, 6R_2, 7m, i$	$6/m\,2/m\,2/m$	
$1\bar{R}_6, 3R_2, 3m$		
$(1\bar{R}_6 = R_3 + m\ \text{perpendicular})$	$\bar{6}\,m\,2$	
$1R_6, 6m$	$6\,mm$	
$1R_6, 6R_2$	622	
$1R_6, 1m, i$	$6/m$	
$1\bar{R}_6\ (= 1R_3 + m\ \text{perpendicular})$	$\bar{6}$	
$1R_6$	6	Hexagonal[a]
$1\bar{R}_3, 3R_2, 3m$		
$(1\bar{R}_3 = 1R_3 + i)$	$\bar{3}\,2/m$	
$1R_3, 3m$	$3m$	
$1R_3, 3R_2$	32	
$1\bar{R}_3\ (= 1R_3 + i)$	$\bar{3}$	
$1R_3$	3	
$1R_4, 4R_2, 5m, i$	$4/m\,2/m\,2/m$	
$1\bar{R}_4 = 2R_2, 2m$	$\bar{4}\,2m$	
$1R_4, 4m$	$4mm$	
$1R_4, 4R_2$	422	Tetragonal
$1R_4, m, i$	$4/m$	
$1\bar{R}_4$	$\bar{4}$	
$1R_4$	4	
$3R_2, 3m, i$	$2/m\,2/m\,2/m$	
$1R_2, 2m$	$m\,m\,2$	Orthorhombic
$3R_2$	$2\,2\,2$	
$1R_2, 1m, i$	$2/m$	
$1m$	m	Monoclinic
$1R_2$	2	
i	$\bar{1}$	Triclinic
R_1	no symmetry	

[a]All crystal classes (point groups) beginning with 6, $\bar{6}$, 3, and $\bar{3}$ are grouped in the hexagonal system.

Isometric

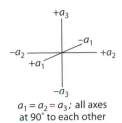

$a_1 = a_2 = a_3$; all axes
at 90° to each other

Orthorhombic

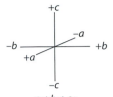

$a \neq b \neq c$;
all axes at 90° to each other

Hexagonal

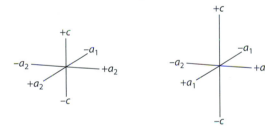

$a_1 = a_2 = a_3$; intersecting at 120°
c perpendicular to plane with a_1, a_2, a_3

Monoclinic

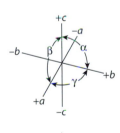

$a \neq b \neq c$
$\beta > 90°$; $\alpha = \gamma = 90°$

Tetragonal

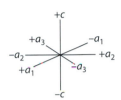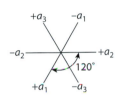

$a = b \neq c$; $a = a_1$; $b = a_2$;
all axes at 90° to each other

Triclinic

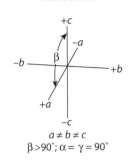

$a \neq b \neq c$
$\alpha \neq \beta \neq \gamma \neq 90°$

Figure 5.4 The choice of crystallographic axes that is compatible with each of the six crystal systems. In the isometric system, all three axes are of equal lengths and are labeled a. In the five less symmetric systems, the vertical axis is always labeled c. When the horizontal axes are of the same length, they are all labeled a. In the orthorhombic, monoclinic, and triclinic systems, all axes are of different lengths and, therefore, are labeled a, b, and c. The notation below each set of axes gives a specific statement about axial lengths and interaxial angles.

5.3.2 Hermann-Mauguin symmetry notation

With the introduction of the concept of crystallographic axes, it is now possible to evaluate the significance of the Hermann-Mauguin notation for crystal classes (or point groups), which is listed in the middle column of Table 5.1. In this table, the left column is simply a listing of the total symmetry content of a specific crystal class. This listing has no relation to any crystallographic aspects of the crystal class. The shorthand of the Hermann-Mauguin notation, however, relates directly to axial directions in crystals. This notation (middle column of Table 5.1), also known as the international notation, was developed from proposals by the German crystallographer Carl Hermann in 1928 that were subsequently modified by the French mineralogist Charles-Victor Mauguin in 1931. Their jointly developed notation system was internationally accepted in the 1935 edition of the *International Tables*. It is explained in Box 5.1, and its use in the notation of what are referred to as space groups is discussed in Section 5.7.

5.3.3 Crystallographic notation for planes in crystals

Now that we have introduced the six different sets of crystallographic reference axes (Fig. 5.4), we can address the question of how planes (external faces of a crystal) and planar sections and directions (inside crystal structures) are defined.

Rather than begin this discussion with the various planes that might be present on the outside of an actual crystal, it is simpler to begin with some imaginary planes inside a regularly

BOX 5.1 | THE HERMANN-MAUGUIN NOTATION EXPLAINED

The understanding of this notation is basic to much of crystallography because it is used in the symbolism for the 32 crystal classes (point groups) but also carries over into the notation for 230 space groups. Let us begin with the external form shown by a cube and evaluate its total symmetry content. The six illustrations depict a cube, its three reference axes, a_1, a_2, and a_3, and show the location of its inherent symmetry elements, except for the presence of a center of symmetry. It locates three axes of 4-fold symmetry (each perpendicular to a cube face and coincident with the three reference axes), four axes of 3-fold rotoinversion (in diagonals between opposite corners of the cube, known as *body diagonals*), and six axes of 2-fold rotation between the centers of opposite edges of the cube (known as *edge diagonals*). It also shows the location of nine mirrors; three of these are at 90° to each other and the other six are in diagonal positions. A center of symmetry is implied because every cube face has an equivalent face of the same size and shape on the opposite side. The symmetry elements that we just listed are the same as those for the first entry in the left column of Table 5.1. The entry in the middle column in that table represents the same information but now directly related to various directions in the cube.

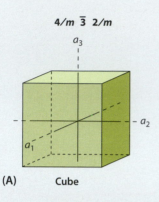

$4/m\ \bar{3}\ 2/m$

(A)   Cube

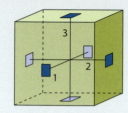

(B) **Three axes of 4-fold rotation**

For a cubic crystal, the first entry refers to the three crystallographic axes, a_1, a_2, and a_3. The second entry refers to the location of the body diagonals, which run from corner to corner. And the third entry refers to the six directions between the centers of edges (edge diagonals).

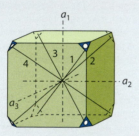

(C) **Four axes of 3-fold rotoinversion**

The digits in the notation represent rotation and/or rotoinversion axes, as shown in Figure 5.2. Mirrors are represented by *m*. If any of the rotational axes are perpendicular to a mirror, a slash is put between the digit (for rotation) and the mirror plane *m*. In the Hermann-Mauguin notation of the symmetry content of the cube, this occurs in two places. The symbol begins with $4/m$ and ends with $2/m$. The $4/m$ means there are 4-fold rotations perpendicular to mirrors, and $2/m$ states

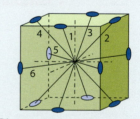

(D) **Six axes of 2-fold rotation**

that there are additional 2-fold rotations with *m*'s perpendicular. We know from inspection of the illustrations that there are three $4/m$ sets, each along one of the three crystallographic axes, a_1, a_2, and a_3. In addition, there are six $2/m$ sets, each between the centers of the 12 edges of a cube. The $\bar{3}$ refers to four body diagonals with 3-fold rotoinversion symmetry. Knowing what the notation means, we can now address the position of each of the symbols ($4/m$, $\bar{3}$, and $2/m$) in the notation. The first entry, $4/m$, in the notation of $4/m\,\bar{3}\,2/m$, refers to the *three* axial directions, a_1, a_2, and a_3. The notation does not state that there are three of these mutually perpendicular $4/m$ sets, but it is implicit in the overall symmetry of a cube. The next entry, $\bar{3}$, refers to four body diagonal directions in the cube. Again, the presence of four such rotoinversion axes is not stated but is implied. The $2/m$ entry refers to six directions between the centers of cube edges. Again, the number of these is not shown but is implied. Because of the high symmetry

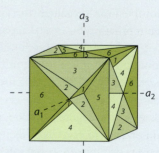

(E) **Three mirrors at 90° to each other**

(F) **Six mirrors in diagonal positions**

content of a cube, and other forms as part of $4/m\,\bar{3}\,2/m$, one must conclude that in this specific notation a lot of information is "collapsed" into each of the three positions of the notation. This highly "condensed" notation is implied in all five crystal classes grouped as isometric. Less and less of this implied multiplicity occurs in crystal systems of lower symmetry content.

Let us look at the significance of the notation in the most symmetric crystal class in the tetragonal system, $4/m$ $2/m\,2/m$, using the drawing of a tetragonal, prismatic crystal (G). In this system (and in the hexagonal system), there is one unique axis of symmetry, $4, \bar{4}$ (in tetragonal) and $6, \bar{6}, 3$, and $\bar{3}$ (in hexagonal). The first entry in $4/m\,2/m\,2/m$ (as seen in the last illustration) refers to the vertical c axis with a horizontal mirror (m) perpendicular to it. The second entry, $2/m$, refers to the two horizontal a_1 and a_2 axes. Because these two axes are equivalent, the $2/m$ notation implies two sets of 2-fold axes, each with a perpendicular mirror. The third $2/m$ entry refers to positions of 45° to the a_1 and a_2 axes (diagonal directions between the two crystallographic axes). There are two sets of these as well. The same system of notation holds for the hexagonal crystal system. In $6/m$ $2/m\,2/m$ the first entry is for the vertical 6-fold axis with a perpendicular mirror. The next $2/m$ relates to the three horizontal axes, a_1, a_2, and a_3. And the third entry is for the diagonal positions between those axes.

In the orthorhombic, monoclinic, and triclinic systems, there is no implied multiplicity in the Hermann-Mauguin notation. For example, $2/m\,2/m\,2/m$ in the orthorhombic system reflects that the a axis is one of $2/m$ symmetry, so is b, and so is c. In the monoclinic system, the 2-fold rotation axis is always the horizontal b axis and m is vertical (in the a-c plane). The triclinic system has no symmetry constrains with respect to the crystallographic axes. We return to some aspects of this notation in Section 5.7, where space groups are discussed.

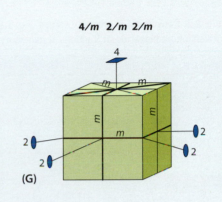

repeated structure referenced to a set of crystallographic axes. Figure 5.5 shows a standard representation of the three perpendicular crystallographic axes in the orthorhombic system. Translation distances along the a, b, and c axes are defined as t_1, t_2, and t_3, where **translation** implies a shift in position without rotation. In Figure 5.5(A), the horizontal plane, marked A, intersects only the vertical c axis at one unit of distance (t) and because it parallels the a and b axes the intercepts along these two axes are stated as ∞. In other words, plane A has the intercepts ∞ (on a), ∞ (on b) and 1 on c, resulting in $\infty, \infty, 1$. The intercepts of plane B in Figure 5.5(B) are ∞ along a, $2(t_2)$ along b, and $3(t_3)$ along c, resulting in $\infty, 2, 3$. Inspection of Figure 5.5(C) shows how intercepts are arrived at for planes C, D, and E. All these intercepts are recorded in Table 5.2. The use of intercept values to represent the orientation of crystal faces was first proposed C. S. Weiss in 1808, and they became known as Weiss intercepts. Subsequently two mineralogists in 1825 and 1829, respectively, recommended a system of representing faces by three numbers that are inversely proportional to Weiss intercepts. These are referred to as Miller indices because W. H. Miller popularized this approach in his book *A Treatise in Crystallography*, published in 1839. **Miller indices** of a face consist of a series of whole numbers that have been derived from intercepts by **inversion** and, if necessary, subsequent clearing of fractions.

The Miller indices listed in the last column of Table 5.2 are the reciprocals (obtained by inversion) of the intercepts of a plane with the axis notation omitted but with the understanding that the first digit refers to a, the second to b, and the third to c. This notation is standard for the definition of the orientation of a crystallographic plane (and/or direction) inside a crystal structure, and for the identification of crystal planes on the outside morphology of crystals. A description of how one obtains a Miller index for a crystal face is given in Box 5.2.

5.3.4 Definition of crystal form

In Box 5.2, the octahedron (an eight-faced form) was described by a Miller index enclosed in braces, {111}. The word *form* has a very special significance in morphological crystallography. Whereas we, in general speech, commonly interchange the words form and shape, in crystallography **form** is used in a special and restricted sense. A form consists of a group of like crystal faces, all of which have the same relation to the symmetry elements. All eight faces of the octahedron shown in Box 5.2 have identical orientations to the symmetry elements contained

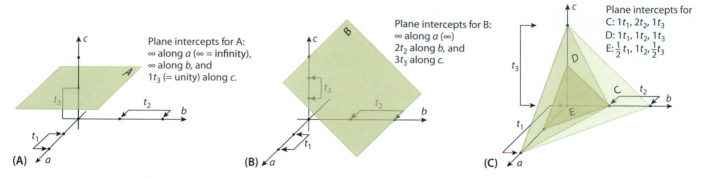

Plane intercepts for A:
∞ along a (∞ = infinity),
∞ along b, and
$1t_3$ (= unity) along c.

Plane intercepts for B:
∞ along a (∞)
$2t_2$ along b, and
$3t_3$ along c.

Plane intercepts for
C: $1t_1$, $2t_2$, $1t_3$
D: $1t_1$, $1t_2$, $1t_3$
E: $\frac{1}{2}t_1$, $1t_2$, $\frac{1}{2}t_3$

(A)   (B)   (C)

Figure 5.5 The intercepts of five planes with different orientations to a set of crystallographic axes. (Adapted from *Minerals and Rocks*, Klein 2008).

Table 5.2 Conversion of intercept values (Fig. 5.5) to Miller indices by inversion and clearing of fractions, if necessary.

Plane	Intercepts	Inversion	Miller index
A	∞a, ∞b, $1c$	$\frac{1}{\infty}$, $\frac{1}{\infty}$, $\frac{1}{1}$	(001)
B	∞a, $2b$, $3c$	$\frac{1}{\infty}$, $\frac{1}{2}$, $\frac{1}{3} \times 6$	(032)
C	$1a$, $2b$, $1c$	$\frac{1}{1}$, $\frac{1}{2}$, $\frac{1}{1} \times 2$	(212)
D	$1a$, $1b$, $1c$	$\frac{1}{1}$, $\frac{1}{1}$, $\frac{1}{1}$	(111)
E	$\frac{1}{2}a$, $1b$, $\frac{1}{2}c$	$\frac{1}{1/2}$, $\frac{1}{1}$, $\frac{1}{1/2}$	(212)

in $4/m\,\bar{3}\,2/m$. The symbol {111} implies the presence of eight equivalent faces in the octahedron. In contrast, using the (111) or (11$\bar{1}$) notation identifies two specific but different faces of the same octahedron. Understanding this distinction is critical because the braces (e.g., {*hkl*}) and parentheses (e.g., (010)) are used to define very different physical entities. In Section 3.4, several prominent cleavage directions were described, and these are universally described by the appropriate Miller index symbol, such as {100} for cubic, {001} for basal, {110} for prismatic, and {111} for octahedral cleavage. These symbols imply that there are one to several cleavage directions as part of the form statement. For example, {100} for cubic cleavage implies three orthogonal cleavage directions, each direction parallel

---

BOX 5.2 | HOW TO ARRIVE AT A MILLER INDEX

The Miller index notation is not easily understood but a lot can be learned from the study of both Figure 5.5 and Table 5.2. The second column in Table 5.2 records the intercepts for the five planes illustrated in Figure 5.5. Whenever an infinity symbol (∞) is used, it means that the plane in question is parallel to one of three crystallographic axes. Plane A has two infinity entries, one for a and the other for b. The inversion of 1/∞ results in 0. Therefore, when zeros (0) appear in the Miller index, it is immediately obvious that the plane in question is parallel to one or two of the crystallographic axes. The third column in Table 5.2, labeled "inversion," contains several fractions. As these are not allowed in Miller index notation, they must be cleared by multiplication with a common factor (e.g., 6 or 2). One great advantage of the Miller index system, over that of the earlier use of intercepts, is that parallel planes are defined by the same Miller index. In Figure 5.5(C), careful inspection shows that planes C and E are parallel to each other. Their intercepts differ but their Miller indices are the same, namely (212) (Table 5.2).

It is conventional to report the Miller index, generally consisting of three digits (not so in the hexagonal system; see Box 5.3), inside parentheses for the identification of a specific external (morphological) face. This is the case in the last column of Table 5.2. Miller indices are read as "oh oh one" for (001), as "oh three two" for (032), and so on. The fourth entry in the last column of Table 5.2 is (111), which is known as the **unit face** (or plane) because it intersects all three axes at unit distances.

In the examples provided in Figure 5.5, we were able to read of the exact intercepts along the three crystallographic axes, which led to specific Miller indices in Table 5.2. This is not always the case, and furthermore, in the evaluation of the position of a specific external face on a crystal (with respect to the location of imagined crystallographic axes), this is generally impossible. For a crystal face that appears to be in a pretty general position, a **general symbol** (*hkl*) is used. The *h*, *k*, and *l* are, respectively, the reciprocals of undefined intercepts along the *a*, *b*, and *c* axes. The (*hkl*) face cuts all three axes without any statement of what the relative units along these axes are. When zeros (0) are used in this general notation, such as in (0*kl*), (*h0l*), and (0*kl*), and (00*l*), it means that the crystal face in question is parallel to one or two axes, with nondefined intercepts on the other.

Until now, we have concerned ourselves only with the position of crystal faces that intersect the positive ends of the crystallographic axes. The illustration depicts the position of a (111) plane with respect to three crystallographic axes in the 4/*m* $\bar{3}$ 2/*m* crystal class in the isometric system (see Table 5.1). This is the class with the highest possible overall symmetry. Being aware of its symmetry content (which is not shown in the figure), one should expect that the one (111) plane is related to seven more, resulting in a solid geometric form known as an **octahedron** {111}. The original (111) plane is repeated by 4-fold axes along a_1, a_2, and a_3, as well as by vertical and horizontal mirrors perpendicular to the 4-fold axes. The resulting eight-faced geometric form consists of the following: (111), (1$\bar{1}$1), ($\bar{1}\bar{1}$1), ($\bar{1}$11), (11$\bar{1}$), (1$\bar{1}\bar{1}$), ($\bar{1}\bar{1}\bar{1}$), and ($\bar{1}$1$\bar{1}$) faces. This illustrates that a bar is placed over the appropriate digit in the Miller index when a face intersects negative ends of crystallographic axes. These overbars on numbers are read as "bar one." Such overbars in Miller indices must not be confused with the overbar notation of rotoinversion axes in the Hermann-Mauguin notation (as in 4/*m* $\bar{3}$ 2/*m*). An interactive animation of Miller index notation is available in Module II of the CD-ROM *Mineralogy Tutorials* that accompanies Klein and Dutrow (2008).

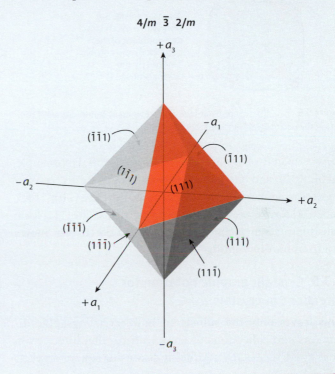

to one of three pairs of parallel faces on a cube, and so on. In Figure 6.10, which gives a synopsis of many significant optical properties, cleavages are noted with the form symbol (e.g., {111}) as well as by angles between specific crystallographic planes, such as (110) Λ (1$\bar{1}$0) = 56°.

Having introduced the concept of form, we must relate it to the concept of the **general form**. The general form is a form in which the faces intersect all the crystallographic axes at different lengths and is identified by the symbol {*h k l*} (or {*h k i l*} in the hexagonal system). The concept of a general form relates to its position with respect to the symmetry elements of a specific point group. Irrespective of the point group (or crystal class) an (*h k l*) face is not parallel or perpendicular to a symmetry plane or axis. For most crystal classes special forms of the same class have fewer faces than the general form. Note that the general form {*h k l*} in 4/*m* $\bar{3}$ 2/*m*, known as a **hexoctahedron**, has 48 faces, whereas the special forms {100}, {011}, and {111} in 4/*m* $\bar{3}$ 2/*m* have 6 faces (a **cube**), 12 faces (a **dodecahedron**), and 8 faces (an **octahedron**), respectively.

In the Miller index notation of forms, it is desirable to choose, if possible, a face symbol with positive digits, such as {011} instead of {0$\bar{1}$1}.

In the first three crystal systems (isometric, hexagonal, tetragonal; Table 5.1) with the higher symmetry contents, the {111} form is not the same as the general form {*h k l*} because the general form is defined as intersecting all crystallographic axes at different lengths. In the orthorhombic, monoclinic, and triclinic systems, the {111} is the general form because in these systems the unit lengths along each of the crystallographic axes are all different ($a \neq b \neq c$).

Figure 5.6 shows examples of two different expressions of the combination of two forms, the octahedron {111} and the cube {100} in crystal class 4/*m* $\bar{3}$ 2/*m*. Form types can be distinguished as **closed forms** and **open forms**. A closed form is one that encloses space. As such, a crystal may consist of only one form (a closed form) such as a cube or an octahedron. An open form is defined as one that does not enclose space. If an open form exists on a crystal, it must be closed off by one or more other open forms. The orthorhombic crystal used in Box 5.4

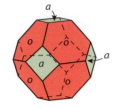

A composite crystal consisting of faces belonging to two forms: {111} octahedron, marked o and {100} cube, marked a

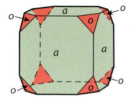

A similar crystal but now with a much larger development of the cube {100} and corners beveled by the octahedron {111}

Figure 5.6 Two different morphological expressions of combinations of two forms, the octahedron {111} and the cube {100}.

consists mainly of a combination of open forms. For example, {100}, {010}, and {001} are all two-faced open forms (consisting of two parallel faces) known as **pinacoids**. The {110}, {101}, and {011} forms are all four-faced open **prisms**. The {111} form consisting of eight inclined corner faces is a closed form known as a **rhombic dipyramid**.

5.3.5 Crystallographic notation for directions in crystals

This leaves only the subject of how crystallographic directions and collections of faces with parallel intersection lines are described by a Miller index inside square brackets

(e.g., [100]). A conspicuous feature of many crystals is the arrangements of groups of faces with parallel intersection edges. Figure 5.7(A) is a crystal drawing of an orthorhombic crystal with symmetry $2/m\ 2/m\ 2/m$. It shows many crystal faces that, in this case, are marked with letters instead of Miller indices, to make the overall illustration simpler. Faces with the same letter belong to the same crystal form. This is standard notation in crystal drawings. All faces marked with the letter m belong to one type of vertical prism, and those marked with s belong to another vertical prism. To enhance the visual perception of the various crystal forms, different colors are used in the two drawings as well as the standard letters. In this text, we generally use Miller indices, but in the case of Figure 5.7, the letters are useful in defining what are known as zones. A **zone** is a collection of crystal faces with parallel edges. In a crystal class with orthogonal axes as in the orthorhombic system, the zone axes, perpendicular to front, side, and top faces, coincide with the crystallographic axes. The front zone axis is perpendicular to the a face (100), the east-west zone axis is perpendicular to the b face (010), and the vertical zone axis is perpendicular to the top c face (001). The standard Miller indices (100), (010), and (001) identify faces, whereas [100], [010], and [001] are directions perpendicular to these faces. In less symmetric systems, such as the monoclinic system, where the a and c crystallographic axes make a β angle with each other that is some angle larger than 90°, the zone axes do not coincide with the perpendiculars to the front and top faces. This is shown in Figure 5.7(B).

BOX 5.3 | **THE MILLER-BRAVAIS INDEX IN THE HEXAGONAL SYSTEM**

We noted earlier that the Miller index notation is different in the hexagonal crystal system because its symmetry is defined with respect to four crystallographic axes instead of three as in the other five crystal systems. These four digits (e.g., $(10\bar{1}0)$), or four letters (e.g., $\{hk\bar{i}l\}$), are known as **Miller-Bravais indices**. The illustrations show two different common forms in crystal class $6/m\ 2/m\ 2/m$ of the hexagonal system. (A) is a **hexagonal prism**, consisting of six vertical faces, with a pair of top and bottom faces known as a pinacoid. The plan view below the drawing of the crystal shows how one vertical plane of the prism intersects two of the horizontal axes ($+a_1$ and $-a_3$) and is parallel to a_2. The three horizontal axes are at 120° to each other, which locates the + end of a_3 to the upper left in this plan view. The intercepts of this plane are shown, as well as the inverted results, thus giving rise to the Miller-Bravais index of $(10\bar{1}0)$. (B) is a **hexagonal dipyramid** consisting of two six-faced pyramids, with the top one reflected into the lower one by the horizontal mirror as stated in the first entry of the Hermann-Mauguin notation, $6/m$. The one shaded face of this dipyramid is redrawn in a perspective image relative to the four hexagonal axes. The bottom drawing allows for a clear view of how the trace of this inclined dipyramidal face cuts the three horizontal axes. It cuts a_1 and a_2 at unit distances along their positive ends. It cuts the minus end of a_3 at one-half the unit distance; it also cuts c at unit distance. Knowing this results in the statement for the intercepts. Below that are the inverted results. The final index is $(11\bar{2}1)$. The form index for the hexagonal prism (consisting of six vertical faces) is $\{10\bar{1}0\}$ and that of the basal pinacoid (consisting of two horizontal faces) is $\{0001\}$. The 12-faced dipyramid has form index $\{11\bar{2}1\}$.

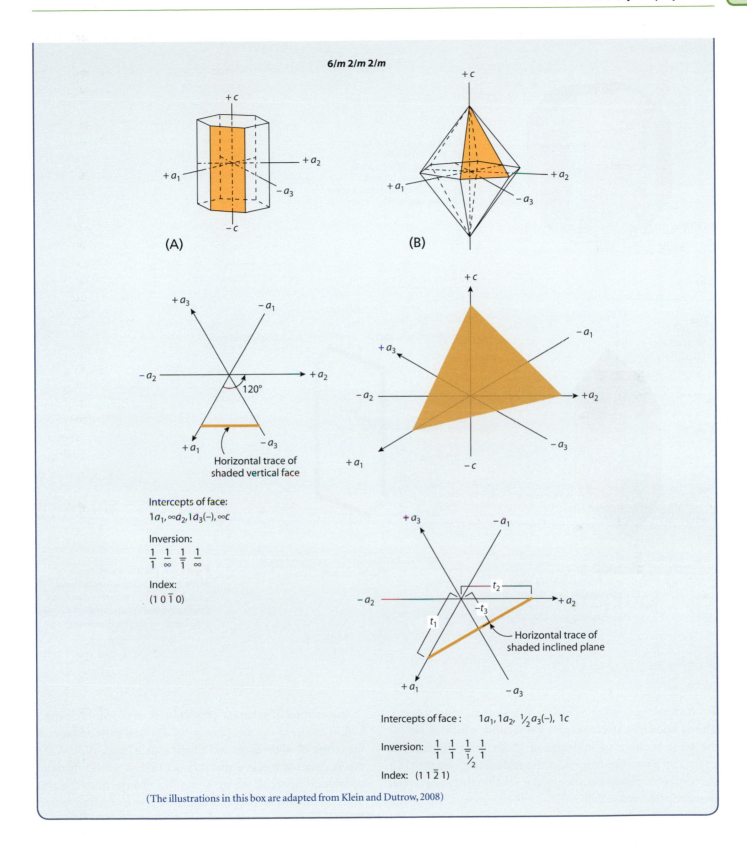

6/m 2/m 2/m

(A)

(B)

Horizontal trace of
shaded vertical face

Intercepts of face:
$1a_1, \infty a_2, 1a_3(-), \infty c$

Inversion:
$\frac{1}{1} \quad \frac{1}{\infty} \quad \frac{1}{1} \quad \frac{1}{\infty}$

Index:
$(1\,0\,\overline{1}\,0)$

Horizontal trace of
shaded inclined plane

Intercepts of face : $1a_1, 1a_2, \frac{1}{2}a_3(-), 1c$

Inversion: $\frac{1}{1} \quad \frac{1}{1} \quad \frac{1}{\frac{1}{2}} \quad \frac{1}{1}$

Index: $(1\,1\,\overline{2}\,1)$

(The illustrations in this box are adapted from Klein and Dutrow, 2008)

5.4 Crystal projections

In prior sections of this chapter, we have used perspective draw-ings of crystal morphology with superimposed symmetry ele-ments to illustrate which elements are inherent in what type of morphology. In Box 5.1, we illustrated the symmetry content of a cube as well as that of a tetragonal prism. Although such perspective sketches are reasonably successful in portraying the relationship between morphology and symmetry, there

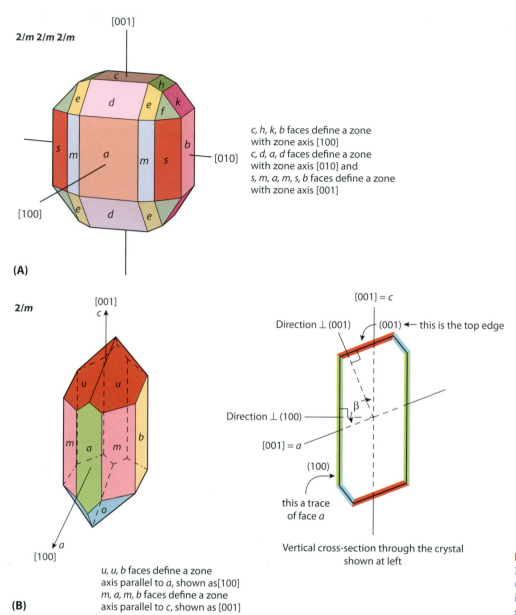

c, h, k, b faces define a zone
with zone axis [100]
c, d, a, d faces define a zone
with zone axis [010] and
s, m, a, m, s, b faces define a zone
with zone axis [001]

(A)

2/m

u, u, b faces define a zone
axis parallel to a, shown as[100]
m, a, m, b faces define a zone
axis parallel to c, shown as [001]

(B)

Vertical cross-section through the crystal
shown at left

Figure 5.7 Illustration of the use of Miller indices inside square brackets (e.g., [100], [010]) to define directions inside morphological crystals and crystal structures. See text for discussion.

is a standard projection technique that does this much better. This is known as **stereographic projection**, which allows for the representation of information about three-dimensional objects on a two-dimensional plane surface (e.g., a page). In a stereographic projection, the angular relations between crystal faces are preserved such that the overall symmetry of the crystal is revealed.

Creation of a stereographic projection of a crystal first involves the **spherical projection** of perpendiculars to each of the crystal faces (known as **face poles**) onto an imaginary sphere that surrounds the crystal; these piercing points are then projected onto a plane. We discuss the spherical projection first (see Box 5.4).

The method of spherical projection is analogous to using a hollow model of the morphology of a crystal with pinholes at the center of all the faces and a light source inside of it. When this is centered inside a much larger hollow sphere made of translucent material, the light rays that emerge from the pinholes produce small points of light on the surrounding sphere. These light points are referred to as face poles (see Box 5.4).

At the beginning of this section, we said that spherical projection (as shown in Box 5.4) would result, through subsequent stereographic projection, in a two-dimensional representation of face poles and symmetry elements. So the next step in creating a stereographic projection is to project the spherical projection points onto a plane (see Box 5.5).

BOX 5.4 | SPHERICAL PROJECTION

2/m 2/m 2/m

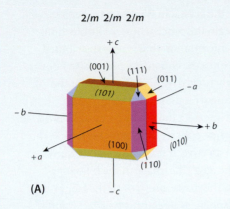

(A)

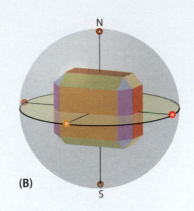

(B)

Let us begin this experiment with a relatively simple crystal of orthorhombic symmetry, $2/m\,2/m\,2/m$. This is shown in the first figure. It consists of a large front, a large right side, and a large top face, each of which has a parallel equivalent as a back, a left side, and bottom face. These represent three pinacoids with form symbols {100}, {010}, and {001}, each of which has one visible face. The vertical corner faces, of which there are four (only two are visible), are part of the {110} form. The forward top and bottom bevels are part of the {101} form, and the side bevels are part of the {011} form. All three forms have four faces, with only two faces visible for each of the forms. The corner bevels, of which there are eight (only four are visible), are part

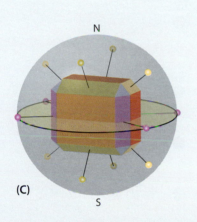

(C)

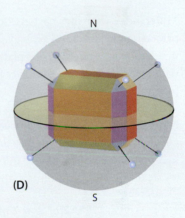

(D)

of the {111} form. All faces belonging to these seven different forms are given different colors for clarity. The first sphere shows the face poles for (100), (010) (two faces), and (001) and (00$\bar{1}$) piercing the surrounding sphere. The second sphere shows the face poles on the surrounding sphere for the {110}, {101}, and {011} forms. The third sphere locates the face poles of {111}. The fourth sphere shows a horizontal equatorial plane (a circle), the upper hemisphere, and the lower hemisphere, both with all of the face poles. Poles normal to vertical faces pierce the sphere along the edge of the equatorial plane. Poles perpendicular to (001) and (00$\bar{1}$) land on the north and south pole, respectively. All other face poles are at various locations in the northern and southern hemisphere. The last sphere shows examples of the assignment of Miller indices to face poles of the three pinacoids {100}, {010}, and {001}. All the other face poles can similarly be indexed.

The illustrations in this box are adapted from the CD-ROM *Mineralogy Tutorials* that accompanies Klein and Dutrow (2008).

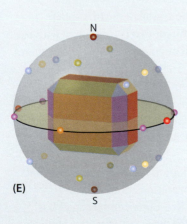

(E)

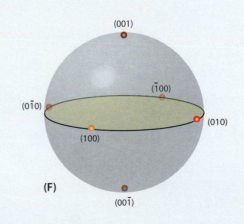

(F)

BOX 5.5 | STEREOGRAPHIC PROJECTION

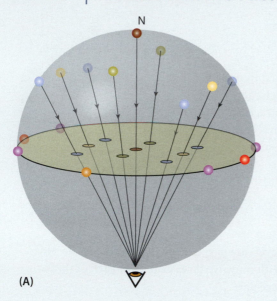

(A)

Now that all the face poles for the crystal in Box 5.4 have been located on the outer sphere, we can begin the process of stereoprojection. This involves bringing all the information that is now distributed over the surface of the sphere onto the equatorial plane, also known as the **primitive circle**. This is achieved by placing the eye at the south pole of the sphere, looking straight up and locating the points where the lines of sight intersect the primitive circle. Those intersections are shown in the first sphere. Until now, we have seen the primitive circle only in perspective. If it is turned 90° toward the observer so it lies flat on the page, the image in the following illustration results. This shows the distribution of all the face poles in the northern (upper) hemisphere as intersection points (along the lines of sight in the first sphere) with the primitive circle. The face poles from the upper hemisphere are shown as small solid dots, and as is customary, they are accompanied by the appropriate

Miller index. (You may have encountered stereographic projections in structural geology, which are constructed in a similar way, but the eye is placed at the north pole and data are projected from the lower hemisphere onto the primitive circle).

To locate the face poles that occur in the southern hemisphere, the exact same procedure is followed by placing the eye at the north pole of the sphere. From that point, looking downward into the sphere, you can locate the intersection points of the sight lines with the

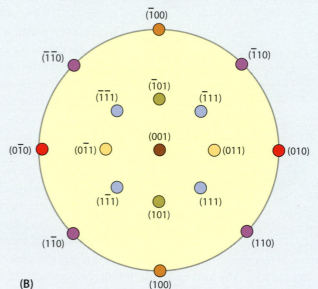

(B)

primitive circle. Because the crystal in the first sphere (in Box 5.4) has a horizontal mirror that is coincident with the primitive circle, all equivalent poles in the two hemispheres intersect at the same point on the primitive circle. The poles from the northern hemisphere are customarily shown as small circular dots and those from the southern hemisphere as small open circles. This results in the image of the second circle with all face poles identified with their respective Miller indices.

At this stage, it is important to realize that our original goal – to reduce all the information about a three-dimensional object onto a two-dimensional page – has indeed been achieved. All the faces on the crystal in Box 5.4

2/m 2/m 2/m

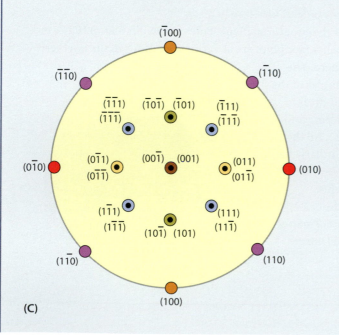

(C)

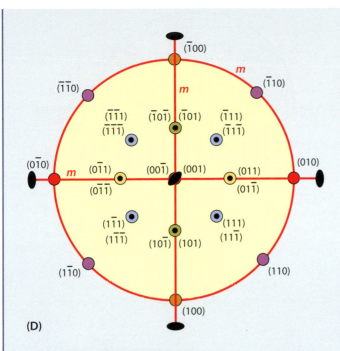

are plotted in their appropriate relative positions on a two-dimensional page. Using the circular diagram, known as a **stereogram**, of all of the face poles of a specific crystal, we can add the symmetry elements implied in the Hermann-Mauguin notation of $2/m\,2/m\,2/m$. That is done in the last circle. All three crystallographic axes are axes of 2-fold symmetry, and each one has a mirror plane perpendicular to it. The vertical mirrors show as a cross at 90° to each other and the horizontal mirror (perpendicular to the c axis) coincides with the edge of the primitive circle (shown in bold).

In mineralogy courses, stereographic projections are commonly made on what are known as **stereographic nets** or **Wulff nets** using specific angles for the plotting of face poles. That is a challenging three-dimensional exercise that we do not pursue in this text. Examples of these stereographic procedures are given in mineralogy texts as well as in the laboratory manual by Klein (2008), *Minerals and Rocks*.

(D)

The illustrations in this box are adapted from the CD-ROM *Mineralogy Tutorials* that accompanies Klein and Dutrow (2008).

5.5 Seven of the thirty-two point groups

We now systematically discuss seven point groups (crystal classes) in which a large number of common minerals crystallize. A listing of how many minerals are part of each of the 32 point groups can be found in Table 2.1 of Bloss (1994). Table 5.1 lists all the 32 crystal classes, of which 19 of the most commonly occurring point groups are treated systematically in Klein and Dutrow (2008). All 32 crystal classes are discussed in Klein and Hurlbut (1993). There is also a powerful software program, titled KrystalShaper, for the interactive development of crystal morphologies in various point groups.

In the treatment that follows, each point group (crystal class) is identified by its Hermann-Mauguin notation. In this text, we use the notation that was introduced in Table 5.1 and discussed in Section 5.3.2. This is considered the easier and more descriptive notation because crystallographers have devised shorter (abbreviated) symbols as well. For five of the seven point groups that follow, here are the longer and the shorter symbols:

$4/m\,\bar{3}\,2/m$ — $m\,3\,m$

$6/m\,2/m\,2/m$ — $6/mmm$

$\bar{3}\,2/m$ — $\bar{3}m$

$4/m\,2/m\,2/m$ — $4/mmm$

$2/m\,2/m\,2/m$ — mmm

The reason for the use of the shorter symbol can be explained by using the last entry. The short symbol for that is *mmm*, which implies that there are three mutually perpendicular mirror planes whose order of listing means that they are perpendicular to the *a*, *b*, and *c* axes of the orthorhombic system. The lines of intersection of these three planes are axes of 2-fold rotation. Therefore, *mmm* is briefer than $2/m\,2/m\,2/m$ but also is much less descriptive. We introduce this duality in notation only because the short symbols are commonly used in the crystallographic literature.

In the illustrations of each of the seven point groups (Figs. 5.11 – 5.17), the Hermann-Mauguin notation is followed by a descriptive name for the point group. This name is derived from the general form of that point group (see Box 5.2). Although we consistently use the Hermann-Mauguin symbol to define a point group or crystal class, we also give the descriptive name for each of the seven point groups (based on the name of the general form) because that name is commonly used in the crystallographic literature and in mineralogy textbooks.

In the seven composite illustrations that follow, the first drawing labeled (A) always gives the standard notation for the crystal axes of the crystal system to which the point group in question belongs. Only in Figure 5.13 is this crystal axis notation omitted because it is the same as that given in Figure 5.12(A). The second illustration (B) is a stereogram of the general form $\{hkl\}$ or $\{hki\bar{l}\}$. Because such a stereogram is given in each of the seven illustrations of the most common crystal classes (point groups) that follow, let us review some aspect of these:

- The graphical symbols used for rotation and rotoinversion axes are those given in Figure 5.2.
- Mirror planes (*m*) are shown as solid bold lines when they are oriented perpendicular to the page and as a bold edge along the primitive circle when they lie in the plane of the page.
- The *c* axis, or in the isometric system a_3, is perpendicular to the page.

Crystal classes in the isometric system have the most complex symmetry. In two of these ($4/m\,\bar{3}\,2/m$ and $\bar{4}\,3m$, but with only the first illustrated in Fig. 5.11), there are four mirror orientations (of the six mirrors between edges of the cube) that are inclined at 45° degrees to the page and appear as arcs when projected onto the plane of the page. Figure 5.8(A) shows the origin of these arcs in the stereogram. It is a representation of the upper hemisphere with three of the four inclined mirrors at their 45° degree inclination to the primitive circle (this is the same upper hemisphere as shown in Box 5.5). When these inclined mirror planes are projected (from a view point at the south pole of the sphere), they appear as four intersecting arcs labeled *m*, as shown in Figure 5.8(B).

Four of the crystal classes depicted with stereograms in the subsequent seven composite figures have mirror planes that coincide with the primitive circle. These are $4/m\,\bar{3}\,2/m$, $6/m\,2/m$ $2/m$, $4/m\,2/m\,2/m$, and $2/m\,2/m\,2/m$. The remaining three do not. When a horizontal mirror is absent, as in $\bar{3}\,2/m$, the stereogram looks distinctly different. This is shown in Figure 5.9(A), with the outline of a **rhombohedron** and the location of the symmetry elements implied by $\bar{3}\,2/m$. The vertical axis is one of the 3-fold rotoinversion and lacks a mirror perpendicular to it. The only mirrors are vertical and are interleaved with the horizontal axes. Figure 5.9(B) is a stereogram of the rhombohedron with the face poles of the three top faces (shown as solid dots) alternating with the three lower face poles (open circles). Because there is no horizontal mirror perpendicular to the *c* axis (along the outer edge of the primitive circle), the face poles of the three top and alternating three bottom faces do not superimpose.

Below the one, two, or three first graphical illustrations of the crystal axes and a stereogram are drawings of the most common morphologies of some rock-forming minerals. These drawings are officially known as **clinographic projections**. These are a type of perspective drawing that yields a portrait-like image of the crystal in two dimensions, and they are much better illustrations of crystals than are photographs. Furthermore, the stylized drawings allow for the quick recognition of different forms on a crystal through the use of color where a specific color identifies a specific face of a form (or faces of the same form) in a designated point group.

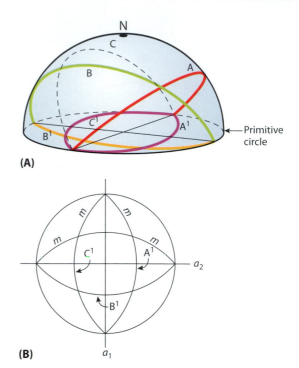

(A)

(B)

Figure 5.8 (**A**) Orientations of the inclined mirror planes in crystal class $4/m$ $\bar{3}2/m$ of the isometric system, as shown in Figure 5.11. Only three of the four possible planes are drawn for clarity. The outer sphere is the same as used in Boxes 5.4 and 5.5. (**B**) The appearance of the four inclined planes as arcs when projected onto the primitive circle.

The crystals are always illustrated as being geometrically perfect, which is rarely seen in nature. However, such idealized illustrations are necessary to convey the overall symmetry content of the various forms. Most naturally occurring crystals are to some extent imperfect or malformed, as shown in Figure 5.10. Perfect crystals or crystal groups are much sought after by curators at natural history museums for public mineral exhibits. It is unlikely that you will have much exposure to such high-quality, expensive crystal specimens in the laboratory that accompanies the course in which this text is used. Nonetheless, some common minerals such as galena, magnetite, fluorite, and garnet (all of which are isometric; see Fig. 5.11) may occur in reasonably well-shaped forms even in the specimens displayed and used for study in the laboratory. In the microscopic study of rock types, in thin section well-crystallized minerals show diagnostic crystal outlines (in cross-sections) that are helpful in their identification.

In the drawings, the crystals are always oriented in a standard position with the *c* axis vertical, the *b* axis east-west, and the *a* axis toward the observer. In the isometric system, the *c* direction is the a_3 axis. The positions of the axes are not shown in the drawings so as to reduce the overall clutter. Different crystal forms shown in a crystal drawing may be identified by different letters, different colors, and Miller index notation.

$\bar{3}\ 2/m$

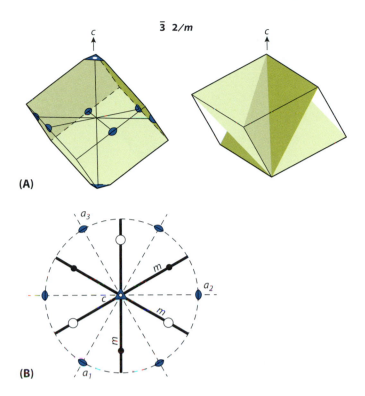

(A)

(B)

Figure 5.9 **(A)** A rhombohedron (in crystal class $\bar{3}\,2/m$) with symmetry axes and mirror planes. It lacks a horizontal mirror perpendicular to the vertical 3-fold rotoinversion axis, which is the same as the crystallographic *c* axis. **(B)** The stereogram of the rhombohedron in which the three upper faces of the form alternate, at 60° separation, with the three lower faces. As noted in Figure 5.2, a $\bar{3}$ is equivalent to a 3-fold rotation axis combined with a center of symmetry (*i*). If an imaginary line is passed from any point on a top face, through the center of the stereogram, and if the same point is found on that line at an equal distance beyond the center, the crystal is said to have a center of symmetry (*i*) (Fig. 5.3(A)). This is shown by the three pairs of parallel faces, with each pair having a top face and an equivalent bottom face (by inversion through the center, and not by a mirror operation).

$4/m\ \bar{3}\ 2/m$

Crystals belonging to this crystal class are referred to three axes of equal lengths that make right angles with each other (see Fig. 5.11(A)). The three crystallographic axes are axes of 4-fold rotation. There are four 3-fold rotoinversion axes that occur between the opposing corners of a cube. There are also six 2-fold rotation axes between the centers of opposing edges of a cube (see Box 5.1). There are nine mirror planes. Three of these are each perpendicular to one of the three crystallographic axes, and six are perpendicular to the six 2-fold rotation axes. There is also a center of symmetry (*i*), the abbreviation for inversion. All this is shown graphically in the stereogram of Figure 5.11(A), as discussed in Box. 5.1.

The general form, $\{hkl\}$, is a hexoctahedron and consists of 48 triangular faces (see Fig. 5.11(B)). All forms in this point group are closed forms on account of its very high symmetry content. The unique geometric shape of these closed forms in the isometric system has resulted in specific names for each of them. Here are those that occur in $4/m\,\bar{3}\,2/m$:

Cube, {001}, consisting of six square faces at 90° to each other. Each face intersects one crystallographic axis and is parallel to the other two.
Octahedron, {111}, consisting of eight triangular faces, each of which intersects all three of the crystallographic axes equally.
Dodecahedron, {011}, is composed of 12 rhomb-shaped faces. Each face intersects two of the crystallographic axes and is parallel to the third.
Tetrahexahedron, $\{0\,k\,l\}$, consists of 24 isosceles triangular faces (isosceles meaning with two equal sides). The most common Miller index is {012}.

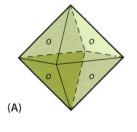

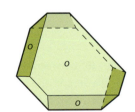

(A)

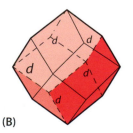

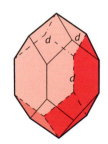

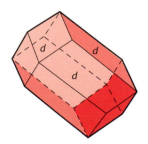

(B)

Figure 5.10 Examples of perfect and distorted crystals. **(A)** Perfect and malformed octahedron. **(B)** Perfect dodecahedron and two malformed ones.

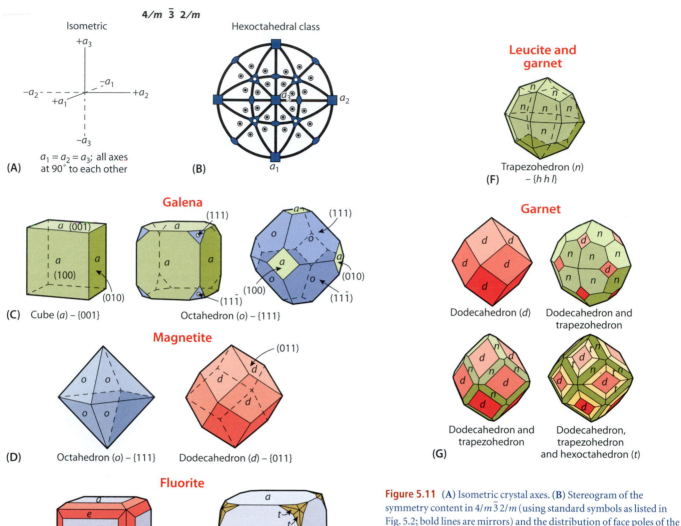

$4/m\,\bar{3}\,2/m$

Isometric — Hexoctahedral class

(A) $a_1 = a_2 = a_3$; all axes at 90° to each other

(B)

Galena

(C) Cube (a) – {001} Octahedron (o) – {111}

Magnetite

(D) Octahedron (o) – {111} Dodecahedron (d) – {011}

Fluorite

(E) Cube (a) and tetrahexadron (e) – {$o\,k\,l$} Cube (a) and hexoctahedron (t) – {$h\,k\,l$}

Leucite and garnet

(F) Trapezohedron (n) – {$h\,h\,l$}

Garnet

Dodecahedron (d) Dodecahedron and trapezohedron

Dodecahedron and trapezohedron Dodecahedron, trapezohedron and hexoctahedron (t)

(G)

Figure 5.11 (**A**) Isometric crystal axes. (**B**) Stereogram of the symmetry content in $4/m\,\bar{3}\,2/m$ (using standard symbols as listed in Fig. 5.2; bold lines are mirrors) and the distribution of face poles of the general form, {$h\,k\,l$}, known as a hexoctahedron. (**C**) Typical galena crystals. (**D**) Common forms on magnetite crystals. (**E**) Common forms on fluorite crystals. (**F**) Leucite and garnet both crystallize commonly as trapezohedra. (**G**) Typical garnet crystals that consist of various combinations of isometric forms.

Trapezohedron, {$h\,h\,l$}, is composed of 24 trapezium-shaped faces. Its most common Miller index is {112}.
Trisoctahedron, {$h\,l\,l$}, consists of 24 isosceles triangular faces. The most common Miller index is {122}.

A large number of the preceding isometric forms are illustrated singly, or in combination, in Figure 5.11.

6/m 2/m 2/m

Crystals whose point group symbols begin with 6, $\bar{6}$, 3, or $\bar{3}$ are classified as part of the hexagonal system, but those with 3 and $\bar{3}$ are part of its rhombohedral division. The hexagonal system is referred to four crystallographic axes, designated a_1, a_2, a_3, and c. The c axis is vertical, and the three a axes lie in a horizontal

plane with 120° between their positive ends (see Fig. 5.12(A)). The vertical axis, c, is one of 6-fold rotation. There are six horizontal axes of 2-fold rotation, three of which coincide with the crystallographic axes and three that lie midway between them. There are seven mirror planes, each perpendicular to one of the symmetry axes. There is also a center of symmetry (i).

The general form, {$h\,k\,\bar{i}\,\ell$}, is a **dihexagonal dipyramid**. This is a closed form consisting of 12 faces at the top and 12 identical faces at the bottom, related by a horizontal mirror plane. The faces of the top and bottom pyramids meet a point, the upper pyramid intersecting the + part of the c axis, the lower pyramid cutting the − part of the c axis at equal distance. The stereogram of this general form is given in Figure 5.12(B).

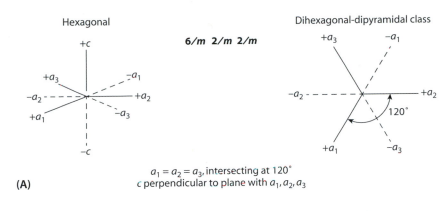

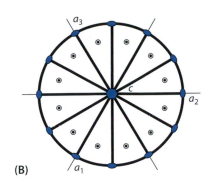

Hexagonal

6/m 2/m 2/m

Dihexagonal-dipyramidal class

$a_1 = a_2 = a_3$, intersecting at 120°
c perpendicular to plane with a_1, a_2, a_3

(A)

(B)

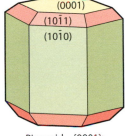

Apatite

(0001)
(10$\bar{1}$1)
(10$\bar{1}$0)

Pinacoid – {0001};
hexagonal prism – {10$\bar{1}$0};
(C) hexagonal dipyramid {10$\bar{1}$1}

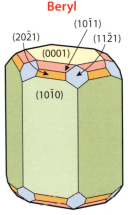

Beryl

(10$\bar{1}$1)
(20$\bar{2}$1)
(11$\bar{2}$1)
(0001)
(10$\bar{1}$0)

Pinacoid; hexagonal
prism; three different
hexagonal dipyramids

Figure 5.12 (A) Hexagonal crystal axes. (B) Stereogram of the symmetry content in 6/m 2/m 2/m and the distribution of face poles of the general form, {h k $\bar{i}$ l}, known as a dihexagonal dipyramid. (C) Typical crystals of apatite and beryl consisting of vertical hexagonal prisms modified by hexagonal dipyramids and with a pinacoid for the top and bottom faces.

In this and subsequent crystal classes, all with symmetry contents less than that of 4/m$\bar{3}$2/m, the form nomenclature includes few specialized names or definitions. Instead, the names of commonly recognized geometric forms predominate, such as **prism** and **pyramid**. In crystallography, a prism is defined as a crystal form consisting of three or more similar faces parallel to a single axis. As such, a vertical hexagonal prism consists of six rectangular faces whose intersection lines are parallel to the vertical c axis (see Fig. 5.12(C)). A pyramid consists of three or more triangular faces that meet at a common point. Hexagonal prisms and pyramids each have, by definition, six faces. A dihexagonal prism is the name for a 12-faced prism, as derived from the Latin root *di*, meaning "two," combined with the word *hexagonal*, from the Greek root *hexa*, meaning "six." *Dihexagonal*, therefore, describes a 12-faced form, be it a prism or a pyramid. A dipyramid consists of two identical pyramids that meet base to base and are the result of a horizontal mirror plane. Common forms in this point group are the following:

Pinacoid, {0001}, consisting of two parallel faces perpendicular to the 6-fold axis. These occur commonly in combination with hexagonal prisms and hexagonal (or dihexagonal) truncated pyramids (see Fig. 5.12(C)).

Hexagonal prism, {10$\bar{1}$0} and {11$\bar{2}$0}. Both prisms consist of six vertical faces that intersect the horizontal axes at different locations. They are identical forms distinguished only in their crystallographic orientation.

Dihexagonal prism, {h k $\bar{i}$ 0}, consists of 12 vertical faces instead of 6.

Hexagonal dipyramid, {h0$\bar{h}$l} and {hh(2$\bar{h}$)l}. Both consist of 12 isosceles faces that intersect in a point along the vertical axis. The two dipyramids differ only in their orientation with respect to the three horizontal axes.

Figure 5.12(C) shows many of the preceding forms in combination in the morphology of apatite and beryl.

$\bar{3}$ 2/m

This crystal class, part of the hexagonal system, is referred to the four crystallographic axes shown in Figure 5.12(A). The vertical crystallographic axis, c, is one of 3-fold rotoinversion, and the three horizontal axes are of 2-fold rotation. The three mirror planes occur midway between the crystallographic axes. The general form, {hk$\bar{i}$l}, is a **scalenohedron**. This is a closed form consisting of 12 scalene triangular faces (*scalene* refers to it having unequal sides). A scalenohedron is distinguished from a dipyramid by a zigzag appearance of the middle edges. The symmetry content of this point group, the distribution of face poles of the scalenohedron, and two morphological illustrations are given in Figure 5.13(A).

A common form in this point group is the rhombohedron, $\{h0\bar{h}l\}$ and $\{0h\bar{h}l\}$. This is a closed form consisting of six rhomb-shaped faces. A rhombohedron can be thought of as a cube deformed (compressed or extended) along one of the

four 3-fold rotoinversion axes between the opposing corners of a cube. A common Miller index for the rhombohedron is $\{10\bar{1}1\}$. A rhombohedron and its stereogram are illustrated in Figure 5.9.

A considerable number of common minerals crystallize in this class. The most common among them is calcite and other members of the calcite group, which includes siderite, magnesite, and rhodochrosite. Figure 5.13(B)–(C) illustrates calcite and corundum in various combinations of hexagonal forms.

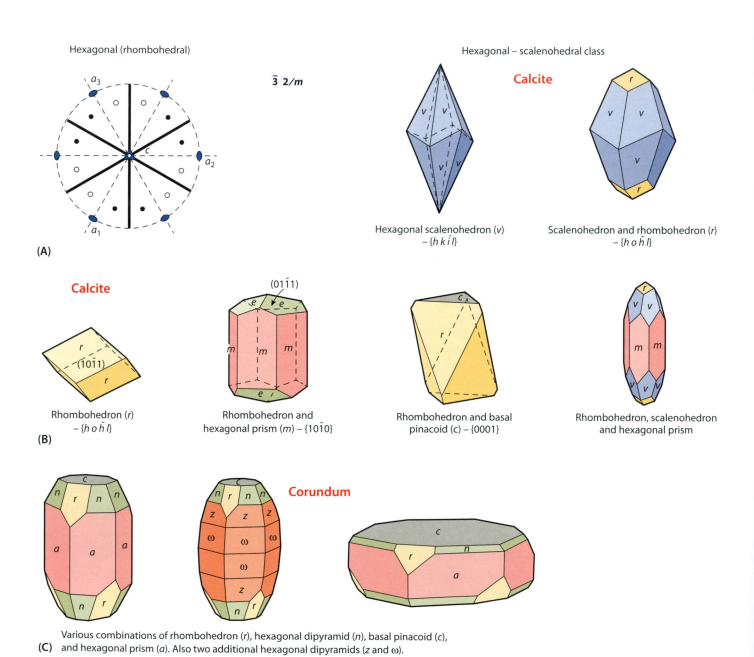

(A) Hexagonal (rhombohedral)

$\bar{3}\,2/m$

Hexagonal – scalenohedral class

Calcite

Hexagonal scalenohedron (v) – $\{hki\bar{l}\}$

Scalenohedron and rhombohedron (r) – $\{ho\bar{h}l\}$

(B) **Calcite**

Rhombohedron (r) – $\{ho\bar{h}l\}$

Rhombohedron and hexagonal prism (m) – $\{10\bar{1}0\}$

Rhombohedron and basal pinacoid (c) – $\{0001\}$

Rhombohedron, scalenohedron and hexagonal prism

(C) **Corundum**

Various combinations of rhombohedron (r), hexagonal dipyramid (n), basal pinacoid (c), and hexagonal prism (a). Also two additional hexagonal dipyramids (z and ω).

Figure 5.13 (**A**) Stereogram of the symmetry content of $\bar{3}\,2/m$ (the crystallographic reference axes are the same as those in Fig. 5.12(A)) and the distribution of face poles of the general form, $\{hki\,l\}$, known as a hexagonal scalenohedron. On the right are two drawings of calcite crystals that exhibit the external form of a scalenohedron. (**B**) Calcite crystals with various common forms. (**C**) Examples of typical corundum crystals. The one in the middle is described as barrel shaped.

4/m 2/m 2/m

Crystals belonging to this crystal class are referred to the system of axes shown in Figure 5.14(A). The vertical axis, c, is one of 4-fold rotation. There are four horizontal axes of 2-fold rotational symmetry, two of which coincide with the crystallographic axes (a_1 and a_2) and the other two are at 45° to them. There are five mirror planes perpendicular to the five rotational axes. There is also a center of symmetry (i). Figure 5.14(B) is a stereogram with all of the symmetry elements as well as the face poles of the general form, {hkl}, known as a **ditetragonal dipyramid**.

The most common forms on three rock-forming minerals are shown in Figure 5.14(C)–(D). These include the following:

Basal pinacoid, {001}, a two-faced form with faces perpendicular to the vertical 4-fold axis and parallel to the horizontal mirror.

Tetragonal prisms, {010} and {110}, both of which are parallel to the vertical c axis but intersect the horizontal a_1 and a_2 axes differently. The {010} prism is parallel to one of the two a axes, whereas the {110} cuts both axes at equal distance. Both prisms consist of four faces.

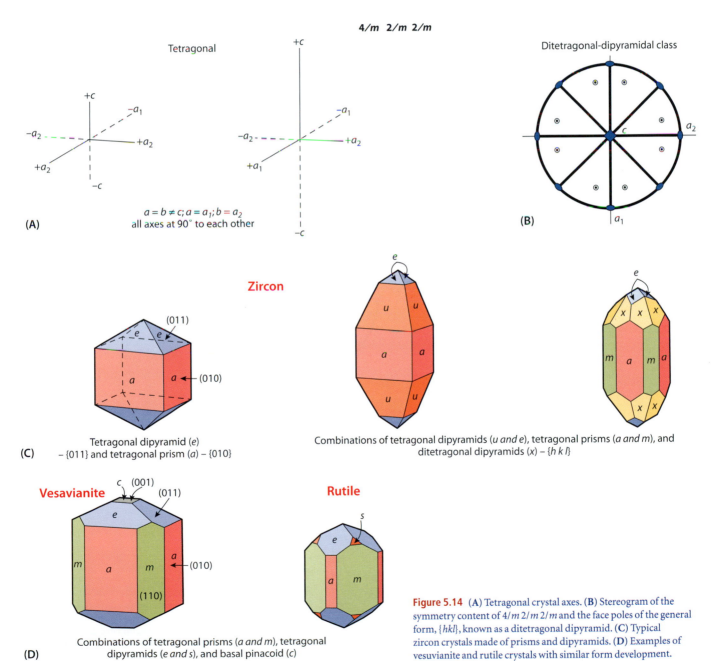

(A) $a = b \neq c; a = a_1; b = a_2$
all axes at 90° to each other

(B)

Tetragonal

4/m 2/m 2/m

Ditetragonal-dipyramidal class

Zircon

(C) Tetragonal dipyramid (e) – {011} and tetragonal prism (a) – {010}

Combinations of tetragonal dipyramids (u and e), tetragonal prisms (a and m), and ditetragonal dipyramids (x) – {h k l}

Vesavianite

Rutile

(D) Combinations of tetragonal prisms (a and m), tetragonal dipyramids (e and s), and basal pinacoid (c)

Figure 5.14 (**A**) Tetragonal crystal axes. (**B**) Stereogram of the symmetry content of 4/m 2/m 2/m and the face poles of the general form, {hkl}, known as a ditetragonal dipyramid. (**C**) Typical zircon crystals made of prisms and dipyramids. (**D**) Examples of vesuvianite and rutile crystals with similar form development.

Ditetragonal prism, {*hk*0}, consists of eight rectangular vertical faces, each of which intersects the two horizontal axes unequally.

Tetragonal dipyramids, {0*kl*} and {*hhl*}, both are closed forms with eight isosceles triangular faces. The {0*kl*} dipyramid intersects one horizontal axis, is parallel to the second, and intersects the vertical axis. The {*hhl*} dipyramid intersects both horizontal axes equally but the *c* axis at a different length. Figure 5.14(C)–(D) shows several combinations of dipyramids and prisms.

2/m 2/m 2/m

Orthorhombic crystals are referred to three crystallographic axes of unequal lengths that make 90° with each other (Fig. 5.15(A)). The three crystallographic axes are of 2-fold rotation, each with a perpendicular mirror. There is also a center of symmetry (*i*). This symmetry content and the face poles of the general form, {*hkl*}, a **rhombic dipyramid**, are shown in Figure 5.15(B). The choice for the location of the vertical *c* axis is mainly based on the habit of the specific mineral. If the mineral shows an elongate habit in one direction, this direction is usually chosen as the vertical *c* axis (see Fig. 5.15(C)). If the mineral is tabular in habit, as shown in Figure 5.15(D), the top face is chosen as a **basal pinacoid**, {001}. Commonly occurring forms in this crystal class are the following:

Pinacoids, {100}, {010}, and {001}, all of which are two-faced forms, each parallel to two different crystallographic axes. The {100} pinacoid is known as the front or *a* pinacoid and cuts the *a* axis but is parallel to *b* and *c*. The {010} or *b* or side pinacoid cuts the *b* axis and is parallel to *a* and *c*. The basal or *c* pinacoid, {001}, cuts the *c* axis and is parallel to *a* and *b*.

Rhombic prisms, {0*kl*}, {*h*0*l*}, and {*hh*0}, consist of four faces that are parallel to one axis and intersect the other two.

Rhombic dipyramid, {*hkl*}, is by definition the general form as noted earlier. Because all three crystallographic axes in the orthorhombic system are of different lengths, the form with Miller index {111} does not represent a specialized form in this crystal system but is one of several indices for the general form. For example, {111}, {113}, and {121} are all examples of the general form, the rhombic dipyramid.

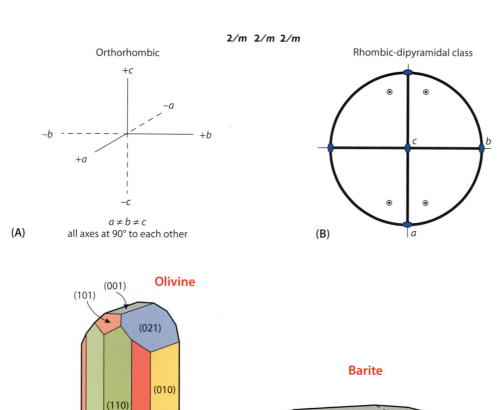

2/m 2/m 2/m

Orthorhombic

+c
−a
−b
+b
+a
−c

a ≠ *b* ≠ *c*
(A) all axes at 90° to each other

Rhombic-dipyramidal class

c
b
a

(B)

Olivine

(001)
(101)
(021)
(010)
(110)
(210)

(C) Combination of four rhombic prisms and two pinacoids (basal and side)

Barite

c
m
d
m
o
o
d

(D) Combination of three rhombic prisms (*m*, *d*, and *o*) and basal pinacoid (*c*)

Figure 5.15 (**A**) Orthorhombic crystal axes. (**B**) Stereogram of the symmetry content of 2/*m* 2/*m* 2/*m* and the distribution of face poles of the general form, {*hkl*}, known as a rhombic dipyramid. (**C**) A vertically elongate olivine crystal showing rhombic prisms and two pinacoids. (**D**) A barite crystal with a distinctly flat habit on account of the pronounced basal pinacoid.

2/m

Monoclinic crystals are referred to three crystallographic axes of unequal lengths. The angles between the a and b axes, and the b and c axes are 90° and the angle between $+a$ and $+c (=\beta)$ is greater than 90° (see Fig. 5.16(A)). The 2-fold rotation axis, which is the direction perpendicular to the mirror plane (in $2/m$), is taken as the b axis. The mirror plane coincides with the $a – c$ plane. There is also a center of symmetry (i). The a axis is inclined downward toward the front and the c axis is vertical. This is shown in the stereogram in Figure 5.16(B). Because the $+$ end of the a axis dips below the equatorial plane (see Boxes 5.4 and 5.5), the $+$ end point of this axis plots inward from the outer edge of the equatorial circle when projected upward (with the eye at the north pole of the spherical projection). The $+$ of the a axis is, therefore, shown as a small open circle inward from the outer edge of the equatorial plane in Figure 5.16(B). The graphical distance between the

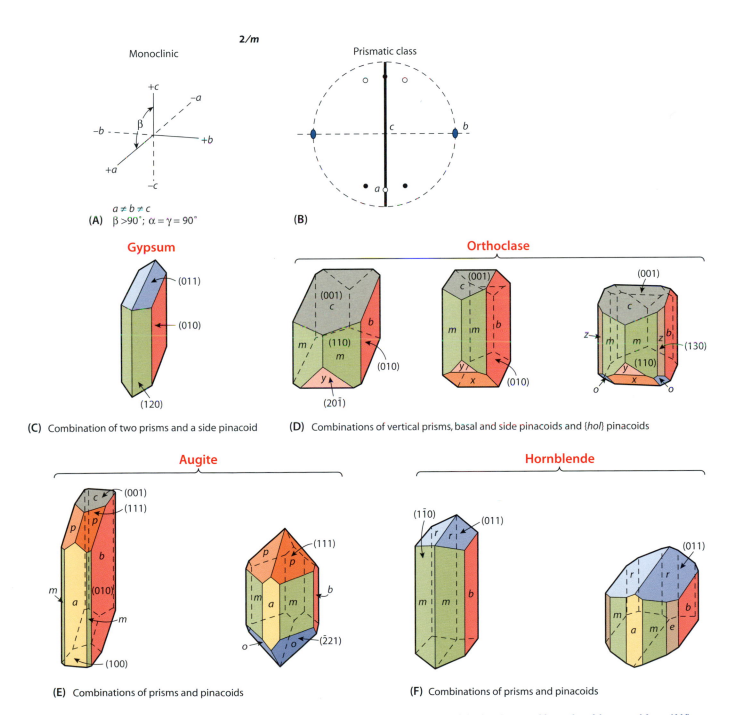

Figure 5.16 (**A**) Monoclinic crystal axes. (**B**) Stereogram of the $2/m$ symmetry in the second setting and the distribution of face poles of the general form, {hkl}, known as a prism. (**C**) Very common habit of gypsum crystals. (**D**) Various typical crystals of orthoclase. (**E**) Augite crystals that show form combinations typical of most of the clinopyroxene group. (**F**) Hornblende crystals showing the development of forms common in clinoamphiboles.

location of the + end of *a* and the outer edge of the equatorial circle is a function of the β angle. If β were 90°, the projection point would lie on the outer perimeter of the circle. With β increasing in size, it moves toward the center of the stereogram.

The previously described orientation for the axes in 2/*m* is most commonly used in mineralogy textbooks. This is known as the second setting. In some of the crystallographic literature, the 2-fold rotation axis is taken as a vertical axis with the mirror coincident with the equatorial plane. This is referred to as the first setting.

The general form, {*hkl*}, in this crystal class, is a four-faced prism, the face poles of which are plotted in the stereogram in Figure 5.16(B). There are also {0*kl*} and {*hk*0} prisms, and several of these are shown in Figure 5.16(C)–(F).

The only other form in this crystal class is the two-sided pinacoid. The most common of these are the {100} or front or *a* pinacoid; the {010} or side or *b* pinacoid; and the {001} or basal or *c* pinacoid. Several of these are shown in Figure 5.16(C)–(F). In addition, there are {$\bar{h}0l$} and {*h0l*} pinacoids.

Many minerals crystallize in this crystal class and several rock-forming minerals are illustrated in Figure 5.16. Orthoclase

is a monoclinic K-feldspar. Augite is a member of the large group known as clinopyroxenes and hornblende is an example of the large group of clinoamphiboles.

$\bar{1}$

Crystals belonging to this crystal class are referred to three crystallographic axes of unequal lengths that make oblique angles with each other (Fig. 5.17(A)). Because of the low symmetry content of this point group, namely *i* (= inversion), which is equivalent to a center of symmetry, the only forms possible are two-sided pinacoids.

The general form, {*hkl*}, pinacoid is shown in the stereogram in Figure 5.17(B). Other pinacoids that occur have Miller indices such as {100}, {010}, {001}, {0*kl*}, {*h0l*}, and {*hk*0}. A number of these are illustrated in Figure 5.17(C)–(D).

There are established rules for the orientation of crystals in the triclinic system. If the crystal is elongate in its habit, that direction is taken as the vertical *c* axis, as in Figure 5.17(C). If it is flat in its habit, then the large pinacoid is chosen as a basal pinacoid, as in Figure 5.17(D).

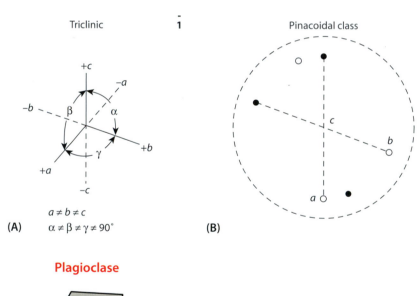

(A) $a \neq b \neq c$
$\alpha \neq \beta \neq \gamma \neq 90°$

(B)

Plagioclase

Rhodonite

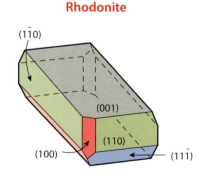

(C) Combination of five pinacoids in plagioclase

(D) Combinations of six pinacoids in rhodonite

Figure 5.17 **(A)** Triclinic crystal axes. **(B)** Stereogram of the location of the three crystallographic axes and the two face poles of the general form, {*hkl*}, known as a pinacoid. These two face poles are related by inversion (*i*). **(C)** and **(D)** typical crystal habits of plagioclase and rhodonite, respectively.

The most common rock-forming minerals that crystallize in this point group are members of the plagioclase series, microcline (a triclinic K-feldspar), rhodonite, and wollastonite.

5.6 Twins

Mineral grains occur over a range of sizes (fine-, medium-, and coarse-grained) as the basic constituents of rocks. These grains are individual crystals that are generally tightly interlocked with neighboring grains. We call these mineral grains and not crystals because most of the time they lack well-developed crystal faces because of their close intergrowth with adjoining grains. We describe such mineral grains as anhedral or subhedral (Sec. 2.4). Euhedral grains, or well-formed crystals, of considerable size (about 3 cm in the long dimension) are relatively rare because they occur mainly in very coarse-grained assemblages found in pegmatites or quartz veins. More rare are well-formed twinned crystals or **twins**. They are not the result of random intergrowths with each other. Instead, they result from a rational, symmetrical intergrowth of two or more crystals of the same substance. The two or more individuals are related by a new symmetry element (known as the **twin element**) that was not already present in the untwinned single crystal. This twin element can be

- A mirror plane or twin plane
- A rotation axis or twin axis
- An inversion about a point or twin center

The previous three operations are known as the **twin laws.**

Let us first illustrate how twinning happens on the atomic scale, before we discuss twin operations as seen in the external form of crystals or under the optical microscope. The bottom right segment of Figure 5.18(A) shows a simple structural arrangement of identical atoms in a rectangular array, an orthorhombic **lattice** (a lattice is an imaginary pattern of points (or nodes) in which every point (or node) has an environment that is identical to that of every other point (or node) in the pattern). It shows the identical lattice on the left side but stacked in an inclined fashion. There is a perfect fit between these two lattices along the inclined plane marked as A-A'. This is the **mirror plane** (or **twin plane**), by which the two identical lattices (but in different orientations) are related.

Figure 5.18(B)–(C) shows another twin operation, on an atomic scale, namely that of a twin related by a rotational axis, the **twin axis**. The lower right-hand side of the lattice is identical to that part of the lattice in Figure 5.18(A). The upper left part shows a segment of that same lattice but rotated 120° from its original right-hand position (Fig. 5.18(B)). Figure 5.18(C) shows the result of 180° rotation of the original lattice segment to its final position. One corner atom is differently colored in these three figures so as to show the distinction between a twin related by a mirror and by a rotational axis. These figures illustrate that the regular arrangement of atoms in the original lattice (lower right side) is interrupted in a nonrandom, very regular fashion. Such twins, known as **growth twins** or primary twins, are considered accidents or nucleation errors that occur during the free growth of a crystal. They are the result of the attachment of atoms, ions, or ionic groups on the outside of a growing crystal in such a way that the original pattern of the crystal structure is interrupted but in a regular way. Processes that modify a crystal after its formation (after its growth is completed) are called secondary. Secondary twinning can result from mechanical

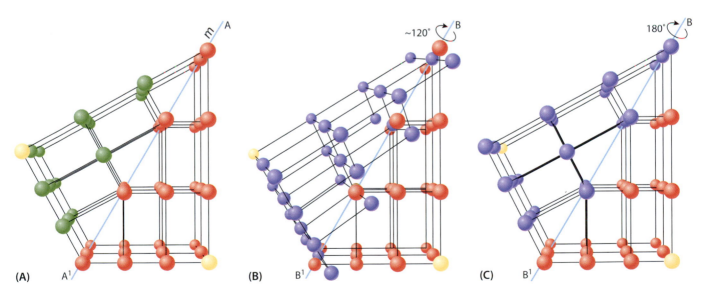

Figure 5.18 **(A)** Two parts of the same orthorhombic lattice related by the inclined mirror plane (A-A'). The twin law, therefore, is a twin plane. One sphere (atom) has a different color to emphasize that the twin operation is indeed a mirror. **(B)** Partial rotation (about 120°) of a segment of the lattice about an axis of rotation (B-B'). **(C)** Complete (180°) rotation of the same segment as in (B) but in a location identical to that shown in (A). Note that (A) and (C) are different, as shown by the locations of the spheres (atoms) at the lower right and their relocation by mirroring (A) or 2-fold rotation (C). (Adapted from the CD-ROM *Mineralogy Tutorials* that accompanies Klein and Dutrow 2008.)

deformation and from the displacive transformation of one polymorph into another (see Sec. 5.8).

Twins are classified as follows:

Contact twins: a regular composition plane joins the two individuals. Such individuals are separated by a twin plane.

Penetration twins: appear to be intergrown with the individuals interpenetrating, and an irregular composition surface. Penetration twins are usually produced by rotation and defined by an axis of rotation.

Multiple or repeated **twins:** three or more individual crystals are twinned according to the same twin law. A **polysynthetic twin** group results if all the composition surfaces are parallel.

We will give some common examples of each of the above twin types beginning with contact twins. But first let us look back at a statement given at the beginning of Section 5.6: "The two or more twinned individuals are related by a new symmetry that was not already present in the untwinned crystal." Figure 5.19(A) shows an octahedron, {111}, a common form in the isometric point group $4/m\,\overline{3}\,2/m$. Diamond and magnetite (a member of the spinel group) crystals commonly occur in such crystals. Figure 5.19(A) also shows an inclined plane that is one of four possible planar directions (all part of {111}) along which the octahedron may be twinned. This is a planar direction that is not a mirror already present in $4/m\,\overline{3}\,2/m$. This is a new symmetry element introduced in the twinning operation. The resultant twin, known as a **spinel twin**, is shown in Figure 5.19(B) and a stereogram of the symmetry content of $4/m\,\overline{3}\,2/m$ as well as the orientation of the new mirror (twin) plane is given in the stereogram in Figure 5.19(C). Additional examples of contact twins in some rock-forming minerals are shown in Figure 5.20.

Penetration twins are generally the result of twinning along a specific direction known as a twin axis. The resulting twinned

intergrowth is intertwined having an irregular composition surface. Figure 5.21 illustrates the twin that results when a crystal of orthoclase (K-feldspar) is rotated about itself by 180° about the vertical c axis (identified by the direction symbol [001]). Figure 5.21(C) shows the resultant twin with the vertical c axis the twin axis. Orthoclase crystals are monoclinic with point group $2/m$. This includes a 2-fold rotation axis parallel to b, but not a 2-fold rotation axis along c. The 180° rotation about c, therefore, is the new symmetry element, or twin element (twin law). This is shown in the stereogram for $2/m$ in Figure 5.21(D). Additional examples of penetration twins are shown in Figure 5.22. Photos of two minerals with contact twins and one with a penetration twin are given in Figure 5.23.

Repeated or multiple twins consist of three of more individuals. When the composition surfaces are closely spaced and parallel, they are known as polysynthetic twins. This twin type is extremely common in triclinic plagioclase and in microcline (a K-feldspar). Figure 5.24(A)–(C) illustrates how a polysynthetic twin comes about. The first thin and planar triclinic unit of the twin with a well-developed {010} pinacoid is shown in Figure 5.24(A). This first basic unit is repeated as a twin by a mirror reflection parallel to (010), as shown in Figure 5.24(B). With many more consecutive parallel twin planes, a polysynthetic twin develops, as shown in Figure 5.24(C). This twin type, known as an **albite twin**, is very common in and highly diagnostic of triclinic feldspars. Note that the newly introduced symmetry element, the twin element, is a mirror plane parallel to (010). Triclinic minerals ($\overline{1}$ or 1) do not have such a mirror already part of their symmetry content. However, monoclinic minerals with point group $2/m$ do have such a mirror in the a-c plane, perpendicular to the b axis. As such, only triclinic feldspars can have this type of twinning. Figure 5.24(D) is a hand specimen of a triclinic feldspar (a member of the plagioclase series) that shows fine lines or striations on a cleavage surface. These lines are the edges of the thin successively twinned parallel plates (as

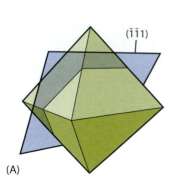

(A)

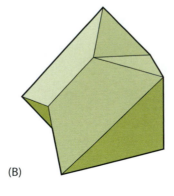

(B)

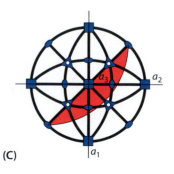

(C)

Figure 5.19 **(A)** An isometric octahedral crystal ($4/m\,\overline{3}\,2/m$) with a shaded plane ($\overline{1}\,\overline{1}1$) that is parallel to one of the four planar directions of the octahedron {111}. If this added plane acts as a mirror, a twin will result. **(B)** The twin, known as a spinel twin, is twinned along one of the four planar directions of {111}. **(C)** A stereogram of the symmetry of $4/m\,\overline{3}\,2/m$ with the location of the newly added twin plane. Note that this mirror plane was not present in the symmetry content of $4/m\,\overline{3}\,2/m$.

Quartz
Hexagonal: **32**

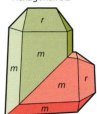

(A) Japan twin on {11$\bar{2}$2}

Staurolite
Pseudo-orthorhombic (monoclinic with β ~90°): **2/m**

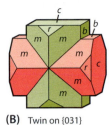

(B) Twin on {031}

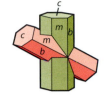

Twin on {231}

Gypsum
Monoclinic: **2/m**

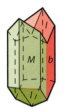

(C) Swallowtail twin on {100}

Orthoclase
Monoclinic: **2/m**

(D) Manebach twin on {001}

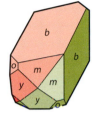

Bareno twin on {021}

Figure 5.20 All drawings of these contact twins are idealized. (**A**) A quartz twin. (**B**) Two types of contact twins common in staurolite. Such twins are uniquely diagnostic in the identification of staurolite in rock types in the field and hand specimen. Staurolite is here referred to as pseudo-orthorhombic because its external form is rectangular in outline along all three axial directions. It is in reality monoclinic, but the β angle is so close to 90° that it appears orthorhombic. (**C**) A swallowtail twin as seen in gypsum. (**D**) Two different types of contact twins in orthoclase.

c axis

[001]

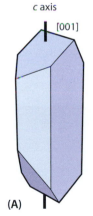

(A)

~80°

[001]

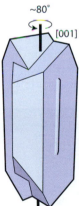

~150°

[001]

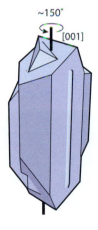

(B)

180°

[001]

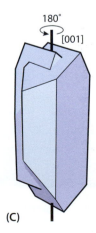

(C)

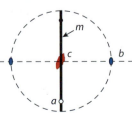

(D)

Figure 5.21 (**A**) A single crystal of monoclinic orthoclase (K-feldspar) with symmetry 2/*m*. (**B**) Two intermediate stages of intergrowth of the two crystals as a result of rotation of crystal in (A) about the vertical *c* axis ([001] direction). (**C**) The final penetration twin, known as a Carlsbad twin. (**D**) Stereogram of the symmetry elements in 2/*m* and the newly introduced 2-fold rotation axis along *c*.

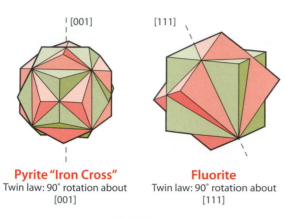

Pyrite "Iron Cross"
Twin law: 90° rotation about [001]

Fluorite
Twin law: 90° rotation about [111]

Figure 5.22 Examples of two penetration twins and their twin laws. The rotations in both twins are 90° about the directions noted. The pyrite iron cross twin results from the regular intergrowth of two pyritohedra. The fluorite twin consists of two intergrown cubes.

in Fig. 5.24(C)). Albite twinning is very easily identified in thin sections, as shown in Figure 3.20(B). The low-temperature polymorph of $KAlSi_3O_8$, known as microcline, is triclinic and always shows albite twinning. However, microcline is also twinned, but according to the pericline law, it results in a **pericline twin**. The lamellae of albite and the pericline twins cross at about 90° on a basal pinacoid, {001}, giving rise to a characteristic tartan pattern on that face, as shown in Figure 3.20(A).

If, in a multiple twin, the twin planes are not all parallel but instead parallel to other faces of the same form, it is called **cyclic twin**.

Yet two other types of twinning are known as **deformation** and **inversion twinning**, both of which represent secondary twinning. Deformation twins form in response to mechanical stress and inversion twins result from the structural change that happens when an earlier-formed crystal transforms to a lower symmetry, as a result of decreasing temperature (discussed in Sec. 5.8).

5.7 Some aspects of space groups

In Section 5.1, we introduced four elements of symmetry (mirrors, rotation axes, center of symmetry, and rotoinversion axes) that in various combinations lead to the 32 crystal classes (or point groups). All these symmetry elements and their combinations are referred to as translation-free elements because they all intersect at one point (32 point groups) in the center of a crystal, or a stereogram thereof. **Space groups**, which represent the various ways in which motifs (e.g., atoms, ions, ionic groups in crystals), can be arranged in a homogeneous array (*homogeneous* meaning that each motif in the pattern is equivalent to every other motif in the pattern) contain additional translational symmetry elements. These are the following:

- **Translation** (*t*)
- **Screw axes** (a rotation axis combined with translation)
- **Glide planes** (a mirror plane combined with a translation)

Space groups and their notation are shorthand for the translational and symmetry elements that define the location of atoms, ions, and/or ionic groups in crystal structures. In subsequent chapters on the systematic treatment of minerals in various rock types (igneous, sedimentary, metamorphic, and ore minerals in Chapters 7, 10, 13, and 15, respectively), each

Figure 5.23 Photos of three twin types. From left to right: a Japan twin in quartz (a contact twin); a Carlsbad twin in orthoclase (a penetration twin); and a butterfly twin in calcite (a contact twin).

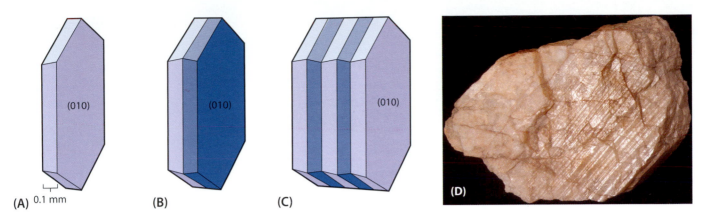

Figure 5.24 **(A)** A thin plate with a large pinacoidal development parallel to {010} is the original unit that is twinned into a pair in **(B)** through a mirror (or composition) plane parallel to (010). **(C)** Successively twinned parallel plates as in albite twinning in triclinic feldspars. **(D)** A hand specimen of plagioclase with albite twin lamellae on a cleavage surface. ((A)–(C) adapted from the CD-ROM *Mineralogy Tutorials* that accompanies Klein and Dutrow 2008).

mineral structure illustration has a headline that consists of mineral name, composition, and space group notation. In the following discussion, we introduce basic aspects of space groups and their notation.

Translation

Three-dimensional order (as inside a crystal structure) results from regular translations (t) in three different directions: x, y, and z. It has been shown that in two-dimensional, regular patterns, there are only five unique planar translational patterns. Using these five patterns as a base, 14 three-dimensional lattice types can be developed that are known as **space lattices**, or the 14 Bravais lattices. A lattice is an imaginary pattern of points (or nodes) in which every point (or node) has an environment that is identical to that of every other point (or node) in the pattern. A lattice has no specific origin, as it can be shifted parallel to itself.

BOX 5.6 | A FEW OF THE MILESTONES IN THE DEVELOPMENT OF CRYSTALLOGRAPHY

Presently, the science of crystallography, through the use of X-ray diffraction techniques, deals with the arrangement of atoms, their bonding, their symmetry, and translations in solid crystalline compounds. This has evolved from the study, in the nineteenth century, of the external (morphological) symmetry of well-formed (euhedral) crystals by refined optical instruments known as reflecting goniometers. These instruments permitted highly accurate and precise measurement of the positions of crystal faces. In the second half of the nineteenth century, Auguste Bravais (1811–1863) made an important contribution to the study of crystal geometry when he introduced a theory of crystal lattices in which he proved that only 14 different lattice types are possible. These are now referred to as the **Bravais lattices**, named in his honor. At the end of the nineteenth century, the Russian mineralogist F. Fedorow and the German mathematician Artur Schönflies independently developed a general theory of internal order and symmetry within crystals. They showed that there could be only 230 different types of structures based on symmetry and translational elements in crystal structures. Schönflies used the mathematical theory of groups and his work was responsible for calling the collective symmetries of spatial patterns space groups. This discovery forms the basis of our present classification of crystal structures determined by X-ray diffraction methods.

An enormous change in the approach to the science of crystallography occurred in 1912 when Max von Laue, together with W. Friedrich and P. Knipping, showed in an X-ray experiment at the University of Munich, Germany, that crystals cause distinct X-ray diffraction patterns due to atoms and that crystals are composed of periodic arrays of atoms. X-rays had previously been discovered by Wilhelm Conrad Röntgen in 1895. In 1914, the first crystal structures were determined by X-ray diffraction techniques by Sir W. L. Bragg (1862–1942). Since then, all of our accurate and precise information about complex crystal structures has been the result of the development of more and more sophisticated X-ray diffraction instrumentation.

All the crystallographic discoveries in the nineteenth century, on the basis of very careful study of the external symmetry of crystals, are now part of the knowledge base that was subsequently developed from X-ray diffraction studies.

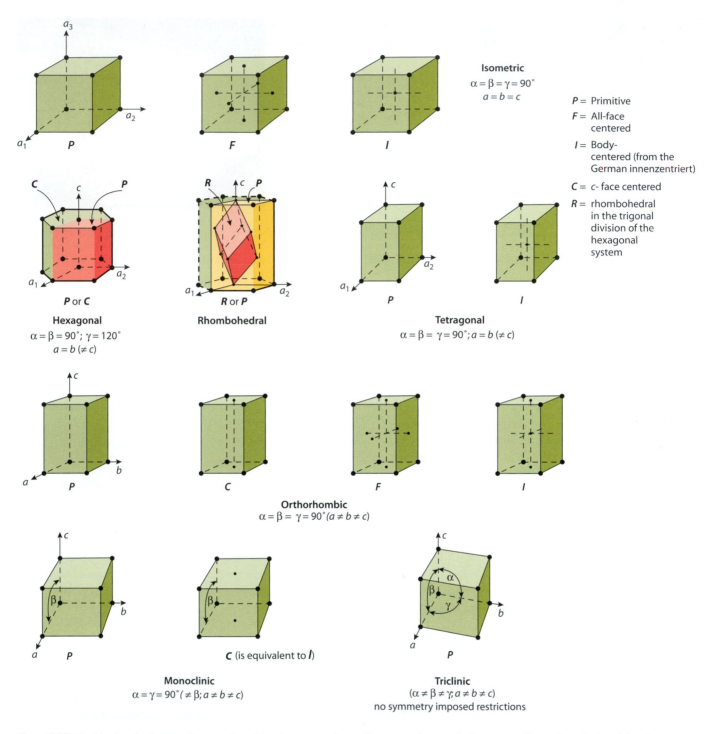

Figure 5.25 The 14 unique lattice types, known as Bravais Lattices, arranged according to crystal system. In the structure illustrations of minerals in subsequent sections on systematic mineralogy (Chapters 7, 10, 13, and 15), each structure drawing is headed by a mineral name, its chemical formula, and its space group notation. The first letter entry in the space group is always the abbreviation for the Bravais lattice type, as shown here. In the hexagonal system, there are three different lattice choices: two for hexagonal structures (*P* or *C* in hexagonal) and *R* in the trigonal division of the hexagonal system. Even though a space group may begin with *R*, it is traditional to describe the unit cell in terms of a primitive hexagonal lattice (*P*). This hexagonal choice of axes is preferred because they are easier to visualize than the rhombohedral ones.

The 14 Bravais lattices represent 14 different ways in which nodes (atoms) can be arranged periodically in three dimensions. They are illustrated in Figure 5.25, and their abbreviations are explained there as well. *Primitive* means that there are nodes at the corners of the lattice only. There is a primitive choice in each of the six crystal systems. Centered lattice choices occur in five of the crystal systems. The various symbols (*P*, *F*, *I*, *C*, and *R*) are the first entry in a space group notation.

When a crystal structure is viewed as a regular repetition of nodes in three-dimensional space, it can be assigned the Bravais lattice appropriate to that crystal structure. This leads to the concept that a crystal structure is really an infinitely extending space lattice made up of an infinite number of unit cells, where a **unit cell** is defined as the smallest unit of structure that can be infinitely repeated to generate the entire structure.

Screw axes

A screw axis is a rotational operation with a translation (t) parallel to the axis of rotation. Figure 5.26 gives three-dimensional representations of two opposite screw motions. The screw axis symbols consist of digits (2, 3, 4, and 6) followed by a subscript, which when inverted represents the fraction of the translation (t) inherent in the operation (Fig. 5.27). For example, 2_1 represents a 2-fold rotation with a translation of ½. The ½ results from placing the subscript over the main axis symbol, 2, to derive the fraction. In the case of 3-, 4-, and 6-fold screw axes, there are additional notations that relate to the handedness of the screw motion (whether the operation is right handed or left handed) or whether the operation is considered neutral. In Figure 5.26, there are drawings of two choices of isogonal screw axes, 4_1 and 4_3, (**isogonal** meaning "rotation through the same angle," from the Greek words *isos*, meaning "same," and *gonia*, meaning "angle"). The translation component in both screw operations is ¼, but a convention allows for a distinction between the two different screw directions, 4_1 and 4_3. When the ratio of the subscript to the digit of the rotation axis (¼ for 4_1) is less than ½, the screw is defined as **right handed**. A right-handed screw operation is one in which the unit of structure (an atom) moves away from the observer in a clockwise direction. When this ratio is more than ½, it is **left-handed**, as in 4_3, where ¾ is larger than ½. Therefore, 4_3 is the notation for a left-handed screw motion. The same holds for 3_1 and 3_2, which are an **enantiomorphous** pair of screw axes. *Enantiomorphous* is derived from the Greek word *enantios*, meaning "opposite." Therefore, 3_1 and 3_2, 4_1 and 4_3, 6_1 and 6_5, and 6_2 and 6_4 are all enantiomorphous pairs of screw axes because each pair represents two screw operations in opposite directions (one right handed, the other left handed). When the ratio is exactly ½, as in 2_1 (fraction is ½), in 4_2 (fraction is ¾ = ½), and in 6_3 (fraction is ⁶⁄₆ = ½), the screw motion is considered **neutral**. This means that the structural unit, the atom, ends up at the same place irrespective of the direction in which the rotation occurs. Figure 5.27 gives the numbers with subscripts as well as the standard symbols used for screw axes in illustrations of crystal structures.

Glide planes

A glide operation is a combination of a mirror reflection with a translation ($t/2$ or $t/4$) parallel to the mirror. Figure 5.28(A) shows how a two-dimensional pattern of human tracks

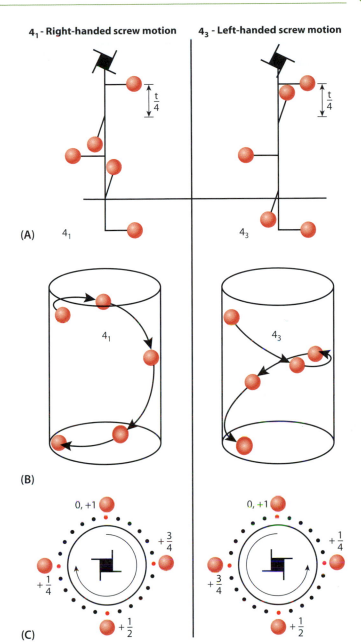

Figure 5.26 (**A**) Vertical views of two screw motions (4_1 and 4_3) with the appropriate graphical symbols for each at the top of the figures. (**B**) Further three-dimensional representation of the two screw motions. (**C**) Two-dimensional projections of both with fractions that reflect the positions of spheres along the vertical axes in (A).

illustrates a glide operation with translation $t/2$. Figure 5.28(B) shows a glide operation in a three-dimensional sketch. Figure 5.28(C) shows three octahedra that are related by a glide operation parallel to the vertical c axis. Figure 5.28(D) shows a more complex structural arrangement of a tetrahedral chain attached to an octahedral chain with a vertical glide operation parallel to the c axis (this is known as a c glide, see Fig. 5.29). Figure 5.28(E) is a polyhedral view of a tetrahedral-octahedral chain in the crystal structure of diopside, $CaMgSi_2O_6$, with a vertical, c glide. Figure 5.29 gives a listing of mirrors and glide

Rotation axis	Isogonal screw axis

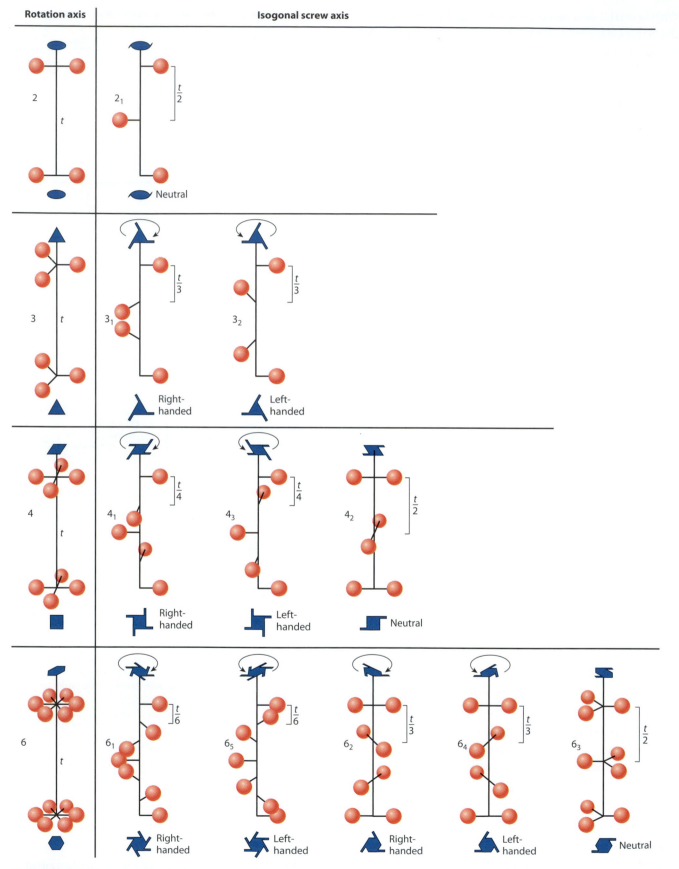

Figure 5.27 This illustrates graphically the operation of rotation and screw motions on motif units, here depicted as small spheres. These may represent atoms, ions, or ionic groups inside structures. The internationally accepted symbols for these operations are shown in perspective sketches at the top of each graphic and flat on the page at the bottom. Circular arrows relate to right- to left-handedness (adapted from *International Tables for Crystallography* 1983, vol. A).

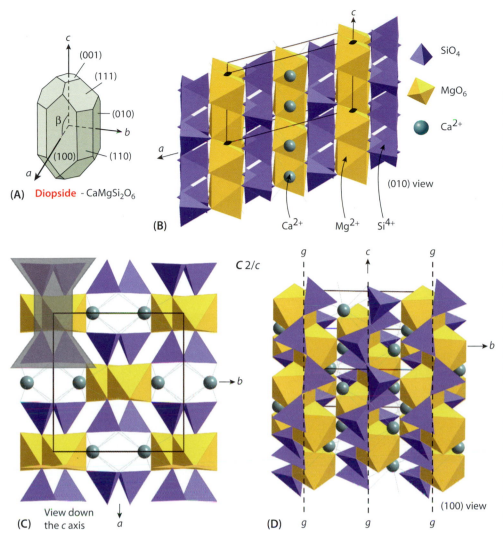

Figure 5.31 Diopside, $CaMgSi_2O_6$, and three different views of its crystal structure. (**A**) A crystal of diopside referred to monoclinic axes showing a β angle between the + ends of the *a* and *c* axes. (**B**) A view down the *b* axis of the structure. A monoclinic unit cell is outlined in the upper part of the structure. Two-fold rotation axes perpendicular to the page are inherent throughout the structure. (**C**) A view down the *c* axis. At the top left, a unit of structure is highlighted in gray and consists of two vertical Si_2O_6 chains that face each other and that coordinate an MgO_6 chain between them. The blue chains are tetrahedral and the yellow chains consist of vertically linked octahedra. This unit of structure is referred to as a *t-o-t* chain (tetrahedral-octahedral-tetrahedral). Note that this structural view reveals five such *t-o-t* packets end on. Four are at the corners of the outlined unit cell, and one is centered inside the unit cell. This leads to the choice of a centered unit cell, in this case *c*-centered, or *C*. (**D**) The front of this perspective view shows the (100) prism face with infinitely extending *t-o-t* chains parallel to the *c* axis. The tetrahedral chains are not perfectly straight but have a slight kinking pattern. The three vertical dashed lines locate vertical glide planes with a glide component parallel to *c*. The resulting space group is $C2/c$.

unit cell. Figure 5.31(C) is a view down the *c* axis of the structure, which reveals the presence of five identical structural packets made of vertical *t-o-t* chains (see caption). The rectangular aspect of the unit cell outline in this view reveals five such *t-o-t* chains, four at the corners and one in the center of the unit cell. This centering results in a unit cell choice known as *c*-centered, or *C*. The third view of the structure (Fig. 5.31(D)) reveals vertical glide planes along the centers of the slightly kinked *t-o-t* chains. The glide component in these is $\frac{1}{2}\,c$, and as such is represented by a small *c*. The appropriate space group choice for the structure of diopside is $C2/c$.

Figure 5.32 gives a synopsis of some structural data for these two structures (low quartz and diopside) that are standard for the structural characterization of any mineral.

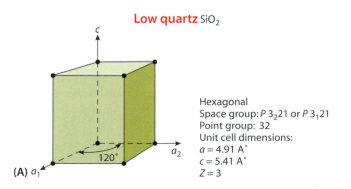

Low quartz SiO_2

Hexagonal
Space group: $P3_221$ or $P3_121$
Point group: 32
Unit cell dimensions:
$a = 4.91$ A°
$c = 5.41$ A°
$Z = 3$

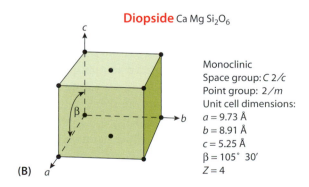

Diopside Ca Mg Si_2O_6

Monoclinic
Space group: $C2/c$
Point group: $2/m$
Unit cell dimensions:
$a = 9.73$ Å
$b = 8.91$ Å
$c = 5.25$ Å
$β = 105°\ 30'$
$Z = 4$

Figure 5.32 (**A**) The hexagonal primitive lattice (and unit cell) choice for low quartz. Its space group and isogonal point group. Unit cell dimensions and Z, the number of formula units of SiO_2 per unit cell. (**B**) The monoclinic *c*-centered lattice choice for diopside. Its space group, isogonal point group, unit cell dimensions and Z(= 4).

5.8 Polymorphism

Some minerals can occur in more than one structural arrangement even though their composition does not change. Polymorphism is the ability of a specific mineral (or chemical compound) to occur in more than one type of structure as a function of changes in temperature, or pressure, or both. The different structures are then referred to as **polymorphs**. The term derives from two Greek words, *poly*, meaning "many," and *morpho*, meaning "shape," resulting in "many shapes" or "many structures."

Table 5.4 gives examples of some common polymorphic minerals, and Figure 5.33 depicts the stability fields (in terms of variable temperature and pressure) of various polymorphs of SiO_2 and Al_2SiO_5. Which polymorph is stable over some specific *T-P* region is determined by energy considerations (see Sec. 14.2.1). The higher-temperature forms have greater lattice energy and as a result have a more expanded structure reflected in lower values for specific gravity (**G**) and refractive index (see Table 5.4). The higher internal energy is a function of increasing temperature, which causes higher frequencies of atomic vibrations and leads to more expanded (open) structure types. Increasing pressure favors structures with closer atomic packing. High-pressure forms of SiO_2 are coesite and stishovite; the high-pressure form of Al_2SiO_5 is kyanite (see Fig. 5.33(A)–(B)). The **G** values for these three minerals and their refractive indices are much larger than those of other respective polymorphs.

There are four types of polymorphism: reconstructive, displacive, order-disorder polymorphism, and polytypism.

Reconstructive polymorphism involves extensive structural rearrangement in going from one structure to the other. It requires the breaking of atomic bonds and a reassembly of structural units into different arrangements. Such transformations require a large amount of energy, they are not easily reversed, and they are sluggish. All the polymorphic mineral names listed for a specific chemical composition in Table 5.4 (except for the first two entries for low and high quartz) are the result of reconstructive transformations (the low to high quartz reaction is displacive, which we discuss later). The higher-temperature forms with greater lattice energy have the more expanded structure, which is generally reflected in lower values for specific gravity and refractive index. In laboratory experiments performed at atmospheric pressure, one can study how SiO_2 transforms via various polymorphs as a function of increasing temperature. The sequence, as depicted on the left side of Figure 5.33(A), is as follows: Low quartz is stable up to about 573°C, at which point it undergoes a reversible, rapid inversion to high quartz; at temperatures between 870°C and 1470°C, the stable polymorph is tridymite; above that, up to its melting point of 1723°C, cristobalite is stable. In general, these polymorphic reactions, as a function of in-

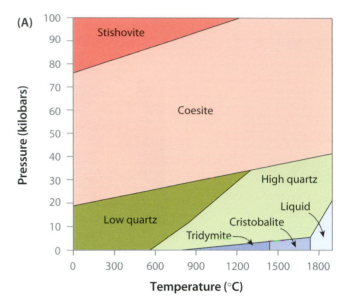

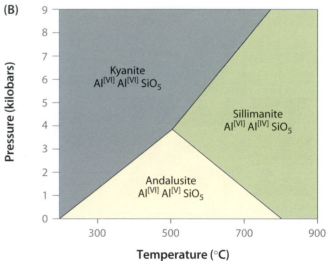

Figure 5.33 (**A**) Experimentally determined stability fields for the polymorphs of SiO_2. (**B**) Experimentally determined stability fields for the three polymorphs of Al_2SiO_5.

creasing temperature, occur as noted because of the increasing thermal energy in the system. However, such responsiveness does not happen when a high-temperature polymorph is cooled to normal crustal conditions. For example, cristobalite and tridymite are formed at high temperature and relatively low pressures in SiO_2-rich lava flows. Because a high activation energy is needed to break atomic bonds and convert the high-temperature forms to low quartz, both high-temperature minerals occur abundantly in very old terrestrial volcanic flows and in Precambrian lunar rocks. These minerals are metastable (see Sec. 14.2.1 and Fig. 14.3) at crustal conditions but transform to the stable SiO_2 form of low quartz only when additional heat (activation energy) is supplied by an external heat source. Examples of the structures of four

Table 5.4 Common polymorphous minerals.

Composition	Mineral Name	Crystal system and space group	G	Average refractive index
SiO_2	Low (α) quartz	Hexagonal – $P3_221$ (or $P3_121$)	2.65	1.55
	High (β) quartz	Hexagonal – $P6_222$ (or $P6_422$)	2.53	1.54
	Tridymite	Monoclinic – Cc	2.27	1.47
	Cristobalite	Tetragonal – $P4_12_12$	2.32	1.48
	Coesite	Monoclinic – $C2/c$	2.92	1.59
	Stishovite	Tetragonal – $P4_2/m\,2_1/n\,2/m$	4.29	1.81
Al_2SiO_5	Andalusite	Orthorhombic – $P2_1/n\,2_1/n\,2/m$	3.15	1.63
	Sillimanite	Orthorhombic – $P2_1/n\,2_1/m\,2_1/a$	3.24	1.66
	Kyanite	Triclinic – $P\bar{1}$	3.65	1.72
$KAlSi_3O_8$	Microcline	Triclinic – $C\bar{1}$	2.58	1.52
	Orthoclase	Monoclinic – $C2/m$	2.57	1.52
	Sanidine	Monoclinic – $C2/m$	2.57	1.52
C	Diamond	Isometric – $F4_1/d\,\bar{3}\,2/m$	3.52	2.42
	Graphite	Hexagonal – $P6_3/m\,2/m\,2/c$	2.23	
$CaCO_3$	Calcite	Hexagonal (rhombohedral) – $R\bar{3}2/c$	2.71	1.57
	Aragonite	Orthorhombic – $P2/m\,2/n\,2_1/a$	2.94	1.63
FeS_2	Pyrite	Isometric – $P2_1/a\bar{3}$	5.02	
	Marcasite	Orthorhombic – $P2_1/n\,2_1/n\,2/m$	4.89	

of the polymorphs of SiO_2 are given in Figure 5.34. (Various views of the low quartz structure are given in Fig. 5.30.) The highest-pressure polymorph of SiO_2 is stishovite, with Si in VI (octahedral) coordination with oxygen, not the normal Si^{IV} (tetrahedral) coordination that exists in all other silicates known from crustal occurrences.

The reconstructive transformations responsible for the three polymorphs of Al_2SiO_5 reflect changes in the coordination numbers of the constituent Al, as shown in Figures 5.33(B) and 5.35. All three are common constituents of medium- to high-grade metamorphic rocks with Al-rich bulk compositions. Andalusite commonly occurs in contact-metamorphosed assemblages; sillimanite is found in regionally metamorphosed terrains; and kyanite, by far the most closely packed of the three structures (G = 3.65; Table 5.4), occurs in regions that have undergone the highest lithostatic pressures.

The structural rearrangement in **displacive polymorphism** is slight when compared with that of reconstructive polymorphism. In displacive reactions, the overall structure is generally left completely intact and no bonds between atoms are broken. There is only a slight displacement of the atoms (or ions) and some readjustment of bond angles between the atoms, which results in a "kinking" of aspects of structure. Only a small amount of energy is required, so transformations occur almost instantaneously and are easily reversed. As such, one polymorph transforms easily into another.

The transformation of low quartz to high quartz, and vice versa, is a prime example of such a displacive polymorphic reaction. Figure 5.33(A) shows that low quartz (the normal structural type known as quartz) is stable from 0°C to a maximum of 573°C, at atmospheric pressure. Above that temperature, quartz instantaneously transforms to the structure displayed by high quartz. The reverse reaction, from high to low quartz, is instantaneous as well as temperature falls below 573°C. The difference between the two forms of quartz is shown in Figure 5.36. Low-quartz, with space group $P3_121$ or $P3_221$ (where 3_1 and 3_2 denote right- and left-handedness, and 1 in the notation means that there is no symmetry operator in that specific part of the Hermann-Mauguin notation) is shown in Figure 5.36(A) with obvious 3-fold (trigonal) open spaces. High quartz, with space group $P6_222$ or $P6_422$ (where 6_2 and 6_4 represent opposite handedness), as shown in Figure 5.36(B), has obvious hexagonal openings in this view of the structure. In the transition from high- to low-quartz, with high-quartz having a slightly higher symmetry than low-quartz, twinning may result. Such **transformation twinning** is known as Dauphiné twinning when the twin elements are obversed and reversed units of structure in low quartz. Such twinning can be seen only on basal, (0001), etched sections of quartz crystals. If low quartz with such twin patterns is heated to above 573°C, the Dauphiné twins instantly disappear.

Order-disorder polymorphism may arise when solid solution of two elements occurs at a specific site in a crystal structure. The degree of ordering in the structure is a function of the temperature at which the mineral originally formed. **Perfect order** in a mineral structure occurs only at absolute zero (0 Kelvin = −273.15°C). An increase in temperature disturbs the **perfect order**, resulting at an intermediate temperature in **partial order** and at some higher temperature in **total disorder**. There is no definite transition point between order and disorder; instead, a continuum of structural states exist.

The best example of different states of order is shown by potassium feldspar, $KAlSi_3O_8$, which occurs as three differently

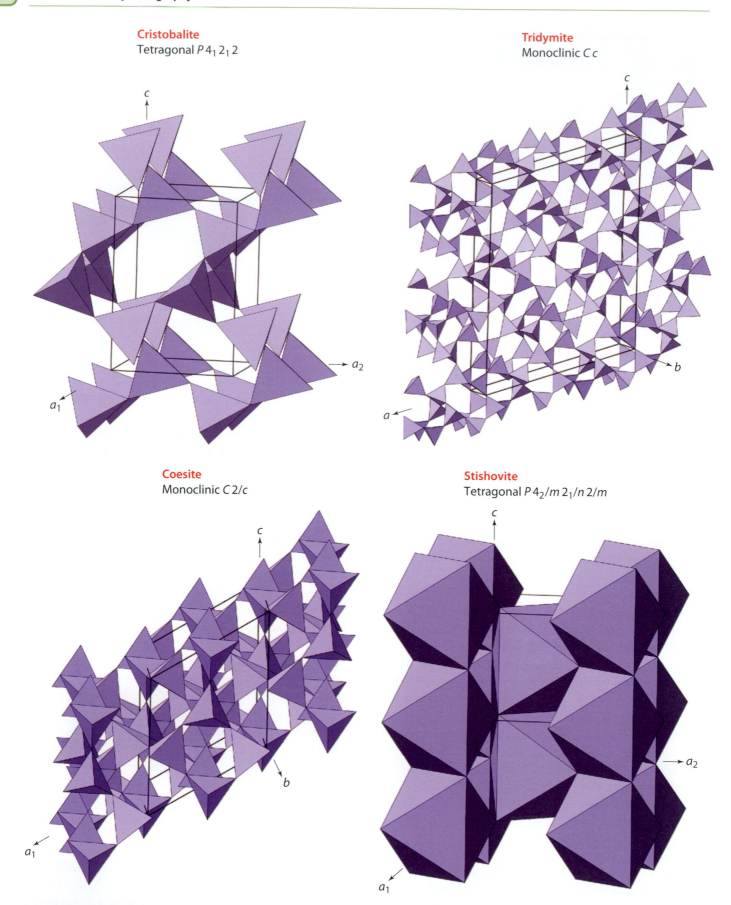

Cristobalite
Tetragonal $P4_12_12$

Tridymite
Monoclinic Cc

Coesite
Monoclinic $C2/c$

Stishovite
Tetragonal $P4_2/m\,2_1/n\,2/m$

Figure 5.34 Structures of four polymorphs of SiO_2 with a unit cell outlined in each. The structure of α quartz (low quartz) is illustrated in Figure 5.30.

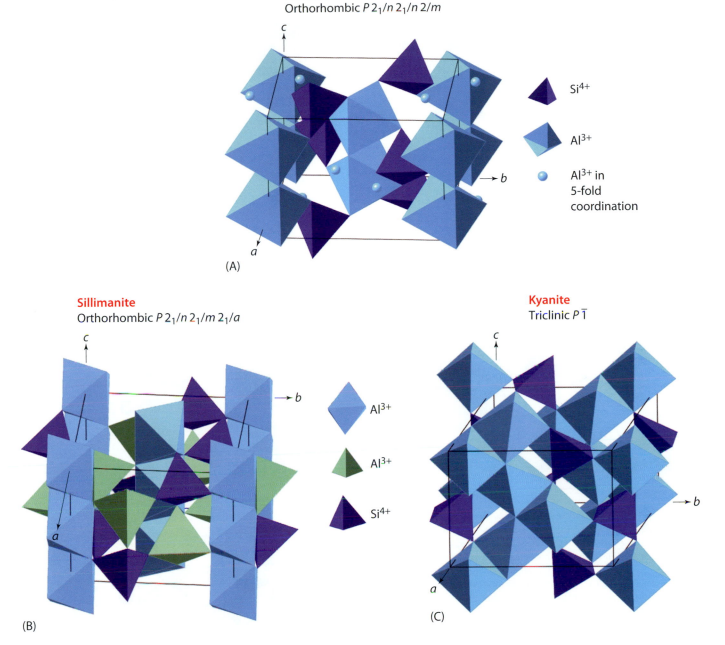

Andalusite
Orthorhombic $P\,2_1/n\,2_1/n\,2/m$

Si^{4+}

Al^{3+}

Al^{3+} in 5-fold coordination

(A)

Sillimanite
Orthorhombic $P\,2_1/n\,2_1/m\,2_1/a$

Al^{3+}

Al^{3+}

Si^{4+}

(B)

Kyanite
Triclinic $P\,\bar{1}$

(C)

Figure 5.35 Structures of the three polymorphs of Al_2SiO_5 with a unit cell outlined in each. (**A**) In the andalusite structure, Al^{3+} is housed in chains of edge-linked AlO_6 octahedra and in 5-fold coordination between the octahedral chains. (**B**) In the sillimanite structure, Al^{3+} occurs in edge-linked chains of AlO_6 octahedra and in AlO_4 tetrahedra. SiO_4 and AlO_4 tetrahedra cross-link the vertical AlO_6 chains. (**C**) The kyanite structure in which all of the Al^{3+} is housed in AlO_6 octahedra; Si^{4+} is in tetrahedra.

named minerals: microcline, orthoclase, and sanidine. All three of these are infinitely extending framework structures in which the Al and Si are distributed over tetrahedral sites. In microcline, the lowest temperature form, Al^{3+} fills one of the four tetrahedral sites and the three Si^{4+} ions (from the formula $K^+ = 1$, $Al^{3+} = 1$, $Si^{4+} = 3$) fill the other three tetrahedral sites (Fig. 5.37(A)), resulting in a triclinic structure ($C\,\bar{1}$). In this structure, there is a 100% probability of finding the Al in only one of the four tetrahedral sites, and such distribution is known as

total order. In the highest-temperature polymorph, sanidine, there is an equal probability of finding the Al^{3+} in any of the four tetrahedral sites. This is known as total disorder and means that statistically, on average, each tetrahedron contains $^1/_4\,Al^{3+}$ and $^3/_4\,Si^{4+}$. It is for this reason that all tetrahedra in the structure illustration of sanidine (Fig. 5.37(C)) are of the same color. Sanidine is monoclinic with space group $C2/m$. The intermediate structure type, known as orthoclase, has the Al^{3+} distributed over two equivalent sets of tetrahedral sites (see Fig. 5.37(B)).

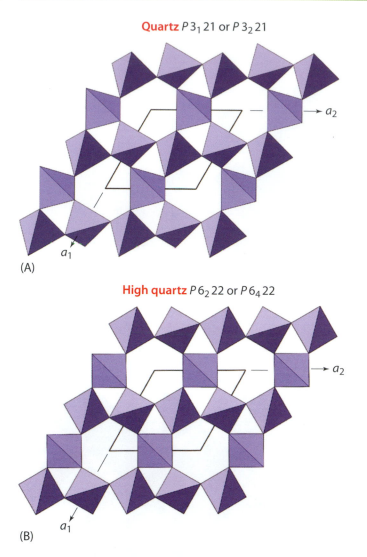

Quartz $P 3_1 21$ or $P 3_2 21$

(A)

High quartz $P 6_2 22$ or $P 6_4 22$

(B)

Figure 5.36 (A) The crystal structure of quartz (low quartz or α quartz) as viewed along the *c* axis, with (0001) the plane of projection. A primitive hexagonal cell is outlined. The large openings have a trigonal outline. Additional views of the quartz structure are given in Figure 5.30. (B) The structure of high quartz (or β quartz) from the same perspective as quartz in (A). Now the large openings are hexagonal in outline.

This structure is referred to as partially ordered and has monoclinic symmetry ($C2/m$). The preceding transformations require little energy and the more ordered structures tend to be stable at low temperatures.

Polytypism is a special kind of polymorphism in which different structural arrangements result from the different ways in which identical sheets (or layers) of structure are stacked in three dimensions. **Polytypes**, also referred to as **stacking polymorphs**, are common in the structures of layer silicates, such as clay minerals and members of the mica group. Polytypism is extensively discussed in the mineralogy textbooks by Dyar et al. (2008) and Klein and Dutrow (2008). Animated stacking sequences of polytypes can be viewed on the CD-ROM, *Mineralogy Tutorials*, which accompanies Klein and Dutrow (2008).

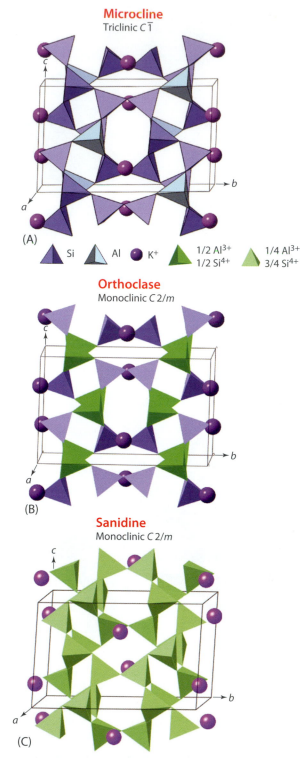

Microcline
Triclinic $C\bar{1}$

(A)

△ Si   △ Al   ● K+   △ 1/2 Al³⁺ 1/2 Si⁴⁺   △ 1/4 Al³⁺ 3/4 Si⁴⁺

Orthoclase
Monoclinic $C2/m$

(B)

Sanidine
Monoclinic $C2/m$

(C)

Figure 5.37 Crystal structures of the three polymorphs of K-feldspar, $KAlSi_3O_8$, with a unit cell outlined in each. (A) The low-temperature structure known as microcline. Here the Al (one Al per four tetrahedral cations; $(Al + Si_3) = 4$) is located in a specific tetrahedral site. This means that the Al distribution is completely ordered. (B) Orthoclase, the medium-temperature polymorph, shows the Al distribution in its structure as partially ordered (or partially disordered). The one Al per formula is distributed evenly over two tetrahedral sites ($1/2$ Al and $1/2$ Si in each of these two sites). (C) Sanidine, the high-temperature polymorph, in which the Al is distributed randomly across all tetrahedral sites. This means that statistically, on average, each tetrahedron contains $1/4$ Al and $3/4$ Si. This is a completely disordered structure.

Summary

This chapter introduced a range of concepts in crystallography. We began with symmetry aspects of the external form of crystals because these are easier to understand (and to locate in well-formed crystals) than the more complex symmetry operations that define the internal crystal structure:

- The four symmetry elements that can be observed in well-formed (euhedral) crystals, or in wooden models that imitate them, are mirrors, rotation axes, rotoinversion axes, and a center of symmetry (*i*).

- Many crystals of minerals show combinations of these symmetry elements. Such combinations can be listed (in terms of how many there are of each) or can be expressed by an internationally accepted notation, known as Hermann-Mauguin notation.

- All the symmetry combinations (or single symmetry elements as in very low-symmetry occurrences) can be grouped into six crystal systems: isometric, hexagonal, tetragonal, orthorhombic, monoclinic, and triclinic.

- These six crystal systems have their own unique reference axes and angles between the axes.

- The orientation of crystal planes in the external form of crystals is described by Miller indices or Miller-Bravais indices (in the hexagonal system).

- The term *crystal form* or *form* is not the same as the commonly used term *shape*. A crystal form is defined as consisting of a group of like crystal faces, all of which have the same relation to the elements of symmetry (contained in the crystal).

- To get away from sketches of three-dimensional crystals, projection schemes were introduced that convey the location of perpendiculars to crystal faces (known as face poles) and symmetry elements in a two-dimensional projection known as a stereogram.

- All the combinations of symmetry elements (or some single symmetry elements) as exhibited by naturally occurring crystals can be expressed in 32 crystal classes.

- We described in some detail only those seven crystal classes in which many crystals of rock-forming minerals can be classified.

- Most well-formed (euhedral) crystals occur as single units or in groups of random intergrowths. However, there are also the less common twinned crystals in which one crystal is related to the other by a symmetry element.

- This chapter ended with a brief introduction to the concept of space groups and their notation. Space group notation is a condensed shorthand for the symmetry and translational elements seen inside crystal structures.

- Crystal structures exhibit, in addition to the four symmetry elements seen in the external form of crystals, translation, screw axes, and glide planes.

- Space group notation, which is internationally accepted as a definition of a particular structure type, includes symbols for the lattice type (Bravais lattice), screw axes, and glide planes, as well as for the nontranslational symmetry elements seen in the morphology of crystals.

- Space group notation can mentally be converted to that of the crystal class (point group) by deletion of all of the translational components. In other words, a point group (or crystal class) is the translation-free derivative of the equivalent space group.

Review questions

1. Give a short definition of a mirror, rotation axis, center of symmetry, and rotoinversion axis.

2. In the evaluation of the symmetry content of a morphological crystal, you can make a list of all of the symmetry elements or you can use an internationally accepted shorthand notation. Which of these two approaches did you use in your course (see Table 5.1)?

3. If you were introduced to the Hermann-Mauguin notation (Box 5.1), what is so useful about this approach?

4. All combinations of possible symmetry elements (and some single elements) are referred to six crystal systems. Which are these six?

5. The concept of crystal axes is basic to the evaluation of the position of a specific crystal face on a morphological crystal. Why is this so?

6. In the isometric system, all three crystal axes are of equal lengths. In the orthorhombic crystal system, the axes are all of unequal lengths. Explain what determines the difference in crystal axis lengths in these two examples.

7. What is meant by *Miller index* and *Miller-Bravais index*? Explain why these two index systems differ.

8. What is the definition of *form*?

9. What is meant by the term *general form*? Give an example of its Miller index.

10. What is meant by a stereogram?

11. How does one arrive at the stereogram of a morphological crystal? Explain the process.

12. Why is a stereogram such a useful illustration?

13. What is meant by the term *crystal class* (or *point group*)?

14. How many crystal classes are there in total, and how many are systematically described in this text.

15. Give an example of a point group (crystal class) notation, using the Hermann-Mauguin system, for an orthorhombic crystal.

16. What is the definition of a *twinned crystal*?

17. What are the three different twin types?

18. What are the three new elements (or operations) encountered in space groups and what are their notations that were not seen in point groups?

19. Pure translation is described by the first letter entry in a space group notation. Give some examples of this purely translational element.

20. Some screw axis pairs are referred to as enantiomorphous. What is meant by that? Give an example of such a pair.

21. In the evaluation of the symmetry content of crystal classes (point groups), mirror operations are represented by *m*. What letters are used in space group notation for glide planes?

22. From Table 5.3 select two space groups from two different crystal systems (e.g., isometric and tetragonal) and derive the appropriate (translation-free) point group therefrom.

23. What is meant by the term *polymorph*?

24. Give the names for three different polymorphic processes.

25. What happens in the polymorphic transition from high to low quartz?

26. How many polymorphs are there for Al_2SiO_5? Give their mineral names.

FURTHER READING

Bloss, F. D. (1994). *Crystallography and Crystal Chemistry: An Introduction.* Reprint of the original text of 1971, Mineralogical Society of America, Chantilly, VA.

Dyar, M. D., Gunter, M. E., and Tasa, D. (2008). *Mineralogy and Optical Mineralogy*, Mineralogical Society of America, Chantilly, VA.

International Tables for X-ray Crystallography (1983). Vol. 1, *Symmetry Groups*; Vol. A, *Space Group Symmetry*, Kynoch Press, Birmingham, UK, and D. Reidel Publishing, Dordrecht, the Netherlands.

Klein, C. (2008). *Minerals and Rocks: Exercises in Crystal and Mineral Chemistry, Crystallography, X-ray Powder diffraction, Mineral and Rock Identification, and Ore Mineralogy*, John Wiley and Sons, New York.

Klein, C., and Dutrow, B. (2008). *Manual of Mineral Science*, 23rd ed., John Wiley and Sons, New York.

Klein, C., and Hurlbut, C. S. Jr. (1993). *Manual of Mineralogy*, 21st ed., John Wiley and Sons, New York.

Perkins, D. (2011). *Mineralogy*, 3rd ed., Prentice Hall, Upper Saddle River, NJ.

Wenk, H. R., and Bulakh, A. (2004). *Minerals: Their Constitution and Origin.* Cambridge University Press, Cambridge.

6 Minerals and rocks observed under the polarizing optical microscope

The study of rocks and minerals was revolutionized in 1850 when Henry Sorby invented the **petrographic microscope**, with which he examined extremely thin (30 µm) slices of rock using **polarized light**. The petrographic microscope is still the most useful instrument for identifying minerals and studying the texture of rocks. It is used by mining and petroleum geologists, environmental scientists, and even forensic scientists. In this chapter, we describe how minerals can be identified using the microscope. The microscope uses polarized light, and we describe how polarized light interacts with the crystal structure of minerals. When light passes through a mineral, it is slowed down, and we measure this slowing using a property called the **refractive index** of the mineral, which provides one of the most useful identifying properties of minerals. When polarized light passes through minerals that do not belong to the isometric crystal system, different refractive indices can be measured in different crystallographic directions. This variation of light velocity with direction in a crystal results in numerous optical effects, all of which provide additional properties that can be used to identify minerals. One of these is the production of **interference colors**, which gives minerals their spectacular colors under crossed polarized light in the microscope. By using special lenses and filters, it is possible to produce interference figures, which allow you to determine which crystal system a mineral belongs to and to give parameters that can be used in mineral identification. By the end of this chapter, you should be able to place a thin section of rock on the microscope stage and identify any of the common rock-forming minerals from their optical properties. Once minerals have been identified, their abundance can be determined under the microscope, and we will have taken the first step toward classifying rocks.

Photomicrograph of orthorhombic pyroxene (gray) and monoclinic pyroxene, augite (brightly colored), in a rock from the world's largest igneous intrusion near Lake St. John, Quebec (width of field is 3 mm). The origin of the lamellar structure in the pyroxene is discussed in Section 7.12.

6.1 Light and the polarizing microscope

In Chapter 3, we saw that the human eye is sensitive to that part of the electromagnetic spectrum known as the visible range. We also saw that the energy transmitted by electromagnetic radiation can be related to its wavelength through a simple equation (Eq. 3.1) involving the Planck constant and the speed of light. The speed of light – 186 000 miles per second – is perhaps one of the most familiar physical constants. This is actually the velocity of any electromagnetic radiation in a complete vacuum, and its precise value, in SI units, is 2.9979×10^8 m/s, or ~300 000 km/s. When light travels through material, it is slowed down, which results in the path of the light bending or refracting. This has two very important consequences:

1. It allows us to construct lenses that can be used to make optical instruments such as telescopes, microscopes, or common reading glasses.
2. When light passes through a mineral, the amount the light is slowed varies from one substance to another. This retardation provides a useful property known as the refractive index, which helps us distinguish and identify minerals.

The history of lenses is long, as Box 6.1 shows. It was in 1850 that Henry Sorby took a significant leap forward when he prepared slices of rock that were so thin that they could be examined under the microscope by light that passed through them. He used a polarized light source, which is light constrained to vibrate in only one plane, similar to the light you observe through a pair of polarizing sunglasses. Sorby then observed the magnified image through another polarizer oriented at 90° to the first one. The interaction of the polarized light with the crystal structure of minerals produced many optical effects that led to major advances in identifying minerals and to the modern polarizing petrographic microscope, such as that shown in Figure 6.1. With this microscope, it is possible to identify many different optical properties of minerals. We are interested in those that allow us to identify minerals. These we refer to as **diagnostic properties** because they allow us to identify or at least narrow the possible choices when trying to identify an unknown mineral.

A modern polarizing petrographic microscope is an expensive precision instrument that has many attachments to help deduce the optical properties of minerals and the textures of rocks. Despite this, it still does essentially the same thing as Henry Sorby's microscope: provide an image of a thin section under crossed polarized light. Figure 6.1 shows a typical petrographic microscope, in this example, a Carl Zeiss research model with the components that are most commonly used and found on all petrographic microscopes labeled. We briefly run through these components, starting with the light source at the base and ending with the eyepiece at the top. The function and use of these components will become apparent when we discuss the optical properties of minerals and their effects on polarized light.

The components of the petrographic microscope

The light source, which is commonly located in the base of the microscope, can have its brightness varied through use of a rheostat or a diaphragm. The light rises into the substage assembly, which consists of the lower polarizer, lower condensing lens, aperture diaphragm, and the upper swing-in condensing lens. The plane of polarization is normally east-west (right-left; we refer to this as E-W) but can be rotated to other orientations. The height of the whole substage assembly can be adjusted by means of a knob.

Next comes the stage on which the thin section is placed. In petrographic microscopes, the stage can be rotated through 360°, with each degree being marked around the circumference of the stage. Most stages have a locking screw on the side that prevents rotation; be sure this is loose before trying to rotate the stage. The stage may have screw holes for attaching accessories, such as the mechanical stage shown in Figure 6.1, which allows a thin section to be moved incrementally in two mutually perpendicular directions.

BOX 6.1 | **HISTORY OF LENSES**

Lenses have been known since the time of the Greeks. The Roman historian Pliny the Elder, who died escaping from the eruption of Pompeii in 79 CE, discussed lenses in his *Naturalis Historia* (Natural History). These early lenses, however, were of poor quality and were probably more often used as a burning glass to light fires than as a magnifying lens. Magnifying lenses were little used until 1300, when the first eyeglasses became available. By the beginning of the seventeenth century, Dutch lens makers had greatly improved the art of shaping, grinding, and polishing lenses, and they put together the first simple telescopes and microscopes. In 1609, Galileo built a telescope with which he observed moons orbiting the planet Jupiter, providing incontrovertible evidence that not all heavenly bodies orbit the Earth. The first microscopes gave magnifications of only several times, but by the middle of the seventeenth century, microscopes with magnifications of several hundred times resulted in remarkable biological discoveries. When the microscope was first used to study minerals and rocks, it served only as a simple magnifying device.

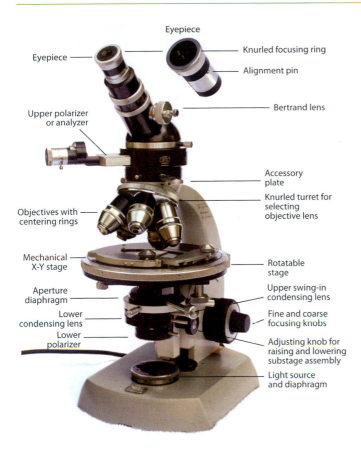

Eyepiece

Eyepiece

Knurled focusing ring

Alignment pin

Upper polarizer
or analyzer

Bertrand lens

Accessory
plate

Knurled turret for
selecting
objective lens

Objectives with
centering rings

Mechanical
X-Y stage

Rotatable
stage

Aperture
diaphragm

Upper swing-in
condensing lens

Lower
condensing lens

Fine and coarse
focusing knobs

Lower
polarizer

Adjusting knob for
raising and lowering
substage assembly

Light source
and diaphragm

Figure 6.1 Polarizing petrographic microscope. The example shown is a Carl Zeiss research model, but most petrographic microscopes have the features that have been labeled.

Above the stage are the objective lenses, which are commonly mounted on a turret that can be rotated so as to select the desired magnification. On some microscopes, only one lens can be mounted at a time, and this has to be removed and replaced with another lens to change magnification. Such lenses are held in place by a quick release clutch mechanism. Most microscopes have at least low (2×), intermediate (10×), and high magnification (40×) objectives, but some may have more, such as an oil immersion lens (100×) for extremely high magnifications. All objective lenses on petrographic microscopes are made so that they can be centered on the axis of rotation of the stage. Centering is done either by rotating the knurled rings on the lenses, as shown in Figure 6.1, or by adjusting centering screws where the lens attaches to the rotating turret. To center a lens, first observe a point under the cross-hairs in a thin section under the microscope as the stage is rotated. If the lens is centered, the point under the cross-hair at the center of the field will remain stationary under the cross-hair. When off-centered, it will sweep out a circle. By adjusting the knurled centering rings or screws on the lens, you can move the apparent position of the cross-hair to where you perceive the center of rotation to be. This may require several attempts before the lens is centered. Under no circumstances should the objective-lens turret be rotated by

grabbing the lenses; this will almost certainly off-center the lenses. Always rotate the turret by means of the knurled wheel.

The upper polarizer, or analyzer, as it is commonly called, is mounted above the objective lenses, with the plane of polarization being oriented N-S, at right angles to that in the lower polarizer. In some microscopes, such as the one shown in Figure 6.1, the plane of polarization in the analyzer can be rotated. Before the light passes through the analyzer, a small slot on the side of the microscope provides a place for insertion of accessory plates into the optical path. As we will see here, these accessory plates allow us to make important optical measurements.

Next a knurled knob on the side of the microscope allows a small lens known as a Bertrand lens to be swung in and out of the optical path to observe interference figures (more about them in Sec. 6.5).

The microscope image is viewed through the eyepiece, or ocular, which typically has a magnification of 8×, 10×, or 12×. The total magnification of the microscope is obtained by multiplying the eyepiece magnification times the objective lens magnification. The eyepiece also houses the cross-hairs, which are used as a reference for all centering and angular measurements made in the microscope. Each individual must focus the cross-hairs to suit his or her eyes. If this is not done correctly, considerable eyestrain can result. When your eyes examine an image in the microscope, they should be focused at infinity; the image of the cross-hairs, which will be superimposed on this image, should also be focused at infinity. The easiest way to ensure that this is the case is to carefully slide the eyepiece out of the microscope, and then, while observing a distant object, move the eyepiece in front of your eye to see whether the cross-hairs are in focus. If they are not, adjust slightly the knurled ring on the eyepiece and check the focus again; make sure you are still looking at a distant object when you do this. When satisfied that the cross-hairs remain in focus when your eye is looking at a distant object, return the eyepiece to the barrel of the microscope making sure that the alignment pin slots into its grove; this aligns the cross-hairs in E-W, N-S directions.

This completes a brief introduction to the various parts of the petrographic microscope. Remember, it is an expensive precision optical instrument and must be treated with care; do not force anything. Before we make use of this instrument, we must understand the behavior of light as it passes from one material to another and, in particular, the effects that crystalline material has on the passage of polarized light.

6.2 Passage of light through a crystal: refractive index and angle of refraction

Light travels through a vacuum at the incredible speed of 300 000 km per second, fast enough to circle the Earth more than seven times in one second. When not in a vacuum, light travels slower, with the velocity depending on the material through which it travels. For example, light is slowed more by traveling through quartz than it is through air. This slowing down provides a useful property that helps identify minerals. It is a difficult task, however, to measure the absolute speed of light, whether in a vacuum or in a mineral. Consequently, rather than measuring the absolute velocity of light in a mineral, we measure its velocity relative to that in air. This ratio is known as the refractive index and is defined as follows:

$$\text{Refractive index (R.I.)} = \frac{\text{velocity of light in air}}{\text{velocity of light in mineral}}$$

$$(6.1)$$

For example, the R.I. of diamond is 2.4, which indicates that light travels 2.4 times slower in diamond than in air. The R.I. of quartz, on the other hand, is only 1.5, which indicates that light is slowed much less in passing through quartz than through diamond. The high R.I. of diamond gives the gem its sparkle.

Equation 6.1 implies that we still need to measure the velocity of light in a mineral if we are to use the R.I. as a means of characterizing minerals. However, light has the peculiar property that it can be treated as traveling either as discrete bundles, or quanta, of energy or as waves (Eq. 3.1). For our purposes, light is most conveniently treated as traveling in waves. One of the immediate consequences of the wave propagation of light is that a light ray passing from one medium to another of different refractive index must bend (refract), and the amount it bends depends on the change of refractive index. The R.I. of a mineral can, therefore, be determined by measuring the angle of refraction rather than absolute velocities.

Figure 6.2 shows light (the blue and white bands indicate wave fronts of light) meeting a boundary between two materials at an oblique angle, where the R.I. of medium 1 is less than that of medium 2; that is, the light travels faster in medium 1 than in medium 2. The path of the light ray is shown by red arrows, whereas the crests and troughs of the light waves are shown in shades of blue. The angle at which the ray impinges on the boundary is known as the angle of incidence (i) and is measured from a line perpendicular to the boundary. On crossing the boundary into the medium of higher refractive index, the light is slowed. At the same time, however, the waves must remain connected across this boundary, and as can be seen in Figure 6.2, this is possible only if the light ray bends toward the perpendicular to the boundary, to create what is known as the angle of refraction (r). The angles of incidence and refraction are related to the R.I. of medium 1 and 2 by the simple relation known as **Snell's law**:

$$\frac{\sin i}{\sin r} = \frac{\text{R.I. 2}}{\text{R.I. 1}}$$

$$(6.2)$$

where sin is the trigonometric function, sine, of the angle of incidence or refraction. These angles, which can be measured with an instrument known as a **refractometer**, provide accurate

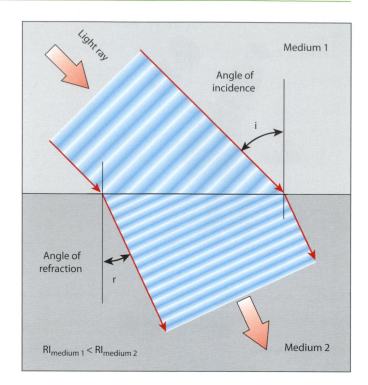

Figure 6.2 Refraction of light. Light waves (blue to white bands indicate wave fronts) are at right angles to the direction in which the light ray (red) travels. When light crosses the boundary between two materials of different refractive index (i.e., different light velocity), the light ray bends so that waves remain together along the boundary.

determinations of refractive index. Even though you may not get to use a refractometer, it is important to remember from Snell's law that if light passes into a substance of higher refractive index, the light will be bent toward the perpendicular to the boundary between the two materials (Fig. 6.2), and if it passes into a substance of lower refractive index, it will be bent away from the perpendicular. You will have many opportunities to apply these general principles in determining relative refractive indexes of grains in thin sections under the microscope.

The example of apatite

Let us consider what happens when light passes through a small grain of apatite that is included in a larger grain of quartz or feldspar, a common occurrence in many igneous rocks. Apatite has a relatively high R.I., around 1.65, whereas quartz and feldspar have a relatively low R.I., near 1.55. Consequently, when light passes into the apatite grain, it is bent toward the perpendicular to the interface between the minerals, but when it leaves the apatite grain and returns to the lower R.I. host mineral, it is bent away from the perpendicular (Fig. 6.3). The result is that light entering the apatite from below tends to focus or concentrate above the grain of higher R.I. Thus, while observing such a grain under the microscope, if the microscope stage is lowered slightly so that the plane of focus is slightly above the mineral, a slightly brighter light will appear just inside the boundary of the grain. This brighter line of light, which is known as the **Becke line**, can

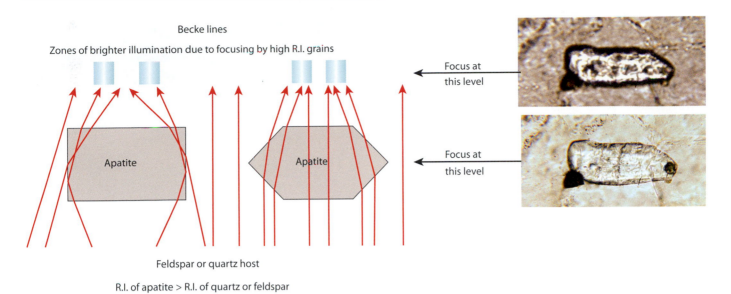

Figure 6.3 When a mineral of relatively high refractive index, such as apatite, is enclosed in one of lower refractive index, such as quartz or alkali feldspar, refraction of the light tends to concentrate the light above the higher refractive index grain. The lower of the two photomicrographs shows a grain of apatite enclosed in quartz and feldspar. In the upper photomicrograph, the plane of focus is just above the apatite grain, so it appears slightly blurred, but the central part of the grain is bright because of light refracted away from the grain boundaries, which appear dark.

be made to move back and forth by slightly lowering and raising the microscope stage, the line moving into the grain of higher R.I. when the working distance is increased. The Becke line can be made more pronounced by closing down the aperture diaphragm or lowering the substage assembly (Fig. 6.1)

The Becke line test provides a simple way of determining relative refractive indexes of adjoining mineral grains, which is one of the most useful optical properties in identifying minerals. Around the edge of thin sections, places can always be found where the R.I. of minerals can be compared with the R.I. of the mounting medium, which is commonly Canada balsam or an epoxy with a similar R.I. of 1.537. Minerals with higher R.I. than 1.537 are said to have **positive relief**, whereas those with lower R.I. have **negative relief**. Minerals such as quartz and feldspar have almost the same refractive index as that of common mounting materials, and so light is not bent in crossing the boundary between the two. This results in the grains having a clean smooth appearance under the microscope, because, although the surface of the ground rock may be quite rough, the glue with matching refractive index smoothes out these irregularities. In contrast, minerals with a much higher or lower refractive index than that of the mounting glue appear stippled. This effect is made more noticeable when the aperture diaphragm is closed down or the substage assembly is lowered.

6.3 Passage of polarized light through minerals

The light we are familiar with coming from the open sky is free to vibrate in any direction. However, light that is reflected off the surface of water, for example, tends to vibrate in a horizontal plane. Figure 6.4(A) shows a view of a pond in which blue sky and white clouds are reflected from the surface of the water. Figure 6.4(B) shows this same pond as viewed through a pair of Polaroid sunglasses. The sunglasses have lenses that allow light to pass vibrating only in a vertical direction. Because the light reflected off the surface of the water is vibrating in a horizontal plane, it is completely eliminated by the sunglasses. One of the features that makes the petrographic microscope different from other microscopes is that it uses a polarized light source; that is, all light that passes up through the microscope is constrained to vibrate in only one plane, which is normally oriented E-W. If you were to turn the light source on in a petrographic microscope and peer into the ocular, you would see a bright light. What your eye would not know is that this light is all vibrating in an E-W plane. If you were to look into the microscope through a pair of Polaroid sunglasses, the field would appear totally black. The N-S plane of polarization of the glasses would eliminate the E-W polarized light rising through the microscope, just as it did with the light reflected off the surface of the pool in Figure 6.4(B). The upper polarizer does exactly the same thing as a pair of Polaroid sunglasses, so that when it is inserted into the optical path, it cuts out all of the light rising from the lower polarizer and the field appears dark.

The refractive index of a mineral was defined in Eq. (6.1) as the velocity of light in air divided by the velocity of light in the mineral. When we measure a refractive index, we find that for most minerals the value obtained depends on the direction in which the light passes through the mineral. It is only minerals belonging to the isometric system and amorphous material such as glass that have only one refractive index independent of the path of the light, and these minerals are said to be **isotropic**.

Plane of polarization of light reflected from surface of water

Plane of polarization
in sunglasses

Figure 6.4 Light reflected off the surface of water tends to be polarized in a horizontal plane. As a result, the reflections of blue sky and white clouds from the surface of the pond in (**A**) are completely eliminated when viewed through a pair of Polaroid sunglasses (**B**) that only allow passage of light that is polarized in a vertical plane (photographs by Doreen Philpotts).

Minerals belonging to all other crystal systems have a refractive index that varies with crystallographic direction and are known as **anisotropic** or **birefringent**. The difference between the maximum and minimum refractive indexes is known as the **birefringence**. For example, the mineral calcite slows the passage of light considerably when it travels perpendicular to the c crystallographic axis, with the result that the R.I. in this direction is 1.66. In contrast, light traveling parallel to the c axis is slowed only slightly, and the R.I. in this direction is 1.49. The birefringence of calcite is, therefore, 0.17 (1.66 – 1.49). Birefringence is responsible for the spectacular colors exhibited by most minerals under crossed polarized light in the microscope, known as **interference colors**, which we learn more of in this chapter.

Relief and birefringence are two of the most readily visible optical properties of minerals under the microscope. Figure 6.5(B)–(C) contains photomicrographs of one of the world's most common types of igneous rock – basalt. The particular sample illustrated is from the Giant's Causeway in County Antrim, Northern Ireland, a UNESCO World Heritage site famous for the spectacular development of columnar joints formed as the lava cooled and shrank (Fig. 6.5(A)). Under plane polarized light (Fig. 6.5(B)), the difference in relief between the plagioclase and pyroxene in this rock is evident. The refractive index of plagioclase (1.56) is almost the same as that of the mounting medium (1.537); its relief is, therefore, low and

the plagioclase appears clear. In contrast, the refractive index of the pyroxene (1.7) is much higher than that of the mounting medium, with the result that light is refracted by irregularities on the surface of grains to create a stippled appearance, and light near grain boundaries is refracted into the grains making their interiors brighter and their margins darker (Becke line). Under crossed polarized light (Fig. 6.5(C)), the birefringence of the grains is evident. Plagioclase has low birefringence (0.012) and hence has gray and white interference colors, whereas pyroxene has high birefringence (0.025) and produces yellow, red, and blue interference colors.

To determine how the R.I. of a mineral varies with crystallographic direction, we need to know in which crystallographic plane the light is vibrating, and so we must use polarized light. The polarized light source in a petrographic microscope gives us a reference plane from which to make measurements. Light rising from the substage illuminator is constrained to vibrate in an E-W plane by the lower polarizer (Fig. 6.6(A)). We can think of this polarized light as a sinusoidal wave rising through the microscope. When we insert the upper polarizer, which allows light to pass vibrating only in a N-S plane, it completely cuts out all of the polarized light rising through the microscope. If, however, we were to rotate the plane of polarization in the upper polarizer to a NE-SW direction (Fig. 6.6(B)), as can be done with some microscopes, the E-W vibrating light coming from

0.1 mm

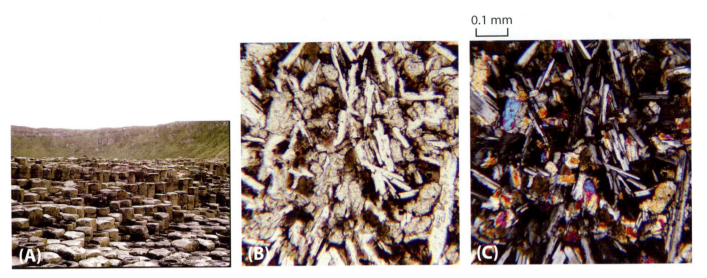

Figure 6.5 (**A**) Giant's Causeway basalt, County Antrim, Northern Ireland, showing columnar joints formed during cooling of the lava. Photomicrographs of a thin section of this basalt are shown under (**B**) plane-polarized light and (**C**) crossed-polarized light. Width of field, 1 mm.

the lower polarizer casts a component of vibration on the NE-SW plane of the upper polarizer, which would then allow light to pass but with a diminished amplitude (the distance from the crest to the trough of the wave would be less).

Looking at anisotropic minerals

What happens when we insert a thin section of an anisotropic mineral into the microscope? First, when plane polarized light enters an anisotropic mineral, it is forced to vibrate in two mutually perpendicular planes, the light in one of these planes

traveling faster than that in the other plane. In Figure 6.6(C), the two planes have been labeled fast and slow. The E-W polarized light rising from the lower polarizer sets light oscillating in the fast and slow vibration planes of the crystal. A crest of a wave arriving at the undersurface of the mineral would immediately create a crest in both the fast and slow vibration planes in the crystal. The amplitude of these waves would be less than that of the E-W polarized wave because only a component of that wave oscillates in the fast and slow vibration planes. The

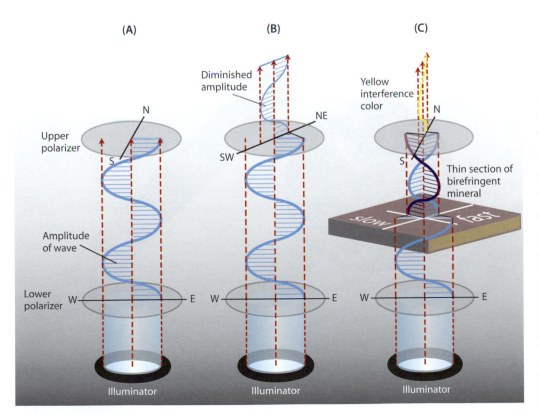

Figure 6.6 Passage of polarized light through the petrographic microscope. (**A**) The E-W polarized light rising from the lower polarizer is eliminated by the N-S plane of polarization of the upper polarizer. (**B**) When the upper polarizer is rotated to a NE-SW direction, a component of the E-W oscillation (black construction line) passes through the upper polarizer. (**C**) When a thin section of an anisotropic mineral is placed on the stage of the microscope, the E-W vibration from the lower polarizer produces vibrations in two mutually perpendicular planes in the mineral that travel with different velocities. When these two components are recombined by the upper polarizer, they are slightly out of phase, which creates an interference color (yellow wave).

amplitude is given by dropping a perpendicular (black construction line in Fig. 6.6(C)) from the E-W sinusoidal wave (blue line) onto the fast and slow vibration planes. Light rising through the crystal sets up sinusoidal waves oscillating in the fast (light blue wave) and slow (dark blue wave) vibration planes. When these waves arrive at the upper polarizer, which will allow light to pass only vibrating in a N-S plane, the fast and slow vibration directions both set up components of oscillation in the N-S plane of the polarizer. However, the waves in the fast vibration plane have traveled faster than those in the slow vibration plane. Thus, a crest of a wave reaching the upper polarizer via the fast vibration plane through the crystal arrives slightly ahead of the same crest that has traveled through the slow vibration plane. Consequently, when the upper polarizer combines these two waves into a single N-S vibrating wave, the fast and slow vibrating waves are slightly out of phase, and their addition causes interference and the generation of an interference color. In Figure 6.6(C), this interference color is shown by the yellow sinusoidal wave rising from the upper polarizer.

When the upper polarizer recombines the fast and slow light waves rising from the mineral, the waves are out of phase by an amount known as the retardation. The retardation is a function of the difference in the velocity of light vibrating in the two planes in the crystal and the thickness of the mineral. Most thin sections are ground to a thickness of 30 μm, so the retardation is usually a function only of the velocity difference between the two planes. If a mineral grain does not go all the way through a thin section, its retardation is less because of its reduced thickness. These interference colors are similar to those formed by

light reflecting off thin films of material, such as a thin film of oil spilled on a wet road (see Fig. 17.7(A)). Light that is reflected from the top and bottom surfaces of a thin film is slightly out of phase because the light that has passed through the thin film has been retarded by the refractive index of the film and by its thickness. The interference colors seen on a patch of oil are reflections of the thickness of the oil (see Fig. 17.7(A)).

Interference colors are a function of the degree of retardation. A retardation of 100 nm, for example, creates a gray interference color. As the retardation increases, the color changes to yellow at 300 nm and then to red at 550 nm (Fig. 6.7(C)). This red is known as **first-order red**, and all the colors resulting from retardations between 0 nm and 550 nm are known as first-order interference colors. As the retardation increases above 550 nm, the interference color changes to **second-order blue**. The second-order colors increase through green and yellow and eventually reach second-order red, with a retardation of 1100 nm. The same sequence of colors repeats for higher orders of interference, with each red being a multiple of 550 nm. Successively higher order colors, however, are less intense, with third-order red appearing a pale pink compared with first-order red (Fig. 6.7(C)).

The interference color created by an anisotropic mineral is one of its most important optical properties when it comes to mineral identification. Figure 6.10, on the inside back cover of the book, shows the interference colors that characterize the common rock-forming minerals in 30 μm-thick thin sections. You will make constant use of Figure 6.10 when identifying minerals.

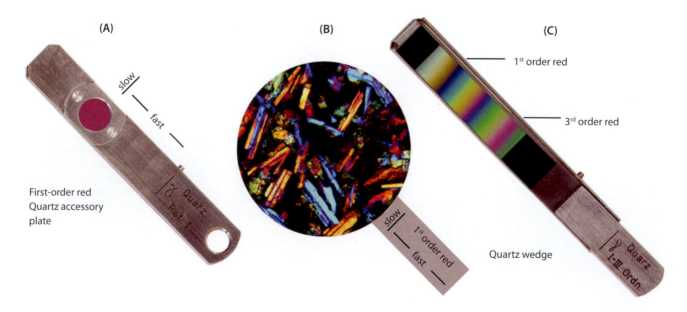

Figure 6.7 Accessory plates for determining the orientation of the fast and slow vibration directions in anisotropic minerals. (**A**) First-order-red interference filter. The color shown is first-order red under crossed polars. (**B**) Plagioclase laths in the Giant's Causeway basalt (Fig. 6.5(C)) showing addition (blue) and subtraction (yellow) of interference colors. (**C**) Interference colors in a quartz wedge. The orientation of fast and slow vibration directions are the same as in the first-order-red plate.

6.4 Accessory plates and determination of fast and slow vibration directions

In Figure 6.6(C), it can be seen that if the mineral grain on the microscope stage were to be rotated, one of the vibration directions in the crystal could parallel the E-W plane of polarization of light rising from the lower polarizer. When this happens, the E-W polarized light vibrates directly in the crystal plane that it parallels, but when this light arrives at the upper polarizer, it is completely eliminated and the mineral appears dark in the microscope. It is said to be in **extinction**. When the stage is rotated through 360°, birefringent crystals go into extinction four times as the fast and slow vibration planes in the crystal coincide with the polarization planes of the lower and upper polarizers.

The angular relation between these extinction positions and prominent crystallographic features, such as cleavage, twinning, or even crystal outline, is a measurable optical property that can be used to identify minerals. Before we can measure such angles, it is necessary to be able to distinguish the fast from the slow vibration direction, and this is done by using an accessory plate inserted into the optical path of the microscope between the thin section and the upper polarizer (Fig. 6.1). The most commonly used accessory plate is a first-order-red interference filter (Fig. 6.7(A)). This is a piece of quartz or gypsum that has been ground to a thickness that creates precisely a retardation of 550 nm; that is, first-order red. In addition, the thin slice of quartz is oriented so that its fast vibration ray parallels the length of the accessory plate and the slow vibration direction, which is commonly designated by the Greek letter gamma (γ), is across the length of the plate. The accessory plate slot on the microscope allows the accessory plate to be inserted at a 45° angle to the E-W plane of polarization of the lower polarizer. Light rising from the lower polarizer sets up vibrations in the fast and slow vibration directions in the quartz accessory plate, which, when recombined into the N-S plane of the upper polarizer, are out of phase by 550 nm and first-order-red results.

If no thin section is present in the microscope, the first-order-red filter simply creates a field of red. When we place a thin section on the stage of the microscope, the interference colors of the various minerals are combined with the red interference color generated by the accessory plate, and several possible conditions can result. For example:

- If the mineral grain is in extinction, it contributes nothing to the interference color of the filter, and consequently such grains have a first-order-red interference color.
- If the fast and slow vibration directions in the crystal coincide with the fast and slow vibration directions in the accessory plate, the retardation is increased and the interference color increases from first-order red to commonly second-order blue.
- If, however, the fast and slow vibration directions in the crystal are opposed to those in the accessory plate, the

amount of retardation is decreased, and the interference color changes from first-order red to first-order yellow.

Figure 6.7(B) shows the thin section of Giant's Causeway basalt (Fig. 6.5(C)) with a first-order-red interference filter inserted in from the SE side of the field of view. The fast vibration direction in the accessory plate is, therefore, oriented NW-SE, and the slow vibration direction is at right angles to this. Note that all plagioclase laths that are oriented NW-SE (or nearly so) are colored blue; that is, the fast vibration direction in these crystals is oriented parallel, or nearly so, to the length of the crystals, and because this coincides with the fast vibration direction in the accessory plate, the retardation is increased from first-order red to second-order blue. In contrast, when the long axis of the plagioclase laths is at right angles to the long axis of the accessory plate (i.e., NE-SW), the retardation decreases and the laths are first-order yellow.

For minerals that have high birefringence, the first-order-red interference filter may not add or subtract from the retardation a sufficient amount to make the effects readily detectable. For these minerals, a quartz wedge can be used (Fig. 6.7(C)). The wedge is similar to the first-order-red filter, but instead of being ground to one thickness that creates a first-order-red interference color, the wedge gives ever increasing degrees of retardation and the interference colors steadily increase, reaching third-order red in the wedge illustrated here. The fast and slow vibration directions in the wedge are oriented the same way as in the first-order-red accessory plate. By inserting the wedge in from the lower right, the fast or slow vibration directions in the crystals can be identified by whether the interference colors of the grains increase (fast and slow directions in crystals and accessory plate coincide) or decrease (fast and slow directions in crystals and accessory plate are opposed).

The quartz wedge can also be used to determine the order of the interference color of a mineral, which is one of its most important diagnostic properties. If the fast and slow vibration directions in the crystal are opposed to those in the wedge, insertion of the wedge causes the interference colors to decrease and when sufficient wedge is inserted to counteract the retardation in the mineral, the mineral will appear black. By counting the orders of colors that change as the wedge is inserted, the interference color of the mineral can be determined. The order of the interference color of a mineral can also commonly be determined without the use of a wedge. Many grains in a thin section taper to a thin edge where they overlap with an adjoining mineral. By finding such thin edges, we can count the number of color fringes (orders) that we go through as the grain thickens to its normal thickness of 30 μm.

In summary, an accessory plate is a useful tool to determine the orientation of the fast and slow vibration directions in a mineral and the order of the interference color. Knowledge of the orientation of the two vibration directions is essential to measuring a mineral's extinction angle, which we consider next.

6.5 Extinction positions and the sign of elongation

Many crystals in thin sections are elongate, and it is often useful when identifying minerals to know how the extinction direction and fast or slow vibration directions are related to this long axis. Measurements can also be made against other prominent crystallographic directions, such as cleavage planes or twin planes. The following are the steps taken in measuring the orientation of an extinction direction:

1. First, orient the long axis of the crystal or other prominent crystallographic direction parallel to the N-S cross-hair in the microscope.

2. Then, with the polarizers crossed, rotate the stage back and forth until the crystal goes into extinction. If the crystal goes into extinction when the long axis of the crystal precisely parallels the N-S direction, the crystal is said to have parallel extinction. This commonly is the case for crystals belonging to the tetragonal or hexagonal systems.

3. Other crystals may go into extinction when the long axis is not N-S. These crystals are said to show inclined extinction, and the angle between the extinction position and the long axis of the crystal (or other reference line) is known as the **extinction angle**.

4. To measure the extinction angle, rotate the crystal to the extinction position and record the angular measure on the microscope stage.

5. Next, rotate the crystal until the long axis (or other reference line) is parallel to the N-S cross-hair, and record the angle on the microscope stage. The difference between these two angles is the extinction angle.

6. Depending on how the crystal is oriented in the thin section, the measured value may fall well short of the true extinction angle. If you were able to rotate the crystal about its long axis on the stage of the microscope, you would find that your extinction measurements would pass through a maximum and this would be the true extinction angle. Because we cannot rotate the crystal about its long axis, this maximum value is determined by measuring numerous grains and choosing the maximum value.

7. To determine whether the extinction angle we have measured is from the fast or slow vibration direction, we need to determine which of the two vibration directions is parallel the N-S cross-hair. Return the crystal to the extinction position that we have just measured, and visualize light vibrating in a N-S plane. Place your index finger on the stage of the microscope pointing in this N-S direction, and rotate the stage of the microscope counter clockwise 45°. Now the vibration direction that previously paralleled the N-S polarizer is oriented in a NW-SE direction and parallel to the fast vibration direction of the accessory plate.

8. Insert the accessory plate and observe how the interference colors of the mineral in question behave. If the interference colors increase, then our finger is pointing in the direction of the fast vibration ray; if the colors decrease, our finger points along the slow vibrating ray.

Some minerals have the fast vibration direction parallel, or nearly parallel, to the long axis of the crystal and are described as **length fast** (also known as **negative elongation**), whereas others have the slow vibration direction parallel to the long axis and are described as **length slow** (also known as **positive elongation**). The plagioclase laths shown in Figure 6.7(B), for example, are length fast, whereas mica flakes are always length slow.

6.6 Anomalous interference colors, pleochroism, and absorption

A small group of minerals exhibits what are known as anomalous interference colors. These minerals have very low birefringence and would be expected to give gray interference colors, but in most of these minerals, the red end of the white light spectrum can be in extinction while the blue end of the spectrum is not. In this case, the mineral exhibits an anomalous blue interference color known as Berlin blue (Fig. 6.8). In some minerals, the blue end of the spectrum can be in extinction while the red end is not. This produces an anomalous brick red to brown interference color. Only a small group of minerals exhibits these anomalous colors including melilite (Fig. 6.8), the iron-poor epidote clinozoisite, and some chlorite, so this is a useful means of deducing the presence of these minerals.

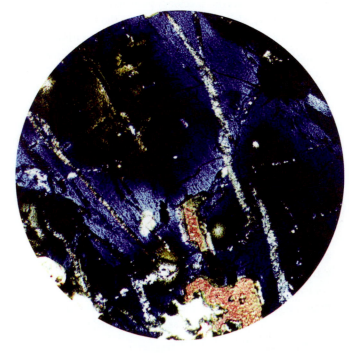

Figure 6.8 Anomalous blue interference color exhibited by zoned crystals of melilite under crossed polars. Width of field is 2.5 mm.

Two other optical properties are readily apparent under plane polarized light in the microscope and these are color and absorption. These are important because they are two of the first properties you see when you look in the microscope, and for many minerals, the color and absorption allow you to narrow your search when identifying unknown minerals. For this reason, color and absorption are included in Figure 6.10 for the common rock-forming minerals.

The causes of color in minerals discussed in Chapter 3 (Sec. 3.3) apply equally to the color of minerals observed in transmitted light under the microscope. The white light generated by the substage illuminator of the microscope may or may not be affected by passage through a mineral. If the mineral has no effect on the light, the mineral will appear colorless (white). In contrast, if the mineral preferentially absorbs certain wavelengths of light, the remaining wavelengths will give the mineral its color. In Section 3.3, the transition elements were shown to be the most important chemical components causing color in minerals. As a result, minerals in which iron is an essential component generally are colored under the microscope.

When polarized light is used, the color of a mineral can change depending on the orientation of the crystal with respect to the plane of polarization of the light. The resulting variability in color is referred to as **pleochroism**. The various colors result from absorption of different wavelengths of light in different directions in the crystals. Commonly the pleochroism of a mineral is a diagnostic property. One such case is the manganese epidote known as piemontite, which is pleochroic from canary yellow in one direction through amethyst to red in the other (Fig. 6.8(A)–(B)). In addition to pleochroism, minerals can show variable absorption of polarized light in different crystallographic directions. For example, the iron-bearing mica biotite shows the maximum absorption when the cleavage is parallel to the lower polarizer and least absorption when it is at right angles (Fig. 6.9(C)–(D)). In contrast, elongate crystals of tourmaline show their maximum absorption when their length is perpendicular to the lower polarizer (Figs. 6.9 (E)–(F)).

6.7 Mineral identification chart

The optical properties we have discussed so far allow us to create a diagnostic scheme for identifying unknown minerals, which is shown in Figure 6.10 (see pp. 534–535). This

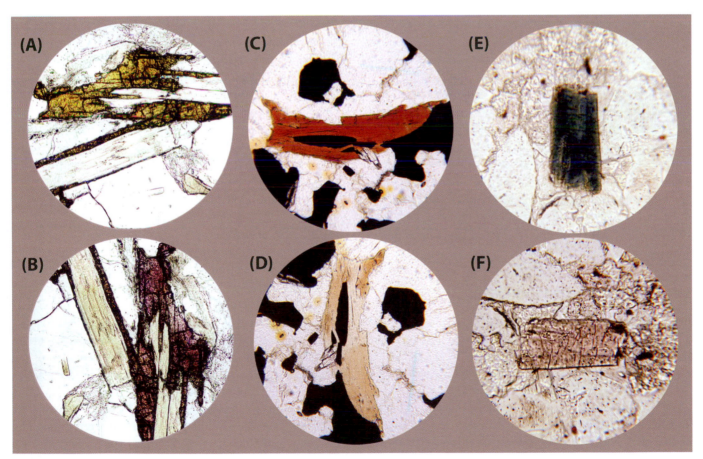

Figure 6.9 Anisotropic minerals may exhibit pleochroism and variable absorption. The manganese epidote piemontite appears yellow in one orientation (**A**) and amethyst color at right angles to this (**B**). The iron-bearing mica, biotite, always absorbs light more strongly when the cleavage direction parallels the polarizer (**C**) than when it is at right angles to the polarizer (**D**). Tourmaline always absorbs more light when the long axis of the crystal is perpendicular to the polarizer (**E**) than it does when parallel to the polarizer (**F**). The polarizer in each of these figures is oriented E-W.

includes almost all of the minerals discussed in this text arranged with increasing refractive index from left to right and increasing birefringence from top to bottom. The only minerals not included in this chart are carbonates and titanite, which have much higher birefringence than the range covered by this chart. But their high birefringence serves to identify them, and titanite is always a dark brown compared with relatively colorless carbonates. The interference colors shown in the central strip of this chart are those that would result for the indicated birefringence in a 30 μm-thick thin section, which is the standard thickness. Figure 6.10 is divided in half by this central strip of interference colors, with colored minerals listed to the left and colorless minerals to the right. Most of the colored minerals are silicates that contain iron and magnesium (ferromagnesian minerals). As the iron content increases, the refractive index also increases. The R.I. of these minerals is, therefore, represented by a bar covering the range of R.I. The general relief of the minerals with respect to a mounting medium with R.I. of 1.537 is shown by the stippled box along the top of the figure. The box indicating the R.I. range is colored to show the mineral's typical color, and if the mineral is pleochroic, these colors are shown as well. The birefringence shown for the mineral is the maximum birefringence it can exhibit. Remember that the birefringence exhibited by a mineral depends on the orientation of a crystal and can be less than the maximum, so it is important to examine numerous grains before deciding on the birefringence of the mineral. The colorless minerals are arranged in a similar manner to the right of the birefringent strip. In general, colorless minerals contain no iron and their R.I. is less than that of the colored minerals, although there are a few exceptions, such as corundum (Al_2O_3) with a R.I. of 1.77.

An unknown mineral can normally be narrowed down to a couple of choices in this figure on the basis of color, relief, and birefringence. In the following sections, we learn of several other optical properties that are listed in the figure and which will almost certainly lead to an identification of an unknown mineral.

6.8 Uniaxial optical indicatrix

In discussing the measurement of extinction angles, we saw that it was very unlikely that any one grain would be lying on the stage of the microscope in an optimal position to give the true extinction angle. We saw that it was, therefore, necessary to measure the extinction angle in numerous grains and to take the maximum value as being closest to the true value. You may have thought that it would be useful to have a device that could be attached to the stage of the microscope that would allow you to tilt the thin section and so get the grain into the optimum position to measure the extinction angle. Such universal stages do exist, but they are very expensive and are not readily available for routine petrographic work. The standard petrographic microscope does have two other features that partially solve our

problem of grain orientation – the upper swing-in condensing lens and the Bertrand lens. Before we can discuss how these two sets of lenses provide a certain degree of three-dimensionality to our measurements, we must introduce the concept of the optical indicatrix.

Crystals belonging to the tetragonal and hexagonal systems have maximum and minimum refractive indexes, and crystals belonging to the orthorhombic, monoclinic, and triclinic systems have maximum, intermediate, and minimum refractive indexes. It is convenient to represent these variations in refractive index with direction by an ellipsoid, which is known as an **optical indicatrix**. The vector from the center of the ellipsoid to its surface gives the magnitude of the refractive index for light vibrating in that particular direction (Fig. 6.11).

The indicatrix for tetragonal and hexagonal crystals is an ellipsoid of revolution, with the axis of revolution being the *c* crystallographic axis. Light that vibrates in a plane including this axis of rotation is referred to as the extraordinary ray and its refractive index is designated by the letter *e*. Light vibrating in a plane at right angles to this (the equatorial plane of the indicatrix) is referred to as the ordinary ray and its refractive index is designated by the letter *o*. The extraordinary ray can be the major refractive index, in which case the indicatrix is prolate (shaped like a tall football or cigar), or the extraordinary ray can be the minimum refractive index, in which case the indicatrix is oblate (short). The prolate indicatrix is described as positive, whereas the oblate indicatrix is said to be negative (Fig. 6.11). For example, the extraordinary ray in the mineral quartz has a refractive index of 1.55, whereas that of the ordinary ray is 1.54, so its indicatrix is prolate and the mineral is optically positive. In contrast, the extraordinary ray in calcite has a refractive index of 1.49, which is significantly less than that of the ordinary ray, which is 1.66; so the indicatrix of calcite is markedly oblate and the mineral is said to be optically negative.

The example of zircon

Let us see how the indicatrix helps us understand the vibration of light passing through a thin section of a grain of zircon, which has tetragonal symmetry. Zircon is optically positive ($e = 2.0$, $o = 1.94$), so the indicatrix is prolate. We can visualize this indicatrix as embedded in the grain and sliced through by the plane of the thin section. The actual orientation of the indicatrix depends on the crystallographic orientation of the zircon grain. Let us first assume that the extraordinary vibration direction (*c* crystallographic axis) lies exactly on the stage of the microscope. Our positive prolate indicatrix would be lying flat on the stage of the microscope and sliced through by the plane of the thin section. Polarized light rising from the lower polarizer and passing into such a zircon grain would set up two rays vibrating in mutually perpendicular planes, one plane paralleling the extraordinary vibration direction and the other the ordinary vibration direction. The refractive indexes of these two rays would be given by the length of the lines from

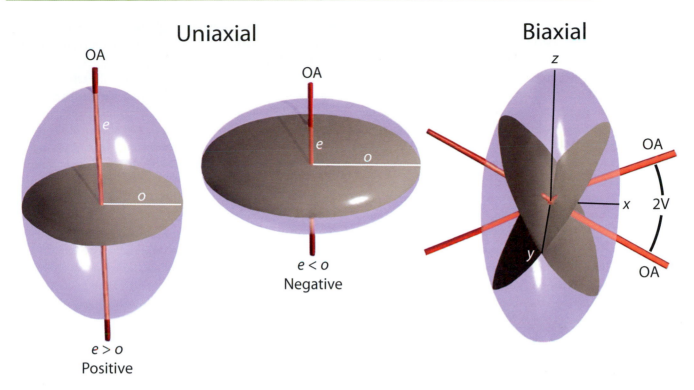

Uniaxial

e > o
Positive

e < o
Negative

Biaxial

Figure 6.11 The optical indicatrix is an ellipsoidal figure indicating how the refractive index in a crystal varies with the orientation of the plane in which light is vibrating. The uniaxial indicatrix is an ellipsoid of revolution that has one optic axis (red) perpendicular to the circular section through the ellipsoid. Light vibrating in a plane containing the axis of rotation is said to be the extraordinary ray (e), whereas that vibrating at right angles to it is the ordinary ray (o). The biaxial indicatrix is a triaxial ellipsoid (R.I. $z > y > x$) with two circular sections at right angles to which are two optic axes.

the center of the indicatrix to its surface in the extraordinary and ordinary vibration directions (i.e., e and o, respectively). As the stage of the microscope is rotated, the extraordinary and ordinary rays would move into parallelism with the polarizers and the grain would go into extinction; this would occur four times in a full rotation.

Next consider a more generally oriented grain of zircon where the indicatrix has some oblique orientation. For example, let us say that the extraordinary axis of the indicatrix pierces the plane of the thin section by rising from the lower right to the upper left as seen in the microscope, as shown in Figure 6.12. In this orientation, the thin section would cut an elliptical section through the indicatrix (see plan view in Fig. 6.12). Again, polarized light rising through this grain would be forced to vibrate in two mutually perpendicular planes, one corresponding to the minimum refractive index and the other to the maximum refractive index. Recall that the refractive index in any given direction is simply the distance from the center of the indicatrix to its surface where cut by the plane of the thin section. With an ellipsoid of revolution, the minimum radius of any section cut through the center of the ellipsoid, regardless of its orientation, is equal to the equatorial radius of the ellipsoid; that is, the ordinary vibration refractive index o. The other vibration direction, though having a higher refractive index than the ordinary vibration direction, is not the full magnitude of the extraordinary refractive index but only a component of it. We, therefore, designate this refractive index as e' (Fig. 6.12).

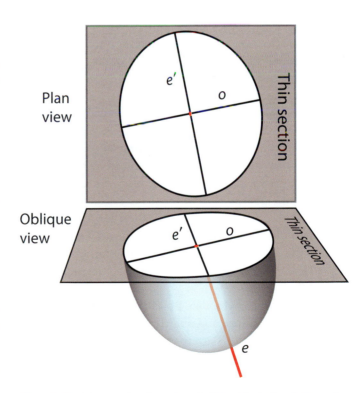

Plan view

Oblique view

Figure 6.12 Oblique section through a uniaxial positive indicatrix shows the two mutually perpendicular vibration directions that polarized light must vibrate in. One of these directions corresponds to the equatorial radius of the ellipsoid and is the refractive index of the ordinary ray (o), whereas the other direction has a greater refractive index but is less than that of the extraordinary ray (e) because e is not lying in the plane of the thin section. Its refractive index is designated e'.

Let us consider one other special orientation of the zircon grain. If the grain were oriented so that its *c* crystallographic axis were parallel to the axis of the microscope, the positive indicatrix would be standing on end, as shown in Figure 6.11, and the plane of the thin section would cut a circular section through the indicatrix. This circular section tells us that the refractive index for light vibrating in any direction would be the same and would be that of the ordinary ray. Polarized light rising through such an oriented grain would be free to vibrate in any two mutually perpendicular planes, and because the refractive indexes in these two planes would be the same, no interference would occur, and the grain would appear as if it were in extinction. This special direction for propagation of light through the indicatrix is designated the **optic axis** (red axis in Figs. 6.11 and 6.12). An ellipsoid of revolution has only one such direction (i.e., perpendicular to the equatorial plane) and, therefore, the indicatrix is described as **uniaxial**.

6.9 Biaxial optical indicatrix

Minerals that belong to the monoclinic, orthorhombic, and triclinic crystal systems have three different refractive indexes. We represent their indicatrix with a triaxial ellipsoid (Fig. 6.11) with the major axis being the maximum refractive index (z), the intermediate axis being the intermediate refractive index (y), and the minimum axis being the minimum refractive index (x). In a triaxial ellipsoid, it is always possible to find two circular sections, which include the intermediate axis. Light traveling normal to either of these planes (shaded gray in Fig. 6.11) has the same refractive index no matter which plane it vibrates in. Consequently, no interference occurs and the grain remains in extinction as it is rotated. The direction normal to the circular section is defined as an optic axis. Because a triaxial ellipsoid has two such circular sections, it has two optic axes, and the indicatrix is said to be **biaxial**. The acute angle between the two axes is referred to as the **optic angle** and designated 2V, and the plane containing the two axes is the optic plane. If the slow vibration direction z (major refractive index) lies in the acute angle, the mineral is said to be biaxial positive. If the fast vibration direction x (minimum refractive index) lies in the acute angle, the mineral is said to be biaxial negative.

A random section through a biaxial ellipsoid is not likely to have the maximum, minimum, or intermediate vibration directions lying on the stage of the microscope. Nonetheless, polarized light on entering a biaxial mineral must vibrate in two mutually perpendicular planes, one being a faster vibration direction (lower refractive index) and the other a slower vibration direction (higher refractive index). If we have no way of knowing the orientation of the indicatrix, the vibration directions can be designated simply as x', y', or z', with the prime indicating that these are only relative differences and do not reflect the true values of x, y, or z. In the next section, we see that we can determine the orientation of the indicatrix through use of what are known as interference figures.

The indicatrix is a representation of how the refractive index varies with direction in a crystal, which is determined by how light interacts with the atoms in a crystal. The indicatrix is, therefore, bound by the same symmetry elements that control all other aspects of the crystal. We have already seen that in tetragonal and hexagonal crystals symmetry requires that the axis of rotation of the uniaxial indicatrix must coincide with the *c* crystallographic axes. In crystals with mirror planes, as for example those with monoclinic and orthorhombic symmetry, the biaxial indicatrix must be symmetrically disposed about the mirror plane. The mineral olivine, for example, is orthorhombic (see Sec. 7.19) and, therefore, the principal axes of the indicatrix must correspond with the crystallographic axes (Fig. 6.13). In addition, the optic plane parallels (001), and the optic axes are equally disposed on either side of crystallographic *a*. Only in the triclinic system is the indicatrix free to have any orientation.

6.10 Uniaxial interference figures

When discussing the measurement of extinction angles, we saw that it was unlikely that any given grain would be lying on the stage of the microscope in exactly the right orientation to give the

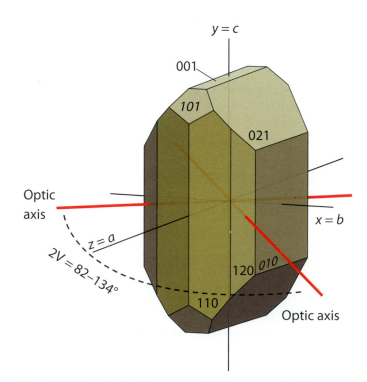

Figure 6.13 The orthorhombic symmetry of olivine requires that the principal axes of the biaxial indicatrix coincide with the crystallographic axes. Most olivine crystals are bounded by the (010), (021), and (110) faces. The *x*, *y*, and *z* axes of the indicatrix correspond to the *b*, *c*, and *a* crystallographic axes, respectively, and the optic axes lie in the mirror plane (001) with the axes equally disposed on either side of *x* and *z*.

correct angle and that it would be desirable to rotate the crystal into the correct orientation. Although we cannot do this on a flat stage (we need a universal stage for that), we can look at light that passes through a crystal at an angle to the axis of the microscope by using some of the additional lenses on the microscope. This gives us the ability to examine optical properties in directions other than those parallel to the axis of the microscope and provides additional optical properties that are useful in identify unknown minerals.

The light we have discussed so far rising through the microscope travels essentially parallel to the axis of the microscope. The lower condensing lens does bend the light slightly toward the axis of the microscope, but we can still think of most of the light as traveling essentially parallel to the axis. When we flip in the upper swing-in condensing lens, a strongly convergent beam of light is focused on the central part of the field of view. Light traveling near the center of the field is still essentially parallel to the microscope axis, but the light in the outer part of the field passes through the thin section at an oblique angle. For example, light in the southeast quadrant rises from the southeast, passes through the thin section and exits toward the northwest. Such a path of light experiences a different transit through a mineral grain than light traveling vertically along the axis of the microscope. If we could view what happens to this oblique light, we would gain knowledge of the optics of the grain in these other directions without having to physically tilt the grain, as might be done using a universal stage. The Bertrand lens allows us to examine, under crossed polars, what are known as interference figures that are produced by the convergent beam of light as it leaves the grain.

The example of quartz

The simplest interference figure is obtained when we look down the optic axis of a uniaxial mineral. Let us consider the mineral quartz, which is uniaxial positive, and if its optic axis were parallel to the axis of the microscope, its indicatrix would be oriented as shown in Figure 6.11. A thin section cut through such an indicatrix would give a circular section; that is, the refractive index of light vibrating in any plane would be the same and would have the value of the ordinary ray, o. As the refractive index is the same in all directions, no interference results, and the grain remains in extinction even when the stage is rotated. In a thin section containing many grains of quartz, a grain with such an orientation is easy to find by looking for one that remains dark under crossed polars as the stage is rotated (Fig. 6.14(A)). Such a grain remains in extinction because the lower polarizer sets up light vibrating in an E–W plane in the mineral that is eliminated by the upper polarizer.

If we flip in the upper swing-in condensing lens, light passes through the indicatrix in many different directions, but when this light exits the mineral and arrives at the objective lens, most of it is lost and only the light traveling parallel to microscope axis is transmitted to the ocular. However, by inserting the Bertrand lens, we can focus this oblique light into the focal plane of the ocular. The same effect can be obtained without the Bertrand lens by simply looking down the microscope without an ocular. This image can be sharpened if we insert a simple pinhole in place of the ocular. Let us examine the property of this convergent light that has traveled through the indicatrix in these different directions.

Light traveling straight up the axis of the microscope in the center of the field is parallel to the optic axis and, therefore, gives a circular section through the indicatrix (Fig. 6.14(B)). This light is in extinction. Convergent light rising from the south side of the field passes through the indicatrix at an angle that creates an elliptical section with the vibration directions being o and e'. However, because these two vibration directions parallel the polarizers, this light is eliminated by the polarizers. The same

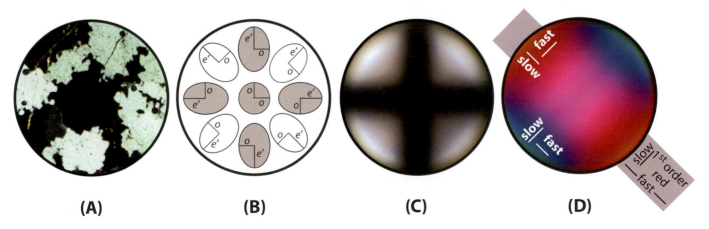

(A)   **(B)**   **(C)**   **(D)**

Figure 6.14 Uniaxial optic axis interference figure. (**A**) Photomicrograph of a thin section showing numerous orientations of quartz grains. The grain in the center is correctly oriented to give an optic axis figure because it remains in extinction when the stage is rotated, so its optic axis must parallel the microscope axis. (**B**) Ellipses show orientation of the o and e' vibration directions for light that has passed obliquely through the crystal. (**C**) We see this oblique illumination by using the Bertrand lens. Where the vibration directions parallel the polarizers (shaded ellipses in (B)), light is in extinction, but between these regions light passes through. The result is a uniaxial optic axis figure consisting of a diffuse dark cross separating quadrants with white interference color. (**D**) When a first-order-red interference filter is inserted, the dark cross turns to first-order red, the NW and SE quadrants turn first-order yellow, and the NE and SW quadrants turn second-order blue, which indicates that $e > o$ and, therefore, quartz is uniaxial positive.

argument applies to light rising from the east, north, and west. These elliptical sections through the indicatrix are shaded gray in Figure 6.14(B) to indicate that this light is eliminated by the polarizers. Light rising from the SE, NE, NW, and SW, however, produces elliptical sections through the indicatrix, in which the vibration directions are oblique to the polarizers, and therefore components of light are transmitted and produce interference colors when recombined in the upper polarizer. Insertion of the Bertrand lens brings these oblique paths of light into the focal plane of the ocular, and we obtain an interference figure.

The N-S and E-W aligned elliptical sections through the indicatrix, which are in extinction, produce two dark diffuse stripes across the field, which are known as **isogyres** (Fig. 6.14(C)). They intersect in the center of the field to produce a cross, which is the uniaxial optic axis interference figure. In the four quadrants between the isogyres, the elliptical sections through the indicatrix show light vibrating in planes that are not parallel to the polarizers, which produces interference colors in these quadrants. In the case of quartz, this interference color is first-order white, but it may reach a cream color on the extreme outside of the field of view. In minerals with higher birefringence, such as muscovite, the interference colors may increase sufficiently to produce what are referred to as isochromatic rings that are concentric about the optic axis (see Fig. 6.16). In the example shown in Figure 6.14, the grain is oriented with the optic axis parallel to the axis of the microscope. If the grain were tilted slightly, an optic axis figure would still be obtained, but the cross would be off centered, and its center would rotate around the center of the field of view as the stage is rotated. If the grain were tilted still more, the center of the cross might be outside the field of view, but as the stage was rotated, the off-centered optic axis figure could still be identified as the two isogyres sweep sequentially across the field of view.

What would a uniaxial mineral look like and what type of interference figure would be obtained if its optic axis were lying on the stage of the microscope? In this orientation, a section through the indicatrix would have light vibrating in the ordinary and extraordinary vibration directions. Consequently, the difference in refractive index between these two directions would be a maximum, and the grain would show the maximum birefringence. If we rotate such a grain into the extinction position and insert the Bertrand lens, a flash figure is obtained. This is a cross that looks just like the optic axis figure, but as soon as the stage is rotated through only a few degrees, the cross splits apart, and two curving isogyres exit the field of view rapidly (hence flash figure). The centered flash figure is useful for proving that a grain exhibits the maximum birefringence, which is what is shown for each of the minerals in Figure 6.10.

Interference figures can be obtained from grains with any orientation between the two extreme cases that we have just considered, but when they are far away from these two orientations, they are more difficult to interpret. In general, the nearer the figure is to that of the optic axis, the isogyres tend to be straight and oriented N-S or E-W. However, as the flash figure is approached, the isogyres become curved and move across the field of view more rapidly. If you are uncertain about which figure you are looking at, it is best to search for another grain with a more identifiable interference figure.

6.11 Determination of optic sign from uniaxial optic axis figure

An optic axis figure provides a convenient way of determining the optic sign of a mineral (i.e., positive or negative), which is one of the important optical properties used to identify minerals. If an accessory plate is inserted into the microscope when we are observing an optic axis figure, its fast and slow vibration directions interact with the fast and slow vibration directions in the different parts of the optic figure and cause the interference colors to change. In Figure 6.14(B), the vibration directions have been labeled in each section through the indicatrix. Take, for example, the elliptical section in the NW quadrant. The ordinary ray (o) vibrates in a NE-SW plane. Light vibrating at right angles to this direction contains a component of the extra-ordinary ray and is designated e'. When the first-order-red accessory plate is inserted, the interference color in this part of the optic figure is lowered from red to yellow. The fast vibration direction in the accessory plate parallels e', which must, therefore, be the slow vibration direction in quartz; that is, light vibrating in e' is slower than light vibrating in the ordinary plane. Therefore, the refractive index of e' must be greater than that of o, making quartz positive. All uniaxial positive minerals show retardation in the NW quadrant of the optic axis figure (creating first-order yellow in the case of quartz) and addition in the NE quadrant (creating second-order blue in the case of quartz). Even when the optic axis figure is badly off center, the sign of the mineral can still be determined as long as you can identify which quadrant you are looking at.

6.12 Biaxial interference figures, optic sign, and optic angle (2V)

Let us now consider the interference figures that we obtain from a biaxial mineral. First, consider a biaxial mineral lying on the stage of the microscope with any two of the principal axes of the indicatrix (x, y, or z) lying on the stage:

- First, we rotate the grain into extinction and insert the Bertrand lens, and the interference figure we obtain resembles the cross of the uniaxial optic axis figure.
- When the stage is rotated, the isogyres split apart and move out into two of the quadrants.
- At 45° of stage rotation, the isogyres reach their maximum separation, which may or may not have the isogyres still in the field of view, depending on the optic angle and the orientation of the grain.
- By 90° of stage rotation, the isogyres come back together and form a cross.

The various interference figures that result from light traveling in different directions through the biaxial indicatrix are shown in Figure 6.15 for a 45° rotation of the stage from the extinction position.

The acute bisectrix figure (BXA) is obtained looking down the axis halfway between the two optic axes. The obtuse bisectrix figure (BXO) is obtained at right angles to this looking down the center of the obtuse angle between the two optic axes. In the extinction position, the BXA forms a centered cross resembling the uniaxial optic axis figure. When the stage is rotated the cross splits apart and curved isogyres separate into two opposing quadrants, which contain the optic axes. At 45° of stage rotation, the isogyres reach their maximum separation (Fig. 6.15(A)). If the optic angle (2V) is less than about 55°, the isogyres remain in the field of view, but if it is larger, the isogyres leave the field of view.

The BXA figure provides a convenient way of determining the optic sign of a mineral:

- First, rotate the figure to the 45° position where the y vibration direction lies between the two isogyres and is tangential to the isogyres at their point of maximum curvature (Fig. 6.15(B)). The vibration direction at right angles to this must be either x or z, depending on whether it is faster or slower than y, which we determine using an accessory plate.
- Next, insert the accessory plate, and if the interference colors between the isogyres increase from first-order red to second-order blue, as shown in Figure 6.15(B), the vibration directions between the isogyres must correspond to the vibration directions in the accessory plate; that is, the y

vibration direction must be slow, and therefore the other vibration direction must be fast and is, therefore, x.

- We can conclude that the z vibration direction must be coming up the axis of the microscope and falling in the acute angle between the optic axes. The mineral is, therefore, biaxial positive.

It will be noticed that if the BXA figure is rotated into the extinction position where it forms a cross and you pretend that it is a uniaxial optic axis figure, the method outlined above for determining signs of uniaxial minerals gives the same result for the biaxial mineral, that is, positive.

The BXO figure is similar to the BXA figure but the optic axes move completely out of the field in the 45° position. When the 2V is large (>55°), it is difficult to distinguish the BXA and BXO figures because the isogyres go completely out of the field. When the 2V becomes 90°, the BXA and BXO are identical.

When the 2V is large, it is preferable to find a centered optic axis figure if you wish to estimate the 2V or determine the sign of the mineral. A centered optic axis figure has an isogyre that passes through the center of the field of view and rotates around like a propeller as the stage is rotated (Fig. 6.15(A)–(C)). When the isogyre is rotated into the 45° position, it shows the maximum curvature, which can be used to estimate the 2V (Fig. 6.15(D)). When the 2V is 90° the isogyre is straight, and as the 2V approaches 0°, the isogyre approaches the right angle bend of a uniaxial optic axis figure. Because of the diffuse nature of isogyres, a 2V of less than 10° is difficult to distinguish from a uniaxial optic axis figure. For example, the feldspar sanidine (Sec. 7.4.2), which is found in some volcanic rocks, has a small 2V (<20°) but often appears uniaxial.

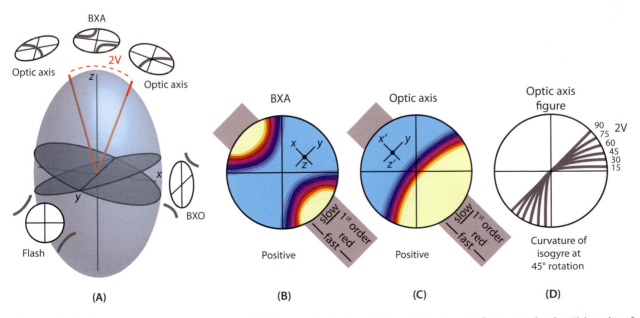

Figure 6.15 (**A**) Principal interference figures associated with the biaxial indicatrix. (**B**) When observing a BXA figure rotated to the 45°, insertion of a first-order-red accessory plate causes the interference colors between the isogyres to increase (as shown here) or decrease depending on the optic sign of the mineral. (**C**) With a centered optic axis figure rotated into the 45° position, insertion of a first-order-red accessory plate causes the interference colors on the convex side of the isogyre to increase (as shown here – blue) or decrease, which allows us to determine the sign (positive in this case). (**D**) Examples of the different degrees of isogyre curvature on a centered optic axis figure rotated to the 45° position for various values of 2V.

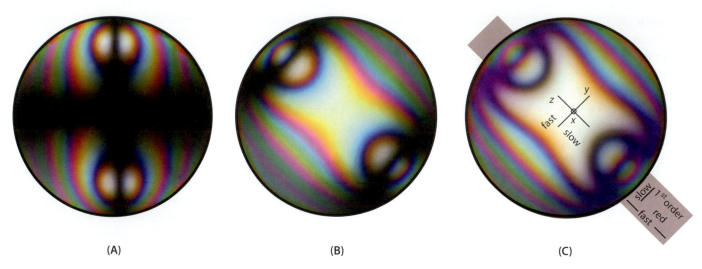

Figure 6.16 Acute bisectrix figure on a flake of muscovite. (**A**) Extinction position showing crossed isogyres, optic axes, and isochromatic lines. (**B**) Rotated into 45° position, with the isogyre separation indicating a 2V of ~40°. (**C**) Insertion of first-order-red accessory plate decreases the interference colors between the isogyres indicating that muscovite is biaxial negative.

The sign of minerals with large 2V can be determined easily from centered optic axis figures (Fig. 6.15(C)). In the 45° position, the convex side of the isogyre points toward the BXA. Knowing this, we apply the same techniques as applied to the BXA in determining the sign. If the isogyre is so straight that no curvature can be detected, the 2V must be very near 90°, at which point distinguishing positive from negative becomes unimportant. Instead, you know the 2V is near 90°, which will help you identify the mineral.

The example of muscovite

Finally, we show you a BXA figure on a flake of muscovite that illustrates some of the important features of the biaxial interference figure (Fig. 6.16). When the grain is rotated into extinction, the isogyres form a cross with the optic axes forming prominent eyes on the north and south segments of the isogyre (Fig. 6.16(A)). A series of concentric lines of increasing interference color surround each optic axis. These are isochromatic lines, reaching, in this case, fourth-order red. When the stage is rotated into the 45° position, the cross pulls apart and the isogyres retreat into the quadrants that contain the optic axes. In this example, the optic plane is, therefore, oriented NW-SE (Fig. 6.16(B)). The 2V of muscovite varies from 30° to 45° depending on composition. This accounts for the separation between the two optic axes. Remember, when the 2V exceeds ~55°, the isogyres exit the field in the 45° position. When the first-order-red accessory plate is inserted with the isogyres in the 45° position, the interference colors between the isogyres are lowered, and hence the fast and slow vibration directions in this part of the field must be opposed to those in the accessory plate. We know that the y vibration direction in the indicatrix is oriented NE-SW and that it is the fast vibration direction. The vibration direction at right angles to this is, therefore, slow and must be z. This indicates that the x vibration direction must be located in the acute angle between the optic axes, and muscovite is, therefore, negative.

6.13 Modal analysis

When identifying rocks, it is necessary to determine not only the minerals present but also their relative abundance. In a two-dimensional thin section under the microscope, we see the area occupied by each mineral. This areal abundance is equal to the volume abundance of the minerals, which is known as the **mode** and is determined by **modal analysis**.

We can perform a quick qualitative modal analysis by comparing what we see in a thin section with a chart showing examples of typical abundances (Fig. 6.17; this figure is shown at the end of the book for quick reference). Although this method is rather subjective and not very accurate, it commonly provides sufficient accuracy to identify rocks. A more accurate mode can be obtained by point counting. This technique makes use of a mechanical stage (Fig. 6.1) that moves the thin section by a metered amount in mutually perpendicular x and y directions. The stage comes to rest on a series of points that form a set of grid points across the thin section. You must identify and record each mineral that is located immediately beneath the cross-hairs before moving the stage to the next grid point. The final modal abundance is calculated from the number of counts on each of the minerals lying beneath the grid points. By calculating the mode as you accumulate data, you can determine how many points are necessary to obtain a statistical significant result. This number can vary considerably, depending on grain size and how evenly distributed the minerals are.

If you have a digital image of a thin section and the minerals in the rock are sufficiently different in color under plain light, the National Institutes of Health (NIH) *ImageJ* software program provides an accurate and rapid way of obtaining a mode (the software can be downloaded for free from http://rsb.info.nih.gov/ij). The instructions in Box 6.2 illustrate how to use this program to determine the modal abundance of minerals in, for example, a basalt. The ImageJ program can be used only if minerals have distinguishable (nonoverlapping) grayscale values. Even when all the minerals cannot be distinguished, the program can often still be used to determine the abundance of

all dark minerals versus all light minerals, which is referred to as the **color index** of a rock. For example, the color index of the rock shown in Box 6.2 is given by the sum of the percentages of pyroxene and magnetite, which is 54.5. The program performs many other useful image analysis operations, such as measuring grain size, grain shape, and grain orientation. This program is well worth downloading and becoming familiar with.

| BOX 6.2 | MODAL ANALYSIS USING NIH IMAGEJ SOFTWARE |

In this photomicrograph of a thin section of basalt under plain light (Fig. A) plagioclase (clear), pyroxene (greenish), and magnetite (opaque) are clearly distinguishable, and their abundances can be measured accurately using the NIH ImageJ software.

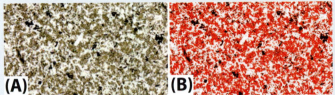

After downloading the NIH ImageJ software from the Web, download the photomicrograph from the text's Web site (http://www.cambridge.org/Kline-Philpotts) under Box 6.2, Figure A.

- Open (NIH) ImageJ, and go to the pull-down menu under File, then Import, to import the photomicrograph using QuickTime (click on button *Convert to 8-bit*). When the file opens, it may fill your computer screen, so after clicking on the magnifying glass in the menu bar, hold down the option key and click on the image a number of times to reduce the image to a manageable size.
- In the pull-down menu under Image, select Type, and 8-bit.
- Next, from the pull-down menu under Analyze, select Set Measurements. This opens a box listing the many operations that can be performed by the program. Click only the box beside Area Fraction.
- Next, from the pull-down menu under Image, select Adjust, followed by Threshold. This opens a box showing a histogram of the grayscale distribution in the image, going from black at the left (0 on the scale below) to white at the right (255). Beneath this scale are two bars that allow you to select a lower and upper threshold limit. The grains in the image falling between these limits are highlighted in red. By adjusting the limits, you can select all the grains of a given mineral.
- Set the lower limit (upper adjusting bar) to 56 and the upper limit (lower adjusting bar) to 205, and the pyroxene grains in the image are selected and will appear red (Fig. B).
- From the pull-down menu under Analyze, select Measure, and a box opens with the result for the fraction of the area occupied by the mineral you have highlighted in red. For the threshold limits given here (56–205), the pyroxene constitutes 50.0% of the sample.
- Next move the upper threshold limit to 255 and the lower one to 190, which does a good job of selecting all of the plagioclase grains. This returns a percentage for plagioclase of 45.5.
- Magnetite is totally opaque, so the lower threshold can be set to 0 and the upper limit to 1. This indicates that the modal abundance of magnetite is 4.5%.
- This rock's modal analysis is, therefore, 50% pyroxene, 45.5% plagioclase, and 4.5% magnetite.
- Its color index (percentage of dark minerals) is 54.5.

Summary

This chapter explained how to use the petrographic polarizing microscope to identify minerals in thin sections of rocks. It describes and tabulates the optical properties of minerals that can be used in their identification. By the end of the chapter, you should be familiar with the techniques necessary to analyze rocks under the microscope. The main points covered in this chapter are the following:

- The petrographic microscope differs from ordinary microscopes in using a polarized light source and another polarizer at right angles to the polarization in the source through which to observe the magnified image. The interaction of polarized light with crystal structures produces optical effects that are useful in identifying minerals.

- Although the speed of light in a vacuum is 300 000 km/s, it is slowed when passing through a mineral, with the level of retardation depending on the particular material.

- The velocity of light in air relative to the velocity of light in a mineral is known as its refractive index. The refractive index is one of the most useful identifying properties of a mineral under the microscope.

- We can measure the refractive index of a mineral by the angle through which light is bent or refracted when passing from air into the mineral. The amount of refraction is related to the refractive indexes by Snell's law.

- The Becke line is a bright fringe of light produced near the margin of a grain by light being refracted into or out of a grain, depending on its refractive index relative to that of the surroundings. The Becke line is used to determine the relative refractive indexes of juxtaposed grains in a thin section. The refractive index of a grain relative to that of the common mounting medium (1.537) is known as its relief, positive if >1.537 and negative if <1.537.

- Polarized light is light that vibrates in a particular plane. Polarized sunglasses, for example, allow the passage only of light vibrating in a vertical plane. The polarizer beneath the stage of a microscope typically constrains light to vibrate in a plane from left to right (E-W), whereas the polarizer above the stage allows passage of light vibrating only in a N-S plane.

- All minerals other than those belonging to the isometric system and amorphous materials, such as glass, have a refractive index that varies with crystallographic direction. These minerals are said to be anisotropic. Polarized light allows us to measure the refractive index in particular directions. The difference between the maximum and minimum directions is known as the birefringence.

- When polarized light enters an anisotropic mineral, it is forced to vibrate in two mutually perpendicular planes, one plane allowing light to travel faster than in the other. When the light waves vibrating in these two directions are recombined into the polarization plane of the upper polarizer, they are out of phase and interference colors are produced. The order of the interference color is determined by how out of phase (retardation) the two waves are, which, in turn, depends on the birefringence.

- Interference filters, with the fast vibration direction parallel to their length, can be inserted into the optical path and will either increase or decrease the retardation in the mineral. If it increases the retardation (higher interference color), the fast and slow vibration directions in the mineral match those in the interference filter; if it decreases the retardation (lower interference color), the directions must be opposed. This allows the orientation of the fast and slow vibration directions in a mineral to be determined. This can provide a diagnostic property for some minerals.

- When the vibration directions in a mineral match those of the polarizers, all light is eliminated and the mineral is said to be in extinction. The orientation of the extinction position relative to crystallographic directions, such as crystal faces, prominent cleavage directions, and twin planes provides another diagnostic property of minerals.

- Absorption of light can vary with direction in anisotropic minerals. Absorption of different wavelengths of light can also vary with crystallographic direction, which leads to pleochroism.

- Interference figures are obtained by using the substage swing-in condensing lens and the Bertrand lens, which allow you to examine the light leaving the grain. The combination of these two lenses essentially allows you to examine the light passing through a mineral grain in three dimensions. Light passing through the grain in a direction where the vibration directions parallel the polarizers is in extinction and produces a dark diffuse line known as an isogyre. Interference figures are produced by isogyres.

- Crystals belonging to the tetragonal and hexagonal crystal systems are said to be uniaxial and produce a diffuse crosslike interference figure when viewed down the c crystallographic axis. This is the uniaxial optic axis figure.

- Crystals belonging to the monoclinic, orthorhombic, and triclinic crystal systems are said to be biaxial because they create two optic axis figures, which consists of a single isogyre that rotates around the center of the field. The angle between the two optic axes is known as the optic angle or 2V.

- Both uniaxial and biaxial minerals can be positive or negative depending on the position of the fast and slow vibration directions with respect to the optic axes.

- By using all the various optical properties of minerals, a definitive mineral identification is almost always possible from observations made under the petrographic microscope.

- Finally, once minerals have been identified, their abundance can be determined to produce what is referred to as a mode – the volume percentage of minerals. This can be done roughly by visual estimate, or more accurately by point counting. Image analysis software can also be used.

Review questions

1. Do longer wavelengths of light have more or less energy than short wavelengths?

2. What was Henry Sorby's contribution to the development of the microscope?

3. What is the refractive index of a mineral?

4. What is the refractive index of air?

5. The refractive index of fluorite is 1.43, whereas that of a garnet is 1.8. How much faster does light travel in fluorite than in garnet?

6. If a light ray with an angle of incidence of 30° passes from air into fluorite and into garnet with refractive indexes of 1.43 and 1.8, respectively, what would the angle of refraction be in both minerals? Through what angle would the light ray be bent in passing into each mineral?

7. Why does the Becke line move into the mineral of higher refractive index when the microscope stage is lowered (i.e., when the distance between stage and lens increases)?

8. When you say a mineral has positive relief, with what refractive index are you comparing the mineral?

9. What materials are isotropic?

10. Calcite is birefringent with maximum and minimum refractive indexes of 1.66 and 1.49, respectively. What are the relative velocities of light traveling in the directions that give these two refractive indexes?

11. Why do birefringent minerals create interference colors under crossed polarizers?

12. Why does the order of the interference color increase as the birefringence increases?

13. What is a first-order-red interference filter and what is the normal orientation of its fast- and slow-vibration directions?

14. When a mineral is in extinction, what can we say about the orientation of its fast- and slow-vibration directions?

15. All micas are length slow. What does this mean, and what test could you apply with an interference filter to check that a grain is length slow?

16. If you have a centered uniaxial optic axis figure (a cross) and you insert a first-order-red interference filter and the northwest and southeast quadrants turn blue and the northeast and southwest quadrants turn yellow, what is the optic sign of the mineral?

17. A centered biaxial optic axis figure has a curved isogyre extending from the northwest to southeast quadrants and is convex toward the northeast quadrant. On inserting a first-order-red interference filter, the region to the northeast of the isogyre turns yellow and to the southwest of the isogyre turns blue. What is the optic sign of the mineral?

ONLINE RESOURCES

The National Institutes of Health's ImageJ software program can be downloaded for free from http://rsb.info.nih.gov/ij

FURTHER READING

More detailed discussions of all aspects of optical mineralogy and the properties of minerals under the polarizing microscope can be found in the following texts.

Klein, C., and Dutrow, B. (2008). *Manual of Mineral Science*, 23rd ed., John Wiley and Sons, New York.

Nesse, W. D. (2000). *Introduction to Mineralogy*, Oxford University Press, New York.

Nesse, W. D. (2003). *Introduction to Optical Mineralogy*, 3rd ed., Oxford University Press, New York.

Philpotts, A. R. (2003). *Petrography of Igneous and Metamorphic Rocks*, Waveland Press, Prospect Heights, IL.

CHAPTER

7

Igneous rock-forming minerals

This chapter is the first of four chapters that present systematic descriptions of common rock-forming minerals. In this chapter these descriptions are preceded by three short sections on some practical chemical aspects of minerals.

The first section introduces the concepts of major and trace elements in mineral and rock analyses. Major elements (usually reported as oxide components) are the most common crustal elements. Trace elements occur in very small amounts. The second section deals with the recalculation scheme of mineral analyses, which leads to chemical formulas with specific subscripts. And the third section deals with the representation of mineral compositions on triangular compositional diagrams. Such diagrams are used to show the extent of solid solution in mineral groups and the coexistence of minerals in specific rock types.

The systematic mineral descriptions in this chapter apply only to igneous rock-forming minerals. The first entry is the mineral name followed by a formula. Then follows a short paragraph on its most common occurrence and a discussion of its chemical variation in terms of its crystal structure. After that follow three headings, "Crystal Form," "Physical Properties," and "Distinguishing Features." These sections are most helpful in hand specimen identification. Each mineral description includes a hand specimen (or, less commonly, a thin section) photograph and concludes with a paragraph on its use in commercial applications.

This image shows a vug filled with coarse-grained granite (commonly referred to as **pegmatitic**) inside a much finer-grained granite matrix. The very center of the photo shows dark gray smoky quartz crystals intergrown with microcline feldspar crystals that have a slightly yellow-beige tinge as a result of oxidation. The whiter rim around that center consists of gray quartz grains intergrown with white microcline and albite. Part of the Conway granite, Moat Mountain, Bartlett, New Hampshire. The size of the field of view is 22 cm by 32 cm. (David Nufer Photography, Albuquerque, New Mexico.)

After three short sections on common elemental abundances in the Earth's crust, the recalculation of mineral formulas from their chemical analyses, and the application of triangular (ternary) diagrams to the depiction of the chemistry of minerals, we start the systematic descriptions of igneous rock-forming minerals. We chose 29 of these as the most representative, of which 19 are silicates, 6 are oxides, 3 are sulfides, and 1 is a phosphate.

Section 7.5 is the first mineral entry of the systematic description of igneous rock-forming minerals that follow. You must read these descriptions to become familiar with the most common minerals; their composition, structure, and physical properties; and their uses in the present-day commercial world. Reading the mineral description in this text while handling one or several specimens of the same mineral in the laboratory that accompanies your course is the best way to familiarize yourself with the mineral's properties. Minerals that are common constituents of sedimentary and metamorphic rocks and of vein deposits are presented after the origin of igneous rocks and their classification, in Chapters 8 and 9.

7.1 Common chemical elements in the Earth's crust and in mineral and rock analyses

In Section 4.1, we mentioned that 13 of the most common chemical elements (Table 4.2) make up about 99% of the Earth's crust. These 13 elements, their coordination numbers, and their ionic radii are listed in Table 4.5. Inspection of the weight percentage (weight %) abundances of these 13 elements leads to the conclusion that only 8 elements account for 98.59 weight % of the crustal composition. These are listed in Table 7.1. When this table is compared with the larger number of elements in Table 4.2, the missing elements turn out to be Mn (with a crustal abundance of 0.09 weight %), Ti (with an abundance of 0.64 weight %), P (with an abundance of 0.06 weight %), and S and C (both in parts per million). Of all the elements in the weight percentage column of Table 7.1, oxygen is the most abundant. This predominance becomes even more obvious when the weight percentage values are recalculated to atomic percentage and subsequently to volume percentage. The volume percentage column is obtained using the radii of the elements involved, and with oxygen being by far the biggest ion in the group of eight, it recalculates to about 86% in volume. This percentage ranges from 86% to 94% in various references as a function of the radius used for oxygen, which varies from 1.26 Å to 1.40 Å. This means that the Earth's crust consists almost entirely of oxygen compounds, especially silicates that also house aluminum, calcium, magnesium, potassium, sodium, and iron. Other major mineral groups that include oxygen are oxides, carbonates, sulfates, and hydroxides. The foregoing mineral groups are the most common constituents of the Earth's crust and are, therefore, referred to as the **rock-forming minerals** (Fig. 7.1).

Table 7.1 The eight most common elements in the Earth's crust.

	Weight[a] percentage	Atom[b] percentage	Ionic radius[c] (Å)	Volume[d] percentage
O	46.60	62.55	1.26	~86
Si	27.72	21.22	$0.40^{[IV]}$	
Al	8.13	6.47	$0.53^{[IV]}$	
Fe	5.00	1.92	$0.92^{[VI]}$	~14
Ca	3.63	1.94	$1.14^{[VI]}$	in total
Na	2.83	2.64	$1.32^{[VIII]}$	
K	2.59	1.42	$1.65^{[VIII]}$	
Mg	2.09	1.84	$0.86^{[VI]}$	
	98.59	100.00		

[a] Data from Mason and Moore, 1982.
[b] Values obtained by dividing the numbers in the first column by the appropriate atomic weights, then normalized to 100.
[c] Radii taken from Table 4.1.
[d] These values fluctuate somewhat depending on the radii used in the calculation of the ionic volume ($V = \frac{4}{3} \pi r^3$).

Let us return briefly to the right-most column in Table 7.1, which shows that about 86% to 94% (depending on the choice of oxygen radius) of the Earth's crust is made up of oxygen by volume. This means that on an atomic scale the Earth's crust is essentially a close packing of oxygen atoms, with all the other cations occupying interstitial spaces. This also means that when one walks over a rock outcrop of, for example, granite, one walks over an infinite package of solid oxygen anions. This concept is well illustrated in the close-packed (space-filling) illustration of the structure of diopside in Figure 4.5(A).

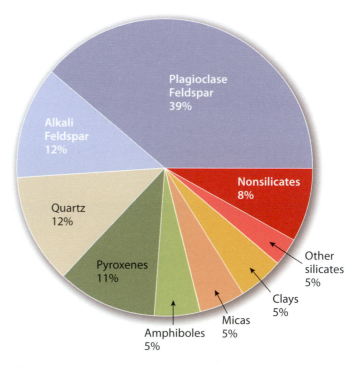

Figure 7.1 Estimated volume percentages for the most common minerals in the Earth's crust. This includes continental and oceanic crust (from Ronov and Yaroshevsky, 1969, Chemical Composition of the Earth's Crust. American Geophysical Union Monograph No. 13, 50).

The chemical composition of minerals is obtained by a professional chemist or mineralogist by using any or several of the analytical techniques discussed in Section 3.8.2. The results of such analyses are most commonly reported as weight percentages of the oxides or as elemental abundances. When chemical elements occur in large amounts (greater than 1 weight %), they are referred to as **major elements**. Elements that have abundances of between 1.0 and 0.1 weight % are referred to as **minor elements**. These are normally reported in weight percentages. Other elements that occur in much smaller amounts (less than 0.1 weight %) are known as **trace elements** and are reported as parts per million (ppm) or parts per billion (ppb). Many of the elements that are of great commercial and economic importance occur in trace amounts when calculated on the basis of crustal abundance. For example, the abundance of Cu is 55 ppm; the abundance of Ni is 75 ppm; and the abundance of Cr is 100 ppm. Only when these trace elements occur in large enough concentrations, as in **ore deposits**, is it feasible to extract these metals for commercial use (see Chapters 15 and 16).

Before we start our systematic descriptions of igneous rock-forming minerals, we must introduce two more subjects that relate directly to the chemical composition of minerals: the calculation of mineral formulas from chemical analyses and the graphical representation of mineral compositions.

7.2 Calculation of mineral formulas

The analytical results of most chemical analyses of minerals and rocks are given in weight percentages of the elements or of the oxides in the material analyzed. Most minerals, except for the native elements, are compounds of two or more elements. A quantitative chemical analysis provides the basic data for calculating a mineral's formula. The analysis may not total exactly 100% because of small analytical errors or some minor (or trace) elements that were not measured in the analysis.

Table 7.2 shows three formula recalculations, beginning in panel A with that of the common and chemically simple mineral quartz SiO_2. To determine the atomic proportions (shown in column 3), the weight of each element (given in column 1) is divided by its atomic weight. From these numbers, the ratio of the two elements in quartz can be determined as 1.0 Si to 2.0 O (column 4).

The chemical analysis of a more complex mineral composition, that of an iron-rich member of the olivine series, is listed in Table 7.2(B). Columns 1 and 2 give the chemical analysis and the molecular weights, respectively. Column 3 gives the results for the division of the weight percentages in column 1 by the appropriate molecular weights in column 2. Column 4 lists the proportions of the metals. These values are the same as the values

Table 7.2 Examples of the recalculation of mineral formulas from their chemical analyses.

(A)

	1 Weight Percent	2 Atomic Weight	3 Atomic Proportions	4 Atomic Ratio
Si	46.74	28.0855	1.6642	1.0
O	53.26	15.9994	3.3287	2.0
Total	100.00			

Formula: SiO_2

(B)

	1 Weight Percent	2 Molecular Weight	3 Molecular Proportions	4 Atomic Proportions Cations	5 Oxygens	6 Oxygen Factor	7 Cations on the basis of 4 oxygens	
SiO_2	31.85	60.084	0.5301	Si^{4+} 0.5301	1.0602	$4/2.0990 =$	1.010	Si^{4+} 1.0
FeO	58.64	71.846	0.8162	Fe^{2+} 0.8162	0.8162	$= 1.90567$	1.555	Fe^{2+}
MnO	0.85	70.937	0.0119	Mn^{2+} 0.0119	0.0119		0.023	Mn^{2+} } 1.98
MgO	8.49	40.299	0.2107	Mg^{2+} 0.2107	0.2107		0.402	Mg^{2+}
Total	99.83				2.0990			

Formula: $(Mg_{0.40} Fe_{1.56} Mn_{0.02})_{1.98} Si_1 O_4$

% of end-member calculation (using data from column 3):

MgO = 0.211   % MgO = 0.211/1.027 × 100 = 20.54 %
FeO = 0.816   % FeO = 0.816/1.027 × 100 = 79.45 %
Total   1.027

(using data from column 7):

Mg = 0.402   % Mg = 0.402/1.957 × 100 = 20.54 %
Fe = 1.555   % Fe = 1.555/1.957 × 100 = 79.45 %
Total   1.957

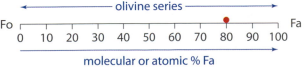

olivine series

Fo — 0 10 20 30 40 50 60 70 80 90 100 — Fa

molecular or atomic % Fa

Table 7.2 cont'd.

(C)

	1 Weight Percent	2 Molecular Weight	3 Molecular Proportions	4 Atomic Proportions Cations		5 Oxygens	6 Oxygen Factor	7 Cations on the basis of 8 oxygens		
SiO_2	66.12	60.084	1.1005	Si^{4+}	1.1005	2.2010	8/3.0197=	2.915	Si^{4+}	
Al_2O_3	21.00	101.944	0.2060	Al^{3+}	0.4120	0.6180	= 2.6493	1.091	Al^{3+}	4.006
Fe_2O_3	0.26	159.676	0.0016	Fe^{3+}	0.0032	0.0048		0.008	Fe^{3+}	
CaO	1.09	56.074	0.0194	Ca^{2+}	0.0194	0.0194		0.051	Ca^{2+}	
Na_2O	9.53	61.969	0.1538	Na^+	0.3076	0.1538		0.815	Na^+	0.986
K_2O	2.14	94.191	0.0227	K^+	0.0454	0.0227		0.120	K^+	
Total	100.14					3.0197				

Formula: $(Na_{0.81} Ca_{0.05} K_{0.12})_{0.99} (Al_{1.09} Si_{2.91})_{4.0} O_8$

% Ab, An, and Or in this composition (using data in column 4):

Na = 0.3076 % Na = 0.3076/0.3724 × 100 = 82.6%
Ca = 0.0194 % Ca = 0.0194/0.3724 × 100 = 5.2%
K = 0.0454 % K = 0.0454/0.3724 × 100 = 12.2%
Total 0.3724

This can be restated as: 82.6% Ab, 5.2% An, and 12.2%, or if the composition is plotted on a composition bar between Ab (albite) and An (anorthite), the two numbers (82.6 and 5.2) must be normalized to 100; 82.6 + 5.2 = 87.8; then 82.6/87.8 × 100 = 94% and 5.2/87.8 × 100 = 6%). This leads to 94% Ab and 6% An.

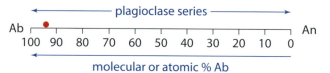

for atomic proportions in column 3. Column 5 shows that the proportion of oxygen in SiO_2 is not the same as in column 4 because divalent metal oxides contribute only one oxygen, but Si^{4+} in the SiO_2 molecule contributes two oxygens. At this stage, one must recall that the formula of olivine is based on four oxygens, $(Mg, Fe)_2SiO_4$. As such, one must recast the total of the number of oxygens in column 5 (total = 2.099) to 4. This is achieved by dividing 4 by 2.099 (= 1.906), and that number, known as the **oxygen factor**, is used as a multiplier for the numbers in column 5. This leads to the values in column 6. When the values in column 6 are inspected, it becomes clear that the resulting formula for this olivine sample turns out to be $(Mg_{0.40} Mn_{0.02} Fe_{1.56})_{1.98}Si_{1.01}O_4$, or essentially $(Mg_{0.4} Fe_{1.6})_{2.0}SiO_4$. This olivine sample is iron-rich and is part of the olivine solid solution series that is known as fayalite (Fa). The series extends from Mg_2SiO_4, or forsterite (Fo), to Fe_2SiO_4, or fayalite (Fa). We expressed its composition as a formula with subscripts, but there is another way to express its composition in terms of the percentage of the two end-member compositions. The amounts of forsterite (Fo) and fayalite (Fa) are directly proportional to the atomic proportions of Mg and Fe (column 4 in Table 7.2(B)) or the molecular proportions of MgO and FeO (column 3 in Table 7.2(B)). For this olivine, using the data in column 7, the number of atoms of Mg and Fe are 0.402 and 1.555, respectively, which sums to 1.957. Therefore, the percentage of Mg is equal to 0.402/1.957 × 100% = 20.54%, and the percentage of Fe is equal to 1.555/1.957 × 100% = 79.45% (these calculations are shown in the lower part of Table 7.2(B) using the data in columns 4 and 7, which produce identical results). The location of this specific composition

is shown in Table 7.2 on a two-component composition diagram (also known as **binary**) for the olivine series. Such composition bars were introduced in Section 4 (see Fig. 4.24(A)). The structure of olivine is shown in Figure 4.23, where the octahedral positions house the Mg^{2+} and Fe^{2+} in its structure.

Table 7.2(C) provides the original chemical analysis of a plagioclase feldspar. The recalculation scheme (in going from columns 2 to 5) is the same as that for olivine in Table 7.2(B), except that the analysis lists different and two more oxide components. The oxygen factor is calculated on the basis of eight oxygens for the generalized formula $(Na, Ca, K) (AlSi)_4O_8$. This recalculation results in the formula $(Na_{0.81} Ca_{0.05} K_{0.12})_{0.99}$ $(Al_{1.09} Si_{2.91})_4O_8$. The calculations underneath the formula allow for the expression of this feldspar composition in terms of three end members, namely albite (Ab), anorthite (An), and orthoclase (Or). Their compositions are $NaAlSi_3O_8$ (albite), $CaAl_2Si_2O_8$ (anorthite), and $KAlSi_3O_8$ (orthoclase, one of three polymorphs of K-feldspar, see Fig. 7.3(A)). Na^+, K^+, and Ca^{2+} substitute for each other in a structural site with large coordination (**coordination number**, or **C.N.**, ranging from VIII to X) and Al^{3+} and Si^{4+} substitute for each other in tetrahedrally coordinated sites. A composition bar in Table 7.2 locates this specific feldspar composition in terms of two end members, Ab and An. This is possible when the Or component (12.2%) is ignored. A better representation of such a composition is on a triangular diagram as is discussed in the following section. Table 7.3 lists the atomic weights of some of the elements most commonly reported in the analyses of rock-forming minerals and is helpful in case you need to recalculate additional analyses.

Table 7.3 Alphabetical listing of the most common elements reported in the chemical analyses of rock-forming minerals, their symbols, atomic numbers, and atomic weights.

Name	Symbol	Atomic Number	Atomic Weight
Aluminum	Al	13	26.98154
Calcium	Ca	20	40.08
Carbon	C	6	12.011
Chromium	Cr	24	51.996
Hydrogen	H	1	1.0079
Iron	Fe	26	55.847
Lead	Pb	82	207.2
Lithium	Li	3	6.941
Magnesium	Mg	12	24.305
Manganese	Mn	25	54.9380
Nickel	Ni	28	58.70
Oxygen	O	8	15.9994
Phosphorus	P	15	30.97376
Potassium	K	19	39.0983
Silicon	Si	14	28.0855
Sodium	Na	11	22.98977
Sulfur	S	16	32.064
Titanium	Ti	22	47.90
Uranium	U	92	238.029
Zinc	Zn	30	65.38

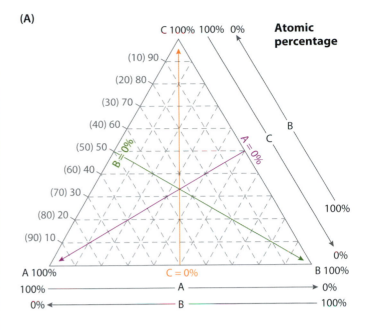

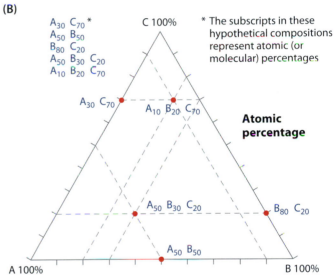

Figure 7.2 (**A**) Some basic aspects of a triangular composition diagram. (**B**) The mechanics of plotting some specific compositions (using capital letters of the alphabet instead of chemical element symbols).

7.3 Triangular diagrams

In the previous section, we used binary compositional bars (between two end-member compositions) to plot the position of a specific olivine and plagioclase composition. In Table 7.2(C), we noted that, to graphically represent the plagioclase on a binary bar, we needed to ignore its K_2O (or Or) content but that its composition would be better represented on a triangular diagram.

Because mineral compositions can be complex when two or more elements in the chemical analysis can substitute for each other in the same structural site, it is useful to have graphical representations of composition that allow for the incorporation of more chemical elements than can be shown in bar diagrams. Here is where **triangular diagrams** come in. These consist basically of three bar diagrams that share corner components in a triangular shape. Figure 7.2(A) shows such a triangular diagram with internal lines that represent percentages at 10% intervals of the three components. The corners are marked as 100% of a specific chemical component (capital letters of the alphabet are used here for simplicity), and three intersecting arrows indicate how the percentage scales start at 0% on the edges opposite the corners of the triangle. Figure 7.2(B) shows the location of several compositions using the letters A, B, and C as the end-member compositions. The compositions that plot along the edges of the triangles can be viewed as locations

in bar diagrams. Three-component compositions plot in the interior of the triangle.

Triangular diagrams (or **ternary diagrams**) are commonly used to display graphically the extent of solid solution that may occur among three end members of a mineral group. The majority of feldspar compositions can be plotted in terms of three feldspar end members: $NaAlSi_3O_8$ (albite), $CaAl_2Si_2O_8$ (anorthite), and $KAlSi_3O_8$ (K-feldspar). If we plot hundreds of analyses of high-temperature feldspars in such a way, the diagram in Figure 7.3(A) results. The shaded region represents the maximal extent of solid solution in feldspars that crystallize at high temperature, mainly in volcanic occurrences. The plagioclase composition calculated from its analysis in Table 7.2(C) is plotted on this diagram as well.

(A)

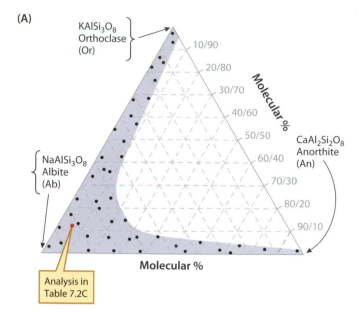

Analysis in Table 7.2C

(B)

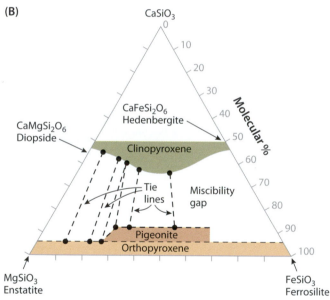

Figure 7.3 (A) A triangular (three-component, also known as a ternary) diagram used to represent the extent of solid solution among feldspars using the three end-member compositions: $NaAlSi_3O_8$, albite; $CaAl_2Si_2O_8$, anorthite; and $KAlSi_3O_8$, K-feldspar with three polymorphs (microcline, orthoclase, and sanidine). Many hundreds of chemical analyses of feldspars that crystallized at high temperature were used to outline the total extent of solid solution in this diagram. The dots represent some of these many analyses. The composition of the plagioclase calculated in Table 7.2(C) is plotted as well. (B) Another example of a triangular composition diagram showing the location of end member compositions ($CaSiO_3$, $MgSiO_3$, $FeSiO_3$, $CaMgSi_2O_6$, and $CaFeSi_2O_6$) of several members of the pyroxene group. The extent of solid solution among these end members in igneous rocks is shaded. Tie lines, across a miscibility gap, connect the locations of the compositions of pyroxenes that occur in pairs, in equilibrium with each other.

Another common use of such diagrams is in the illustration of minerals that may coexist with each other in a rock. Figure 7.3(B) is an example of a triangular composition diagram that not only depicts the end-member compositions of four pyroxenes and their high-temperature solid solution extent but also shows the

chemistry of specific pyroxene compositions that may coexist in igneous rocks. Such pairs of compositions are joined by **tie lines** across a miscibility gap.

7.4 Systematic mineralogical descriptions of common igneous minerals

The following subject matter consists of systematic descriptions of mineral groups, mineral series, and individual mineral species that are common constituents of igneous rocks. The reason for restricting the remainder of this chapter to minerals common in igneous rocks is that Chapters 8 and 9 deal solely with igneous rock formation processes and classification. Because minerals are the basic constituents of rocks, a discussion of igneous minerals before that of igneous processes is necessary.

In this text, the systematic descriptive aspects of mineralogy are spread over four different chapters. Igneous minerals are presented in this chapter; sedimentary rock-forming minerals in Chapter 10; metamorphic minerals in Chapter 13; and economic minerals in Chapter 15. This approach is different from what is normally done in standard mineralogy texts, where all minerals are discussed according to the classification outlined in Section 2.2. As such, mineralogy texts commonly start with native elements, then sulfides, then oxides, and so on, and end with silicates. In this text, which deals with minerals and rocks, we decided that close integration of the subjects of mineralogy and petrology would lead to a more efficient and better learning process. That approach brings us now to igneous minerals, which include mainly silicates with subordinate oxides and sulfides. The predominance of silicates is clear from Figure 7.1, which shows that 92% of the Earth's crust consists of silicate minerals.

Most mineralogy texts group and discuss minerals on the basis of their chemistry and structural type (see, e.g., Dyar et al., 2008; Klein and Dutrow, 2008; Nesse, 2000). Because there are so many silicates (about 1,140 silicates among the 4,150 known minerals; see Sec. 2.2), their systematic treatment is generally based on structure types. These are shown in Figure 7.4. Cristobalite (Fig. 7.4(A)) is an example of an infinitely extending tetrahedral framework structure in which each tetrahedral corner is linked to another tetrahedron via a bridging oxygen. These **framework** structures (also known as **tectosilicates**, from the Greek word *tecton*, meaning "builder") are found in all the polymorphs of SiO_2 (except stishovite), in all feldspars, nepheline, leucite, and sodalite. Figure 7.4(B) illustrates an infinitely extending tetrahedral single chain that is basic to all members of the pyroxene group. Figure 7.4(C) shows an infinitely extending tetrahedral double chain that characterizes the structure of all members of the amphibole group. Both types of **chain silicates** are also known as **inosilicates**, from the Greek word *inos*, meaning "thread." Figure 7.4(D) shows an infinitely extending tetrahedral sheet that is present in layer silicates such as talc, kaolinite, members of the mica group, chlorite,

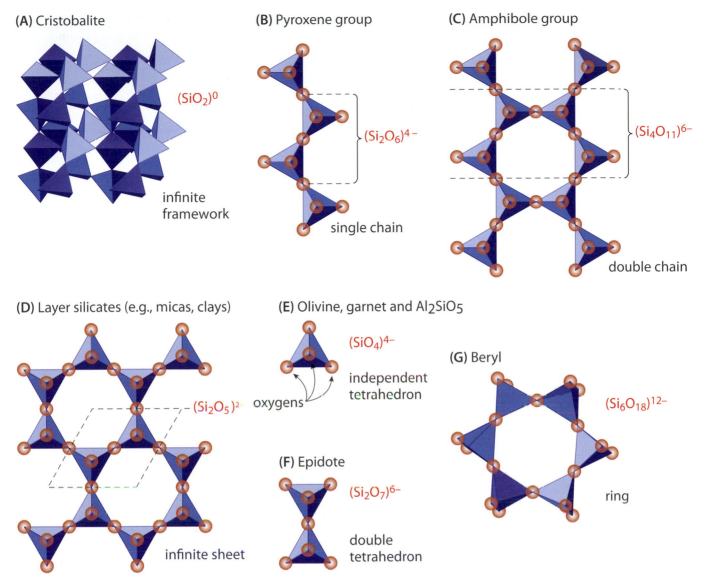

(A) Cristobalite

$(SiO_2)^0$

infinite framework

(B) Pyroxene group

$(Si_2O_6)^{4-}$

single chain

(C) Amphibole group

$(Si_4O_{11})^{6-}$

double chain

(D) Layer silicates (e.g., micas, clays)

$(Si_2O_5)^{2-}$

infinite sheet

(E) Olivine, garnet and Al$_2$SiO$_5$

$(SiO_4)^{4-}$

independent tetrahedron

oxygens

(F) Epidote

$(Si_2O_7)^{6-}$

double tetrahedron

(G) Beryl

$(Si_6O_{18})^{12-}$

ring

Figure 7.4 Classification of silicate structures according to type. (**A**) An infinite framework as in cristobalite, SiO$_2$. (**B**) and (**C**) infinitely extending single and double tetrahedral chains as in pyroxenes and amphiboles. (**D**) An infinitely extending tetrahedral sheet as in sheet silicates. (**E**) A silicate structure with independent tetrahedra as in olivine. (**F**) Double tetrahedral groups as in epidote. (**G**) Tetrahedra forming a ring as in the structures of beryl and tourmaline.

and serpentine. These **sheet silicates** are also referred to as **phyllosilicates**, from the Greek word *phyllon*, meaning "leaf." Independent (SiO$_4$)$^{4-}$ tetrahedra, as shown in Figure 7.4(E), are referred to as independent because they occur in silicate structures in which they do not link with other silica tetrahedra. Instead, all their corner oxygens link to other cation polyhedra. Examples of such structures are olivine; garnet; and the three polymorphs of Al$_2$SiO$_5$, andalusite, sillimanite, and kyanite. These are also known as **nesosilicates**, from the Greek word *nesos*, meaning "island," or as **orthosilicates**, from the Greek word *orthos*, meaning "normal." Double tetrahedral units, as in Figure 7.4(F), occur in epidote. These are also referred to as **sorosilicates**, from the Greek word *soros*, meaning "heap." Another name is **disilicates**, in reference to the double tetrahedral groups. Ring structures (Fig. 7.4(G)) are characteristic

of beryl, tourmaline, and cordierite. They are also known as **cyclosilicates**, from the Greek word *kyklos*, meaning "circle."

In standard mineralogy texts, all the silicates that belong to one of the previously mentioned seven structure types, shown in Figure 7.4, are grouped together, and each group is discussed in an order that commonly begins with independent tetrahedral structures (E) and ends with infinitely extending framework structures (A). This text does not follow this order. Instead, the order reflects how common a specific mineral, or mineral group, is in crustal rocks. Thus, in this chapter, in accordance with the abundance numbers given for various silicates, or silicate groups, in Figure 7.1, we begin with plagioclase, followed by K-feldspar, quartz, pyroxenes, and so on.

The layout of the mineral descriptions that follow is designed so that the most pertinent information appears in an informative, readable, and user-friendly format. The information provided is considered directly relevant to the goals of this text, namely (1) familiarity with the names, chemistry, structure, and physical properties of about 85 rock-forming minerals; (2) a good understanding of the mineralogical makeup of a range of rock types and their origin; (3) the ability to recognize minerals and rocks in hand specimens; (4) some familiarity with the study of rocks in thin sections (under the microscope) if aspects of optical mineralogy are part of the course; and (5) some understanding of what minerals and rocks are used for in commercial and technical applications. With those goals in mind, some of the more encyclopedic aspects of mineral descriptions, as are commonly listed in mineralogy textbooks and other reference volumes, are excluded from the descriptions. Such omitted descriptive aspects include extensive discussion of morphological crystal forms, the listing of unit cell measurements, and optical and X-ray diffraction data. All this information is readily available in standard mineralogy textbooks (e.g., Dyar et al., 2008; Klein and Dutrow, 2008; Nesse, 2000) and in reference works (e.g., Deer et al., 1993, 1982–2009; Gaines et al., 1997; Anthony et al., 1995–2003). In addition, several Web sites provide exhaustive mineralogical databases. Three of these are listed as part of "Further Reading" at the end of this chapter (Mineralogical Society of America and two mineralogy databases, webmineral and mindat).

Optical parameters are not listed because all of these, for most of the minerals covered in this text, are given in Figure 6.10 (see pp. 534–535 and discussed in Chapter 6). The only mineral group that is not covered in that chart is the carbonates, which have such high birefringence that they do not plot in the range covered by the chart. Their high birefringence, however, serves to identify them easily.

7.5 Plagioclase feldspar: $\underset{\text{albite (Ab)}}{NaAlSi_3O_8} - \underset{\text{anorthite (An)}}{CaAl_2Si_2O_8}$

Occurrence: Plagioclase feldspars are the most common rock-forming minerals in the Earth's crust (see Fig. 7.1), occurring in almost all igneous and most metamorphic rocks, and they may be major constituents of some sedimentary rocks.

Chemical composition and crystal structure: A complete solid solution exists at higher than about 800°C between the two end members $NaAlSi_3O_8$ and $CaAl_2Si_2O_8$, as outlined in Figures 7.3(A) and 7.5. Both diagrams show the extent of solid solution at the highest temperature (marked as 900°C in Fig. 7.5(A)). At under ~800°C, there is a lack of continuum of homogeneous feldspars between the two end-member compositions. These inhomogeneities are the result of three types of exsolution features (caused by Na^+ and Ca^{2+} distributing themselves in submicroscopic lamellae of slightly different compositions) that are not visible to the naked eye but cause iridescence. One occurs between An_2 and An_{15} (known as the **peristerite gap**) and is responsible for the iridescence shown by

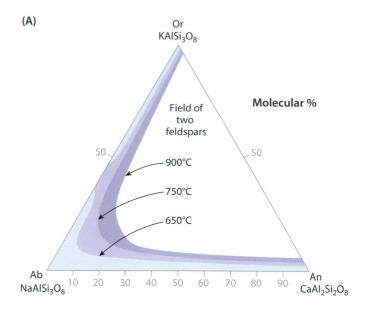

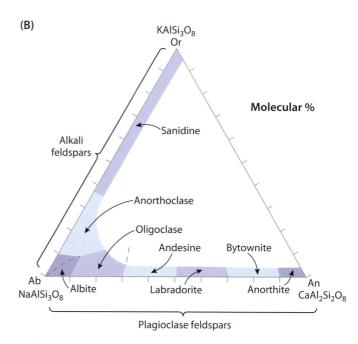

Figure 7.5 **(A)** Experimentally determined extent of solid solution in the system orthoclase (Or) – albite (Ab) – anorthite (An) at $P_{H2O}= 1$ kilobar (after Ribbe, P. H., 1983, *Chemistry, Structure, and Nomenclature of Feldspars*, Reviews in Mineralogy 2, Mineralogical Society of America, Chantilly, VA). **(B)** Nomenclature for the plagioclase feldspar series and the high-temperature alkali feldspars.

moonstone. Labradorescence (a pronounced play of colors) is seen between about An_{40} and An_{65}. A third compositional gap occurs between An_{60} and An_{85}. The overall compositional range of plagioclase has been subdivided into six compositional regions with different names, as shown in Figure 7.5(B).

The crystal structure of common albite, illustrated in Figure 7.6, is a framework structure in which Al^{3+} is housed in a specific tetrahedral site (referred to as **completely ordered**) with the Si^{4+} in the other tetrahedral sites. In both the albite and the anorthite end members, the Na^+ and Ca^{2+} are in 6-fold coordination.

Albite, NaAlSi$_3$O$_8$
Triclinic, C $\bar{1}$

Perspective view along the *c* axis

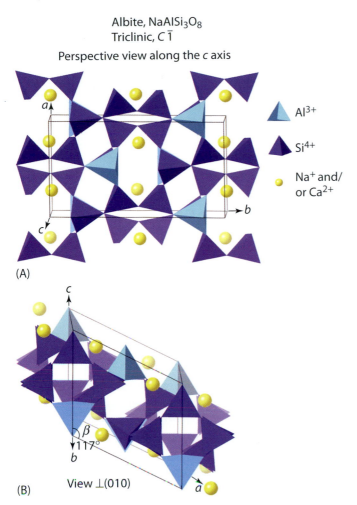

(A)

(B)   View ⊥(010)

Figure 7.6 Two different views of the framework structure of albite, NaAlSi$_3$O$_8$, with a triclinic unit cell superimposed. The light-blue-colored tetrahedra house Al^{3+}. This means that one specific tetrahedron of four ((AlSi$_3$) = 4 tetrahedral sites) is occupied by Al^{3+}, and the structure is, therefore, described as totally ordered. (**A**) This perspective view shows the locations of the Na$^+$ and/or Ca^{2+} ions in relatively large openings. (**B**) This view outlines a triclinic unit cell and shows four-membered tetrahedral rings that are linked in crankshaft-like chains parallel to the *a* axis.

The structure of anorthite is almost identical to that of albite; it is triclinic with space group $P\bar{1}$. The overall structure of plagioclase is very similar to that of microcline (Fig. 5.37(A)).

Common albite (also known as low albite) may contain as much as a few weight percentage of K$_2$O, resulting in only slight solid solution toward the KAlSi$_3$O$_8$ end member (Fig. 7.7). However, at temperatures of about 700°C, a complete solid solution series exists between albite and KAlSi$_3$O$_8$ (with end-member names orthoclase and sanidine; see Fig. 7.7). The higher-temperature albite species are labeled "intermediate albite," "high albite," and "monalbite" in Figure 7.7 and have structures that are more disordered (in terms of the Al distribution among tetrahedra) than the common variety of albite. Upon cooling, these high-temperature structures commonly invert to low albite.

Crystal form: Triclinic, $\bar{1}$. In igneous rocks, when the crystals are euhedral, they tend to be flattened on {010}, or rectangular, or blocky in cross-sections because the γ angle between the *a* and

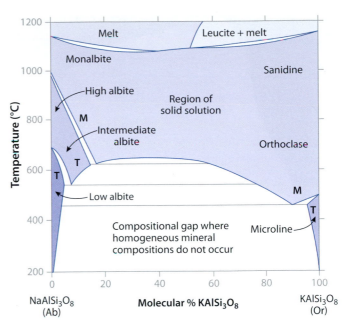

Figure 7.7 Schematic phase diagram for the system NaAlSi$_3$O$_8$ (Ab) – KAlSi$_3$O$_8$ (K-feldspar) showing a large miscibility gap at temperatures less than about 650°C. M and T mean monoclinic and triclinic, respectively (modified after J. V. Smith and W. L. Brown, 1988, *Feldspar Minerals*, Springer Verlag, New York, fig. 1.2; with kind permission of Springer Science and Business Media).

b axes (in the triclinic system) is very close to 90° (ranging from 88° to 91°). In metamorphic rocks, plagioclase grains tend to be anhedral, and in sedimentary rocks they occur as detrital grains. (A detrital mineral grain is the result of the mechanical disintegration of the parent rock; see Sec. 11.2.)

Physical properties: Perfect {001} and good {010} cleavage. **H** = 6; **G** = 2.62 (for Ab) to 2.76 (for An). Color ranges from white (mainly in Ab) to various shades of gray toward the An end member. Albite twinning parallel to {010} (see Fig. 3.20(B)) occurs in all members of this series. It shows as finely spaced parallel striations in hand specimens on a {001} crystal or cleavage face (Fig. 7.8). Carlsbad twins, also parallel to {010}, are common in igneous plagioclase but almost never occur in metamorphic plagioclase. They generally divide a crystal into two equal parts with different {001} cleavage directions in each half.

Distinguishing features: The ubiquitous presence of albite twin lamellae, best seen as striations on {001}. In thin sections, such twin lamellae are instantly recognized and highly diagnostic (Fig. 3.20(B)). Blocky cleavage. Change in color from white in Ab to darker gray tones toward An is helpful in identification. Labradorite commonly shows iridescent color play (see Fig. 3.12(A)).

Uses: Feldspars, being the most common minerals in igneous rocks, constitute a large component of the industrial aggregate used to make concrete and asphalt-based pavement. It is a major constituent in rocks that are cut and polished for ornamental use: countertops (see Fig. 2.20), facings for exteriors of buildings and the interiors of lobbies, and tombstones. Finely powdered plagioclase serves as a filler in paint, plastic, and rubber. It is also used in the manufacture of glass and ceramics.

Figure 7.8 Hand specimen of gray plagioclase with albite twin lamellae.

7.6 K-feldspar: KAlSi$_3$O$_8$ with three polymorphs (microcline, orthoclase, and sanidine)

Occurrence: Microcline and orthoclase, the lowest- and medium-temperature polymorphs (Figs. 7.7 and 5.37), are common as anhedral and euhedral grains in many igneous and metamorphic rocks. Both are found in granite, granodiorite, syenite, granitic pegmatites, and related low- to medium-temperature plutonic rocks. Both are common constituents of sedimentary rocks such as sandstone, arkose, and greywacke. Sanidine, the high-temperature polymorph, occurs as phenocrysts in volcanic rocks (rhyolites and trachytes) that have cooled rapidly from the high temperature of the original eruption.

Chemical composition and crystal structure: The composition of microcline is generally close to KAlSi$_3$O$_8$ (see Fig. 7.7), with only minor substitution of Na$^+$ for K$^+$. At about 600°C (Fig. 7.7), a region of almost complete solid solution exists between orthoclase, as well as sanidine, toward albite, NaAlSi$_3$O$_8$. At temperatures greater than about 1000°C, solid solution between sanidine and albite (referred to as monalbite; Fig. 7.7) is complete.

The structures of the three polymorphs are illustrated in Figure 5.37. All three are framework silicates. In microcline, the Al is housed in a specific tetrahedral site (known as **ordered**). In orthoclase, the one Al atom is statistically distributed between two tetrahedral sites (known as **partial order**), and in sanidine the Al is statistically distributed among all four tetrahedral sites (known as **total disorder**).

Crystal form: Microcline triclinic, $\bar{1}$; orthoclase monoclinic, 2/m; sanidine monoclinic, 2/m. Both microcline and

orthoclase occur as anhedral and euhedral grains, typically elongate along the *a* or *c* axis, in felsic igneous rocks. Sanidine occurs as phenocrysts in volcanic rocks with a tabular habit parallel to {010}.

Physical properties: Both microcline and orthoclase have perfect {001} and good {010} cleavage, with the angle between these at about 90°. **H** = 6; **G** ~2.55. Color ranges from white to light pink. A green variety of microcline is known as amazonite (see Fig. 2.6), with the coloring agent being trace amounts of lead and water. Sanidine is generally colorless to white and may range from translucent to transparent.

Distinguishing features: For microcline – blocky cleavage fragments. Commonly with a **perthitic** texture that consists of a parallel intergrowth of microcline (or orthoclase) and albite-rich plagioclase in which the K-feldspar is the host (Fig. 7.9). This texture is the result of the formation of exsolution lamellae when a Na-containing K-feldspar cools from higher temperatures (see Fig. 7.7). Green feldspar (amazonite) is microcline. Under the microscope, the presence of pericline twinning (tartan twin pattern; see Fig. 3.20(A)) is highly diagnostic.

Orthoclase – it is generally impossible to distinguish orthoclase from microcline in hand specimens. Commonly exhibits good cleavage in two directions resulting in blocky fragments (Fig. 7.9). Carlsbad twinning is common and is best seen under the microscope, where it lacks the tartan twin pattern of microcline, and it differs from sanidine in having a large optic angle (~50°).

Sanidine – occurs commonly in good crystal form as phenocrysts in silicic volcanic rocks. Optical and/or X-ray measurements are needed for unambiguous identification. Sanidine is distinguished from orthoclase under the microscope by its small optic angle (less than 20°); indeed, most sanidine appears almost uniaxial. Sanidine is shown in Figure 7.10.

Uses: Alkali feldspar is commonly added to ceramic mixes as a flux. Feldspar and feldspathic rocks are also used in making glass and in enamel and glaze formulations. Finely ground feldspar is extensively used in scouring and cleaning compounds.

Figure 7.9 On the left, a hand specimen of green perthitic microcline (known as amazonite); on the right, a perfect, blocky cleavage fragment of orthoclase.

and extrusive rocks such as granite, granodiorite, rhyolite, and granitic pegmatites. It is also a common vein-forming mineral. Because it is stable in the weathering environment, it is a major component of clastic sedimentary rocks, and it may be a cementing agent in the form of chalcedony. It occurs in many metamorphic rocks as well. **Chert** (**flint**), a hard, dense, microcrystalline to cryptocrystalline variety of quartz is common in marine sedimentary rocks (see Sec. 10.15).

Tridymite and cristobalite, the high-temperature polymorphs of silica, may occur as phenocrysts or as part of the groundmass of felsic extrusive igneous rocks, and also line cavities and vesicles.

Coesite and stishovite are indicative of a high-pressure origin. Both are found in meteorite impact craters and in high-pressure metamorphic assemblages (see Fig. 5.33(A)).

Chemical composition and crystal structure: Quartz is essentially pure SiO_2. Tridymite and cristobalite are very close to SiO_2 as well, but their slightly more open structures allow for the housing of trace to small amounts of Na^+, K^+, and Ca^{2+}, with some Al^{3+} substituting for Si^{4+} for electrostatic charge balance.

Figure 7.10 Close-up of a small vug in rhyolite with colorless to white, transparent sanidine crystals intermixed with black hornblende.

7.7 Quartz: SiO_2 and polymorphs tridymite, cristobalite, coesite, and stishovite

Occurrence: Quartz is a very common mineral in continental crustal rocks. It is a major constituent of felsic intrusive

Figure 7.11 Hand specimens (clockwise from left to right) of rose quartz, citrine, smoky quartz, amethyst, and rock crystal (clear transparent quartz).

The silicate framework structure of quartz is illustrated in Figure 5.36 and those of the four polymorphs are given in Figure 5.34.

Crystal form: Low quartz (referred to as quartz) hexagonal (trigonal), 32. High quartz, hexagonal, 622. Quartz crystals are commonly prismatic with top (and/or bottom) terminations a combination of two rhombohedra. These may be so equally developed as to give the impression of being a hexagonal dipyramid. Figure 2.2(B) illustrates a group of quartz crystals, and Figure 2.2(E) shows a synthetic crystal used in the manufacture of tiny quartz plates that serve as oscillators in quartz watches.

Physical properties: H = 7; **G** = 2.65. Conchoidal fracture. Usually colorless or white, with a glassy luster also referred to as **vitreous luster** (from the Latin word *vitrum*, meaning "glass"). Colored varieties have specific names (see Fig. 7.11):

Amethyst – various shades of violet; commonly in crystals

Rose quartz – rose red or pink; rarely in crystals; usually massive

Smoky quartz – smoky yellow, brown to dark gray

Citrine – light yellow to brownish orange (may resemble topaz)

Rock crystal – colorless, transparent, and commonly in distinct crystals

Distinguishing features: Glassy luster, hardness, conchoidal fracture, and crystal form.

Uses: The main uses of SiO_2 are in glassmaking and the production of ceramics. It is also used as foundry sand for the casting of steel and ductile iron. It is a major component of the sand and gravel used in construction, and it is used as an abrasive in sandpaper and sandblasting. Quartz is the prime source of silicon that is chemically extracted therefrom. Silicon is used in the manufacture of silicon chips (used in the computer industry) and solar cells.

7.8 Nepheline: $(Na,K)AlSiO_4$

Occurrence: A common mineral in silica-deficient intrusive and extrusive rocks. A major constituent of nepheline syenite and occurs in nepheline gabbro and its volcanic equivalent, basanite.

Chemical composition and crystal structure: The end-member composition is $NaAlSiO_4$, but there is generally considerable K^+ substitution for Na^+, which leads to the common composition of $(Na_3K)Al_4(SiO_4)_4$ in which there is one K^+ for every three Na^+ ions. At high temperature there is a complete solid solution series from $NaAlSiO_4$ to $KAlSiO_4$, kalsilite.

An alumino-silicate framework structure that is considerably more open than that of feldspar. The structure of $(Na_3K)Al_4(SiO_4)_4$ is shown in Figure 7.12, with K^+ inside the large 6-fold rings.

Nepheline is a member of the **feldspathoids**, which also include leucite and sodalite. Feldspathoids are similar to feldspar in their structures and physical properties, but they contain less

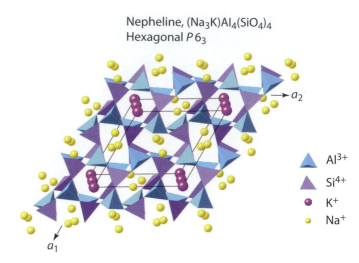

Nepheline, $(Na_3K)Al_4(SiO_4)_4$
Hexagonal $P6_3$

△ Al^{3+}
▲ Si^{4+}
● K^+
● Na^+

Figure 7.12 Perspective view of the crystal structure of nepheline. A primitive hexagonal unit cell is outlined. The sites in which K^+ are housed are nearly hexagonal in symmetry. The Na^+ ions are in more irregular sites.

silica than the feldspars relative to the amounts of Na and K. The framework structure of feldspathoids is slightly more open than that of feldspars.

Crystal form: Hexagonal, 6. Rarely in crystals, almost invariably compact, massive, and in embedded grains (see Fig. 7.13). Forms euhedral phenocrysts in some silica-poor igneous rocks.

Physical properties: H = 5½–6; **G** ~2.62. Distinct $\{10\bar{1}0\}$ cleavage. Luster vitreous to greasy. Generally colorless, white to gray. Difficult to distinguish from quartz but has lower hardness than quartz (with **H** = 7) and lacks conchoidal fracture.

Figure 7.13 Massive nepheline.

Distinguishing features: As noted already, difficult to distinguish from quartz but usually appears greasy by comparison. Easily distinguished on the weathered surface because nepheline is highly susceptible to weathering and hence weathers recessively, whereas quartz and feldspar stand out in relief. If blue sodalite is part of the assemblage, this reflects a silica-poor bulk composition and quartz will not be present. If quartz is not a possibility, then you may be dealing with nepheline.

Uses: Nepheline, extracted with feldspar from nepheline syenite, is used in the manufacture of glass and ceramics. It is also used as a filler in paints and plastics.

7.9 Leucite: KAlSi₂O₆

Occurrence: Leucite is a rare mineral that occurs as phenocrysts in K-rich volcanic rocks (Fig. 7.15). It is commonly associated with plagioclase, nepheline, and aegirine, a Na-rich pyroxene, $NaFe^{3+}Si_2O_6$. These are silica-deficient assemblages that never contain quartz.

Chemical composition and crystal structure: Most leucites are close to $KAlSi_2O_6$, but small amounts of Na^+ may replace K^+. Leucite is a member of the feldspathoids, which also include nepheline and sodalite. Feldspathoids are similar in composition to feldspars but contain less silica relative to the amounts of Na and K.

The structure, shown in Figure 7.14, is that of a framework silicate with 4- and 6-member rings of Al/Si tetrahedra in which K^+ occupies large openings. Al^{3+} and Si^{4+} are randomly distributed (disordered) over the tetrahedral sites.

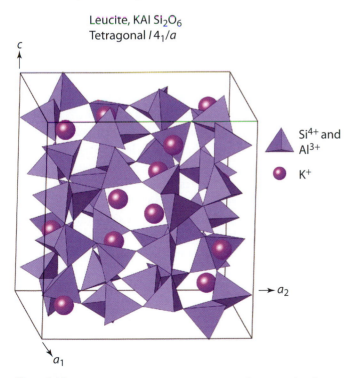

Leucite, KAl Si₂O₆
Tetragonal I4₁/a

Si⁴⁺ and Al³⁺

K⁺

Figure 7.14 Perspective view of the crystal structure of leucite, within the outline of a tetragonal unit cell. The K^+ ions are in 12-coordination with oxygen in large cavities of this silicate framework structure.

Figure 7.15 Trapezohedral leucite crystals in a volcanic rock.

Crystal form: Leucite crystallizes from magma with isometric symmetry but on cooling below 605°C, becomes tetragonal, $4/m$. Usually in trapezohedral (Fig. 7.15) or dodecahedral crystals. On cooling below 409°C, leucite becomes unstable and changes into an intergrowth of feldspar and kalsilite, which retains the external shape of the high-temperature crystal. This intergrowth is known as pseudoleucite.

Physical properties: $H = 5\frac{1}{2}$–6; $G = 2.47$. White to gray and translucent. Conchoidal fracture and brittle.

Distinguishing features: The trapezohedral crystal form of phenocrysts in silica-deficient extrusive rocks. Under the microscope its crystals are characterized by multiple twins in different sectors.

Uses: No commercial uses.

7.10 Sodalite: Na₄Al₃Si₃O₁₂Cl

Occurrence: Sodalite is a comparatively rare rock-forming mineral that occurs in Na-rich, silica-deficient (undersaturated), felsic rocks such as nepheline syenite and sodalite syenite.

Chemical composition and crystal structure: Generally close in composition to the formula $Na_4Al_3Si_3O_{10}Cl$, with some substitution of K^+ for Na^+.

The structure is that of an alumino-silicate framework that is more open than that of feldspar. This allows for the housing of the relatively large Na^+ and Cl^- ions. Al^{3+} and Si^{4+} are housed in tetrahedra that alternate throughout the structure (Fig. 7.16). Sodalite is classified with nepheline and leucite as a member of the feldspathoid group.

Crystal form: Isometric, $\overline{4}3m$. Usually massive or in foliated cleavable masses. Also as embedded grains. Rarely in good crystals but does form euhedral phenocrysts in sodalite syenite.

Physical properties: $H = 5\frac{1}{2}$–6; $G \sim 2.2$. Cleavage {011} poor. Color a deep azure blue (Fig. 7.17); rarely white to gray.

Sodalite, $Na_4Al_3Si_3O_{12}Cl$
Isometric $P\bar{4}3n$

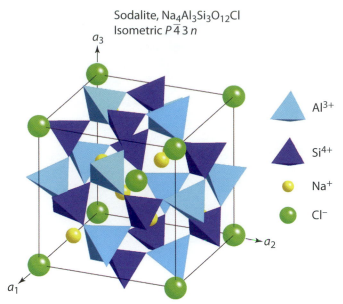

- Al^{3+}
- Si^{4+}
- Na^+
- Cl^-

Figure 7.16 Perspective view of the crystal structure of sodalite. A primitive isometric unit cell is outlined. The structure consists of a framework of linked SiO_4 and AlO_4 tetrahedra in which large cavities are occupied by Na^+ and Cl^-.

Distinguishing features: Identified by its blue color. Closely resembles lazurite, a complex Na-Ca-Al silicate, which also has a dark blue color. Lazurite (also known as lapis lazuli) is commonly associated with pyrite; sodalite is not.

Uses: Sodalite-containing rocks are used as valuable dimension stone. In massive form, sodalite is used for the carving of ornamental objects. Also used as a gemstone in en cabochon cuts.

Figure 7.17 Hand specimen of sodalite with nepheline and muscovite.

7.11 Enstatite: $MgSiO_3$ – $(Mg, Fe)SiO_3$

Occurrence: A major rock-forming silicate in mafic and ultramafic igneous rocks such as basalt, gabbro, norite, and peridotite. Common associations include augite, olivine, and plagioclase.

$CaSiO_3$ - Wollastonite (Wo) - a pyroxenoid

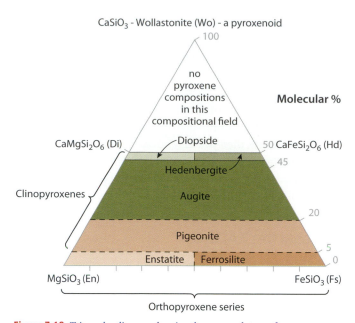

Figure 7.18 Triangular diagram showing the nomenclature of common pyroxenes in terms of the compositional space in the end-member compositions of Wo (wollastonite) – En (enstatite) – Fs (ferrosilite). All pyroxene compositions lie below the line that connects the diopside and hedenbergite compositions. Wollastonite is a pyroxenoid.

It also occurs in high-grade metamorphic rocks and in stony meteorites.

Chemical composition and crystal structure: The end-member composition of enstatite is $MgSiO_3$, but the name extends all the way to $(Mg_{0.5}Fe_{0.5})SiO_3$, as shown in Figure 7.18. The iron-rich part of this series is known as ferrosilite, ranging from $(Mg_{0.5}Fe_{0.5})SiO_3$ to the end-member $FeSiO_3$. This series is referred to as **orthopyroxenes** because they are members of the pyroxene group with orthorhombic symmetry (commonly abbreviated as "opx"). The compositional extent of this series toward ferrosilite, the Fe-rich end member, is limited at crustal pressures. Ferrosilite becomes stable only at high pressures in the mantle. The compositional extent of enstatite toward the $CaSiO_3$ end member (Fig. 7.18) is limited to 5 molecular % of $CaSiO_3$. (This end member, known as wollastonite is a member of a group of silicates known as **pyroxenoids**, with a chain structure different from that in pyroxenes). All pyroxene compositions above that dividing line are more Ca-rich (ranging from 5 to 50 molecular % Wo), and they are all part of the **clinopyroxene** group with monoclinic symmetry (commonly abbreviated as "cpx").

The crystal structure of an orthopyroxene with enstatite composition is shown in Figure 7.19, in which 7.19(A) is an almost frontal (though perspective) view showing the infinitely extending vertical, single chains of silicate tetrahedra of (SiO_3) composition. Two opposing tetrahedral chains link a central octahedral chain. This is best seen in Figure 7.19(B). In enstatite compositions with Fe^{2+} and Mg^{2+}, the larger Fe^{2+} ions tend to concentrate in atomic positions outside the regular octahedral chain, as

Enstatite, (Mg,Fe)SiO$_3$
Orthorhombic $P\,2_1/b\,2_1/c\,2_1/a$

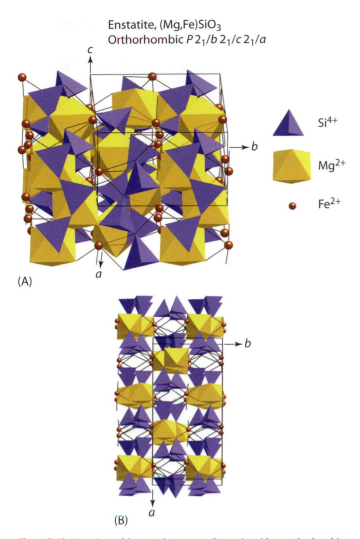

Si^{4+}

Mg^{2+}

Fe^{2+}

(A)

(B)

Figure 7.19 Two views of the crystal structure of enstatite with an orthorhombic unit cell superimposed. **(A)** A perspective view showing the presence of kinked tetrahedral and octahedral chains parallel to the c axis. The smaller Mg^{2+} ions are preferentially housed in the yellow octahedra. The larger Fe^{2+} ions occur in a somewhat larger site shown as a sphere. **(B)** A view down along the c axis showing the t-o-t (tetrahedral-octahedral-tetrahedral) infinite chains along c.

Figure 7.20 Hand specimen of bronze-brown enstatite.

7.12 Pigeonite: ~Ca$_{0.25}$(Mg, Fe)$_{1.75}$Si$_2$O$_6$

Occurrence: A common Ca-poor clinopyroxene in subalkaline rocks. It is stable only at high temperature and preserved only in rapidly cooled igneous rocks such as lava flows and dikes. It crystallizes in mafic intrusive rocks but inverts, on relatively slow cooling, by reconstructive polymorphism, to orthopyroxene. In this inversion process, it first exsolves augite lamellae parallel to {001} and subsequently inverts to the orthorhombic pyroxene known as enstatite. On further cooling, this may exsolve augite lamellae parallel to {100} (see Figs. 7.21 and 7.22 and compare with the chapter-opening photomicrograph in Chapter 6). Pigeonite never occurs in silica-undersaturated rocks. It forms in some high-temperature metamorphic rocks but inverts to enstatite on cooling.

Chemical composition and crystal structure: Figure 7.18 shows the compositional range of pigeonite in a ternary diagram, where its Ca-component (CaSiO$_3$, wollastonite) varies from 5 to 20 molecular percent. Most pigeonite ranges between 30 and 70% of the Fe component. At more magnesian compositions, orthopyroxene (enstatite) is common.

The structure of pigeonite is that of a clinopyroxene. This monoclinic group of pyroxenes includes augite, hedenbergite (see Fig. 7.18), and aegirine as well. The structure shown in Figure 7.23(A) is that of the clinopyroxene diopside, CaMgSi$_2$O$_6$. The pigeonite structure (Fig. 7.23(B)) is almost identical except for the coordination of the Ca site. In diopside, this is an irregular 8-coordinated site (**C.N.** = VIII); in pigeonite, this same site is an irregular 7-coordinated site (**C.N.** = VII). This difference in coordination is because in diopside this site is completely filled by Ca^{2+} (ionic radius 1.26 Å), whereas in pigeonite, with a composition Ca$_{0.25}$(Mg, Fe)$_{1.75}$Si$_2$O$_6$, it is, on average, occupied by Ca$_{0.25}$ (Mg, Fe)$_{0.75}$. With Mg^{2+} and Fe^{2+} considerably smaller than Ca^{2+}, the site becomes smaller and exhibits a lower coordination

shown by the brown spheres in Figure 7.19. These positions are somewhat larger than the regular 6-fold octahedra that house Mg^{2+}, and they represent a distorted 6-fold coordination site.

Crystal form: Orthorhombic; $2/m\,2/m\,2/m$. Prismatic habit but crystals rare. Usually massive.

Physical properties: **H** = 5½–6; **G** ~3.2–3.6, and is a function of increasing iron content. {210} cleavage good at about 90° (see Fig. 3.15). Color ranges from light brown to bronze (Fig. 7.20) to olive green and brown, depending on the composition.

Distinguishing features: Characterized by good pyroxene cleavage and bronze to brown color. When Fe-rich (ferrosilite), it may be almost black and is difficult to distinguish from augite.

Uses: Some orthopyroxene-rich rocks are used for attractive dimension stone or decorative purposes as countertops. Also for carvings and occasionally as a minor gemstone.

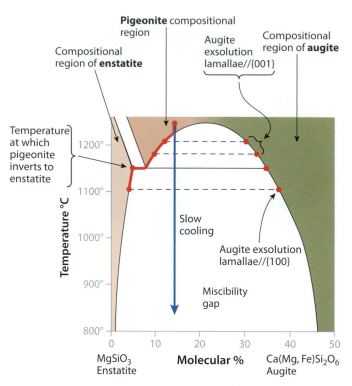

Figure 7.21 Temperature-composition diagram between two members of the pyroxene group: enstatite, $MgSiO_3$ (can also be written as $Mg_2Si_2O_6$), and augite, $Ca(Mg,Fe)Si_2O_6$. At temperatures below about 1250°C, a wide miscibility gap exists between the two compositions, but above this temperature there is an almost complete solid solution series. The compositional field of the high-temperature clinopyroxene pigeonite is located at the upper left of the diagram. A specific pigeonite composition is represented by a red dot. When this composition is cooled slowly along the vertically downward path (see arrow), two sets of exsolution lamellae of somewhat different augite compositions may develop inside the mineral grains of originally homogeneous high-temperature pigeonite. The compositions of the pigeonite host and the first set of exsolved lamellae are shown at about 1200°C. At about 1150°C pigeonite, inverts to the lower-temperature structure of enstatite (an orthopyroxene) and at that inversion (and at somewhat lower temperatures) further exsolution lamellae (with a different crystallographic orientation) may develop. The final cooled product is an enstatite host with two possible sets of augite lamellae (see Fig. 7.22 and opening photograph to Chapter 6). Such a crystal is known as inverted pigeonite.

number. This same site in enstatite, $Mg_2Si_2O_6$, is completely filled by Mg^{2+} and accordingly has a **C.N.** = VI.

Crystal form: Monoclinic, $2/m$. Rarely as euhedral phenocrysts with a prismatic habit parallel to c. May be rimmed by augite.

Physical properties: H = 6; G = 3.30–3.46. Good {110} cleavage. Color brown, greenish brown, to black. When cooled slowly, it inverts to enstatite with augite exsolution lamellae, as shown in Figure 7.22. Such an occurrence, which can be seen only under the microscope, is commonly referred to as inverted pigeonite, because the host is an orthopyroxene, enstatite, with augite exsolution lamellae, and not the original high-temperature, homogeneous pigeonite (see the chapter-opening photomicrograph to Chapter 6).

Distinguishing features: In hand specimens, cleavage and color may suggest it to be a clinopyroxene. However, in hand specimens, it cannot be distinguished from enstatite or augite.

Uses: None.

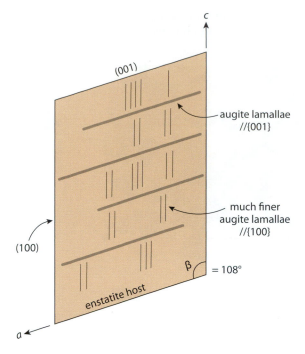

Figure 7.22 Schematic drawing of a (010) section through an inverted pigeonite as can be seen under a petrographic microscope. The outline of the grain is clearly monoclinic, but the host is an orthopyroxene, enstatite. The two sets of exsolution lamellae are the result of exsolution from an original pigeonite host.

7.13 Augite: $(Ca, Na)(Mg, Fe, Al)(Si, Al)_2O_6$

Occurrence: Augite is a major constituent of most mafic igneous rocks such as basalt and gabbro. It is also common in ultramafic rocks including pyroxenite and the type of peridotite known as lherzolite, which is thought to be an important rock type in the upper mantle. It also occurs in andesite and its intrusive equivalent diorite. It occurs in metamorphic rocks such as amphibolites and granulites.

Chemical composition and crystal structure: The best way to visualize the compositional range of augite is to begin with the end-member composition of the clinopyroxene diopside, $CaMgSi_2O_6$. The formula of augite (given previously) differs in that there may be some substitution in the Ca position by Na, and there is always considerable Fe^{2+} substitution for Mg^{2+}, as well as Al^{3+} substitution not only in the tetrahedral (Si^{4+}) site but also in the octahedral sites normally occupied by Mg^{2+} and Fe^{2+}. The schematic extent of solid solution in augite is shown in Figure 7.18, a triangular composition diagram between the end members $MgSiO_3$ (enstatite), $FeSiO_3$ (ferrosilite), and $CaSiO_3$ (wollastonite). This diagram does not show the extent of solid solution of Al^{3+} or Na+ components in augite.

The structure of diopside is shown in Figure 7.23(A). Augite's structure is very similar with some Na^+ in the Ca^{2+} site, a wide range of Fe^{2+} substitution in the octahedral sites occupied by Mg^{2+}, and additional Al^{3+} in octahedral as well as

Diopside, CaMg (Si_2O_6)
Monoclinic, C 2/c

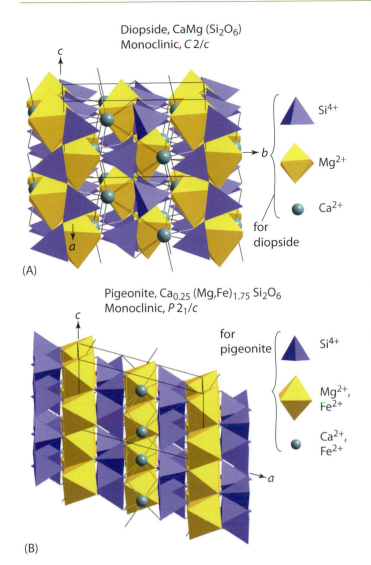

(A)

Pigeonite, $Ca_{0.25}$ $(Mg,Fe)_{1.75}$ Si_2O_6
Monoclinic, P 2_1/c

for pigeonite

Si^{4+}

Mg^{2+}, Fe^{2+}

Ca^{2+}, Fe^{2+}

(B)

Figure 7.23 Two different views of the crystal structure of the clinopyroxene diopside. **(A)** A frontal view (looking essentially perpendicular to (100)) shows the vertical, infinitely extending tetrahedral and octahedral chains in the structure. Mg^{2+} is preferentially housed in the 6-coordinated (octahedral) sites, whereas Ca^{2+} is located in irregular, 8-coordinated sites. **(B)** A view along the *b* axis (looking almost perpendicular to (010)) that shows the outline of the monoclinic unit cell and the vertically extending tetrahedral and octahedral chains.

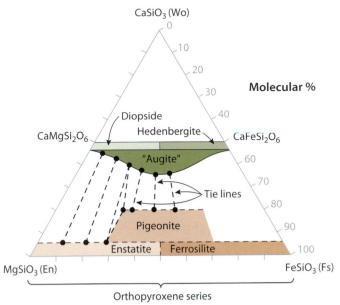

Figure 7.24 Common compositional fields among various pyroxenes. The white space, with crossing tie lines, outlines the approximate field of a miscibility gap between the Ca-rich clinopyroxenes colored in green and the orthopyroxene and pigeonite fields, at the bottom of the diagram, colored in shades of brown. Representative tie lines that connect the compositions of coexisting minerals (i.e., mineral pairs) across the miscibility gap are shown.

common coexistences of augite and pigeonite, as well as augite and enstatite. A temperature-composition section through this miscibility gap is shown in Figure 7.21.

Crystal form: Monoclinic; 2/*m*. In stubby, prismatic crystals showing square and eight-sided cross-sections (Fig. 7.25). Also in granular masses and anhedral grains.

tetrahedral sites. The Al^{3+} substitution in both the octahedral and the tetrahedral sites requires a coupling with another cation because the charge of Al^{3+} is different from that of Mg^{2+} (or Fe^{2+}) as well as Si^{4+}. An example of such limited coupled substitution is $Ca^{2+} Mg^{2+} \rightleftarrows Na^+ Al^{3+}$.

Figure 7.24 is very similar to Figure 7.18, but it shows the extent of the compositional fields of augite, pigeonite, and enstatite at relatively high temperatures. The white space in the lower half of the triangle outlines the extent of a miscibility gap between the three named clinopyroxenes in the upper field and pigeonite, and the orthopyroxenes series. "Augite" is labeled between quotation marks because the Al^{3+} content of many common augites cannot be shown in this figure, or in Figure 7.18. Tie lines connect the

Figure 7.25 Dark green to black grouping of prismatic augite crystals.

Physical properties: Good to fair cleavage on {110} at 87° and 93°. May show parting on {001}. **H** = 5–6; **G** = 3.2–3.3. Augite color is very dark green to black (see Fig. 7.25). Diopside color ranges from white to light green.

Distinguishing features: Characterized by square cross-sections of crystals and relatively good prismatic cleavage at 87° and 93°. The color of augite is most commonly black, but it may be very dark brownish green as well.

Uses: There is no specific use for augite, but because it is a major constituent of basalt, diabase, and gabbro, it is a major component of crushed stone used in highway construction (see Fig. 2.9).

7.14 Aegirine: $NaFe^{3+}Si_2O_6$

Occurrence: Aegirine and aegirine-augite (an intermediate composition between aegirine and augite) occur in igneous rocks with substantial amounts of Na. Examples are alkali granite, syenite, and nepheline syenite. Commonly in association with nepheline and sodalite in nepheline syenite.

Chemical composition and crystal structure: Aegirine shows a wide range of composition and in most occurrences there is atomic substitution according to $Na^+ Fe^{3+} \rightleftarrows Ca^{2+} (Mg^{2+}, Fe^{2+})$, which results in a complete solid solution series toward augite. Intermediate compositions are referred to as aegirine-augite.

The structure of aegirine is similar to that of other monoclinic pyroxenes such as diopside (Fig. 7.23(A)). Na^+ substitutes for Ca^{2+} in a large, irregular 8-coordinated site, and Fe^{3+} substitutes in the octahedral positions normally occupied by Mg^{2+} and Fe^{2+}.

Crystal form: Monoclinic; $2/m$. Slender prismatic crystals with steep terminations. Also fibrous and in anhedral grains.

Figure 7.26 Photomicrograph of a thin section of nepheline syenite from Mount Johnson, Quebec, Canada, containing pyroxene crystals that are zoned from aegirine-augite (pale green) in the core to aegirine (bright green) on the rim. The darker crystals, which are zoned from brown cores to dirty green rims, are hornblende. Two high-relief clear crystals show hexagonal cross-sections through apatite prisms. Smaller apatite crystals can be seen included in the aegirine crystal at the left. The clear low-relief minerals are alkali feldspar and nepheline. Width of field is 1 mm.

Physical properties: Cleavage {110} imperfect at 87° and 93°. **H** = 6–6½; **G** = 3.40–3.55. Color dark green to greenish black.

Distinguishing features: In hand specimens, aegirine can be identified as a pyroxene by cleavage, hardness, and color. It may be mistaken for hornblende, which has better cleavage at 56° and 124°. In hand specimens, it is commonly very hard to distinguish aegirine from aegirine-augite or augite. Occurrence in Na-rich and SiO_2-poor rocks is characteristic.

Best identified in thin sections on the basis of color and pleochroism. Aegirine and aegirine-augite are clear (Fig. 7.26) compared with the murky color of hornblende.

Uses: None.

7.15 Hornblende: $(Na, K)_{0-1}Ca_2(Mg, Fe, Al)_5$ $(Si, Al)_8O_{22}(OH)_2$

Occurrence: Hornblende is one of the most common rock-forming minerals. It is the main ferromagnesian mineral in many igneous rocks such as diorite, granodiorite, andesite, and dacite. It also occurs in gabbro and norite.

It is one of the major minerals in metamorphosed mafic igneous rocks that have undergone the metamorphic conditions of the **amphibolite facies**. Some specific rock types are amphibolite and hornblende gneiss.

Chemical composition and crystal structure: Hornblende, a member of the **amphibole group** of silicates, is chemically one of the most complex of all major rock-forming silicates. This was discussed in Section 4.4 under the heading of "Rule 5: The Principle of Parsimony." The chemical formula for hornblende given at the beginning of this section lists seven different cations distributed over four different cationic sites. In addition, the (OH) group has a unique atomic location. This, therefore, is the first hydrous silicate among the minerals we have discussed that are common in igneous rocks.

The formula for hornblende is best understood if we begin with the simpler formula of a closely related clinoamphibole, tremolite, $Ca_2Mg_5Si_8O_{22}(OH)_2$. This can be considered a relatively simple end member composition in the amphibole group. Hornblende contains several additional cations that may range over considerable atomic percentages. In pure tremolite, Na^+ and K^+ are absent and so are Fe^{2+} and Al^{3+}. This means that in tremolite a large atomic site (referred to as the "A" site with 10-fold coordination) is empty, whereas in hornblende, this may contain variable amounts of Na^+ and minor K^+. The very first entry in the hornblende formula $(Na, K)_{0-1}$ reflects this. When the "A" site is empty, the subscript is zero, but when some Na^+ and/or K^+ are present, this site may show a fractional occupancy such as $(Na^+, K^+)_{0.35}$. The atomic site that is occupied mainly by Ca^{2+} is in 8-fold coordination, and variable amounts of Mg^{2+}, Fe^{2+}, and Al^{3+} are distributed over the regular octahedral sites in the structure. Al^{3+} substitutes in the regular octahedral sites as well as the tetrahedral sites that are mainly occupied by Si^{4+}. Additional chemical elements that may be present in a chemical analysis of hornblende are Ti^{4+} and Mn^{2+}.

Two different views of a **clinoamphibole** structure were given in Figure 4.13, and additional illustrations appear in Figure 7.27. A perspective view of a clinoamphibole structure is given in Figure 7.27(A) and a close-up view of the double-chain structure is given in Figure 7.27(B). The various atomic sites are identified in the legend of both illustrations. The overall structure of amphiboles is more open than that of pyroxenes (see Fig. 7.23), which allows for the greater extent of atomic substitutions in amphiboles.

Clinoamphibole: hornblende,
$(Ca,Na)_{0-1}Ca_2(Mg,Fe,Al)_5(Si,Al)_8O_{22}(OH)_2$
Monoclinic, $C\,2/m$

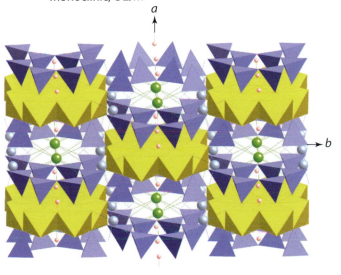

(A)

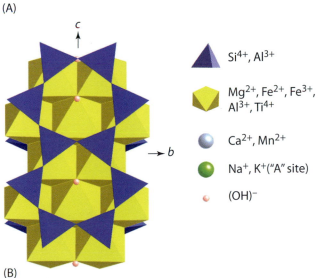

▲ (blue triangle)	Si^{4+}, Al^{3+}
⬡ (yellow)	Mg^{2+}, Fe^{2+}, Fe^{3+}, Al^{3+}, Ti^{4+}
● (light blue)	Ca^{2+}, Mn^{2+}
● (green)	Na^+, K^+("A" site)
● (pink)	$(OH)^-$

(B)

Figure 7.27 **(A)** Perspective view of the structure of a clinoamphibole. A monoclinic unit cell is outlined. The various atomic sites are labeled in the legend. **(B)** A close-up view of the vertical double tetrahedral chain that is coordinated to a chain of regular octahedra. The "A" site is centered inside the hexagonal outline of the double chain.

Figure 7.28 Cluster of black, euhedral hornblende crystals.

Crystal form: Monoclinic; $2/m$. Commonly in prismatic crystals (Fig. 7.28). Also columnar and fibrous.

Physical properties: Commonly in prismatic crystals or showing perfect {110} cleavage at 56° and 124° (see Fig. 3.15). H = 5–6; G = 3–3.4. Color various shades of dark green to black.

Distinguishing features: Crystal form and cleavage angles distinguish hornblende from pyroxenes such as augite. The dark color (dark green to black) distinguishes hornblende from other types of amphiboles.

Uses: No specific use, except for it being a constituent of decorative or dimension stone, as well as rock types used as aggregate in road building.

7.16 Muscovite: $KAl_2(AlSi_3O_{10})(OH)_2$

Occurrence: Muscovite is an important constituent of aluminous granite (commonly associated with biotite), granodiorite, aplite, and other related felsic rocks. It is an extremely common constituent of a range of metamorphic rocks from slate to schist and gneiss.

Chemical composition and crystal structure: Muscovite, biotite, and phlogopite are all members of the mica group of **layer silicates** (also known as **phyllosilicates**). Muscovite with a most common composition of $KAl_2(AlSi_3O_{10})(OH)_2$ consists of infinitely extending tetrahedral sheets of composition $(AlSi_3O_{10})$ that are linked to infinitely extending octahedral sheets of $Al_2(OH)_6$ composition (see Fig. 7.29(A)). The geometries of these two sheets of very different composition are compatible so that they can link together. The structure of a gibbsite sheet ($Al(OH)_3$ or $Al_2(OH)_6$) is shown in Figure 7.29(B)–(C). Because Al^{3+} is housed in an octahedral site (**C.N.** = VI), the electrostatic valency (e.v.) of each of the six Al-(OH) bonds = $^3/_6$ = ½. Each $(OH)^-$ group can, therefore, be shared only between two adjacent octahedra. This results in

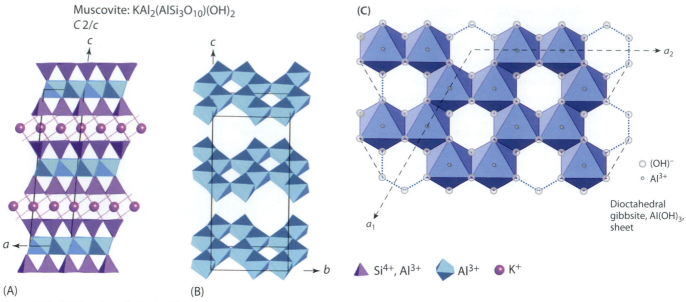

Muscovite: $KAl_2(AlSi_3O_{10})(OH)_2$
$C\,2/c$

(A)   (B)   (C)

Si^{4+}, Al^{3+} Al^{3+} K$^+$

(OH)$^-$
Al^{3+}
Dioctahedral gibbsite, Al(OH)$_3$, sheet

Figure 7.29 (**A**) View along the b axis of the muscovite structure. It consists of infinitely extending tetrahedral and octahedral sheets (when combined known as *t-o-t* layers) with interlayer K$^+$ ions. A monoclinic unit cell is outlined. (**B**) A perspective view of only the octahedral sheets showing the regularity of vacant sites in these sheets. (**C**) A horizontal view of a gibbsite (Al(OH)$_3$ or Al$_2$(OH)$_6$) sheet showing the regularity of two occupied octahedral sites alternating with one vacant one. This is known as a dioctahedral sheet.

Figure 7.30 Randomly intergrown cluster of well-formed tabular crystals of muscovite with pseudohexagonal outlines. These are referred to as pseudohexagonal because the underlying structure of muscovite is monoclinic.

an octahedral sheet in which one octahedron is vacant for every two octahedra filled with Al^{3+}. This geometry is known as **dioctahedral** because only two octahedra (of three available) are occupied. When the alumino-silicate sheet and the gibbsite sheet are joined, the composition $Al_2(AlSi_3O_{10})(OH_2)$ results, which, if we check all the charges of the ions involved, is not a neutral, stable compound. That is, $Al_2^{3+}(Al^{3+}Si_3^{4+}O_{10}^{2-})(OH)_2^-$ has a residual charge of -1. This results from the 6^+, 3^+, 12^+ (totaling 21^+ on the cations) versus the charges on the anions of 20^-, and 2^- (totaling 22^-). This residual negative charge is balanced by a large interlayer cation, K^+, between successive tetrahedral-octahedral-tetrahedral (*t-o-t*) layers in the structure of muscovite (see Fig. 7.29(A)).

Crystal form: Monoclinic; $2/m$. Distinct crystals, as in Figure 7.30, rare. Usually tabular with pseudohexagonal outlines referred to as "books" because of the stacking of thin cleavable sheets.

Physical properties: Perfect {001} cleavage, resulting in very thin sheets that can be peeled off (see Fig. 3.13). The folia are flexible and elastic. $H = 2–2½$; G ~2.8. Colorless and translucent in thin sheets but in thicker sheets blocks light. Color ranges from shades of light yellow to brown. Pearly luster. Sericite is the name for white mica (commonly muscovite) that is a hydrothermal alteration product of feldspar and other K-Al-rich minerals.

Distinguishing features: Characterized by its micaceous habit, excellent cleavage, pearly luster, and light color.

Uses: An early use of muscovite was as **isinglass** in furnace and stove doors. Because of its high dielectric and heat-resisting properties, cleavage plates are used as insulating material in many aspects of the electronics industry. Other industrial applications of ground muscovite include use as a filler in paint, in plastic, and in wallboard. Also as coatings on wallpaper to produce a silky sheen. It is also a constituent of drilling mud in the recovery of oil and gas. Cosmetic products such as nail polish, lipstick, and eye shadow may contain finely ground muscovite for the production of luster.

7.17 Phlogopite: $KMg_3(AlSi_3O_{10})(OH)_2$

Occurrence: Phlogopite occurs in some K-rich ultramafic rocks, especially those containing leucite. It is also a major component of the groundmass of kimberlites (Fig. 9.38), which occur as diatremes and dikes and are important because they bring diamonds up from the mantle. It is also present in many carbonatites, an igneous rock composed mainly of calcite (Fig. 9.39). The most common occurrence in metamorphic rocks is in marbles and related carbonate-rich rocks.

Chemical composition and crystal structure: In Section 7.16, we discussed the structure of muscovite (see Fig. 7.29). Phlogopite (and biotite) have overall structures that are similar, the only difference being the composition of the octahedral sheet that is sandwiched between two opposing tetrahedral sheets (Fig. 7.31). In muscovite, this octahedral sheet has the

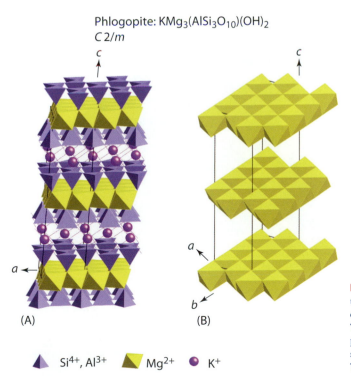

Phlogopite: $KMg_3(AlSi_3O_{10})(OH)_2$
$C\,2/m$

(A)   (B)

▲ Si^{4+}, Al^{3+} ◼ Mg^{2+} ● K^+

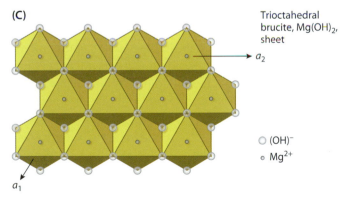

(C)

Trioctahedral brucite, $Mg(OH)_2$, sheet

○ $(OH)^-$
· Mg^{2+}

Figure 7.31 (**A**) Perspective view of the structure of phlogopite. A monoclinic unit cell is outlined. It clearly shows, as does muscovite (Fig. 7.29), the presence of tetrahedral-octahedral-tetrahedral (*t-o-t*) layers, and K^+ as interlayer cations. The octahedral sheet in this case is of $Mg(OH)_2$, brucite composition. (**B**) A perspective view of only the octahedral sheets in the phlogopite structure. These sheets are continuous without vacancies. (**C**) A horizontal view of a brucite sheet. This is known as a trioctahedral sheet.

Figure 7.32 View of the {001} cleavage in a large book of phlogopite. Note the pseudohexagonal outline. The pink mineral in the back is calcite; the blue, apatite.

gibbsite, Al(OH)$_3$, composition, but in phlogopite, it is composed of Mg(OH)$_2$, brucite. The structure of this sheet is shown in Figure 7.31(B)–(C). With Mg^{2+} housed in an octahedral sheet, the electrostatic valency (e.v.) of the six Mg-OH bonds = $\frac{2}{6}$ = $\frac{1}{3}$. Each (OH)$^-$ group can, therefore, be shared by three adjacent octahedra, giving rise to an octahedral sheet without vacancies. This continuous sheet is known as **trioctahedral** because all three adjacent octahedral sites are occupied. In contrast, in a dioctahedral sheet, as in muscovite, two out of three octahedral sites are occupied, and the third is empty. Except for the composition and architecture of the octahedral sheets, the structures of muscovite and phlogopite are otherwise alike. The K$^+$ ions occur as interlayer cations between *t-o-t* layers.

Crystal form: Monoclinic; 2/*m*. Usually in six-sided plates and/or books (Fig. 7.32). Crystals may be large and coarse.

Physical properties: Cleavage {001} perfect. Folia flexible and elastic, **H** = 2½–3; **G** = 2.86. Color yellowish brown, commonly with copperlike reflections from the cleavage surface.

Distinguishing features: Characterized by its micaceous habit and yellowish brown (bronze) color. May be difficult to distinguish from biotite.

Uses: Phlogopite is used in commercial applications similar to those listed for muscovite (Sec. 7.16). Phlogopite is less common in large commercial deposits than is muscovite. The use of dry ground mica (muscovite as well as phlogopite) continues to increase in the manufacture of automotive plastics.

7.18 Biotite: K(Mg, Fe)$_3$(AlSi$_3$O$_{10}$)(OH)$_2$

Occurrence: One of the most common rock-forming silicates. In many granites, it may be the major ferromagnesian mineral (Fig. 9.44). It occurs also in granodiorite, quartz diorite, syenite, nepheline syenite, and pegmatites. In many mafic igneous rocks, it is a late crystallizing product, commonly forming rims around magnetite and ilmenite grains. It is also a common constituent of metamorphic rocks such as schists and gneisses.

Chemical composition and crystal structure: The composition of biotite is similar to that of phlogopite but with considerable substitution of Fe^{2+} for Mg^{2+}. There is also substitution of Fe^{3+}, Ti^{4+}, and Al^{3+} for Mg^{2+}, with coupled substitution of Al^{3+} in the tetrahedral site occupied mainly by Si^{4+}. The trioctahedral structure of biotite is the same as that of phlogopite shown in Figure 7.31.

Crystal form: Monoclinic; 2/*m*. Usually in foliated masses, rarely in tabular or short prismatic crystals with pronounced {001} and pseudohexagonal outline (Fig. 7.33).

Physical properties: Usually in irregular foliated masses. Cleavage {001} perfect. Folia flexible and elastic. **H** = 2½–3; **G** = 2.8–3.2. Color dark green, brown, to black.

Distinguishing features: Black color and excellent micaceous cleavage.

Uses: No commercial application. However, vermiculite, a hydrated alteration product of biotite or phlogopite, is used in a wide variety of thermal and acoustical insulation materials because of its low bulk density. It is also used as a soil additive. When added to soil, vermiculite helps aerate and improve the workability of the soil.

Figure 7.33 Perfect basal cleavage {001} shown by a book of black biotite, with an irregular pseudohexagonal outline.

7.19 Olivine: (Mg, Fe)₂SiO₄

Occurrence: A common mineral in dark-colored (mafic) igneous rocks such as gabbro, basalt, and peridotite (see Figs. 8.8, 8.26, and 9.45). Dunite is essentially a monomineralic rock composed of olivine (see Fig. 3.2). Olivine is considered a major constituent of the Earth's mantle (Fig. 8.1). Relatively pure forsterite (Mg_2SiO_4) occurs in high-temperature metamorphosed siliceous carbonate rocks.

Chemical composition and crystal structure: A complete solid solution series exists between Mg_2SiO_4, forsterite (Fo) and Fe_2SiO_4, fayalite (Fa), with homogeneous intermediate compositions in between (Fig. 8.9). Most common olivines are richer in Mg^{2+} than Fe^{2+}. The structure consists of layers of octahedra (housing Mg^{2+} and Fe^{2+}) cross-linked by independent SiO_4 tetrahedra (Fig. 7.34).

Crystal form: Orthorhombic; $2/m\ 2/m\ 2/m$. Crystals are a combination of various prisms and pinacoids (Fig. 5.15(C)).

Physical Properties: Usually as embedded grains or in granular masses (see Fig. 3.2) but also occurs as euhedral phenocrysts in basaltic rocks (Fig. 8.32). Conchoidal fracture. $H = 6\frac{1}{2}–7$; $G = 3.27–4.37$, increasing with Fe content. Color pale yellow green in forsterite, darker brownish green toward the fayalite end member. A gem-quality, yellow-green crystal of forsterite, known as peridot, is illustrated in Figure 7.35. In peridot, ~10% of the total (Mg + Fe) is Fe.

Distinguishing features: Its granular nature, green color, conchoidal fracture and commonly glassy luster. In thin sections, it shows high birefringence and high relief (see variation diagram for refractive indices, Fig. 4.25).

Uses: A traditional use is as foundry sand for the shaping of metals because of its high melting temperature. Also used as a slag conditioner in iron making and steelmaking, as a raw material in the manufacture of refractories, and as a blast cleaning

Figure 7.35 Large, yellow-green, euhedral crystal of gem-quality forsterite known as peridot.

agent replacing quartz in sandblasting. Peridot is the gem variety of the mineral olivine.

7.20 Zircon: ZrSiO₄

Occurrence: Zircon is a common accessory mineral in granitic igneous rocks. It is also found as detrital grains in sedimentary rocks and as an accessory in metamorphosed sedimentary rocks that originally contained detrital zircon.

Chemical composition and crystal structure: The composition is close to $ZrSiO_4$, with variable but small amounts of hafnium, uranium, thorium, lead, and rare earth elements (REEs). The structure, shown in Figure 7.36, consists of Zr^{4+} in irregular

Olivine: (Mg,Fe)₂SiO₄
$P2_1/n\ 2_1/m\ 2_1/a$

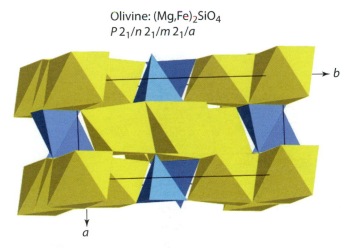

Figure 7.34 Perspective view, looking down the *c* axis, of the structure of olivine. A rectangular cross-section of an orthorhombic unit cell is shown. The octahedra, housing Mg^{2+} and Fe^{2+}, occur in layers parallel to {100}. These layers are cross-linked by independent SiO_4 tetrahedra.

Zircon, ZrSiO₄
$I4_1/a\ 2/m\ 2/d$

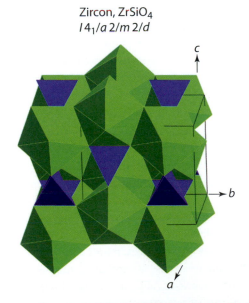

Figure 7.36 Crystal structure of zircon, $ZrSiO_4$. A tetragonal unit cell is outlined. The Zr^{4+} ions are in irregular polyhedra with **C.N.** = 7–8.

Figure 7.37 Small zircon crystals consisting of prisms terminated by distorted dipyramids.

7- to 8-fold coordination polyhedra cross-linked by independent (SiO_4) tetrahedra.

Although the structure of zircon is very resistant to chemical disintegration during the weathering cycle, it is commonly in a **metamict** state. This reflects internal structure damage caused by self-irradiation from the radioactive decay of U and Th that are commonly present in small amounts in the structure. The end result of this continuous structural disintegration is an isotropic glass with a reduction in density of about 16% and a lowering of the original refractive index. In thin sections, under the polarizing microscope, the radiation from uranium and thorium in zircon crystals that are included in other colored minerals, such as biotite and hornblende, damages the structure of the surrounding mineral and produces a halo that changes color as the thin section is rotated. These **pleochroic haloes** are often much larger than the zircon crystal from which the radiation comes.

Crystal form: Tetragonal; $4/m\,2/m\,2/m$. Crystals commonly consist of a combination of a simple prism terminated by a dipyramid (Fig. 7.37). Metamict grains may be bounded by crystal faces, but these reflect the crystal form of the original nonmetamict mineral. Such grains are amorphous pseudomorphs after the earlier crystalline zircon (*pseudomorph* from the Greek words *pseudo*, "false," and *morpho*, "form").

Physical properties: Commonly in small crystals exhibiting a combination of a tetragonal prism with a dipyramid (see Fig. 5.14(C)). Also in irregular grains; cleavage {010} poor. **H** = 7½; **G** = 4.68. Color generally some shade of brown; also colorless, and gray. High luster. A nonmetamict grain may show some cleavage, but the metamict product will show conchoidal fracture instead.

Distinguishing features: Recognized by its characteristic crystal form, color, luster, and high hardness.

Uses: Zircon is an ore of zirconium and hafnium as well as rare earth elements (REEs). Zirconium is used in the construction of nuclear reactors. As an industrial mineral, it is used in the manufacture of refractory bricks and as a foundry sand for the making of castings.

When crystals are of a brownish to red-orange color, and transparent, zircon is used as a gem. Because of its high refractive index and high dispersion, a faceted stone will show exceptional brilliance and fire. Cubic zirconia (CZ) with composition ZrO_2 is a substitute for diamond in inexpensive jewelry. This is a synthetic product made from zirconium extracted from zircon.

7.21 Tourmaline: $(Na, Ca, K)(Fe^{2+}, Mg, Al, Mn, Li)_3\,(Al, Fe^{3+})_6(BO_3)_3(Si_6O_{18})(OH)_3(O, OH, F)$

Occurrence: The most common occurrence of tourmaline is in granite pegmatites and in rocks immediately surrounding them. It occurs as an accessory mineral in granite, granodiorite, and related felsic rocks. It is also a common accessory mineral in metamorphic rocks such as schists and gneisses.

Chemical composition and crystal structure: Tourmaline is a very complex mineral, both structurally and chemically. The structure (Fig. 7.38) consists of six-membered tetrahedral (Si_6O_{18}) rings and three-membered borate (BO_3) rings interconnected by large Na octahedra and Al octahedra.

The most commonly occurring variety known as schorl, is black and has a general formula, $NaFe_3Al_6(BO_3)_3(Si_6O_{18})(O, OH, F)_4$. All the other cations listed in the formula that heads this section substitute for one another in groups that are listed inside parentheses. The great variability in the chemical composition leads to many differently colored varieties that, in the jewelry industry, have resulted in a large number of varietal names based on color (Sec. 15.18).

Crystal form: Hexagonal (trigonal); $3m$. Commonly in crystals with a pronounced trigonal prism outline. May be vertically striated, or may show columnar habit (see Fig. 3.1(B)).

Tourmaline (variety schorl):
$NaFe_3Al_6(BO_3)_3(Si_6O_{18})(O,OH,F)_4$
$R\,3\,m$

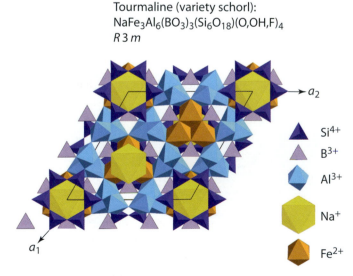

Figure 7.38 Crystal structure of tourmaline in a *c* axis projection, looking down the *c* axis. A primitive hexagonal unit cell is outlined for this rhombohedral structure. See caption for Figure 5.25 for further explanation.

Allanite, $(Ca,Ce)_2(Al,Fe^{2+},Fe^{3+})_3(SiO_4)(Si_2O_7)(OH)$
$P\,2_1/m$

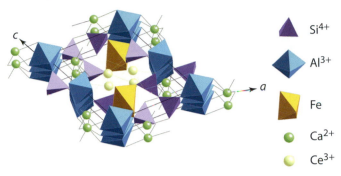

Si^{4+}	
Al^{3+}	
Fe	
Ca^{2+}	
Ce^{3+}	

Figure 7.40 Perspective view, down the *b* axis, of allanite. The SiO$_4$ tetrahedra occur as single tetrahedral (SiO$_4$) units as well as double units (Si$_2$O$_7$). A monoclinic unit cell is outlined.

Figure 7.39 Euhedral black tourmaline crystals, known as schorl, on a quartz matrix.

Physical properties: Crystals commonly show trigonal (3-fold) outline and cross-section. $\mathbf{H} = 7–7½$; $\mathbf{G} = 3.0–3.25$. May be massive, compact, and/or in radiating crystal groups. Conchoidal fracture. Color highly variable. Most commonly black and known as schorl (Fig. 7.39).

Distinguishing features: Usually recognized by the characteristic rounded, triangular cross-section of prismatic crystals, as well as by its conchoidal fracture.

Uses: Tourmaline in clear, transparent, and deeply colored varieties make beautiful gemstones (see Figs. 15.30(C) and 15.32(A)). Examples of several names applied to colored gem varieties by jewelers are as follows: rubellite, red to pink; indicolite, dark blue; verdelite, green; and so on.

7.22 Allanite: $(Ca, Ce)_2(Al, Fe^{2+}, Fe^{3+})_3(SiO_4)$ $(Si_2O_7)(OH)$

Occurrence: Allanite is a common accessory mineral in many granites, granodiorites, syenites, and pegmatites. It is also found in metamorphic rocks such as gneisses and in some limestone **skarn deposits**. Skarn deposits are contact metamorphosed rocks lying in the aureole of an intrusive igneous body.

Chemical composition and crystal structure: The main compositional variation is due to the substitution of Ce, La, Y, and other rare earth elements for Ca^{2+}. Significant amounts of Th and U may also be present substituting for Ca^{2+}. Because of the radioactive decay of these two elements, the allanite structure may become metamict (see Sec. 7.20 for discussion of the metamict state).

The crystal structure of allanite, which is essentially the same as that of epidote (allanite may be referred to as an REE-epidote) is shown in Figure 7.40. It contains both independent (SiO$_4$) tetrahedra and double tetrahedral (Si$_2$O$_7$) groups.

Crystal form: Monoclinic; $2/m$. Crystals may be tabular (Fig. 7.41), prismatic, to acicular. May also be metamict.

Physical properties: $\mathbf{H} = 5½–6$; $\mathbf{G} = 3.5–4.2$. Color light brown to black. Fracture conchoidal to uneven. May be somewhat radioactive.

Distinguishing features: Dark color and occurrence in granitic rocks are suggestive. For positive identification, optical or other analytical techniques may be required.

Uses: If allanite's cerium content is high, it may be mined for this metal. Cerium oxide is used as a decolorizing agent in glass manufacture. Cerium is also used in automotive catalytic converters.

Figure 7.41 Tabular crystal of allanite.

Melilite: $(Ca,Na)_2(Mg,Al)(Si,Al)_2O_7$
$P\bar{4}2_1m$

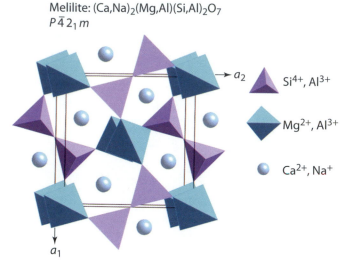

Figure 7.42 Perspective view, along the c axis, of the crystal structure of melilite. A tetragonal unit cell is outlined.

7.23 Melilite: $(Ca, Na)_2(Mg, Al)(Si, Al)_2O_7$

Occurrence: Melilite is found in highly silica-undersaturated alkaline igneous rocks. It also occurs in high-temperature metamorphosed impure dolomites.

Chemical composition and crystal structure: Melilite is a name given to a range of minerals with solid solution from gehlenite, $Ca_2Al(Al, Si)_2O_7$ through åkermanite, $Ca_2MgSi_2O_7$. The structure of a synthetic sodium-rich melilite is given in Figure 7.42.

Crystal form: Tetragonal, $\bar{4}\,2m$. Equant to short prismatic crystals.

Physical properties: H = 5–6; G = 2.9–3.1. Colorless to pale yellow and light brown. Translucent. Conchoidal fracture. Best identified by optical microscopy.

Distinguishing features: In a thin section, it may show anomalous (first order) interference colors (Fig. 7.43).

Uses: None.

Figure 7.43 Photomicrograph of a thin section with two zoned crystals of melilite under crossed polarized light, where the interference colors range from black at the core to anomalous blue on the rim. Width of field is 3.75 mm.

7.24 Magnetite: Fe_3O_4

Occurrence: A common accessory in igneous, sedimentary, and metamorphic rocks. Massive magnetite segregations are found in mafic magmatic bodies. In alkaline and calcalkaline igneous rocks with significant amounts of Fe^{3+}, magnetite tends to crystallize early, forming octahedral phenocrysts, whereas in tholeitic igneous rocks, it crystallizes late and is anhedral or skeletal. Magnetite may be concentrated in the heavy mineral fraction of clastic sedimentary rocks. It is a major constituent of Precambrian banded iron-formation, where it represents a chemical precipitate from an anoxic ocean.

Chemical composition and crystal structure: The composition of most magnetite is close to Fe_3O_4 with only minor substitution of Mg^{2+} and Mn^{2+} for Fe^{2+}, and Al^{3+}, Cr^{4+}, and Ti^{4+} for Fe^{3+}. However, at magmatic temperatures, magnetite forms a complete solid solution series with ulvöspinel, Fe_2TiO_4, so that most igneous magnetite crystallizes with a composition between these two end members. However, on slow cooling, the ulvöspinel component undergoes oxidation to ilmenite and magnetite ($3Fe_2TiO_4 + 0.5O_2 \rightarrow 3FeTiO_3 + Fe_3O_4$). As a result, most igneous magnetite, when examined under the microscope in reflected light, shows ilmenite lamellae in a relatively pure magnetite, Fe_3O_4, host.

Magnetite is one of several members of the oxide structure type referred to as the **spinel group**, with a general composition XY_2O_4 (the mineral spinel is $MgAl_2O_4$). Ulvöspinel is the titanium member of this group. The magnetite structure is illustrated in Figure 7.44. The magnetite formula can be rewritten as $Fe^{3+}(Fe^{2+}, Fe^{3+})O_4$, where Fe^{3+} occupies all of the tetrahedrally coordinated sites and (Fe^{2+}, Fe^{3+}) together are distributed among the octahedral sites. Octahedra are joined along edges to form rows and planes parallel to {111} of the structure, and tetrahedra provide cross-links between layers of octahedra.

Magnetite: Fe_3O_4 or $(Fe^{2+}Fe_2^{3+}O_4)$
$F4_1/d\bar{3}2/m$

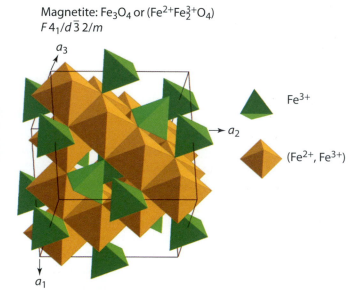

Figure 7.44 Perspective view of the crystal structure of magnetite. A cubic unit cell is outlined.

Figure 7.45 Octahedral magnetite crystals.

Figure 7.47 Massive black chromite with a submetallic luster.

Crystal form: Isometric; $4/m\,\overline{3}\,2/m$. Very commonly in octahedral crystals (Fig. 7.45), more rarely in dodecahedrons (Fig. 5.11(D)).

Physical properties: Commonly in octahedral crystals (Fig. 7.45). $H = 6$; $G = 5.2$. Color is iron black with metallic luster. Strongly magnetic. Black streak.

Distinguishing features: Strong magnetism, black color, hardness, and octahedral crystal form taken together are highly diagnostic.

Uses: Magnetite is commonly a major constituent of iron ores derived from banded iron-formation that are mined in the Precambrian cratonic regions of many continents. It may also occur as a segregation from a silicate magma, as in Kiruna, Sweden. It is a major source of iron for the iron and steel industry.

7.25 Chromite: $FeCr_2O_4$

Occurrence: Chromite occurs as an accessory mineral in mafic igneous rocks such as gabbro, dunite, peridotite (Fig. 8.8), and

Chromite: $FeCr_2O_4$
$F\,4_1/d\,\overline{3}\,2/m$

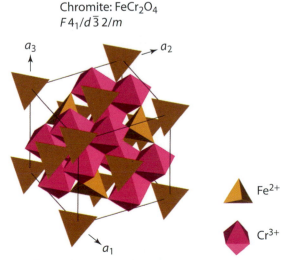

Figure 7.46 Perspective view of the crystal structure of chromite. A cubic unit cell is outlined. This structure is the chromium analogue of magnetite, $Fe^{2+}Fe_2^{3+}O_4$.

pyroxenite and may be concentrated into layers that are mined as chromite ore (Fig. 9.8(B)). Also in serpentinites derived from ultramafic rocks.

Chemical composition and crystal structure: The composition, $FeCr_2O_4$, can be viewed as the chromium analogue of magnetite, $Fe^{2+}Fe_2^{3+}O_4$. Both minerals are members of the spinel group of oxides. The structure of chromite is illustrated in Figure 7.46. Fe^{2+} is housed in the tetrahedral and Cr^{3+} in the octahedral sites.

Crystal form: Isometric; $4/m\,\overline{3}\,2/m$. Octahedra are the most common crystal form.

Physical properties: Crystals are rare; most commonly massive (Fig. 7.47), granular, and compact. $H = 5½$; $G = 4.6$. Color is iron black to brownish black. Luster metallic to submetallic. Streak dark brown.

Distinguishing features: Resembles magnetite and/or ilmenite but is nonmagnetic. Brown streak and submetallic luster are helpful in identification. In a thin section, it is brown on thin edges.

Uses: Chromite is the only ore mineral for chromium metal. Most of its production is used in the making of stainless steel and high-strength alloys. It is also extensively used in the chemical industry in catalysts, corrosion inhibitors, metal plating, pigments, printing chemicals, and tanning compounds. For example, bright yellow pigment, a lead-chromate base, is used in the yellow stripes found on all major streets and highways.

7.26 Hematite: Fe_2O_3

Occurrence: Hematite is a relatively rare constituent of igneous rocks but is an important iron ore extracted from sedimentary banded iron-formations and their metamorphosed equivalents. In igneous rocks, it may be a minor constituent when the magma from which the rock crystallized is rich in Fe^{3+} and poor in Fe^{2+}. Examples of such rock types are granite, syenite, and trachyte.

Chemical composition and crystal structure: Hematite is generally of constant composition, $Fe_2^{3+}O_3$. Its structure, illustrated in Figure 7.48, is identical to that of corundum, Al_2O_3. The Fe^{3+}

Hematite, $Fe_2^{3+}O_3$
$R\bar{3}2/c$

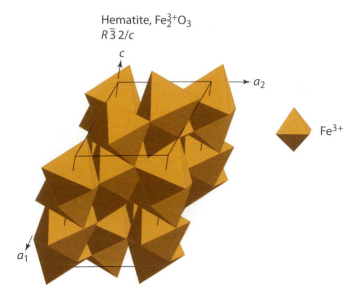

Fe³⁺

Figure 7.48 Perspective view of the crystal structure of hematite, consisting of Fe³⁺ in octahedral coordination. A primitive hexagonal unit cell is outlined for this rhombohedral structure. See the caption for Figure 5.25 for further explanation.

is housed in octahedra that are stacked vertically such that each octahedron shares a face between two adjoining layers. The Fe³⁺ ions in the octahedra will tend to move away from the shared face because of the repulsive electrical forces between them. The oxygens

are in hexagonal closest packing (HCP), with the octahedra occupied by Fe³⁺ in a dioctahedral arrangement (see Sec. 7.16).

Crystal form: Hexagonal; (trigonal) $\bar{3}\,2/m$. Crystals are commonly platy, tabular on {0001}. Also in botryoidal masses (Fig. 7.49).

Physical properties: Crystals may be tabular with rhombohedral modifications. Also platy, micaceous, and botryoidal. When foliated known as specular hematite, or specularite (see Fig. 3.5). **H** = 5½–6½; **G** = 5.3 (for crystals). Color reddish brown to black. Metallic luster in crystals but may be dull in earthy varieties. Red streak.

Distinguishing features: Recognized by its reddish brown to black color. Metallic luster in crystals, hardness, and red streak. In thin sections, under the microscope, it is red, in contrast with magnetite, which is opaque, and ilmenite and chromite, which are brown.

Uses: Hematite is the most important iron ore for the iron and steel industry. It is also a major source for mineral pigments and polishing powder. Black crystals may be cut as gems, which are misleadingly called black diamonds.

7.27 Ilmenite: FeTiO₃

Occurrence: A common accessory mineral in many igneous and metamorphic rocks. Large concentrations may be associated

Figure 7.49 Hematite in two different forms: as fine platy crystals with a metallic luster and as a reddish-brown botryoidal mass. The whitish translucent, glassy crystals are quartz.

Ilmenite: FeTiO₃
$R\overline{3}$

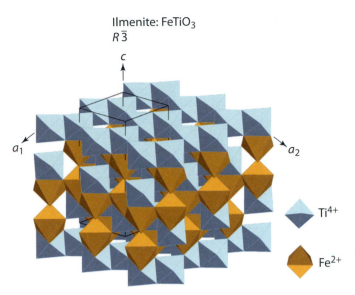

Figure 7.50 Perspective view of the crystal structure of ilmenite. A primitive hexagonal unit cell is outlined for this rhombohedral structure. See caption for Figure 5.25 for further explanation.

with gabbro, norite, anorthosite, and ultramafic rocks. Detrital sediments may also contain ilmenite, and are referred to as "black sands."

Chemical composition and crystal structure: Although an idealized chemical formula is given in the heading here, a more realistic formula would be $(Fe, Mg, Mn)TiO_3$ with limited Mg, and Mn substitution. The structure (Fig. 7.50) is similar to that of hematite (Fig. 7.48) but with Fe and Ti arranged (ordered) in separate, alternating, octahedrally coordinated layers perpendicular to the c axis. Because the alternating layers house two different elements (Fe^{2+} and Ti^{4+}) and not just one element throughout (as in hematite with Fe^{3+}) the overall symmetry of ilmenite ($\overline{3}$) is lower than that of hematite ($\overline{3}\,2/m$).

Crystal form: Hexagonal; (trigonal) $\overline{3}$. Crystals are commonly thick and tabular with prominent basal planes (Fig. 7.51).

Physical properties: Commonly not in good crystals. May be in thin plates. May be massive or compact. Occurs as grains

in sand. **H** = 5½–6; **G** = 4.7. Color iron black. Metallic luster, streak black to reddish brown. May show slight magnetism (due to association with magnetite grains). In thin sections, under the microscope, it has a deep brown color on thin edges.

Distinguishing features: Ilmenite is distinguished from magnetite by its lack of strong magnetism and from hematite by not having a red streak. In thin sections, it can be distinguished from magnetite by its crystal form (thin plates versus octahedra) and on thin edges by being translucent brown, whereas magnetite is always opaque.

Uses: Ilmenite is a major ore mineral for the extraction of titanium. This is used as metal and in alloys because of its high strength-to-weight ratio and great corrosion resistance. Titanium is a major component of aircraft and space vehicles in both engines and frames. Total hip- and knee-joint replacements are made of titanium and/or titanium alloys. A very large use is as a bright white pigment (TiO_2) used in paints and plastics. It has replaced the use of lead in earlier paints. Ilmenite is also used as a filter material in wastewater treatment plants. Much ilmenite is mined from heavy mineral sands, also known as **placer deposits**.

7.28 Rutile: TiO₂

Occurrence: A common accessory mineral in many igneous and metamorphic rocks. Commonly in small grain sizes, which makes identification difficult. Its occurrence in larger crystals is limited to granite pegmatites and quartz veins. May occur as fine, needlelike inclusions in quartz crystals.

Chemical composition and crystal structure: Most commonly the chemical composition is TiO_2. The structure of rutile is shown in Figure 7.52 and consists of Ti^{4+} in octahedral coordination. The structure can be visualized as edge-sharing octahedra that are arranged parallel to the c axis. Octahedra in neighboring chains share corners.

Figure 7.51 Ilmenite in warped plates in a matrix of quartz.

Rutile: TiO₂
$P\,4_2/m\,2_1/n\,2/m$

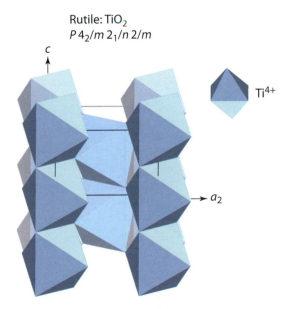

Figure 7.52 Perspective view of the structure of rutile with the c axis vertical. A tetragonal unit cell is outlined.

Figure 7.53 Striated crystal of rutile in the shape of an elbow twin.

Crystal form: Tetragonal; $4/m\,2/m\,2/m$. Commonly as prismatic crystals with vertically striated faces. May be twinned along {011} forming elbow twins (see Fig. 7.53).

Physical properties: Although commonly in small crystals, also massive and compact. Cleavage {110} distinct. $H = 6-6\frac{1}{2}$; $G = 4.2$. Color red, reddish brown to black. Streak pale brown. Submetallic to adamantine luster (adamantine means "brilliant," from the Greek word *adamas*, meaning "diamond").

Distinguishing features: Brilliant luster and red color.

Uses: Rutile is a source for titanium and is used in the applications listed under the ilmenite section. It is mined mainly as detrital grains from placer deposits.

7.29 Uraninite: UO_2

Occurrence: Uraninite may occur as a minor constituent of granite pegmatites with the following mineral association: microcline, quartz, muscovite, zircon, garnet, and allanite. Its most abundant occurrence is in hydrothermal veins and in bedded sedimentary deposits, chiefly conglomerates and sandstones that are essentially unmetamorphosed.

Chemical composition and crystal structure: Uraninite may contain variable amounts of thorium and a complete solid solution series exists between UO_2 and ThO_2, thorianite. Uraninite is also always partially oxidized with a composition ranging from UO_2 to U_3O_8 ($= U^{4+}O_2 + 2\,U^{6+}O_3$).

The structure of uraninite is illustrated in Figure 7.54, with the oxygen ions at the center of a tetrahedral group of metal ions. The structure of thorianite is the same.

Crystal form: Isometric; $4/m\,\overline{3}\,2/m$. Rarely in good crystal form (Fig. 7.55) as octahedra with subordinate dodecahedral and cube faces. Most commonly in the massive variety known as pitchblende.

Physical properties: Rare crystals are octahedra. Most commonly massive, pitchblende. $H = 5\frac{1}{2}$; $G = 7.6-9.7$ (for crystals); $G = 6.5-9$ (for pitchblende). Color black; luster metallic to dull; streak brownish black. Radioactive.

Distinguishing features: A combination of properties such as black color, high specific gravity, and radioactivity (as measured with a Geiger or scintillation counter; Sec. 3.7) may lead to its proper identification.

Uses: Uraninite and pitchblende are major sources of uranium used as a fuel in nuclear reactors that produce a considerable part of global electricity. Also used in nuclear warheads.

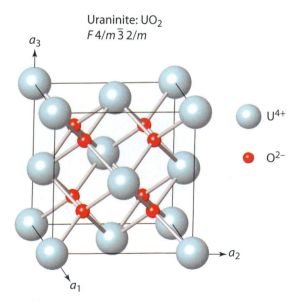

Uraninite: UO_2
$F\,4/m\,\overline{3}\,2/m$

a_3

a_2

a_1

U^{4+}

O^{2-}

Figure 7.54 Perspective, ball and stick representation of the crystal structure of uraninite. A cubic unit cell is outlined.

Figure 7.55 Octahedral crystal of uraninite on quartz. The four small modifying faces at the front tip of the octahedron are those belonging to a dodecahedron.

7.30 Pyrite: FeS$_2$

Occurrence: Pyrite is an accessory mineral in many igneous, sedimentary, and metamorphic rocks. In mafic intrusive complexes, it may form magmatic segregations. It is common in shales and in contact metamorphic deposits. It is abundant in many hydrothermal ore deposits.

Chemical composition and crystal structure: Pyrite's composition is generally FeS$_2$ with only minor amounts of Ni and/or Co. Its structure is shown in Figure 7.56, in which each iron atom is surrounded by six sulfur atoms in octahedral coordination. This type of structure is very similar to that of halite, NaCl (Fig. 10.22). FeS$_2$ also occurs as the less common polymorph marcasite.

Crystal form: Isometric; $2/m\,\overline{3}$. Commonly in crystals with the most common crystal forms the cube, pyritohedron (Fig. 7.57), and octahedron. Also massive.

Physical properties: Frequently in crystals. Cubes may show striated faces. **H** = 6–6½; **G** = 5.0. Brittle. Conchoidal fracture. Color pale brass yellow. Metallic luster.

Figure 7.57 Pyritohedron of pyrite with encrustations of small euhedral, cubic crystals of pyrite on two crystal faces.

Distinguishing features: May be confused with chalcopyrite, but pyrite has a paler yellow color and higher hardness (**H** = 6–6½ is unusually hard for a sulfide). Although it is commonly referred to as fool's gold, it is distinguished from gold by its brittleness and much greater hardness. Marcasite is somewhat paler in color and has a different crystal form.

Uses: Pyrite, because of its high sulfur content (about 53 weight % S) is a source for the production of sulfuric acid, H$_2$SO$_4$, which is used in a variety of industrial and medicinal products and in the production of fertilizer. Most sulfur is mined from native sulfur deposits in volcanic regions and salt domes. In some countries, pyrite is mined as a source of iron as well, if iron oxide deposits are not available. Pyrite may also be mined for the gold and copper associated with it.

7.31 Pyrrhotite: Fe$_{1-x}$S

Occurrence: Commonly present in mafic and ultramafic igneous rocks. Also in pegmatites and contact metamorphic rocks. A very common constituent of hydrothermal sulfide deposits.

Chemical composition and crystal structure: Most pyrrhotites have a deficiency of Fe with respect to S, as shown by the formula Fe$_{1-x}$S, with x ranging from 0 to 0.2. This is known as **omission solid solution** (see Sec. 4.6.2 for specific discussion of the structural vacancies in the pyrrhotite structure).

The structure of pyrrhotite is shown in Figure 7.58. The sulfur atoms are arranged in hexagonal closest packing (HCP) with Fe atoms occupying octahedral interstices between the sulfur layers. In a composition such as Fe$_7$S$_8$, one-eighth of the octahedral sites are vacant.

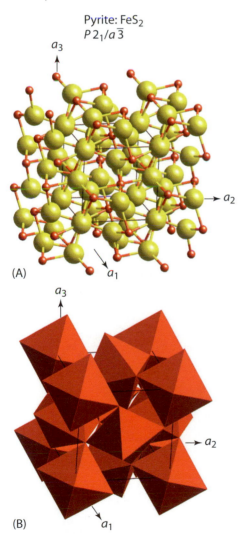

Pyrite: FeS$_2$
$P2_1/a\,\overline{3}$

(A)

(B)

Figure 7.56 Two perspective views, with the a_3 axis vertical, of the crystal structure of pyrite. A cubic unit cell is outlined in both illustrations. **(A)** A ball and stick representation and **(B)** a polyhedral image of the pyrite structure.

Pyrrhotite: $Fe_{1-x}S$
$C\,2/c$

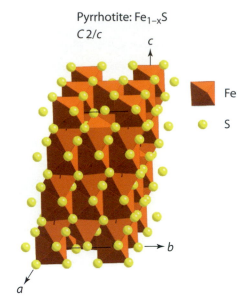

Figure 7.58 Perspective view of the crystal structure of pyrrhotite. A monoclinic unit cell is outlined.

Chalcopyrite: $CuFeS_2$
$I\bar{4}2d$

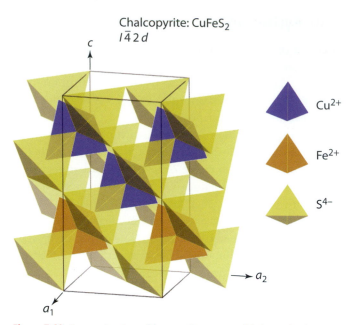

Figure 7.60 Perspective view of the crystal structure of chalcopyrite. A tetragonal unit cell is outlined.

Crystal form: Monoclinic; $2/m$ (pseudohexagonal). Crystals commonly pyramidal (Fig. 7.59). Generally massive.

Physical properties: Generally massive. $H = 4$; $G \sim 4.6$. Color brownish bronze. Metallic luster. Somewhat magnetic.

Distinguishing features: Recognized by its commonly massive occurrence, bronze color, and low-intensity magnetism.

Uses: It may be mined for the copper, nickel, and platinum contents that are housed in associated sulfide minerals.

7.32 Chalcopyrite: $CuFeS_2$

Occurrence: Chalcopyrite is an accessory mineral in some mafic igneous rocks. It is the most common copper-bearing mineral in hydrothermal ore deposits.

Chemical composition and crystal structure: The chalcopyrite composition is essentially $CuFeS_2$ with little variation. Its crystal structure is shown in Figure 7.60. The Cu^{2+} and Fe^{2+} ions are in tetrahedral coordination with S, and the S^{4+} ions are tetrahedrally coordinated by two Cu^{2+} and two Fe^{2+} ions.

Crystal form: Tetragonal; $\bar{4}2m$. Occurs as tetragonal disphenoidal crystals (Fig. 7.61). Most commonly massive.

Physical properties: When in crystals, they are tetrahedral in appearance. $H = 3\frac{1}{2}-4$; $G = 4.1-4.3$. Color brass yellow. Metallic luster. Streak greenish black. Brittle.

Figure 7.59 Dipyramidal crystals of pyrrhotite among smaller, almost black, crystals of Fe-rich sphalerite, and some white quartz overgrowths.

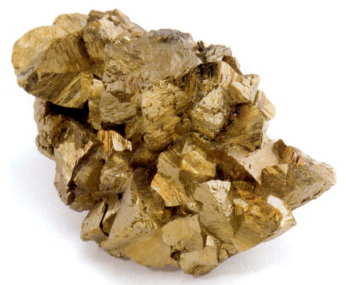

Figure 7.61 Brass yellow tetragonal disphenoidal crystals of chalcopyrite.

Distinguishing features: Recognized by its brass-yellow color. Distinguished from pyrite by being softer and from gold by its brittleness.

Uses: Chalcopyrite is the major ore mineral of copper that is used in most electrical wiring and in the construction industry. Copper is essential to the global infrastructure for industrial manufacturing as well as social and commercial activities.

7.33 Apatite: $Ca_5(PO_4)_3$ (OH, F, Cl)

Occurrence: Because apatite is the only common phosphorus-bearing mineral, it is a ubiquitous accessory mineral in igneous, sedimentary, and metamorphic rocks. Its abundance is simply a function of the abundance of phosphorus in the bulk rock. It is rare in most ultramafic rocks but may be present in felsic

Figure 7.63 Green, hexagonal prismatic apatite crystal with pink calcite.

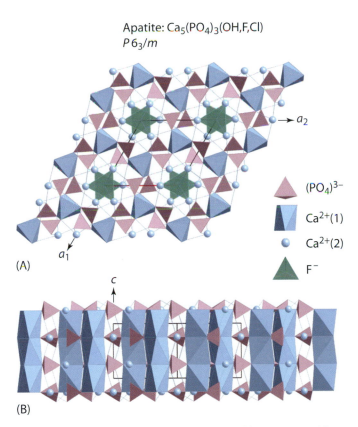

Apatite: $Ca_5(PO_4)_3(OH,F,Cl)$
$P6_3/m$

(PO₄)³⁻ ... $(PO_4)^{3-}$
Ca²⁺(1) ... $Ca^{2+}(1)$
Ca²⁺(2) ... $Ca^{2+}(2)$
F⁻ ... F^-

(A)

(B)

Figure 7.62 Crystal structure of apatite. **(A)** A view of the structure straight down along the c axis. The outline of a primitive hexagonal unit cell is shown. **(B)** A view straight down the direction of the a_3 axis with the c axis vertical.

igneous rocks and pegmatites. Hydroxyapatite, $Ca_5(PO_4)_3(OH)$, is the main constituent of bones and teeth.

Chemical composition and crystal structure: The main compositional variation in apatite is in the proportion of OH, F, and Cl. The crystal structure of apatite (Fig. 7.62) consists of tetrahedral $(PO_4)^{3-}$ groups bonded laterally by Ca^{2+}. Such calcium ions occur in two different structural sites. Half are in distorted 6-fold coordination, and the other half are in 7-fold coordination with oxygen from the phosphate groups and additional (OH, F, Cl). The irregularly 6-fold coordinated calcium ions appear as vertical columns in Figure 7.62(B). Each fluorine atom is surrounded by three Ca atoms at the same level, which results in triangular coordination.

Crystal form: Hexagonal; $6/m$. Commonly in crystals with long prismatic habit, as seen in Figure 7.63. Also in granular to compact masses.

Physical properties: Common in long prismatic, hexagonal crystals. **H** = 5; **G** ~3.2. Color usually some shade of green, brown, or blue; also colorless. Vitreous luster.

Distinguishing features: Apatite is usually recognized by its crystal shape (Fig. 7.63), color, and hardness (**H** = 5 can just be scratched by a knife).

Uses: Crystallized apatite has been a major source of phosphate for fertilizer, but nowadays phosphorite deposits supply most of the phosphate mined for fertilizer production (see Sec. 10.16).

Summary

This chapter concentrated mainly on the mineralogical makeup of igneous rocks. But before we began the systematic descriptions of igneous minerals, we introduced three concepts that apply to all minerals and rock types: (1) a discussion of the distinction between major, minor, and trace elements; (2) the recalculation scheme for arriving at a mineral formula from its chemical analysis; and (3) the plotting of mineral compositions on triangular compositional diagrams. Aspects of these three concepts are used throughout this text.

We also introduced the standard, universally accepted structural classification of silicates. This allows for the organization of silicates into subgroups that have distinctly different atomic structural arrangements, as follows:

- Framework structures (or tectosilicates)

- Chain structures (inosilicates or chain silicates)

- Sheet (or layer) structures (phyllosilicates or layer silicates)

- Structures with independent (SiO_4) tetrahedra (also known as nesosilicates or orthosilicates)

- Structures with double tetrahedral (Si_2O_7) groups (also known as disilicates)

- Ring structures (also known as cyclosilicates).

These different atomic arrangements can be seen in the various illustrations of crystal structures of silicates that appear throughout this text and in CrystalViewer.

In the systematic treatments of the 29 igneous minerals (and in subsequent chapters on systematic mineralogy), after the mineral name and formula, the first entry is "occurrence." This lists various rock types in which the mineral under discussion is commonly found.

This is followed by "chemical composition and crystal structure," which gives a short synopsis of the range of composition and the crystal structure of the mineral at hand.

After that come three entries: "crystal form," "physical properties," and "Distinguishing features." These three entries combined introduce physical properties as seen in hand specimens (with a few observations that relate to thin sections) that are basic to mineral identification.

The last section is that of "use." This describes the usage of a specific mineral in commercial applications. This reflects the large role of Earth materials in the economics of the world. In case you wish to pursue in greater depth the occurrence and application of a specific industrial mineral, consult one of the two editions of *Industrial Mineral and Rocks*, 1994 and 2006, listed in "Further Reading."

Review questions

1. What are the differences between major, minor, and trace elements?

2. In what components are the chemical analyses of silicates most commonly expressed?

3. Recalculate the following chemical analysis of an alkali feldspar into its formula: $SiO_2 = 65.67$, $Al_2O_3 = 20.84$, $CaO = 0.50$, $Na_2O = 7.59$, and $K_2O = 5.49$ weight %. Remember to look up the number of oxygens per formula unit for feldspar.

4. On a triangular diagram, corners represent the feldspar end members $KAlSi_3O_8$ (at the top), $NaAlSi_3O_8$ (at the lower left), and $CaAl_2Si_2O_8$ (at the lower right). On this diagram, plot the following compositions: $Or_{90}Ab_{10}$; $Or_{20}Ab_{80}$; $Ab_{40}An_{60}$; and $Ab_{50}Or_{10}An_{40}$.

5. On a triangular diagram having corners representing $CaSiO_3$ (top), $MgSiO_3$ (lower left) and $FeSiO_3$ (lower right), plot the end-member compositions of diopside and hedenbergite. Also plot the following: $En_{80} Fs_{20}$; $Ca(Mg_{0.5} Fe_{0.5}) Si_2O_6$; $Ca(Mg_{0.8} Fe_{0.2}) Si_2O_6$; also $Wo_{48} En_{44} Fs_8$ and $Wo_{50} En_{20} Fs_{30}$.

6. What are the three polymorphs of K-feldspar?

7. How do these three differ in atomic structure, and what is the main factor that controls these structures in the first place?

8. What are the end-member formulas of the plagioclase feldspar series?

9. What is a highly diagnostic physical property of the plagioclase series?

10. The most common occurrence of SiO_2 is as quartz. What are the names of the other four polymorphs?

11. Which of these represent high-temperature forms of SiO_2, and which represent high-pressure forms?

12. What are three colored gem varieties of quartz?

13. How do the crystal structures of nepheline and leucite differ from those of the feldspars?

14. To what silicate group do nepheline and leucite belong?

15. What are three common igneous pyroxenes?

16. How do the structures of pyroxenes and amphiboles differ?

17. What are three common igneous silicates that tend to be dark green in color?

18. What are three common igneous members of the mica group?

19. What distinguishes the crystal structures of micas from those of the pyroxenes and amphiboles?

20. The *t-o-t* layer in muscovite is described as dioctahedral; that in phlogopite is described as trioctahedral. What is the difference?

21. Give the end-member compositions of the olivine series. What is meant by the gem name *peridot*?

22. Give the chemical formula for zircon. Many zircons are described as *metamict*. What is meant by that?

23. The mineral tourmaline is most commonly black (schorl), but several gem varieties have very different colors. Why do tourmalines have so many different colors?

24. What is a unique aspect of the crystal structure of tourmaline?

25. What are the names of three metallic oxides that may be present in igneous rocks as accessory minerals?

26. What are the names of two sulfides that occur as accessory minerals in igneous rocks?

27. What is the name of the symbolic notation that is printed above the illustrations of the various crystal structures?

28. How do you derive the notation of the crystal class (or point group) from what you answered in question 27?

FURTHER READING

The following eleven books give many useful mineralogical references in which you can locate additional mineral data that are not recorded in this text. Three Web site addresses lead to vast mineral databases. The *Industrial Minerals and Rocks* are two references on industrial minerals and rocks. These are extremely helpful for data on geological occurrence, mining, markets, and applications for a specific industrial mineral.

Anthony, J. W., Bideaux, R. A., Bladh, K. W., and Nichols, M. C. (1995–2003). *Handbook of Mineralogy*, 5 vols., Mineral Data Publishing, Tucson, AZ.

Carr, D. M., ed. (1994). *Industrial Minerals and Rocks*, 6th ed., Society for Mining, Metallurgy, and Exploration, Littleton, CO.

Deer, W. A., Howie, R. A., and Zussman, J. (1993). *An Introduction to the Rock-Forming Minerals*, 2nd ed., Longman Group, UK Limited, Essex, U.K.

Deer, W. A., Howie, R. A., and Zussman, J. (1982–2009). *Rock-Forming Minerals*, vol. 1A, 1B, 2A, 2B, 3A, 3B, 4B, and 5B, Geological Society, London.

Dyar, M. D., Gunter, M. E., and Tasa, D. (2008). *Mineralogy and Optical Mineralogy*, Mineralogical Society of America, Chantilly, VA.

Gaines, R. V., Skinner, H. C., Foord, E. E., Mason, B., and Rosenzweig (1997). *Dana's New Mineralogy*, John Wiley and Sons, New York.

Klein, C., and Dutrow, B. (2008). *Manual of Mineral Science*, 23rd ed., John Wiley and Sons, New York.

Kogel, J. E., Trivedi, N. C., Baker, J. M., and Krukowsk, S. T., eds. (2006). *Industrial Minerals and Rocks: Commodities, Markets, and Uses*, 7th ed., Society for Mining, Metallurgy, and Exploration, Littleton, CO.

Mason, B., and Moore, C. B. (1982). *Principles of Geochemistry*, 4th ed., John Wiley and Sons, New York.

Mineralogical Society of America, http://www.minsocam. org/MSA/ (lists on its home page *Handbook of Mineralogy* by J. W. Anthony, R. A. Bideaux, K. W. Bladh, and M. C. Nichols, a five-volume set with encyclopedic data for all known mineral species; also the MSA Crystal Structure Database compiled by Bob Downs and Paul Heese).

Mineralogy Database, http://webmineral.com.

Mineralogy Database, http://www.mindat.org.

Nesse, W. D. (2000). *Introduction to Mineralogy*, Oxford University Press, New York.

Nesse, W. D. (2004). *Introduction to Optical Mineralogy*, 3rd ed., Oxford University Press, New York.

8 How do igneous rocks form?

Igneous rocks are the most abundant rocks in the Earth's crust, formed by the crystallization of melts that have risen from the planet's interior. The crust itself has formed over an extended period of time, in fact billions of years, by just such crystallization. The outer half of the planet is normally solid, so it is natural to wonder where magmas come from. Where does this melting take place, and why do normally solid rocks become molten? What composition do rocks have in the source region? Is melting partial or total? What magma compositions are formed when these rocks melt? What makes magma rise toward the surface, and how fast does it travel? As magma ascends and begins to cool, it becomes a mixture of liquid and crystals that can separate from one another and change the composition of the magma. Eventually, magma cools and solidifies to form igneous rock. The processes leading to the formation of this rock are of great importance because they have controlled the differentiation of our planet. The composition of the crust is very different from that of the planet as a whole, and life as we know it would be very different if these processes had not taken minor elements from the Earth's interior and concentrated them in the crust. In this chapter, we look into each of these igneous rock-forming processes. We leave a discussion of the diversity and classification of igneous rocks to Chapter 9.

Heat from the Earth's mantle is evident in this photograph of basaltic lava issuing from the front of a flow on the south coast of the Big Island of Hawaii. The molten lava cools rapidly by radiation to produce a ropy surface. (Photograph courtesy of Brian R. Elms.)

Igneous rocks are those formed by the solidification of molten rock. This molten material, which we call magma, is formed at depth in the Earth and rises toward the surface, where it cools and solidifies, either beneath the surface, where it usually has time to crystallize, or on the surface as volcanic rocks, where cooling may be rapid enough to form glass. We classify igneous rocks on the basis of the minerals they contain, which are determined by the composition of the magma. Magma compositions are determined in the source region by the nature of the rock that undergoes partial melting, but they can be modified during ascent and solidification, especially in large magma chambers, where solidification can take thousands of years. Throughout geologic time, the rise of magmas from the mantle has slowly generated the Earth's crust, whose composition is, therefore, determined by the composition of magmas.

We can think of an igneous rock as the end product of a lengthy series of processes, all of which play important roles in determining the rock's composition, its appearance and mineral makeup, and the shape and position of igneous bodies it forms. It is these processes that we discuss in this chapter, leaving the mode of occurrence and classification of igneous rocks to Chapter 9. Many things can happen to magma between its formation as a partial melt in the source region and its eventual solidification to form a rock. In Box 8.1, we trace the three main types of magma from their sources to their final resting places. These magmas solidify to form the important rock types, basalt, andesite, and granite, which were introduced in Chapter 2. The first two of these are commonly associated with divergent and convergent plate boundaries, respectively. Most granitic magmas are generated near the base of the continental crust, where

BOX 8.1 | FROM MAGMA SOURCE TO IGNEOUS ROCK

In this cross-section through the lithosphere, we trace the three main types of magma, basaltic, andesitic, and granitic, from their source regions to their final place of solidification.

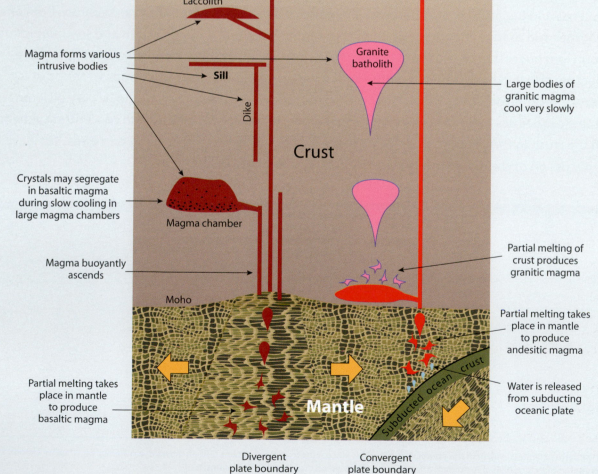

they are heated by deep burial and by the intrusion of mantle-derived magma. They are common at convergent boundaries involving continental crust.

Most magmas have their source in the upper mantle or lower crust, which normally is completely solid but under certain circumstances can undergo partial melting (there is insufficient heat to cause total melting). Most partial melting is caused by tectonic plate divergence or convergence, but some can occur at hot spots over mantle plumes (Fig. 1.7(E)). At divergent plate boundaries, the rise of the asthenosphere causes melting as a result of decreasing pressure to form basaltic magma, and at convergent plate boundaries, the release of water from subducting plates causes melting, with formation of andesitic magma. Most granitic magma is formed in the lower continental crust, where it is heated by magma rising from the mantle.

All magmas are less dense than the solid rock from which they form, and consequently they tend to rise toward the surface. At first, the partial melt is dispersed throughout the source rock and must segregate to form larger bodies before it can rise at significant rates. In the plastic mantle and lower crust, these segregation bodies may have lens or inverted drop shapes, but on rising into the brittle crust, sheetlike bodies known as dikes become common, especially near divergent plate boundaries, where they are oriented parallel to the plate boundary. Near the surface, horizontal sheetlike intrusions known as sills may form, and very near the surface these may dome upward to form mushroom-shaped laccoliths. Once granitic magma has segregated from the partially melted lower crust, it tends to rise in large diapiric domes, which form huge bodies known as batholiths. When magma erupts on the surface, it forms volcanoes whose shapes are determined by how fluid the lava is and how much explosive activity occurs. Fluid basaltic magma forms flat shield volcanoes, whereas less fluid andesitic magma forms steep conical volcanoes that contain a considerable amount of ash produced by explosive release of gases.

As magma rises into the crust, it encounters colder rocks and begins to cool and solidify. In thin sheets, this may occur rapidly (hours), but in large magma chambers, it may take thousands of years. When magma slowly solidifies, crystals can separate from the liquid, which can produce igneous rocks that are either richer in early crystallizing minerals or enriched in the residual liquid. This leads to fractional crystallization and adds to the great diversity of igneous rocks.

In this chapter, we examine all of these processes that govern the composition of magma and lead to the diversity of rock types that we deal with in Chapter 9.

8.1 Why, and how, does solid rock become molten?

Although the Earth's interior is hot (Sec. 1.5), the crust and mantle are not normally molten. The passage of seismic shear waves (S waves) through the crust and mantle indicates that this part of the Earth is mostly solid, as S waves cannot pass through a liquid.

Only below the mantle-core boundary are S waves eliminated, which indicates that the outer part of the core is liquid (Fig. 1.5). The liquid in the outer core is composed largely of iron and nickel and cannot be the source of magmas, which are predominantly silicate melts. The relatively low seismic velocity of the **asthenosphere**, which extends from depths of 70 to 250 km, is likely due to the presence of a thin film of melt along grain boundaries. This melt is trapped in the rock by surface tension and cannot rise as magma toward the surface, but it does weaken the rock and creates the zone on which lithospheric plates move.

Extrusion of lava from volcanoes, however, indicates that magma does exist at certain places at certain times. If the crust and mantle are normally solid, what causes melting? In the following sections, we answer this question by discussing what rocks are present in the source region and at what temperatures and pressures they melt.

8.1.1 Composition of the upper mantle

What types of rock occur in the region where magma is generated? Magma has two major sources, the upper mantle and the lower crust. **Basaltic magma** comes from the upper mantle, and most, but not all, **granitic magma** comes from the lower crust (see Sec. 2.7 for definitions of granite and basalt). For the moment, we restrict our discussion to the Earth's most abundant magma, basalt, and leave to Section 8.6.2 the origin of granitic magma.

Beneath active basaltic volcanoes, earthquake tremors are associated with the ascent of magma from depths of ~60 km to the surface. On rare occasions, when eruptions are rapid or even explosive, fragments of wall rock are brought to the surface and provide samples of the rocks through which the magma ascended (Fig. 8.1(A)), in some cases bringing up diamonds from depths of at least 200 km. The fragments of wall rock are known as **xenoliths**, from the Greek word *xeno*, meaning "foreign."

The most common type of mantle xenolith is a rock known as **peridotite**, which consists essentially of three minerals, olivine, and ortho- and clinopyroxene (Fig. 8.1(B)–(C)). Mantle peridotite differs from crustal peridotite in that it does not contain plagioclase feldspar, which is commonly a minor constituent of crustal peridotites. Instead, it contains small amounts of either spinel ($MgAl_2O_4$) or garnet ($Mg_3Al_2Si_3O_{12}$). Experiments show that plagioclase in peridotite reacts to form spinel and then garnet with increasing pressure. We conclude that the upper mantle in the region from which basaltic magma is derived is composed of either spinel or garnet peridotite, depending on the depth of the source region.

8.1.2 Melting range of upper mantle peridotite

Mantle peridotite, being a rock consisting of several minerals, melts over a temperature range of about 600°C. We are interested mainly in the temperature at which melting begins, which is known as the **solidus**, because it is there that the first melt appears from which magma can form. Figure 8.2 shows the experimentally determined solidus for a typical mantle peridotite under anhydrous (no water present) and water-saturated (excess water

Figure 8.1 (**A**) Breccia containing a mixture of fragments derived from the mantle (dark) and the crust (light colored) in an explosive volcanic vent known as a **diatreme**, from Île Bizard, near Montreal, Quebec. (**B**) Photomicrograph under plane polarized light of a thin section of the dark mantle peridotite nodules shown in (A) consisting of relatively colorless olivine (Ol), pale green clino- and orthopyroxene (Py), and deep brown chromium-bearing spinel (Sp). (**C**) Same field of view as in (B) but under crossed polars.

present) conditions. It also shows the **liquidus**, which is a line indicating the conditions under which the rock becomes totally liquid. At atmospheric pressure and under anhydrous conditions, melting begins at about 1130°C, but with increasing pressure (and hence depth) the solidus rises rapidly, and at a depth of 100 km it is at 1500°C. The two cusps on the solidus at depths of 20 km and 70 km result from the change from plagioclase-, to spinel-, to garnet-bearing peridotite with increasing pressure.

Melting experiments in the presence of excess water indicate melting also begins at 1130°C at atmospheric pressure, but as pressure increases, the melting point decreases at first, because the water dissolves in the melt but then rises steadily when the melt can dissolve no more water (long dashed line in Fig. 8.2). If water is present in excess (water saturated), the solidus temperature at a depth of 100 km is only 1000°C, which is 500°C lower than under anhydrous conditions.

Latent heat of fusion

The beginning-of-melting curves shown in Figure 8.2 are the curves that must be crossed if mantle peridotite is to melt and form magma. However, when solid is converted to liquid at constant temperature, an amount of heat known as the **latent heat of fusion** is required, which for rocks is about 400 kJ/kg (kilojoules/kilogram). Figure 8.3 illustrates the magnitude of this hidden heat sink. Approximately 1200 kJ is needed to heat a kilogram of crystalline basalt from room temperature to the molten state. Fully one-third of this amount of heat is needed not for heating but for converting the solids to liquid.

If a limitless supply of heat were available, mantle peridotite would continue melting as its temperature rose above the solidus and would become completely molten at the liquidus. The liquid composition at this temperature would be the same as that of the whole rock, a peridotite liquid. However, the latent heat of fusion prevents the temperature from rising much above the solidus, and the rock only partially melts. The liquid formed at the solidus is compositionally very different from peridotite. Peridotite contains about 40 weight % silica, but the liquid that is formed at the solidus at a depth of 150 km is basaltic in composition and contains almost 50 weight % silica. **Partial melting**, therefore, gives rise to magma of very different composition from the source rock.

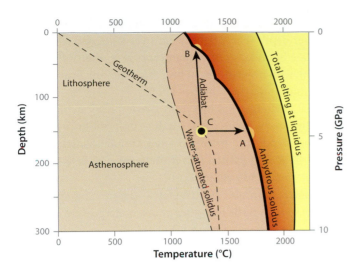

Figure 8.2 The beginning of melting (solidus) and total melting (liquidus) of mantle peridotite as a function of pressure and depth. The thick solid line indicates the solidus of mantle peridotite when no water is present (anhydrous), and the long dashed line is the solidus when water is present in excess (water saturated). The short dashed line represents a typical geotherm. The arrows labeled A and B represent paths by which a point on the geotherm at a depth of 150 km might reach the solidus and begin melting. Point C does not change pressure or temperature but begins melting when the solidus is lowered by the addition of water. Drawn from data presented by Takahashi (1986) and Kushiro et al. (1968).

The latent heat of fusion of rocks is so large that it has played a pivotal role in the cooling and differentiation of the planet. We can emphasize this with three important conclusions:

- The latent heat of fusion of rocks is so large that melting of rock rarely allows the temperature in the Earth to rise far above the solidus.
- Because the temperature rarely rises far above the solidus, the liquids formed are generally the **lowest possible melting fraction**.
- The Earth's crust has formed from these low-melting fractions, which have risen as magma to form igneous rocks.

8.1.3 Geothermal gradient and the geotherm

The temperature distribution in the Earth is known as the **geotherm** (Fig. 8.2). Near-surface temperature gradients are about 25°C/km. Below about 100 km, the lithosphere passes into the asthenosphere, where convection flattens the gradient to about 0.6°C/km. The important conclusion we draw from the position of the normal geotherm is that it does not intersect the anhydrous solidus of mantle peridotite, which is why this part of the Earth is normally solid. If the mantle were water saturated, melting could begin between a depth

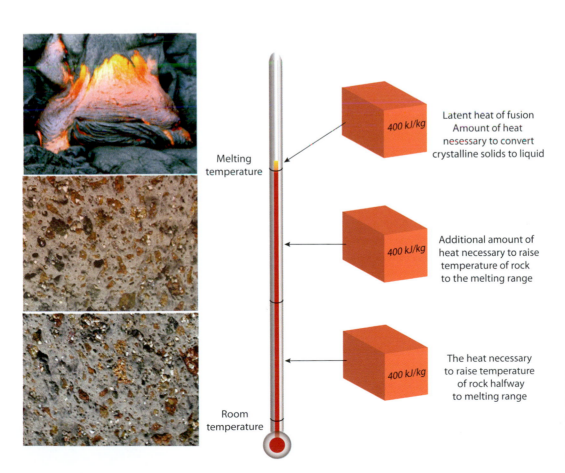

Figure 8.3 The latent heat of fusion of most rocks (Hawaiian basalt in this example) is approximately one-third of the heat needed to raise a rock from room temperature to a completely molten state.

of 100 km and 300 km. Very little water is present in the mantle, so it is unlikely that we could ever have water-saturated melting in the mantle. However, if small amounts of water are present, some lowering of the solidus could occur, which would explain the low seismic velocity of the asthenosphere. Carbon dioxide, another gas in the mantle, may also affect the properties of rocks at this depth, but it lowers melting temperatures far less than does water.

8.2 Three primary causes of melting and their plate tectonic settings

In the previous section, we saw that the normal geotherm is not high enough to reach the anhydrous solidus of mantle peridotite, and consequently magma cannot be generated. For example, at a depth of 150 km, the mantle is about 400°C below the anhydrous solidus of peridotite. However, beneath Hawaii, midocean ridges, and any number of large volcanoes above subduction zones melting is clearly taking place today. How do the conditions in these regions differ from the normal conditions portrayed in Figure 8.2? Melting can be achieved in three different ways labeled A to C in Figure 8.2: by raising the temperature, by lowering the pressure, and by lowering the melting temperature by the addition of water. All three of these changes occur and are brought about by plate tectonic processes. We discuss each of these in order.

8.2.1 Raising the temperature of mantle peridotite to the melting range over hot spots

If we were asked to melt a rock in the laboratory, we would heat it in a furnace, but in nature no heat sources are normally available to raise the temperature above the geotherm. However, if mantle peridotite were moved by plate motion over a mantle plume bringing heat up from the core-mantle boundary (Sec. 1.6.4), its temperature might rise sufficiently to cross the solidus and cause melting (path A in Fig. 8.2). This, in part, is what happens beneath Hawaii, for example. The heating of the mantle by the plume does not cause the temperature to continue rising once it crosses the solidus, because the change in state from solid to liquid consumes all the excess heat (latent heat of fusion).

8.2.2 Decompression melting at divergent plate boundaries

When the lithosphere at divergent plate boundaries is stretched, rift valleys form at the surface, and the asthenosphere rises from below. The rising asthenosphere does not have time to lose heat (**adiabatic**), but the pressure decreases as it approaches the surface (path B in Fig. 8.2). Because of the slope on the solidus, the rising asthenosphere eventually

intersects the solidus and melting takes place but at a much shallower depth than might occur over a hot spot. At divergent plate boundaries, melting occurs as a result of decompression, with no addition of heat. This melting process is by far the most important on the planet.

Partial melting resulting from decompression of mantle peridotite produces liquids of basaltic composition, but at this shallower depth, the basaltic liquids contain slightly higher silica (~52 weight %) than do those formed at greater depth over hot spots (<50 weight %).

Decompression also contributes to the melting associated with hot spots. Mantle rising in a plume undergoes decompression, and if the mantle is already hot, the chances of encountering the solidus are increased by the decompression.

8.2.3 Fluxing with water at convergent plate boundaries (subduction zones)

In Section 8.1.2, we saw that the water-saturated solidus of mantle peridotite is much lower than the anhydrous solidus (Fig. 8.2). At a depth of 150 km, it is 500°C lower, which is sufficient for the solidus to intersect the geotherm and cause melting. The mantle, however, normally contains very little water, so that if water is going to lower the solidus significantly, there must be a way of introducing it at depth. This happens at convergent plate boundaries.

Enormous volumes of ocean water circulate through newly formed igneous rocks at midocean ridges. This water not only cools the rocks but also alters many of the primary igneous minerals to hydrous minerals. At convergent plate boundaries, these hydrothermally altered rocks are subducted back into the mantle (see Fig. 2.21), and as the pressure on them rises, metamorphic reactions take place that release the water, which rises into the overlying mantle wedge. This water acts as a flux and allows melt to form at the water-saturated solidus (Fig. 8.2(C)).

The melting process consumes all the available water. If the melt does not remain water saturated, the solidus temperature rises and further melting ceases. The production of melt, consequently, depends on the rate of release of water from the subducting slab, which in turn depends on the rate of subduction. The periodicity of volcanic eruptions above subduction zones is probably in large part determined by the quantity of fluids released from the subducting plate. It is interesting that such volcanoes are commonly located where the subducting plate reaches a depth of 100 km, as long as the subduction angle exceeds 25°. This depth appears critical to bringing about the metamorphic reactions that release the water needed for melting.

Partial melting at the water-saturated solidus of peridotite produces magma that is richer in silica (58–60 weight %) than that formed at the anhydrous solidus as a result of decompression melting or heating by mantle plumes. This more silica-rich magma produces the volcanic rock known as **andesite** (see Sec. 9.3.2 for igneous rock classification).

In summary, magma forms in the upper mantle by partial melting of peridotite following three different schemes. The most voluminous production occurs as a result of decompression associated with the rise of the asthenosphere at divergent plate boundaries. This produces relatively silica-rich basaltic magma (~52 weight % SiO_2). The second most productive source of magma occurs in the mantle wedge above subducting oceanic plates, which release water that lowers the melting point of peridotite. The magma formed under these conditions is more silica rich and is of andesitic composition (~58–60 weight % SiO_2). The third method of producing magma is by heating associated with mantle hot spots. Here the addition of heat and decompression from the rise of mantle plumes results in the production of basaltic magma, which typically contains slightly less than 50 weight % SiO_2.

8.3 Melting processes in rocks

In Section 8.2.2, we saw that because rocks are made of an assemblage of minerals they melt over a range of temperature. We might expect that if a rock were heated, the lowest melting mineral would melt first and the highest melting mineral would melt last, but one simple fact shows that this cannot be the case. Basalt begins melting around 1000°C, well below the melting points of its constituent minerals, which have melting points ranging from 1300°C to 1800°C. Rocks melt at much lower temperatures than do any of their constituent minerals.

In this section, we examine how and why a mixture of minerals melts at a lower temperature than do individual minerals. The melting processes are shown graphically in **phase diagrams**, which show the composition of minerals that coexist with liquids in terms of temperature (also pressure). These diagrams are created from carefully controlled experiments in which a mixture with a specific chemical composition is held at a desired temperature and pressure for sufficient time to achieve equilibrium, and then all the phases are identified and analyzed. These diagrams are useful in accounting for the composition and temperature of liquids that form when rock melts or the sequence in which minerals appear when magma crystallizes. The diagram is the same for melting or crystallizing, only the direction in which temperature changes is different. Phase diagrams are important to understanding igneous rocks and interpreting their textures. You will be well rewarded for taking a little time to understand them when you come to examine igneous rocks more closely.

8.3.1 Melting of a mixture of minerals

If you wanted to study how a rock begins to melt, you might heat it in the laboratory until melting begins and then, after cooling it rapidly, study thin sections of the partially melted rock under the microscope. Robert Bunsen, who invented the Bunsen burner, used his burner in 1851 to try exactly such experiments. He managed to melt Icelandic basalt but could not melt granite, so he erroneously concluded that granite had a higher melting temperature than basalt. His experiments, unfortunately, did not last long enough to produce detectable liquid in the slowly melting granite. We start this section by examining the results of an experiment in which a sample of basalt was partially melted in a furnace. This allows us to appreciate how melting and crystallization occur in a rock.

Figure 8.4(A) shows a 1-cm cube of basalt that was placed over a hole in the base of a graphite crucible so that it was supported only at its corners. When heated to the beginning of melting, liquid began draining from the sample to form the large drip on its lower surface. The sample was rapidly cooled, which froze the molten part to glass, and then cut and polished. Figure 8.4(B) shows a photomicrograph of this polished surface after etching with hydrofluoric acid fumes. The important result of this experiment is that melting occurred only where plagioclase and pyroxene grains came in contact. This experiment reveals one of the most important facts of igneous petrology; that is, two different minerals touching one another have a lower melting point along their common boundary than either mineral has separately. This explains why basalt begins melting at ~1000°C, well below the melting points of plagioclase or pyroxene.

What we have witnessed in this experiment is the same phenomenon that we see when we sprinkle salt on an icy pathway. The melting point of ice at atmospheric pressure is 0°C and that of NaCl is 800°C. Allow the two materials to come in contact so that molecules jumping back and forth between the two solids have the opportunity to form brine, which has a freezing point well below 0°C, and the ice starts melting, but so does the salt, as evidenced by the salty taste of the liquid formed. Because salt dissolves in water, it lowers the freezing point of ice. Similarly, water dissolves in molten NaCl, so its freezing point is lowered by water. When minerals can mix in the molten state, their melting points are lowered and magma is the result.

The melting point lowering that results from mixing in the liquid state is quantified through the **cryoscopic equation**, which shows that the lowering of temperature is dependent only on the amount to which a mineral can be diluted by a solvent and the latent heat of fusion of the mineral (Box 8.2). This equation is important to igneous petrology because it explains why mixtures of minerals have lower melting points than single minerals, and these *low-melting mixtures are the common magmas.*

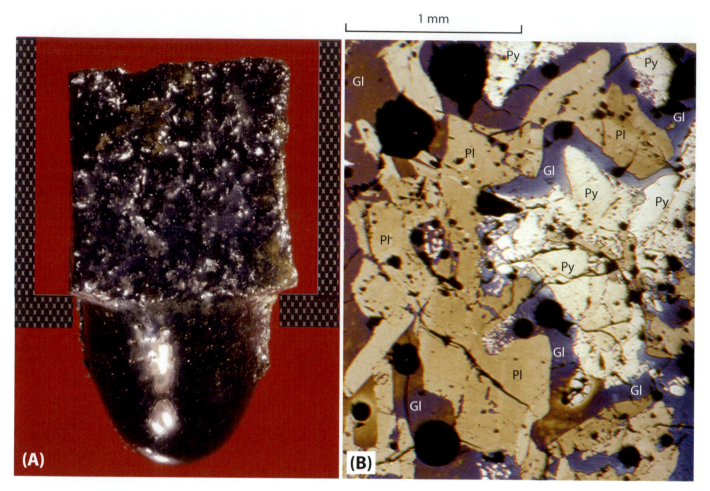

1 mm

(A)

(B)

Figure 8.4 **(A)** Partially melted 1-cm cube of basalt positioned over a hole in a graphite crucible so that liquid could drain through the hole. The cube, which is 70% melted, is supported only by its corners, whereas the partial melt forms the large drip hanging from its lower surface. Although the cube is basalt, the partial melt forming the drip is andesitic in composition. **(B)** Photomicrograph of a polished section of the experimental charge in (A) seen under reflected light. Melt, which was quenched to glass (Gl) formed only where pyroxene (Py) and plagioclase (Pl) were in contact.

Lowering of the melting point of ice

Before examining the lowering of melting points of minerals and its application to the formation of magma, we discuss the melting of ice-salt mixtures, because this is familiar to us if we live where icy conditions occur during the winter. The behavior of an ice-salt mixture has many parallels to those in magmas.

The graph in Figure 8.5, which was calculated using the cryoscopic equation (Box 8.2), shows how the melting point of ice is lowered by the addition of salt to the solution. Pure water ($X = 1.0$) melts at 0°C (273 K). As salt is added to the water, the mole fraction of H_2O is decreased, and when it gets to 0.95, for example, the melting point of ice has been lowered to −5°C. As more salt is added to the solution, the mole fraction of H_2O decreases, and the melting point of ice is lowered still more. Eventually the solution becomes so salty that salt starts precipitating, and because it no longer is dissolving in the liquid and lowering the mole fraction of H_2O, the melting point is no

longer lowered. The line in the graph indicating the temperature at which ice crystallizes from a salty solution is known as the liquidus of ice, because above this line, H_2O exists as liquid water.

Figure 8.5 illustrates another important effect when crystallization takes place from a solution such as the brine in this figure or magma in nature. Consider cooling brine whose mole fraction of H_2O is 0.95 (a in Fig. 8.5). When the temperature reaches −5°C (b in Fig. 8.5) ice begins to crystallize. If you remove and taste the crystals, you find they are not salty; in fact, they are pure H_2O (c in Fig. 8.5). If pure ice crystals form from the solution that initially had a mole fraction of H_2O of 0.95, the solution must become saltier, but this requires the temperature to drop if ice is to keep crystallizing. The result is that as the temperature is lowered, ice keeps crystallizing as the liquid changes composition down the liquidus (red path b to e in Fig. 8.5), becoming saltier as it does so. For example, by the time the temperature has dropped to −16°C, the liquid has reached point e on the liquidus. The ability of a liquid to change its composition as a result of crystallization is extremely important in nature and leads to *igneous differentiation* and variation in the composition of igneous rocks. We can see the results of this process in the experiment shown in Figure 8.4, where the liquid dripping from the bottom of the cube of basalt is an-

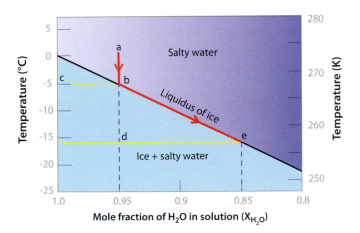

Figure 8.5 Lowering of the melting point of ice brought about by the solution of salt, according to the cryoscopic equation (Box 8.2). Letters refer to crystallization paths discussed in the text.

desitic in composition. Thus, andesite can form by separation of plagioclase and clinopyroxene from basalt.

8.3.2 Melting of a pair of minerals: the eutectic

We are now in a position to explain why melting occurred between pyroxene and plagioclase crystals in the basalt shown in Figure 8.4. Just as salt lowers the melting point of ice, plagioclase lowers the melting point of pyroxene and vice versa. For simplicity, let us treat these minerals as pure diopside, $CaMgSi_2O_6$, and pure anorthite, $CaAl_2Si_2O_8$. At atmospheric pressure, diopside melts at 1392°C and anorthite at 1557°C. When molten, these minerals can dissolve in each other and lower each other's melting points. Figure 8.6(A) shows a temperature-composition phase diagram in which diopside is plotted at the left, anorthite at the right, and mixtures of the two are expressed in weight percentages along the horizontal axis; temperature is plotted on the vertical axis. Descending from the melting point of each pure mineral is a liquidus line indicating the lowering of melting point of that mineral when diluted by the other mineral. Where these lines intersect, both diopside and plagioclase crystallize, and no more lowering of the melting point is possible. This point is, therefore, the lowest possible melting mixture of the two minerals and is known as a **eutectic** after the Greek word *eutektos*, meaning "easily melting."

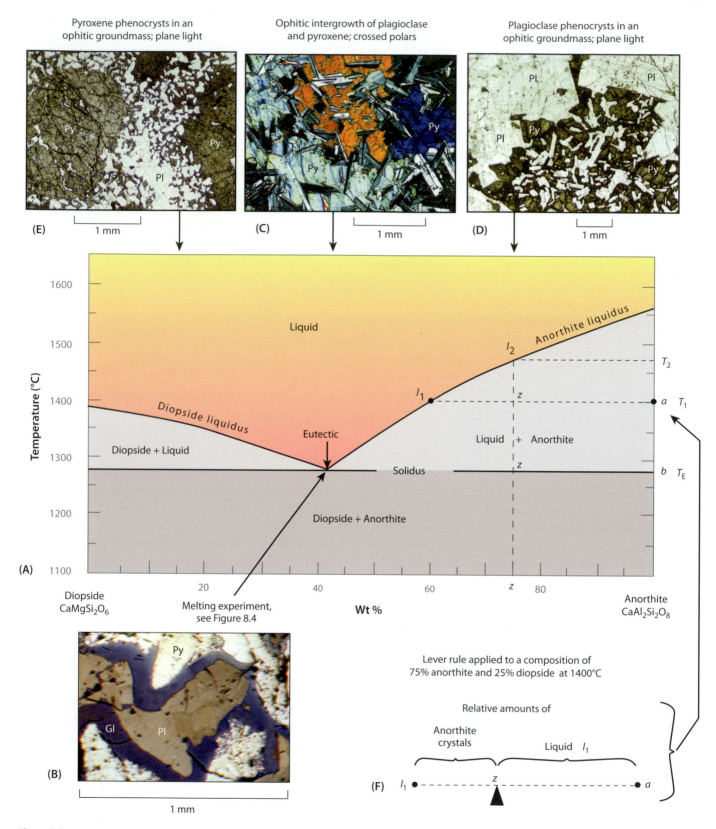

Pyroxene phenocrysts in an ophitic groundmass; plane light

Ophitic intergrowth of plagioclase and pyroxene; crossed polars

Plagioclase phenocrysts in an ophitic groundmass; plane light

Figure 8.6 (**A**) Melting relations for mixtures of diopside and anorthite at one atmosphere pressure (after Weill et al., 1980). (**B**) Photomicrograph of polished section of the partially melted basalt sample shown in Figure 8.4 with melt (now glass, Gl) between pyroxene (Py) and plagioclase (Pl) formed at the eutectic. (**C**) Photomicrograph under crossed polars showing the ophitic texture. (**D**) Photomicrograph of a rock containing plagioclase phenocrysts in a groundmass with ophitic texture. (**E**) Photomicrograph of a rock that falls on the pyroxene side of the eutectic, with pyroxene phenocrysts surrounded by groundmass with ophitic texture. (**F**) The lever rule indicating the proportions of liquid and crystals of anorthite in composition z at 1400°C.

Eutectics are the most important physical-chemical relation in igneous petrology. We have seen that the Earth does not have extra heat and that melting is a rare occurrence. When it does occur, the lowest possible melting mixtures form, which are usually eutectic mixtures (two other possibilities are considered in the following two sections). Most magmas, therefore, have near-eutectic compositions Basalt, for example, typically contains ~60 weight % pyroxene and ~40 weight % plagioclase, which we see from Figure 8.6(A) is close to the eutectic between diopside and anorthite (58 weight % and 42 weight %, respectively). In Section 8.3.3, we see that granite is a eutectic mixture of quartz and alkali feldspar.

Another important line in Figure 8.6(A) is drawn horizontally through the eutectic and extends to diopside and anorthite on either side. Below this line, which is known as the **solidus**, everything is solid. Above the liquidus, everything is liquid. Between the two lines, solids and liquids coexist, and this is the region where magmas plot.

Melting and crystallization in a simple eutectic phase diagram

Regardless of the proportions of diopside and anorthite in a rock, when it is heated to the solidus, melting begins with formation of the eutectic liquid, because this is the only composition that can form a liquid at this low temperature. This explains the restricted range of magma compositions and the relatively small number of igneous rock types.

Phase diagrams such as that in Figure 8.6(A) can be used to illustrate what happens during melting of solid rock and crystallizing of magma. Let us consider what happens if we heat a rock containing 75 weight % anorthite and 25 weight % diopside. This composition is indicated by the letter z in the diagram. On reaching the solidus (T_E), melting occurs where the diopside and anorthite grains are in contact to produce a liquid of eutectic composition (Fig. 8.6(B)). If we remove and cool this liquid, it would simultaneously crystallize diopside and anorthite in eutectic proportions and produce a **eutectic intergrowth**. In basaltic rocks, this intergrowth creates the **ophitic texture**, with small laths of plagioclase embedded in larger host grains of pyroxene (Figs. 8.6(C) and 2.10(A)).

If the eutectic liquid is not drained away from the sample, melting continues at the eutectic temperature with production of more and more eutectic liquid. Even though we add heat, the temperature remains fixed because all the added heat is used for the latent heat of fusion. Eventually, when all the diopside has melted and only anorthite crystals remain, the temperature begins to rise as the liquid changes its composition along the anorthite liquidus. The liquid eventually reaches a composition containing 75% anorthite and 25% diopside, which was the starting composition of the solids, and

thus the sample becomes completely melted. In nature, however, melting never gets this far. The Earth does not have large quantities of excess heat, so melting is rare, and when it does occur, the latent heat of fusion prevents melts rising far above the eutectic temperature.

Igneous textures related to a eutectic

If we cool the liquid we have just discussed from temperature T_2 in Figure 8.6(A), it starts crystallizing anorthite, which enriches the liquid in diopside as it descends the liquidus. When the liquid reaches the eutectic, diopside begins crystallizing as well. Because the anorthite crystallized first without competition from clinopyroxene, it forms large euhedral crystals known as **phenocrysts** after the Greek word *phainein*, meaning "to show" (Fig. 8.6(D)). These conspicuous crystals sit in a **groundmass** of later-forming smaller crystals. A rock containing abundant phenocrysts, for example of feldspar, is known as feldspar **porphyry**. Porphyry is a very old term, which the Roman historian Pliny the Elder used to describe a purplish building stone in Egypt; *porphura* is Greek for "purple." Phenocrysts give a rock a porphyritic texture. If a melt has a composition on the diopside side of the eutectic, it first crystallizes along the diopside liquidus and forms diopside phenocrysts (Fig. 8.6(E)). Regardless of whether anorthite or diopside crystallizes first, the last liquid to crystallize does so at the eutectic where the ophitic intergrowth forms the groundmass (Fig. 8.6(D)–(E)).

The large latent heat of fusion of rocks (~400 kJ/kg) prevents magmas from being heated to temperatures much above the eutectic. In 1928, N. L. Bowen noted that almost all rapidly cooled volcanic rocks are porphyritic and, therefore, could not have had temperatures above the liquidus. From this he concluded that *magmas are rarely superheated*, which in terms of Figure 8.6(A) means that they rarely enter the field labeled liquid. If they did and were erupted from a volcano, they would quench to a glass that contains no phenocrysts, and such rocks are rare.

In this section, we have seen that because rocks are composed of a mixture of minerals, they melt over a temperature range. When heated, they begin to melt where different minerals come in contact. The lowest melting mixture in a rock is known as the eutectic. Many igneous rocks have near-eutectic compositions, because the Earth has only enough heat to melt these lowest melting fractions. Crystallization of several minerals at a eutectic produces a eutectic intergrowth, for example, the ophitic texture in basalt. Porphyritic textures form from magmas with compositions that deviate from the eutectic composition, with the mineral in excess of the eutectic mixture forming larger more euhedral crystals known as phenocrysts. Although many igneous rocks are porphyritic and hence have formed from magmas that were not superheated, compositions generally deviate only slightly from eutectic compositions.

BOX 8.3 | THE LEVER RULE

Phase diagrams contain information about the proportions of crystals and liquid. The **lever rule** allows us to read these proportions. It is called this because it applies the same principle that is used in a lever or the balancing of a seesaw. Let us apply it to a rock containing 75% anorthite and 25% diopside that has been heated to temperature T_1 in Figure 8.6(A). At this temperature, the bulk composition (z) is made of a mixture of liquid of composition l_1 and anorthite crystals at a, and we wish to determine the proportions of each.

First, we join l_1 and a with a horizontal line, known as a **tie line** because it ties together coexisting phases (i.e., they are in equilibrium with each other). This line is the lever that allows us to determine the proportions of liquid and crystals present at this temperature. Think of this line as a seesaw that has a quantity of liquid sitting at one end and a quantity of anorthite crystals at the other, and that it is balanced about a fulcrum at z.

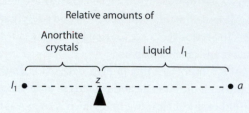

If two people are seated on a seesaw, the fulcrum has to be closer to the heavier person to achieve a balance, and the same applies to the quantities of liquid and crystals on either end of our tie line. We can determine the relative weights of two people on a balanced seesaw from their distances from the fulcrum, and the same applies to the weights of liquid and anorthite crystals on each end of the tie line.

The fulcrum on the seesaw represents the bulk composition of the rock (z). If the bulk composition were half way along this line, it would indicate equal weights of liquid and anorthite, but point z is closer to the liquid end of the line in our example, so the weight of liquid must be greater than that of anorthite crystals. The actual amounts of these two can be calculated knowing that on a balanced seesaw the product of the weight and distance to the fulcrum (moment) must be the same on both sides. Applying this to our tie line, we have the following:

weight of l_1 × length(l_1z) = weight of anorthite × length(az).

On rearranging, we have the following:

weight of l_1 / weight of anorthite = az / l_1z.

In other words, the weight proportions of phases can be read from the ratios of the lengths of the lines on either side of the bulk composition. The lines in our particular example indicate 64% liquid and 36% anorthite.

8.3.3 Congruent melting and the granite and nepheline syenite eutectics

In this section, we consider the melting relations between two important rock-forming minerals, quartz, SiO_2, and nepheline, $NaAlSiO_4$. These relations are important because they provide an explanation for the composition of one of the most abundant igneous rocks, **granite**, and a less abundant rock known as **nepheline syenite**.

The phase diagram for quartz and nepheline differs from that of diopside and anorthite because the mineral albite, $NaAlSi_3O_8$, plots between them. Nepheline reacts with quartz to form albite:

$$2SiO_2 + NaAlSiO_4 \rightarrow NaAlSi_3O_8$$
quartz   nepheline   albite

The composition of albite plots very near the center of the nepheline-quartz diagram (Fig. 8.7). On heating at atmospheric pressure, albite melts at 1100°C to form a liquid of albite composition. Because the solid and liquid have the same composition, melting is said to be **congruent**. In the following section, we discuss minerals that melt to form a liquid of different composition, and these are said to melt **incongruently**. The diagram also contains several extra fields that mark the stability ranges of the high-temperature polymorphs of quartz and nepheline. Quartz changes to tridymite at 867°C and then to cristobalite at 1470°C, and nepheline changes to carnegieite around 1280°C. We are not concerned with these polymorphs here and refer to them simply as quartz and nepheline, because these are the minerals that are normally found in rocks.

The presence of albite between quartz and nepheline creates two eutectics in this diagram, one on the quartz side of the diagram and one on the nepheline side. Both halves of the diagram are read in exactly the same way as we read the simple

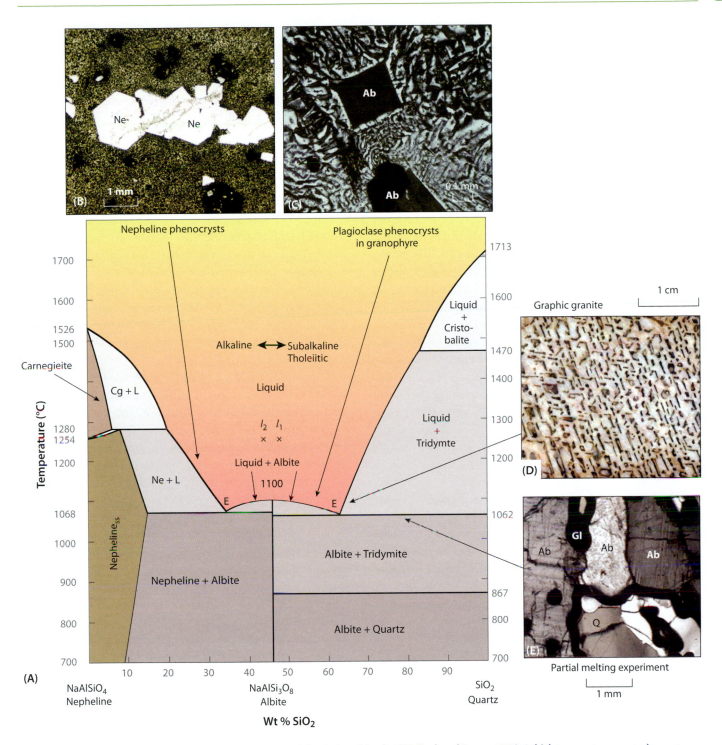

Figure 8.7 **(A)** Melting relations in the system nepheline-quartz (after Greig and Barth, 1938; Tuttle and Bowen, 1958). At high temperature, quartz changes to tridymite and then cristobalite, and nepheline (Ne) to carnegieite (Cg). **(B)** Photomicrograph of rock containing nepheline phenocrysts. **(C)** Photomicrograph of phenocrysts of albite-rich plagioclase (in extinction) in a granophyric intergrowth of quartz and albite-rich feldspar. **(D)** Hand specimen of graphic granite shows the end-on view of quartz rods in a large alkali feldspar host grain. **(E)** Photomicrograph under crossed polars of an experimentally melted sandstone in which melt (Gl) formed between quartz (Q) and albite (Ab).

anorthite-diopside diagram. Crystallization of any liquid in this diagram leads to one or the other of these eutectics. It is helpful to think of the liquidus in this diagram as if it were a topographic cross-section through a continent, with the melting point of albite creating a thermal maximum that is equivalent

to the continental divide. Just as rain falling to the east of the continental divide drains to the east, so a liquid with a composition on the quartz side of albite (l_1 in Fig. 8.7(A)) would, on crystallizing albite, change its composition toward the eutectic with quartz. In contrast, a liquid on the nepheline side of albite

(l_2 in Fig. 8.7(A)) would, on crystallizing albite, change its composition toward the eutectic with nepheline.

The thermal maximum produced by albite divides all igneous rocks into two major groups, with those plotting on the nepheline side of albite referred to as **alkaline**, because they tend to be richer in alkalis than the other group that plots on the quartz-rich side of albite and are known as **subalkaline** or **tholeiitic**. (Tholey is a village in the Saarland of Germany and the type locality for tholeiitic basalt.) Tholeiites are the world's most abundant igneous rocks, forming most of the ocean floor (midocean-ridge basalt, MORB) and **large igneous provinces** (**LIP**). In terms of Figure 8.2, they are formed by relatively large-scale decompression melting at shallow mantle depths. Alkaline rocks, such as the one shown in Figure 8.7(B), are less abundant and are formed by relatively small degrees of melting at greater depths than tholeiites. They are common along rift valleys and during the waning stages of magmatism over hot spots.

The two eutectics in Figure 8.7 correspond in composition to the rocks granite and nepheline syenite. The lever rule (Box 8.3) indicates that the eutectic liquid between albite and quartz would crystallize to form a rock containing two-thirds albite and one-third quartz. Most granite contains approximately two-thirds alkali feldspar and one-third quartz. The proportions of nepheline and albite in the eutectic between the minerals is similar to that in nepheline syenite.

Quartz and albite crystallizing together at the "granite eutectic" commonly form an intergrowth known as **graphic granite** because of its resemblance to cuneiform writing (Fig. 8.7(D)). The quartz forms long parallel rods embedded in large single crystals of feldspar. The rods, when sliced perpendicular to their length, form small hexagonal or wedge-shaped sections that resemble cuneiform characters. Graphic granite textures are common in the extremely coarse-grained igneous rock known as **pegmatite**. The groundmass of many tholeiitic basalts contains an extremely fine-grained, slightly less regular graphic intergrowth of quartz and alkali feldspar, which is described as a **granophyric intergrowth** (Fig. 8.7(C)).

Any mixture of quartz and albite, regardless of its bulk composition, will start melting at the eutectic temperature to form a liquid of eutectic granite composition. The photomicrograph shown in Figure 8.7(E) illustrates sandstone containing quartz and albite grains that was heated in the laboratory until melt began to form, which was rapidly quenched to glass. The melt forms only where quartz and albite grains are in contact. *Most large bodies of granite are formed by the segregation of low-temperature eutectic melts of granitic composition formed by partial melting in the lower crust.*

8.3.4 Incongruent melting and the peritectic

We now examine part of the phase diagram for the Mg-rich olivine, forsterite, and cristobalite, the high-temperature polymorph of quartz. Only the magnesium-rich half of the diagram is shown in Figure 8.8(A) and for temperatures that cover the lowest part of the liquidus. This diagram looks as if it should be similar to the nepheline-quartz diagram, because forsterite, Mg_2SiO_4, and cristobalite, SiO_2, react to form enstatite, $Mg_2Si_2O_6$, according to the following reaction:

$$\underset{\text{cristobalite}}{SiO_2} + \underset{\text{forsterite}}{Mg_2SiO_4} \rightarrow \underset{\text{enstatite}}{Mg_2Si_2O_6}$$

However, when enstatite begins to melt at 1557°C, it breaks down into a pair of phases that straddle the enstatite composition; a solid, which is forsterite, and a liquid, which is more silica rich than enstatite. Because the composition of the liquid is different from that of the initial solid (enstatite), this type of melting is described as incongruent. The point on the liquidus indicating the composition of the liquid is known as a **peritectic** (P in Fig. 8.8(A)). In contrast to a congruent melting mineral, such as albite, that creates two eutectics separated by a thermal maximum (Fig. 8.7), an incongruent melting mineral does not introduce a second eutectic.

If a mixture of quartz and enstatite is heated, melting begins at 1543°C with formation of an enstatite-cristobalite eutectic liquid. In contrast, if we heat a mixture of forsterite and enstatite, melting does not begin until the peritectic temperature of 1557°C is reached. Regardless of the proportion of forsterite to enstatite, the liquid formed is of peritectic composition. From the position of the peritectic in Figure 8.8(A), we see that a liquid of this composition (plots to the right of enstatite) must eventually crystallize at the enstatite-cristobalite eutectic. Thus, from an initial rock containing no cristobalite (or quartz at low temperature), partial melting at the peritectic produces a liquid that eventually crystallizes at the enstatite-cristobalite eutectic. If, for example, the mantle nodule in Figure 8.1, which contains forsterite-rich olivine and enstatite-rich pyroxene, were heated, a peritectic liquid would form that would eventually crystallize cristobalite. Next to eutectics, peritectics are the most important phase relation controlling the initial composition of magmas.

Peritectics also play an important role in the crystallization of igneous rocks. For example, when a basaltic liquid with composition x in Figure 8.8(A) cools, it first crystallizes forsterite, which would form phenocrysts, such as those in the basalt that filled the Kilauea Iki lava lake (Fig. 2.11(A)). When the cooling liquid reaches the peritectic, enstatite starts to crystallize, forming a reaction rim around the olivine. Many basaltic igneous rocks, especially slowly cooled intrusive ones, preserve **reaction textures** formed at peritectics. The photomicrograph in Figure 8.8(B) is of a coarse-grained sample from the Stillwater Complex in Montana that shows forsterite-rich olivine grains (bright interference colors) surrounded by reaction rims of enstatite-rich orthorhombic pyroxene (gray). The original outlines of the olivine crystals before the peritectic reaction are

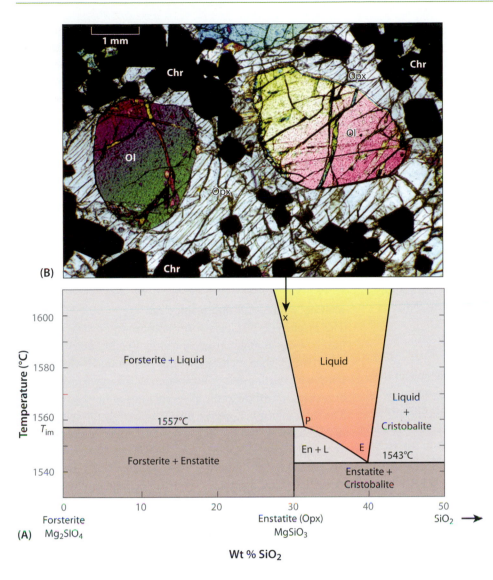

Figure 8.8 (**A**) Melting relations in part of the system forsterite-silica, which contains the important intermediate mineral, enstatite (after Bowen and Anderson, 1914). Only the magnesium-rich half of the diagram is shown for temperatures ranging from 1530°C to 1610°C. At these temperatures, cristobalite is the stable polymorph of silica. Enstatite (En) melts incongruently at 1557°C (T_{im}) to form a liquid of peritectic composition (P) and solid forsterite. (**B**) Photomicrograph under crossed polars showing chromite grains (Chr) that initially surrounded olivine (Ol) crystals that reacted with magma to produce orthorhombic pyroxene (Opx) in the Stillwater Complex, Montana.

marked by the distribution of small opaque grains of chromite, which crystallized early with the olivine.

8.3.5 Melting relations of solid solutions

Many igneous rock-forming minerals belong to solid solution series (Sec. 4.7) whose melting relations have two distinct ways of behaving. In one, melting temperatures decrease smoothly from the high-temperature to the low-temperature end member, and in the other, melting temperatures decrease to a minimum between the two end members. We illustrate these two styles of melting with olivine and plagioclase as examples of the first and alkali feldspars as an example of the second.

Melting and crystallizing olivine

By substituting iron for magnesium, olivine changes its composition continuously from forsterite, Mg_2SiO_4, to fayalite, Fe_2SiO_4, forming an almost ideal **solid solution**, which in phase diagrams is indicated by the subscript *ss* after the name of the mineral. The ionic radii of Mg^{2+} and Fe^{2+} in octahedral

coordination are 0.71 Å and 0.77 Å, respectively (Table 4.1). This results in the ionic bond between magnesium and oxygen being significantly shorter and, hence, stronger than that between iron and oxygen. Forsterite, consequently, melts at higher temperature (1890°C) than does fayalite (1205°C). Other minerals that are solid solutions between Mg and Fe end members (e.g., pyroxenes and amphiboles) show similar behavior.

The melting relations between forsterite and fayalite are shown in Figure 8.9. This diagram differs from the previous diagrams in having no eutectic. Instead, the liquidus falls along a gently convex upward curve from the high-melting forsterite to the low-melting fayalite, and the solidus is a gently concave upward line that also descends from the high-melting forsterite to the low-melting fayalite but at more forsterite-rich compositions than the liquidus at any given temperature.

Melting in this system takes place incongruently. For example, on heating olivine containing 50 weight % fayalite, melting begins at s_3 in Figure 8.9 with formation of a liquid of composition

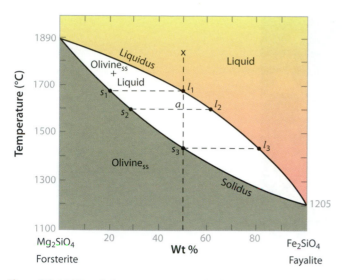

Figure 8.9 Melting relations at 1 atm pressure for the olivine solid solution series (Olivine$_{ss}$) between forsterite and fayalite (after Bowen and Shairer, 1935).

l_3; that is, a liquid much richer in fayalite than the starting composition. As heating continues, olivine changes composition along the solidus, steadily becoming richer in forsterite. Meanwhile, the first relatively fayalite-rich liquid (l_3) changes its composition along the liquidus, steadily becoming richer in forsterite. For example, when the temperature reaches 1600°C, the olivine has a composition of s_2 and the coexisting liquid has a composition of l_2. The lever rule indicates that the proportion of liquid to solid at this temperature is given by the lengths s_2a/al_2. When the temperature reaches ~1680°C, the liquid has a composition of 50 weight % fayalite (l_1), which is the same as that of the starting solid (s_3), so all the solid must have been converted to liquid. The last solid to melt has a composition of s_1. Any further heating does not change the liquid's composition but simply causes its temperature to rise, reaching point x, for example, at 1850°C.

Cooling of a liquid of composition x in Figure 8.9 results in the reverse sequence of steps encountered in the melting of olivine of this composition, as long as equilibrium is maintained. Equilibrium requires that all crystals have the same composition from core to rim at any given temperature (see Sec. 14.2.1 for fuller discussion of equilibrium). When liquid of composition x cools, the first olivine to crystallize has a composition of s_1. With continued cooling, the olivine changes its composition along the solidus from s_1 to s_3 as the liquid changes its composition along the liquidus from l_1 to l_3. This requires that the first-formed olivine (s_1) continuously reacts with the liquid to make its composition progressively more fayalitic. This type of reaction is, therefore, referred to as **continuous** in contrast to the **discontinuous reaction** at a peritectic. As the crystals grow larger, however, reactants find it increasingly difficult to diffuse through the crystals. Unless cooling is very slow, diffusion often has insufficient time to go to completion, and crystals

have cores that retain some of the earlier crystallizing more forsteritic olivine. This type of zoning is common in minerals belonging to solid solution series and is referred to as **normal zoning**; that is, the core is enriched in the early crystallizing high-temperature component and the rim is enriched in the low-temperature component.

The failure of diffusion to eliminate the forsterite-rich cores of olivine grains means that **equilibrium** is not achieved. The result is that the liquid continues farther down the liquidus under **disequilibrium** conditions than it does under equilibrium conditions, and in the case of extreme disequilibrium, liquids can descend the liquidus all the way to pure fayalite. Disequilibrium can also result from the first-formed olivine crystals sinking and being removed from the liquid with which they should react. This also results in the residual liquid becoming enriched in the low-melting fayalitic component.

Melting and crystallizing plagioclase

The plagioclase feldspars form another important solid solution series ranging from anorthite ($CaAl_2Si_2O_8$) to albite ($NaAlSi_3O_8$). Their melting relations are identical to those for olivine, with the liquidus-solidus loop decreasing from the higher melting anorthite to the lower melting albite (Fig. 8.10(A)). The solid solution, however, involves not only the substitution of Na^+ for Ca^{2+} but Si^{4+} for Al^{3+} (see coupled substitution in Sec. 4.7.1). The Si^{4+} for Al^{3+} substitution involves exchanges in the tetrahedral site with its strong covalent bond. Diffusional exchange of $Na^+ + Si^{4+}$ for $Ca^{2+} + Al^{3+}$ is, therefore, a slow process, and equilibrium crystallization is difficult to achieve. As a result, plagioclase crystals, especially in volcanic rocks where cooling rates are relatively rapid, are normally zoned. Because of slow diffusion, many plagioclase crystals preserve complex zoning patterns. For example, some have cores that are richer in lower-melting albite and rims that are richer in higher-melting anorthite. Such **reverse zoning** probably reflects changes in magma temperature resulting from influxes of fresh hot magma, or possibly the mixing of two completely different batches of magma. Some plagioclase crystals have complex **oscillatory zoning**, where compositions oscillate to higher and lower anorthite content from core to rim of a single grain (Fig. 8.10(B)). This zoning pattern might result from the cooling and heating of crystals as they move around in a convecting body of magma or from complex diffusion processes.

Melting and crystallizing alkali feldspars

The melting relations in the alkali feldspar solid solution are very different from those in the plagioclase solid solution in that both the temperature of the liquidus and solidus decrease from the melting points of the pure end members to a **minimum** at about 30 weight % orthoclase and 70 weight % albite. Figure 8.11(A) shows these melting relations at a pressure of 0.2 GPa in the presence of excess water. Two liquidus-solidus loops can be seen on either side of the minimum. Crystallization

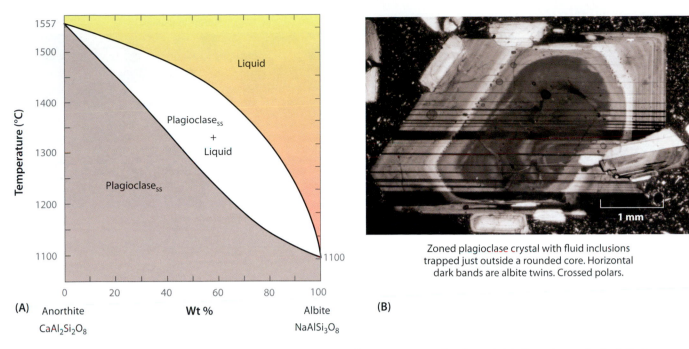

Figure 8.10 (**A**) Melting relations in the plagioclase feldspars at 1 atm pressure (after Bowen, 1913; Weil et al., 1980). (**B**) Photomicrograph of a plagioclase phenocryst in andesite from Mount St. Helens, Washington, showing complex zoning under crossed polarized light.

Zoned plagioclase crystal with fluid inclusions trapped just outside a rounded core. Horizontal dark bands are albite twins. Crossed polars.

of any liquid in this diagram causes its composition to change toward the minimum. For example, cooling of liquid x results in crystals of s_1 composition forming from the liquid at l_1. Crystallization of s_1, which is relatively potassium-rich compared with the starting composition, enriches the remaining liquid in albite components, and the liquid travels down the liquidus to l_2, while the solid continuously reacts with the liquid to change its composition along the solidus to s_2. When the solid reaches s_2, which is identical to the starting composition x, all the liquid is consumed, and the system becomes solid. The residual liquid, therefore, does not reach the minimum under equilibrium conditions. However, if the early formed crystals do not react completely with the liquid, the liquid would continue to change its composition until reaching the minimum, at which temperature the liquid would finally solidify. Similar arguments apply to liquids approaching the minimum from the albite-rich side of the minimum.

Although an alkali feldspar solid solution crystallizes at the solidus in Figure 8.11(A), on cooling, this solid solution becomes unstable and, given time, separates into orthoclase-rich and albite-rich solid solutions. The line in the phase diagram indicating the temperature and composition below which this separation should occur is known as the **solvus**. The separation of one mineral from another is known as **exsolution**. In this case, it requires sodium and potassium ions to diffuse through the crystal structure. Diffusion in silicates is slow, so exsolution is more common in slowly cooled intrusive rocks than in rapidly cooled volcanic rocks. When exsolution does

occur, the two minerals commonly are intergrown along crystallographic planes that minimize strain. In the alkali feldspars, this results in the **perthite** texture (Fig. 8.11(C)), where exsolution lamellae typically parallel the b crystallographic axis and make an angle of ~73° with (001). For example, when the alkali feldspar crystals of composition s_2 formed from the equilibrium crystallization of liquid of composition x in Figure 8.11(A) are cooled to the solvus (s_3), albite-rich exsolution lamellae of composition s'_3 are formed. With continued cooling, the orthoclase and albite solid solutions descend either side of the solvus adjusting their compositions, for example, to s_4 and s'_4, respectively, at 750°C.

If the pressure in Figure 8.11(A) is increased to 0.5 GPa, more water dissolves in liquid, which lowers the water-saturated liquidus and solidus so much that they intersect the solvus, and then the minimum changes to a eutectic (Fig. 8.11(B)) from which two separate alkali feldspar minerals crystallize, one an orthoclase solid solution (s_2) and the other an albite solid solution (s'_2). If we consider the cooling of liquid of composition x, we see that orthoclase solid solutions with compositions of s_1 to s_2, crystallize first as the liquid changes composition from l_1 to the eutectic (E), whereupon an albite solid solution of composition s'_2 starts crystallizing. On further cooling, both the orthoclase solid solution and the albite solid solution has to exsolve the other phase as they descend the solvus, reaching, for example, compositions of s_3 and s'_3 at 600°C.

Magmas crystallizing under the conditions shown in Figure 8.11(A)–(B) produce very different-looking rocks.

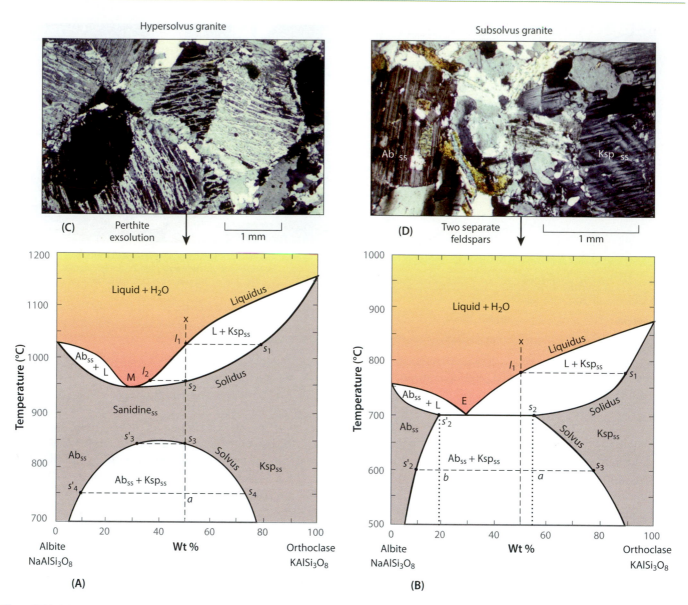

Figure 8.11 (A) Melting relations in the alkali feldspars in the presence of H_2O at a pressure of 0.2 GPa (after Bowen and Tuttle, 1950). This system exhibits a minimum (M) on the liquidus. (B) Melting relations in the same system as in (A) at a pressure of 0.5 GPa (after Morse, 1970). At this pressure, the minimum changes to a eutectic (E). (C) Photomicrograph (crossed polars) of hypersolvus granite that initially crystallized a single feldspar, which later exsolved to form perthite. (D) Subsolvus granite that initially crystallized two different alkali feldspars, an albite solid solution (Ab_{ss}) and an orthoclase solid solution (Ksp_{ss}).

Magma in (A) would crystallize to a single alkali feldspar whereas magma in (B) would crystallize two different alkali feldspars. All of these feldspars may develop exsolution lamellae on cooling slowly below the solvus, but the texture of the rock clearly shows whether one or two different feldspars crystallized from the magma. Because liquids in (A) crystallize above the solvus, rocks formed under these conditions are said to be **hypersolvus**, whereas those that crystallize under the conditions in (B) are said to be **subsolvus**. Figure 8.11(C)–(D) shows examples of hypersolvus and subsolvus granites respectively. This distinction is important, because subsolvus granites tend to be lower temperature but are richer in water, which is released during crystallization to

form hydrothermal fluids that can lead to the formation of ore deposits (Chapter 16).

8.4 Effect of pressure on melting

Pressure increases rapidly with depth in the Earth and has a profound effect on melting. In this section, we investigate the effect of pressure on melting, being careful to distinguish between anhydrous and hydrous environments. The effect of pressure in these two environments is completely opposite, causing melting points to rise in anhydrous environments and to fall in hydrous environments. First, we need to know what pressures to expect in the Earth.

Pressure in the Earth

When you dive into the deep end of a swimming pool, you feel the pressure build on your eardrums as you descend. Water is a liquid, so it has no shear strength; that is, it conforms to the shape of the container it is in. The pressure we experience is therefore due only to the weight of the overlying column of water. This is known as the **hydrostatic pressure**. Rocks near the surface do have shear strength, but at depth they become plastic and flow, so we can, as a first approximation, calculate the pressure in the Earth the same way we would in water. We call this the **lithostatic pressure**, *lithos* being Greek, for "stone." The pressure at any given depth is determined from the product of the density of the overlying rocks, the acceleration of gravity (9.8 m/s^2), and the depth. If a 35 km-thick continental crust has an average density of 2800 kg/m^3, the pressure at its base would be given by $2800 \times 9.8 \times 35 \times 10^3 = 0.96 \times 10^9$ (kg/m^3) $\times$ (m/s^2) $\times$ m. But a kg·m/s^2 is the unit of force known as the **newton** (N). The units, therefore, reduce to N/m^2, which is the unit of pressure known as a **pascal** (Pa). The pressure at the base of a 35 km-thick crust is, therefore, 1×10^9 Pa or 1 GPa (G is short for *giga* = 10^9). In cgs (centimeter gram second) units, this would be 10 kilobars (10^5 Pa = 1 bar = 10^6 dyne/cm^2 = 0.9869 atmosphere).

Below the Moho discontinuity (base of crust), density jumps to 3300 kg/m^3, so pressure rises more steeply in the mantle. Diagrams in this book showing depth also show pressure, which has been calculated in this manner by inserting appropriate densities. For example, in Figure 8.2, we see that basaltic magma rising to the surface of the Earth from a depth of 150 km would decrease its pressure by 5 GPa.

8.4.1 Effect of pressure on the anhydrous melting of rock

When any rock melts, its volume increases by ~10%; that is, the liquid rock is 10% less dense than the solid rock. If we were to first melt a rock and then increase the pressure on the liquid, we might expect the liquid to change back into a solid because it would occupy a smaller volume. This is precisely what happens, so at higher pressures, higher temperatures are needed to cause melting. It is important to point out that this applies only to anhydrous rocks. If water is present, increasing pressure lowers the melting point (see Sec. 8.4.2).

Pressure has a similar effect on the anhydrous melting of all rocks, which we illustrate with granite (Fig. 8.12). At atmospheric pressure, this granite begins melting at 950°C. If the granite had a eutectic composition, its liquidus and solidus temperatures would be the same (see Fig. 8.7), but this granite contains some feldspar phenocrysts, so its liquidus is 100°C higher than the solidus. The temperatures of the solidus and liquidus both rise linearly with increasing pressure so that melting begins at 1100°C at 1.5 GPa, which corresponds to a depth of ~50 km.

Let us consider what might happen to a eutectic melt formed from this granite at a depth of 30 km (point A in Fig. 8.12).

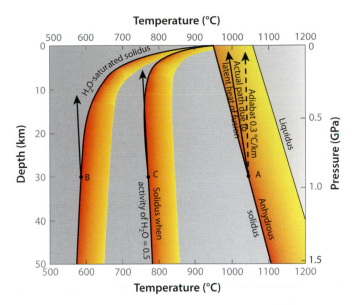

Figure 8.12 Melting of granite as a function of pressure (depth) under anhydrous conditions, water-saturated conditions, and under an activity of water (relative humidity) of 0.5. Synthesized from various sources including Tuttle and Bowen (1958), Burnham (1979), and Wyllie (1977).

Magmas typically rise fast enough so that they do not have time to lose significant heat; that is, they rise nearly adiabatically (no heat loss). The decrease in pressure during ascent results in adiabatic cooling by about 0.3°C/km (Fig. 8.12). Magma formed just above the solidus, however, is likely to contain many crystals, so as the magma rises, these crystals dissolve and the magma experiences additional cooling as a result of the latent heat of fusion of the crystals. The path followed by the magma toward the surface, therefore, falls below the adiabatic path (Fig. 8.12). The magma would, however, steadily rise above the solidus toward the liquidus but would probably not reach the liquidus. If erupted on the surface, it would contain phenocrysts that were rounded by resorption during ascent (see crystal core in Fig. 8.10(B)).

In summary, anhydrous melts formed at depth (high pressure) have higher melting points than at low pressure, and during ascent they can resorb crystals. This melting path is typical of the decompression melting that occurs at divergent plate boundaries.

8.4.2 Hydrous melting of rock and the solubility of water in magma

In Section 8.2.3, we saw that the melting point of mantle peridotite was lowered in the presence of water. This is caused by the solution of water in the melt. For water to dissolve in magma, however, elevated pressures are required. The same is true of CO_2 in carbonated beverages, and when you remove the cap from a bottle and release the pressure, the CO_2 comes out of solution and forms bubbles. Similarly, as the pressure on magma rising toward the surface decreases, volatiles in the magma are exsolved to form gas bubbles (see Fig. 8.16(A)). At atmospheric pressure, the solubility of water and most other gases in magma is essentially zero.

Experiments on a wide range of different magmas show similar increases in solubility of water with pressure (Fig. 8.13). The solubility increases rapidly at first to about 4 to 5 weight % at pressures near 0.1 GPa (depths of 3–5 km) and then increases linearly, but at a lesser rate at greater pressure. When the pressure reaches 0.5 GPa (16 km) most magmas contain about 10 weight % water, with midocean-ridge basalt (MORB) being slightly less (8.5 weight %).

At low pressures (<0.1 GPa), water dissolves in magma either as hydroxyl ions (OH^-) or as molecular water (H_2O), but at higher pressures, it dissolves only as molecular water. Water that dissolves by forming hydroxyl groups does so by first dissociating into separate hydroxyl and hydrogen ions ($H_2O \rightarrow OH^- + H^+$). The hydrogen ion attaches to an oxygen that is shared between two silica tetrahedra in the melt, breaking away one complete tetrahedral grouping while the hydroxyl ion completes the other tetrahedron (Fig. 8.14). This process takes place until the pressure reaches ~0.1 GPa and the melt contains about 4 weight % water. Above this pressure, water dissolves mainly as molecular water by fitting into holes in the structure of the melt, which is almost a linear function of pressure.

An extremely important result of water dissolving in magma by the hydrolysis reaction is that it breaks up, or depolymerizes, the framework structure of the melt. We see in Section 8.5.2 that granitic magmas (especially) tend to be extremely viscous; that is, they do not flow easily. The high viscosity is due to silica tetrahedra linking together through shared oxygen atoms to form polymerized groups such as those encountered in the structure of quartz and feldspar (Chapter 7). When water depolymerizes magma, the viscosity decreases significantly, and magma flows more rapidly.

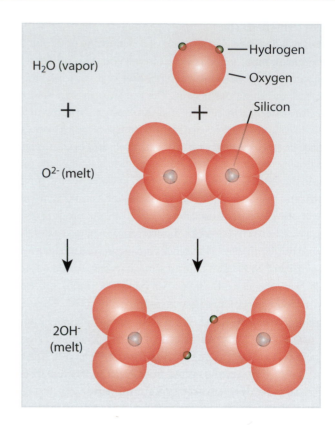

Figure 8.14 A mechanism by which water can dissolve in a silicate melt by hydrolysis of a bridging oxygen in the SiO_4^{4-} tetrahedral framework leading to breakage of the framework structure across the bridging oxygen.

The temperature at which rocks begin to melt is lowered by water that dissolves in the melt. Figure 8.12 shows that the water-saturated beginning of melting of granite drops from 950°C at atmospheric pressure to 600°C at 0.5 GPa. Most of this temperature drop occurs at low pressures, because this is the range over which water shows the greatest increase in solubility with pressure. Above about 0.5 GPa, the lowering of melting point with increased pressure is only slight, and at extremely high pressures (above those shown in Fig. 8.12), the melting point begins to increase, because water becomes less soluble. Other rocks show similar lowering of melting points as a result of solution of water at elevated pressures.

Let us consider what happens to a water-saturated granite magma that forms in the lower crust at a depth of 30 km (B in Fig. 8.12). As soon as the magma starts rising, the pressure decreases, water starts coming out of solution, and the magma drops below the solidus and solidifies. Although a *water-saturated magma* may form near the base of the crust, *it can never rise toward the surface without solidifying.*

Effect of water undersaturation on the melting of rocks
Analyses of volcanic fumaroles indicate the presence of many gases, but water and carbon dioxide (CO_2) are by far the most

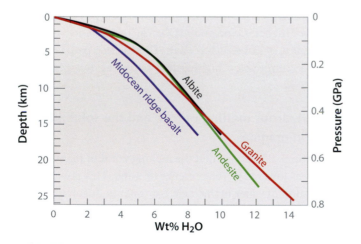

Figure 8.13 Solubility of water as a function of pressure in three common magmas and molten albite. Data from Dixon et al. (1995); Holtz et al. (1995); Silver and Stolper (1989); Moore et al. (1995); Burnham (1979).

abundant. Let us consider what happens if carbon dioxide is present along with water in the region where magma is generated. Experiments indicate that CO_2 is relatively insoluble in magmas at pressures less than 2 GPa and as a result has little effect on the melting temperature of rocks at low pressure. When CO_2 is mixed with water, it behaves as an inert component that lowers water's concentration in the fluid, which decreases its ability to lower melting points. To quantify this effect, we need to know the concentration of water in the H_2O-CO_2 mixture. We do this by using a concept with which we are familiar from weather forecasts, the relative humidity. When the relative humidity is 100%, air is saturated in water and it is probably raining. If the relative humidity is 50%, air contains only half the amount of water it can at the particular temperature. When discussing magmas, instead of relative humidity, we use the **activity** of water, which uses fractions rather than percentages, so that 100% humidity would be an activity of 1 and 50% humidity would be an activity of 0.5.

The anhydrous solidus given in Figure 8.12 is for a relative humidity of 0% (partial pressure of $H_2O = 0$) or an activity of 0. The water-saturated solidus is for a relative humidity of 100% or an activity of 1. A whole range of melting curves between the water-saturated and anhydrous curves could be shown, but we show only one, the solidus for an activity of 0.5.

Again, let us consider what happens to magma that forms at a depth of 30 km in an environment where the activity of water is 0.5 (C in Fig. 8.12). First, we see that its temperature is considerably higher than the water-saturated magma at the same depth. Second, the slope on the solidus is in the opposite direction from the water-saturated solidus, at least at depths greater than 10 km. This means that as the magma starts to rise, its path falls above the solidus, and so it remains molten. In fact, it is able to rise 15 km (halfway through the crust) before it crosses the solidus and solidifies. If we had chosen a magma with a still-lower activity of water, it could have ascended still farther through the crust before solidifying. Magmas formed under an activity of water of only 0.25 are able to rise almost to the Earth's surface before encountering the solidus. Only when these magmas get very near the surface do they start to crystallize, and at that shallow depth, exsolving gas may cause a volcanic explosion (see Sec. 8.4.4)

8.4.3 Solubility of other gases in magma

Water and carbon dioxide are by far the most abundant gases in volcanic emanations, constituting from 30–80 and 10–40 molecular %, respectively, of the vapor phase. However, anyone who has been near an active volcano can testify to the strong smell of hydrogen sulfide gas (Fig. 8.15(A)), but our noses are very sensitive to this gas (because of its toxicity – exposure to 800 ppm (parts per million) for five minutes can be fatal) and the amount is actually very small. Most volcanic gases are compounds of hydrogen, oxygen, carbon, and sulfur, and include

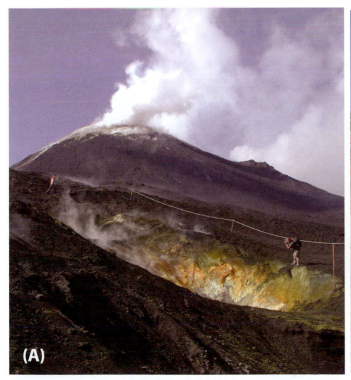

Figure 8.15 (**A**) Hot volcanic gases deposited sulfur and altered the rocks surrounding this fumarole near the summit of Etna, Sicily. (**B**) Similar caustic solutions emanating from magma bodies produced the gray and red hydrothermal alteration around the rich copper deposits of Bisbee, Arizona (after Philpotts and Ague, 2009).

H_2O, CO_2, CO, SO_2, H_2S, H_2, S, and O_2. Other minor constituents include nitrogen, argon, hydrochloric acid, hydrofluoric acid, and boron.

Not all gases coming from volcanoes are derived from magma. A large fraction of the steam is recycled groundwater. Large bodies of magma near the Earth's surface set up large groundwater circulation cells in the surrounding rocks, which can cause large amounts of hydrothermal alteration (Fig. 8.15(B)). This is particularly true of magma chambers beneath mid-ocean ridges. It is estimated that the entire volume of water in the Earth's oceans cycles through midocean ridges every 8 million years. Groundwater circulation cells are capable of transferring large quantities of elements held in solution, especially those classified as **chalcophile** (Greek for "copper-loving"). Chalcophile elements, such as copper, lead, zinc, silver, and arsenic, do not bond readily with silicates and prefer to bond with sulfur. These sulfides may be disseminated in extremely small amounts throughout large bodies of rock, but groundwater cells associated with bodies of magma can transport and concentrate them in hydrothermal veins of economic value.

8.4.4 Exsolution of magmatic gases and explosive volcanism

As magma approaches the surface of the Earth, the decrease in pressure causes dissolved gases to exsolve and form bubbles, which in rocks are known as **vesicles** (Fig 8.16(A)). Vesicles can later be filled with minerals, such as quartz, carbonates, and zeolites, to form **amygdales** (diminutive size = **amygdules**)

because of their resemblance to almonds, the Latin word for which is *amygdala* (Fig. 8.16(B)). Almost all volcanic rocks contain vesicles (or amygdules), and their abundance can increase dramatically during eruption if the magma contains significant amounts of dissolved gases. This can produce extremely vesicular rock, which in the case of basaltic lava is known as **scoria** (Fig 8.26(A)), and in rhyolitic lava, as **pumice** (see Fig. 8.17(C)).

When the percentage of bubbles in magma becomes large, their continued growth is impeded by the rate at which the intervening liquid can be squeezed aside. When bubbles are far apart, this presents little problem, but flow in the film between bubbles becomes progressively more difficult the thinner the film becomes. In basaltic magma, which is relatively fluid (see Sec. 8.5.2), this is not too difficult, and bubbles eventually coalesce and rise to the surface of the lava where they may pop and form small explosions (see opening photo to Chapter 9). In this way, basaltic magma rids itself of excess gas. Granitic magma is at least a thousand times more viscous than basaltic magma, so flow in the thin film of liquid between bubbles often cannot keep pace with bubble growth. When this happens, the thin film ruptures, and the gas forms a continuous phase in which are suspended small concave tetrahedral patches of liquid. This disrupts the magma, changing it from a highly viscous, bubbly mass to an extremely fluid gas carrying small particles of melt in suspension (Fig. 8.17(A)). This dramatic change in properties normally triggers a violent volcanic explosion, with the suspended particles of melt forming a large component of what we refer to as **volcanic ash** (see Fig. 9.23).

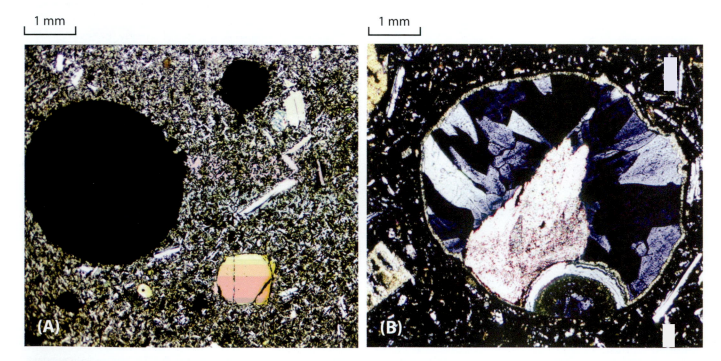

Figure 8.16 (A) Photomicrograph under crossed polars of vesicular Hawaiian basalt, showing spherical vesicles (black). (B) Photomicrograph under crossed polars of an amygdale in a Tahitian basalt filled with calcite (pale pink) and zeolite (anomalous blue interference color).

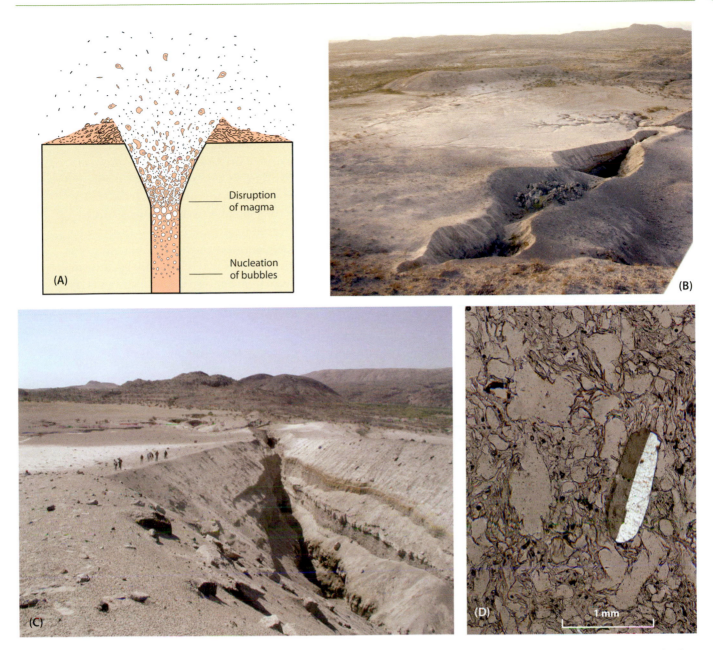

Figure 8.17 (**A**) When exsolved gas bubbles grow large enough to connect, they disrupt magma and form a rapidly expanding gas carrying a suspension of molten particles. This occurs commonly in highly viscous rhyolitic magma. (**B**) A 500 m-long, 60 m-deep vent formed by explosive eruption of rhyolitic magma from a north-south trending fault at Da'ure in the Afar region of Ethiopia (see Box 8.4). White volcanic ash and blocks of rock ripped from the walls of the vent litter the surroundings. The eruption ended when a small pumice dome rose and plugged the center of the vent. (Photograph courtesy of Asfawossen Asrat.) (**C**) View looking south from the pumice dome. (**D**) Photomicrograph under partially crossed polars of a sample from the pumice dome. The relatively light-colored areas are gas bubbles separated by thin films of colorless glass. The rounding of the sanidine phenocryst with a Carlsbad twin is evidence of heating of the magma before eruption. (Modified from Philpotts and Ague, 2009.)

The important conclusions to draw from Section 8.4 are that pressure either increases or decreases the beginning of melting temperature of rocks depending on whether the environment is anhydrous or water-saturated, respectively. Anhydrous magmas undergo decompression melting on rising, and hence the proportion of liquid to crystals increases as the magma ascends. In contrast, water-saturated magma solidifies as soon as it starts to rise. All possible conditions exist between anhydrous and water-saturated melting. In general, relatively wet magmas are restricted to the lower part of the crust because they solidify during ascent. Relatively drier magmas have a greater chance of rising through the crust. Solution of water in magma breaks the framework structure of the liquid, thus dramatically lowering its viscosity. Finally, all gases tend to exsolve as magma approaches the surface, explaining why most volcanic rocks are vesicular. If the fraction of vesicles reaches the critical value where they connect and disrupt the magma, explosive volcanism results.

BOX 8.4 | VOLCANIC BIRTH THROUGH EXPLOSIVE RELEASE OF DISSOLVED GAS

On September 26, 2005, a violent explosion in the Afar region of Ethiopia (Fig. (A)) gave rise to a new volcano where previously there had been only desert. Twelve days earlier, the first of 163 earthquakes with a magnitude of more than 3.9 on the Richter scale started between two older volcanoes, Dabbahu and Gabho (D and G, respectively, in Fig. (B)), and progressed southward for 60 km. The earthquakes were accompanied by normal faulting, which extended in a southerly direction from the Red Sea, a direct consequence of the movement of the Arabian plate away from the Nubian plate (Fig. (A)). Satellite images recorded this motion in almost live time. By comparing images before and after the disturbance, Wright and others (2006) were able to create the **Interferometric synthetic aperture radar** (**InSAR**; see Sec. 17.4) image shown in Figure (B) where each color fringe corresponds to a change in distance to the satellite of 40 cm (roughly elevation change). The results show that the region extended by 6 m in an east–west direction, and that up to 1.5 m of uplift occurred toward the center of the region, except along a 2–3 km-wide zone that subsided ~2 m (white line in (B)). They were able to show that this disturbance could be modeled by intrusion of a thin 60 km-long vertical sheet of magma (dike) that intruded southward from between the Dabbahu and Gabho volcanoes. This magma would have to have been low-viscosity basalt to intrude that distance in little more than a week. At its point of origin between the Dabbahu and Gabho volcanoes, the basalt must have reheated an older chamber full of granitic (rhyolite) magma. It was this magma that exploded violently from one of the normal faults on September 26 (Fig. 8.17(B)–(C)), throwing out pumice and older rock fragments. The eruption lasted for several days and then ended when a viscous mass of pumice plugged the vent.

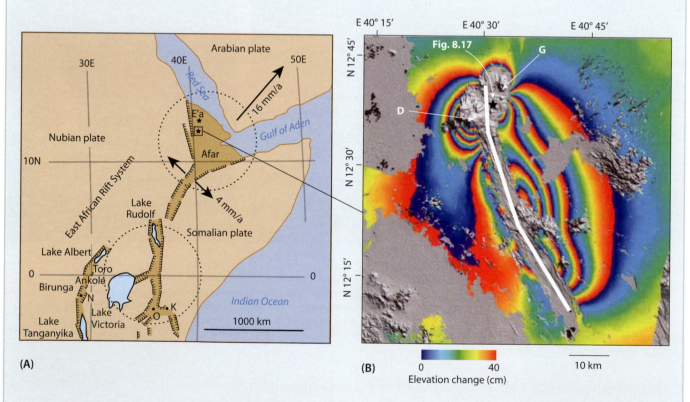

(A)

(B)   Elevation change (cm)

The crustal extension and accompanying intrusion of magma along a dike is the way in which new crust is created at divergent plate boundaries. Thanks to satellite imagery, the Afar event provided us with the first opportunity to witness, almost in live time, the formation of new crust.

8.5 Physical properties of magma

So far in this chapter we have seen how magma is formed from solid rock and why it tends to have eutectic or peritectic compositions. We have also seen how pressure affects melting in both hydrous and anhydrous environments. Now we must address the question of how magma, once formed, rises toward the Earth's surface. What forces cause magma to rise, and what forces resist this movement? To answer these questions, we must know the physical properties of magma, in particular its density and viscosity.

8.5.1 Magma density

The **density** of magma relative to the density of its surroundings is of importance because density contrast in a gravitational field is one of the major forces in nature causing things to move; it drives tectonic plates, ocean circulation, weather systems, and most certainly magma. Magma density is also important, because when crystallization begins, the density contrast between crystals and liquid can redistribute these phases and change the composition of the magma.

We have already seen that when rocks melt, their volume increases by ~10%, and hence the density of the liquid must be 10% less than that of the equivalent solid. However, we have also seen that the latent heat of fusion of rocks is so large that total fusion is unlikely, in which case the density contrast between the partial melt and the solid residue are of greater importance, and the compositions of these two fractions may be very different.

Rocks can be melted in the laboratory and rapidly quenched to glass, which can have its density measured in the same way we measure the density of a mineral (e.g., using the Jolly balance; see Sec. 3.6). Although glass is a supercooled liquid and its structure is essentially the same as that of the high-temperature liquid, its room-temperature density must be adjusted to take care of the shrinkage involved in cooling from magmatic temperatures; this requires knowledge of the coefficient of thermal expansion.

The density of magma is ultimately determined by its composition, which, if known, can be used to calculate the density, as is done in the computer program known as MELTS (see "Online Resources" at the end of this chapter). Direct measurements and calculations from chemical composition indicate that magmas have densities ranging from 2300 to 3000 kg/m^3, with most basaltic magma being between 2600 and 2750 kg/m^3, and most granitic magma being around 2400 kg/m^3. These values are of interest when compared with the densities of common rock-forming minerals. For example, pyroxene and olivine crystals, which are major constituents of basalts, have densities of 3000 to 3700 kg/m^3 and would tend to sink in basaltic magma. The density of plagioclase feldspar, however, ranges from 2630 to 2760 kg/m^3, which is similar to the density

of basaltic magma, and, therefore, plagioclase may be neutrally buoyant in this magma. If basaltic magma is generated by partial melting in mantle peridotite, the magma would be buoyant with respect to the surrounding olivine and pyroxene. Granitic magma is less dense than any of the rock-forming minerals and will, therefore, always try to rise toward the surface. In Section 8.6.1, we see that the density contrast between magma and wall rocks provides the main driving force for the ascent of magma.

8.5.2 Magma viscosity

Viscosity is the property of a liquid to resist flow when a shear stress is applied. Imagine placing a surfboard on a calm body of water and giving it a push. It would glide for some considerable distance before coming to a halt. Now imagine placing the same surfboard on a pool of honey and giving it the same push. It would not glide as far, because honey is much more viscous than water. Viscosity is defined as the ratio of the applied shear stress (e.g., the push you give to the surf board) to the rate at which the fluid deforms. It has units of pressure × time, which in SI units are pascal × seconds (Pa·s).

The viscosity of magmas varies by more than six orders of magnitude depending on the composition of the magma (Table 8.1). In general, the higher the silica content of magma, the larger the number of bridging oxygens forming a framework structure and the higher the viscosity. Basaltic magma, with ~50 weight % silica, has a viscosity of ~50–100 Pa·s, which is about the same as that of ketchup (Fig. 8.18), whereas rhyolitic magma, with ~73 weight % silica, has a viscosity of more than 10^5 Pa·s at 1200°C but is 10^8 Pa·s at 800°C. The addition of water to magma decreases its viscosity (Sec. 8.4.2).

Table 8.1 Viscosities of magmas and common substances.

Material	Viscosity (Pa s)	Weight % SiO$_2$	Temp. (°C)
Water	1.002×10^{-3}	–	20
ASE 30 motor oil	2×10^{-1}	–	20
Kimberlite	$10^{-1} - 10$	30–35	~1000
Komatiite	$10^{-1} - 10$	40–45	1400
Ketchup	$\sim 5 \times 10$	–	20
Basalt	$10 - 10^2$	45–52	1200
Peanut butter	$\sim 2.5 \times 10^2$	–	20
Crisco shortening	2×10^3	–	20
Andesite	$\sim 3.5 \times 10^3$	~58–62	1200
Silly Putty	$\sim 10^4$		
Tonalite 6% H$_2$O	$\sim 10^4$	65	950
Rhyolite	$\sim 10^5$	~73–77	1200
Granite 6% H$_2$O	$\sim 10^5$	75	750
Rhyolite	$\sim 10^8$	~73–77	800
Average mantle	10^{21}	–	–

Note: Magma viscosities from Dingwell (1995) and references therein. Granite and Tonalite viscosities from Petford (2003). Mantle viscosity is from King (1995).

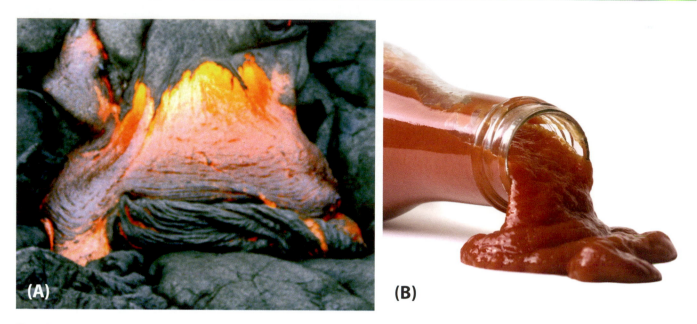

(A)

(B)

Figure 8.18 Basalt (**A**) and ketchup (**B**) have about the same viscosity. The rapid increase in viscosity as lava cools causes it to pile up and form a ropy surface.

As magma cools, its viscosity increases exponentially. The basaltic lava issuing from the front of the Hawaiian flow in the chapter-opening photo and Figure 8.18(A) is at about 1160°C. It flows rapidly on escaping from beneath the crust, but its surface cools rapidly by radiation, reaching a temperature of ~700°C where it is a dull cherry-red color. This drop in temperature causes a huge increase in viscosity, and the lava slows and piles up with the surface becoming wrinkled into a ropy

Figure 8.19 Ropy crust of a pahoehoe flow that was domed up on the side of a large blister, where lava flowing beneath the crust forced its way to the surface on the south coast of the Big Island of Hawaii. The later more viscous lava, formed loaf-sized blobs extruded from the upper left side of the blister.

Figure 8.20 A younger pahoehoe flow lying on top of an older aa flow on the south coast of the Big Island of Hawaii.

mass. This is particularly common on the surface of smooth lava flows to which the Hawaiian name **pahoehoe** is given. As these flows begin to solidify, their surface is often deformed into wrinkles that range in scale from fine string to coarse rope, with the coarser wrinkles forming on more viscous lava (Fig. 8.19).

Pahoehoe flows cool as they descend the flanks of volcanoes and become more oxidized from interaction with air. This increases their viscosity, which eventually leads to a change in the nature of the flow from the fluid pahoehoe to a slowly moving pile of clinkerlike rubble, to which the Hawaiian name **aa** is given (Fig. 8.20). Many andesitic lavas also have a rubbly surface, but the particles are larger (~20–40 cm) than those in aa, and these flows are referred to as **blocky**. At a completely different scale, **rhyolite** flows can show a ropy structure when seen at a great distance from the air, but the wrinkles have wavelengths of tens of meters (Fig. 8.21(A)). This structure is not easily visible on the ground where the surface of the flow is composed of large blocks of glassy rhyolite known as **obsidian** (Fig. 8.21(B)).

Some liquids do not begin to flow until the applied shear stress exceeds a minimum value known as the **yield strength**. Paint is an example of such a liquid. When a thin layer of paint is applied to a wall, it does not flow and drip, but application of a thicker coat increases the shear stress on the paint and it begins to drip. Magmas can have a yield strength, especially when they cool into the crystallization range. One consequence of such behavior is that in the central part of a magma conduit where the shear stress drops to zero, all the magma may flow with the same velocity to produce what is known as plug flow. Toothpaste squeezed from a tube exhibits plug flow because it has a yield strength. Another important consequence for magmas is that crystals that are denser than the magma may not be able to sink unless they can grow large enough to overcome the yield strength.

8.5.3 Diffusion in magma, crystal growth, and grain size of igneous rocks

We have seen that magma viscosity is related to its silica content because silica tends to form a tetrahedral framework whose

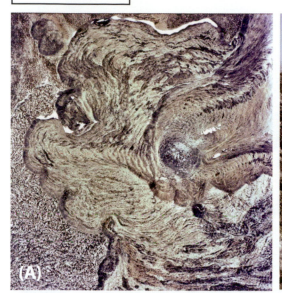

1 km

(A)

(B)

Figure 8.21 **(A)** Aerial photograph of Glass Mountain, Medicine Lake Highlands, California. (Photograph courtesy of U.S. Forestry Service.) Eruption of rhyolitic lava from the dome just to the right of center of the photograph produced thick lava flows whose surfaces are wrinkled into ridges with wavelengths of ~50 m. **(B)** On the ground, the surface of these flows are seen to consist of large blocks of obsidian (glassy rhyolite).

strong bonds are difficult to break. These strong bonds also impede diffusion of ions through magma. We consequently find that magmas with high viscosity have low diffusion rates. Diffusion is important for the growth of crystals and bubbles in magma, and it determines the **grain size** of igneous rocks.

For a crystal (or bubble) to grow in magma, the necessary ions must first form a large enough group to remain stable; small groups are unstable and break apart. This smallest stable group is known as a nucleus. Growth of the nucleus then takes place by ions diffusing through the magma to its surface. The magma surrounding a growing nucleus becomes depleted in the necessary ions for growth, and then ions have to diffuse from greater distance, which takes progressively more time. If the magma is cooling slowly, the ions may have sufficient time to diffuse to the growing crystal, but if cooling is rapid, as in a lava flow, the ions do not have time to diffuse great distances, so instead they form new nuclei. Consequently, if cooling is slow, few nuclei need form because there is sufficient time for ions to diffuse significant distances (~1 cm), but if cooling is rapid, diffusion has time to move ions much shorter distances (<1 mm) and many more nuclei form. The number of nuclei formed is the number of crystals the rock eventually contains when crystallization is complete. If few nuclei form, the rock will be *coarse grained*; if many nuclei form, it will be *fine grained*.

The grain size of igneous rocks is classified as *fine* if the average diameter of grains is less than 1 mm; *medium*, if it is between 1 and 5 mm; and *coarse*, if it is greater than 5 mm. These divisions correspond approximately to whether a microscope, hand lens, or unaided eye is needed to identify the minerals.

Diffusion rates, like viscosity, are affected by the presence of water in the melt, because of its ability to break the tetrahedral framework. Granitic magma with a high water content has much higher diffusion rates than anhydrous granitic magma; consequently, it produces coarser-grained granite. In extreme cases, where a separate vapor phase may be present in which diffusion rates are orders of magnitude greater than in the granitic magma, so few nuclei form that giant crystals with dimensions of meters may result (Fig. 8.22). Growth of such large crystals obviously requires diffusion over great distances. This produces the rock known as pegmatite.

8.6 Magma ascent

As with any fluid, magma flows only in response to pressure gradients. What pressure gradient can cause magma to rise from its source toward the Earth's surface? The answer is buoyancy driven by the density contrast between magma and the surrounding rock; that is, magmas float toward the Earth's surface. Near the Earth's surface, exsolution of gas can cause magma to rise rapidly and produce volcanic explosions.

8.6.1 Buoyancy

If mantle peridotite has a density of 3300 kg/m³ and basaltic magma a density of 2700 kg/m³ (Sec. 8.5.1), the density contrast is 600 kg/m³. Multiplying this by the acceleration of gravity (9.8 m/s²) gives a buoyant pressure gradient of 5880 Pa/m. This is not a huge gradient (recall that 1 atm = 10^5 Pa), but it is sufficient to make magma rise buoyantly through the mantle.

Figure 8.22 Granite pegmatite with meter-size crystals of cream-colored microcline in dark gray quartz (geological hammer with meter-long handle for scale). Hale pegmatite, Middletown pegmatite district, Connecticut.

When basaltic magma enters the lower crust where the rock density is ~3000 kg/m³, the density contrast between magma and surrounding rock drops to 300 kg/m³ and the buoyant pressure gradient decreases to ~2940 Pa/m, which means that magma rises more slowly than through the mantle below and causes a "traffic jam," which is probably compensated for by magma spreading laterally. The traffic jam becomes even more serious when basaltic magma tries to enter the upper crust where the average rock density is ~2750 kg/m³. Here the buoyant pressure gradient drops to only 490 Pa/m, which may be insufficient to overcome the viscous drag of the wall rock, and the magma stops rising. As new magma arrives from the mantle below, it has to spread laterally. This, in fact, is one popular theory for how the lower crust has formed, by **underplating** with basaltic magma.

As basaltic magma rises from the mantle, it expands slightly and decreases its density. It might, therefore, rise buoyantly into the upper crust. In any case, we do observe basaltic magma erupting from volcanoes with densities as low as 2600 kg/m³, which would make such magma remain buoyant through the entire crust. Magma that is denser than the upper crust might still make it to the surface if its conduit remained continuous all the way down into the mantle, where the large density contrast between wall rock and magma could provide the extra force necessary to get the magma through the upper crust. This might happen at divergent plate boundaries where opening fractures could create vertically continuous bodies of magma from the source to the surface.

Most magma probably rises as buoyant lenses that open a fracture and force aside the rock as it passes through; think of a boa constrictor swallowing a large meal. Such lenses do not remain connected to the source region, and their continued rise depends entirely on the density contrast between the magma and the surrounding rock. In general, magmas are thought to rise to a level in the crust where they match the density of their surroundings and then spread laterally. This is known as the **level of neutral buoyancy**. Whether magma spreads laterally as a vertical (**dike**) or horizontal sheet (**sill**) depends on the local stresses. For example, at a divergent plate boundary, a vertical dike might form, but if magma spread laterally into a sedimentary basin, it might intrude as a horizontal sill between the layers of sedimentary rock. In either case, it would be spreading at a level of neutral buoyancy.

8.6.2 Buoyant rise of magma

Buoyancy is the main driving force causing magma to ascend. In the magma source region, liquid is formed where different minerals come in contact. The liquid, therefore, forms an interconnected network of thin films and tubes along grain boundaries. Its flow through this porous mixture of crystals and liquid is identical to the flow of groundwater through soil and is governed by **Darcy's law**, which states that the flow rate is proportional to the pressure gradient. We saw in the previous section that this pressure gradient is produced by the density contrast between the solids and liquid. If a volume of magma is to rise buoyantly, an equal volume of solids must descend. Because the amount of melt in the source region is always small (only a few percent), the rate at which it can rise is determined by the rate at which the solids can deform; that is the solids have to recrystallize and compact downward to expel the buoyant liquid upward. Deformation of the solids is slow and, therefore, the ascent rate of magma out of the source region is slow.

Experiments show that if partially melted rock is deformed, the liquid on grain boundaries tends to segregate into lenses, which greatly increases the velocity with which it can flow. Field evidence indicates that such lenses form in magma source regions in the mantle beneath divergent plate boundaries and in the lower crust in orogenic zones. These lenses coalesce to form still larger veins, which eventually converge to form thick steeply dipping sheets up which magma can travel with significant velocity. On cooling and solidifying, these sheets of liquid form what are known as dikes. Near divergent plate boundaries, for example, swarms of dikes parallel the boundary.

Dikes vary in width from millimeters (Fig. 2.12) to tens of meters, and their width may change as magma passes through them. They do not form by filling open fractures. Instead, magma's buoyant force initiates a fracture and then pushes apart the rock to form the dike. The flow velocity of magma in a dike is proportional to the square of the width of the sheet, and consequently a dike need only become slightly wider to allow much larger volumes of magma to pass through.

To illustrate the velocity with which magma can buoyantly rise in a dike, let us consider basaltic and granitic magma with viscosities of 300 and 10^6 Pa·s, respectively, rising in a 4 m-wide dike. If the density contrast between the magma and wall rock is 100 kg/m³ for the basalt and 300 kg/m³ for the granite, the average velocity in the dike would be 4.4 m/s for the basaltic magma and 4 mm/s for the granitic magma. Clearly, viscosity makes a huge difference to the magma's flow velocity, with basaltic magma moving very rapidly and granitic magma moving extremely slowly.

Because granitic magma flows so slowly through thin sheets, it is believed that this highly viscous magma finds it easier to rise through the lower crust in the form of large domes. Granitic magma formed in the lower crust may start as a laterally extensive sheet, but because of its low density (2400 kg/m³), the sheet becomes gravitationally unstable relative to the overlying dense rocks (>2700 kg/m³) and tries to rise. It cannot rise everywhere at the same time, so instead it develops a series of evenly spaced domes on its upper surface in which the buoyant material rises and between which the denser material sinks. Such gravitational instabilities are common in nature and are known as **Rayleigh-Taylor instabilities**. The rise of thunderheads, for example, is a common example of such a phenomenon (Fig 8.23(A)). Figure 8.23(B) shows a series of photographs of domes forming on a sheet of oil as it rises through an overlying layer of denser honey. The photographs, taken over a period of a number of seconds,

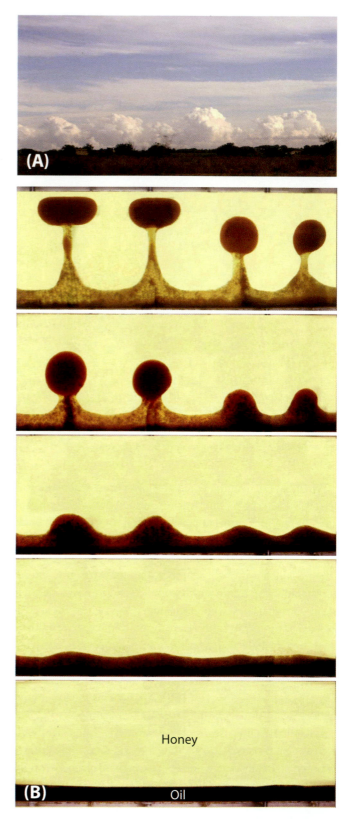

Figure 8.23 **(A)** Regularly spaced thunderclouds rising through cooler denser air is an example of a gravitational instability. **(B)** A series of photographs taken over a period of seconds (bottom to top) showing a layer of oil rising through a denser overlying layer of honey as a series of regularly spaced domes.

show a wave developing on the boundary between the two liquids that grows to form domes and eventually mushroom-shaped bodies as the oil nears the top of the container. Bodies of granitic magma are thought to rise through the crust in such domes at a rate of possibly a couple of meters per year, and when they solidify they form what are known as **batholiths** (Fig. 9.9). The spacing of the domes, which is determined by the viscosities of the two materials and the thickness of the buoyant layer, may account for the regular spacing of batholiths and volcanoes along convergent plate boundaries.

8.7 Processes associated with the solidification of magma in the crust

During magma's ascent through the crust, it begins to cool and crystallize and numerous processes may separate crystals from liquid. This can change the composition of the magma, a process known as **magmatic differentiation**. Magma that does not make it to the surface may cool very slowly in **intrusive bodies**, thus allowing time for differentiation processes to effect dramatic changes in magma composition. Most of the compositional range of igneous rocks is generated at this stage. Differentiation can produce rocks that are compositionally extremely different from the magma from which they form. Some important ore bodies, for example, form when economically important minerals are concentrated by differentiation. Box 8.5 provides a flow chart of the processes that may affect magma as it rises toward the surface. Most of these processes are extremely slow, and only in magma bodies cooling at depth is there sufficient time for them to function. The cooling of magma sets the time scale in which differentiation can operate, so we consider it first before discussing differentiation processes.

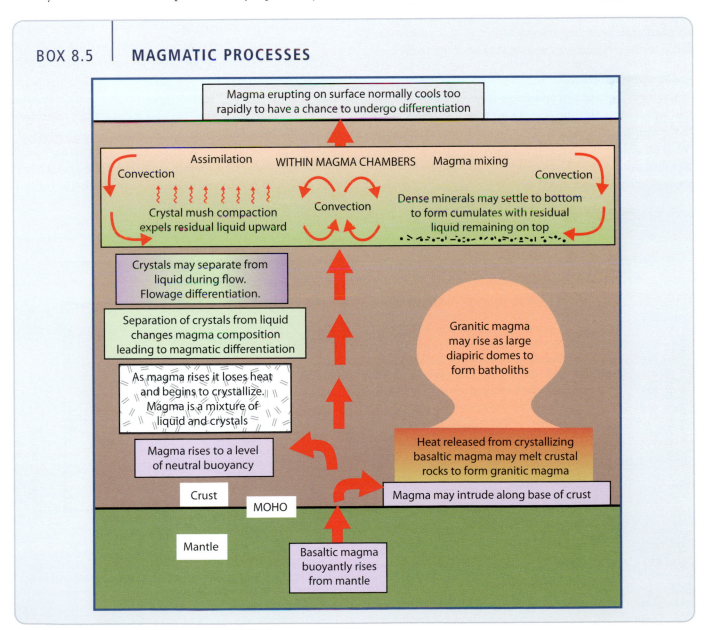

BOX 8.5 | MAGMATIC PROCESSES

Magma erupting on surface normally cools too rapidly to have a chance to undergo differentiation

WITHIN MAGMA CHAMBERS

Assimilation   Magma mixing

Convection   Convection

Convection

Crystal mush compaction expels residual liquid upward

Dense minerals may settle to bottom to form cumulates with residual liquid remaining on top

Crystals may separate from liquid during flow. Flowage differentiation.

Separation of crystals from liquid changes magma composition leading to magmatic differentiation

As magma rises it loses heat and begins to crystallize. Magma is a mixture of liquid and crystals

Granitic magma may rise as large diapiric domes to form batholiths

Magma rises to a level of neutral buoyancy

Heat released from crystallizing basaltic magma may melt crustal rocks to form granitic magma

Crust

MOHO

Magma may intrude along base of crust

Mantle

Basaltic magma buoyantly rises from mantle

8.7.1 Cooling of bodies of magma by heat conduction

A detailed understanding of how igneous bodies cool involves considerable mathematics, which is beyond the scope of this book (for a survey, see Philpotts and Ague, 2009). We can, however, summarize a number of the important principles.

For magma to cool and crystallize, it must transfer heat from the hot magma to the cold surrounding country rocks or to the air in case of the surface of lava. This takes place mainly through **conduction**, where atoms transfer thermal vibrational energy to adjoining atoms. When magma is still hot and fluid, **convection** may transfer heat by bodily moving hot magma nearer to contacts, but heat transfer into the country rock must still be by conduction. Hot lava on the Earth's surface can cool by **radiation** (infrared electromagnetic radiation; see Sec. 3.3) but as soon as a centimeter-thick crust forms, cooling again is controlled by conduction. Conduction, therefore, plays the dominant role in cooling igneous bodies. We discuss it first and touch on the other two processes at the end of this section.

Cooling by conduction

The transfer of heat by conduction is governed by **Fourier's law**, which states that the rate of heat transfer is proportional to the temperature gradient. The heat loss from a body of magma is greatest at first when the temperature gradient is steep and then decreases as the gradient flattens. Heat is transferred through magma (and rock) very slowly, because its **thermal**

conductivity is very low and of the order of 2 to 3 W/m°C (1 watt (W) = 1 joule/second), which compares with 430 W/m°C for a silver spoon with which you might stir coffee.

Fourier's law allows us to calculate the temperature distribution across an igneous contact at any time following the injection of magma. Figure 8.24 shows such temperature distributions across a planar igneous contact, where the temperatures in the magma and in the country rock (T) are expressed as a fraction of the initial magma temperature T_0. The initial country rock temperature is taken to be 0°C; if it is not, we adjust its temperature to zero and subtract the same number of degrees from the magma temperature, because it is only the temperature difference, not the absolute temperatures, that cause cooling. The calculated temperature also assumes that magma at the initial temperature T_0 and country rock at 0°C always exist at some distance from the contact. Also, no account is taken of the latent heat of crystallization, which we will see has an enormous effect on the cooling time.

When magma is first injected, its temperature is T_0, say 1200°C for a basaltic magma, and the country rock is at 0°C. The temperature profile across the contact at time zero steps sharply from 0°C to T_0. This steep temperature gradient causes heat to diffuse rapidly from the magma near the contact into the country rock. Following the initial heat transfer, which affects only the region near the contact (see gradients for 1 week and 1 month), the temperature gradient begins to make heat start flowing at greater and greater distances from the contact.

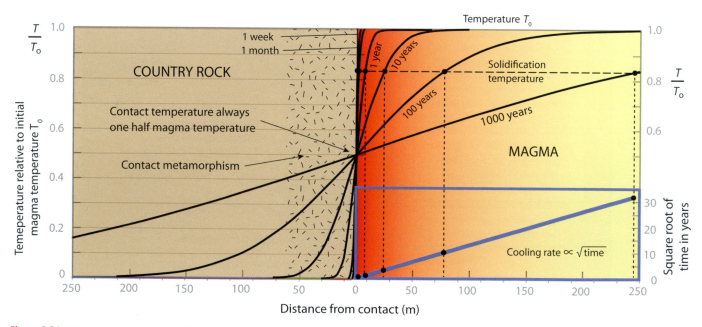

Figure 8.24 Calculated temperature gradients across a planar igneous contact at various times following injection of magma at temperature of T_0 into country rock at 0°C. Temperatures are represented as fractions of the initial magma temperature (T/T_0). No account is taken of the latent heat of crystallization. The inset graph in the lower right shows that the advance of the solidification front, which has a temperature of $T/T_0 = 0.833$, is a linear function of the square root of the cooling time.

With continued cooling, this gradient becomes flatter and extends farther and farther from the contact. The gradients rotate clockwise with time and are symmetrical about the contact; that is, the heat lost from the magma, matches the heat gained by the country rock.

Generalizations about the cooling of igneous bodies

We can draw a number of important conclusions from the calculated cooling curves in Figure 8.24:

- The temperature at the contact of an igneous intrusion is one-half the temperature of the magma immediately following intrusion and remains at this temperature as long as hot magma remains in the center of the body. This result may come as a surprise, because we might expect contact temperatures on huge bodies of magma to be higher than those on small bodies, but size makes no difference.

- Cooling becomes slower as the thermal gradients become flatter, which is a direct consequence of Fourier's law. Comparison of the thermal gradients after 1, 10, and 100 years shows that considerably more time is required to cause cooling as the igneous body cools. This increase can be quantified by tracking the progress of a particular temperature into the body, its solidification temperature for example. Let us assume that country rock at 0°C is intruded by 1200°C basaltic magma that solidifies at 1000°C. We track the solidification front into the magma by following the progress of a value of T/T_0 of 0.833 (i.e., 1000°C/1200°C). The long-dashed horizontal line in Figure 8.24 shows this solidification temperature. After 1 year of cooling, the solidification front advances into the igneous body a distance of 7.66 m; after 10 years, it reaches 24.3 m; after 100 years, it reaches 76.6 m; and after 1000 years, it reaches 243 m. The rate at which it advances slows significantly with time.

- *The cooling rate is proportional to the square root of time.* If the distance from the contact reached by the solidifying basalt in the previous example is plotted against the square root of the cooling time, a straight line is obtained (see bottom of Fig. 8.24). This is one of the most important consequences of Fourier's heat-flow equation, and we find many processes in nature at all scales that are controlled by it.

In late 1959 and early 1960, an eruption of Kilauea Iki, Hawaii, formed a 100 m-deep lava lake whose cooling and solidification was carefully monitored by the U.S. Geological Survey (Fig. 8.25(A)). As soon as the crust on the lake was strong enough to support the weight of a helicopter that brought in a drilling rig, the thickening of the crust was monitored over the following 20 years by drilling through it. The base of the crust was taken to be the depth at which the drill broke through into the magma. This occurred where the percentage of crystals dropped below 50% and the temperature was 1065°C,

which was 85°C below the initial magma temperature. A plot of the thickness of the crust, which tracked the advance of the 1065°C isotherm, against the square root of cooling time is linear (Fig. 8.25(B)). This indicates that cooling of the lava lake was proportional to the square root of time, as predicted by Fourier's law.

Figure 8.25(C) shows a linear plot of the ocean-floor depth versus the square root of its age. This linear relation is explained by the fact that the ocean-floor depth depends on the density of the oceanic crust, which in turn is determined by the cooling time since formation at a midocean ridge.

Effect of latent heat of crystallization on cooling of magma

In Figure 8.25(B), the cooling of the Kilauea Iki lava lake has been calculated using the same equation as that used to construct Figure 8.24; that is, the latent heat of crystallization was ignored. As magma cools, it crystallizes and liberates the latent heat of crystallization (~400 kJ/kg), which is extra heat that must be gotten rid of and that greatly prolongs cooling. The measured cooling times for the Kilauea Iki lava lake are at least three times longer than the calculated cooling times ignoring latent heat (Fig. 8.25(B)). We can conclude, therefore, that the latent heat of crystallization prolongs cooling of magma by at least three times those shown in Figure 8.24, but the cooling rate is still proportional to the square root of time.

8.7.2 Cooling of bodies of magma by convection and radiation

Cooling near igneous contacts causes magma density to increase, which develops a gravitational instability that causes magma to convect. This is particularly true when a magma's viscosity is low. As cooling proceeds and the viscosity increases, convection becomes less likely. Convection dramatically increases cooling rates, because cool magma near contacts is replaced with hot magma that maintains a high temperature gradient across the contact; this, in turn, increases the conduction of heat into the country rock.

The surface of lava flows and particles of volcanic ash ejected into the atmosphere are able to cool by radiation. Radiation cooling is extremely rapid (proportional to the fourth power of the absolute temperature), and as a result the surface of lava flows and particles of ash cool so rapidly that they do not have time to crystallize but are quenched to glass. The rate of radiation cooling can be seen in the chapter-opening photograph, where lava exiting from the front of the flow at 1160°C cools to 700°C (dull red color) in a matter of seconds.

In summary, intrusive bodies of magma cool exceptionally slowly because of very low thermal conductivities and large latent heats of crystallization. Although convection may speed the process early in the cooling of a body, the transfer of heat

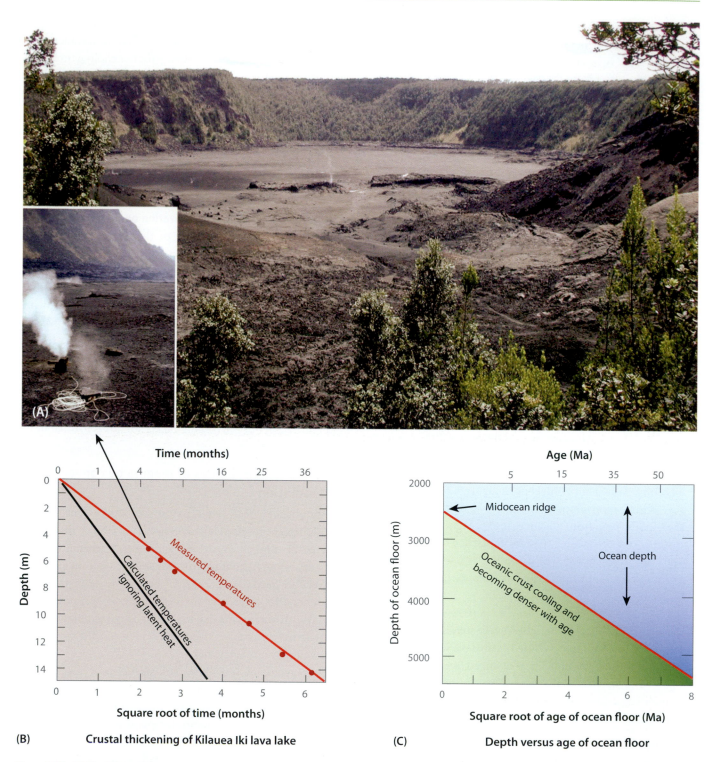

Figure 8.25 (A) The Kilauea Iki crater, Hawaii, filled with a 100 m-deep lava lake during an eruption late in 1959 and early 1960. The inset photo shows the head of one of the U.S. Geological Survey's drill holes with the thermocouple cable used for measuring down-hole temperatures. (B) Thickness of the crust on the Kilauea Iki lava lake plotted against the square root of time since the eruption (data from Peck et al., 1964). Also shown is the calculated cooling time if the latent heat of crystallization is ignored. (C) Linear plot of ocean depth versus the square root of the ocean floor age (after Sclater et al., 1981.)

by thermal conduction through the country rock eventually controls the cooling and crystallization of magma. Except near contacts where cooling may be rapid, bodies of magma remain molten or partially molten for extended periods of time, which provides magmatic differentiation processes ample time to effect changes in magma compositions. The Bushveld Complex, South Africa, for example, took 200 000 years to go from its liquidus to solidus temperatures.

8.7.3 Magmatic differentiation by crystal settling

Lavas from a single volcano commonly show a compositional range that is interpreted to result from magmatic differentiation. Most of this variation can be explained by the separation of crystals from liquid. For example, the compositional variation in Kilauean lavas is due to variable amounts of olivine, which forms prominent phenocrysts in these lavas. Separation of crystals from liquid is undoubtedly the main process of magmatic differentiation, but how does this separation take place?

In 1915, Norman Bowen published an important paper titled "Crystallization Differentiation in Silicate Systems," in which he demonstrated through experimental studies that igneous differentiation could result from *crystal settling*, a process that Charles Darwin had previously postulated. While determining phase diagrams such as those shown in Section 8.3, Bowen noticed crystals of forsterite and diopside had sunk to the bottom of his crucibles in hour-long experiments. If such rapid separation occurs in the laboratory, then surely it must occur in nature, where cooling times are much longer. Following publication of this paper, crystal settling became the favored explanation for differentiation of igneous rocks and remained such during most of the twentieth century.

Crystal settling was well documented in the Kilauea Iki lava lake referred to in Section 8.7.1. Lava erupted from a vent on the side of the crater and produced a large cinder cone of scoriaceous basalt, the edge of which can be seen on the right side of the photograph in Figure 8.25(A). These cinders provided a continuous sample of the erupting lava, which was a porphyritic olivine basalt (Fig. 8.26(A)). Drill holes that later penetrated the entire 100 m-thickness of the lava lake show that olivine crystals had time to sink before the magma solidified. The upper half of the lava lake lost most of its olivine phenocrysts (Fig. 8.26(B)), which gives it a significantly different composition from that of the original magma. The olivine that sank from the upper part of the lava lake accumulated in the lower part to create a rock that is much richer in olivine (Fig. 8.26(C)) than the original magma. Crystal settling, therefore, differentiated the magma in the lava lake during its solidification.

How rapidly can a crystal sink in magma? As we might expect, it depends on how viscous the liquid is and on the density contrast between the crystal and liquid. What is not immediately obvious, however, is that the sinking velocity is proportional to the diameter of the crystal squared; that is, doubling a grain's size quadruples its settling velocity. The crystal size is, therefore, the most important factor. The settling velocity in terms of viscosity, density, and grain size is given by **Stokes' law** (Box 8.6).

Consider how grain size would affect the settling velocity of olivine phenocrysts in the Kilauea Iki lava lake. We can show that the diameters of olivine crystals in Figure 8.26(A) vary by a factor of three, which would equate to a difference in settling

1 mm

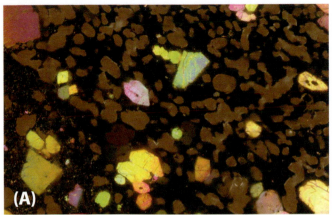
(A)

(B)

(C)

Figure 8.26 Photomicrographs of thin sections under crossed polars of samples from the 1959–1960 Kilauea Iki lava lake (Fig. 8.25(A)) demonstrating crystal settling of olivine (large birefringent crystals). Sample **(A)** is from the cinder cone formed at the feeder vent and contains olivine phenocrysts and many vesicles in a black glass. Samples **(B)** and **(C)** are from U.S. Geological Survey drill core at depths of 33 and 81 m in the 100 m-deep lava lake.

velocity of nine times (3^2). In the sample from the lower part of the lava lake shown in Figure 8.26(C), the one very large olivine crystal, which is almost in extinction, would have sunk from a much greater height in the lava lake than the surrounding smaller crystals.

BOX 8.6 | CRYSTAL SETTLING AND STOKES' LAW

The terminal velocity of a spherical particle sinking (or floating) through a liquid is given by **Stokes' law**:

$$\text{Terminal velocity} = \frac{\text{gravitational acceleration} \times \text{density contrast} \times (\text{grain diameter})^2}{18 \times \text{viscosity}}$$

The equation is valid as long as the flow of liquid around the particle is laminar (not turbulent), which is the case for slowly sinking crystals in viscous magma, and the liquid does not have a yield strength.

From inspection of the equation, we see the following:

- The sinking velocity is proportional to the acceleration of gravity. Thus, sinking would be six times slower on the Moon than on Earth, because the Moon's gravitational acceleration is one-sixth that of the Earth's (1.62 versus 9.8 m/s^2).
- The sinking velocity is proportional to the density contrast between the solid and the liquid. Thus, denser minerals sink faster than less dense ones.
- The sinking velocity is inversely proportional to the viscosity. Thus, a crystal would sink faster in basaltic magma than it would in more viscous granitic magma.
- The sinking velocity is proportional to the square of the diameter of the grain. Thus, if a grain's diameter doubles, it sinks four times as fast.

Stokes' law shows that grain size plays a very important role in crystal settling. In Section 11.6.2, we see that it plays an equally important role in determining what grain-size material can be carried in suspension in rivers.

An olivine-rich layer near the base of the Palisades Sill, which forms the prominent cliff along the Hudson River across from New York City, was one of the first bodies of rock that was interpreted to have formed through crystal settling. We can use Stokes' law (Box 8.6) to calculate the velocity at which olivine crystals might have sunk in this sill. The basaltic magma that formed the sill is estimated to have had a density of 2620 kg/m^3 and a viscosity of 400 Pa·s. If the olivine crystals had a diameter of 2 mm and a density of 3500 kg/m^3, the crystals would have sunk at 4.8×10^{-6} m/s, which is 1.7 cm/hour, or 151 m/year. The sill is 300 m thick and would have taken many years to solidify, so this settling velocity is more than adequate to have allowed olivine crystals to sink and form the olivine-rich layer near the base of the sill.

Igneous cumulates

When dense crystals sink and accumulate on the floor of a body of magma, they can form rocks that have significantly different compositions from the magma from which they formed (differentiation). For example, the rock in the lower part of the Kilauea Iki lava lake is enriched in olivine relative to the basalt that filled the lava lake. Rocks that are formed by accumulation of crystals are called **cumulates**, with the name of the mineral accumulating being added as an adjective; thus, the rock in the lower part of the lava lake is an olivine cumulate. The mineral that accumulates is known as the *cumulus phase* and the minerals that crystallize from the surrounding liquid are known as the *intercumulus phases*. Identification of the intercumulus

material in the lower part of the Kilauea Iki lava lake is easy, because it looks no different from the basaltic rock formed from the liquid that was left behind in the upper part of the lava lake; it consists of a fine-grained mixture of pyroxene, plagioclase, and opaque grains. However, in more slowly cooling intrusive bodies, the intercumulus liquid can crystallize to an extremely coarse-grain size that forms a cement surrounding the cumulus grains. For example, Figure 8.27 shows a pyroxene cumulate from the Great Dike of Zimbabwe (the Great Dyke of Rhodesia in older literature), in which a single large intercumulus grain of plagioclase surrounds 12 euhedral cumulus grains of augite and orthopyroxene. Rocks that have smaller grains embedded in larger grains are said to have a **poikilitic texture**.

Igneous layering

Many igneous rocks exhibit layering due to variations in the abundance of minerals (**modal layering**). Slowly cooled bodies of basaltic magma commonly exhibit prominent layering, such as that in the Skaergaard Intrusion of Southeast Greenland (Fig. 8.28(A)–(B)). Wager and Deer, who first studied this intrusion, attributed this layering to different rates of crystal settling from magma currents that convected down the walls and across the floor of the magma chamber. They postulated that the denser minerals, such as olivine and pyroxene, settled first and the less dense plagioclase settled last, creating graded layers (Fig. 8.28(D)). Figure 8.28 shows features found in layered rocks that indicate deposition from magma currents, such as trough structures (B), cross-bedding (C), rip-up clasts (E),

1 mm

Figure 8.27 Photomicrograph under crossed polars of a pyroxene cumulate from the Great Dyke of Zimbabwe (Rhodesia), containing cumulus crystals of augite (Aug) and orthopyroxene (Opx) surrounded by intercumulus plagioclase (Pl).

settling. Moreover, if the basaltic magma has a yield strength, as many do, even olivine and pyroxene crystals could not sink unless they had diameters well in excess of a centimeter, which is unusual. Other factors that may play a role in forming layers in slowly cooled igneous rocks include, different rates of nucleation and crystal growth on solidifying surfaces and redistribution of minerals by diffusion through the liquid in piles of crystal mush.

8.7.4 Compaction of crystal mush

We have seen that crystal settling has difficulty explaining the differentiation of basaltic magma because plagioclase tends to be neutrally buoyant and dense minerals may not separate if the magma has a yield strength, but the chemical evidence from differentiated suites of rocks indicates that all these minerals must be able to separate from the melt. One mechanism that solves both of these problems is **crystal mush compaction**. A mush forms when crystals are abundant enough to make contact with each other. Almost two-thirds crystallization is necessary before equidimensional crystals, such as olivine, form a mush, but platy plagioclase crystals can form a mush with as little as one-third crystallization. Once a three-dimensional network of crystals forms, it behaves as a deformable porous medium through which the remaining liquid can flow, in much the same way water might be squeezed from a wet paper towel. It is the bulk density of the crystal mush relative to that of the interstitial liquid that causes compaction of the mush and expulsion of the interstitial liquid and not the density of individual crystals. Even if plagioclase is slightly buoyant in the liquid, as long as denser minerals such as olivine and pyroxene are present to raise the bulk solid density above that of the liquid, compaction takes place. In addition, a thick pile of compacting crystals is able to overcome the yield strength of magma that might prevent individual crystals from sinking. The resulting change in the composition of the liquid will consequently indicate subtraction not only of olivine and pyroxene but also of plagioclase. In large igneous intrusions, where cooling rates are slow, compaction of crystal mush is a likely mechanism of igneous differentiation. It is also the mechanism by which basaltic magma is extracted from the partially molten mantle beneath divergent plate boundaries.

8.7.5 Assimilation and fractional crystallization

Magma from the mantle may come in contact with rocks of strikingly different composition in the crust, which can contaminate the magma. Near the margins of many igneous bodies, xenoliths in various stages of **assimilation** are common (Fig. 8.29(A)). Most magmas do not have enough heat to totally melt xenoliths, but heat liberated by crystallization can partially fuse them, creating a low-melting fraction that is added to the magma. This process of differentiation is known as assimilation and fractional crystallization (AFC).

and slump structures (F). These structures resemble features seen in sedimentary rocks formed in fluvial environments and are strong evidence that layering in igneous rocks can form through similar sedimentary processes, albeit from a far more viscous liquid at much higher temperatures.

The sedimentary-like appearance in the gently dipping layers on the floor of basaltic intrusions was used as evidence that crystal settling was the main mechanism by which these magmas differentiated. However, identical structures are found in steeply dipping layers on vertical walls where gravity could not possibly be the cause. Also, although olivine and pyroxene are dense enough to settle from basaltic magma, plagioclase has about the same density as the liquid and would not sink. Calculated settling velocities for olivine and pyroxene in basaltic magma are orders of magnitude less than the velocity with which the magma convects, giving the crystals little chance of

Figure 8.28 (**A**) Gently dipping compositionally graded layers in gabbro of the Skaergaard Intrusion, southeast Greenland. Graded layers, pass from olivine rich (dark) at the base, through pyroxene rich to plagioclase rich (light color) at the top. (**B**) A large trough structure trending to the upper left of the photograph is filled with layered gabbro. (Photographs in (A) and (B) courtesy of Christian Tegner, Aarhus University, Denmark.) (**C**) Gently dipping plagioclase- and orthopyroxene-rich layers in the Bjerkreim-Sokndal Intrusion, southwestern Norway exhibiting channel and cross-bedding structures. (**D**) Detail of layers in (C) showing grading from pyroxene-rich at right (bottom) to plagioclase-rich at left (top). (**E**) Mafic layers in the Main Zone of the Bushveld Complex, South Africa (Fig. 9.8), which were ripped up by a magmatic disturbance prior to complete solidification. (Photograph courtesy of Grant Cawthorn.) (**F**) Layers of olivine gabbro (gray) slumped and draped over a block of yellowish-brown peridotite (olivine + pyroxene) on the Isle of Rum, Scotland.

Figure 8.29 (A) Igneous breccia with xenoliths of metamorphosed sedimentary rocks stoped from the contact (lower right of photograph) of the Brome Mountain gabbroic intrusion, Quebec. Many xenoliths of the metamorphosed sedimentary rocks that strike into the intrusion in the lower right of photograph show signs of partial melting. (B) Detail of slate xenolith showing veins of granite formed by partial melting.

Most xenoliths of country rock do not melt completely. Instead, only a low-temperature eutectic fraction forms. The amount of this melt depends on the composition of the xenolith. For example, in the igneous breccia at the margin of the Brome Mountain gabbroic intrusion, Quebec (Fig. 8.29(A)), a fragment of pure quartz shows no signs of melting, because it does not contain any other minerals with which to form a eutectic mixture. In contrast, fragments of metamorphosed sandstone, which contain both quartz and alkali feldspar, are almost totally melted and form ghostlike xenoliths. Fragments of slate, which contain abundant refractory alumina-rich minerals, produced only a small amount of eutectic granite melt, which forms thin light-colored veins throughout the xenolith (Fig. 8.29(B)).

Fractional crystallization of tholeiitic basaltic magma eventually produces small amounts of granitic residual liquid (see Sec. 8.3.3 and Fig. 8.7(C)). The effect of assimilation is simply to increase the amount of this granitic fraction.

Assimilation by Reaction

Some crustal rocks change the composition of magma by reacting with it rather than having to melt. This is particularly true of carbonate sedimentary rocks. For example, when magma intrudes sedimentary rock containing dolomite, the following reaction can occur:

$$\underset{\text{dolomite}}{CaMg(CO_3)_2} + \underset{\text{silica in magma}}{SiO_2} \Rightarrow \underset{\text{diopside}}{CaMgSi_2O_6} + \underset{\text{vapor}}{2CO_2}$$

This reaction removes silica from the magma is known as a **desilication reaction**. Desilication may lead to the formation of nepheline from feldspar through the following reaction (note the minus sign in the reaction indicating subtraction of silica):

$$\underset{\text{albite}}{NaAlSi_3O_8} - \underset{\text{silica}}{2SiO_2} \Rightarrow \underset{\text{nepheline}}{NaAlSiO_4}$$

Desilication reactions, like fusion, require heat, which is provided by the latent heat of crystallization of magma, so they normally cause magmas to solidify.

8.7.6 Liquid immiscibility

When some magmas cool, they split into two liquids, a process known as **liquid immiscibility**. Magma that becomes iron rich through fractional crystallization of ferromagnesian minerals, such as olivine and pyroxene (Sec. 8.3.5 and Fig. 8.9), can intersect a field of liquid immiscibility, where the liquid splits into silica-rich and silica-poor fractions that form droplets of one in the other (Fig. 8.30). The silica-rich liquid consists of a network of shared silica tetrahedra having a structure similar to that of feldspar but less regular (Sec. 6.4.2), whereas the silica-poor

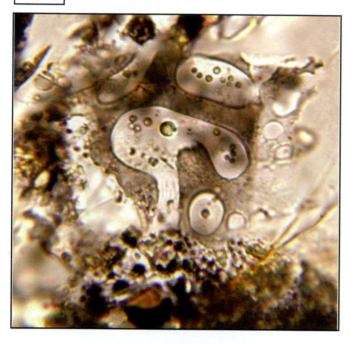

10 μm

Figure 8.30 Photomicrograph under plane polarized light of droplets of silica-rich (clear) and silica-poor (brown) immiscible glasses in the groundmass of a basalt from Southbury, Connecticut.

liquid consists of isolated tetrahedra bonded together by Mg, Fe, and Ca. Because of the structural difference between the two liquids, elements may prefer to enter one liquid more than the other. This is known as element partitioning. Iron, for example, has a strong preference for the silica-poor liquid, which explains its brown color. As a result of element partitioning, the silica-rich liquid has the composition of granite and the silica-poor liquid the composition of iron-rich pyroxene. These liquids have very different densities and may separate, with the less dense granitic fraction floating upward, just as oil does in salad dressing. Lenses of granite toward the top of slowly crystallizing bodies of basaltic magma may have separated in this way.

A rare type of calcite-rich igneous rock known as **carbonatite** probably forms from an immiscible carbonate-rich melt that separates from silica-poor alkaline magma. Although rare, carbonatites are extremely important in being the host rock for ore deposits of niobium and **rare earth elements**. These elements have many high-tech uses in devices such as cell phones, computers, and catalytic converters in cars. The high concentration of these elements in carbonatite results from the preferential partitioning of these elements into the carbonate, rather than the silicate, immiscible liquid.

Liquid immiscibility also plays an important role in the formation of certain copper and nickel sulfide ore deposits. If basaltic magma is sufficiently rich in sulfur, a dense immiscible sulfide liquid may separate and sink to the bottom of an intrusion. Elements such as copper, nickel, platinum, and palladium partition strongly into the sulfide liquid and can form ore deposits of these elements.

Summary

In this chapter, we described how magma is generated, what determines its initial composition, and how that composition can be modified as the magma rises toward the surface. The following are the main points addressed in this chapter:

- The temperature in the Earth normally lies well below the beginning of melting of peridotite, the rock that forms the upper mantle. Melting occurs only near hot spots, at divergent plate boundaries where the mantle undergoes decompression, or at subduction zones where water rising from subducting plates lowers the solidus in the overlying mantle wedge.

- The latent heat of fusion of rock is so high that only partial melting ever occurs in the Earth. Partial melting of peridotite, the olivine-pyroxene rock in the upper mantle, produces basaltic magma under anhydrous conditions and andesitic magma under hydrous conditions. These partial melts have risen from the mantle to produce the Earth's crust.

- A mixture of minerals melts at a lower temperature than do individual minerals. The lowest melting mixture is known as a eutectic. Because the Earth does not have excess heat, most magmas have compositions close to eutectic compositions. Most basalt has a composition close to the plagioclase-pyroxene eutectic, and the composition of granite is close to the quartz-alkali feldspar eutectic.

- Minerals that melt to form liquid of the same composition melt congruently. Minerals that melt by breaking down into a liquid of a different composition and a new solid melt incongruently. The liquid formed from incongruent melting is known as a peritectic. Peritectics are also important in determining magma compositions.

- Crystals belonging to solid solution series continuously react with liquid as they crystallize. If equilibrium is not maintained, which is commonly the case, compositionally zoned crystals result.

- Although increasing pressure raises the melting point of anhydrous rocks, it lowers the melting point if water is present, because at high pressure, water dissolves in magma. When pressure decreases as magma approaches the surface, gas exsolves and can cause violent volcanic explosions.

- The density of magma is ~10% less than that of the equivalent solid rock.

- Magma viscosities vary over six orders of magnitude, from fluid basalt to highly viscous granite. Viscosity increases with silica content because bridging oxygen atoms form strong framework structures, and it decreases with water content, which breaks the framework structure.

- The grain size of igneous rocks depends on the number of nuclei that form. Rapid cooling promotes rapid nucleation, which results in fine grain size.

- Magma rises toward the surface because of buoyancy and stops when it reaches a level of neutral buoyancy.

- In the source region, magma flows along grain boundaries by porous flow. As it rises, it segregates into veins that eventually feed into large sheets, which at divergent plate boundaries form dikes parallel to the plate boundary.

- Highly viscous granitic magma formed in the lower crust rises slowly in large mushroom-shaped domes known as batholiths.

- Magma and rock are such poor conductors of heat that magma cools very slowly.

- Magmatic differentiation refers to any process by which magma compositions can change. Most differentiation is caused by separation of crystals from the liquid in magma (fractional crystallization).

- Dense crystals may sink to the bottom of magma chambers to form rocks known as cumulates. These rocks are commonly layered with the proportions of different minerals varying gradationally through a layer.

- Crystal mush on the floor of a magma chamber undergoes compaction with upward expulsion of interstitial liquid.

- Magma can change its composition by assimilating intruded rocks but only with heat supplied by crystallization of the magma. Consequently, assimilation is commonly accompanied by fractional crystallization.

- Liquid immiscibility, which splits magma into contrasting liquid fractions, plays a role in the formation of some granite formed from basaltic magma, carbonate-rich igneous rocks, and some magmatic sulfide ore bodies of nickel and copper.

Review questions

1. What is the evidence that Earth's crust and mantle are normally solid?

2. How do we know that the upper mantle is composed of peridotite?

3. If the geotherm is normally well below the beginning of melting temperature of anhydrous mantle peridotite, what three mechanisms cause melting of this peridotite?

4. What plate tectonic settings provide the conditions for the three different melting mechanisms of mantle peridotite, and what composition magmas are generated at each?

5. What is the latent heat of fusion, and why does it limit the amount of melting that takes place in the Earth?

6. Why is the eutectic of importance in determining the composition of magmas?

7. What is a porphyritic rock, and what does it tell you about the composition of magma relative to the eutectic composition?

8. Using the lever rule, measure the proportion of liquid to anorthite in the magma of composition z in Figure 8.6 at a temperature of T_1.

9. What evidence indicates that magmas are rarely superheated?

10. Using the phase diagram for nepheline-quartz (Fig. 8.7), explain how igneous rocks can be classified as either alkaline or tholeiitic.

11. If peridotite containing a 50/50 weight % mixture of forsterite and enstatite (Fig. 8.8) were heated just to the beginning of melting and the liquid was removed to form a magma that rose toward the surface, what mixture of minerals would this magma eventually crystallize on cooling?

12. If an olivine liquid containing 50 weight % forsterite (Fig. 8.9) were cooled slowly until it was 50% crystallized, and then all of the crystals were separated from the liquid (perhaps as a result of crystal settling) what composition would the remaining liquid have (use the lever rule)?

13. In light of your answer to question 12, what general conclusion can you draw about the effect of fractional crystallization on the composition of magmas?

14. Can you think of a reason why olivine phenocrysts show very little compositional zoning, whereas plagioclase phenocrysts are almost always zoned, despite the similarity in the phase diagrams for these two solid solution series (Figs. 8.9 and 8.10)?

15. Discuss the differences between hypersolvus and subsolvus granites, and explain why subsolvus granites are less likely to rise to the surface than are hypersolvus granites.

16. Calculate the pressure at the base of an 11 km-thick oceanic crust with an average density of 3000 kg/m³. Make the calculation for an oceanic ridge that just reaches the ocean surface (e.g., Iceland) so no account need be taken of the density of water.

17. Why does the melting point of anhydrous rock increase with increasing pressure?

18. Why does the melting point of hydrous rock decrease with increasing pressure?

19. Explain explosive volcanism in terms of the solubility of volatiles in magma.

20. Why are the viscosities of basaltic and rhyolitic magma so different?

21. What determines the grain size of an igneous rock?

22. What determines the rate at which basaltic magma can escape from partially molten mantle peridotite?

23. Why does buoyant material segregate into regularly spaced domes when it begins to rise?

24. You have been asked to walk out onto a newly crusted over lava lake in Hawaii 7.5 days (648 000 s) after the eruption. It is planned to bring in a helicopter with a large drilling rig as soon as the crust is 2 m thick. You carry a small portable drill, which allows you to determine that the crust is 1 m thick. How long will it be before the helicopter and drilling rig can land on the surface of the lava lake?

25. If plagioclase is neutrally buoyant in basaltic magma, how can compaction of crystal mush lead to its concentration in the lower part of an igneous body?

26. If magmas do not have excess heat, where does heat come from to cause melting of included blocks of country rock?

ONLINE RESOURCES

Copies of historic (1565–1835) papers expressing ideas on volcanoes, basalt, and geologic time can be found at http://www.lhl.lib.mo.us/events_exhib/exhibit/exhibits/vulcan/index.shtml.

The MELTS program of Ghiorso and Sack (1995) uses thermodynamic data to calculate, from a chemical analysis, the physical properties of a magma and the sequence of minerals that will crystallize from it. The program can be used on the Web or can be downloaded for free from http://melts.ofm-research.org.

For Smithsonian's world map of volcanoes, earthquakes, impact craters, and plate tectonics, see http://www.minerals.si.edu/tdpmap.

For Smithsonian's list of volcanoes as an overlay on Google Earth, see http://www.volcano.si.edu/world/globallists.cfm?listpage=googleearth.

U.S. Geological Survey volcanic hazard Web site: http://www.usgs.gov/hazards/volcanoes

The VolatileCalc program of Newman and Lowenstern (2002) for calculating the solubility of H_2O and CO_2 in magmas can be downloaded from http://volcanoes.usgs.gov/yvo/aboutus/jlowenstern/other/software_jbl.html.

FURTHER READING

Bowen, N. L. (1928). *The Evolution of the Igneous Rocks,* Princeton University Press, Princeton, NJ (reprinted in 1956, Dover Publications, New York). This book is of historic importance because it marks the beginning of the use of physical-chemical principles for the study of igneous rocks. Be warned, however, that it is difficult reading.

Philpotts, A. R., and Ague, J. J. (2009). *Principles of Igneous and Metamorphic Petrology,* 2nd ed., Cambridge University Press, Cambridge. This book presents a detailed and quantitative coverage of all topics presented in this chapter.

Sources of data referenced in this chapter include the following:

Bowen, N. L. (1913). The melting phenomena of the plagioclase feldspars. *American Journal of Science,* 34, 577–599.

Bowen, N. L., and Anderson, O. (1914). The binary system MgO-SiO_2. *American Journal of Science,* 37, 487–500.

Bowen, N. L., and Schairer, J. F. (1935). The system MgO-FeO-SiO_2. *American Journal of Science,* 29, 151–217.

Bowen, N. L. and Tuttle, O. F. (1950). The system $NaAlSi_3O_8$-$KAlSi_3O_8$-H_2O. *Journal of Geology,* 58, 489–511.

Burnham, C. W. (1979). The importance of volatile constituents. In *The Evolution of the Igneous Rocks: Fiftieth Anniversary Perspectives,* ed. H. S. Yoder, Jr., Princeton University Press, Princeton, NJ, 439–482.

Dixon, J. E., Stolper, E. M., and Holloway, J. R. (1995). An experimental study of water and carbon dioxide solubilities in mid-ocean ridge basaltic liquids. Part I: Calibration and solubility models. *Journal of Petrology,* 36, 1607–1631.

Greig, J. W., and Barth, T. F. W. (1938). The system $Na_2O \cdot Al_2O_3 \cdot 5SiO_2$ (nephelite, carnegieite)–$Na_2O \cdot Al_2O_3 \cdot 6SiO_2$ (albite). *American Journal of Science,* 35A, 93–112.

Holtz, F., Behrens, H., Dingwell, D. B., and Johannes W. (1995). H_2O solubility in haplogranitic melts: Compositional, pressure, and temperature dependence. *American Mineralogist,* 80, 94–108.

King, S. D. (1995). Models of mantle viscosity. In *Mineral Physics and Crystallography: A Handbook of Physical Constants, American Geophysical Union Reference Shelf 2,* ed. T.J. Ahrens, American Geophysical Union, Washington, DC, 227–236.

Kushiro, I., Syong, Y., and Akimoto, S. (1968). Melting of a peridotite nodule at high pressures and high water pressures. *Journal of Geophysical Research,* 73, 6023–6029.

Moore, G., Vennemann, T., and Carmichael, I. S. E. (1998). An empirical model for the solubility of water in magmas to 3 kilobars. *American Mineralogist,* 83, 36–42.

Peck, D. L., Moore, J. G., and Kojima, G. (1964). Temperatures in the crust and melt of Alae lava lake, Hawaii, after the August 1963 eruption of Kilauea volcano–a preliminary report. *U.S. Geological Survey Professional Paper* 501D, 1–7.

Petford, N. (2003). Rheology of granitic magmas during ascent and emplacement. *Annual Review of Earth and Planetary Sciences,* 31, 399–427.

Sclater, J. G., Parsons, B., and Jaupart, C. (1981). Oceans and continents: similarities and differences in the mechanisms of heat loss. *Journal of Geophysical Research,* 86, 11535–11552.

Silver, L., and Stolper, E. M. (1989). Water in albitic glass. *Journal of Petrology,* 30, 667–709.

Takahashi, E. (1986). Melting of a dry peridotite KLB-1 up to 14 GPa: implications on the origin of peridotitic upper mantle. *Journal of Geophysical Research,* 91, 9367–9382.

Tuttle, O. F., and Bowen, N. L. (1958). Origin of granite in the light of experimental studies in the system $NaAlSi_3O_8$–$KalSi_3O_8$–SiO_2–H_2O. *Geological Society of America Memoir,* 74.

Weill, D. F., Hon, R., and Navrotsky, A. (1980). The igneous system $CaMgSi_2O_6$-$CaAl_2Si_2O_8$-$NaAlSi_3O_8$: variations on a classic theme by Bowen. In *Physics of Magmatic Processes,* ed. R. B. Hargraves, Princeton University Press, Princeton, NJ, 49–92.

Wright, T. J., Ebinger, C., Biggs, J., et al. (2006). Magma-maintained rift segmentation at continental rupture in the 2005 Afar dyking episode. *Nature,* 442, 291–294.

Wyllie, P. J. (1977). Crustal anatexis: an experimental review. *Tectonophysics,* 13, 41–71.

CHAPTER

9

Igneous rocks

Their mode of occurrence, classification, and plate tectonic setting

The aim of this chapter is to introduce the main types of igneous rock and their modes of occurrence. We classify igneous rocks according to their mineral content as well as by their texture, which varies according to whether the magma solidifies slowly beneath the surface or rapidly on the surface. We accordingly start the chapter by discussing where magmas cool and solidify and the types of igneous bodies formed. We then discuss the classification of igneous rocks using the scheme proposed by the International Union of Geological Sciences. Although many igneous rocks have been found and named, the vast majority are described by only a few names, and these rocks form common associations that can be related to their plate tectonic settings. We discuss and illustrate examples of the main rock types in each of these associations and explain why each forms in its particular setting. We end the chapter by discussing three unusual rock associations, two were formed only in the Precambrian, and the third was formed by large meteorite impacts.

Anthony Philpotts in October 2005 beside the 60 m-wide lava lake on the Erte'ale shield volcano in the Afar region of Ethiopia (see E'a in Box 8.4(A) for location). Convecting magma rises where the large bubble is bursting, traverses the lava lake, and descends the opposite wall. The dark crust, which forms almost immediately by radiation cooling, splits apart as it moves radially away from the ascending plume, exposing molten lava immediately beneath the crust and producing the irregular red lines emanating from the bubble burst.

So far in this book we have classified rocks into igneous, sedimentary, and metamorphic, and we have discussed the origin of magmas and igneous rocks using only a few common rock names, such as basalt, andesite, and granite. In this chapter, we consider the classification of igneous rocks in more detail, and we see, for example, that some rock types commonly occur together, whereas others are never associated. In Chapter 8, we learned about igneous rock-forming processes – why rocks melt, where they melt, and how they melt. We saw that the melting process controls the composition of the liquid, which determines the composition of magmas that ascend into the crust. We also learned about the physical properties of magma and of processes that can change its composition and, hence, the spectrum of igneous rocks it can form. We are in a position to use this knowledge, first, to describe where igneous rocks form; second, to learn how igneous rocks are classified; and finally, to relate igneous rocks to plate tectonics.

9.1 Why an igneous rock classification is necessary

In the natural sciences, the development of nomenclature to describe and classify the natural world has been a necessary first step to understanding natural phenomenon. For example, in biology, the classification of plants and animals led to understanding the principles behind the diversity of life, as laid out in Darwin's *Origin of Species*. Similarly, the classifications of minerals and rocks led to an understanding of the processes that govern their diversity.

The historical literature contains vast numbers of igneous rock names, not all of which have been used consistently. Fortunately, the internationally accepted rock classification that we introduce in this chapter helps eliminate many of the problems associated with rock nomenclature.

Igneous rocks are classified using three different criteria:

- Mode of occurrence
- Mineralogical makeup
- Chemical composition

Igneous rocks that are formed by magma erupting on the Earth's surface appear strikingly different from those that cool slowly at depth, even if they have the same mineralogical makeup. On the Earth's surface, the appearance of rocks can vary considerably depending on the type of volcano or type of eruption involved. Much of the nomenclature of igneous rocks relates to textural differences resulting from the environment in which magma solidifies. Second, magmas have a wide compositional range, which results in the crystallization of many different minerals and provides the criteria for the second mode of classification. Some igneous rocks are so fine grained or even glassy that minerals may not be identifiable, in which case chemical criteria must be used to classify them.

Because the textures and mineralogical makeup of igneous rocks, which are the basis for their classification, are so dependent on their mode of occurrence, we start by discussing where magmas solidify and the types of bodies they form.

9.2 Mode of occurrence of igneous rocks

In Chapter 8, we saw that as soon as magma is generated, it buoyantly rises toward the surface and must, therefore, intrude overlying rocks. If magma reaches the surface, it extrudes as lava or erupts as volcanic ash. Igneous rocks and the bodies they form are, therefore, classified as being either **intrusive** or **extrusive**. The intrusive rocks and bodies are further subdivided into **hypabyssal** and **plutonic** (derived from *Pluto*, the Greek god of the underworld), depending on whether they crystallize near the surface or at great depth, respectively. The distinction, which does not imply a specific depth, is based mainly on grain size. Plutonic magma cools and crystallizes so slowly that few crystal nuclei form, which results in coarse-grained rock. Hypabyssal rocks crystallize relatively rapidly and have large numbers of nuclei and are, therefore, finer grained. Indeed, most hypabyssal rocks resemble volcanic rocks more than they do plutonic rocks.

The type of body in which an igneous rock forms determines many of the properties that are used for classification. An important step in classifying an igneous rock is, therefore, to appreciate where the rock comes from. The most important igneous bodies are shown in Figure 9.1. Although we have detailed knowledge about extrusive bodies through eyewitness accounts of volcanic eruptions, many intrusive bodies are poorly understood, especially large bodies. Nowhere is a complete cross-section through the Earth's crust exposed, and thus the shape of large igneous bodies must be pieced together from multiple exposures of different bodies that have been eroded to different depths, and considerable controversy still surrounds the precise shape of large bodies.

The formation of intrusive bodies is a necessary first step in the ascent of magma, so we deal with these first and follow this with a discussion of extrusive bodies.

9.2.1 Shallow intrusive igneous bodies: dikes, sills, laccoliths, cone sheets, ring dikes, and diatremes

Intrusive bodies vary in size from millimeter-thick sheets (Fig. 2.12) to huge bodies covering thousands of square kilometers (Fig. 9.9). Large igneous bodies almost certainly are created from multiple injections of small pulses of magma. We have seen that the latent heat of fusion of rocks prevents large-scale melting in the Earth, and once formed, melts buoyantly rise toward the surface. Magma can, therefore, never be stored in large quantities in the source region; instead, it probably rises in small batches. The volume of lava released during a single eruptive episodes gives some idea of the volume of these batches. For example, the

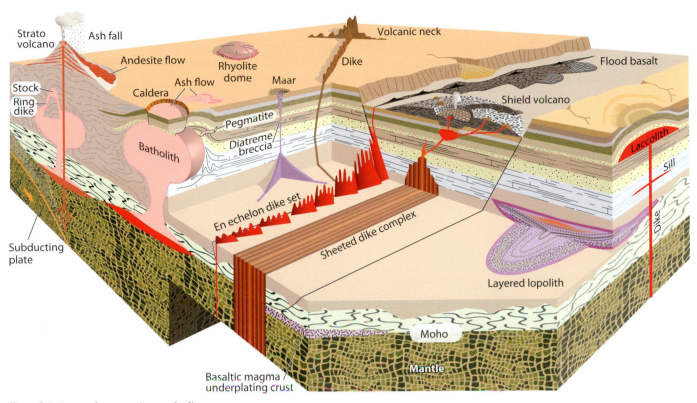

Figure 9.1 Forms of common igneous bodies.

largest eruption of lava in historic times is 14 km³ of basalt at Lakagigar, Iceland (see Sec. 9.2.3). Granitic magma would need to travel in larger batches to overcome the much greater viscous forces of that magma. We can conclude that large intrusive bodies of igneous rock are formed from multiple pulses of magma.

Intrusive bodies are divided into two broad categories based on their relation to the surrounding rocks. If an igneous body parallels the layering in the surrounding rock, it is said to be **concordant**, and if it cuts across the layering, it is **discordant**. Magma always intrudes in a direction that enables it to force aside the rock in the minimum stress direction. The orientation of the minimum stress direction relative to layering in surrounding rocks, therefore, determines whether an intrusion is concordant or discordant.

Dikes, sills, and laccoliths

The simplest intrusive bodies are sheetlike intrusions, which have a length to breadth ratio on the order of 100 to 10 000, the ratio increasing with confining pressure and depth. A discordant sheet is a **dike** (also spelled dyke), and a concordant one is a **sill** (Figs. 9.1 and 9.2). They can form at any depth in the crust, but horizontal sills are more common near the Earth's surface where the minimum stress direction is vertical. Although dikes and sills can have any dip depending on the attitude of the layering in the host rock, large dikes tend to have steep dips, and large sills tend to

Figure 9.2 Cretaceous dikes and sills in gently dipping Ordovician limestone on Mount Royal in Montreal, Quebec.

have shallow dips. The rocks in these sheets are commonly more resistant to weathering than are the surrounding rocks, which results in their standing out in relief. Many steeply dipping dikes erode to form walls (Fig. 9.3(A)–(B)), hence their name, from the Dutch word *dyke*. Gently dipping sills erode to form ridges with prominent escarpments on their up-dip side.

Dikes vary in width from millimeters to tens of meters, reaching, in rare cases, 150 m. Dikes that are tens of meters wide can be hundreds or even a thousand kilometers long (see Fig. 9.9). Some long dikes, rather than forming a continuous sheet, form a series of shorter segments that overlap with adjoining segments to form en echelon dike sets (Fig. 9.1).

Figure 9.3 (**A**) Shiprock volcanic neck, New Mexico, rises 335 m above the surrounding plane. The prominent ridge extending to the left of the neck is one of a number of dikes that radiate from the neck, as shown in (**B**). (**C**) Satellite image showing dikes radiating from Shiprock. (Satellite image from NASA Earth Observatory, www.visibleearth.nasa.gov/17587/shiprock.)

Figure 9.4 Thick Jurassic sill in the Dry Valleys region of Antarctica. (Photograph courtesy of Jean H. J. Bédard of the Geological Survey of Canada.)

Sills can be much thicker than dikes because they make room for themselves by lifting the overlying rocks. For example, the 300 m-thick Palisades sill intruded near the base of the Mesozoic sedimentary rocks in the Newark Basin, New Jersey, exploiting the density difference between lighter overlying sedimentary rocks and denser underlying crystalline rocks. Thick sills can be laterally extensive, the Palisades sill being traceable along strike for 80 km. The Whin sill, although only 75 m thick, can be traced 125 km across northern England. The best-exposed large sills are in the Dry Valleys region of Antarctica, where they intrude sediments deposited in Mesozoic basins that formed during the breakup of Gondwanaland (Fig. 9.4).

Dikes and sills are commonly formed by multiple injections of magma, with each new batch being intruded near the center of the sheet. Dike and sill margins cool rapidly and are finer grained than interior parts. These **chilled margins** are commonly found inside wide dikes and sills where later pulses of magma froze against earlier pulses. When a dike or sill contains more than one injection of magma, it is described as multiple if successive injections are of essentially the same composition, or composite, if they are significantly different.

Dikes are particularly common in zones of crustal extension, where they form in large numbers known as **dike swarms**. Dike swarms form at divergent plate boundaries (Fig. 9.1) where magma, formed by decompression melting in the ascending asthenosphere (Sec. 8.3.2), intrudes as a dike along the plate boundary. Chilled margins develop along both contacts of the dike, but as plate divergence continues, the initial dike is split in half and a new dike intrudes along its center, again producing chilled margins on both contacts. Subsequently, this dike is split, and another dike intrudes along its center. The result is a whole series of dikes, with those on one side of the spreading axis showing chilled margins on only one side and those on the other side of the axis having chilled margins on the opposite side (Fig. 9.1). Such dike swarms are known as **sheeted dike complexes**. They are important because they form new crust at divergent plate boundaries (see Sec. 9.4.1, Figs. 9.31 and 9.32(C), and Box 8.4).

Dikes can form **radial swarms** around prominent igneous features, such as volcanic necks (Fig. 9.3(C)). They can also occur on a huge regional scale, such as those radiating southward for more than 1000 km from the Mackenzie region of

(A)

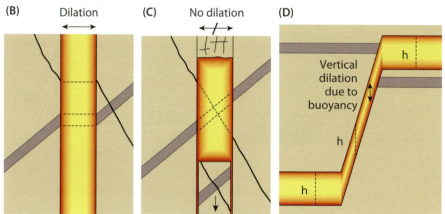

(B) Dilation (C) No dilation (D)

Vertical dilation due to buoyancy

h

h

h

Figure 9.5 (A) Quartz-rich layer (light gray) in metamorphic rock offset by intrusion of a granite pegmatite. (B) Simple dilation resulting from intrusion of a sheet of magma offsets structures in the wall rocks perpendicularly across the sheet. (C) A sheetlike body showing no dilation, as evidenced by the alignment of structures on either side of the sheet. (D) The thickness of sills joined by sloping dikes is commonly the same as the vertical distance across the dike, which indicates that magma intruded buoyantly (see, e.g., Fig. 9.4).

northwestern Canada. These huge radiating patterns are probably related to the heads of mantle plumes (Fig. 1.6(E)).

The intrusion of a dike or sill normally involves magma simply forcing the wall rocks apart. This process of dilation offsets any layering or veins in the wall rock that intersect the intrusion at angles other than 90° and were present prior to the intrusion (Fig. 9.5(A)–(B)). The dilation associated with sills that are connected by dikes commonly provides evidence that the force causing intrusion of magma is **buoyancy**. Figure 9.5(D) illustrates two sills connected by an obliquely dipping dike. The dilation associated with the intrusion of the sills is indicated by their thickness, h. The dike connecting the two sills is much narrower than the sills, but its width measured in a vertical direction is the same as the thickness of the sills. This indicates that magma was emplaced buoyantly and that the country rock in the upper left side of the figure floated vertically in the magma. Figure 9.4 shows an example of such relations where a large slab of sandstone has floated upward from the base of the major sill to produce a series of steplike sills, each connected by a dike whose width in the vertical direction is the same.

Pressure generated by buoyancy is able to propagate fractures through solid rock. The intrusion of magma concentrates enormous stress on the tip of a fracture, which causes the fracture to propagate in the same way that it would if we were to drive a wedge into the fracture. The same principle is used in the engineering practice known as **hydraulic fracturing** (**hydrofracking** or **fracking**), where water pumped into a well under high pressure fractures the surrounding rock, which increases the permeability of the region surrounding the well.

When magma pressure causes fractures to extend, magma typically lags behind the tip of the fracture because of its viscosity. As sills extend laterally, viscous drag on the magma makes it increasingly difficult for the sill to propagate. Eventually, the magma finds it easier to lift the overlying rocks than to continue spreading laterally. This is particularly true near the Earth's surface. When a sill's tendency to lift the overlying rocks exceeds its tendency to spread laterally, the overlying rocks dome upward and form a mushroom-shaped intrusion known as a **laccolith**, from the Greek words *laccos*, "cistern," and *lithos*, "stone" (Figs. 9.1 and 9.6). Laccoliths typically form within 3 km of the Earth's surface. Deeper laccoliths must lift more overlying rock, and, therefore, their diameter must be greater to generate the leverage necessary to dome the overlying rocks.

Figure 9.6 Maverick Mountain, western Texas, is a partially eroded laccolith intrusive into less resistant limestone.

Cone sheets and ring dikes

Changes in magma pressure in chambers beneath volcanoes can fracture the overlying rock (Fig. 9.7(A)). When pressure increases, a conical fracture propagates from the top of the chamber toward the surface, into which magma may intrude to form a **cone sheet**. When pressure decreases, steeply outward dipping fractures form, which allows large segments of the roof to sink into the chamber, with magma welling up into the fracture to form a **ring dike**. Cone sheets are rarely more than a couple of meters thick (Fig. 9.8(B)), but ring dikes can be much thicker and, therefore, contain coarser-grained rocks. Ring fractures may penetrate through to the Earth's surface, in which case a **caldera** forms (see the following section). The alternation between over- and underpressured magma chambers commonly results in successive generations of cone sheets and ring dikes.

Diatremes

Another type of near surface intrusion is the **diatreme** (Fig. 9.1), which we mentioned when discussing how samples of the Earth's mantle are brought rapidly to the surface (Sec. 8.2.1 and Fig. 8.1(A)). Diatremes are elongate carrot-shaped bodies filled with broken rock fragments known as **breccia**. The fragments include rocks through which the diatreme passes on its way to the surface. These fragments, which can include diamond-bearing rocks from depths of more than 200 km, are thoroughly mixed with fragments from other depths, some of which may be from above the present depth of exposure of the diatreme. The breccia matrix consists of finer particles abraded from larger fragments. In rare diatremes, the breccia can have an igneous matrix that contains the magnesium mica, phlogopite (Fig. 9.38). This matrix, which is present in diatremes that contain diamonds, is called **kimberlite**, named after Kimberly, South Africa, where diamonds, were first found in bedrock (previously diamonds had been found only in sands and gravels). The surface expression of a diatreme is referred to as a **maar**, a shallow explosive volcanic crater surrounded by debris from the crater but from which no lava erupts.

(A)

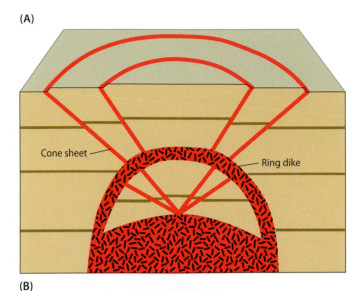

(B)

Figure 9.7 **(A)** Inward-dipping cone sheet and steeply outward-dipping ring dike formed above a shallow magma chamber. **(B)** Cone sheet on the Isle of Mull, Scotland, dipping toward the volcanic center, located to the upper right of the photograph.

9.2.2 Plutonic igneous bodies: lopoliths, batholiths, and stocks

Lopoliths

A **lopolith** (*lopas* is Greek for "shallow basin") is a large saucer-shaped intrusion, which can have a diameter of hundreds of kilometers. The Bushveld Complex in South Africa, the largest known lopolith, has a diameter of about 300 km and a thickness of 8 km (Fig. 9.8(A)). Almost all large lopoliths are Precambrian, and the Bushveld is 2.0 billion years old. Although lopoliths are generally concordant, they are so large that their relation to the intruded rock may be difficult to ascertain. This is particularly true of the overlying roof rocks, which are commonly melted and difficult to distinguish from rocks that are generated by differentiation from the magma that fills the lopolith. Intrusion of the lopolith at Sudbury, Ontario, was triggered by the impact of a large meteorite. The roof rocks to this lopolith

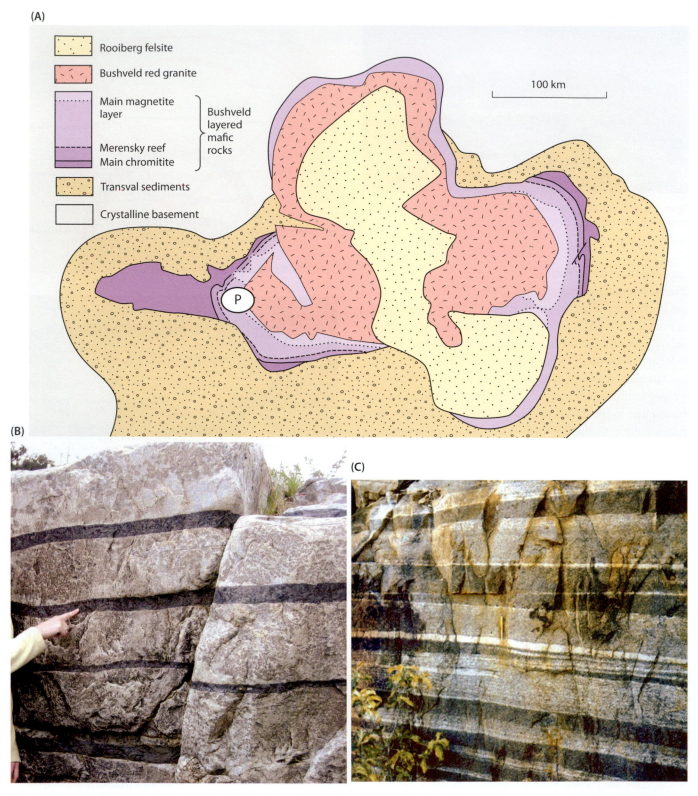

Figure 9.8 (**A**) Map of the Bushveld Complex, South Africa, the world's largest lopolith. The Pilanesberg Mountain intrusion (P) is a younger body (simplified from maps in Wager and Brown, 1967). (**B**) Chromitite layers in plagioclase-rich rock in the Dwars River section in the eastern part of the Bushveld. (Photograph courtesy of Alan Boudreau.) (**C**) Variable amounts of cumulus plagioclase, orthopyroxene, and clinopyroxene creating layers in the Main Zone of the Bushveld. (Photograph courtesy of Grant Cawthorn.)

consist of impact debris that was deposited back in the meteorite explosion crater (see Sec. 9.5.3). Perhaps other large Precambrian lopoliths fill meteorite explosion craters, in which case they would not strictly be intrusive bodies. However, lopoliths are so large that they crystallize slowly (200 000 years for the Bushveld) and contain extremely coarse-grained rocks.

Lopoliths contain rocks formed from basaltic magma, but because of differentiation during crystallization, extreme rock types can form, ranging from rocks composed almost entirely of olivine (**dunite**), or chromite (**chromitite**) through to granite. Many of the rocks exhibit meter- to millimeter-scale layering (Fig. 8.28), which parallels the saucerlike shape of the intrusion. The layering results from variations in the abundance of minerals, variation in the composition of minerals, or alignment of elongate grains (Fig. 9.8(B)–(C)). Many economic mineral deposits occur in lopoliths, including ores of platinum, chromium, copper, nickel, and vanadium. The Merensky Reef (Fig. 9.8(A)) and the Upper Group 2 chromitite layer in the Bushveld Complex, for example, host two of the world's most important platinum deposits.

Because of their enormous size, lopoliths are not formed from a single intrusion of magma. Instead, multiple pulses continuously recharge the chamber and mix with whatever magma remains from previous pulses, which may have undergone differentiation by the time of the new magma influx. Mixing of magmas at different stages of fractionation can change the sequence of crystallization and can produce some of the mineral layering seen in lopoliths. For example, in the Bushveld Complex, a large volume of new magma was injected close to the level of the Merensky Reef, and in the 2.7 billion-year-old Stillwater Complex of Montana, an influx of new magma marks the horizon where platinum mineralization occurs in that intrusion. New batches of magma in these two cases appear to have triggered platinum mineralization, but the exact nature of the mechanism is still hotly debated.

Batholiths

Batholiths are large bodies of granitic rock that extend to considerable depth in the crust and have exposures on the Earth's surface exceeding 100 km^2 (Figs. 9.1 and 9.9); if smaller than this, they are called stocks (see "Stocks" section herein). Detailed mapping of many batholiths indicate that they are composed of numerous large intrusions. The name *batholith* is derived from the Greek word *bathos*, which means "depth," which alludes to the fact that they have deep roots. Because of their enormous size, their extension to depth is uncertain and has to be inferred from different bodies that have been eroded to different depths. Some batholiths are exposed at shallow depths, where they intrude their own volcanic capping. Others are exposed at midcrustal levels and still others at deep-crustal levels. Near the surface, batholiths intrude brittle rock and appear to intrude by stoping (see Sec. 8.7 and Fig. 8.29). In the lower crust, where the surrounding rocks behave plastically, magma makes room for itself by pushing aside the surrounding rock, as shown in Figure 9.11. Although the rocks in most batholiths are massive, they commonly exhibit a foliation due to alignment of feldspar phenocrysts or streaky concentrations of dark minerals. This foliation, which generally parallels the margins of the batholith, has also been used to show that magma in some batholiths has flowed horizontally for considerable distances.

50 km

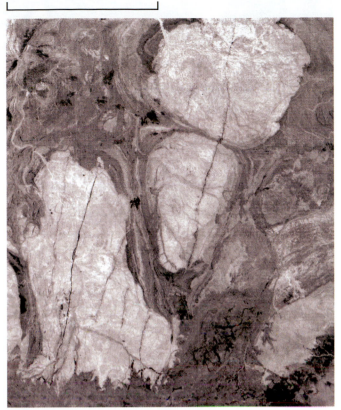

Figure 9.9 Satellite image of deeply eroded sections through granite batholiths in the Pilbara district, Western Australia. Thin north-northeasterly trending dark lines that cut the batholiths are basaltic dikes. (Photograph courtesy of NASA.)

Batholiths typically form in orogenic belts and are intimately related to the melting process that accompanies the release of water from subducting plates. However, some batholiths may be related to mantle plumes beneath continental crust, whereas others are associated with continental rifting. In orogenic belts, batholiths commonly have a regular spacing (Fig. 9.9), which is similar to the periodicity of volcanic centers along island arcs. The spacing could be governed by the same principles that led to the regular spacing of the domes in the oil-and-honey model shown in Figure 8.23 (see Rayleigh-Taylor instability in Sec. 8.6.2).

Batholiths are always composed of rocks that we refer to as *granitic* in the broadest sense of the word, but as we show in Section 9.4.5, rocks containing smaller amounts of quartz and larger amounts of dark minerals than typical granite are also abundant. Isotopic studies reveal that most rocks in batholiths have been derived from melting of the continental crust, but in some, a significant input of mantle-derived magma can be identified. A characteristic feature of granitic rocks in most batholiths is inclusions of basalt. Figure 9.10(A) shows a quarry face in the granite batholith at Vinalhaven, on the coastal islands of Maine, in which one large blob and several smaller pieces of basalt can be seen enclosed in the granite. Because blobs of basalt, whether solid or liquid, have a higher density than granitic magma, the granite must have been a crystal mush at time of entrainment of the basalt, otherwise the basalt would have sunk to the bottom of

Figure 9.10 (**A**) Basalt inclusions in the Vinalhaven granite coastal Maine, United States. Sheet joints parallel the surface of the outcrop. (**B**) Large pillowlike blobs of basalt separated by granite in a deeper section of the Vinalhaven granite.

the intrusion. In a deeper exposure of the Vinalhaven batholith, still larger blobs of basalt form pillowlike bodies in the granite (Fig. 9.10(B)). Similar basaltic pillows are found near the base of other batholiths around the world. The ubiquity of basaltic inclusions in granitic batholiths points to an important role for an input of mantle-derived magma in the generation of batholiths.

Observations from many batholiths allow us to construct a general model of a batholith (Fig. 9.11(A)). First, their overall

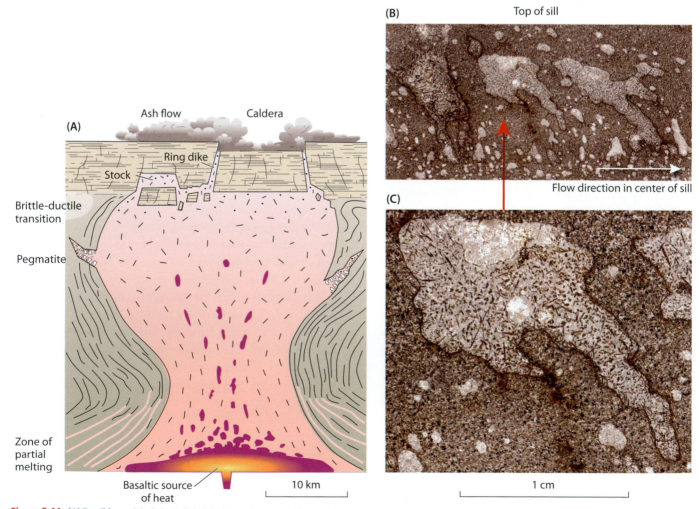

Figure 9.11 (**A**) Possible model of a batholith. (**B**) Photomicrograph of a large thin section from the upper part of a basaltic sill near Montreal, Quebec, showing diapirs of alkali feldspar-rich rock that buoyantly rose through the basalt. (**C**) Enlarged view of the central diapir reveals features similar to those found in granite batholiths.

shape is probably that of a diapir, which buoyantly rises through the lower plastic part of the crust and then mushrooms out on nearing the brittle upper crust. Such a diapir would probably resemble the shape of those formed in the oil-and-honey model (Fig. 8.23). A batholith's extension to depth would be via the neck, which tends to narrow and pinch off as the diapir rises; some may even completely separate from their root zone. On nearing the surface, the diapir would spread laterally, which would explain the evidence for horizontal flow in some batholiths. Near the surface where rocks are brittle, further rise of the batholith would be through stoping. Large sections of the crust might sink into the magma chamber, producing a caldera on the surface. The thermal energy required to form and maintain a batholith as it rises through the crust is enormous and in most cases requires input of heat from the mantle. The most likely source of heat would be from intrusion of basaltic magma into the lower crust. If this magma is injected rapidly, it could fountain into the bottom of a batholith, forming the large pillowlike blobs of basalt found near the floor of many batholiths.

Figure 9.11(B) shows bodies of rock that were formed by very similar processes to those that form batholiths but on a centimeter scale rather than the kilometer scale of batholiths. As described in Box 9.1, they provide a remarkable analogy to the very much larger batholith.

Stocks

If the surface exposure of a pluton is less than 100 km², it is called a **stock**. Although this distinction from a batholith (>100 km²) appears arbitrary and indicates only the area of exposure of an intrusion at its present level of erosion, large numbers of intrusions fall into this category and are clearly of very different origin from batholiths. First, they have steep contacts and are commonly cylindrical in form. Second, they are not limited to rocks of granitic composition. Indeed, many stocks are composed of rocks derived from basaltic magma, including rocks with extreme compositions, such as peridotite (pyroxene and olivine) and syenite (mainly alkali feldspar). The isotopic composition of many of these rocks indicates derivation directly from the mantle. Alkaline rocks associated with rift valleys and hot spots commonly form stocks. Some stocks form in the roof zone of granite batholiths where a cylindrical piece of the roof sinks into the magma chamber and magma buoyantly rises to fill the space (Fig. 9.11(A)).

9.2.3 Extrusive igneous bodies: flood basalts, shield volcanoes, composite volcanoes, domes, calderas, ash-fall and ash-flow deposits

Once magma extrudes onto the Earth's surface, the shape of the body is no longer determined by how magma makes room for itself but by its viscosity. As viscosity increases, **lava** flows more

---

| BOX 9.1 | **CENTIMETER-SCALE DIAPIRS THAT MIMIC BATHOLITHS** |

The sample shown in Figure 9.11(B) is from the upper part of a meter-thick basaltic sill near Montreal, Quebec. The sill was formed by multiple injections of basaltic magma that differentiated to produce a low-density alkali feldspar-rich liquid that collected to form sheets beneath the downward solidifying roof. With intrusion of a new batch of magma in the center of the sill, the roof rocks were reheated and softened. The sheet of low-density differentiated liquid developed a Rayleigh-Taylor instability and rose into the denser overlying basaltic crystal mush as a series of regularly spaced diapirs, which solidified to form the light-colored rock. This spacing would be analogous to the regular spacing of batholiths along an orogenic belt. As the diapirs rose, their ascent was terminated by the cooler upper part of the sill, which formed an impervious brittle barrier against which they flattened out. This would be analogous to the mushroomlike head that forms on some batholiths on reaching the brittle crust. The diapirs floated vertically, but their tails all point to the lower right in the photograph because magma in the center of the sill (below in the image) continued to flow to the right as the diapirs rose. The relative motion of tectonic plates similarly shears granite batholiths rising through the crust at convergent plate boundaries.

Figure 9.11(C) shows an enlarged view of the central diapir in Figure 9.11(B), which can be compared with the batholith model in Figure 9.11(A). In this photomicrograph, the diapir can be seen to vary in composition from bottom to top. Although it is composed largely of alkali feldspar (gray), dark needles of augite and hornblende are more abundant in the lower part of the diapir than in the top. Many batholiths show similar compositional zoning. Along the top of the diapir and in one central patch, the rock is significantly lighter color because of a much lower content of mafic minerals and the presence of abundant analcime (white). Analcime is a hydrous mineral, so clearly the upper part of this diapir was enriched in water. Granite batholiths are also enriched in volatiles toward the top, which can lead to formation of pegmatites or can result in violent explosive volcanism accompanying caldera collapse.

slowly, but it also decreases the magma's ability to rid itself of gas bubbles, which leads to more explosive eruptions and formation of fragmental material, which is referred to as **tephra** (from the Greek word *tephra* for "ashes"). When tephra solidifies, it forms a rock called **tuff**.

How explosive an eruption is and the proportion of tephra to lava created is one way to describe the style of a volcanic eruption. **Hawaiian eruptions** involve fluid lava with little associated explosive activity. **Strombolian eruptions** liberate more gas in the form of bursting bubbles that eject blobs of magma tens of meters into the air (see photograph opening this chapter). These ejected particles can become streamlined as they fly through the air and develop twisted spindle-shaped tails. Large blobs (>64 mm) are known as **bombs**. When a bomb lands, its interior is commonly still molten, and the bomb flattens to form a shape resembling a cow pancake (Fig. 9.12(A)). Particles between 64 mm and 2 mm are referred to as **lapilli** and less than 2 mm as **ash**. Blobs that accumulate around the vent build

Figure 9.12 Volcanic eruptive styles. (**A**) Strombolian eruptions eject volcanic bombs, which may flatten into a pancake shape on landing, as seen in this bomb from Haleakala, Maui, Hawaii. (**B**) Spatter cones on Bartholomew Island, Galapagos, formed from the buildup of bombs around the vent. (**C**) Plinian eruption of Mount St. Helens on June 9, 1982, formed a 3.5 km-high column of ash. (Photograph courtesy of James Zollweg.) (**D**) Pliny the Younger described the column of ash rising from Vesuvius in 79 CE as having the shape of a Roman pine tree, as seen here behind the Forum in Rome.

steep-sided **spatter cones** Fig. 9.12(B)). **Vulcanian eruptions** occur when lava in the vent solidifies and prevents gas from escaping until the pressure builds sufficiently to cause a more violent explosion. The fragments ejected are predominantly solid and angular. If the amount of gas released is extremely large, particles can be ejected thousands of meters into the atmosphere. As the column of ash rises, it typically spreads out into a large mushroom-shaped cloud (Fig. 9.12(C)). This type of eruption was first described by Pliny the Younger (a Roman lawyer whose many letters have survived and tell us much about this period in history) when Vesuvius erupted in 79 CE, burying the town of Pompeii and killing his famous uncle, Pliny the Elder. He likened the shape of the cloud to the umbrella shape of the pine trees of Rome (Fig. 9.12(D)). This type of eruption is referred to as a **Plinian eruption**. Eruptions of high-temperature ash flows, such as those described by Lacroix from Mont Pelée in Martinique in the eastern Caribbean, are referred to as **Peléan eruptions**. A continuous spectrum exists between these various types of eruption, and some volcanoes can simultaneously erupt ash from the summit and lava from the flanks.

The **volcanic explosivity index (VEI)**, quantifies the explosive power of volcanoes. The index, like the Richter scale for earthquakes, is logarithmic, so each successive number is a 10-fold increase in explosivity. The explosivity is gauged from the volume of tephra produced and the height of the explosive column. Eruptions such as those in Hawaii where mostly lava is erupted have values near zero. The 1980 eruption of Mount St. Helens in Washington State had a VEI of 5, and the 1883 eruption of Krakatoa in Indonesia had a VEI of 6. The largest explosive eruptions in the geologic record are estimated to have had a VEI of 8. Fortunately, no such eruption has occurred in historical times, but they will occur in the future and will have a significant effect on our atmosphere (see the section "Calderas" here on Yellowstone).

Volcanic eruptions are also classified by the shape of the volcanic structures they form. These structures are determined largely by magma viscosity, which in Section 8.5.2 we saw can vary by six orders of magnitude in going from basalt to rhyolite. The shape of the volcanic structure also depends on whether the eruption occurs from a central vent or a long fissure, and whether or not it is explosive. We now consider each of the main volcanic edifices.

Figure 9.13 (**A**) View north from Krafla volcano, Iceland, along the rift valley separating the North American (high ground to left) and European (high ground to right) plates. Between 1975 and 1984 magma flowed northward in dikes from Krafla and erupted from fissures to flood across the rift valley forming the black basaltic flow. Small spatter and cinder cones developed along some of the fissures. (**B**) Location of Iceland relative to Mid-Atlantic Ridge. (**C**) Map of Iceland showing the Mid-Atlantic ridge rift system (shaded), and the locations of Lakagigar (L) and Krafla (K). The red rectangle is the region photographed in (A).

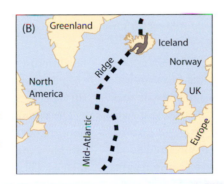

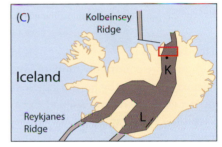

(A)

(A)

(B)

Figure 9.14 (**A**) Columnar cooling joints in the 15 m-thick basaltic lava flow at Aldeyjarfoss, Iceland. At the left of the photograph, columns fan around where the lava cooled against the bank of the former riverbed in which it flowed. (**B**) "Chisel marks" on the side of a column in (A).

Flood basalts

Although we commonly think of lava erupting from conical-shaped volcanoes, by far the largest volumes of volcanic rock in the geologic record erupted from long fissures that have very little topographic relief. This type of eruption involves fluid basaltic magma that erupts from long fissures and floods the countryside to form a **flood basalt**. They occur in regions of crustal extension and are commonly the precursors to the breakup of tectonic plates. For example, flood basalts erupted at the beginning of the Jurassic Period in eastern North America and Morocco when Pangaea broke apart to form the North Atlantic Ocean. A little later, the Parana basalts in South America and the Kirkpatrick basalts in Antarctica erupted with the breakup of Gondwanaland. Still later, Tertiary flood basalts of the Brito-Arctic, or Thulean, Province accompanied the opening of the northern part of the North Atlantic. Flood basalts are also associated with rift valleys and mantle plumes, such as those in the Afar region of Ethiopia (Box 8.4(A)).

No large flood basalt has formed in historical times. The world's largest historic eruption is, nonetheless, a small flood basalt, which erupted 14 km^3 of magma in 1783 from a 25 km-long fissure at Lakagigar, Iceland (L in Fig. 9.13(C)). This fissure formed at the divergent boundary where the Mid-Atlantic Ridge crosses Iceland. Smaller flood basalt flows have formed more recently to the north of Lakagigar near the Krafla Volcano (Fig. 9.13(A)), which is situated in the rift valley on the north side of Iceland (Fig. 9.13(C)). Here, magma intruding

northward along dikes from the main volcanic center of Krafla erupted from fissures to produce small flood-basalt flows in the rift valley separating the European and North American plates.

The volume of these eruptions pales in comparison with flood-basalt flows in the geologic record, which individually can have volumes of hundreds or thousands of cubic kilometers. They occur in regions where multiple flood-basalt flows add up to staggering volumes of lava. For example, in the Columbia River area of Washington and Oregon, Miocene flood basalts cover an area of 120 000 km^2 and have a total thickness of 1500 m (Fig. 2.21(E)). Around Lake Superior, the Keweenawan basalts form a 1.1 billion-year-old flood-basalt province. One of these flows, the Greenstone, forms the backbone of the Keweenaw Peninsula on the south side of Lake Superior and is up to 600 m thick. Such large eruptions and the gases they emit can have severe environmental consequences. The eruption of the Deccan flood basalts in western India, which cover an area of 500 000 km^2 and have an average total thickness of 600 m, have been blamed for the extinction of dinosaurs at the end of the Cretaceous (meteorite impact has also been blamed; see Sec. 9.5.3). Eruption of the 251 million-year-old Siberia flood basalts, which cover an area of at least 1 500 000 km^2, have been blamed for the mass extinction at the end of the Paleozoic era.

When flood-basalt flows cool and shrink, sets of **columnar joints** propagate into the flow from their lower and upper surfaces. Those that extend up from the base form regular five- to six-sided columns perpendicular to the surface from which they cool and are referred to as the **colonnade** because of their resemblance to the columns in a Greek temple, and those that propagate down from the top, which are known as the **entablature**, are smaller and form more irregular patterns that fan out from master joints down which water percolates and promotes cooling (Fig. 9.14(A)). The sides of columnar joints are marked by decimeter-wide horizontal segments (**chisel marks**), which indicate the distance the joint propagated on each successive fracture episode (Fig. 9.14(B)). The cracking sound of these successive fracture events has been recorded on cooling Hawaiian lava lakes.

Divergent tectonic plate boundaries with which flood basalts are associated are topographic lows that may be occupied by lakes or arms of the ocean, so eruption commonly occurs into water. When this happens, the surface of lava rapidly forms a crust, and to advance, lava breaks through the crust and issues as sacklike blobs that pinch off and roll down the front of the flow. These blobs, which are called **pillows**, are still molten inside a thin glassy crust when they come to rest on the upper surface of previously deposited

Figure 9.15 Vertically dipping basaltic pillow lava on the Ungava Peninsula, northern Quebec. The convex upper surfaces of the pillows face toward the right.

Figure 9.16 (**A**) Mauna Loa, the world's largest shield volcano on the Big Island of Hawaii, rises 4200 m above sea level and 8000 m above the ocean floor. (**B**) Satellite image showing the 165 m-deep caldera at the summit of Kilauea, a younger shield volcano to the southeast of Mauna Loa. The Halemaumau crater is a dormant lava lake. Eruption dates of individual flows on the caldera floor are shown. HVO is the U.S. Geological Survey Hawaiian Volcano Observatory. (Image from NASA's Visible Earth catalog: http://visibleearth.nasa.gov.) (**C**) View from Kilauea's caldera rim (point C in (B)) toward the Halemaumau crater in the floor of the caldera.

(A)

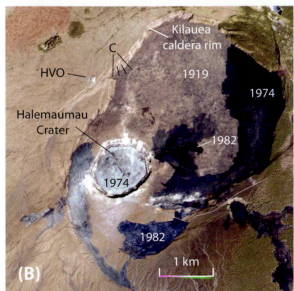

(B)

(C)

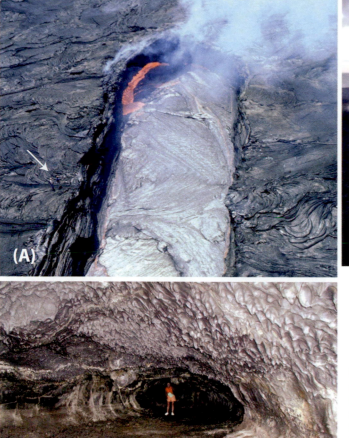

Figure 9.17 (**A**) Lava draining into a tube from the lava lake at Pu'u 'Ō'ō to the east of Kilauea in 1987. Note two geologists at lower left for scale (arrow). (**B**) Lava that descended from Pu'u 'Ō'ō issues from beneath the surface of the flow into the sea along the south shore of the Big Island of Hawaii. (Photograph courtesy of Brian Elms.) (**C**) Lava tube at Pu'uhonua o Hōnaunau National Historic Park, Hawaii.

pillows to which they mold their shape. Pillows consequently develop a characteristic convex upper surface and concave lower surface where they have molded themselves to the topography of the underlying pillows. This shape can be used to determine the top of the pillows in folded pillow lava successions, such as the vertically dipping lava shown in Figure 9.15. Much of the basalt erupted at midocean ridges forms pillows (Fig. 2.21(A)).

Shield volcanoes

Basaltic lava erupted from a central vent builds a broad gently sloping volcano, which, because of its shape, is called a **shield volcano** (Fig. 9.16(A)). Although fed from a central conduit, shield volcanoes tend to be leaky and have many points on their flanks from which eruptions may occur (Fig. 9.1). This tends to destroy their symmetry. Flank eruptions commonly occur along prominent fissures or **rift zones**. Shield volcanoes typically have a large down-dropped area at their summit known as a caldera (Fig. 9.16(B)–(C)). Seismic surveys reveal that calderas collapse into near-surface magma chambers (Fig. 9.1). As a

chamber is replenished with new magma, the volcano expands, only to be deflated when magma erupts onto the floor of the caldera (Fig. 9.16(B)) or moves into the flanks of the volcano, where it can erupt from rift zones. Calderas are distinct from smaller volcanic **craters** whose form is determined by the expansion of gas escaping from the volcanic vent.

Most of the material erupted from shield volcanoes consists of basaltic lava flows, but small cones of scoriaceous cinders can form above active vents. Small volumes of differentiated magma may also erupt. Basaltic lava forms either **pahoehoe** or **aa** flows (Sec. 8.5.2; Fig. 8.20), both of which are able to travel great distances before solidifying. When eruption rates are high, lava flows in open channels, but when the rate is moderate or low, flows crust over, but the lava continues to move inside the flow (Fig. 9.17(A)). Lava that flows beneath an insulating crust can travel many kilometers (Fig. 9.17(B)). When the supply of magma is cut off, lava drains from the flow's interior, leaving a **lava tube** (Fig. 9.17(C)). The flanks of shield volcanoes may have many such tubes. Whether eruption is rapid or slow, magma on shield volcanoes normally descends the flanks a considerable distance. This, along with the small amount of ash produced, gives shield volcanoes their characteristic shape.

Figure 9.18 **(A)** The Osorno composite volcano, Chile. **(B)** Alternating layers of ash and lava in the crater wall of the Masaya composite volcano in Nicaragua. **(C)** A lahar, formed during the 1980 catastrophic eruption of Mount St. Helens in Washington State, descended the Toutle River valley, forming the dark material in the foreground, which has since been cut into by the river. **(D)** This lahar consists of various size rock fragments suspended in a muddy matrix.

Composite volcanoes

When magma is more viscous than basalt, which happens with increasing silica content (Sec. 8.5.2), magma cannot flow as far before solidifying, and it cannot rid itself of gas bubbles as easily, so the amount of explosive activity increases. This changes the shape of the volcano to what most people think of as the typical conical form (Fig. 9.18(A)). This type of volcano is characteristic of those that erupt andesitic magma, which is an order of magnitude more viscous than basalt. They are called **composite volcanoes** because they erupt both lava and ash, which form very stratified deposits, and so the name **strato-volcano** is also used (Fig. 9.18(B)). In contrast to shield volcanoes, their slopes are concave upward, with slopes increasing to ~36° near the summit. They are normally topped by a single crater and rarely produce flank eruptions, which preserves their symmetric shape (Fig. 9.18(A)).

Eruptions from composite volcanoes commonly alternate between explosion of tephra and extrusion of lava. This probably results from bubbles rising in the volcano's feeder conduit to form a frothy, potentially explosive magma at the top (Sec. 8.4.4) and a degassed magma below. Upon eruption of the gas-rich magma, tephra is ejected from the vent to produce a **pyroclastic deposit** (*pyro* means "fire," and *clastic* means "broken"). This material can consist of particles of freshly disrupted magma or fragments of older rocks ripped from the vent walls. During powerful explosions, small particles may be ejected high enough into the atmosphere to circle the Earth. The transport of smaller particles is also affected by wind, which can give tephra deposits an asymmetric distribution elongated downwind.

The largest particles ejected during eruptions settle quickly, causing pyroclastic deposits to thin away from the vent. Tephra would soon be eroded away from the steep slopes of these volcanoes if the tephra were not covered by more resistant lava flows. These flows commonly have a blocky surface with fragments considerably larger than those in aa flows.

Volcanic glass in pyroclastic deposits is rapidly converted to clay minerals, which on steep slopes in wet climates can lead to mudslides, which are known as **lahars**. These mudslides are potentially dangerous because they can involve huge volumes of material that flow rapidly (Sec. 17.4). The eruption in 1980 that removed the top of Mount St. Helens also triggered a lahar that descended the Toutle River valley and caused serious sedimentation problems as far as 100 km downstream (Fig. 9.18(C)–(D)).

Very large explosive eruptions can eject ash high into the atmosphere, which may take years to settle. This ash blocks sunlight and causes global cooling. The effect of large volcanic eruptions on the Earth's atmosphere is discussed in Section 17.4.

Major eruptions can spread layers of ash over large parts of a continent, and even though these layers may be only millimeters thick, they can often be identified by specific compositions of glassy fragments or phenocrysts. Because deposition of such a widespread layer is essentially instantaneous on the geologic time scale, they make excellent markers for correlating the age of rocks from one area to another. Dating rocks by means of layers of ash is known as **tephrochronology**.

Although volcanoes can be active over long periods of time, possibly going through extended periods of dormancy, they do eventually become extinct. During the life of many composite volcanoes, lavas become progressively more silica rich. Some volcanoes start with a basaltic shield before developing an andesitic composite cone and then finally, during old age, erupting rhyolitic lava. Mount St. Helens, for example, is built largely of andesite, but following the catastrophic eruption of 1980, a dome of a more silica-rich rock known as dacite (Fig. 9.25) has been slowly growing in its crater (Fig. 9.12(C)). The increase in silica content with age of composite volcanoes could result from differentiation in magma chambers at depth, but in most cases it probably results from increasing degrees of assimilation of crustal rocks (Sec. 8.7.5).

Domes

As magma becomes more viscous, it flows more slowly and tends to cool and solidify before traveling any great distance. Viscous lavas, such as rhyolite, need to be tens of meters thick to flow. Because rhyolite has such high viscosity, it does not erupt rapidly enough to form a typical volcanic cone. Instead, it develops a **dome** whose surface cools and cracks as additional rhyolite inflates the dome from within. These domes can be from tens of meters to kilometers in diameter (Fig. 9.19(A)).

For example, the small dome that formed in the fissure vent of the Dabbahu Volcano in Ethiopia is ~30 m in diameter (Fig. 8.17(B)). The Puy de Sarcoui in Auverne, France, is 400 m in diameter and 150 m high, and Lassen Peak in California is 2 km in diameter and 600 m high. If a dome develops on a sloping surface, flow may occur if it has sufficient thickness (Fig. 8.21). The rhyolite in many domes does not have time to crystallize and consequently forms glassy **obsidian** (Fig. 9.43). Most obsidian is layered because of the presence of minute crystals (**microlites**) that develop in layers parallel to the walls of the conduit up which the magma rises. On reaching the surface, this layering becomes distorted into flow folds as the lava spreads laterally (Fig. 9.19(C)).

As fresh magma inflates a dome, its solid crust continuously cracks. The surface of most domes is, consequently, covered with blocks of obsidian (Fig. 8.19(B)). On some domes, ridges are pushed upward by inflation. Cracks develop along these ridges, and as the ridge continues to rise, the crack propagates downward. Because the obsidian on either side of the crack is still soft, the surfaces of the crack become curved (Fig. 9.19(B)). Inflation can also push spines of rock out of the top of domes in a pistonlike fashion (Fig. 9.19(D)). As long as the fracturing and expansion of the crust is able to cope with the supply of new magma, domes are able to grow peacefully. However, many domes destroy themselves in violent explosions. For example, the small dome that is growing in the crater of Mount St. Helens has destroyed itself numerous times, exploding and throwing ash high into the atmosphere (Fig. 9.12(C)).

Calderas

Based on the geologic record, rhyolitic magma is erupted more often in the form of ash than as lava flows. Indeed, thick rhyolitic pyroclastic deposits come a close second to flood basalts in forming the largest volumes of volcanic rock on Earth. Before discussing these rocks, we must first describe one other source of tephra, because it is from this source that most huge pyroclastic deposits have come. During historic times, most tephra has erupted from composite volcanoes and has not involved huge eruptive volumes. However, in the geologic record, eruptions with a VEI of 8 have been associated with the collapse of huge calderas into the top of magma chambers (Fig. 9.1). As the caldera sinks into the chamber, magma rises along the ring fracture and, with decreasing pressure, is disrupted by exsolving gas, which results in an explosive eruption. The sinking caldera acts like a piston and pushes the buoyant magma out of the chamber, but the expansion associated with the exsolution of gas adds to this expulsive force by increasing the pressure and decreasing the density of the magma. Consequently, once the eruption starts, it tends to sustain itself through the related feedback processes. This is similar to the effect you get after shaking a bottle of carbonated beverage and then removing the cap. The bubbling drink starts

Figure 9.19 (A) Rhyolite dome in the Big Bend district, West Texas, showing columnar cooling joints. (B) Curving fractures on the surface of the Little Glass Mountain obsidian dome, Medicine Lake Highlands, California. (C) Folded layers in a thick rhyolite lava flow from the Big Bend district of West Texas. (D) Obsidian spine extruded from the surface of the Little Glass Mountain obsidian dome, Medicine Lake Highlands, California.

to foam out of the top, but as the bottle begins to empty, the pressure on the remaining liquid in the bottle decreases and the gas keeps exsolving until the bottle is essentially empty.

The volcanic activity at Yellowstone National Park in northwestern Wyoming provides an excellent example of this type of eruption (Fig. 9.20). Yellowstone is at the eastern end of a hot-spot trail that crosses the Snake River plain in Idaho. To the west of Yellowstone, earlier silicic volcanoes have been almost completely buried by the younger flood basalts of the Snake River plain. Yellowstone is the most recent of these silicic volcanic centers to form over the hot spot. Two million years ago, a caldera measuring 75 km in diameter subsided into the magma chamber and expelled more than 2500 km³ of magma, which now forms the yellow tuff of Yellowstone National Park (caldera 1 in Fig. 9.20(A)–(B)). To give an idea of the magnitude of this eruption, this quantity of magma would bury the states of Connecticut and Rhode Island in the United States or Wales in the United Kingdom to a depth of 125 m (410 feet). Eruptions were not to end with this one event, and 800 000 years later, a smaller caldera formed at the western end of the first caldera, erupting another

280 km³ of ash. Again, 600 000 years later, another large caldera formed to the east (caldera 3 in Fig. 9.20(A)), erupting another 1000 km³ of magma. Since that eruption, the floor of the latest caldera has developed two domes. The Old Faithful geyser (Fig. 9.20(C)), located on the western dome, bears testimony to the continued presence of hot magma not far below the surface. As the North American plate continues moving westward, the igneous activity associated with this hot spot can be expected to continue to generate volcanic centers to the east of Yellowstone. When these occur, they may again have a VEI of 8 (see Sec. 17.4).

Ash falls and ash flows

Volcanic ash erupted from giant calderas such as Yellowstone and from composite volcanoes has two very different modes of transport. In one, the ash is ejected high into the air and is cool by the time it settles to form an **ash-fall** deposit. The other mode involves a sideways blast of hot ash that hugs the ground and flows down valleys at great speed and forms an **ash-flow** deposit (nuée ardente). Ash falls blanket the topography with a layer that thins gradually away from the source (Fig. 9.21(A)).

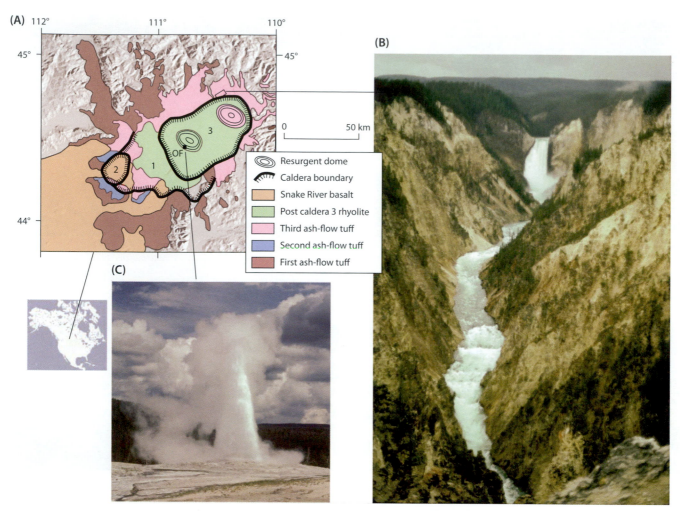

Figure 9.20 (A) Map of Yellowstone volcanic center, Wyoming. Eruptions of volcanic ash occurred during the subsidence of three calderas (1, 2, 3) (map simplified from Christiansen, 1979). (B) The Grand Canyon of the Yellowstone River is cut into ash-flow tuff. (C) Old Faithful geyser (OF).

Ash flows, by contrast, flow to topographic lows such as valleys, and thus their thickness reflects the underlying topography (Fig. 9.21(B)). Nowhere is this difference between ash falls and ash flows better illustrated than in the two ash layers deposited on a farmer's plowed field on the slopes of Vesuvius during the 79 CE eruption that destroyed Pompeii (Fig. 9.24(D)–(E)). The eruption began with an ash fall that deposited a several-centimeter-thick layer of white pumice ash evenly over the furrows of the plowed field. Later, an ash flow filled in all of the furrows as it raced down the slope toward Pompeii. The top of the ash-flow layer is even, but its base follows the surface of the ash fall that was deposited on the furrows.

Ash flows move as dense hot suspensions of particles, so they do not cool as rapidly as ash falls. Indeed, their rapid turbulent flow can even generate frictional heat, so that they are still hot when they finally come to rest. The eruption of ash flows was first described by the French geologist Alfred Lacroix who witnessed their eruption from Mont Pelée in Martinique, where one flow in 1902 killed all 30 000 residents of the nearby city of St. Pierre. Lacroix described them as **nuées ardentes** (French for "glowing clouds"), but they are commonly referred to as

ash flows or **pyroclastic flows**. When they come to rest, their suspended particles are commonly still hot enough to weld themselves together to form rock known as **welded ash-flow tuff** or **ignimbrite**. The welding can be so complete that a solid mass of glass is formed that develops columnar joints as it cools (Fig. 9.22(A)–(B)). Considerable compaction goes on during the welding, and fragments of glass and pumice can become extremely flattened (Fig. 9.22(C)–(D)).

Ash-fall tuffs and welded ash-flow tuffs appear very different under the microscope (Fig. 9.23(A)–(B), respectively). Ash-fall tuffs commonly contain small glassy particles known as **shards**, which form from the quenched magma trapped between expanding gas bubbles. Shards are bounded by concave surfaces that mark the outlines of the surrounding bubbles; they consequently commonly have a Y shape. Fragments of pumice and phenocrysts, some of which may be broken, are also common. Similar particles can be found in ash-flow tuffs, but their appearance is quite different because of the flattening during welding. Glass shards may still have three terminations, but their Y shape becomes

Figure 9.21 **(A)** Layers of ash-fall tuff near Bend, Oregon. **(B)** The Valley of Ten Thousand Smokes is filled by this ash-flow tuff, which gave off steam for 50 years following its eruption in 1912 near Katmai, Alaska (see K in Fig. 9.41(A)).

completely flattened. Bubbles in pumice fragments also become extremely flattened, giving the pumice a flamelike appearance and consequently is referred to as **fiamme** from the Italian word for "flame" (Fig. 9.23(B)). Glassy particles drape themselves around rigid phenocrysts, which gives the rock a very fluidal appearance.

Ash flows are far more hazardous to humans than ash falls because of the velocity with which they travel; those from Mont

Figure 9.22 **(A)** Columnar joints in a welded ash-flow tuff in the Big Bend district of West Texas (centimeter and inch scale at lower right). **(B)** Giant columnar joints in the thick welded ash-flow tuff of Chiricahua National Monument, Arizona. Weathering along columns has left some as freestanding pillars. **(C)** The welded ash-flow tuff in (A) contains red glassy fragments that were flattened during compaction and welding of the ash. **(D)** The Chiricahua welded ash-flow tuff shown in (B) contains light-colored pumice fragments, which would have been roughly spherical before flattening during the compaction that accompanied welding.

1 mm

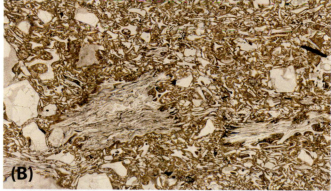

Figure 9.23 Photomicrographs of thin sections of ash-fall (**A**) and ash-flow (**B**) tuffs from the Bishop Tuff, Long Valley caldera, California.

Pelée have been clocked at 100 m/s. Ash flows were responsible for most of the deaths at Pompeii (Fig. 9.24) during the 79 CE eruption of Vesuvius. The eruption began with ash falling on roofs, which collapsed as the ash thickened, but many residents were able to escape during this stage of the eruption. The ash falls, however, were followed by a number of ash flows that raced down the slopes of Vesuvius and sheared off the tops of buildings that protruded above the earlier ash-fall deposits (Fig. 9.24(B)). Most of the fossil corpses at Pompeii are found in the ash-flow deposits and are presumably of people who chose not to escape during the early eruptive phase (Fig. 9.24(C)).

9.3 International Union of Geological Sciences classification of igneous rocks

In 1964, an international group of scientists began working on a systematic approach to the nomenclature of igneous rocks, taking into account all of the different usages of existing terms and trying to find common ground among diverse classifications. In 1972, the International Union of the Geological Sciences (**IUGS**) approved a systematic classification and nomenclature for igneous rocks that has since become accepted worldwide.

The IUGS classification of igneous rocks is based on mineral content as determined from modal analysis (volume percentage; see Sec. 6.13). Obtaining modal analyses from coarse-grained plutonic rocks is easy but difficult from fine-grained volcanic

rocks and impossible in the case of glassy rocks. Fine-grained and glassy rocks must be classified using a chemical analysis, which can be recalculated into standard igneous minerals so that the IUGS mineralogical classification can be applied.

9.3.1 Mode and norm

Before presenting the IUGS classification, we discuss how a chemical analysis of a rock is recast into a set of hypothetical igneous minerals known as a **norm** to contrast it with a **mode**, which is the abundance of actual minerals observed in a rock. Four petrologists, Cross, Iddings, Pirsson, and Washington, devised a method by which chemical analyses of rocks could be recast into common anhydrous rock-forming minerals. In recognition of their contribution, this calculated mineral composition is known as a **CIPW norm**. We do not spell out the CIPW norm calculation rules in this book, because they are tedious and are best done by computer. A computer program for calculating the norm can be downloaded free from the U.S. Geological Survey's Volcano Hazards Program Web site (http://volcanoes.usgs.gov/yvo/aboutus/jlowenstern/other/software_jbl.html).

In Table 9.1, CIPW norms are given along with the chemical analyses of some common plutonic igneous rocks. Because the CIPW norm calculation recasts the weight percentage of the oxides in the chemical analysis into normative minerals, the CIPW norm is a weight norm. To compare this with the mode, which is in volume percentage (what you actually see in the microscope), the CIPW normative mineral abundances must be recalculated to volume percentage by dividing the weight percentage of each normative mineral by its density (see Chapter 7) and then recalculating the total to 100%. This does not make much difference to the percentages of most normative minerals except for dense minerals, such as magnetite and ilmenite, whose volume percentages are far less than their weight percentages.

9.3.2 IUGS classification of igneous rocks

The IUGS classification divides igneous rocks into two large classes based on their mafic (ferromagnesian) mineral content, with the division set at 90%. Those with less than 90% mafic minerals (dark) constitute the largest group and include almost all volcanic rocks. Rocks with more than 90% mafic minerals are restricted almost completely to plutonic occurrences.

Rocks containing less than 90% mafic minerals are classified on the basis of their content of the following groups of minerals:

Q = quartz
A = alkali feldspar (orthoclase, microcline, sanidine, perthite, anorthoclase, albite ($<An_5$))
P = plagioclase (An_5 to An_{100}), scapolite
F = feldspathoids, including nepheline, sodalite, leucite (pseudoleucite), analcite
M = mafic minerals including olivine, pyroxenes, amphiboles, micas, opaque minerals

Figure 9.24 (**A**) View of Vesuvius from the Forum in Pompeii. All walls and pillars still standing were buried in the early Plinian eruptive stage on August 24, 79 CE. All taller structures were removed by the ash-flow surge that arrived early on the morning of August 25. (**B**) Pyroclastic deposits in the Necropolis (burial chambers) immediately south of Pompeii's city wall. The lower part of the tomb's wall is buried in ash-fall deposits, but the upper part was sheared off by an ash-flow surge. (**C**) Casts of a man and pregnant woman found in the first ash-flow deposit on top of the ash-fall deposit. (**D**) Quarry wall on the flanks of Vesuvius showing a vertical section through the undulating surface of what was a plowed field buried by the eruption in 79 CE. The close-up of this wall (**E**) shows a couple of centimeters of white ash-fall deposit covered by a grayer ash-flow deposit.

Rocks are commonly plotted in either a QAP or FAP triangular diagram for classification purposes. However, it is easy to think of the feldspathoid content of a rock as negative quartz (or silica deficiency). This can be illustrated by writing simple chemical reactions between alkali feldspar and negative quartz. For example:

$$NaAlSi_3O_8 - 2SiO_2 \Rightarrow NaAlSiO_4$$

albite   quartz   nepheline

$$KAlSi_3O_8 - SiO_2 \Rightarrow KAlSi_2O_6$$

orthoclase quartz   leucite

As magmas become increasingly deficient in silica, more and more sodium, potassium, and alumina form feldspathoids rather than feldspar. If we think of feldspathoids in terms of silica deficiency or negative quartz, we can plot the minerals used for the IUGS classification on a simple orthogonal graph (Fig. 9.25), with quartz increasing upward, feldspathoids increasing

downward, and the composition of the feldspar (A to P) indicated on the horizontal axis.

Figure 9.25 is divided at its center by a horizontal heavy line labeled A to P; that is, alkali feldspar to plagioclase. Above the line, rocks contain increasing amounts of quartz; below the line, they contain increasing amounts of feldspathoids. Recall from our discussion of the binary phase diagram for nepheline – quartz (Fig. 8.7) that albite creates a thermal maximum on the liquidus, which divides the diagram in two, with **alkaline igneous rocks** plotting on the nepheline side and **tholeiitic igneous rocks** on the quartz side of the maximum. This same division we see is the horizontal line A-P in the IUGS classification. The phase diagram therefore presents a physical-chemical justification for this major division in the IUGS classification.

The fields above and below line A-P in Figure 9.25 are subdivided into regions to which common igneous rock names

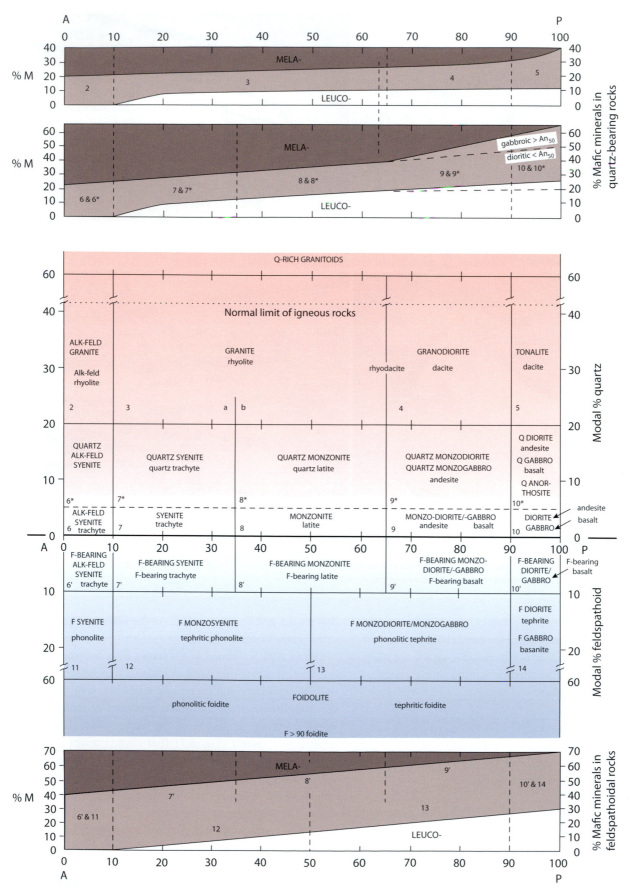

Figure 9.25 IUGS classification of igneous rocks containing less than 90% mafic minerals. The heavy horizontal line goes from alkali feldspar (A), which includes plagioclase <An₅, to plagioclase (P, An₅–An₁₀₀). Rocks containing quartz plot above this line, and those containing feldspathoids plot below it. Plutonic rock names are given in upper case and volcanic rocks in lower case. The normal mafic mineral contents of these rocks are indicated above and below the main diagram. Q and F stand for quartz and feldspathoid, respectively.

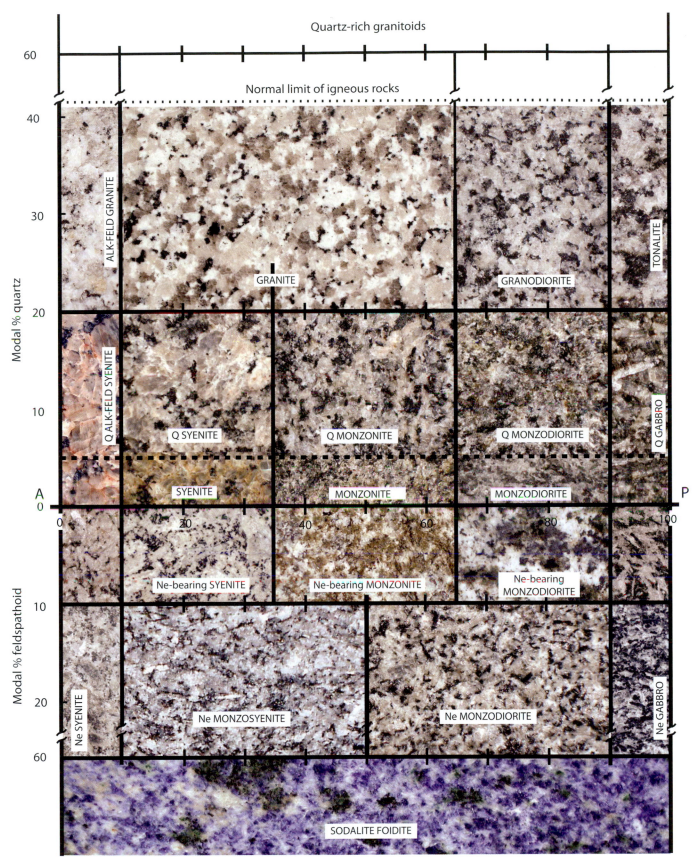

Figure 9.26 Typical hand specimens of the plutonic rocks in the IUGS classification. Fields are the same as in Figure 9.25. Q and Ne stand for quartz and nepheline, respectively.

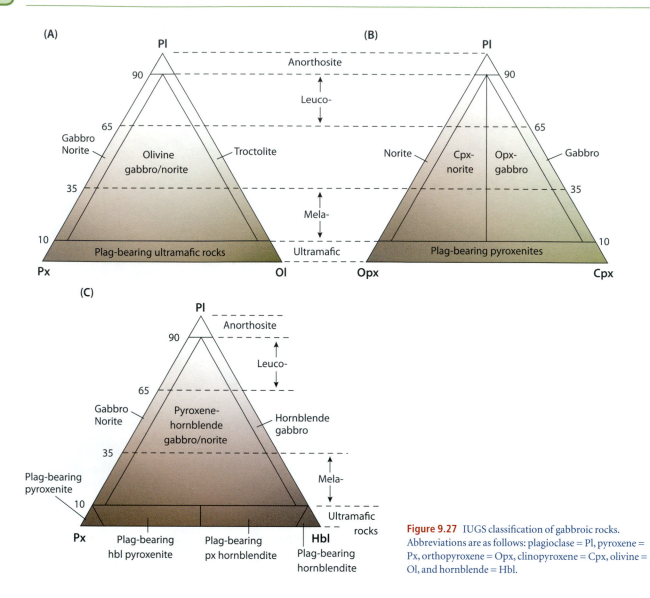

Figure 9.27 IUGS classification of gabbroic rocks. Abbreviations are as follows: plagioclase = Pl, pyroxene = Px, orthopyroxene = Opx, clinopyroxene = Cpx, olivine = Ol, and hornblende = Hbl.

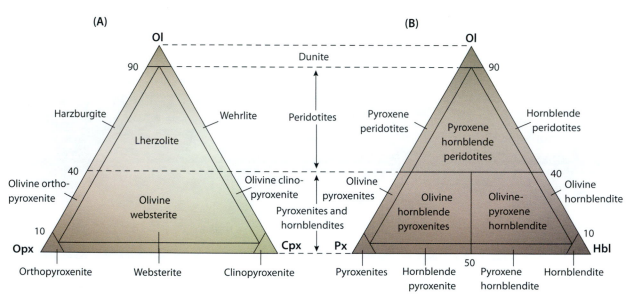

Figure 9.28 IUGS classification of ultramafic rocks (M > 90%). Abbreviations are the same as in Figure 9.27.

should be applied. Although the boundaries between the fields are arbitrary, they were positioned so that rocks with a particular name plot near the center of each field rather than near a boundary. The coarse-grained plutonic rock names are given in each field in upper case, and equivalent volcanic rock names are given in lower case. For example, **granodiorite** is the plutonic equivalent of the volcanic rock **dacite**. For convenience, each of the fields is numbered. If you wish to indicate additional mineralogical information about a rock, especially if it is transitional between fields, you can do so by adding mineral names to the rock name, with the most abundant mineral closest to the root name. A biotite-hornblende granodiorite, for example, would contain more hornblende than biotite.

Figure 9.25 does not uniquely define all rocks. For example, as rocks become more plagioclase rich (toward the right in Fig. 9.25), plagioclase tends to become more anorthite rich, but in some rocks it becomes more anorthite rich than in others. For example, **gabbro** and **diorite** both plot in field 10, but plagioclase in gabbro is more anorthite rich and in diorite less anorthite rich than An_{50}. In addition, as rocks plot farther to the right in Figure 9.25, they become more mafic. Separate graphs indicating the typical mafic mineral content of rocks in the different fields are plotted above and below the diagram for quartz-bearing and feldspathoid-bearing rocks, respectively. For rocks containing exceptionally high or low contents of mafic minerals, the prefix *mela-* (dark) and *leuco-* (light), respectively, can be added to the rock name. Another term used to indicate the mafic mineral content is the **color index**, which is simply the total percentage of dark minerals in a rock.

Figure 9.26 shows typical hand specimens of each of the plutonic rock types in the IUGS classification. The colors of the rocks shown are typical, but it should be emphasized that granites and quartz syenites could just as easily have been illustrated with pink samples. Alkali feldspar is commonly pink because of the presence of small amounts of ferric iron (Sec. 7.4.2).

The IUGS classification subdivides gabbroic rocks on the basis of their major mafic minerals, which include clino- and orthopyroxene, olivine, and hornblende. Most gabbroic rocks consist of plagioclase, pyroxene, and olivine (Fig. 9.27(A)), but in many plutonic bodies, compositional layering can give rise to rocks that are enriched in any of these minerals. If the rock consists of more than 90% pyroxene or olivine, it is designated an **ultramafic rock**. If it contains more than 90% plagioclase, it is named **anorthosite**. Names of rocks consisting of intermediate mixtures of these minerals are indicated in Figure 9.27(A). Gabbroic rocks are also subdivided on the basis of the type of pyroxene (Fig. 9.27(B)). The pyroxene in normal gabbro is the clinopyroxene augite (Sec. 7.4.9), but in some, orthopyroxene predominates, in which case the rock is called **norite**. In some gabbroic rocks, especially those we classify as alkaline (see Sec. 9.3.5), hornblende can be a major constituent. Figure 9.27(C) shows the rock names that should be applied to hornblende-bearing gabbroic rocks.

Ultramafic rocks are also subdivided on the basis of mafic minerals (Fig. 9.28). **Peridotites**, for example, are composed predominantly of olivine and pyroxene. If ortho- and clinopyroxenes are present in approximately equal amounts, the peridotite is called **lherzolite**. This is the type of peridotite brought up in diatreme breccias from great depth along with diamond-bearing nodules that we believe are representative of the upper mantle (see Sec. 8.1.1 and Fig. 8.1). The mantle peridotite immediately beneath midocean ridges from which basaltic magma has been extracted is composed of olivine and orthopyroxene and is called **harzburgite**. In many lopoliths, layers consisting almost entirely of one mafic mineral can be found. For example, if the layer is composed of olivine, it would be referred to as dunite.

No volcanic equivalents are given for the ultramafic rocks, because for the most part, no magmas of these compositions ever existed. Ultramafic rocks are formed by the accumulation of early crystallizing minerals from magma or by the redistribution of minerals in crystal mush. However, one important exception, which is discussed in Section 9.5.1, is the ultramafic lava known as **komatiite**, named from its occurrence in the Komati River of South Africa. It has the composition of a lherzolite, but its texture and mode of occurrence indicate that it crystallized in lava flows.

9.3.3 Composition of common plutonic igneous rocks

Table 9.1 lists average chemical analyses of some common plutonic igneous rocks compiled by Le Maitre (1976). The averages are based on large numbers of analyses, as indicated in the table. The abundances of the major (>1.0%) and minor (1.0–0.1%) elements in rocks are normally presented in analyses as weight percent oxides. Trace elements (<0.1%), however, are presented as parts per million (ppm) of the element by weight. Table 9.1 gives only major and minor elements.

The CIPW norms of each of the average rock types are also presented in Table 9.1. All but the nepheline syenite contain normative quartz and, therefore, plot above the A-P line in Figure 9.25. The granite and granodiorite both contain small amounts of normative corundum. These rocks most likely do not contain corundum, but some of the rocks used in calculating the average probably contain muscovite. Because the CIPW norm does not calculate hydrous minerals, any muscovite in a rock appears as normative orthoclase and corundum, as indicated from the following reaction:

$$KAl_3Si_3O_{10}(OH)_2 \Rightarrow KAlSi_3O_8 + Al_2O_3 + H_2O$$

$$\text{muscovite} \qquad \text{orthoclase} \quad \text{corundum} \quad \text{water}$$

Note that the abundance of normative albite and anorthite in the gabbro and diorite are consistent with the plagioclase in these rocks being respectively more or less anorthite rich than An_{50}.

Table 9.1 Average compositions (weight %) of some common plutonic igneous rocks (after Le Maitre, 1976) and their CIPW norms.

	Granite	Nepheline syenite	Granodiorite	Monzonite	Tonalite	Diorite	Gabbro
Number of analyses averaged	2485	115	885	336	97	872	1451
SiO_2	71.30	54.99	66.09	62.60	61.52	57.48	50.14
TiO_2	0.31	0.60	0.54	0.78	0.73	0.95	1.12
Al_2O_3	14.32	20.96	15.73	15.67	16.48	16.67	15.48
Fe_2O_3	1.21	2.25	1.38	1.92	1.83	2.50	3.01
FeO	1.64	2.05	2.73	3.08	3.82	4.92	7.62
MnO	0.05	0.15	0.08	0.10	0.08	0.12	0.12
MgO	0.71	0.77	1.74	2.02	2.80	3.71	7.59
CaO	1.84	2.31	3.83	4.17	5.42	6.58	9.58
Na_2O	3.68	8.23	3.75	3.73	3.63	3.54	2.39
K_2O	4.07	5.58	2.73	4.06	2.07	1.76	0.93
P_2O_5	0.12	0.13	0.18	0.25	0.25	0.29	0.24
H_2O+[a]	0.64	1.30	0.85	0.90	1.04	1.15	0.75
Total	99.89	99.32	99.63	99.28	99.67	99.67	98.97

CIPW norms

Normative mineral[b]	Granite	Nepheline syenite	Granodiorite	Monzonite	Tonalite	Diorite	Gabbro
Q	28.93		22.11	13.86	15.96	10.15	0.73
C	0.80		0.07				
Or	24.05	32.98	16.13	23.99	12.23	10.40	5.50
Plag	39.48	29.45	49.56	45.58	53.27	54.35	48.99
Ne		22.37					
Di		5.58		4.09	2.20	5.17	13.94
Hy	3.36		7.46	6.01	10.35	12.34	22.03
Ol		0.29					
Mt	1.75	3.26	2.00	2.78	2.65	3.62	4.36
Il	0.59	1.14	1.03	1.48	1.39	1.80	2.13
Ap	0.28	0.30	0.42	0.58	0.58	0.67	0.56
Mg/Mg+Fe	0.593	0.679	0.645	0.690	0.673	0.683	0.715
Plag An%	20.2	11.1	34.6	29.5	40.9	43.4	57.3

[a] H_2O+ is water driven off by heating above 105°C.

[b] Abbreviations for the normative minerals are as follows: quartz = Q, corundum = C, orthoclase = Or, plagioclase = Plag, nepheline = Ne, diopside (cpx) = Di, hypersthene (opx) = Hy, olivine = Ol, magnetite = Mt, ilmenite = Il, and apatite = Ap. Mg/Mg+Fe is the ratio of magnesium to magnesium + iron in the diopside, hypersthene, and olivine, and Plag An% is the percentage of anorthite in plagioclase.

9.3.4 IUGS classification of volcanic igneous rocks

The IUGS recommends a simple chemical classification based on total alkalis ($Na_2O + K_2O$) and silica contents for volcanic rocks that are too fine grained for modal analysis or are glassy. Figure 9.29 shows a plot of these two parameters, with the diagram being divided into 15 fields that indicate the compositional ranges covered by common volcanic rocks. To plot a rock in this diagram, its analysis is recalculated to 100% on a H_2O- and CO_2-free basis. Rocks classified using this diagram are almost totally consistent with those using QAPF in Figure 9.25. Along the bottom of the diagram are some general rock names that indicate the silica content of the rock. Early in the science, silica was thought to combine with water in magmas to form siliceous acid (H_2SiO_3), and magnesium and iron were thought to form bases (e.g., $Mg(OH)_2$). Rocks rich in silica were, therefore, referred to as acidic, and rocks poor in silica and rich in magnesium and iron were referred to as basic. Rocks that were exceptionally poor in silica due to an abundance of mafic minerals were referred to as ultrabasic. Although rocks are no longer thought of in terms of this acidity scale, the names persist for general classification purposes. However, instead of acid, basic, and ultrabasic, the descriptive terms felsic, mafic, and ultramafic are preferable.

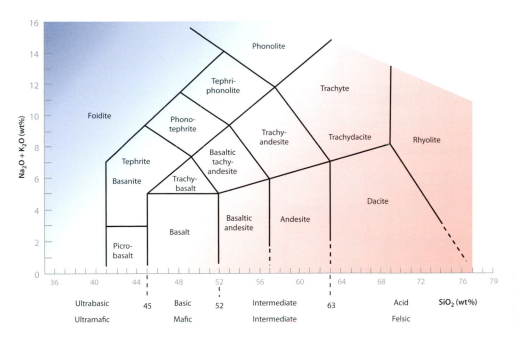

Figure 9.29 IUGS chemical classification of volcanic rocks. The pink and blue areas correspond to the same colored areas in Figure 9.25.

9.3.5 Irvine-Baragar classification of volcanic rocks

The IUGS classification is not a genetic classification. Rock names are based purely on the modal or normative abundance of minerals, or in the case of volcanic rocks, on their total alkalis and silica contents. With the advent of plate tectonics, it became apparent that certain rock types commonly occur together and are genetically related and are characteristic of specific plate tectonic settings. For example, lavas covering a wide range of compositions can be erupted from a single volcano and must, therefore, be genetically related through processes taking place beneath that volcano. Lavas covering a similar range of composition can be found erupted from other volcanoes that have similar plate tectonic settings. Recognition of such relations lead to classifications that tie rock names to genetic lineages that have plate tectonic significance. One such classification, by Irvine and Baragar (1971), has gained wide acceptance. It is based on a few simple chemical parameters. Most of the rock names are the same as those in the IUGS classification, but they are grouped into genetically related series.

Irvine and Baragar distinguish three main series of igneous rocks referred to as **subalkaline**, **alkaline**, and **peralkaline** (Fig. 9.30). As the names imply, these divisions are based on alkali contents. The subalkaline rocks contain the least alkalis. They can be separated from alkaline rocks on the basis of normative contents of olivine, nepheline, and quartz, with the subalkaline rocks containing normative quartz and the alkaline rocks containing normative nepheline. This corresponds to the division we saw in the phase diagram for nepheline and quartz created by the thermal maximum at the composition of albite (Fig. 8.7) and the main division in the IUGS classification. The subalkaline group is subdivided into tholeiitic and **calcalkaline igneous rock series** on the basis of iron content, with

tholeiitic rocks being more iron-rich. In addition, calcalkaline basalts are slightly richer in alumina than tholeiitic basalts and are referred to as **high-alumina basalt**. The alkaline group is subdivided into the **alkali olivine basalt** series, which has sodic and potassic branches, and the **nephelinitic-leucitic-analcitic** series. Finally, the relatively rare **peralkaline igneous rock** series have molecular amounts of $(Na_2O + K_2O) > Al_2O_3$, which means that the rock contains more alkalis than can be accommodated in feldspar or nepheline; therefore, they also contain the sodic pyroxene aegirine or a sodic amphibole.

At the top of each of the series in Figure 9.30 is a rock type formed from magma that can be thought of as parental to the other rocks in the series. The subsequent rocks in the series are formed through differentiation of the parental magma, mainly as a result of crystal fractionation, but assimilation of crustal rocks and even liquid immiscibility may play a role at late stages. In general, rocks at the top of each series are dark colored because of abundant mafic minerals and contain relatively anorthite-rich plagioclase. Rocks become lighter colored toward the bottom of each series as the content of quartz or feldspathoid increases, and the plagioclase becomes progressively more albitic and is accompanied by increasing amounts of alkali feldspar.

Rocks belonging to each of these series occur in definite plate tectonic settings. Tholeiitic rocks are characteristic of active divergent plate boundaries where considerable melting takes place in response to adiabatic decompression of the mantle. Basalts erupted from midocean ridges and covering most of the ocean floor are tholeiitic. Divergent plate boundaries on continents are also characterized by tholeiites, as long as spreading rates are significant. Volcanism over hot spots and hypothesized mantle plumes is also tholeiitic until the waning stages, when it can become alkaline. Calcalkaline rocks are characteristic of

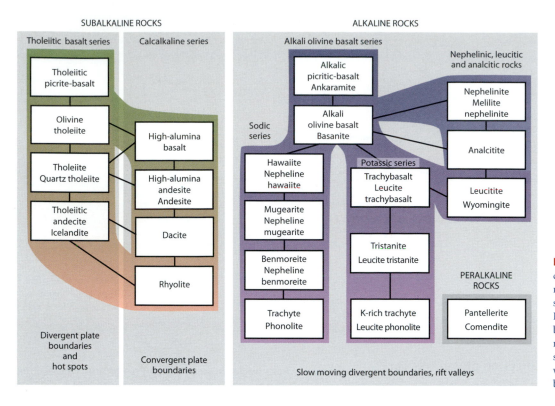

SUBALKALINE ROCKS   ALKALINE ROCKS

Figure 9.30 General classification scheme for volcanic rocks based on genetically related series (modified from Irvine and Baragar, 1971). Lines joining boxes link commonly related rocks. Each main series has a specific plate tectonic association, which is indicated in the background gray boxes.

convergent plate boundaries. The sequence of basalt, andesite, dacite, rhyolite (BADR) is typical of lavas erupted from composite volcanoes generated above the subducting plate when it reaches a depth of 100 km, and the plutonic equivalents – gabbro, diorite, granodiorite, and granite – are the rock types found in the batholiths of orogenic belts. Alkaline rocks form when the degree of melting in the mantle is far less than in the case of tholeiitic rocks and the source is deeper. Where plate divergence rates are low, such as rift valleys in continents, the rocks formed are alkaline. As a tholeiitic shield volcano moves off a hot spot, as has happened to each of the volcanoes in the Hawaiian chain, the degree of melting in the source decreases and moves to greater depth, and the magma changes from tholeiitic to alkaline. In the following section, we describe typical examples from each of these plate tectonic settings.

9.4 Igneous rocks and their plate tectonic setting

In Section 8.2, we saw that the Earth's crust and mantle are normally solid, and only on rare occasions do conditions allow them to partially melt. These conditions result from plate tectonic processes. Magma is generated in response to heating over mantle plumes, decompression at divergent plate boundaries, or fluxing with water above subducting oceanic plates. Each of these processes involves different ways of crossing the solidus and generating magmas of slightly different compositions (Sec. 8.2). These magmas then rise toward the surface and, through

differentiation and assimilation, generate the wide range of rock types found in the Earth's crust. It is not surprising, therefore, that these rocks reflect the plate tectonic setting in which they form. We now discuss the major rock types that typify the various plate tectonic settings where rocks are created.

9.4.1 Igneous rocks formed at midocean-ridge divergent plate boundaries

Midocean ridges are the most productive rock factories on Earth. More than 10 km³ of tholeiitic lava (MORB) is erupted worldwide each year from the 65 000 km-long midocean ridge system, and still more magma solidifies at depth. Despite extensive sampling with dredges and shallow drill holes, we know little about these deeper rocks because most of them are subducted into the mantle at convergent plate boundaries. On rare occasions, however, fault-bounded slices of ocean crust have been thrust onto continents as a result of **obduction** to form what are known as **ophiolite** suites. The rocks making up these suites can be correlated with seismic studies of modern ocean floor to give the clearest picture of the nature of the oceanic crust.

Figure 9.31 shows a cross-section of oceanic crust in the vicinity of a midocean ridge based largely on knowledge gained from ophiolites, which typically consist of three major rock types:

- An ultramafic rock composed largely of serpentine
- Pillowed basalt (often badly altered) and its plutonic equivalent gabbro
- A silica-rich sedimentary rock known as **chert** (Sec. 11.3.1)

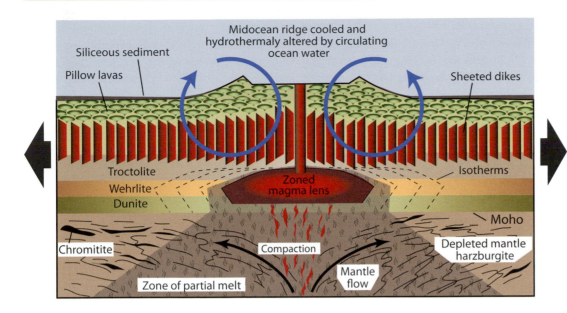

Figure 9.31 Cross-section through the oceanic lithosphere near a midocean ridge based on seismic data and oceanic lithosphere that is obducted onto continents to form ophiolites.

We illustrate this sequence with rocks from the Troodos massif, in the core of the Mediterranean island of Cyprus, which has brought to the surface a complete section through the oceanic crust (Fig. 9.32(A)).

The sedimentary rocks surrounding the Troodos massif (light color in Fig. 9.32(A)) were formed from deepwater marine sediments. Some of these consist of **radiolaria**, which on compaction was converted to the silica-rich rock known as

Figure 9.32 **(A)** The dark area in this NASA satellite image of the island of Cyprus in the eastern Mediterranean is the Troodos massif, which exposes a complete section through the oceanic crust. *S* indicates location of the Skouriotissa mine. **(B)** Pillow lavas exposed on the north side of the Troodos massif. **(C)** Sheeted dikes with chilled contacts only on their left side (see Fig. 9.33(F)). **(D)** Copper mine at Skouriotissa (*place of slag*). (Photograph courtesy of Colin Wilkins.) **(E)** Roman slag from smelting copper ore at Skouriotissa.

chert, a cryptocrystalline variety of quartz (Fig. 9.33(A)–(C)). Below the sediments are the MORBs, which are olivine tholeiites (Fig. 9.30 and field 10 in Fig. 9.25). Because they are erupted onto the seafloor, they are commonly pillowed (Fig. 9.32(B)). Rapid quenching by water produces an extremely fine-grained rock (Fig. 9.33(D)–(E)). Ocean water circulating through this basalt hydrothermally alters the plagioclase to albite and the mafic minerals to hydrous minerals. This altered sodium-rich basalt is known as **spilite**.

Immediately beneath the pillows is a zone composed almost entirely of dikes intruding dikes to form a sheeted dike complex (Fig. 9.32(C)). These dikes typically have only one chilled contact, with the other chilled contact occurring on the other side of the divergent plate boundary, having been separated by sequential intrusions of new magma in the center of a previous dike. Figure 9.33(F) shows a photomicrograph of the chilled margin of one of the dikes in Figure 9.32(C), which contains small vesicles that were sheared into ellipses by the flow of magma. Figure 9.33(G) shows how the ellipticity of the vesicles indicates the magma flow direction. The sheeted dikes pass down into layered gabbroic to ultramafic rocks (Fig. 9.33(H)–(M)), grading from troctolite (plagioclase + olivine) to wehrlite (olivine + clinopyroxene) to dunite (olivine) at the base (Fig. 9.31). These layers can be laterally extensive, which might suggest formation in huge magma chambers, but seismic surveys across active oceanic ridges, such as the East Pacific Rise and the Galapagos Spreading Axis, reveal chambers only 1 km wide and 100 m thick. The size of the chambers probably remains relatively small at all times, but continuous spreading from the axis produces layers that are laterally far more extensive (see isotherms in Fig. 9.31).

All these igneous rocks are hydrothermally altered, but their primary igneous textures are still clearly visible (Fig. 9.33(H)–(J)). However, the rocks beneath the dunite appear quite different in that they are highly deformed and are composed largely of the mineral serpentine; they are, therefore, referred to as **serpentinites**. Before serpentinization, these rocks were harzburgites (olivine + orthopyroxene), as can be identified from their textures (Fig. 9.33(K)–(L)). Olivine, when converted to serpentine, shows curving fractures that are commonly decorated with magnetite grains. The serpentine along these fractures is the fibrous chrysotile variety. Serpentine formed from orthopyroxene has a very different appearance and in many respects still resembles the original orthopyroxene grains. For example, the original cleavage of the pyroxene may still be preserved, and the serpentine shows the same parallel extinction under crossed polars that characterizes orthopyroxene. The serpentinized harzburgite is believed to be the refractory mantle residue from which the basaltic magma was extracted during decompression melting. Its deformation is thought to result from flow in the mantle associated with the rise of the asthenosphere beneath the divergent plate boundary. According to this interpretation, the boundary between the layered gabbroic rocks and the underlying harzburgite is the **Mohorovičić (Moho) discontinuity**.

Several economic mineral deposits are associated with ophiolites. The name Cyprus is from the ancient Greek word for "copper," *kypros*, which has been mined on the Island for more than 4000 years (Fig. 9.32(D)–(E)). The copper deposits are associated with the pillow lavas and formed on the ocean floor from hydrothermal vents. Modern equivalents of these vents are the **black smokers** that have been identified along midocean ridges where superheated steam rising from newly formed ocean floor precipitates iron, copper, and zinc sulfides on entering the cold ocean water (Fig. 9.34). The sulfides build meter-high chimneys from which plumes of dark, suspended sulfide particles rise through the cold, dense ocean water until they find a level of neutral buoyancy, whereupon they spread laterally and slowly settle out the suspended sulfide particles onto the ocean floor. In this way a **strata-bound sulfide deposit** is formed, which can be contrasted with hydrothermal vein deposits, which cut across the strata.

Ophiolites also host important economic chromite deposits. Rocks composed largely of chromite are known as chromitites. They occur both as layers in the dunite and as lenses and pods in the harzburgite. The chromitite in the layered rocks appears to have formed as crystal cumulates from the magma beneath the spreading axes. They form laterally continuous layers and have textures consistent with having accumulated through crystal settling.

The world's main source of **asbestos** comes from the chrysotile in the serpentinites of ophiolites, as in the Eastern Townships of Quebec. This mining industry was dramatically reduced after some forms of asbestos were linked to lung cancer (see Sec. 17.3.2).

Figure 9.33 Photographs of hand specimens and thin sections of rocks through the oceanic crust as exposed in the Troodos ophiolite suite, Cyprus. Samples are arranged in stratigraphic order from sediment at the top to mantle peridotite at the bottom. (A) Radiolarian chert. (B) and (C) Photomicrograph of radiolaria in chert under plane and crossed polarized light, respectively. (D) and (E) Photomicrograph of pillowed basalt under plane and crossed polarized light, respectively. (F) Chilled margin (left) of one of the dikes in Figure 9.32(C). The white spots are vesicles, whose elliptical shape indicates that magma flowed upward. (G) Laminar flow of magma past a contact shears a spherical vesicle into an ellipsoidal shape. (H) Hydrothermally altered gabbro from beneath the sheeted dike complex. (I) Photomicrograph of sample (H) under plane light showing that most of the original pyroxene has been altered to green amphibole. (J) Same as (I) but under crossed polars. (K) Hand sample of serpentinized harzburgite from Mount Olympus at the core of the Troodos massif. Serpentine formed from olivine is stained red and that formed from orthopyroxene is green. (L) Photomicrograph of (K) showing green serpentine, with that formed from olivine being full of small opaque magnetite grains, whereas that formed from orthopyroxene is a smooth green color. (M) Same as (L) but under crossed polars. The small veins in the serpentinized olivine are filled with fibrous chrysotile asbestos (white).

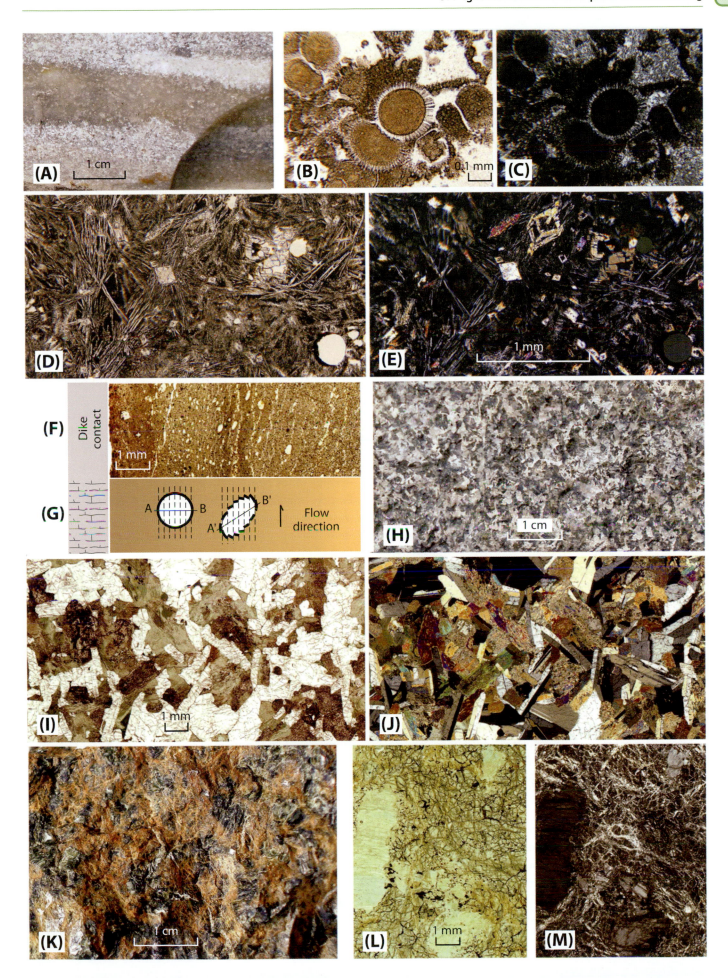

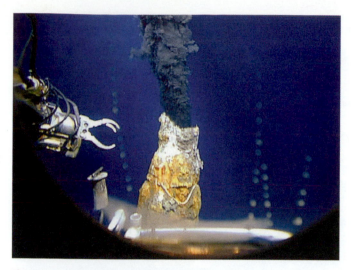

Figure 9.34 A black smoker hydrothermal vent on the East Pacific Rise (17° 37.2′ S, 113° 15.1′ W) at a depth of 2591 m below sea level, as seen from the window of the research submersible *Alvin*. The 353°C hydrothermal fluids are depositing anhydrite ($CaSO_4$) and chalcopyrite ($CuFeS_2$) on the wall of the chimney. (Photograph courtesy of Woods Hole Oceanographic Institution (WHOI); M. Lilley and K. Von Damm, chief scientists.)

9.4.2 Igneous rocks of oceanic islands formed above hot spots

Before research vessels sampled the ocean floor, rocks of oceanic regions were thought to be predominantly alkaline. This was because so many oceanic islands (excluding those of island arcs; see Sec. 9.4.5) are made of alkaline rocks. We now know that the ocean floor is underlain predominantly by MORB, which is tholeiitic, and even large oceanic islands are mainly tholeiitic, although they often have a late capping of alkaline rocks. As plate motion moves a volcano off a hot spot, cooling and densification cause the volcanic island to sink, and it is not surprising, therefore, that the rocks found on these islands are late-stage alkaline rocks.

Tholeiitic rocks of oceanic islands have a wider compositional range than do MORBs. This is in part because of differentiation of magma in chambers beneath volcanoes, but it is also because the basalts come from a greater range of depth. The tholeiitic magmas result from relatively large-scale partial melting at shallow depths, whereas alkaline magmas are the result of smaller-scale melting at greater depth. Clearly, large volumes of magma must be generated and erupted just for a volcano to build from the ocean floor to sea level. This is normally achieved through the eruption of olivine tholeiite. We have already seen what this rock looks like in our discussion of the eruption of Kilauea Iki (Fig. 8.26). The only mineral that commonly forms phenocrysts in these basalts is olivine. The **groundmass** contains plagioclase, augite, and pigeonite, and magnetite typically crystallizes very late. In slowly cooled intrusive bodies, quartz may form in the final residue.

The alkaline rocks of oceanic islands are strikingly different in both hand specimens and thin sections from the tholeiites. The main rock type is **alkali olivine basalt** (AOB), which belongs to the sodic series in Figure 9.30. As the name implies, olivine is abundant and forms phenocrysts, but this basalt also contains phenocrysts of augite and plagioclase. Moreover, AOB never contains pigeonite or orthopyroxene, and the augite differs from that in tholeiites in that it is titanium rich. The titanium content of titanaugite is a reflection of the silica deficiency in the magma, which is, of course, why these rocks are classified as alkaline (Sec. 9.3.2). In coarser-grained intrusive rocks, this silica deficiency causes nepheline to form as a late-crystallizing mineral (Fig. 9.35). Differentiation of alkali olivine basalt magma can lead to the eruption of less mafic lavas on oceanic islands. One of the most common of these is known as **trachyte**, which is the volcanic equivalent of **syenite** (Fig. 9.25).

9.4.3 Continental flood basalts and large igneous provinces

At various times throughout geologic history, parts of continents have been covered by vast outpourings of flood-basalt flows that were the precursor to continental rifting and formation of new ocean floor. For example, eruption of the early Triassic flood basalts of eastern North America was the harbinger to the breakup of Pangaea to form the North Atlantic Ocean, and the eruption of Tertiary flood basalts across Greenland, Iceland, and the northwestern coast of Britain led to the opening of the North Atlantic Ocean between Greenland and Europe. In many cases, flood basalts on continents have been traced via oceanic ridges to oceanic islands located over hot spots. This led to the idea that these vast flood-basalt provinces formed over possible mantle plumes (Fig. 1.6(E)) that supplied the heat necessary for such large-scale melting. Associated with flood basalts are large volumes of intrusive rocks. Feeder dikes to the flows can form radial patterns around the plume head, such as those of the McKenzie region of northwestern Canada. Sedimentary basins form during rifting, and when magma enters these, it commonly spreads laterally to form sills (Fig. 9.4). Because all of these related igneous rocks can cover large areas, the regions are referred to as **large igneous provinces** (**LIPs**). We should add that some LIPs form on oceanic crust, thus creating large plateaus such as those of Kerguelen in the southern Indian Ocean and Ontong-Java in the southwestern Pacific. The Ontong-Java Plateau was formed by 100 million km^3 of flood basalt that erupted about 120 million years ago. The origin of these vast outpourings on the ocean floor is still poorly understood, but they represent some of the largest igneous events in the Earth's history.

The flow of the mantle away from a plume head leads to rifting of the lithosphere in much the same way that magma convecting beneath the crust of a lava lake rifts its surface (see photograph at beginning of this chapter). The rifts intersect at a triple junction above the plume (Fig. 1.6(E)). Normally, two of

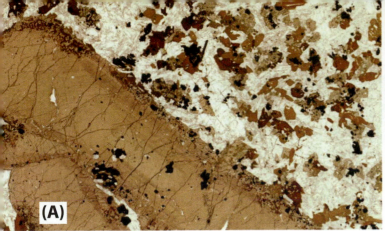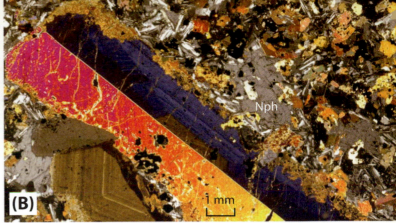

(A)

(B)

1 mm

Figure 9.35 Photomicrograph of a thin section of porphyritic nepheline basalt from Tahiti. (A) Under plane light, a phenocryst of titanaugite, showing titanium-rich pink sectors, is surrounded by groundmass containing light-colored plagioclase and nepheline, deep-brown hornblende, and opaque magnetite. (B) Under crossed polars, the titanaugite exhibits anomalous blue and brown interference colors and oscillatory zoning. Large areas with gray interference color are nepheline (Nph).

the rifts become dominant and open to form new ocean floor, whereas the third becomes a failed arm and either stops moving or moves very slowly. Such a plume and rift system is thought to be centered in the Afar region of Ethiopia (Box 8.4), where the Red Sea and Gulf of Aden rifts are opening to form new ocean floor, with the Arabian plate moving northeastward away from the African continent at 16 mm/year. The East African Rift is spreading more slowly at only 4 mm/year.

The geologic record shows that igneous activity associated with the development of a new plume peaks early, with huge initial outpourings of flood basalt. The first basalts erupted are what are referred to as **continental flood basalts**. This rock type, which has remained remarkably constant in composition from the Precambrian to the present, is a quartz tholeiite. Although it may contain phenocrysts of olivine, its silica content is high enough that olivine should react to form orthopyroxene or pigeonite if given sufficient time. In Figure 8.8, continental flood basalts plot to the right of enstatite, whereas MORBs plot to the left of it. Some flood basalts can be alkaline, but these are not common.

Figure 9.36(A)–(B) shows photomicrographs of a thin section of typical continental flood basalt. It has plagioclase phenocrysts in a groundmass of clusters of small plagioclase laths, augite, and pigeonite. Dark patches of mesostasis, formed from the residual liquid, consist of dendritic magnetite crystals and immiscible iron-rich and silica-rich glassy droplets (Fig. 8.30).

This same magma, when cooling more slowly in sills, such as those of the Karoo region of South Africa (Fig. 9.36(C)–(D)), crystallizes the same minerals as in the flood basalts, but they have time to develop the ophitic texture, where plagioclase laths are enclosed in larger crystals of pyroxene or olivine (Sec. 8.3.2). This coarser-grained basaltic rock is referred to as **diabase** (**dolerite** in Britain). The photomicrograph under crossed polars (Fig. 9.36(D)) shows that augite and pigeonite crystals are much larger than the enclosed plagioclase crystals.

These same magmas can form other intrusive bodies that cool more slowly and undergo significantly more fractionation during solidification. The Skaergaard intrusion in East Greenland, formed during the initial breakup of Greenland from Europe, is one of the world's most strongly differentiated intrusions. Most of its gabbroic rocks show prominent layering that point to active magma convection during solidification (Fig. 8.28(A)–(B)). The grain size is considerably coarser than in sills, and even though the same minerals are present, slower cooling allows minerals to equilibrate to lower temperatures. Figure 9.36(E)–(F) shows a Skaergaard gabbro that initially crystallized plagioclase, augite, and pigeonite, with a few olivine grains preserved in the pigeonite. On cooling slowly, the pigeonite inverted to orthopyroxene (pale yellow interference color) and at the same time exsolved small patches of augite (red dots throughout the yellow) to form **inverted pigeonite** (see photomicrograph opening Chapter 6).

Slower cooling also allows for differentiation processes to segregate late-stage liquids that contain significantly more silica than the initial basaltic magma. Both quartz and alkali feldspar crystallize from these late liquids, forming an intergrowth called a **granophyre** (Fig. 9.36(G)–(H)). Some granophyres consist almost entirely of quartz and alkali feldspar and are light colored, but those in the Skaergaard contain significant amounts of iron-bearing minerals and are dark. The pyroxene shown in Figure 9.36(G) is rich in the iron end member hedenbergite, and the olivine (top center) is rich in the fayalite end member. Note that fayalitic olivine can coexist with quartz, whereas forsteritic olivine reacts to form orthopyroxene (Fig. 8.8).

9.4.4 Alkaline igneous rocks associated with continental rift valleys

Igneous rocks associated with failed arms of triple junctions (Fig. 1.6) or along slowly diverging continental plates tend to be alkaline. For example, tholeiitic flood basalts were erupted near the Afar triple junction, where rifting successfully opened the Red Sea and Gulf of Aden, but farther south along the

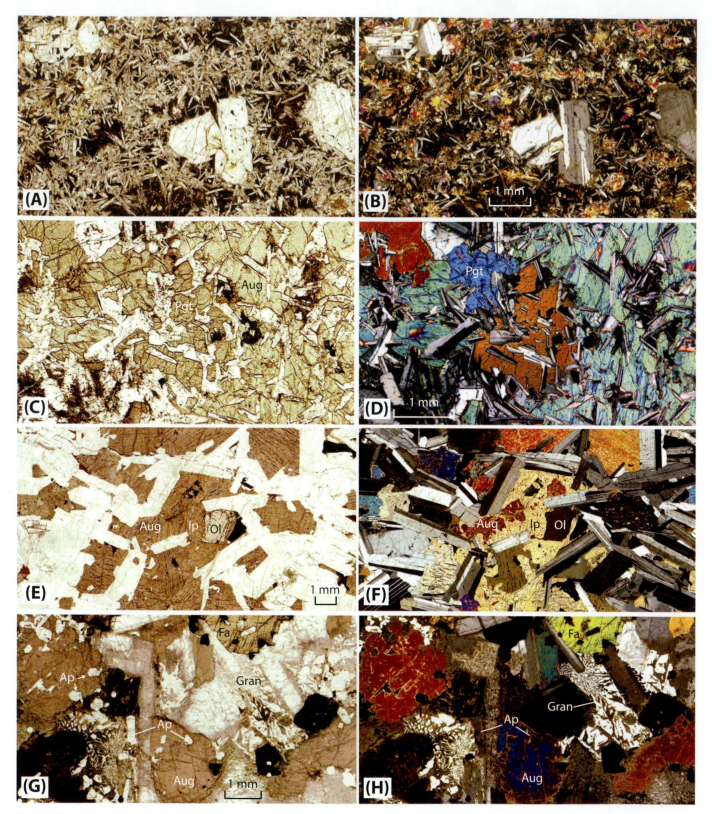

Figure 9.36 Photomicrographs of thin sections of rocks associated with large igneous provinces as seen under plane and crossed polarized light. (**A**) and (**B**) Holyoke flood-basalt flow from the Mesozoic Hartford Basin, Connecticut. (**C**) and (**D**) Karoo diabase from the Birds River sill, South Africa. Plagioclase laths are ophitically enclosed in large crystals of augite (Aug) and pigeonite (Pgt). (**E**) and (**F**) Gabbro from the Skaergaard intrusion (Fig. 8.28), East Greenland, composed of augite and olivine (Ol), which is surrounded by a reaction rim of pigeonite that has changed to inverted pigeonite (Ip). (**G**) and (**H**) Granophyre from the Skaergaard under plane light containing intergrown quartz and alkali feldspar (Gran) between crystals of iron-rich augite (Aug), fayalitic olivine (Fa), plagioclase, magnetite, and apatite (Ap).

East African Rift system igneous activity is distinctly alkaline. Mount Kilimanjaro (K in Box 8.4(A)), for example, is a large volcano built of alkali basalt, trachyte, and phonolite. The alkaline nature of these magmas is due to smaller degrees of partial melting in the source region, which is also at greater depth than tholeiitic sources.

The most common mafic intrusive rock of this alkaline association is nepheline gabbro, the plutonic equivalent of alkali olivine basalt. Olivine, titanaugite, and deep reddish-brown hornblende and biotite are the main mafic minerals. Magnetite is an abundant accessory mineral and contains a large percentage of ulvöspinel ($Fe_2^{+2}TiO_4$). Ilmenite ($FeTiO_3$) may also be present and is distinguished from magnetite by its elongate crystal shape (magnetite is equant). Titanite, $CaTiSiO_5$ (also called sphene), is another common titanium-bearing mineral in these rocks. It is identified by its wedge-shaped crystals, which, because of extremely high birefringence, appear the same brownish color under plane or cross-polarized light. Alkaline rocks have much higher phosphorus contents than do tholeiitic rocks, and consequently apatite is an abundant accessory.

The main felsic rock of this series is nepheline syenite. It is composed largely of nepheline and alkali feldspar but may also contain sodalite. The main mafic mineral is the sodic pyroxene aegirine, which in thin sections is bright green (Fig. 7.26). Greenish-blue sodic amphiboles are also common. Apatite is an extremely abundant accessory and may even be present as phenocrysts. Alkaline rocks tend to be rich in zirconium, and consequently zircon is a common accessory, as are other rarer zirconium minerals.

Mafic dike and sill rocks associated with alkaline intrusions commonly contain large phenocrysts of hornblende or biotite. These rocks are known as **lamprophyres** (Fig. 9.37(A)–(C)). They have essentially the same composition as alkali olivine basalt. The presence of hydrous phenocrysts testifies to the high volatile content of the magma from which they form.

Some alkaline rocks are so silica-poor that nepheline, $NaAlSiO_4$, forms in place of albite, $NaAlSi_3O_8$, calcium melilite, gehlenite, $Ca_2Al_2SiO_7$, forms in place of anorthite, $CaAl_2Si_2O_8$, and calcium magnesium melilite, åkermanite, $Ca_2MgSi_2O_7$, forms instead of pyroxene, $CaMgSi_2O_6$. The magnesium-rich mica, phlogopite, can also be present, forming a major mineral in the rock known as kimberlite (Fig. 9.38), which hosts mantle xenoliths that bring diamonds to the surface.

Another economically important alkaline rock is **carbonatite** (Fig. 9.39), which is composed largely of calcite with only small amounts of silicate minerals, including diopsidic pyroxene, phlogopitic mica, forsteritic olivine, and monticellite, $CaMgSiO_4$. The high Fe^{3+}/Fe^{2+} in these rocks causes the silicate minerals to be magnesium rich, with most of the iron entering magnetite. Magnetite, apatite, and perovskite are common accessories. Many carbonatites contain

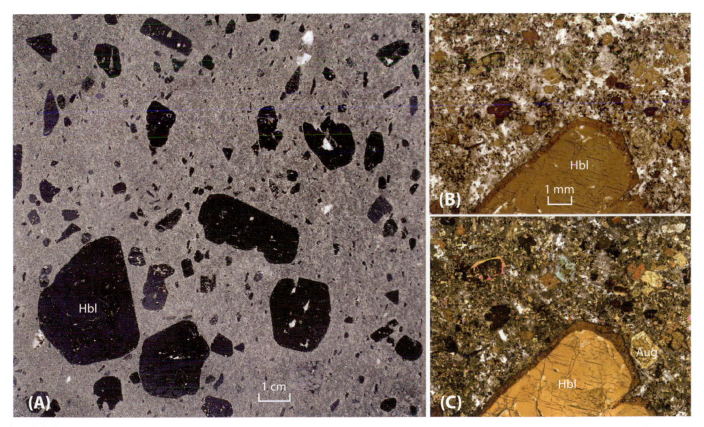

Figure 9.37 (A) Lamprophyric dike rock from Montreal, Quebec, containing large black phenocrysts of hornblende (Hbl). (B) Photomicrograph of a thin section under plane light showing the upper tip of the large hornblende phenocryst in the lower left of (A) in a groundmass of titanium-rich hornblende, titanaugite (Aug), magnetite, and late crystallizing plagioclase. (C) Same as (B) under crossed polars.

1 mm

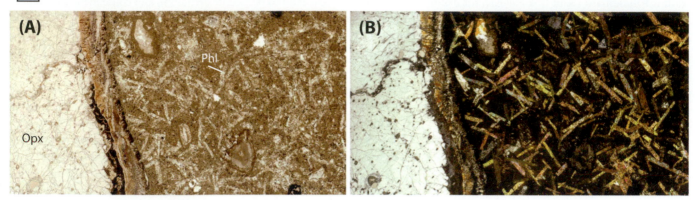

Figure 9.38 **(A)** Photomicrograph of a thin section under plane light of kimberlite, which reacted with an orthopyroxene-rich (Opx) mantle nodule (left) from a diatreme pipe on Isle Bizard, Quebec (see Fig. 8.1(A)). The kimberlite has light-colored millimeter-size phlogopite (Phl) crystals in an extremely fine-grained matrix. **(B)** Same as (A) under crossed polars.

pyrochlore, a complex calcium, sodium, niobium, tantalum oxide, which forms small red octahedra (Fig. 9.39(A)). This mineral, which provides the main source of niobium and tantalum, is also a valuable source of **rare-earth elements**, which substitute into the mineral's structure.

9.4.5 Igneous rocks formed near convergent plate boundaries

Convergent boundaries are the second-largest rock factory on Earth. However, in contrast to the more productive midocean ridge system whose rocks are subducted back into the mantle, rocks produced at convergent plate boundaries have slowly built the continents throughout geologic time. They are, therefore, of immense importance, because they not only provide the primary rocks of continents but, through their weathering, also create the soils on which our agriculture depends and the sediment to form sedimentary rocks. The rise of magma into the crust near convergent boundaries also brings heat with it

that causes metamorphism and circulation of fluids that form ore deposits. Without the igneous rock production at convergent plate boundaries and the formation of continents, life on our planet would have remained largely marine.

Let us briefly review the fundamental processes discussed in Section 8.2.3 that lead to formation of igneous rocks at convergent plate boundaries. New ocean floor created at midocean ridges is cooled and hydrothermally altered by circulating ocean water. By the time these rocks reach a subduction zone, they contain hydrous minerals and are overlain by a thin veneer of sediment, which also contains considerable water (Fig. 9.40). The pressure increase that accompanies subduction compacts the sediment and drives off water and eventually causes metamorphic reactions that convert hydrous minerals to anhydrous minerals, releasing still more water. This water rises into the overlying mantle wedge, where its fluxing action causes partial hydrous melting of mantle peridotite. The critical release of water that cause melting occurs at a depth of about 100 km.

5 mm

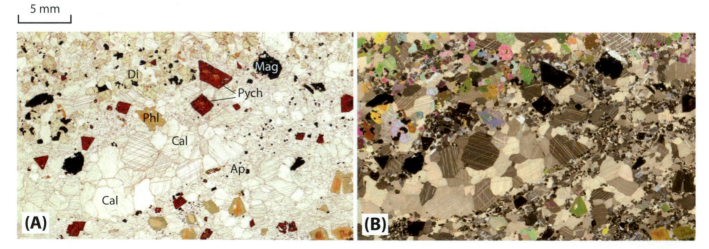

Figure 9.39 **(A)** Photomicrograph of a thin section under plane light of carbonatite from Oka, Quebec, containing clear calcite (Cal) with intersecting twin lamellae, small colorless high-relief crystals of apatite (Ap), opaque magnetite (Mag), pale green diopside (Di), orange to green zoned phlogopite (Phl), and red pyrochlore (Pych). **(B)** Under crossed polars, small apatite grains (low birefringence) outline original coarse grains of calcite that subsequently recrystallized to smaller polygonal grains.

We know this because the **Benioff seismic zone** of earthquakes in the descending slab always occurs ~100 km beneath the volcanic arc that forms above the subducting plate (Fig. 9.40). However, if the Benioff zone dips at less than 25°, as occurs along part of the Andes in South America, no volcanism occurs. Perhaps at these shallow angles water is released over too great a distance or the temperature in the mantle wedge is too low to cause melting.

Volcanic arcs

The line of volcanoes above a subducting plate can form an **island arc** on oceanic plates, or a **volcanic arc** atop a mountain range on continental plates. Along these arcs, volcanoes are regularly spaced. For example, those along the Aleutian arc are spaced ~70 km apart, whereas those along the adjoining Kamchatka arc are ~30 km apart (Fig. 9.41). This spacing is best explained as a result of a Rayleigh-Taylor instability (Sec. 8.6.2 and Fig. 8.23). Magma generated above the subducting plate would segregate to form a sheet parallel to the boundary that would be gravitationally unstable as a result of buoyancy of the magma. Diapirs would develop along this sheet that would determine the spacing of the volcanoes along the arc. Based on typical volcano spacings (30–70 km) and the viscosities of the magma and the lithosphere, we can calculate that the thickness of the magma sheet that segregates above the subducting plate is only 100–500 m thick.

Subduction can involve convergence of oceanic, oceanic and continental, or continental plates. When oceanic plates converge, the older, cooler, and hence denser ocean floor subducts beneath the younger less dense ocean floor. When oceanic and continental plates converge, the denser oceanic plate subducts beneath the continent. Continental plates can be subducted beneath continental plates if carried along on a subducting oceanic plate that transports the continent as if it were on a conveyor belt. India, for example, is following the ocean floor that is being subducted beneath Tibet. The convergence of continents, however, presents problems because the buoyant continental rocks do not sink as easily as dense metamorphosed ocean-floor rocks, so continent-continent collisions result in considerable thickening of the crust.

A subducting slab consists of cold dense rocks, and the metamorphic reactions that accompany subduction increase its density still more and provide the main driving force for subduction. Coupling between the subducting plate and the overlying mantle wedge sets up convection in the wedge (Fig. 9.40), which not only drives plate convergence but also brings in a constant flux of new hot mantle from beneath the overriding plate. This same convection creates tensional forces in the lithosphere behind the arc, which can result in rifting and formation of a **back-arc basin** and possibly creation of new ocean floor (Fig. 9.40). Not all convergent plate boundaries, however, have back-arc spreading. For example, none is seen in the Bearing Sea behind the Aleutian arc (Fig. 9.41).

Calcalkaline magma production above a subducting plate

The release of the hydrous fluids from the subducting plate, the convection in the mantle wedge, and the possible back-arc spreading all lead to perturbations that generate a wide variety of igneous rocks near convergent boundaries. These rocks are generally referred to as calcalkaline, but they show a wide range of composition depending on the type of plates involved (oceanic versus continental), the distance behind the primary volcanic arc, and whether back-arc spreading has played a role in their generation. The main rocks in the volcanic arc

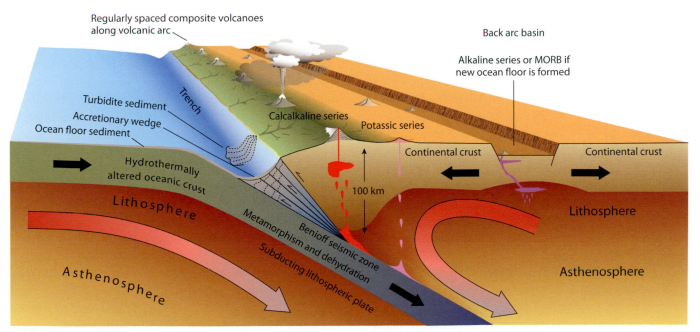

Figure 9.40 Cross-section showing magmatic activity associated with subduction of an oceanic plate beneath a continental plate.

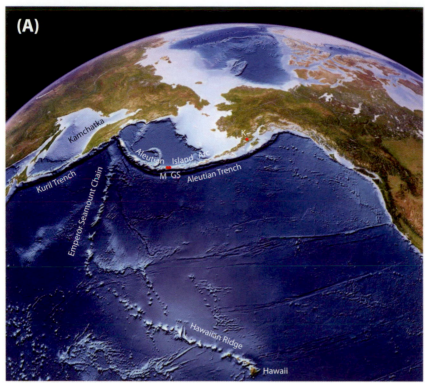

Figure 9.41 (**A**) The 2500 km-long Aleutian volcanic arc, formed by subduction of the Pacific plate beneath the North American plate. Volcanoes shown are Katmai (K), Moffett (M), and Great Sitkin (GS). The Hawaiian Hot Spot track follows the Hawaiian Ridge and Emperor Seamount Chain. (Image from Planetary Visions Ltd./ Science Photo Library.) (**B**) View looking east from Moffett volcano toward Great Sitkin, which is 45 km away. (Photograph courtesy of Bruce Marsh.)

are high-alumina basalt ($Al_2O_3 > 16.0\%$), andesite, dacite, and rhyolite, and their plutonic equivalents gabbro, diorite, granodiorite, and granite. However, the proportions of these rocks vary considerably depending on whether continental plates are involved in the convergence. Island-arc volcanoes produced at the convergence of oceanic plates are composed predominantly of high-alumina basalt, basaltic andesite, and andesite, with only minor amounts of more felsic rocks. When a continental plate is involved, the proportion of felsic rocks increases. This variation in the proportion of rock types is well documented along convergent boundaries that change from ocean – ocean convergence to ocean – continent convergence along their length, such as the Aleutians passing into Alaska at their eastern end (Fig. 9.41). Along the western half of the Aleutian arc where oceanic plates converge, basalts and andesites predominate, but toward the eastern end where oceanic plate converges with continental plate, larger quantities of felsic rocks occur. For example, the 1912 eruption near Katmai (K in Fig. 9.41(A)) produced 7 km³ of rhyolitic pumice, the world's largest eruption of the twentieth century (Fig. 9.21(B)). As magma rises through thicker and thicker sections of continental crust, the proportion of felsic rocks increases. This implies that the felsic magmas are derived largely from the melting of crustal rocks, a conclusion supported by isotopic studies.

Volcanoes continue to form as a subducting plate goes deeper than 100 km, but with diminishing size and frequency. One feature found worldwide is that as the Benioff zone deepens, magmas become progressively more potassium rich. The increase might relate to mica playing a more significant role at greater depth or to the magma's ability to scavenge potassium from the rocks through which it passes on route to the surface.

When a **back-arc basin** develops, the lithosphere is thinned, and decompression melting may occur in the ascending asthenosphere beneath the basin (Fig. 9.40). If the scale of melting is small, alkaline magmas form, but if the extension leads to rifting and the generation of new ocean floor, the increased degree of melting leads to the eruption of tholeiitic MORBs.

Calcalkaline volcanic rocks

The earliest rock erupted in volcanic arcs is subalkaline basalt that tends to have higher alumina (>16 weight %) than do tholeiitic flood basalts of LIPs, which typically contain about 15% Al_2O_3. They are, therefore, called high-alumina basalt. A complete gradation exists from high-alumina basalt to andesitic basalt and then basaltic andesite to andesite. The basaltic members of this series form a major component of the calcalkaline series of island arcs but are less abundant on continental arcs, or if they are initially abundant, later, more siliceous, lavas obscure them. The typical high-alumina basalt contains abundant plagioclase phenocrysts, which can be as calcic as An_{90} and account for the rock's high alumina content. These phenocrysts typically show complex zoning patterns and commonly contain zones of melt inclusions, which are trapped during periods of rapid crystal growth (Fig. 9.42(A)–(B)). These features are consistent with repeated fluctuations in magma temperature, which probably result from surges of magma entering and

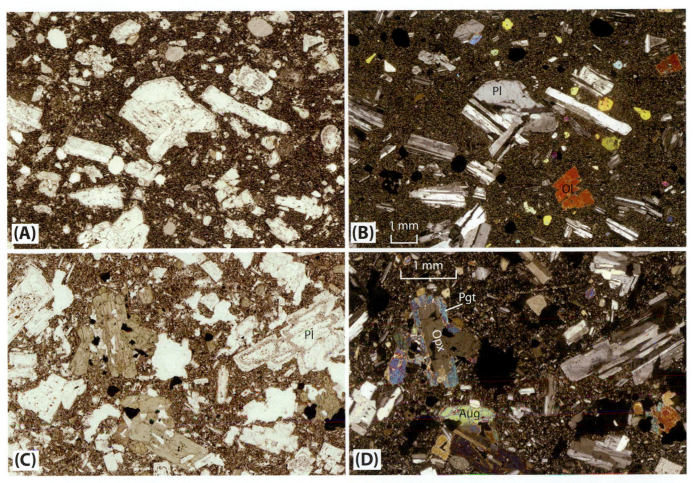

Figure 9.42 (A) and (B) Photomicrographs of a thin section of vesicular porphyritic high-alumina basalt from Mount Fuji, Japan, under plane and crossed polarized light, respectively. Phenocrysts of olivine (Ol) and plagioclase (Pl), occur in a brown glassy matrix containing small crystals of plagioclase, pyroxene and magnetite. (C) and (D) Photomicrographs of a thin section of vesicular porphyritic andesite from the Hakone Volcano, Japan, under plane and crossed polarized light, respectively. Phenocrysts of plagioclase (Pl), augite (Aug), orthopyroxene (Opx), and magnetite (opaque) occur in a groundmass of small crystals of plagioclase, pyroxene, magnetite, and brown glass. The orthopyroxene phenocrysts are rimmed by highly birefringent pigeonite (Pgt).

mixing with older magma in chambers beneath the volcanoes. The main mafic phenocryst is olivine, but augite and magnetite may also occur. The groundmass contains augite as well as pigeonite, magnetite, feldspar, and residual siliceous glass.

Andesite is the most abundant volcanic rock of the calcalkali series (Fig. 9.42(C)–(D)). Most andesite is strongly porphyritic, with abundant plagioclase phenocrysts, which show both normal and oscillatory zoning. They contain many glassy melt inclusions, which are trapped in zones when the crystals grew rapidly. Plagioclase phenocrysts commonly show evidence of having been melted along cleavage planes to produce what is called a **sieve texture**. Although the average plagioclase composition in andesites is andesine (An_{50-30}), the cores of some crystals can be much more calcic. Olivine phenocrysts may be present, but they are typically replaced by orthopyroxene, which may have olivine inclusions in its core. Orthopyroxene phenocrysts, in turn, are rimmed by pigeonite (Fig. 9.42(C)–(D)). Augite and magnetite also form phenocrysts. The groundmass of andesites is commonly glassy, which indicates

its high silica content. Many andesites contain as much as 50% phenocrysts, which approaches the limit that allows magma to move. Complex phenocryst assemblages, zoning patterns, and sieve texture indicate that magma mixing has played an important role in the generation of andesites.

Magmas more silica rich than andesite erupt to form progressively more glassy rocks. **Dacite** and **rhyolite**, for example, commonly form the glassy rock known as obsidian (Figs. 9.19 and 9.43(A)). Most obsidian exhibits layering, which is produced by concentrations of extremely small crystals that are barely visible, even under the microscope. In Figure 9.43(B), layering in the Glass Mountain obsidian flow from the Medicine Lake Highlands of California can be seen in a thin section to produce slightly darker bands in a clear glassy host. Small laths of feldspar have been rotated into parallelism with this flow layering, thus giving the rock a **trachytic texture**. The layers wrap around the phenocryst of sanidine. Close inspection of the sanidine phenocryst reveals that it is surrounded by a rim of brown glass, which, because of its color, must be richer in iron than the clear glass in

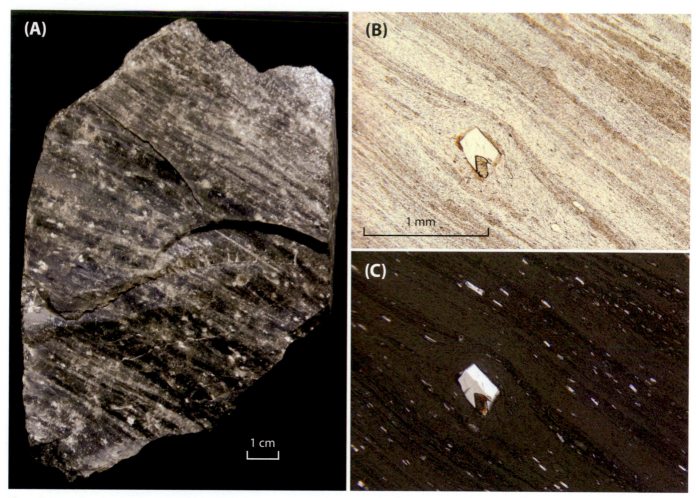

Figure 9.43 (**A**) Hand specimen of layered obsidian from Glass Mountain, Medicine Lake Highlands, California (see Fig. 8.21), showing conchoidal fracture. White crystals are sanidine phenocrysts. Photomicrographs of a thin section of this obsidian under plane (**B**) and partially crossed (**C**) polarized light. Layering is due to concentrations of minute crystals of pyroxene in clear glass.

the rest of the sample. Compositionally different glasses in many obsidians indicate magma mixing. It is likely that the intrusion of basalt into a rhyolitic magma chamber provides the influx of thermal energy necessary to trigger eruption of the obsidian. In this respect, obsidians share the same feature that characterizes most plutonic bodies of granite; that is, the ubiquitous presence of basaltic inclusions (see Sec. 9.2.2 and Fig. 9.10).

As rocks become more siliceous, increasingly larger proportions of them erupt as either ash falls or ash flows. The rocks formed from these deposits are known as tuffs. The differences between the ash-fall and ash-flow tuffs were discussed in Section 9.2.3 and illustrated in Figure 9.23.

As calcalkaline rocks become more potassium rich with increasing depth to the Benioff zone, potassium feldspar becomes more abundant and biotite appears. At the basic end of the series, biotite-bearing lamprophyres may be present. In intermediate andesitic compositions, phenocrysts may include biotite and hornblende in addition to plagioclase, augite, and magnetite. At the felsic end, extremely potassium-rich volcanic rocks may contain phenocrysts of leucite, as at Vesuvius.

Calcalkaline plutonic rocks

Erosion of volcanic arcs on continental plates exposes the underlying stocks and batholiths, which contain the plutonic equivalents of the overlying volcanic rocks. This includes the typical calcalkaline suite of rocks, ranging from diorite through granodiorite to granite. A complete compositional gradient exists between the most mafic and most felsic members. In the Sierra Nevada (California) and Idaho batholiths, for example, rocks vary from predominantly diorite on the western side through granodiorite to mainly granite on the eastern side (Fig. 9.44). This variation most likely reflects the type of rocks through which the magmas rose and the amount of felsic material assimilated. We describe here a few of the main rock types.

The most mafic rocks are typically gabbro and diorite, the distinction between which is based on whether the plagioclase is more or less calcic, respectively, than An_{50}. However, many diorites whose average plagioclase is less than An_{50} contain phenocrysts whose cores are extremely anorthite rich. This distinction is, therefore, sometimes difficult to apply, unless you have a chemical analysis and CIPW norm. Diorite tends to be more

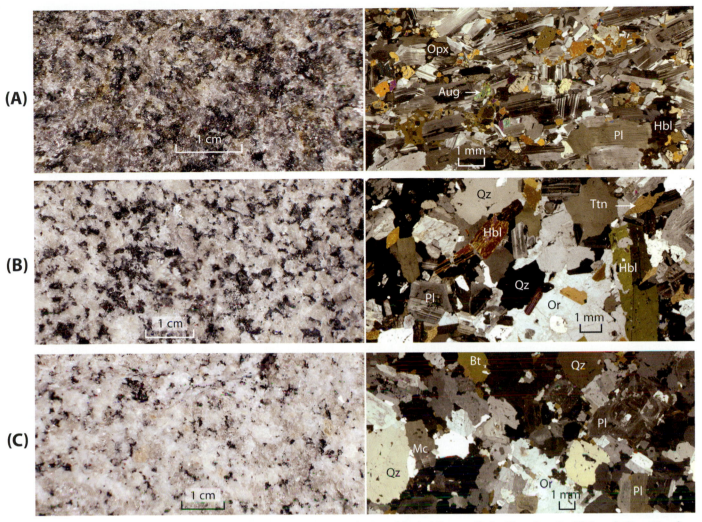

Figure 9.44 Hand specimens and photomicrographs of thin sections under crossed polars of diorite (**A**), granodiorite (**B**), and granite (**C**) from the Sierra Nevada batholith, California (plagioclase (Pl), orthoclase (Or), microcline (Mc), quartz (Qz), orthopyroxene (Opx), augite (Aug), hornblende (Hbl), biotite (Bt), and titanite (Ttn)).

plagioclase rich than gabbro and is, therefore, a lighter color (Fig. 9.44(A)). Plagioclase crystals in diorite have complex zoning patterns, such as that shown at the bottom of the photomicrograph in Figure 9.44(A). The zoning records a long history of magma recharge and mixing. The main mafic minerals in diorite are olivine, orthopyroxene, and augite. Olivine reacts with magma at an early stage to form orthopyroxene, which is, therefore, more common than olivine in diorite. In many diorites, late-stage liquid reacts with the earlier crystallizing pyroxene to form rims of hornblende. Quartz and biotite crystallize from the residual liquid to fill interstices between the earlier-formed minerals. Accessory minerals include titanite and apatite.

The most abundant member of the calcalkaline series is granodiorite. For example, it is the major rock type in the Sierra Nevada batholith of California (Fig. 9.44(B)). This rock contains both zoned plagioclase (An_{50-25}) and orthoclase, which are easily distinguished in thin sections but may require a hand lens to identify in a hand specimen. The plagioclase

crystals tend to be euhedral and are surrounded by orthoclase and quartz. Orthoclase, therefore, crystallizes later than plagioclase, which is the normal sequence in these rocks, but is different from the sequence in **rapakivi granite**, where the plagioclase rims the orthoclase (Fig. 2.13). Pleochroic green hornblende is usually the most abundant mafic mineral in granodiorite, but in potassium-rich varieties, biotite is a major mafic mineral. Several percent of titanite, with its characteristic highly birefringent wedge-shaped crystals, can be present (Fig. 9.44(B)), which indicates the relatively high titanium content of the rocks. Apatite and magnetite are common accessories.

Granite forms the felsic end member of the calcalkaline series and is, therefore, the lightest colored. The feldspar and quartz grains tend to be of the same size. Some feldspar grains, especially plagioclase, if present, develop crystal faces, but other grains are anhedral. The feldspar is rich in the alkali components and can be present as single alkali feldspar crystals in hypersolvus granite or as separate potassium-rich (orthoclase or microcline) and sodium-rich (albite) crystals in subsolvus granite (Sec. 8.3.6 and Fig. 8.11). In the example shown in Figure 9.44(C) from the Sierra Nevada batholith, two separate feldspars are present, a subhedral zoned albite-rich plagioclase and an anhedral microcline. Biotite is the main mafic mineral, but hornblende can also be present. Accessories include apatite, magnetite, and zircon.

9.5 Special Precambrian associations

Geology, as a modern science, began when James Hutton (1726–1797) broke with eighteenth-century ideas on the origin of rocks through catastrophic processes and introduced the uniformitarian approach, which interpreted rocks as having formed through the same processes operating today but acting over extended periods of time. The uniformitarian approach, "the present is the key to the past," formed the main principle expounded in Sir Charles Lyell's (1797–1875) *Principles of Geology*, the first geology textbook. The principle of uniformitarianism has served us well in unraveling the geologic record. However, in the last three sections of this chapter, we deal with rocks that appear to be at odds with this principle. We discuss rocks that were formed only in the Precambrian or were formed through truly catastrophic processes. These are komatiites, which formed only during the Archean; anorthosites, which formed only during the Proterozoic; and impact rocks produced by catastrophic meteorite impacts.

9.5.1 Komatiites

In Section 9.3.2, we saw that most plutonic igneous rocks have volcanic equivalents, but ultramafic rocks (i.e., >90% mafics) do not. This has a simple explanation. Most ultramafic rocks never existed as a liquid; instead, they form by the accumulation of mafic minerals from less mafic magma. For example, in the Kilauea Iki lava lake, olivine crystals sank to produce an olivine-rich rock in the lower part of the lake (Fig. 8.26(C)). No rapidly quenched glassy lava having the composition of the olivine-rich rock is found in Hawaii. We can melt the olivine-rich rock in the laboratory, but this requires temperatures well above 1200°C, typically the highest temperature recorded during volcanic eruptions. We conclude, therefore, that ultramafic rocks are cumulates, formed typically from basaltic magma, and they have never existed as liquids.

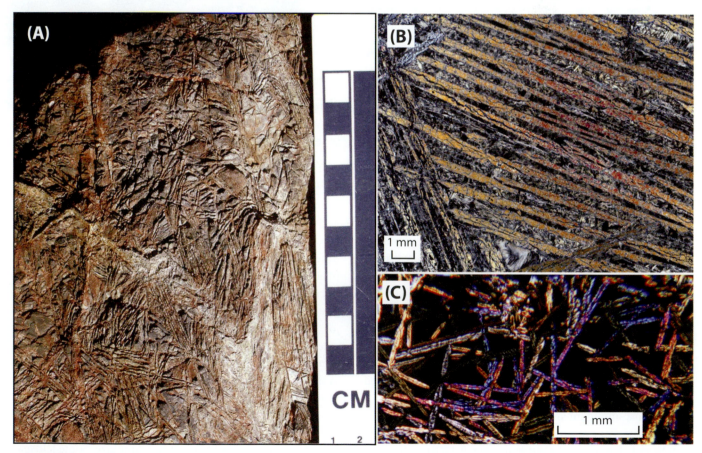

Figure 9.45 **(A)** Long blades of olivine formed by rapid crystallization in the top of a komatiite lava flow, Monroe Township, Ontario. **(B)** Photomicrograph of a thin section of komatiite in (A) showing skeletal crystals of olivine under crossed polars. **(C)** Skeletal crystals of olivine in Roman slag produced during smelting of copper ore at Skouriotissa, Cyprus (Fig. 9.32(E)).

Komatiites are an exception. These remarkable rocks, which were first described in detail from the Komati River in South Africa, are ultramafic lavas. They have been found worldwide but are restricted almost entirely to Archean greenstone belts, where they occur along with normal tholeiitic basalt. Komatiites contain more than 18% MgO, but many have as much as 33% MgO. Basaltic lava, by contrast, rarely contains more than 10% MgO. Basaltic lava can contain exceptionally high contents of olivine phenocrysts, which can elevate its MgO. These are known as picrites. In thin sections, they resemble the rock from the lower part of the Kilauea Iki lava lake (Fig. 8.26(C)). The olivine crystals are euhedral and clearly had time to grow to a large size and accumulate in a groundmass that quenches to normal basalt. In contrast, the upper rapidly quenched part of komatiite lava flows has no euhedral olivine crystals. Instead, olivine forms long thin needles or blades up to a meter in length that radiate downward in clusters from the upper surface of the flow (Fig. 9.45(A)). Because these splays resemble tufts of the Australian grass known as spinifex, they form what is known as the **spinifex texture**. In thin sections, this olivine is seen to form skeletal dendritic crystals (Fig. 9.45(B)), which is the typical morphology of crystals that grow rapidly from a liquid. Compare their shape with those that must have grown in only seconds in slag formed when the Romans smelted Cypriot copper ore (Fig. 9.45(C)). Although regularly shaped olivine crystals are found in the lower part of komatiite flows where cooling was slower, the thin quench crystals in the upper part of the flow are what show that these magnesium-rich lavas existed as liquids.

Melting experiments at one atmosphere indicate that komatiites become liquid around 1650°C (basaltic lava becomes liquid about 1200°C). Such high temperatures imply that the Archean geotherm must have been higher than in later times, which is consistent with a cooling Earth. As the pressure on mantle peridotite is increased, the first-formed melts become progressively more magnesium rich, reaching greater than 30% MgO at 7 GPa, which corresponds to a depth of 200 km. It is concluded, therefore, that komatiite magmas formed at depths greater than 150 km and basaltic magmas form at depths less than 100 km.

Komatiites are of economic importance because of associated sulfide ore deposits of nickel, copper, and platinum group elements (PGE), such as those at Kambalda, Western Australia. These deposits occur at the base of komatiite flows, where the high-temperature lava appears to have thermally eroded a channel in the underlying sedimentary rocks. Whether komatiite assimilated sulfur from sulfide-bearing sediments or derived it from other sources, they became so rich in sulfur that an immiscible sulfide liquid formed, into which the nickel, copper, and PGE partitioned (Sec. 8.7.6). Because the immiscible sulfide liquid was denser than the silicate liquid, it sank and accumulated near the base of the flow to form massive concentrations of nickel, copper, and PGE sulfide ore.

9.5.2 Massif-type anorthosites

The IUGS classifies rocks containing more than 90% plagioclase as anorthosite (Fig. 9.27). Anorthosite occurs as layers in large lopolithic intrusions, such as the Bushveld Complex of South Africa (Fig. 9.8), but it also forms huge intrusive bodies of batholithic proportions, which are referred to as **massif-type anorthosites**. It also forms a large part of the lunar highlands (the light-colored areas on the moon, in contrast to the dark areas, or maria, which are underlain by basaltic rock). Most plagioclase in anorthosite layers in lopoliths is more calcic than An_{70}, as are lunar anorthosites, whereas that in massif-type anorthosites is between An_{65} and An_{40}. Another feature of massif-type anorthosites is that they formed only in the Proterozoic between 1.45 billion and 0.9 billion years ago. In this section, we discuss massif-type anorthosites.

Massif-type anorthosites form large diapiric bodies of low-density rock that have risen in Proterozoic terranes through denser high-temperature, high-pressure metamorphic rocks. The most prominent belt of these massifs extends northeastward from the Adirondack Mountains in New York State, through the Grenville and Nain geological provinces of the Canadian Shield to the coast of Labrador (Fig. 9.46). If we

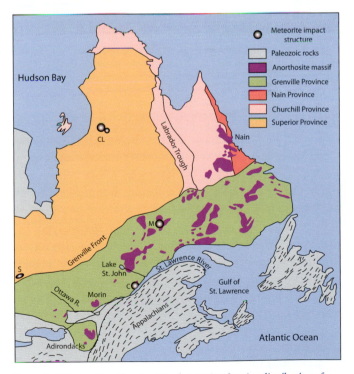

Figure 9.46 Map of northeastern North America showing distribution of Proterozoic anorthosite massifs. Also shown are four large meteorite impact structures: Manicouagan (M), Charlevoix (C), Sudbury (S), and Clearwater Lakes (CL).

Figure 9.47 (**A**) Large single meter-size crystal of labradorescent plagioclase from Nain, Labrador (for location, see Fig. 9.46). (**B**) Single crystal of orthopyroxene showing kink bands from the Lake St. John anorthosite massif, Quebec. (**C**) Ophitic intergrowth of decimeter-long plagioclase laths in a single crystal of pyroxene (dark) from the Lake St. John massif.

reassemble Pangaea, this belt continues through southern Greenland into Scandinavia, and to the west, it probably extends beneath the younger sedimentary rocks of the central United States to reappear in Wyoming. Other belts can be traced through Pangaea, such as the one from India through Madagascar to South Africa.

One of the most striking features of massif-type anorthosites is their coarse grain size. At the type locality for the mineral **labradorite** at Nain, Labrador (Fig. 9.46), plagioclase crystals are up to a meter in diameter (Fig. 9.47(A)). The main mafic mineral in anorthosites is orthopyroxene, which can form decimeter-size crystals (Fig. 9.47(B)). Where plagioclase and pyroxene are intergrown, they form an ophitic intergrowth, which resembles that seen in diabase (Fig. 9.36(D)), but the grain size, instead of being millimeters, is decimeters (Fig. 9.47(C)). These coarse grain sizes characterize the primary igneous rock in the core of the massifs. Toward the margins these rocks are deformed and recrystallized to finer-grain size. Recrystallization also changes the color of

anorthosite from dark to light as a result of the expulsion of minute exsolved grains of magnetite, ilmenite, and hematite. Whiteface Mountain in the Adirondack Mountains of New York State, for example, is made of recrystallized anorthosite. Also found in the recrystallized margins of many massifs are intrusive bodies of rock composed largely of ilmenite. These rocks, and their weathering products, provide the world's main source of titanium ore.

Another remarkable feature of anorthosite massifs is the limited compositional range of plagioclase in any one body. Plagioclase rarely changes by more than a few percent anorthite across an entire massif, but each massif can be characterized by a different plagioclase composition. The Nain anorthosite on the coast of Labrador, for example, is a labradorite anorthosite, whereas the Morin anorthosite north of Montreal (Fig. 9.46) is an andesine anorthosite. In thin sections, crystals lack zoning, which is in striking contrast to the complex zoning seen in plagioclase crystals in diorites (Fig. 9.44(A)). The large plagioclase crystal in Figure 9.47(A) exhibits iridescence (Secs. 3.3.4 and

7.4.1) in shades of green to blue, which indicates that its anorthite content is 55 ±0.04%, a remarkably constant composition for such a large crystal.

The lack of zoning in the plagioclase crystals is difficult to explain given the nature of the plagioclase phase diagram (Fig. 8.10(A)) and the common occurrence of zoning in other slowly cooled intrusive rocks such as diorites (Fig. 9.44(A)). Homogenizing a crystal as large as the one shown in Figure 9.47(A) would take more than all of geologic time, given the low diffusion rates in plagioclase. We can conclude only that these crystals must have grown initially with a constant composition. Why crystals would not show some compositional variation across a huge massif remains a puzzle.

There are no lavas of anorthosite composition, so we must conclude that the plagioclase in anorthosites is a cumulus mineral. Unfortunately, we do not know the composition of the magma from which they crystallized. If it was basaltic, where are all the mafic minerals that must also have crystallized? Gravity surveys indicate that no large bodies of mafic rock underlie anorthosite massifs, so this mafic fraction would have to have been left behind at considerable depth. However, anorthosites cannot have been formed at too great a depth, because plagioclase is not stable at high pressure. The magma may have been closer to granodiorite in composition, which would explain the common occurrence of quartz monzonite and granite capping anorthosite massifs. Although the contact of these rocks with anorthosite is gradational, isotopic studies show that a large part of these more quartz-rich rocks formed by partial melting of the surrounding metamorphic rocks and did not form by differentiation of the magma from which the anorthosite crystallized.

Finally, there remains the problem of why massif anorthosites formed only between 1.4 billion and 0.9 billion years ago. The origin of anorthosites remains one of the biggest unsolved problems in geology, and it is unlikely that evidence from the "present" will provide "the key to the past." The conditions and processes must have evolved as the Earth has aged.

9.5.3 Rocks associated with large meteorite impacts

One of the most dramatic events in the history of life on our planet was the extinction of dinosaurs at the end of the Cretaceous Period (66 Ma). Although many other species disappeared at this time, the extinction of dinosaurs is what has captured the popular imagination. How could gigantic beasts such as *Tyrannosaurus rex* suddenly become extinct? The discovery of an anomalously high concentration of the rare element iridium on the Cretaceous-Tertiary boundary (K-T boundary), led Luis and Walter Alvarez to postulate that the extinction resulted from a large meteorite impact. Meteorites are known to contain elevated concentrations of iridium, an element that on Earth sank to the core early in the planet's history. Correlating the extinction with this catastrophic event added to the popu-

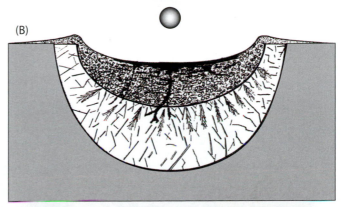

Figure 9.48 **(A)** Meteor Crater, Arizona. (Photograph courtesy of U.S. Geological Survey.) **(B)** Hypothetical cross-section of a meteorite crater, with relative dimensions based on the meteorite being a stony variety with a density of 3500 kg/m³ traveling at 20 km/s. **(C)** Shatter cones in fine-grained Ordovician limestone at Charlevoix, Quebec (C in Fig. 9.46).

lar mystique surrounding dinosaurs. At first, no appropriately aged meteorite impact structure was known, but eventually the smoking gun was found at Chicxulub on the northern tip of Mexico's Yucatán Peninsula. Unfortunately, younger sedimentary rocks cover this 100 km-diameter impact structure, and its presence can be demonstrated only through geophysical surveys and samples recovered from oil-exploration drill holes. Its lack of exposure is disappointing considering the dramatic

Figure 9.49 (A) Generalized geological map of the Sudbury lopolith and meteorite impact structure, Ontario (S in Fig. 9.46). (B) Sudbury impact breccia. (C) Photomicrograph of a thin section of Onaping tuff under plane light, showing small glassy fragments of impacted rock that were flash melted. (D) Quartz grains in Onaping tuff, showing shock-induced planar features.

effect it had on life on the planet. However, numerous other impact structures of similar size are well exposed on land and provide samples of rocks that allow us to gauge the magnitude of such a catastrophic event.

Most large meteorite impacts are preserved on Precambrian cratons. Impacts that occur in younger orogenic belts are destroyed by deformation. It is appropriate, therefore, to discuss rocks formed by meteorite impact in this section dealing with Precambrian associations, even though impact-generated rocks are not necessarily confined to the Precambrian period.

Small meteorites enter the Earth's atmosphere every day and burn up as shooting stars. Occasionally, some are large enough to survive this fiery transit and produce small impact craters on reaching the surface. More rarely, larger meteorites, which are not slowed perceptibly by their passage through the atmosphere, impact the Earth with the velocity they had in space, about 20 km/second. The kinetic energy of a meteorite only a meter or two in diameter traveling at this velocity is enormous (kinetic energy = $\frac{1}{2}$ mass × velocity^2) and is sufficient to completely evaporate the

meteorite. This is why you will never see a meteorite larger than a meter or two in diameter in museums. On impacting the Earth, a meteorite's velocity suddenly drops to zero, and its kinetic energy is converted to heat and shock waves, which produce an explosion crater and evaporate the meteorite (Fig. 9.48).

We have learned much about explosion craters from underground nuclear tests, which generate similar amounts of energy (Fig. 9.48(B)). The depth of an explosion crater is approximately one-tenth its diameter. The crater is surrounded by a blanket of ejecta, with the size of particles decreasing rapidly away from the crater. Rocks in the crater rim may actually be turned upside down from the force of the explosion, as seen in the rim of Meteor Crater in Arizona (Fig. 9.48(A)). Upon impact, about 20% of the kinetic energy of a meteorite is converted to heat, which evaporates the meteorite and melts the impacted rock. Particles of melt may be ejected into space by the explosion to later fall back to Earth as **tektites**, small centimeter-scale glassy droplets whose surfaces indicate the effects of frictional heating and ablation during reentry through the atmosphere. Beneath the melt zone, rock is totally crushed to a depth of approximately one-third the crater diameter. Beneath this, rock is intensely fractured to a depth of two-thirds the crater diameter. A characteristic type of fracture that is unique to meteorite impact sites is known as **shatter cones** because of the conical sets of fractures that develop with their apex pointing in the direction from which the shock wave came (Fig. 9.48(C)).

Meteor Crater in Arizona (Fig. 9.48(A)) is the world's best-preserved impact structure. It has suffered little erosion since its formation 49 000 years ago. With a diameter of only 1.2 km, however, it is extremely small compared with older impact structures that might correlate with mass extinctions. Erosion, however, has removed most of the upper part of these older craters, and all that remains is their lower part, whose diameter may be only a third of the original crater diameter. For example, the 100 km-diameter structure at Chicxulub, with which dinosaur extinction has been associated, may originally have had a diameter of 300 km, which would dwarf the crater in Arizona.

We illustrate the rocks formed during these cataclysmic events by discussing the impact structure at Sudbury, Ontario, which is of particular interest because it is associated with the world's largest nickel deposit. The impact occurred 1.85 billion years ago, leaving a circular structure that was later deformed into an ellipse (Fig. 9.49(A)) during the Grenville orogeny, which occurred just to the south of the structure (S in Fig. 9.46). Shock waves from the impact created shatter cones that are generally radial and point back toward the center of the explosion, except in a few places where rocks have been overturned. The transient shock wave also formed irregular veins of breccia, which contain fragments of country rock up to meters in diameter in a matrix of smaller fragments and a black aphanitic rock formed by flash melting (Fig. 9.49(B)).

Following the impact, ejecta from the explosion were deposited back into the crater to form what was originally mistakenly mapped as a volcanic rock, the Onaping tuff. Thin sections of this rock (Fig. 9.49(C)) reveal that it contains many rock fragments, most of which have been melted and then quenched to a glass that has subsequently devitrified to extremely fine feathery crystals. It also contains quartz grains that contain closely spaced planar features produced by the passage of the shock wave. These are made visible by minute inclusions on the planes or by differences in refractive index or crystallographic orientation (Fig. 9.49(D)). The tuff also contains microscopic diamonds, which formed during the passage of the shock wave, where pressures would have momentarily exceeded 10^3 GPa, which is well into the diamond stability field.

We do not know the diameter of the original Sudbury crater, but it was probably on the order of 200 km. Rocks would, therefore, have been fractured all the way to the base of the crust, which likely triggered melting through decompression to produce magma that may have been as hot as 2000°C. This melt then differentiated to produce a lower mafic part (norite) and an upper felsic part (granophyre), which intruded into the Onaping tuff (Fig. 9.49(A)). An immiscible sulfide melt also formed, probably from crustal rocks, and this dense liquid sank and accumulated along the base of the lopolith to form the nickel ore bodies.

Many other large meteorite impact structures have been found worldwide. Three additional ones, Manicouagan (Fig. 17.9), Charlevoix, and Clearwater Lakes, are seen in the area covered by the map in Figure 9.46. Although an iridium anomaly was first identified on the K-T boundary, others have subsequently been found at other boundaries marked by mass extinctions. It is likely, therefore, that catastrophic meteorite impacts have played an important role at many times during the evolution of life on our planet.

Summary

This chapter first dealt with where intrusive and extrusive rocks are formed. We then saw that igneous rocks are classified on the basis of their mineralogy. We learned that groups of rocks form common associations that are characteristic of the plate tectonic setting in which they form. Finally, we discussed two rock types that were formed only in the Precambrian and rocks that were formed by large meteorite impacts. The following list presents the main points in this chapter:

- Igneous rocks are classified as extrusive or intrusive, with the later being subdivided into plutonic (formed at great depth) and hypabyssal (formed near surface). Generally, extrusive and hypabyssal rocks are fine grained or glassy, whereas plutonic rocks are coarse grained.

- Magma rises buoyantly and pushes aside the intruded rock to form bodies ranging from narrow sheets to large batholiths.

- Extrusion of magma produces a variety of volcanic rocks, ranging from lava flows to volcanic ash, depending on how explosive the eruption is. As the silica content of magma increases from basalt (~50%) to rhyolite (73%), viscosity increases, which makes it more difficult for gas bubbles to escape from the magma and increases the likelihood of explosive eruptions.

- The largest eruptions are from long fissures to form flood basalts. Magma extruded from a central vent forms gently sloping basaltic shield volcanoes, steep-sided andesitic composite cones, and rhyolitic domes.

- Collapse of the roof of a batholith produces a caldera with expulsion of magma commonly as ash flows that become welded during solidification.

- The International Union of Geological Sciences (IUGS) classifies igneous rocks on the basis of their mineral content (mode).

- Rocks fall into three main series that have distinct plate tectonic settings: tholeiitic rocks are associated with divergent plate boundaries; calcalkaline rocks, with convergent plate boundaries; and alkaline rocks, with continental rift valleys and late stages of volcanism on oceanic islands.

- The largest production of igneous rocks occurs at midocean ridges, where tholeiitic midocean ridge basalt (MORB) is erupted. Where ocean crust has been obducted onto continents (ophiolite suites), MORBs are underlain by sheeted dike complexes, layered gabbro, and serpentinized ultramafic rocks.

- Ocean-floor rocks are hydrothermally altered by circulating ocean water that cools them when they are formed at midocean ridges. Economic mineral deposits of iron, copper, and zinc sulfides, chromite, and asbestos are found with these rocks.

- Oceanic islands formed over hot spots are mostly tholeiitic, but as they move off the hot spot, they are commonly capped by alkaline rocks.

- At numerous times during Earth's history, huge eruptions of tholeiitic flood basalt have occurred over extensive areas to form large igneous provinces (LIPs).

- At convergent plate boundaries, a volcanic arc forms as soon as the subducting plate sinks to ~100 km, as long as the subduction angle is greater than 25°. Water released at this depth from the hydrothermally altered ocean-floor rocks acts as a flux to cause melting in the overlying mantle wedge.

- Igneous rocks formed above subducting plates include high-alumina basalt, andesite, dacite, and rhyolite, as well as their plutonic equivalents gabbro, diorite, granodiorite, and granite. The felsic rocks are more abundant where convergence involves a continental plate. Calcalkaline rocks have slowly built the continents over geologic time.

- Two igneous rock types are exclusively of Precambrian age, the ultramafic lava komatiite, and large massif anorthosites (An_{65-40}).

- Large meteorite impact structures contain rocks that bear testament to huge explosions that appear to have played an important role in the evolution of life on the planet.

Review questions

1. How do intrusive igneous bodies make room for themselves?

2. From the crosscutting relations in Figure 9.2, deduce the chronology of intrusive events seen in this outcrop.

3. How do magma density and viscosity affect the shape of intrusive bodies?

4. Contrast the eruptive styles of Hawaiian, Strombolian, Vulcanian, Plinian, and Peléan volcanoes.

5. How does magma viscosity affect the shape of volcanoes?

6. What is the difference between a mode and a norm?

7. Using the normative compositions of rocks in Table 9.1, plot each rock in Figure 9.25 and determine whether it has been correctly named according to the IUGS classification of igneous rocks.

8. What name would you give to a rock composed of 50% olivine, 25% orthopyroxene, and 25% clinopyroxene? Much of the Earth's upper mantle is thought to have this composition.

9. What are the main rocks in an ophiolite suite, and how are they formed?

10. Why are igneous rocks at midocean ridges so altered to hydrous minerals?

11. How are sulfide ore deposits formed near midocean ridges?

12. Why are so many oceanic islands generated over hot spots capped with alkaline rocks, even though the main volcano-building period erupts tholeiitic rocks?

13. What is a large igneous province, and what rock type characterizes such regions?

14. What is inverted pigeonite, and how does it form?

15. What are some titanium-bearing minerals that form in alkaline rocks as a result of their high titanium content?

16. What is kimberlite, and why is it of economic interest?

17. What is carbonatite, and what economic mineral deposits are associated with it?

18. What are the main calcalkaline rocks erupted in the volcanic arc above subduction zones, and what are their plutonic equivalents?

19. What difference is there between volcanic rocks erupted on island arcs involving convergent oceanic plates and volcanic arcs developed on continental plates, and what is the likely cause of the difference?

20. What can you conclude about the history of an andesite (or diorite) from the complex zoning patterns in its plagioclase phenocrysts?

21. What importance is attached to the ubiquitous occurrence of basaltic inclusions in rhyolite and granite?

22. What is a komatiite, and what textural evidence shows that it was derived from an olivine-rich liquid?

23. What distinguishes plagioclase in massif-type anorthosites from plagioclase in most other plutonic rocks?

ONLINE RESOURCES

The National Institutes of Health's **ImageJ** software program can be downloaded for free from http://rsb.info.nih.gov/ij. This is a very useful program for doing a variety of measurements on digital images of photomicrographs. It can be used for measuring grain size and orientation and the modal abundance of minerals that have distinctive gray-scale values.

The CIPW norm calculation found on the U.S. Geological Survey's Volcano Hazards Program site (http://volcanoes.usgs.gov/yvo/aboutus/jlowenstern/other/software_jbl.html) gives reliable results, whereas many others on the Web do not.

NASA's Moderate Resolution Imaging Spectroradiometers (MODIS), carried on the Earth Observing System (EOS) satellites, surveys the entire surface of the planet every 48 hours at a resolution of 1 km/pixel and indicates regions of elevated temperature, such as new volcanic eruptions and forest fires. This near-real-time satellite coverage is available at http://modis.higp.hawaii.edu.

FURTHER READING

Christiansen, R. L. (1979). Cooling units and composite sheets in relation to caldera structure. In *Ash Flow Tuffs*, ed. C. E. Chapin and W. E. Elston, Geological Society of America Special Paper, 180, 29–42.

de Boer, J. Z., and Sanders, D. T., 2002. *Volcanoes in Human History*, Princeton University Press, Princeton, NJ. This entertaining and instructive book discusses nine major historic volcanic eruptions and the effects they had on humans, both positive and negative.

Irvine, T. N., and Baragar, W.R.A. (1971). A guide to the chemical classification of the common volcanic rocks. *Canadian Journal of Earth Science*, 8, 523–548.

Le Maitre, R. W. (1976). The chemical variability of some common igneous rocks. *Journal of Petrology*, 17, 589–598. This paper presents the average compositions of common igneous rock types based on 26,373 analyses.

MacKenzie, W. S., Donaldson, C. H., and Guilford, C. (1982). *Atlas of Igneous Rocks and Their Textures*, Longman Group Ltd., Harlow, Essex, UK.
This book contains beautiful colored illustrations of igneous rocks and their textures as seen under the microscope.

Wager, L. R., and Brown, G. M. (1967). *Layered Igneous Rocks*, Oliver & Boyd, Edinburgh.
This book gives a survey of all major layered igneous intrusions in the world.

CHAPTER

10 Sedimentary rock-forming minerals and materials

This chapter presents the systematic descriptions of uniquely sedimentary rock-forming minerals and materials. These are mainly the newly formed minerals that result from interactions of preexisting minerals with the atmosphere or precipitation from the oceans. Detrital minerals such as quartz, feldspar, and the micas, which were discussed in Chapter 7, are the major constituents of sedimentary rocks such as sandstones, arkoses, and graywackes. By studying hand specimens of these minerals in the laboratory while referring to the appropriate specimen descriptions in this chapter, you will further develop your mineral identification skills.

A deep gorge showing cliffs of well-banded, brown (oxidized) Precambrian iron-formations in the Hamersley Range of Western Australia. These consist mainly of finely banded chert-magnetite-hematite and are the result of widespread deposition of iron-rich chemical sediments during a long time span in the Precambrian when the atmosphere-ocean system was essentially anoxic. This iron-rich formation is the host to many vast open pit mines in northwestern Western Australia producing enormous amounts of iron ore for the steel industry worldwide (see Fig. 16.7).

In the prior three chapters (Chapters 7–9), we addressed many aspects of minerals and rocks of igneous origin. They are the result of crystallization from a liquid magma over a wide range of high temperatures (~1400 to 600°C) in an environment in which oxygen gas (O_2) is nonexistent and in which H_2O is generally a minor constituent. This chapter deals with minerals that (1) are newly formed as a result of chemical reactions (under atmospheric conditions) that slowly destroy earlier minerals (e.g., feldspars) and recombine various ions in solution in water to form **clay minerals**, and **oxides** or **hydroxides**, and (2) those that are **chemical precipitates** in depositional basins such as carbonates, evaporite minerals, and Precambrian iron-formations. However, we must also mention those minerals that have survived the physical and chemical weathering processes while exposed to the atmosphere; these are known as **detrital** minerals.

We systematically treat 13 sedimentary rock-forming minerals. These include ice, the solid form of H_2O; a hydroxide; a clay mineral (representative of the large clay mineral group); two polymorphs of $CaCO_3$; four more carbonates; two halides; and two sulfates. Clearly, this group of common sedimentary minerals is totally different from the igneous rock-forming minerals discussed in Chapter 7.

We close with discussions of three sedimentary materials: chert and agate, phosphorite, and soil.

10.1 The interaction of the Earth's atmosphere with minerals

The Earth's atmosphere is composed mainly of nitrogen (78%), oxygen (21%), argon (0.93%), and carbon dioxide (CO_2; 0.039%) and variable amounts of water vapor, ranging from 1 to 5%. This mixture of gases, combined with the processes of physical weathering, leads to the disintegration of earlier-formed igneous, metamorphic, and sedimentary rock constituents. Some primary (igneous and/or metamorphic) minerals are more susceptible than others to chemical disintegration under atmospheric conditions. Table 10.1 lists the relative stabilities of some common igneous rock-forming minerals discussed in Chapter 7. This table shows that quartz is one of the most chemically and physically stable minerals, whereas olivine is easily altered chemically, mainly through oxidation and hydration (see Equations (10.1) and (10.2)). The positions of the minerals between quartz and olivine represent intermediate stabilities during weathering. This order shows that, in general, igneous minerals that form at the highest temperature (e.g., olivine) are the most unstable, whereas minerals formed at lower temperatures (e.g., quartz, muscovite, K-feldspar) are more stable under atmospheric conditions. Quartz, K-feldspar, and muscovite are common constituents of detrital sedimentary rocks such as arkose and greywacke (see Sec. 12.1.1).

Table 10.1 Relative stabilities of some igneous rock-forming minerals during weathering.

High stability	Quartz	
	Muscovite	
	K-feldspar	
	Biotite	Albite
	Hornblende	Intermediate plagioclase compositions
	Augite	Anorthite
Low stability	Olivine	

(Increasing stability ↑)

The position of olivine in Table 10.1 indicates that it is the least stable of the ferromagnesian minerals listed. Figure 10.1 shows a high-resolution transmission electron microscopy (HRTEM) image of a partially oxidized olivine grain in a basalt. The oxidation of the Fe-containing olivine results in a

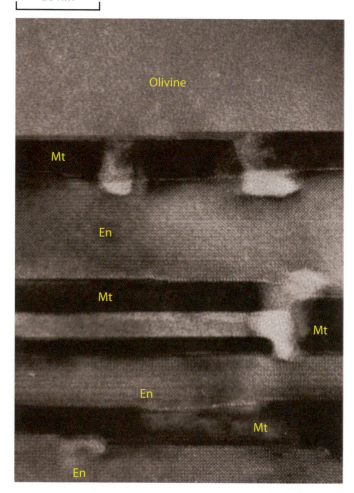

20 nm

Figure 10.1 A high-resolution transmission microscopy (HRTEM) image of a partially oxidized iron-containing forsterite (an olivine) in a basalt flow. The oxidation products are enstatite (En) and magnetite (Mt) in a crystallographically controlled intergrowth. For the equation that describes this chemical reaction, see Reaction (10.1). (Photograph courtesy of Hiromi Konishi and Huifang Xu, University of Wisconsin-Madison.)

crystallographically oriented intergrowth of enstatite and magnetite, according to the following reaction:

$$15(Mg_{0.8}Fe_{0.2})_2 SiO_4 \quad + \quad 13O_2$$
iron-containing forsterite   oxygen
$$\rightarrow 24MgSiO_3 \quad + \quad 2Fe_3O_4 \quad + \quad 3SiO_2 \qquad (10.1)$$
enstatite   magnetite   amorphous silica

Olivine is also highly susceptible to hydration, by hydrothermal fluids, as follows:

$$2Mg_2SiO_4 + 3H_2O \rightarrow Mg_3Si_2O_5(OH)_4 + Mg(OH)_2$$
forsterite   serpentine   brucite   (10.2)

New mineral formation, as shown on the right-hand side of the two above equations, is the result of the chemical breakdown of the more unstable minerals in Table 10.1 as a result of oxidation and hydrolysis. A hydration reaction for the breakdown of K-feldspar to form kaolinite is as follows:

$$2KAlSi_3O_8 + 11H_2O \rightarrow Al_2Si_2O_5(OH)_4 +$$
K-feldspar   kaolinite
$$4H_4SiO_4 + 2(OH)^- + 2K^+ \qquad (10.3)$$
silicic acid

The reaction products are kaolinite and several ions and compounds that go into solution (silicic acid, hydroxyl ion, and K^+). In similar breakdown reactions of Ca^{2+}- and Na^+-containing silicates, such as members of the plagioclase feldspar series, the Ca^{2+} and Na^+ go into solution as ions as did the K^+ in reaction (10.3). These ions may become so concentrated in ocean water that new minerals are precipitated out, such as halides (NaCl and KCl) and carbonates ($CaCO_3$, $CaMg(CO_3)_2$), as a result of reactions with anions such as Cl^- and $(CO_3)^{2-}$.

Pyrite, FeS_2, a common accessory mineral in some igneous rocks and a major constituent of sulfide-rich ore deposits, is highly unstable under atmospheric conditions and will react as follows:

Figure 10.2 The weathered surface of an aa lava flow on the south coast of Molokai, Hawaii, which was covered by a later flow. Although the structure and texture of the older flow top is still visible, with clinkerlike fragments containing white plagioclase phenocrysts, all the primary igneous minerals have been converted to soft red iron oxides and hydroxides and clay minerals.

$$FeS_2 + 3.5O_2 + H_2O \rightarrow Fe^{2+} + 2H^+ + 2(SO_4)^{2-}$$
$$\text{pyrite} \qquad\qquad\qquad \text{ions in solution} \qquad (10.4)$$

The Fe^{2+} in solution is immediately oxidized to Fe^{3+} and may be precipitated as the Fe^{3+}-containing hydroxide, goethite, $FeO(OH)$, or it may form a mixture of hydrous iron oxides called limonite, which is a field term that refers to an amorphous mixture of iron hydroxides, of yellow to brown color, and of uncertain identity. Such yellow-brown weathering products commonly overly sulfide deposits and are known as **gossan**. Such occurrences are of much interest to prospectors and exploration geologists because they help locate possible underlying sulfide deposits of economic value. A brownish hue on the weathered surface of a granite is due to a microscopically thin layer of limonite. Many soils exhibit such a color as well. Trace amounts of Fe^{2+} in minerals such as feldspar may oxidize in an environment of oxygen-rich groundwater to produce a microscale coating of hematite, which results in the red hue of many granites and sandstones in rock exposures.

In addition to the oxides (and hydroxides) formed as a result of Equation (10.4), H^+ and $(SO_4)^{2-}$ produce acidic and sulfate-rich waters. These are generated in the waste dumps of sulfide-rich mining districts and coal mines, which commonly contain pyrite. This, then, is the main reason for what is known as **acid mine drainage (AMD)**, which can cause extensive pollution that requires expensive and complex remediation procedures. Major highway construction commonly involves leveling the topography by filling valleys with material cut from hilltops (cut and fill). If the cut rocks contain sulfides, as is the case in many igneous and metamorphic rocks, the freshly broken rock provides a ready source of extremely acid **leachate**, which can cause major environmental problems in the water system adjoining the highway.

The result of the foregoing reactions acting on a fresh igneous rock is to convert it to an assemblage of clay minerals, iron oxides and hydroxides, a few chemically resistant minerals (e.g., quartz, if present), and soluble products (Na^+, K^+, Ca^{2+}, and Mg^{2+}) that are carried away in groundwater and eventually reach the ocean. The effect of these reactions on a relatively young igneous rock can be seen in the example of an **aa** lava flow on the Hawaiian Island of Molokai (Fig. 10.2). The prominent weathering profile in the lower (older) flow that grades from gray at the base of the road to red at the top of the flow is the result of the production of iron oxides and hydroxides from the primary dark ferromagnesian minerals in the basalt.

10.2 Ice: H_2O

We begin with ice, the crystalline form of water, because it fits all the criteria included in the definition of mineral, as discussed in Section 2.1.

Occurrence: Most of the natural ice on Earth is located at extreme latitudes: in the Greenland ice sheet and sea ice at the North Pole and in the Antarctic ice sheet at the South Pole. Also found in high-mountain glaciers and present as permafrost in very large polar and subpolar regions of the globe.

Chemical composition and crystal structure: Pure H_2O occurs in many polymorphs, as shown in Figure 10.3. The crystal structures of two of these ice polymorphs are illustrated in Figure 10.4. The first is the structure of ice that forms naturally under freezing conditions. This is ice I, also known as I_h, because it has a hexagonal ring structure, with a low $G = 0.915$. The second structure is that of ice VIII, a cubic structure with $G = 1.6$, which is the densest form of ice.

Crystal form: Normal ice, ice I, exhibits hexagonal crystal form, $6/m\,2/m\,2/m$, as shown in snowflakes (see Fig. 10.5).

Physical properties: $H = 1.5$, but the hardness varies with temperature. At $-70°C$, $H = 6$, but in most glaciers, the temperature is $-30°C$. The G of ice at $0°C$ is 0.915, which is extremely

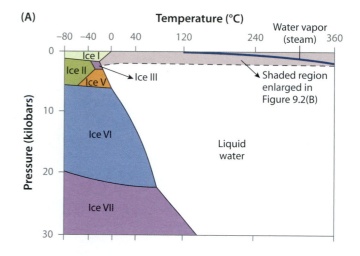

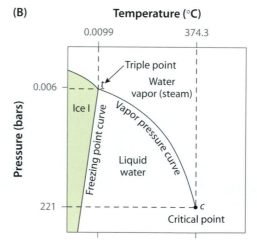

Figure 10.3 Stability diagrams for H_2O. **(A)** Stability fields of six polymorphs of ice (ice I, II, III, V, VI, and VII) and that of liquid water (modified after Figs. 1915 and 1916 in *Phase Diagrams for Ceramists*, 1964, and reprinted with permission of the American Ceramic Society, http://www.ceramics.org; all rights reserved). **(B)** An enlargement, not to scale, of the shaded region in (A).

light for a mineral (ice floats in water). The **G** of pure water at 0°C is 0.999. At −30°C the **G** of ice is 0.920.

Ice is generally colorless and transparent, but it may also be bluish white and opaque as in glaciers. Some ice appears white because of the inclusion of air bubbles. Transparent. Crystals are hexagonal (Fig. 10.5). Fracture is conchoidal.

Distinguishing features: Melts quickly, upon the touch of a hand, to make water.

Uses: Until the second half of the nineteenth century, ice harvesting was a big business, using block ice, mainly for the cooling of food and food products. Since the advent of artificial refrigeration, the harvesting and delivery of block ice has become obsolete. Artificial ice production is a large business because of the use of ice in food storage and processing, in the curing and mixing of concrete, and in the use of packaged ice by consumers.

Ice I, H$_2$O
Hexagonal, $P\,6/m\,2/m\,2/m$

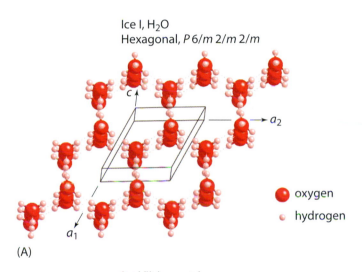

oxygen

hydrogen

(A)

Ice VII; Isometric,
$P\,4_2/n\,\bar{3}\,2/m$

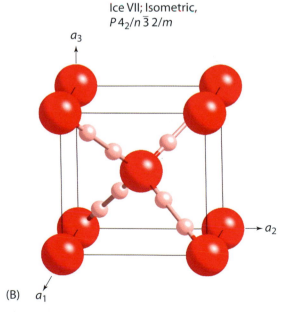

(B)

Figure 10.4 **(A)** A ball-and-stick representation of the crystal structure of ice I, the normal, everyday type of ice. A hexagonal unit cell is outlined. **(B)** A ball-and-stick representation of the crystal structure of ice VII. A cubic unit cell is outlined.

Figure 10.5 Two examples of hexagonal snowflakes. There are seemingly endless variations in the hexagonal shape of snowflakes resulting from constantly changing temperature and humidity when a snowflake forms. Many of these can be seen on http://www.snowflakes.com. (These photographs courtesy of Kenneth G. Libbrecht, California Institute of Technology; from http://snowflakes.com.)

10.3 Goethite: FeO(OH)

Occurrence: Goethite is a common oxidation product formed during the weathering of iron-bearing minerals. It may be present in bogs and springs as a result of inorganic or biogenic precipitation from groundwater. Together with hematite it constitutes the most common commercially mined iron ore. Such iron-rich ores are the product of oxidation of earlier iron-rich minerals (e.g., carbonates, silicates) and the leaching away of silica, which is a major constituent of Precambrian banded iron-formation.

Chemical composition and crystal structure: Closely approaches FeO(OH) with variable Mn content of up to about

Goethite: FeO(OH)
$P\,2_1/n\,2_1/m\,2/a$

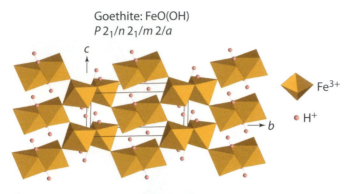

Figure 10.6 Polyhedral representation of the structure of goethite with Fe^{3+} in octahedral coordination with oxygen and $(OH)^-$ groups. An orthorhombic unit cell is outlined.

5 weight %. The crystal structure is shown in Figure 10.6 with $(OH)^-$ groups in hexagonal closest packing (HCP) and Fe^{3+} in octahedral coordination between them.

Crystal form: Orthorhombic; $2/m\,2/m\,2/m$. Is rarely in crystals. Commonly massive, stalactitic with radiating fibrous texture (Fig. 10.7). So-called bog ore is porous and loose.

Physical properties: $H = 5–5\frac{1}{2}$; $G = 4.37$, but may be as low as 3.3 for impure and/or porous varieties. Color is yellowish brown with a silky luster in fibrous varieties (Fig. 10.7). Streak is yellow brown.

Limonite is a field term that refers to a yellow-brown to brown oxidation and/or hydration product consisting of amorphous hydrous iron oxides of uncertain identity.

Figure 10.7 Goethite with a radiating, fibrous texture.

Distinguishing features: Distinguished by its yellow-brown streak and its common occurrence in massive to fibrous textures. Hematite, by contrast, has a red-brown streak.

Uses: Goethite, together with hematite, is the major source of iron worldwide from oxidized portions of Precambrian iron-formations.

10.4 Kaolinite: $Al_2Si_2O_5(OH)_4$

Occurrence: Kaolinite is a common mineral and is the main constituent of **kaolin** or clay. It forms at low temperature and pressure as an **authigenic** mineral (*authigenic* meaning that the mineral formed in place and was not transported there from elsewhere; from the Greek words *authi*, meaning "in place," and *genesis*, meaning "origin"). Also a secondary weathering product of aluminosilicates such as feldspar (see Equation 10.3). A common constituent of soils, and as massive clay deposits. Two other clay minerals, montmorillonite and illite, with compositions and structures different from that of kaolinite, are also common in soils (see Sec. 10.17).

Chemical composition and crystal structure: The composition of most kaolinite is essentially $Al_2Si_2O_5(OH)_4$. The crystal structure is shown in Figure 10.8, in which the octahedral sheet is dioctahedral (see Sec. 7.16) with one out of three tetrahedra in the sheet being vacant.

Crystal form: Triclinic; 1. Occurs as minute (submicroscopic), thin plates. Usually massive in claylike masses. Can be pseudomorphous after K-feldspar (Fig. 10.9).

Physical properties: H = 2; G = 2.6. Perfect {001} cleavage. Dull, earthy luster. Color most commonly white. Other colors are due to impurities. Becomes plastic when wetted.

Figure 10.9 Very fine-grained kaolinite pseudomorphous after two intergrown Carlsbad twins of orthoclase. One twin is horizontal, the other inclined at a high angle.

Distinguishing features: Very low hardness and claylike character. Without X-ray powder diffraction tests, kaolinite cannot be uniquely identified or distinguished from several other clay minerals.

Uses: Commercial clays, or clays used as raw materials in manufacturing, are among the most important nonmetallic mineral resources (see Sec. 16.3). Major commercial products are common brick, paving brick, sewer pipe, and draining tile. High-grade clays known as **china clay** or kaolin are major constituents of pottery and chinaware. A very large use is as filler in paper. It is also used in the manufacture of refractories and in the rubber industry.

Other related clay minerals such as montmorillonite and vermiculite have very specific industrial applications. Montmorillonite is used in drilling muds and as a catalyst in the petroleum industry. It is also used as a bonding clay in foundries and in bonding iron-ore pellets. Vermiculite is used extensively in agriculture for soil conditioning. When vermiculite is rapidly heated, it becomes a lightweight expanded product used in potting soil and thermal insulation.

10.5 Calcite: $CaCO_3$

Occurrence: Calcite is the main constituent of limestone, the most abundant chemically precipitated sedimentary rock. Limestone is also the result of organic precipitation. Many organisms that live in the oceans extract calcium carbonate from the water to build their protective shells, and on death their hard parts accumulate on the sea floor. **Coquina** is a loosely cemented, fragmental limestone made of fossil shells and fragments. **Chalk** is a soft, fine-grained limestone mainly made of foraminiferal shells. **Marble** is the metamorphic equivalent of limestone. Hydrothermal veins commonly contain calcite. It is also the main constituent of the igneous rock **carbonatite** (see Sec. 9.4.4, Fig. 9.50).

Kaolinite: $Al_2Si_2O_5(OH)_4$
C 1

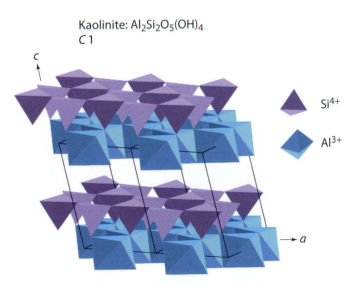

Figure 10.8 A perspective, polyhedral view of the structure of kaolinite showing infinitely extending tetrahedral and octahedral sheets bonding together to make tetrahedral-octahedral (*t-o*) layers. Two identical monoclinic unit cells are outlined.

Calcite: $CaCO_3$
$R\bar{3}2/c$ also   siderite: $FeCO_3$
                  magnesite: $MgCO_3$
                  rhodochrosite: $MnCO_3$

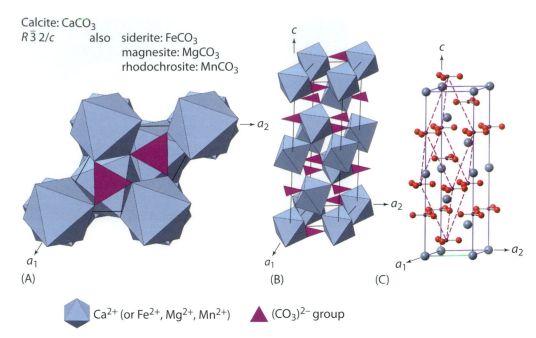

(A)   (B)   (C)

⬡ Ca^{2+} (or Fe^{2+}, Mg^{2+}, Mn^{2+})   ▲ $(CO_3)^{2-}$ group

Figure 10.10 Three different views of the crystal structure of calcite. (**A**) A view down the vertical *c* axis. (**B**) A perspective, vertical view of the same structure. In both illustrations a primitive hexagonal unit cell is outlined for this rhombohedral structure. (**C**) A third view of the calcite structure in a ball-and-stick representation. This shows the two different Bravais lattice choices: *P* for the hexagonal and *R* for the rhombohedral lattice choice. See caption for Figure 5.25 for further explanation.

Chemical composition and crystal structure: Most calcites are close to the formula $CaCO_3$. The structure of calcite is shown in Figure 10.10, with triangular $(CO_3)^{2-}$ groups in planes at right angles to the vertical *c* (3-fold) axis. The Ca^{2+} ions, in alternate planes, are in 6-coordination with oxygens of the (CO_3) groups. Magnesite ($MgCO_3$), siderite ($FeCO_3$), and rhodochrosite ($MnCO_3$) exhibit the same structure. These four common carbonates, therefore, are referred to as **isostructural**. A polymorph of calcite, aragonite, which is stable at high pressures (Fig. 10.11), is discussed in Section 10.6.

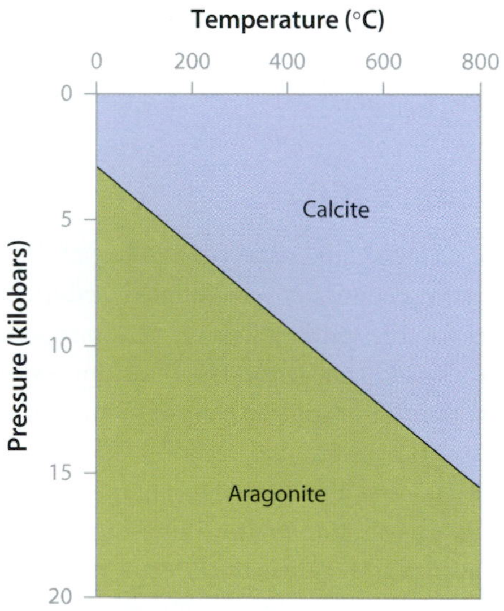

Crystal form: Commonly occurs in good crystals of varied habit. Rhombohedral and scalenohedral crystals (Fig. 10.12) are most common. Also occurs in coarsely crystalline masses, as well as in fine-grained, compact, and stalactitic forms.

Physical properties: H = 3 (on cleavage surface); **G** = 2.71. Perfect rhombohedral $\{10\bar{1}1\}$ cleavage. Color is usually white to colorless but may show other colors. Transparent to translucent.

Distinguishing features: Effervesces readily in cold, dilute HCl. Characterized by its **H** = 3, excellent rhombohedral cleavage, and light color.

Uses: The main use for calcite is in the manufacture of cements and mortars. Limestone is the chief source, which when heated to about 900°C forms CaO, **quicklime**, by the reaction $CaCO_3 \rightarrow CaO + CO_2$. The CaO, when mixed with water, forms one of several CaO hydrates (**slaked lime**), which swell, give

Temperature (°C)

Figure 10.11 The approximate locations of the experimentally determined stability fields of two polymorphs of $CaCO_3$: calcite and aragonite.

Figure 10.12 Three intergrown scalenohedral crystals of calcite.

off heat, and harden. Quicklime when mixed with sand forms common mortar. The most widely used cement is known as **Portland cement**, which consists of about 75% calcium carbonate with the remainder being silica and alumina.

Limestone is the main raw material for the production of various inorganic Ca-containing compounds. When finely crushed, it is a soil conditioner, and it is a component in whiting and white wash. It is also a flux in smelting various metallic ores.

Limestone and marble are used extensively in the building industry as dimension stone.

10.6 Aragonite: CaCO₃

Occurrence: Aragonite is less common than calcite because it is less stable at atmospheric conditions. It is an inorganic precipitate in near-surface conditions of relatively warm ocean waters as found in the Bahamas. CaCO₃, as aragonite, is secreted by mollusks, but with time it changes to calcite, resulting in calcite pseudomorphous after aragonite. It is also present in metamorphic rocks that have been subjected to high pressure. Figure 10.11 shows the relative stabilities of calcite and aragonite.

Chemical composition and crystal structure: Most aragonite is close to the composition CaCO₃. The structure of aragonite is somewhat denser than that of calcite. Figure 10.13 shows the orthorhombic crystal structure of aragonite. In the aragonite structure, the $(CO_3)^{2-}$ groups occur, as in calcite, in planes perpendicular to the c axis, but now in two different structural planes, with the triangular carbonate groups pointing in opposite directions to each other. Ca^{2+} is in 9-fold coordination with oxygens from adjoining $(CO_3)^{2-}$ groups.

Crystal form: Orthorhombic; $2/m\ 2/m\ 2/m$. May occur in acicular pyramidal crystals, as well in tabular crystals. Usually in radiating groups of large to very small crystals. Also as

Figure 10.14 A cluster of white pseudohexagonal twins of aragonite forming a hexagonal-like prism terminated by a basal plane. This results from the intergrowth of three individual aragonite crystals twinned on {110} forming cyclic twins. The yellow mineral in the background is sulfur.

pseudohexagonal twins (Fig. 10.14) showing a hexagonal-like prism terminated by a basal plane. These result from the twinned intergrowth (a cyclic twin) of three individuals twinned on {110} with {001} forms in common.

Physical properties: H = 3½–4; G = 2.94. Usually occurs in radiating groups of smaller acicular crystals. Also in reniform, columnar, and stalactitic aggregates. Generally colorless or white.

Distinguishing features: Effervesces in cold HCl. Distinguished from calcite by higher specific gravity and lack of rhombohedral cleavage.

Uses: Because of the extensive availability of CaCO₃ in the form of calcite in limestones, and because of the lack of large deposits of aragonite, this polymorph of calcite has little to no commercial use.

10.7 Dolomite: CaMg(CO₃)₂

Occurrence: Dolomite occurs as extensive sedimentary rock sequences and as its metamorphic equivalent, dolomitic marble. Dolomite (also referred to as **dolostone**, a rock type) is commonly the result of replacement of preexisting limestone with relict calcite grains, calcitic fossils, and sedimentary structures still preserved in the dolomitic matrix. Also found in hydrothermal veins.

Chemical composition and crystal structure: An essentially complete solid solution exists between dolomite and ankerite, CaFe(CO₃)₂ (see Fig. 10.15). The dolomite composition is halfway between calcite and magnesite (Fig. 10.15). The

Aragonite: CaCO₃
$P\,2_1/n\,2_1/m\,2_1/a$

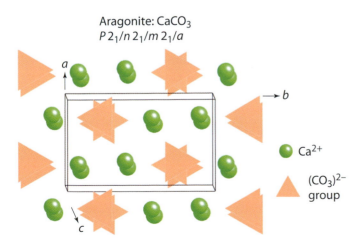

Figure 10.13 View down the vertical c axis of the structure of aragonite. As in calcite, the triangular $(CO_3)^{2-}$ groups lie perpendicular to the c axis. Here, however, the triangular groups in one plane point in opposite directions to those in the other. An orthorhombic unit cell is outlined.

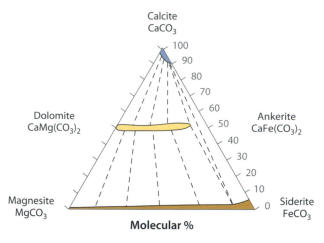

Figure 10.15 Graphical illustration of the extent of solid solution among carbonates in the system CaO-MgO-FeO-CO₂. Tie lines connect commonly coexisting carbonates. Calcite-dolomite pairs occur in Mg-containing limestones. Ankerite-siderite pairs are common in Precambrian banded iron-formations (adapted from Essene, E. Z., 1983, Solid solutions and solvi among metamorphic carbonates with applications to geologic thermobarometry, *Reviews in Mineralogy*, 11, 77–96, Mineralogical Society of America, Chantilly, VA).

Figure 10.17 A hand specimen of dolomite showing the intergrowth of many, slightly pink, curved rhombohedral crystals.

compositional field of dolomite-ankerite is halfway between that of calcite and the solid solution field of magnesite-siderite. These compositional fields are separated by miscibility gaps, as shown by pairs of coexisting compositions in Figure 10.15

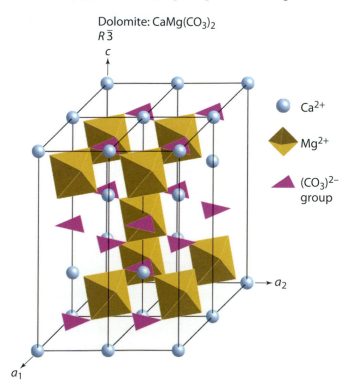

Figure 10.16 Perspective view of the crystal structure of dolomite. In the calcite structure (see Fig. 10.10) all octahedra are occupied by Ca²⁺. In dolomite, the Ca²⁺ and Mg²⁺, on account of their large difference in size (about 33%), are ordered, appearing in alternating levels of the structure. A cluster of four primitive hexagonal unit cells is outlined for this rhombohedral structure. See caption for Figure 5.25 for further explanation.

connected by tie lines. The crystal structure of dolomite is similar to that of calcite but with separate layers of Ca²⁺ and Mg²⁺ alternating along the vertical c axis (Fig. 10.16). The large difference in size between Ca²⁺ and Mg²⁺ is the cause for the cation ordering into separate layers of the structure.

Crystal form: Hexagonal; $\bar{3}$. Crystals are commonly curved rhombohedra ("saddle shaped"; see Fig. 10.17). Generally in coarse, cleavable masses.

Physical properties: **H** = 3½–4; **G** = 2.85. Shows perfect rhombohedral cleavage, {10$\bar{1}$1}. Color ranges from colorless to flesh colored to pink; also white, gray, and black.

Distinguishing features: Crystals commonly show curved rhombohedra and flesh color. Dolomite reacts only very slowly in cold HCl but effervesces in hot HCl.

Uses: In the manufacture of some cements. Also in the extraction of magnesia, MgO, for refractory linings used in the steelmaking process. Also used in the glass industry and in metallurgical fluxes. Agricultural applications include soil amendment and feed additive. Used as dimension stone in the construction industry.

10.8 Magnesite: MgCO₃

Occurrence: In sedimentary rocks as a primary precipitate or as a replacement of limestone by Mg-rich solutions, forming dolomite as an intermediate product. Also a common constituent of hydrothermal veins. It also occurs in metamorphosed igneous rocks, such as peridotite, pyroxenite, and dunite.

Chemical composition and crystal structure: The most common magnesites are rich in magnesium, with only minor substitution of Fe²⁺ for Mg²⁺. However, a complete solid solution exists between magnesite and siderite, FeCO₃ (see

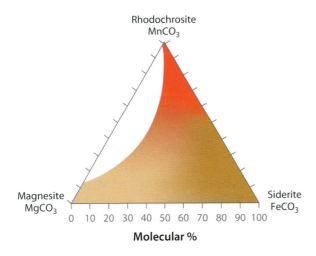

Figure 10.18 The approximate extent of solid solution among three isostructural carbonates: magnesite ($MgCO_3$)–siderite ($FeCO_3$)–rhodochrosite ($MnCO_3$) (adapted from Essene, E. Z., 1983, Solid solutions and solvi among metamorphic carbonates with applications to geologic thermobarometry, *Reviews in Mineralogy*, 11, 77–96, Mineralogical Society of America, Chantilly, VA).

Figs. 10.15 and 10.18). The iron-rich compositions are common in Precambrian iron-formations. Magnesite is isostructural with calcite (see Fig. 10.10).

Crystal form: Hexagonal; $\bar{3}\,2/m$. Rhombohedral crystals are rare (Fig. 10.19). Usually compact and fine grained.

Physical properties: H = 3½–5; G = 3.0–3.2. Perfect rhombohedral $\{10\bar{1}1\}$ cleavage. Color ranges from white to gray, yellow, and brown (the latter two colors result from variable substitution of Fe^{2+} for Mg^{2+}).

Distinguishing features: Reacts only very slowly to cold HCl but dissolves with effervescence in hot HCl. Cleavable varieties

may resemble dolomite but have higher specific gravity. The massive white variety resembles chert but has a lower hardness.

Uses: Magnesite may be mined for the production of Mg metal and other Mg compounds. However, most of the magnesium produced is from seawater brines (Mg being the third most common element in seawater, after chlorine and sodium). A major use of Mg is as MgO, which is a basic component of some refractories.

10.9 Siderite: $FeCO_3$

Occurrence: Siderite may be a major primary constituent of Precambrian banded iron-formation, in association with other iron-rich minerals such as magnetite, pyrite, and Fe-rich silicates. It is also present in metamorphosed iron-formations. Hydrothermal veins commonly contain coarsely crystalline siderite.

Chemical composition and crystal structure: Siderite that occurs in iron- formations is generally low in Mg, although a complete solid solution series exists between siderite and magnesite (Fig. 10.18). There is also a complete series toward rhodochrosite, $MnCO_3$. Siderite is isostructural with calcite, magnesite, and rhodochrosite (see Fig. 10.10).

Crystal form: Hexagonal; $\bar{3}\,2/m$. Crystals are usually rhombohedral in form (Fig. 10.20). Commonly cleavable, granular.

Physical properties: H = 3½–4; G = 3.96 (for pure $FeCO_3$). Shows perfect rhombohedral $\{10\bar{1}1\}$ cleavage. May be compact or botryoidal. Color usually light to dark brown.

Distinguishing features: Distinguished from other carbonates by its brownish color and high specific gravity. Soluble in hot HCl with effervescence.

Uses: In only a few countries is siderite mined as an iron-ore mineral. The bulk of all global iron is extracted from Precambrian iron-formations rich in magnetite, hematite, and/or goethite. A source of brown pigments used in paints.

Figure 10.19 A cluster of light beige-colored tabular (discoid) crystals of magnesite. These crystals are combinations of almost flat rhombohedra terminated by basal pinacoids. Most magnesite is commonly massive.

Figure 10.20 A group of yellowish-brown rhombohedral crystals of siderite with composite and somewhat curved faces.

10.10 Rhodochrosite: MnCO₃

Occurrence: Sedimentary, low-temperature rhodochrosite, of **diagenetic** origin, occurs in finely laminated sediments or concretionary aggregates in manganese formations (**diagenesis** refers to all the chemical, physical, and biological changes that take place in a sediment after its initial deposition). It is also present in some manganese-rich iron-formations. It may be of hydrothermal origin, commonly in association with sulfide minerals. There is a unique occurrence of rhodochrosite at Capillitas in Argentina, where it is present as stalagtites and stalagmites in a former silver mine.

Its metamorphic, high-temperature reaction product is commonly rhodonite, $MnSiO_3$.

Chemical composition and crystal structure: Most rhodochrosites are close to $MnCO_3$ in composition, but there is a complete solid solution series between rhodochrosite and siderite (Fig. 10.18). Rhodochrosite is isostructural with calcite (Fig. 10.10).

Crystal form: Hexagonal; $\bar{3}\,2/m$. Occurs only rarely in well-developed rhombohedral crystals. Generally massive, granular.

Physical properties: $H = 3\frac{1}{2}–4$; $G = 3.5–3.7$. Color is usually some shade of rose red (Fig. 10.21). It cleaves with rhombohedral cleavage.

Distinguishing features: Pink color, rhombohedral cleavage, and $H \approx 4$. Rhodonite resembles rhodochrosite in color but is much harder ($H = 5\frac{1}{2}–6\frac{1}{2}$). Rhodochrosite effervesces in warm HCl and rhodonite does not.

Uses: Because of its rose-red color, rhodochrosite is a desirable ornamental stone. On account of its low hardness and good cleavage, it does not make a good material for jewelry. However, the banded variety (Fig. 10.21) is popular in beads and pendants.

10.11 Halite: NaCl

Occurrence: Halite is dissolved in the waters of oceans and salt lakes. It is a common constituent of **evaporites**, which are the result of the evaporation of restricted bodies of seawater or saline lakes, where it is associated with gypsum, sylvite, calcite, and anhydrite. It is a major constituent of playa deposits. It also occurs in **salt domes**, which are nearly vertical pipelike masses of salt that have risen upward to the surface from an underlying salt bed (salt is less dense than the overlying rock).

Chemical composition and crystal structure: Generally close to NaCl in composition but may contain inclusions of other minerals and saline water. The crystal structure of halite is shown in Figure 10.22. Each Na^+ is in 6-fold coordination with Cl^-, and similarly each Cl^- is surrounded by six Na^+ neighbors.

Crystal form: Isometric; $4/m\bar{3}\,2/m$. Most commonly occurs in cubes (Fig. 10.23), with other crystal forms rare.

Physical properties: $H = 2\frac{1}{2}$; $G = 2.16$. Cubic {001} cleavage is perfect. Readily soluble in water. Generally transparent to translucent. Salty taste. Color is most commonly colorless or white, but when impure may be yellow, blue, or purple.

Distinguishing features: Characterized by its salty taste, cubic cleavage, and low hardness. Distinguished from sylvite by its less bitter taste.

Uses: The most extensive use of halite is in the chemical industry for the production of sodium and chlorine as well as hydrochloric acid. In its natural state, it is used in tanning hides, in stock feeds, in fertilizers, in salting icy highways, in

Halite: NaCl
and sylvite: KCl
$F\,4/m\,\bar{3}\,2/m$

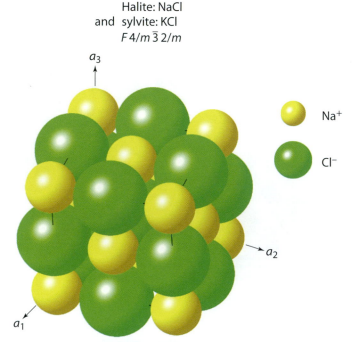

Figure 10.21 Section through a concentrically banded stalactite of rhodochrosite. This material originates from Capillitas, Catamarca Province, Argentina. This locality supplies the world market with ornamental rhodochrosite.

Figure 10.22 Perspective view of a close-packed model of the structure of halite, NaCl. The three crystallographic axes outline a cubic unit cell.

Figure 10.23 Translucent, cubic crystals of halite.

pharmaceuticals, and as a weed killer. It is used at home in meal preparation, and it finds extensive use in the preservation of prepared foods, as well as of cheese, butter, fish, and meat.

10.12 Sylvite: KCl

Occurrence: A constituent of marine evaporite deposits but less common than halite. The precipitation of sylvite from a brine requires a great deal more evaporation than does halite.

Chemical composition and crystal structure: Because the radii of Na^+ (VIII coordination, 1.28Å) and K^+ (VIII coordination, 1.65Å; see Table 4.4) are so far apart, there is little substitution of these elements for each other. Sylvite is isostructural with halite (Fig. 10.22).

Crystal form: Isometric; $4/m\,\bar{3}\,2/m$. Commonly found in crystals that combine cube and octahedron (Fig. 10.24). Usually granular.

Figure 10.24 Cubo-octahedral crystals of sylvite.

Physical properties: H = 2; **G** = 1.99. Perfect {001} cubic cleavage. Commonly transparent. Color is colorless to white. Readily soluble in water. Has a salty taste but is more bitter than halite.

Distinguishing features: Distinguished from halite by its more bitter taste.

Uses: It is the chief ore mineral for the production of potassium compounds. Soluble potassium is a major constituent of fertilizers. A mixture of sylvite and halite, known as **sylvinite**, is the highest-grade potash ore.

10.13 Gypsum: CaSO₄·2H₂O

Occurrence: Gypsum is a common mineral in evaporite deposits, in association with halite, sylvite, calcite, dolomite, anhydrite, and clays. It also occurs in saline lakes and as an efflorescence on desert soils.

Chemical composition and crystal structure: Almost all gypsum has the composition $CaSO_4 \cdot 2H_2O$. A mainly polyhedral image of the structure is given in Figure 10.25. It consists of strongly bonded layers of $(SO_4)^{2-}$ and Ca^{2+} parallel to {010} alternating with layers of H_2O molecules. The Ca^{2+} in octahedral coordination shares edges with the tetrahedral $(SO_4)^{2-}$ groups. Because the bonds between H_2O molecules and the neighboring layers are weak, it has excellent cleavage parallel to {010}.

Crystal form: Monoclinic; $2/m$. Commonly in crystals of simple habit, tabular on {010}. Somewhat more complex crystals of selenite are shown in Figure 10.26. The name *selenite* is used for colorless and transparent varieties.

Physical properties: H = 2; **G** = 2.32. Perfect {010} cleavage. When not in crystals it is massive and cleavable. Color is most commonly colorless, white, or gray. May have other colors due to impurities.

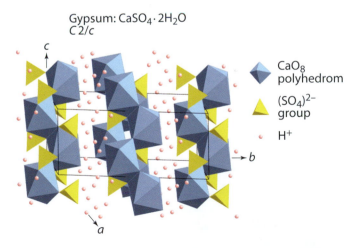

Figure 10.25 Perspective view of the crystal structure of gypsum with the *c* axis vertical. The location of water molecules is that of the H⁺ ions. Successive layers of $(SO_4)^{2-}$ tetrahedra bonded to Ca^{2+} are separated by sheets of the H_2O molecules. The bonds between the H_2O molecules and the neighboring ions are weak giving rise to excellent {010} cleavage. A monoclinic unit cell is outlined.

Figure 10.26 Several monoclinic crystals of colorless and transparent gypsum also known as selenite. The crystals show {010} side pinacoids, {hk0} prismatic bevels, and {001} basal pinacoids.

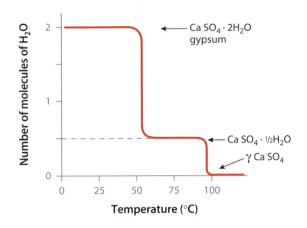

Figure 10.29 A cluster of equant anhydrite crystals each with three sets of different pinacoidal faces.

Figure 10.27 Dehydration curve of gypsum illustrates the formation of a metastable phase, $CaSO_4 \cdot \frac{1}{2}H_2O$ at about 65°C. At about 100°C, with complete dehydration, a metastable γ polymorph of $CaSO_4$ appears.

Distinguishing features: Characterized by its low hardness and excellent cleavage.

Uses: Gypsum is used mainly in the production of gypsum board or wallboard, also known as Sheetrock or drywall. In the manufacture of wallboard, gypsum is heated to drive off a large part of the water, resulting in the production of the hemihydrate of calcium sulfate, $CaSO_4 \cdot \frac{1}{2}H_2O$ (Fig. 10.27). This is subsequently ground to form a material called stucco (similar to material sold as plaster of Paris). Stucco mixed with water, reinforcing fibers, aggregates, and some additives produces a watery slurry that is cast into a narrow paper envelope. This produces a gypsum core with specially prepared paper bonded to both sides, known as wallboard.

10.14 Anhydrite: $CaSO_4$

Occurrence: Anhydrite is a common constituent of evaporite deposits in association with halite, calcite, dolomite, and gypsum. Also found in salt deposits in the cap rock of salt domes and in limestone. Also present as filling of vesicles in basalt flows.

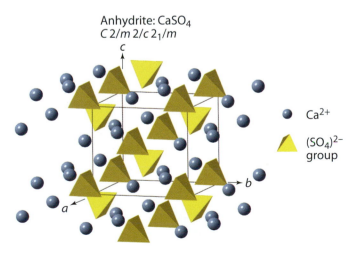

Figure 10.28 Perspective view of the crystal structure of anhydrite. An orthorhombic unit cell is outlined.

Chemical composition and crystal structure: Essentially $CaSO_4$. The crystal structure of anhydrite is given in Figure 10.28, which consists of $(SO_4)^{2-}$ tetrahedra and distorted CaO_8 polyhedra. A metastable polymorph of anhydrite, $\gamma CaSO_4$ (Fig. 10.27), is formed as the result of slow dehydration (during heating) of gypsum.

Crystal form: Orthorhombic; $2/m\, 2/m\, 2/m$. Crystals are rare but appear blocky (Fig. 10.29) when present.

Physical properties: H = 3–3½; G = 2.89–2.98. {010} cleavage perfect; {100} cleavage nearly perfect; {001} good. Color is generally colorless to gray to bluish white.

Distinguishing features: The three good cleavages result in equant, blocky cleavage fragments. It has a higher **H** than gypsum. It may appear similar to calcite but does not effervesce in HCl.

Uses: In ground form as a soil conditioner. In some parts of the world, anhydrite is used in the production of cement as well as sulfuric acid.

10.15 Chert and agate: both made of very fine-grained SiO_2

Occurrence: Most chert is of marine origin. It is abundant in Precambrian banded iron-formations. In Phanerozoic rocks, it forms beds in ophiolite suites overlying pillow lavas. It may also occur in limestone, especially chalk. Agates commonly occupy **vugs** (cavities) in volcanic rocks and cavities in other rocks as well. They may be the result of silica deposition from groundwater percolating through the host rock slowly filling the original cavities with banded chalcedony (the main constituent of agate). They may also originate from silica deposition that is simultaneous with the crystallization of the matrix (host) rock.

Chemical composition and crystal structure: The composition is that of quartz, SiO_2. Grain size ranges from microcrystalline to cryptocrystalline with the structure of low quartz (Fig. 5.36(A)). *Cryptocrystalline* (from the Greek word *krypto*, meaning "hidden") means that it consists of crystals that are too small to be recognized even under an optical microscope,

although the crystallinity can be proved by X-ray diffraction techniques or electron microscopy.

Shape: Chert occurs in bedded deposits and as nodules in sedimentary sequences. Agates are commonly in rounded shapes that reflect the original cavity that was filled with chalcedony. Fine banding tends to parallel the outline of the cavity in which the silica was deposited (Fig. 10.30).

Physical properties for chalcedony: H = 6½–7; **G** = 2.60 ± 0.10. For microcrystalline quartz, these are: **H** = 7; **G** = 2.66. Chert is usually grayish white, occasionally dark gray. In agates, color banding is commonly in thin parallel bands, in irregular clouds, or in mosslike forms. It can be found in many natural colors, but the very vivid blues, greens, and pinks seen commercially in shops that sell agate for ornamental purposes is due to artificial coloring.

Distinguishing features: Hardness essentially that of quartz. In most agates, the banding is concentric and the chalcedony completely fills the cavity. In others, the banded portion only partially fills the cavity and the central portion is void, or filled completely, or in part, by coarsely crystalline quartz (Fig. 3.2(F)).

Uses: Cut and polished agates, as well as polished slices thereof, are sold as ornaments. Uniquely textured material is used in the jewelry industry.

10.16 Phosphorite

Phosphorite is a sedimentary rock with a high enough phosphate mineral content to be of economic interest. Phosphorites are defined as having more than about 20% P_2O_5. Most commonly, it is a marine rock composed of microcrystalline fluorapatite, $Ca_5(PO_4)_3(F)$, in the form of pellets; laminae; oolites;

Figure 10.31 On the left, a black, granular hand specimen of phosphorite (the dark color is probably due to inclusions of organic matter); on the right, a phosphatic (apatite-rich) limestone.

nodules; and skeletal, shell, and bone fragments. The mineral apatite was discussed in Section 7.33. Figure 10.31 illustrates (on the left) a black, granular hand specimen of phosphorite and on the right a specimen of phosphatic limestone. Other sedimentary rocks with less than 15% P_2O_5 but still more than the average phosphorus content of sediments are referred to as phosphatic, as in phosphatic limestone.

The total volume of sedimentary phosphates in the world is small. But like very iron-rich sedimentary rocks (iron-formations), phosphorites are of great economic importance (see Sec. 17.1). They contribute more than 80% of the world's phosphate production and make up about 96% of the world's total resources of phosphate rock.

The origin of marine sedimentary phosphate-rich formations is not yet clearly understood. It is generally believed that such sedimentary rocks are the result of upwelling of phosphorous-rich waters from deeper parts of the ocean and that the (PO_4) was biologically used to form soft body tissue. When such organic materials are deposited and decomposed on the seafloor, the (PO_4) is released to the seawater. When this is re-precipitated on or replaces limestone, it becomes part of the precursor materials of the present-day phosphorites. Apatite may also have precipitated diagenetically from phosphate-rich pore waters, forming phosphate grains and cement, and replacing skeletal and carbonate grains. Yet another view of the origin of phosphorites is that they are the result of direct precipitation from concentrated phosphorus solutions that originated from deep-ocean hydrothermal activity admixed with seawater. This would represent an inorganic chemical instead of biogenic source. Precambrian iron-formations are considered the chemical sedimentary result of such deep-ocean hydrothermal sources as well.

Uses: About 90% of phosphate rock production is used for fertilizer and animal feed supplements, with the balance used for industrial chemicals (e.g., phosphoric acid). A typical fertilizer used for garden soil might have the three bold digits printed on the outside of the container or bag: 16-8-8. This means that 16% of the material is nitrogen, 8% is phosphoric acid (P_2O_5), and 8% is soluble potassium (K_2O) (see Sec. 17.2).

The United States is the leading producer of phosphate fertilizers, but Morocco may have the largest reserves of phosphate raw material. Presently, the United States, China, and Morocco together produce about 75% of the world's phosphorus.

10.17 Soil

Soil is mentioned in this text because it is an extremely important Earth material. Soil is essential to our survival. Engineers may define soil as the unconsolidated materials above the bedrock, but to soil scientists it is the medium for growth of land plants.

Figure 10.30 An agate consisting of translucent, cryptocrystalline quartz known as chalcedony. The delicate pattern of banding suggests the shape of a flame at the lower part.

Soils exhibit subtly colored, usually horizontal layers that are related to the weathering of surficial deposits or bedrock. Plants and soil are closely linked. Plants grow up from the surface, but their roots penetrate down into soil. Decaying organic matter accumulates on top and is mixed by organisms into the top layers of the soil. Air and water, also mixed into the soil, aid in plant growth.

As shown in Figure 10.32, soils can be described in terms of **soil horizons** that differ from each other in particle size, chemical composition, and other soil characteristics. The upper layer of the **A horizon** is usually enriched in organic matter relative to other horizons. This organic matter, also referred to as humus, produces the dark brown to gray color and earthy odor of moist soil. The lower, main part of the A horizon lacks minerals that are most susceptible to dissolution.

The **B horizon** is typically richer in clays, such as kaolinite as well as montmorillonite and illite. This horizon may have distinctive red, orange, and yellow hues caused by the presence of oxides and hydroxides of iron, such as microcrystalline hematite and goethite, as well as amorphous limonite. This soil horizon is one of accumulation of clays and soluble suspended material brought down by water into the B horizon.

The **C horizon** is the lowermost horizon of the soil and consists generally of the essentially unmodified parent material for soil formation. In Figure 10.32, this is the horizon that consists of bedrock.

Uses: In agriculture soil is normally the primary nutrient base for plants, by providing minerals as well as water. In construction projects soil serves as the foundation; massive amounts of soil (usually called fill) are used in road building and dam construction. Soil absorbs rainwater, thus slowing runoff, and releases it later, thus preventing floods and drought. Soil also cleans water as it percolates through. Organic soils such a peat are used as a fuel resource in many countries.

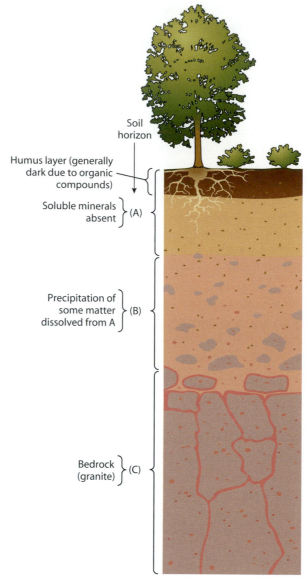

Figure 10.32 Typical soil profile developed in sediments overlying granite bedrock under temperate climatic conditions. Soil characteristics depend on (1) climate, (2) organisms, (3) relief, (4) parent material, and (5) age of the soil.

Summary

This chapter introduced 13 common sedimentary minerals and 3 sedimentary materials. It mentions the importance of several detrital minerals that were systematically described in Chapter 7, namely, quartz, feldspars, and members of the mica group. The majority of minerals described in this chapter are carbonates, halides, and sulfates. Only one, kaolinite, is a silicate, but this is part of a large mineral group known as the clay minerals, which, among others, includes montmorillonite, illite, and vermiculite.

- Oxidation and hydration reactions are common under atmospheric conditions.

- Soluble products (Na^+, K^+, Ca^{2+}, and Mg^{2+}) build up in the oceans and subsequently lead to precipitation of carbonates, halides, and sulfates that incorporate the originally soluble cations.

- Ice is a mineral but is not commonly thought of as such.

- Goethite and limonite are common oxidation and/or hydration products of all rocks, even those that carry only traces of iron.

- Kaolinite is a common alteration product of originally high-temperature feldspars.

- Calcite and dolomite are the most common rock-forming carbonates.

- Aragonite is secreted by mollusks, but with time aragonite converts to calcite. Aragonite also occurs in metamorphic rocks that have been subjected to high pressures.

- Halides and sulfates are common constituents of evaporite sequences.

- Agates, consisting of extremely fine-grained (microcrystalline to cryptocrystalline) quartz, are the result of silica deposition by groundwater in open spaces.

- Phosphorites are the main global source of phosphorus used in fertilizer.

- Soil is a complex material on which all global agriculture is dependent.

Review questions

1. What is the number of polymorphs exhibited by ice?

2. What is the crystallographic system of ice that forms normally under freezing conditions?

3. What is the crystal form of ice and how is it exhibited?

4. What is the chemical formula of goethite?

5. What is the oxidation state of iron in goethite?

6. Kaolinite can be pseudomorphous after what feldspar? Give the chemical formulas of both minerals.

7. Kaolin is a major constituent of what commercially produced products?

8. Calcite and three other carbonates are isostructural. What does that mean, and which are the other three? What are their names and chemical formulas?

9. What is the main commercial use of limestone or calcite?

10. What is the chemical formula of dolomite?

11. What is different in the dolomite crystal structure as compared with that of calcite?

12. Halite is also known as table salt. What other major commercial uses does halite have?

13. Agates can be very aesthetically appealing and are commonly on exhibit in museums of natural history and for sale in rock and mineral shops. What is so appealing about them?

14. Phosphorites are mined for their content of what mineral? Give that mineral's name and chemical formula.

15. Soils commonly show various horizontal soil horizons. How do these various layers differ?

FURTHER READING

Please refer to the listing at the end of Chapter 7.

11 Formation, transport, and lithification of sediment

Sedimentary rocks form through the compaction and cementing together of loose sediment, which can be derived from the weathering of preexisting rocks, the hard parts of organisms, or chemical precipitates. Most sediment is formed from weathered rock, and this detrital material must be transported by water, wind, or ice to depositional sites before it can begin the process of turning into solid rock. Grain size is an important property of sediment because it determines how easily sediment can be moved. During transport, sediment is sorted by grain size. In rivers, which are the main agent of sediment transport, the coarser fraction travels on the bottom, whereas the finer fraction travels in suspension, which leads to separate deposits of coarser sand and gravel and finer mud, respectively. Most detrital sediment is transported to the sea, but some accumulates in basins on continents. Along shorelines, wave action further abrades and sorts sediment. In warm climates, coastal sediments may consist almost entirely of the hard parts of calcareous organisms. Sediment eventually accumulates in basins, most of which have a specific plate tectonic setting, such as forearc basins at convergent plate boundaries, subsiding passive continental margins, and rift and pull-apart basins on continents. Sediments accumulating in these basins begin to compact under their own weight as pore fluids are expelled. These fluids are commonly saturated in the minerals with which they are in contact, and as the fluids rise, they precipitate minerals that cement the detrital grains together to form a rock. In this chapter, we discuss the formation, transport, deposition, and eventual solidification of sediment to form sedimentary rock.

Gorges du Verdon in the south of France cuts through Jurassic limestones that were thrust out of the sea by the Alpine orogeny onto the low terrain seen in the upper left of the photograph. As the mountains grew, the Jurassic rocks were exposed to weathering, and rivers, such as the Verdon, moved the weathering products back to the sea and created the gorge in doing so. The milky turquoise color of the Verdon River is testament to the large quantity of solids it carries in suspension, to say nothing of the amount carried as bed load and in solution.

11.1 Importance of sediments in understanding the history of the Earth

Sedimentary rocks, which are the topic of Chapter 12, are formed by the accumulation and burial of sediment in depositional basins. Most sediment is formed from the weathering of preexisting rocks as a result of physical, chemical, and biological processes. Some sediment is formed directly by organisms, and some may be a direct chemical precipitate. Sedimentary rocks have a diverse compositional range, not only because of differences in the type and source of the sediment but also because of processes associated with the transport and burial of sediment. Before we discuss sedimentary rocks, we must, therefore, understand how sediment is formed, transported, deposited, and turned into rock (Fig. 11.1).

The formation of sediment, its transport, and deposition are dependent on topographic relief, which in large part is produced by plate tectonic processes. High relief at convergent and divergent plate boundaries promotes weathering and transport of sediment, so many sedimentary rocks are related to these regions.

When rocks weather, some minerals, such as quartz, remain stable, but most become unstable and are changed into new minerals characteristic of the weathering zone. Disintegration of the primary minerals is promoted by biological activity, with organisms feeding on the nutrients provided by the primary minerals. Because most of the weathering products of rock are silicate minerals, this type of sediment is referred to as **siliciclastic sediment**. It is the most abundant type of sediment, but sediment derived from organisms and chemical precipitates is also important.

These various types of sediment are transported to sites of deposition by water, wind, and ice. Water also carries the soluble ions produced during weathering. During transport, sedimentary particles tend to separate on the basis of size and density. Most sediment eventually makes its way to the sea, where wave action and currents on beaches may further process sand-sized and coarser particles. The sorting of grains during transport leads to formation of the main types of siliciclastic sedimentary rocks (conglomerates, sandstones, and mudrocks). The soluble ions are removed from seawater by precipitation, especially as a result of evaporation in restricted basins, and by organisms that build shells and other hard parts of $CaCO_3$ and SiO_2. Along some shorelines, carbonate grains formed from the breakup of shells and other animal hard parts form calcareous gravel, sand, and mud. These accumulate to form limestones.

On reaching a site of accumulation, sediment is buried by more sediment and slowly, through compaction, recrystallization, and cementation, undergoes **lithification**, to form sedimentary rock.

Topographic relief is necessary to provide sources of sediment and basins in which it can be deposited. The importance of uplift and subsidence to the formation of sedimentary rocks was one of the pillars of the modern science of Geology as laid out

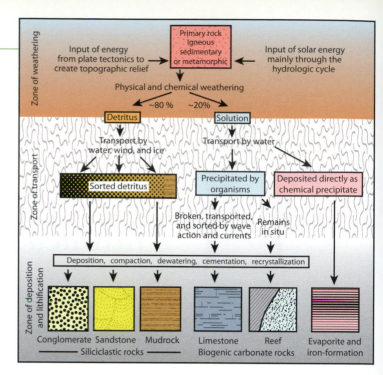

Figure 11.1 Flow chart showing the various paths that sediment can take from its source in the zone of weathering to its final site of deposition where it is turned into solid sedimentary rock.

in James Hutton's *Theory of the Earth* (1788). At Siccar Point, on the east coast of Scotland near the border with England, Hutton discovered an angular unconformity between underlying vertically dipping Lower Silurian sandstones and overlying gently dipping Upper Devonian "Old Red Sandstone" (Fig. 11.2). By identifying such features as ripple marks and cross-bedding in the rocks, he correctly interpreted that they formed from sediment deposited by water in near horizontal layers. The underlying sandstone was formed from deepwater marine sediment, whereas most of the Old Red Sandstone was deposited by rivers on a coastal plain. Although Hutton did not have the benefit of

Figure 11.2 Anthony Philpotts standing on the angular unconformity discovered by James Hutton at Siccar Point on the east coast of Scotland. Vertically dipping Lower Silurian sandstone is overlain by gently dipping Upper Devonian Old Red Sandstone.

our knowledge of plate tectonics, he understood that the unconformity recorded periods of crustal deformation and uplift followed by erosion, and then subsidence, more deposition, and finally uplift again. Based on his observation of the rates of these processes in nature and applying the principle of uniformitarianism, he concluded that the unconformity proved that the geologic record must cover an enormous time span. And so began the modern science of geology.

Today, we know that plate tectonics is the main cause of topographic relief, and without it, continents would soon be eroded away. The elevation of a mountain belt represents a balance between constructive and destructive processes. The rate of uplift depends on plate convergence rates, which are on the order of 50 mm/year (about as fast as our fingernails grow) and the types of plates involved (oceanic or continental). The rate the surface is lowered by erosion depends on many factors, including climate (wet or dry, cold or warm), latitude, and elevation, but is on the order of 0.3 mm/year. If the terrain undergoing weathering consists of granite with a density of 2750 kg/m^3, this amount of surface lowering corresponds to removal of 1000 tonnes of weathered granite per square kilometer each year.

Such large amounts of sediment being eroded from mountain ranges may sound excessive, but if, for example, you look at the Verdon River flowing out of the Alps through the Gorges du Verdon in the chapter-opening photograph, you can see from the river's milky color that it is carrying a considerable suspended load. In addition, sediment is moving along the bed of the river and the water is carrying ~20% of the sediment load as dissolved ions. This part of France is underlain predominantly by Mesozoic shallow marine limestones that were raised out of the sea to form the Alps when the African plate collided with the Eurasian plate beginning at the end of the Mesozoic era. As the mountains rose, rivers, such as the Verdon, kept removing the weathering products from the newly elevated terrains and transporting them to the sea, where they accumulate as sediment that is ultimately buried and converted to new sedimentary rock.

In this chapter, we examine the history of sediment from its source to its final site of accumulation and conversion to sedimentary rock. The processes of weathering and sediment transport lead to major redistributions of the elements in the crust. The resulting sedimentary rocks are, therefore, of great importance in adding to the diversity of Earth materials.

11.2 Sediment formed from weathering of rock

Most sediment is formed from the physical and chemical weathering of igneous, sedimentary, and metamorphic rocks. Physical weathering involves the breaking apart of rocks along fractures such as joints that release stresses generated by tectonism, cooling of igneous rocks, or simply exhumation during erosion. Even the boundaries between mineral grains or the cleavage planes in individual grains are potential surfaces along which physical weathering can occur. Freezing and thawing of water or the growth of salt crystals in such fractures can break rock apart. The growth of plant roots in fractures can augment this breakage. Physical weathering is important because it provides avenues for the access of water, which is the chief chemical weathering agent. Many minerals react with rainwater, which contains dissolved oxygen and carbon dioxide, to produce hydration, oxidation, or carbonation reaction products, whereas other minerals, such as quartz, remain stable. This mixture of new minerals and stable minerals forms the **detritus** that becomes sediment that is transported to sites of deposition to form the majority of sedimentary rocks.

In Section 10.1, we discussed hydration and oxidation reactions and their role in forming soils. Before discussing the weathering of rock to form sediment, we must first consider the role played by carbon dioxide in chemical weathering.

11.2.1 Role of carbon dioxide in weathering

In reactions (10.2) and (10.3), we saw how olivine and feldspar (respectively) can react with water in the weathering zone to form new hydrous minerals. In these reactions, water was taken to be pure H_2O, but in nature, water is rarely pure. Rainwater, for example, contains many dissolved gases that make it acidic.

When droplets of rain condense and fall through the atmosphere, they equilibrate with the carbon dioxide (and other gases) in the atmosphere, which reacts with the water to form carbonic acid according to the reaction:

$$\underset{\text{carbon dioxide}}{CO_2} + \underset{\text{rain}}{H_2O} \rightarrow \underset{\text{carbonic acid}}{H_2CO_3} \qquad (11.1)$$

At normal atmospheric concentrations of CO_2 (~390 ppm by volume), this reaction makes rainwater slightly acidic, with a pH of about 5.3, which means that the concentration of hydrogen ions (H^+) in rainwater is more than ten times greater than in pure water (pH = 7; pH is a negative logarithmic scale of the hydrogen ion concentration).

The presence of CO_2 in the atmosphere and the acidity it creates in rainwater accelerates the weathering of minerals through carbonation and hydration reactions. The following is a simple **carbonation reaction** that would destroy olivine:

$$\underset{\text{forsterite}}{Mg_2SiO_4} + 2CO2 \rightarrow \underset{\text{magnesite}}{2MgCO_3} + \underset{\text{amorphous silica}}{SiO_2} \qquad (11.2)$$

If we combine water with this reaction, olivine breaks down as follows:

$$\underset{\text{forsterite}}{Mg_2SiO_4} + \underset{\text{rainwater}}{4CO_2 + 4H_2O} \rightarrow \underset{\text{Mg ion}}{2Mg^{2+}} +$$
$$\underset{\text{bicarbonate ion}}{4HCO_3^-} + \underset{\text{silicic acid}}{H_4SiO_4} \qquad (11.3)$$

All of these reaction products are removed in solution. We can write a similar reaction for rainwater's effect on potassium feldspar:

$$2KAlSi_3O_8 + 2CO_2 + 11H_2O \rightarrow Al_2Si_2O_5(OH)_4 +$$
$$\text{K-feldspar} \quad\quad \text{rainwater} \quad\quad\quad \text{kaolinite}$$
$$2K^+ \quad + \quad 2HCO_3^- \quad + \quad 4H_4SiO_4 \quad\quad (11.4)$$
$$\text{K ion} \quad\quad \text{bicarbonate ion} \quad \text{silicic acid}$$

This reaction produces both solid kaolinite and soluble products. Most of the soluble products eventually find their way to the sea, where they may be precipitated as chemical sediment or used to build the shells and tests of organisms that eventually form biogenic sediment.

Reactions (11.3) and (11.4) are examples of **hydrolysis reactions** because the water molecule is split into hydrogen and hydroxide ions ($H_2O \rightarrow H^+ + OH^-$). They are similar to reactions (10.2) and (10.3), but the inclusion of CO_2 dramatically increases reaction rates. It is such reactions, therefore, that are primarily responsible for weathering of rocks.

Because these reactions remove CO_2 from the atmosphere, weathering plays an important role in determining the concentration of this **greenhouse gas** in the atmosphere. It is estimated that at normal weathering rates, all CO_2 in the atmosphere would be removed by weathering in slightly more than 3000 years if it were not for other processes that restore CO_2 to the atmosphere. This rate of consumption of CO_2 during weathering is so high that weathering plays an important role in buffering the CO_2 content of the atmosphere. For example, if the temperature of the planet were to increase, weathering rates would increase, and hence CO_2 would be consumed more rapidly, and this greenhouse gas would be removed from the atmosphere and cause global cooling. However, if large amounts of limestone are exposed to weathering, as for example, in the rise of the Tibetan Plateau, CO_2 is released into the atmosphere, which would cause warming. This illustrates some of the complex interactions that must be taken into account in modeling the carbon cycle to predict climate change. The fact that weathering reactions can remove CO_2 from the atmosphere has also led to serious thought being given to sequestering CO_2 by pumping it into rocks where it could be stored as carbonate reaction products (Sec. 17.3.6).

11.2.2 Weathering products of rock

Early in the Earth's history, igneous rocks were the dominant source of sediment, but with time, sedimentary and metamorphic rocks have made ever-increasing contributions. The weathering products from these different rock types are essentially the same; only their proportions differ. We chose to illustrate the weathering process by focusing on granite.

In Section 9.3.2 (Fig. 9.25), we saw that granite is composed essentially of one-third quartz, two-thirds alkali feldspar, and small amounts of a mafic mineral, usually biotite or hornblende (Fig. 11.3(B)). It also contains accessory amounts of

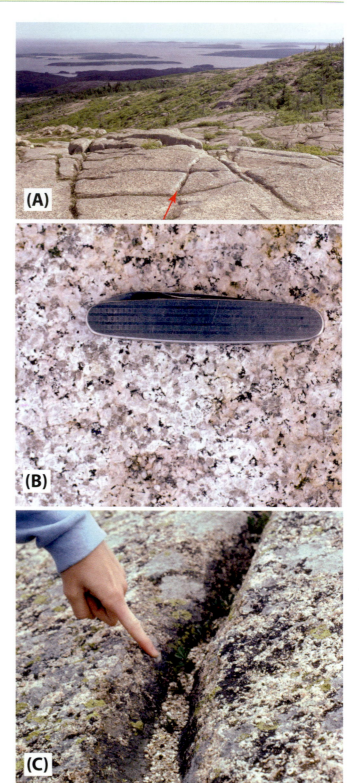

Figure 11.3 (**A**) View from the top of Cadillac Mountain, near Bar Harbor, Maine. This granite mountaintop was scraped clean of weathering products during the Wisconsin Ice Age. Since becoming free of ice 16 000 years ago, the fresh granite has begun to weather, especially along joints. (**B**) Cadillac Mountain granite is composed of white to pink weathering feldspar, gray quartz, and black hornblende. (**C**) Quartz, feldspar, and hornblende grains weathering out of the granite form detritus that accumulates in the differentially weathered joints. The outcrop surface is covered with gray, green, and black lichens that help weather the granite.

minerals such as magnetite, apatite, and zircon. On weathered surfaces, these minerals undergo oxidation, hydration, and carbonation reactions. Let us illustrate the development of these weathering products by examining the Devonian granite of Cadillac Mountain near Bar Harbor, Maine, on the northeastern coast of the United States. During the Wisconsin Ice Age, all of the weathered granite on this mountain was removed by glacial erosion. When ice left the area 16 000 years ago, the mountain was a smooth polished outcropping of fresh granite rising 470 m above the sea (Fig. 11.3(A)). Since then, the granite has begun to weather. Its alkali feldspar has turned chalky white to pink on the weathered surface as a result of the formation of kaolinite (Sec. 10.4) and the oxidation of trace amounts of iron (Fig. 11.3(B)). These are examples of hydrolysis and oxidation reactions. The granite is relatively impervious, so water is able to penetrate only short distances below the surface except along joints where it can descend hundreds of meters, thus allowing reactions to occur well below the present surface. As a result, the granite is more weathered along joints than in the massive part of the outcrop (Fig. 11.3(A)). On the outcrop surface, lichens help accelerate weathering (Fig. 11.3(C)). Lichens are a symbiotic association of a fungus with a photosynthesizing green or blue-green algae (cyanobacteria), the former giving the lichen a green color and the latter a blue-gray color, but other pigments can be present. Lichens are capable of deriving nutrients directly from the rock, which in the granite is mainly potassium from the feldspar. They penetrate into the cleavage planes in feldspar and other minerals. Lichens also release acids that help weather rocks.

Despite the evidence for weathering on the surface of the Cadillac Mountain granite, weathering has not progressed far beneath the surface in the 16 000 years since the ice left the region. On breaking open a piece of this granite, perfectly fresh rock is found less than a centimeter beneath the surface. This is in striking contrast with the depth of weathering of granite in regions that have not experienced recent glaciation. For example, a recent road cut through granite near the gold-mining town of Sumpter, Oregon, shows that weathering extends to a depth of more than 10 m (Fig. 11.4). This region did not experience continental glaciation, and so the weathering zone is much older and hence deeper. The small amount of weathering on Cadillac Mountain shows that granite weathering is a slow process, which probably requires on the order of hundreds of thousands of years to develop a thick weathering zone, at least in this temperate climate. Weathering rates in tropical climates are noticeably faster.

Weathering products consist of grains of chemically resistant minerals and products of hydrolysis, hydration, oxidation, and carbonation reactions, which include clay minerals, iron hydroxides, and soluble ions. In Figure 11.3(C), we see the detritus formed from weathering of the Cadillac Mountain granite accumulating in the depressions along differentially weathered

Figure 11.4 Road cut through deeply weathered granite near Sumpter, Oregon. Although original joints are still visible, weathering is so advanced that the mineral grains lack cohesion and are forming detritus (granite grus) along the base of the outcrop.

joints. This detritus obviously consists of grains of the minerals in the granite; that is, quartz, alkali feldspar, hornblende, and magnetite (too small to be seen in the figure). Similarly, at the base of the outcrop of granite near Sumpter, Oregon (Fig. 11.4), detritus consists of the weathered grains in the granite. However, if you pick up a handful of this detritus, you find that your hands are covered with fine dust; this is the kaolinite formed from the weathering of the feldspar. Despite its fine grain size, it too is part of the detritus.

11.2.3 Detrital grain size

The grain size of detritus produced from weathering of the Cadillac Mountain and Sumpter granites is clearly related to the grain size of the granites, except for the newly formed kaolinite, which is extremely fine. As this detritus is transported to lower elevations and moved by water and wind, its grain size is reduced by abrasion and actual breakage of grains. The finer a particle, the more easily it is moved by water or wind. However, different minerals abrade at different rates, creating grain-size variations that ultimately result in sorting of grains on the basis

Table 11.1 The Udden-Wentworth detrital grain-size scale.

Particle name		Grain diameter (mm)
Gravel	Boulders	
		256
	Cobbles	
		64
	Pebbles	
		4
	Granules	
		2
Sand	Very coarse sand	
		1
	Coarse sand	
		0.5
	Medium sand	
		0.25
	Fine sand	
		0.125
	Very fine sand	
		0.0625
Mud	Silt	
		0.0039
	Clay	

of size, and consequently type of mineral, during transport. The grain size of detrital sediment is, therefore, one of its most important properties, and we must have a definite way of quantifying this property.

The most widely used scale for measuring the grain size of sediment is the **Udden-Wentworth scale** (Table 11.1). This is a simple geometric scale, in which a grain diameter of 1 mm is taken as the boundary between *coarse* and *very coarse sand*, and other boundaries are found by multiplying or dividing by 2.

For example, very coarse sand includes particles with diameters from 1 mm to 2 mm, and particles from 2 mm to 4 mm are referred to as *granules*. On the finer side of the scale, coarse sand includes grains down to a diameter of 0.5 mm, below which grains are referred to as *medium sand* down to a diameter of 0.25 mm, below which they are referred to as *fine sand*, and so on. The names and sizes of particles are given in Table 11.1. Coarser particles in the granule to boulder range are grouped together as *gravel*, and fine particles in the *silt* to *clay* size range are called *mud*. Note that the term *clay* also refers to a mineral group, and this can cause some confusion. However, nearly all clay-size grains in detrital sediments consist of clay minerals (Sec. 10.4).

One way of measuring the grain size of sediment is to pass the sediment through a set of sieves of different mesh size (i.e., different spacing between wires making up the sieve). A sieve of a given mesh size traps larger particles and allows finer particles to pass through. Grain sizes can also be estimated qualitatively by visually comparing the sample with grains of a known grain size. Figure 11.5 shows particles that were sorted by sieving into various size ranges from silt to very coarse sand. Above the boundary between each size bin, a scale bar is shown in fractions of a millimeter to help in judging grain size. Note that because particles in the very coarse sand are not very rounded some grains have maximum diameters greater than 2 mm (this is one of the problems with using sieves to determine grain size).

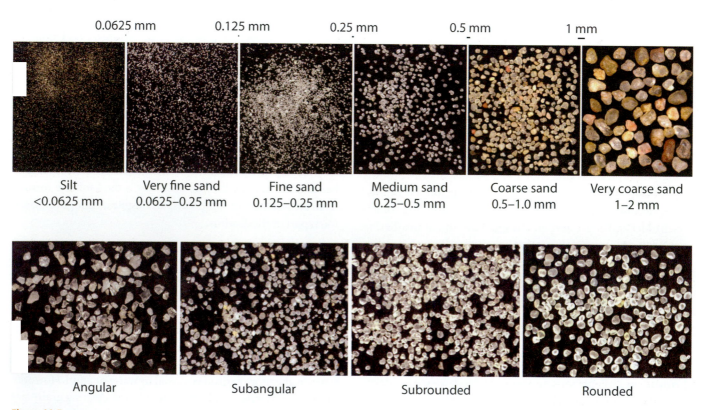

0.0625 mm	0.125 mm	0.25 mm	0.5 mm	1 mm

| Silt <0.0625 mm | Very fine sand 0.0625–0.25 mm | Fine sand 0.125–0.25 mm | Medium sand 0.25–0.5 mm | Coarse sand 0.5–1.0 mm | Very coarse sand 1–2 mm |

Angular	Subangular	Subrounded	Rounded

Figure 11.5 Photographs of detrital particles ranging from silt to very coarse sand and particles ranging in shape from angular to rounded.

In a sample of detrital sediment, grains usually show a range of sizes. For example, if we were to place a sample of beach sand in the top of a stack of sieves with decreasing mesh size going downward and then shake the sieves until the sample had a chance to sort itself, we might find particles in many different sieves. By weighing the sediment collected in each sieve, we could determine what percentage each fraction represents of the total sample. This gives a measure of the *grain-size distribution* of the sample. As sediment is transported by water or wind, grains separate on the basis of size. As a result, sediment that has undergone considerable transport will tend to have grains of a uniform size and is said to be *well sorted*, in contrast to sediment that is *poorly sorted*, which has a wide range of grain sizes. For example, most beach and desert sands are well sorted, whereas most glacial sediment is poorly sorted.

11.2.4 Detrital grain roundness and resistance to abrasion

As detritus makes its way from the source to its site of deposition, grain size decreases and grains become more rounded as a result of physical abrasion and further chemical weathering. Grains of feldspar weathering from granite, for example, tend to be blocky because of their subhedral shape in the granite, whereas quartz, which is usually interstitial in granite, forms irregularly shaped detrital grains. As soon as these grains are transported, either by moving down slope or by being carried by water or wind, their sharp corners are knocked off and grains become more rounded. This process continues as long as a grain keeps moving. If grains have a long transport history, they are more rounded than grains that are deposited quickly. Some rounding may occur by solution after grains have been deposited and stopped moving. Indeed, this process, which

involves solution and redeposition, is partially responsible for turning unconsolidated sediment into solid rock (Sec. 11.9.2). The roundness of grains can be estimated by comparing them with grains of known degrees of roundness. For this purpose, Figure 11.5 shows detrital grains ranging from *angular* through *subangular* and *subrounded* to *rounded*.

In 1951, Folk introduced the concept of **textural maturity** for detrital sediment, which takes into account the degree of rounding and sorting of particles. An **immature** sediment consists of angular, poorly sorted grains in which a considerable amount of mud-size particles still remain mixed with sand-size particles. Such sediment would not have been exposed to significant transport. In **submature** sediment, the mud-size fraction has been removed, but grains remain angular and poorly sorted. **Mature** sediment is well sorted, and grains are subangular. **Supermature** sediment, which consists of well-sorted rounded grains, is indicative of considerable transport or a high-energy environment in which sediment has been strongly agitated (e.g., beach with strong wave action).

As detrital sediment matures, grain size is reduced by abrasion, breakage of grains, solution, and chemical weathering, most of which cause rounding, although grain breakage can produce new sharp edges. The rates of these processes vary from mineral to mineral and on the climate. The medium in which grains are transported also plays a role. The viscosity of water tends to cushion collisions between grains, whereas the low viscosity of air provides no such protection. As a result, the surface of wind-blown grains is commonly pitted with minute percussion marks, giving them a frosted appearance (see the rounded grains in Fig. 11.5). The lack of cushioning by air makes wind abrasion orders of magnitude more effective than river abrasion.

1 mm   1 mm   1 mm   1 mm

A   B   C   D

Figure 11.6 Examples of sand from different source regions. **(A)** Beach sand from Long Island Sound; source glaciated metamorphic and igneous terrain. Light-colored grains are quartz and feldspar, black are magnetite. **(B)** Garnet-rich river sand from a metamorphic terrain in Connecticut. Garnet grains are red, quartz are clear to rusty, feldspar are white, and magnetite are black. **(C)** Sand from Black Sand Beach, Hawaii, composed of grains of black glass and green olivine. **(D)** Carbonate beach sand from the Bahamas composed of shell fragments. Note that the scale in (D) is twice the scale in the other photographs.

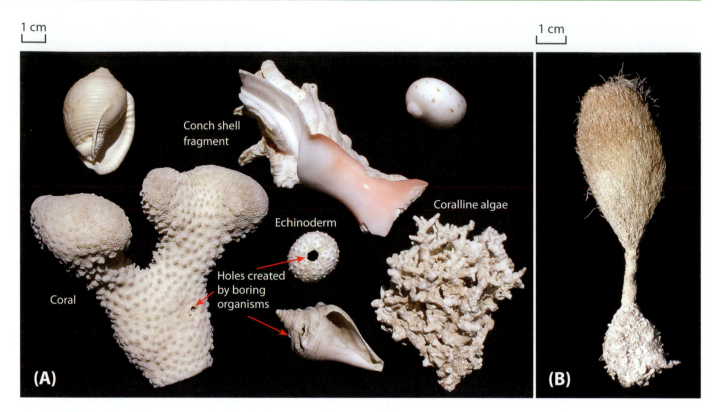

Figure 11.7 **(A)** Some of the larger particles found on a beach in the Bahamas, including coral, echinoderm (sea urchin), coralline red algae, various shells and large conch fragment. Wave action breaks these particles down to create beach sand (Fig. 11.6(D)). Some particles have holes created by boring organism, which produce carbonate mud. **(B)** Calcareous green algae, *Penicillus*, secretes 1–4 μm aragonite needles in its tissue, which upon decay are released to the environment to form carbonate mud. Bulbous mass at base is where it was "rooted" in the sediment.

A detrital grain's resistance to abrasion depends on a number of factors. Minerals with good cleavage tend to break more easily, soft minerals are abraded more easily, and more soluble and chemically reactive minerals dissolve more easily. In Section 10.1 and Table 10.1, minerals are listed in order of relative stabilities during weathering. Of the common rock-forming minerals, quartz is the most resistant. As a result, the longer detritus is exposed to abrasion, the richer it becomes in quartz as the less resistant minerals are winnowed out. If the detritus is from a metamorphic terrain, garnet is another rock-forming mineral that resists abrasion (Fig. 11.6(B)). It is not surprising, considering the abrasion resistance of quartz and garnet, that most sandpaper is made from these minerals. Other less abundant rock-forming minerals that are resistant to abrasion, and therefore persist in detritus, are magnetite, ilmenite, zircon, and tourmaline.

A noted exception to hardness being important in preserving detrital grains is gold, which has a Mohs hardness of only 2.5–3. However, it is extremely malleable and is able to absorb impacts with other detrital grains without breaking. Because of the resistance of some minerals to abrasion, they can become concentrated in detrital residues. If the mineral is of economic importance, the deposit is known as a **placer deposit**. Resistant detrital minerals that are mined from placers include gold, ilmenite, diamonds, and the tin oxide cassiterite (SnO_2).

11.3 Organically produced sediment

Sediment not only is produced as the detritus from the weathering of preexisting rocks but also can be of organic or chemical origin. In places such as the Bahamas, most of the sediment on beaches is formed from the breakup of marine organisms by wave action (Fig. 11.6(D)). Because the majority of these organisms have hard parts made of aragonite or calcite, they produce sediment that consists of carbonate grains, which eventually form limestone (or dolostone). Some organisms build hard parts of silica, extracted from the silicic acid in seawater, which ultimately is derived from the weathering of silicate minerals on continents (Reaction (11.4)). The hard parts of these silica-secreting organisms form much of the sediment on the deep ocean floor (Fig. 9.38(B)). Other sediment is formed by the direct accumulation of vegetation in situ, and this eventually produces coal.

11.3.1 Formation of carbonate and siliceous sediment

Carbonate sediment can be of chemical or biogenic origin. During the Precambrian, most carbonate sediment was a chemical precipitate, but the Cambrian faunal explosion saw the evolution of a large number of fauna with calcareous hard parts, and throughout the Phanerozoic, biogenic carbonates

1 mm

1 mm

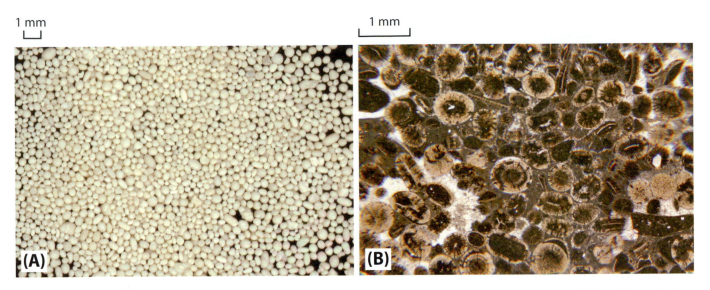

Figure 11.8 **(A)** Ooids from the Joulters Cays area, Bahamas, where currents across shallow shoals keep the ooids in constant motion. **(B)** Photomicrograph of a thin section of ooids in the Middle Ordovician Laysburg Formation of Central Pennsylvania.

became the dominant type of carbonate sediment. This explosion also dramatically increased the abundance of carbonate rocks in the Phanerozoic.

Biogenic carbonate sediment is formed from the breakup of the hard parts of a wide variety of organisms, including algae, corals, mollusks, sponges, echinoderms, and bryozoans (Fig. 11.7(A)). Although we might expect these hard parts to be made of calcite, which is thermodynamically the stable form of $CaCO_3$ under the conditions found in ocean water, they are composed of either aragonite or calcite containing several percent magnesium (high-magnesium calcite). Aragonite is the stable form of $CaCO_3$ at high pressure and is found in some high-pressure, low-temperature metamorphic rocks (Sec. 14.10.2). The aragonite and high-magnesium calcite that are grown by organisms tend to change with time into the stable calcite form with less than 5 mole percent magnesium, so that most limestone is made of calcite.

The calcareous organisms are broken by wave and current action and by other organisms that feed on them. Because the grain size of carbonate sediment is reduced by both transport and attack from other organisms, it commonly has a bimodal distribution. Grains the size of gravel to sand are broken skeletal remains (Fig. 11.6(D)), whereas the finer mud fraction is produced from the breakage of these skeletal parts or from calcareous green algae, such as *Penicillus*, shown in Figure 11.7(B), which secretes 1–4 μm aragonite needles in its tissue. When these organisms die and decay, the needles are released to the environment to form carbonate mud.

Another type of carbonate sedimentary grain that is most likely of inorganic origin is an **ooid** (also referred to as **oolith**), a subrounded sand-size grain consisting of concentric layers

of carbonate (Fig. 11.8(A)). The argonite or calcite precipitates around a preexisting grain, which forms the nucleus (Fig. 11.8(B)). Most layers are continuous around an ooid, rather than being on just one side, which indicates that the particle must have been continuously moved. This is consistent with ooids occurring where strong currents agitate the water. In thin sections, the layering often shows minor unconformities, where earlier layers have been abraded and then covered by subsequent layers. Also visible in thin sections is the radial growth pattern of aragonite crystals. Although modern ooids are made of aragonite, following burial and lithification, they change to calcite, which may destroy the layering and radial structure (Fig. 11.8(B)). Most **oolitic limestone** is made of calcite.

Other sand-size particles of carbonate sediment are formed by fecal pellets, which are produced by many different organisms. These are held together by organic material that may eventually decay, and the pellets often fall apart to form carbonate mud, but some survive to become **pellets** (or **peloids**) in limestone.

Many of the organisms that form carbonate sediment live together and form reefs, especially in warm, clear water (Fig. 11.9). Although these are commonly referred to as **coral reefs**, algae are an important component, and reef-forming coral itself contains symbiotic photosynthetic protozoans. Because photosynthetic organisms require sunlight for growth, they are normally restricted to water depths of less than 20 m. Reefs grow upward until they reach a level where they are exposed to wave action at low tide. The wave action is important to life in the reef because it brings in nutrients, but at the same time, it is constantly destroying the reef. The existence of a reef, therefore, depends on the balance between the two opposing

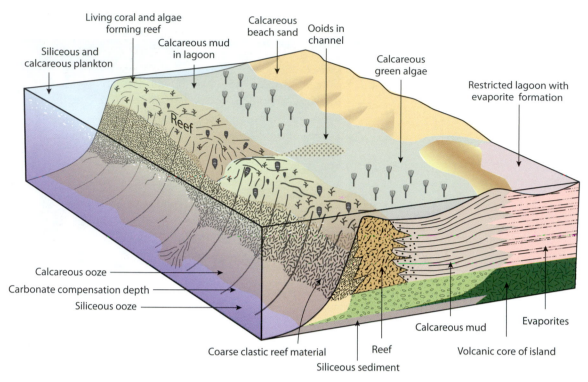

Figure 11.9 Sites of carbonate sedimentation in the vicinity of a reef.

processes of growth and destruction. As with other systems in dynamic equilibrium, reefs are sensitive to environmental change. A healthy reef, however, is a constant source of carbonate sediment. On the offshore side, where a reef normally descends into deeper water, material broken from the reef accumulates as a wedge of coarse bioclastic material. Between the reef and the shoreline is commonly a shallow lagoon in which finer carbonate sand and mud accumulate. Reefs have been important in the geologic record since the lower Paleozoic.

Deepwater calcareous and siliceous sediment

In addition to organisms that live near the shoreline, oceans contain an enormous quantity of **plankton**; that is, plants and animals that are essentially free floating and are carried wherever ocean currents take them. When these die they sink toward the bottom to form deep-sea sediment. Many of these have calcareous hard parts, which form calcareous ooze on the ocean floor. One of the most prolific forms of calcareous plankton in the oceans for the last 150 million years is the foraminifera, *Globigerina*. Another important contribution comes from planktonic algae that secrete small plates of calcite known as coccoliths. These form a major part of many chalk deposits, such as that in the White Cliffs of Dover. Both the tests of *Globigerina* and coccoliths are made of magnesium-bearing calcite rather than aragonite.

Although the plankton in oceans ties up an enormous quantity of calcium carbonate, not all of this finds its way to the ocean floor. As we know from opening a carbonated beverage, CO_2 is more soluble in water at high pressure than it is at low pressure. In addition, water can dissolve more CO_2 at low temperatures than at high temperatures. As a result, with increasing depth in the ocean, pressure increases, temperature decreases, and the solubility of CO_2 increases. This, in turn, makes more carbonic acid according to Reaction (11.1), which dissolves calcite. Thus, if calcareous plankton sinks deep enough in the ocean, it dissolves. This depth is known as the **carbonate compensation depth**. In the Atlantic Ocean, the compensation depth for calcite is ~5000 m, whereas in the Pacific Ocean it is ~1000 m less, the difference being due largely to higher concentrations of CO_2 in the Pacific deep waters, which are older than deep waters in the Atlantic and, therefore, have had time to increase their CO_2 content. The compensation depth for aragonite is less than that for calcite.

Other plankton such as diatoms and radiolaria (Fig. 9.33(B)) build tests of silica. When these die and sink, they form siliceous ooze on the ocean floor. Unlike the calcareous plankton, the siliceous tests have no restriction on depth. As a result, since the Jurassic, the deepest ocean floor, where calcite is not stable, has been covered with siliceous ooze.

Sponges are animals with a soft jellylike body that is strengthened by small rods or **spicules** made of silica or calcite. They live in shallow water, but when they die and decay, the spicules become part of the sediment, which may be transported to deeper water by currents. Sponge spicules are common in **chert** nodules found in limestone.

11.3.2 Formation of hydrocarbons in sediment

Since the Silurian (420 Ma), when the first land plants evolved, plant detritus has been a potential component of sediment. Normally when plants die, they soon decay (oxidize) and are not likely to contribute much sediment. However, if sedimentation is rapid enough to bury plant material before it oxidizes, it can be a major component of sediment, reaching 100% in the case of that which forms **coal**. The percentage of plant material in land-derived sediment is highly variable. Even a small amount, however, can impart a gray color to the sediment. Because decaying plant material consumes oxygen, even a small amount in sediment is sufficient to reduce the oxygen content of the sediment to a level at which it prevents ferric iron from forming. The result is that the sediment tends to be gray or even black if the concentration of plant material is high enough. River deltas provide an environment where the influx of large quantities of sediment commonly results in burial and accumulation of organic material. Many of the coal deposits of Pennsylvania are believed to have accumulated in such environments.

The burial of large amounts of plant material to form coal decreases the CO_2 content of the atmosphere. Formation of large quantities of coal during the Carboniferous resulted in significant lowering of the atmosphere's CO_2 content.

Though not normally considered sediment, oil is a direct product of the burial of animal and plant remains in sediment. It forms from the breakdown of kerogen from the organic matter at slightly elevated temperatures and pressures known as the **oil window** (60–120°C, 2–4 km depth). Petroleum is a viscous liquid that, once formed in a source rock, is able to move through permeable sedimentary rocks and, if we are lucky, accumulates in a porous reservoir rock beneath an impermeable cap to form an oil reservoir (Fig. 12.36). Natural gas is formed in the same way but at slightly higher temperatures (100–200°C) and pressures (depth 3–6 km). If these hydrocarbons are heated beyond this range, they change to graphite.

11.4 Chemically produced sediment

Approximately 20% of the weathering products of rocks are removed in solution, most of which eventually make it to the sea, but some are trapped in saline lakes in large basins that have no connection to the sea (e.g., Death Valley, the Dead Sea). The concentration of these soluble ions (e.g., Ca^{2+}, Na^+, K^+, Mg^{2+}, $(SO_4)^{-2}$, Cl^{-1}, $(HCO_3)^{-1}$) would steadily increase in seawater if there were nothing to remove them. We have seen that calcareous organisms provide a sink for calcium and, to a much lesser extent, magnesium ions, but what happens to the other ions? These are eventually removed as chemical precipitates in restricted basins where evaporation raises their concentration to super-saturation levels. This forms **evaporites**.

The sequence in which minerals precipitate from water as it evaporates is determined by the composition of the water and the solubilities of the minerals that precipitate. The first minerals to precipitate from seawater are normally aragonite or calcite and dolomite, followed by gypsum and anhydrite, and then halite. If the seawater were to completely evaporate, halite would be by far the most abundant mineral formed, but halite is one of the most soluble minerals, and in basins where evaporation does not go to completion, halite may not even precipitate. When it does, it is deposited on top of the gypsum and anhydrite. Because of their much lower solubility, gypsum and anhydrite are far more common than halite in evaporite deposits. Whether gypsum or anhydrite precipitates depends on temperature and salinity, with anhydrite being favored by higher temperatures and salinities. In saline lakes in arid climates, evaporite minerals can include borates, the most common of which is borax, $Na_2B_4O_5(OH)_4 \cdot 8H_2O$.

Unlike detrital sediment from the weathering of rock and calcareous sediment from the breakage of organisms or precipitation from seawater, evaporite minerals undergo little transport. Indeed, most evaporite minerals are precipitated at the site in which they accumulate to eventually form rock. These rocks are commonly layered due to variations in mineralogy, grain size, or concentrations of impurities, which may be transported into the basin of deposition as fine wind-blown particles. In some modern evaporites, each layer can be seen to represent an annual deposit and is, therefore, called a **varve**. Rhythmic layers in older evaporite deposits, such as that of the Castile Formation of New Mexico (Fig. 11.10), are interpreted to be varves.

Another type of chemical sediment is **banded iron-formation** (BIF), which is important because it is the host

1 cm

Figure 11.10 Rhythmic layers of gypsum in the Castile Formation (Permian), New Mexico. The dark layers contain small amounts of calcite and organic matter.

10 cm

Figure 11.11 Finely banded 2500 million-year-old Brockman Iron-Formation of the Hamersley Range, Western Australia. Brownish-red layers are predominantly quartz with some ankerite and an iron-silicate named stilpnomelane, whereas the darkest thin bands contain small amounts of magnetite and/or hematite.

rock for the world's major source of iron ore (Sec. 16.2). This sediment was formed only during the Precambrian, and, therefore, we have no modern analogues to help us interpret how it formed. Iron-formations range in age from 3.8 billion to about 1.8 billion years ago, with some additional occurrences between 0.8 and 0.6 billion years ago. The most extensive occurrences, which form belts hundreds of kilometers long, range in age from 2.8 to 1.8 billion years. Those between 2.8 and 2.2 billion years of age show prominent centimeter-size banding (hence its name), with alternating silica- and iron-rich layers (Fig. 11.11). The silica-rich layers are composed of chert,

or quartz, or red jasper (a chert admixed with fine hematite inclusions), and the iron-rich layers consist mainly of magnetite or hematite but also include siderite; ankerite; and several hydrous iron silicates, the most common of which is greenalite, $Fe_3Si_2O_5(OH)_4$.

These well-banded iron-rich chemical sediments have been deposited in a mainly anoxic marine environment, and because the layers show great consistency over large distances, they must have formed below wave base. Iron-formations that occur in the Early Proterozoic (2.2 to 1.8 billion years ago) show less well developed banding but also exhibit extensive oolitic textures with finely banded magnetite-hematite ooids in a chert matrix. These BIFs are considered the result of deposition in shallow water platforms. During the Phanerozoic, some hematite-rich sediments, with oolitic textures, were deposited and are known as ironstones.

11.5 Sediment produced by glacial erosion

Much of the northern part of the Northern Hemisphere is presently covered by surficial deposits of sediment that were laid down during the retreat of the last continental ice sheet. Similar deposits are found preserved in ancient rocks, which testifies to the existence of much earlier periods of continental-scale glaciation. Glacial sediment is either deposited directly from ice, in which case it is referred to as **till**, or is deposited from meltwater, in which case it becomes stratified.

Moving ice is capable of dislodging and transporting large pieces of the bedrock over which it moves (Fig. 11.12(A)). The base of an ice sheet is typically full of such blocks, which gouge the underlying bedrock, leaving glacial striations. The blocks themselves become abraded and commonly faceted as the ice

Figure 11.12 (A) Large glacial boulder left by the Wisconsin ice sheet with glacial striations on bedrock in foreground showing two directions of ice movement in Glastonbury, Connecticut. (B) Polished glaciated surface developed on basalt at North Branford, Connecticut. White arrows in both photos indicate direction of ice movement.

pushes them across the bedrock surface. The base of the ice soon contains not only large fragments of the bedrock but also fine material produced by their abrasion. While large particles gouge the bedrock surface, finer ones tend to polish it and can produce an almost mirror finish on hard fine-grained rock such as basalt (Fig. 11.12(B)). When ice movement is impeded by drag along the base of a glacier, the ice can be thrust up along shear planes over the impeded ice. This motion takes sediment from the base of a glacier up into the ice sheet. As a result, glacial ice typically has sediment distributed throughout the ice sheet. This sediment, which is released from the ice sheet at its terminus or from wasting masses of stagnant ice, is called till. Till formed at the base of an ice sheet tends to be extremely compact because of the weight of the overlying ice. This is known as **lodgement till** (Fig. 11.13(A)). Sediment that is distributed throughout the ice sheet, if not washed away by meltwater, is lowered onto the land surface as the ice slowly ablates. Meltwater typically removes the finer sedimentary particles, leaving only the coarser fraction. This residual or **ablation till** is not as compact as lodgement till and lacks the finer-grain fraction. The feature that characterizes glacial till from most other types of detrital sediment is its almost complete lack of sorting; giant boulders can be juxtaposed with clay-size particles (Fig. 11.13(A)). Many of the boulders are faceted, with striations indicating where they were ground against the bedrock.

As continental ice sheets reach lower latitudes or mountain glaciers reach lower elevations, they eventually melt, and much of the sediment in the ice is washed away and deposited by meltwater as **stratified meltwater deposits**. Some may be deposited by water flowing along the sides of glaciers or in tunnels within the ice forming **kames** and **eskers**, respectively (Fig. 11.13(B)). Meltwater may carry sediment out of the glacier into **outwash plains** and from here the finer particles may be transported great distances away from the terminus of the ice. The amount and grain size of sediment released from ice sheets depends on the amount of melting, which varies seasonally. Figure 11.13(B) shows ice-contact stratified deposits in an esker where grain size ranges from fine sand to large boulders. The sand would have been deposited during the winter when little meltwater was flowing, whereas the large boulders would have been moved by the large flux of meltwater during the summer. Sediment that accumulates in standing bodies of water that are fed by melting ice typically have rhythmic layers of coarser and finer sediment which reflect the seasonal changes of sediment influx. These annual deposits are known as varves (Fig. 11.13(C)).

11.6 Transport of sediment

Detrital sediment can move by simply falling in a gravitational field, which certainly happens on cliff faces and talus slopes, but most detritus is moved by water, wind, or ice, with water being by far the most important agent. These three agents move as fluids, albeit with very different viscosities (see Sec. 8.6.2 for definition of viscosity). Before we can understand how fluids transport sediment, we need to know how fluids move.

11.6.1 Laminar and turbulent flow

Fluids have no shear strength and consequently must flow in response to stresses imposed on them. The stresses could result from the Earth's gravitational field in the case of water or ice flowing down hill, or from atmospheric pressure differences in the case of wind. The speed with which fluids react to stresses is measured by viscosity (Sec. 8.6.2). Fluids move in two different

Figure 11.13 (**A**) Compact lodgement till exposed in an excavation at Haverill, Massachusetts. The cobbles are held in place so firmly by the surrounding clay that some were broken during excavation. (**B**) Stratified meltwater deposits in an esker from North Windham, Connecticut. (**C**) Centimeter-scale varved silt and clay layers deposited in Glacial Lake Hitchcock, South Windsor, Connecticut. The lighter colored clay-rich layers stand out in relief. (Photographs (A) and (C) courtesy of Janet Radway Stone, U.S. Geological Survey.)

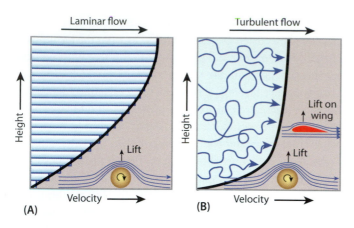

Figure 11.14 Fluid flow is either laminar (**A**) or turbulent (**B**). Turbulent flow occurs when the viscosity of the fluid is low and the velocity is high. Fluid flow over a grain resting on the bottom decreases the pressure above the grain and causes it to lift the same way that air passing over an airplane wing provides lift. This movement causes grains to jump along the bottom.

ways that we refer to as laminar or turbulent. Fluids that flow in a **laminar** manner can be thought of as consisting of thin sheets or laminae that slide over each other (Fig. 11.14(A)). Just as cards remain in the same position in a deck of cards when sheared sideways, the fluid laminae maintain their same positions relative to the surface over which the fluid flows. For example, if we were to place a drop of dye in a fluid at 1 cm from the boundary over which it is flowing, that dye would remain at this distance as long as the flow remained laminar. In contrast, if the fluid were in **turbulent flow** (Fig. 11.14(B)), we would not be able to predict where the dye would go, because it would follow an irregular flow path, and if we were to insert another drop of dye, it could follow a completely different path.

When you pour honey slowly from a jar, it is in laminar flow, whereas a mountain stream is in turbulent flow. What determines the type of flow? In the case of honey, our intuition tells us that the fluid is extremely viscous and does not flow fast, whereas the mountain stream has low viscosity and flows rapidly. Our intuition would have led us to the two most important factors in determining whether flow is laminar or turbulent; that is, viscosity and velocity. For fluid flowing in a channel, such as a river, we can define a term that involves the viscosity and density of the fluid and its average velocity that indicates whether flow should be laminar or turbulent. This term is known as the **Reynolds number** (**Re**), which is defined as

$$\mathrm{Re} \equiv 4 \times \text{hydraulic radius} \times \text{density} \atop \times \text{ average velocity/viscosity} \qquad (11.5)$$

where the **hydraulic radius** is the cross-sectional area of the channel divided by the wetted perimeter (i.e., 2 × depth + width). All of the units cancel in this expression, so we are simply left with a number. When the Reynolds number is less than 500, flow is laminar, and when it exceeds 2000, it is turbulent. Between these two values, the roughness of the channel bed determines whether the flow is laminar or turbulent.

Let us calculate the Reynolds number for flowing water. The viscosity of water is ~10^{-3} Pa·s and its density is 10^3 kg/m³. Substituting these values into Equation (11.5), the Reynolds number for flowing water becomes

$$\mathrm{Re} = 4 \times 10^6 \times \text{hydraulic radius} \atop \times \text{ average velocity} \qquad (11.6)$$

If the hydraulic radius of a stream were 1 m, as soon as the water velocity reached 0.125 mm/s (7.5 mm/minute), the Reynolds number would reach 500 and flow could become turbulent. If the hydraulic radius were still larger, as it would be in a river, the velocity required for turbulent flow would be still less. We can conclude that water almost always flows turbulently. Air having a still lower viscosity than water always flows turbulently. Glacial ice has such high viscosity that its flow is always laminar.

11.6.2 Movement of particles by fluid flow

Rolling

When fluid passes over a particle at rest, the shear force applied by the fluid tends to roll the particle (Fig. 11.14). Consequently, particles at the bottom of a river, on a beach, or on a desert floor are rolled along by the movement of the fluid. Their velocity depends on the fluid's velocity and on how spherical the particles are.

Saltation

As fluid flows over a stationary particle, flow lines are deflected and forced closer together. This causes the fluid momentarily to move a little faster over the top of the grain. According to Bernoulli's principle, this velocity increase decreases the pressure above the grain relative to the pressure where the flow lines are undisturbed. This pressure decrease can lift the particle, as long as it is not too dense or too large (Fig. 11.14). This is exactly the same effect an airplane wing experiences when air flows more rapidly over its curved top side than it does on the flat bottom side. Sedimentary particles that are lifted by this force rise into the flowing fluid and are carried along for a short distance. Once a particle leaves the bottom, flow lines above and below the particle become the same, so the Bernoulli effect is eliminated and the particle drops back to the bottom. This results in the particle's motion being a series of short jumps, which is known as **saltation**, from the Latin word *saltare*, which means "to dance or leap." Saltation is an efficient means of moving sediment along the bottom of rivers or across sand dunes.

Suspension of sedimentary particles and Stokes' law

When flow is turbulent, which is the case for wind and most moving water, turbulent eddy currents can lift particles and carry them in **suspension**, as long as the particles are not too large or dense. When a turbulent eddy lifts a particle, gravity tries to pull the particle downward. Whether the particle is lifted by the current or falls back depends on the current's velocity and the

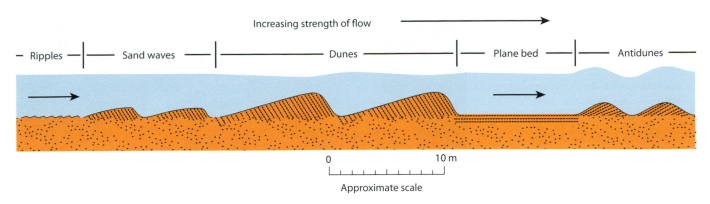

Increasing strength of flow

— Ripples — | — Sand waves — | — Dunes — | — Plane bed — | — Antidunes —

0   10 m

Approximate scale

Figure 11.15 Ripples, sand waves, dunes, plane beds, and antidunes are bedforms that can form as a result of flowing water (or wind).

settling velocity of the particle through the fluid. In Section 8.7.3, we saw that **Stokes' law** (Box 8.6) gives the terminal velocity at which a particle settles through a fluid. It is this velocity that the fluid must exceed if it is to lift the particle. Stokes' law shows that the settling velocity is proportional to the square of the diameter of the particle, directly proportional to the density contrast between particle and fluid, and inversely proportional to the viscosity of the fluid. Because the grain diameter is squared, it is the most important factor in determining settling velocities, and it explains why only small clay- and silt-size particles are normally carried in suspension. Also, because the density contrast of a particle with air is so much greater than with water, and air's viscosity is much lower than water's, only very small particles can be carried in suspension in air (e.g., dust storm).

Fluid, then, has two ways of transporting detrital grains, either as particles rolling or jumping along the surface or as particles suspended in the fluid. In the case of a river, sediment traveling along or near the bottom is referred to as **bedload**, and that which is in suspension as **suspended load**, and both contribute to the formation of **alluvial deposits**. Usually the suspended load is greater than the bed load. Because the motion of particles in the bed load is discontinuous, the bed load moves slower than the suspended load, which moves with the same velocity as the water (or wind). The result is that the silt- and clay-size fraction (mud) in suspension is effectively separated from the coarser fraction in the bedload. In addition, clay-size particles take a very long time to settle, even when water has essentially stopped moving. The clay fraction, therefore, travels faster and farther than the sand-size fraction. Sediment transport by water or wind leads to effective sorting on the basis of grain size. The separate deposits of coarse- and fine-grained sediment, when lithified give us the three main types of detrital sedimentary rock, the coarser **conglomerates** and **sandstone** and the finer **mudrocks**. In the case of wind blowing across deserts, the fine-grained particles are removed and can be deposited to form what is known as **loess**, from the German word for *loose*. These deposits lack prominent bedding and commonly weather to form vertical cliffs. Loess forms extremely fertile soil (Sec. 17.2.1).

Bedforms

Bedload moving along a river bottom rarely moves as a planar sheet. Instead, it develops **bedforms** that are characteristic of the flow strength (Fig. 11.15). At the lowest velocities, **ripples** are formed that have wavelengths from centimeters to decimeters. As stream velocity increases, larger **sand waves** develop, eventually forming **dunes**, which can have wavelengths up to 10 m. The wavelike shape of dunes is out of phase with the wave that develops on the surface of the water. At still greater velocity, irregularities on the bottom disappear and the bedload moves along as a planar sheet. At the greatest velocities, antidunes develop where the wave on the surface of the water is in phase with the undulating riverbed. These bedforms are preserved in sedimentary rocks, except for antidunes, which are rare. Figure 11.16, for example, shows ripple marks preserved in the Cambrian Potsdam Sandstone of northern New York State, where we see that water flowed from left to right (toward the steeper lee side of the ripples).

Ripples, sand waves, and dunes all move in the same manner, with erosion taking place on their gentler upstream side and deposition on their steeper lee side. All the sediment in these structures is consequently deposited on the lee surface of the

Figure 11.16 Ripple marks in the Cambrian Potsdam Sandstone exposed in the Ausable River, New York.

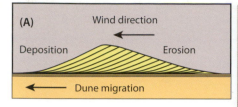

Figure 11.17 (**A**) Cross-bedding in dunes is formed when sand eroded from the windward side of the dune is deposited on the lee side. (**B**) Cross-bedding in the Jurassic Navajo Sandstone, Zion National Park, Utah, formed by sand dunes that migrated from right to left.

structures as they migrate. This produces an internal stratification that is at an angle to the main stratification of the deposit and dips in the downstream direction. This cross-stratification is one of the ways in which the sedimentary structure known as **cross-bedding** can form.

Animations of the growth of bedforms and cross-bedding can be seen at the U.S. Geological Survey's Web site (http://walrus.wr.usgs.gov/seds/bedforms/index.html). Videos showing the growth of similar features in flume experiments is given at Paul Heller's Web site (http://faculty.gg.uwyo.edu/heller/sed_video_downloads.htm).

Wind-blown sand dunes show the same type of cross-bedding as do ripples, because they migrate in the same way with the sediment eroded from the windward side being deposited on the lee (Fig. 11.17(A)). Figure 11.17(B) shows cross-bedding in

the Jurassic Navajo Sandstone in Zion National Park, Utah. The cross-bedding in this sandstone was formed by sand dunes that migrated from right to left. Notice how the bottom of each cross-bed thins and curves to become tangential with the underlying stratification. The top of each cross-bed, however, is truncated where the sand was eroded away and transported downwind to form a new layer on the lee surface of the dune. This difference between the lower and upper terminations of cross-beds provides a useful criterion for determining tops in ancient folded rocks.

Cross-bedding can form in other environments. For example, as a stream channel meanders back and forth across a floodplain, it continuously cuts into earlier stratified deposits and then redeposits the sediment farther downstream (Fig. 11.18). When a channel, scoured out during a period of high flow, is later filled with sediment during a period of

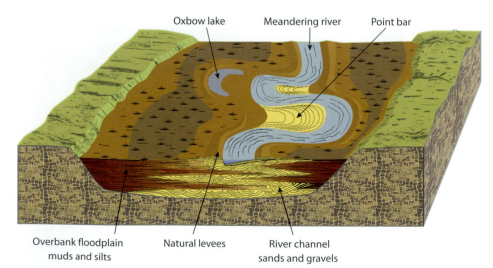

Figure 11.18 Alluvial deposits formed by a meandering river.

Figure 11.19 Cross-bedding formed by water flowing left to right in the Jurassic feldspathic sandstone at Portland, Connecticut. Alkali feldspar grains are white and quartz grains are coated with a film of reddish-brownish hematite.

lower flow, the new layers of sediment parallel the walls of the channel but generally thicken toward its center. This alternate scouring and filling of channels can be seen to have occurred in the esker deposit shown in Figure 11.13(B). A meandering river moves laterally across a floodplain by cutting into the banks on the outside of meanders where the water velocity is greatest, and then depositing the sediment on the inside of the next meander where the water velocity is less. The result is the formation of a point bar on the inside of the meander (Fig. 11.18). Point bars continuously grow with deposition of new layers, which commonly show a low angle cross-stratification. In some cases, the newly arriving sediment builds a small delta out into the deeper water. Figure 11.19 shows cross-bedding that was formed in such a fluvial setting in the Jurassic Portland feldspathic sandstone, in the Hartford Basin of Connecticut.

When rivers flood, they not only carry more sediment but also often overflow their banks. Once water leaves the main channel, its velocity decreases and its suspended load is deposited across the floodplain (Fig. 11.18). Because these deposits are formed from the suspended load, they are always fine-grained and carry considerable clay. When the flood subsides, the wet clay begins to dry and shrink, which results in the formation of **mud cracks** (Fig. 11.20). These cracks are initiated at irregularities in the mud (e.g., a small stick) and typically form a set of three fractures roughly at 120° to each other. As the cracks widen and propagate from these triple junctions, they eventually reach fractures emanating from other triple junctions against which they terminate. This results in the formation of polygonal-shaped mud cracks. Should another flood occur at this stage of desiccation, the cracks are filled in by the first sediment to settle from the new flood. In this way, mud cracks can be preserved and eventually converted to solid rock, as seen in the example from the Cambrian Potsdam Sandstone in northern New York State (Fig. 11.21). If desiccation proceeds farther, mud polygons commonly curl up because of the greater degree of shrinkage on their upper side (Fig. 11.20). When next the river floods, these curled-up mud polygons are washed away as detrital fragments known as **rip-up clasts**. Rip-up clasts are also produced in carbonate deposits by erosion of semilithified muddy bottom sediment (Fig. 12.22). Mud cracks can also form in marine sediment along coastal planes.

When rivers are in flood, they commonly flush out organic material that may have accumulated since the last flood. Fallen trees, for example, are carried along in floods and may form logjams that are eventually pushed onto the bank where they

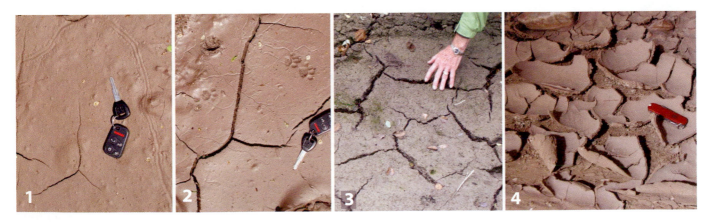

Figure 11.20 Successive stages (1–4) in the development of mud cracks.

Figure 11.21 Mud cracks in the Cambrian Potsdam Sandstone, Ausable Chasm, New York.

remain until the next flood. Floodplain deposits commonly contain fossils of such material, such as the fossilized pieces of wood that were pushed into a subparallel stack in the Mesozoic mudrock from the Hartford Basin in Massachusetts (Fig. 11.22(A)). Before drying out, the soft mud on a floodplain provides an ideal medium in which to preserve fossils. Tracks of various insects and animals can be seen in the modern mud, which is just beginning to form mud cracks shown in Figure 11.20. Dinosaurs roaming Mesozoic floodplains left their imprints in this soft mud (Fig. 11.22(B)).

11.6.3 Movement of particles in turbidity currents

Sediment has another important way of moving in water, and that is in **turbidity currents** in lakes and especially ocean basins. In this case, water does not actually move the particles but acts simply as a lubricant. A turbidity current involves a dense suspension of sediment that moves down slope under its own gravitational force. It has low viscosity because of the water between the particles, but it has a high density because of the suspended particles, and consequently flows rapidly, at velocities of as high as 10 m/s.

Turbidity currents are formed when unconsolidated sediment is disturbed as a result of either slumping triggered by an earthquake or the pounding of storm waves on a nearby shore. Once the sediment is dislodged and starts moving, the turbidity current takes on a life of its own and continues moving as long as there is a slope down which it can travel. They can travel hundreds of kilometers across the continental slope in oceans. They have great erosive power and are thought to be responsible for the erosion of submarine canyons on the continental slope. Much of the sediment that eventually makes it into oceanic trenches is delivered as turbidity currents.

A turbidity current flows as a turbulent mass of sedimentary particles, and consequently sorting has little chance to occur while rapid flow continues. However, once the current begins to slow, the coarser particles make their way toward the bottom, so that when the current finally comes to rest these larger particles are deposited first and the sediment becomes finer upward; the clay-size fraction may take considerable time to settle. The result is a layer of sediment with a gradation in particle size from coarse at the bottom to fine at the top. This is known as

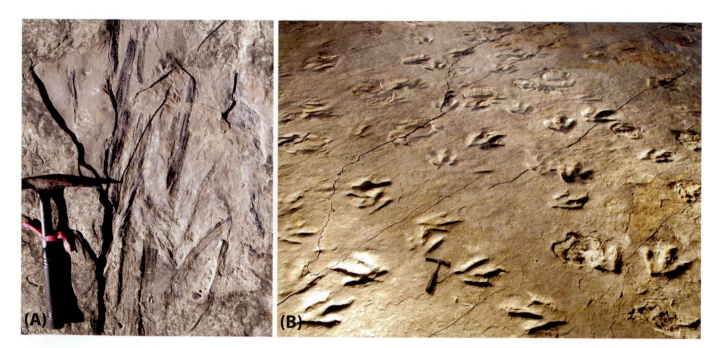

Figure 11.22 (**A**) Fossilized wood and (**B**) dinosaur footprints in Mesozoic floodplain deposits in the Hartford Basin of Massachusetts and Connecticut.

graded bedding and the deposit itself is known as a **turbidite**. Figure 11.23 shows a sample of turbidite from a sequence of marine sandstones in southeastern Quebec that were deposited in deep water but later thrust onto the North American craton during the Appalachian orogeny. The base of the layer consists of coarse detrital grains of quartz, feldspar, mafic minerals, and even rock fragments. This grades upward through fine sandstone to siltstone at the top of the photograph. The gradation continues above this to produce a fine dark maroon mudrock similar to the one shown at the bottom of the photograph, which is the top of the underlying graded layer.

11.6.4 Movement of sediment in debris flows

On steep hillsides, sediment can become unstable especially if it becomes water saturated. Soil and **regolith** (unconsolidated material above bedrock) normally move slowly downhill through creep and solifluction. The velocity of these processes is usually less than 1 cm/year. If the regolith becomes water saturated, however, slope instabilities can generate **debris flows**, where velocities can range from as little as 1 m/year to as much as 1 km/hour, the later often causing catastrophes when they descend into built-up areas (Sec. 17.4). Such debris flows are common around volcanoes where they are known as lahars (Fig. 9.18(D)). Debris flows are also common on alluvial fans.

The sediment in debris flows is poorly sorted and can range in size from boulders to clay. When the flow consists largely of particles finer than sand-size, it is called a **mudflow**, and these can have velocities of up to 100 km/hour. Once a debris flow or mudflow starts moving, it can travel considerable distances and can descend into valleys or across alluvial fans and out into sedimentary basins. They tend to have lobate fronts, and their upper surfaces are marked by concentric ridges that are similar to the ridges seen on rhyolite lava flows (Fig. 8.21(A)).

11.7 Layering in sediments and sedimentary rocks

Most sediments and sedimentary rocks are characterized by layering, which can vary in thickness from hundreds of meters to fractions of a millimeter. Contrast, for example, the thick layers of flat lying Paleozoic sedimentary rocks in the Grand Canyon (Fig. 11.24(A)) with the thin layers in the mudrock formed on the floor of a deep lake in the Hartford Mesozoic Basin (Fig. 11.24(B)).

The main sedimentary unit that produces layering in sedimentary rocks is defined as a bed. A **sedimentary bed** is a layer that is distinguishable from layers above and below it on the basis of rock type, grain size, or some physical property, such as **fissility**, which is the splitting apart of a bed into thin sheets (Fig. 11.25). Beds are normally thicker than 1 cm; if less, they are referred to as **laminae**. The boundaries of a bed can be sharp

1 cm

Slaty cleavage

Figure 11.23 Ordovician sandstone from the Appalachians of southeastern Quebec, showing graded bedding from coarse-grained green sand at the base to fine-grained dark maroon mudstone at the top. The dark mudstone at the bottom of the photograph is the top of the underlying graded bed. The sample is bounded by slaty cleavage planes.

Permian Kaibab
Limestone
Permian Coconino
Sandstone
Mississippian Redwall
Limestone

1 mm

Precambrian
Vishnu Schist

(A)

(B)

Figure 11.24 (A) The Grand Canyon in Arizona exposes a spectacular 1.5 km-thick section of flat-lying, prominently layered Paleozoic sedimentary rocks lying unconformably on top of the Precambrian Vishnu Schist, which can be seen in the deepest part of the canyon. (B) Photomicrograph of thinly bedded shale from the East Berlin Formation, formed by accumulation of mud on the floor of a deep Jurassic lake in the Hartford Basin, Durham, Connecticut.

or gradational, and the grain size within a bed can vary, as in a graded bed (Fig. 11.23). A bed can also have internal layering or stratification as seen in the case of cross-bedding (Fig. 11.19). A bed should also have some lateral continuity extending beyond the outcrop scale. If layers are shorter than this, they are referred to as **lenses** (shorter than a few meters). As you can see, the definition of a bed is rather flexible but in general refers to the prominent layering visible in an outcrop.

A sedimentary bed reflects a certain supply rate of sediment and a set of depositional conditions. If these change, a new bed is likely to form. The rate of supply of sediment can vary considerably from several meters per year in an active delta to a fraction of a millimeter in a thousand years in the deep ocean far from land. Normally, coarser sediments are deposited more rapidly than fine sediment. The coarse boulder beds in the esker (Fig. 11.13(B)) may have been deposited during surges in

Figure 11.25 Black shale weathering to form thin flakes due to fissility parallel to the bedding.

the river below the ice that perhaps lasted no more than a day, whereas each of the fine laminae in Figure 11.24(B) may have taken a year to be deposited. Depositional environments can change as a result of climatic changes or tectonic processes. Seasonal changes in rainfall can dramatically alter the supply rate of sediment. Episodes of down faulting of sedimentary basins increase relief and the supply of sediment.

11.7.1 Law of superposition

The **law of superposition**, first clearly stated by Nicolas Steno in 1669, states that sedimentary layers are deposited one on top of the other in a time sequence, with the oldest at the bottom and the youngest on top. The important point is that a sequence of sedimentary beds records relative ages. It is natural, therefore, to ask how much time an individual bed represents. This is often a difficult question to answer. A general idea can be obtained from rates of accumulation mentioned in the previous paragraph, but these vary widely. Coarse beds of detrital sediment may be deposited during a single storm, so they could represent a period of as little as minutes. Similarly, turbidity currents travel rapidly to their site of deposition, so a single turbidite bed may represent a period on the order of a day or less. Rhythmically layered sediments, with alternating coarser and finer layers, may represent annual deposits. Varved glacial lake sediments (Fig. 11.13(C)) are definitely formed this way, as are some layers in evaporites (Fig. 11.10).

11.7.2 Milankovitch cycles

Sequences of beds that reflect climatic changes on the scale of tens of thousands of years have been related to **Milankovitch cycles**, which are driven by variations in the amount of solar energy received as a result of perturbations in the Earth's orbital

Figure 11.26 (**A**) A sequence of sedimentary rocks in the Jurassic East Berlin Formation of the Hartford Basin, Connecticut, passing from thick red sandstone at the base (lower left) through thin black shale to thick red sandstone at the top reflects climatic changes brought about by the ~22 000-year-precession cycle of the Earth's axis (a Milankovitch cycle). (**B**) Above the black shale shown in (A), which is located at the left end of this photograph, two other black shales occur in the overlying rocks (2 and 3), which mark subsequent 22 000 year Milankovitch cycles when the climate was wetter.

eccentricity around the Sun (~100 000 years), its tilt (~40 000 years), and its precession (19 000–23 000 years). Figure 11.26(A) shows a sequence of sedimentary rocks that were deposited in the Hartford Mesozoic Basin of Connecticut during one complete precession cycle of the Earth's axis and hence represents ~21 000 years of the Earth's history. At the base of the outcrop is a red sandstone with ripple marks and mud cracks. It was deposited in a fluvial setting in a relatively arid climate. Successively higher beds become finer grained and pass into a black shale that was deposited in a deep lake at the bottom of which conditions were anoxic, so that organic material that sank to the bottom (e.g., dead fish) were preserved. The change from red fluvial sandstone to black lacustrine deposits indicates that the climate became wetter. Still higher in the outcrop, thin beds of silt and then sand start appearing in the shale, and eventually the beds become fluvial sandstone similar to the ones at the base of the outcrop. Over this same interval, the rock's color changes from black to red, which indicates that the anoxic conditions in the lake changed to oxidizing in the fluvial environment. These changes record the shrinkage and drying up of the lake as the climate became relatively arid again. The continuation of this outcrop (Fig. 11.26(B)) shows two more overlying black shale units that were formed during subsequent wet periods.

11.7.3 Sediments related to tectonic processes

Still thicker sequences of sedimentary beds, such as the major formations exposed in the Grand Canyon (Fig. 11.24(A)) are

the product of tectonic processes. Most of these formations in the canyon, except the Coconino Sandstone, Supai, Moenkopi, and Surprise Canyon Formations were formed during incursion of shallow seas into the area. The top of the Grand Canyon is now more than 2000 m above sea level and has clearly risen since the deposition of the Paleozoic marine rocks. Moreover, each of the marine deposits was uplifted and eroded before deposition of the next marine deposit. At least 14 major unconformities in the canyon testify to these periods of uplift and erosion. The land surface must have changed relative to sea level at least that many times.

Geochemical evidence indicates that the volume of seawater has remained essentially constant throughout most of geologic time, and, therefore, this is not a variable that we can use to explain marine incursions. The volume of seawater decreases during periods of continental glaciation (because of ice storage on continents), but this lowers sea level and would not explain rises in sea level into the Grand Canyon area. We must conclude, therefore, that the elevation of the land has moved up and down, or something has caused sea level to change. Both of these processes have operated in the Grand Canyon, and both are the result of tectonic processes.

Changes in land elevation are to be expected near convergent plate boundaries, where thrusting and folding build mountain chains. The Paleozoic rocks in the Grand Canyon, however, show no signs of such deformation, and the beds still have almost the same horizontal attitude they had when deposited. We

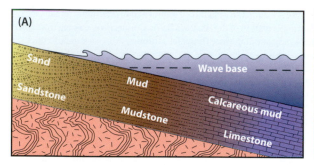 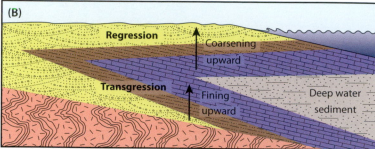

Figure 11.27 (A) Variation in sediment grain size away from a marine beach. (B) Sequence of sediment formed during the rise (transgression) and fall (regression) of sea level.

have to resort to the principle of **isostasy** (buoyancy) to explain the changes in elevation. To make the lithosphere lighter it is necessary to heat it. This can result from the rise of a **mantle plume**, but the Grand Canyon does not show the widespread igneous activity and rifting that characterize plume heads. It can also be heated as a result of **delamination** of the base of the crust. It is believed that parts of the base of the crust can become so dense as a result of metamorphic reactions that it sinks into the mantle and is replaced by hot material rising from the asthenosphere. This mechanism has been invoked to explain the rise of the Colorado Plateau and the downcutting of the Colorado River to produce the Grand Canyon.

Worldwide changes in sea level also result from changes in seafloor spreading rates. The midocean-ridge system rises above the deep ocean floor because its rocks are buoyant as a result of residual heat left over from their formation at divergent plate boundaries. As the new crust cools, it becomes denser and the ocean floor sinks. During periods of exceptionally rapid ocean-floor spreading, ocean ridges become larger (more hot crust created) and the displaced water causes a worldwide rise in sea level. During these periods, continents can be flooded with shallow seas (**epeiric seas**). For example, rapid seafloor spreading during the Cretaceous produced epeiric seas, from which many of the chalk deposits were formed on continents. Many of the formations in the Grand Canyon were deposited in epeiric seas, but no ocean floor of this age remains with which to correlate fluctuations in ocean-floor spreading rates.

Marine transgressions and regressions

Sedimentary rocks formed during incursions of epeiric seas have a characteristic sequence that records, first, deepening water as sea level rises and then progressively shallower water as sea level falls. Deposits formed during the rise of sea level are said to be **transgressive** because the shoreline advances or transgresses the continent, whereas those formed during the fall in sea level are described as **regressive**. If we were to take sediment samples across a modern ocean beach, we would find a decrease in grain size from coarse sand at the top of the beach, through finer sand in the surf zone, to mud below wave base, and, if the climate were warm enough, we might find carbonate mud beyond that (Fig. 11.27(A)). If sea level were to rise, this lateral succession of sedi-

ments would move landward, and where sand was previously being deposited mud would now be deposited (Fig. 11.27(B)). This produces a vertical succession of sedimentary layers that is identical to the lateral variation found at any instant of time. This is known as **Walther's law**. In the case of a transgressive sequence, the rocks become finer grained upward, whereas in a regressive sequence, they become coarser grained (Fig. 11.27(B)).

11.8 Sites of deposition and tectonic significance

We have seen that as sediment is transported from its source to final resting place, it may form deposits in river valleys, lakes, alluvial fans, beaches, and continental shelves. Considerable thicknesses of sediment must accumulate before the sediment is converted to rock. Sediment, therefore, needs a depression in which to accumulate, and unless that depression is very deep, such as a forearc basin, it is soon filled, unless it is able to subside and make room for more sediment. Here is where tectonics plays an important role. There are three main sites for the deposition and accumulation of thick sequences of sediment, two of which are marine and the other continental. At convergent plate boundaries, deep forearc basins receive sediment from erosion of the mountain range on the overriding plate. Divergent plate boundaries that split continents apart create passive margins that receive sediment from the continent to build a continental shelf. Basins are formed on continents by extensional rifting and strike-slip movement.

11.8.1 Convergent plate boundaries

Convergent plate boundaries have the greatest relief on Earth, passing from high mountain chains to deep oceanic trenches. The mountains are a source of an enormous amount of sediment, but little of this finds its way to the trench, because at most convergent plate boundaries it is trapped in a forearc basin, which is separated from the trench by a forearc ridge (Fig. 11.28). An **accretionary wedge**, composed of highly deformed sediments from the ocean floor, the trench slope, and older forearc basins, underlies the forearc ridge. This mixture of rocks is referred to as a **mélange** from the French word for *mixture*.

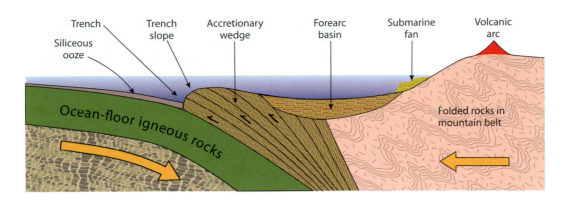

Figure 11.28 Sedimentary depositional sites at a convergent plate boundary.

Rivers flowing from the mountain ranges toward forearc basins tend to be short, so their drainage basins are relatively small. What they lack in drainage area, however, they make up for in stream velocity and energy, which results in large quantities of sediment being supplied to these basins. Because transport is rapid, sediment does not have time to mature; that is, grains tend to be angular, poorly sorted, and only slightly weathered; detrital grains of volcanic rock and igneous and metamorphic rocks from the mountain range are common. On reaching the sea, the sediment is deposited and accumulates rapidly, often forming submarine fans off the mouths of major rivers. Commonly, this sediment becomes unstable and slumps, forming turbidity currents that take the sediment into deeper parts of the basin.

The trench receives very little terrigenous sediment and contains mainly siliceous ooze formed from pelagic organisms such as radiolaria (Fig. 9.33(B)). The side of the trench bounded by the accretionary wedge is steeper than the ocean side, and slumping does occur and canyons can develop that allow sediment from the forearc basin to spill into the trench, where it forms submarine fans on the floor of the trench.

11.8.2 Passive continental margins

Passive continental margins result from the rifting apart of a continent with formation of new ocean floor between the plates. At the time of initial rifting, the land elevation is high because of the rise of hot asthenosphere beneath the rift. The Red Sea provides an example of newly formed ocean floor that is bounded by high mountains on both sides of the rift (Box 8.4). As the ocean widens and the crust cools and becomes denser, the passive margins continuously subside, providing sites for deposition of siliciclastic sediment eroded from the continent and for carbonate sediment generated along its shore. Rivers draining into passive margins tend to be long and have gentle slopes (e.g., Mississippi, Amazon), but their drainage basins are large, and hence they supply enormous quantities of sediment to passive margins. Because the rivers tend to flow slowly, sediment has time to mature and become better sorted than in rivers on convergent plate boundaries. Some rivers, such as the Mississippi, build large deltas, whereas others, such as the Amazon, have their sediment removed by ocean currents and tidal

activity. Much of the fluvial sediment, on reaching the sea, is transported along coastlines by longshore currents and drift, where it mixes with sediment formed by coastal erosion. Wave action along beaches further rounds and sorts the sediment, with the finer sediment being transported to deeper water away from shore (Fig. 11.27(A)).

Deltas are one of the major constructional sedimentary structures formed along passive margins. These can become so large that their load on the crust causes subsidence, which makes room for still more deposition. Sediment is supplied to the delta through a system of distributaries, which are constantly changing. As soon as a distributary becomes inactive, wave action goes to work on the sediment. The result is a sequence of sediments with complex cross-stratification. In addition, the surfaces of deltas are commonly covered with marshes that contribute considerable organic material to the sediment. Indeed, it was this organic material on deltas and coastal swamps that formed many of the coal deposits in Pennsylvania (Fig. 2.18).

Waves and tidal currents both play roles in transporting sediment away from coasts. On coasts where tidal action is small, waves generated during storms play the dominant role. In marine basins where tides are large, tidal currents are the dominant process. Both waves and currents transport the sediment out onto the continental shelf. If the shelf is on a passive margin that is still cooling, the shelf sediments continuously subside, making room for more. Calculations of the amount of subsidence resulting from cooling and from isostatic adjustment to sediment load indicate that a thickness of about 10 km of sediment can accumulate on a passive margin.

11.8.3 Rift and pull-apart basins

Crustal extension and strike-slip motion both lead to the formation of fault-bounded basins, which are known as **rift basins** and **pull-apart basins**, respectively (Fig. 11.29(A)–(B)). These basins are major depositional sites on continents. Their steep walls result in rapid erosion, with sediment being transferred to the floor of the basin, which in many cases has no drainage outlet. As long as the extension or strike-slip movement continues, the basins continuously deepen, making room for more sediment. Sedimentary thicknesses of many kilometers can accumulate in these basins.

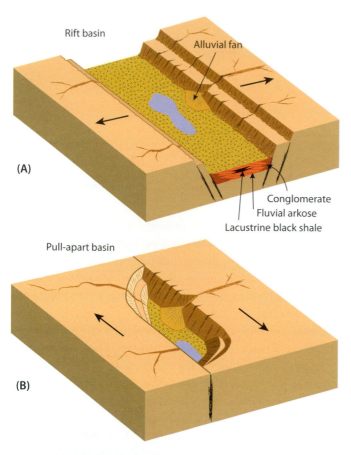

Figure 11.29 (A) Rift basins, formed by crustal extension, and (B) pull-apart basins, formed by strike-slip movement, are important sedimentation sites on continents.

Figure 11.30 Coarse conglomerate of the Jurassic Portland Formation formed in an alluvial fan along the eastern border fault of the Hartford Basin, near Durham, Connecticut.

Rapid erosion of fault scarps bounding these basins commonly develops **alluvial fans**, especially in arid climates where rivers are ephemeral and may flow only during thunderstorms. Sediment transported onto fans can be coarse grained with even boulders being moved during periods of rapid flow. As basins subside, the alluvial fan deposits are buried and turned into rock known as conglomerate (Fig. 11.30). The finer sediment is transported across fans onto the floor of the basin, where rivers transport it toward lakes, in the case of enclosed basins, or to the sea if the rift connects to the ocean. Transport distances are short, and sediment has little time to mature. Sands consequently contain a considerable amount of feldspar. When this sediment is converted to rock, it forms **feldspathic sandstone** (Fig. 11.19), which is one of the most common rock types in continental basins.

Another characteristic feature of most siliciclastic sedimentary rocks deposited in continental settings is that it is usually red, a result of staining with iron oxide. Unlike sand that is transported to deep water in marine environments where there is little oxygen (iron is reduced and sands tend to be gray to green), the fluvial deposits in rift and pull-apart basins remain in oxygen-rich environments, even after deposition when oxygen-rich ground water circulates through them. As a result, grains become coated with a layer of limonite that later dehydrates to form hematite, which gives the rock its red color (Fig. 11.26). This is one of the most common ways in which rocks that are referred to as red beds can form.

In the center of closed basins, lakes may form, especially when the climate is wet. Here the finest sediment settles out as layers of mud on the floor of the lake. Vegetation around the lake can contribute organic material to this fine sediment. If a lake becomes deep enough, the lack of oxygen in the bottom waters allows organic material to be preserved, which makes the mud black. This is one of the environments in which the rock known as **black shale** can form (Fig. 11.26(A)).

11.9 Conversion of unconsolidated sediment to sedimetary rock: lithification

The accumulation of successive layers of sediment results in compaction and expulsion of fluids from the lowest layers. At the same time grains become cemented together and the unconsolidated sediment is converted into solid rock. The initial stage of this process is known as lithification. As burial continues, pressures and temperatures rise, and still other processes help lithify the sediment. The sum of all these processes is known as **diagenesis**. At still higher temperatures, diagenetic processes grade into metamorphic processes, with the boundary between the two being arbitrarily set at about 150°C (Chapter 14).

11.9.1 Porosity and compaction

When sediment is first deposited, it has a high **porosity**; that is, void spaces constitute a large percentage of the total volume. If the

Figure 11.31 Fossilized tree replaced by silica in Triassic volcanic-ash-rich mudrock of the Chinle Formation in the Painted Desert National Park, Arizona.

only if pore fluids are able to escape, so the rate of dewatering determines the rate of compaction. As fluids rise through a compacting pile of sediment, the decrease in pressure on the fluid may lead to supersaturation and precipitation of minerals in the pores, which cements the detrital particles together and helps lithify the sediment. The related processes of compaction and dewatering are the first important steps in lithification.

11.9.2 Cementation of sediment

The pore fluid in most sediment is mainly water, but it contains considerable amounts of dissolved ions. If the sediment is marine, the pore fluid starts out as trapped seawater, and if the sediment is lacustrine it may contain large amounts of dissolved ions if the lake is saline. After deposition, these pore fluids soon become saturated in the minerals with which they are in contact. As the pore fluids rise through the compacting pile of sediment, the decreasing pressure results in a decrease in the solubility of most minerals, which causes them to precipitate and, in some cases, replace minerals that are less stable. This precipitate forms a **cement** that binds the detrital grains together. The composition of the cement depends to a large extent on the composition of the sediment undergoing compaction. For example, where sediment contains a large amount of volcanic ash, the pore fluids can become rich in silica, which cannot only cement the detrital grains together but, as shown in Figure 11.31, can completely replace and fossilize trees that may be buried in the sediment. In some cases, escaping pore fluids can dissolve cement and increase the porosity of sediment.

sediment consists of spherical grains of quartz all of the same size, its porosity is ~33%. More poorly sorted sediment would have lower porosity, because smaller grains could fit in the spaces between larger grains. If sediment consists primarily of clay particles, these small flakes, when having a random orientation, can have a porosity of up to 80%, and 60% is common in many marine sediments. Clay minerals also attract water molecules to their surface, which further helps maintain a high porosity.

The pressure toward the base of a thick pile of sediment causes **compaction**. If the sediment is composed of rounded quartz grains, only limited compaction is possible at first, but if large amounts of clay are present, the compaction can be as much as 80%, as the clay particles rotate into parallelism with the bedding planes. Compaction is possible, however,

Calcite is one of the most common cements. This is to be expected in carbonate sediments, but it is also common in quartz-rich sandstones (Fig. 11.32(A)). Calcite is one of

1 mm

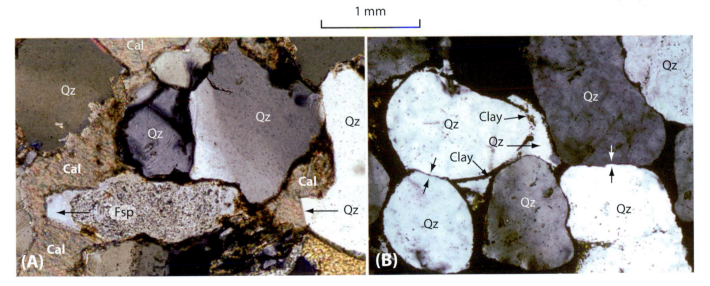

Figure 11.32 Photomicrographs of thin sections under crossed polarized light of cement in sandstone. **(A)** Feldspathic sandstone containing angular grains of quartz (Qz) and feldspar (Fsp) cemented by calcite (Cal). Both feldspar and quartz grains, where marked by arrows, have overgrowths of new clean feldspar and quartz. **(B)** Quartz-rich sandstone, in which grains of quartz had begun dissolving where they push into each other (opposing arrows), with the dissolved quartz being precipitated in the pores, where it grows onto existing grains in crystallographic continuity (single arrow). The original detrital grain boundary is marked by clay deposits on the grain surface.

the most soluble of the rock-forming minerals, and we have already seen how pressure affects its solubility (see carbonate compensation depth in Sec. 11.3.1). As pore fluids migrate upward through sediment and the pressure decreases, the solubility of CO_2 in water decreases and calcite precipitates to form cement. Quartz also forms cement in sandstones, and as can be seen in Figure 11.32(B), it grows on detrital quartz grains in crystallographic continuity. In many cases, the outline of the initial grain boundary can be seen beneath these overgrowths as inclusions of clay or iron oxide that coated the original grain. These are commonly referred to as *dust rings*.

11.9.3 Pressure solution

As sediment is buried deeper, pressure rises still higher, temperatures also begin to rise, and other processes become active that help lithify the rock. As long as the sediment remains permeable and the pore fluid is able to escape, we can think of a pile of sediments as consisting of two parts: a mass of detrital grains that rest on one another, and a pore fluid that is able to move freely between the detrital grains (high **permeability**). At the base of such a pile, there are two different pressures that can be measured. One is the pressure felt by the detrital grains resting one on top of another, and the other pressure exists in the pore fluid. These two pressures depend on the densities of the material in the solid and fluid columns. If we assume the detrital grains are quartz with a density of 2650 kg/m^3 and the fluid is brackish water with a density of 1060 kg/m^3, the pressure on the solid particles would be 2.5 times greater than that on the adjoining fluid. In addition, the pressure on the solids is not evenly distributed but is focused on those points where detrital grains contact one another. At first when grains might be relatively round, these points of contact represent a very small fraction of the total horizontal area of the sediment, and thus the pressure they experience must be multiplied by this factor. The result is that grains experience pressures that are many times larger than 2.5 times the pore pressure.

High pressures at grain-to-grain contacts cause solids to dissolve in the pore fluid and precipitate where the pressure is low in the adjoining pore spaces. This process, which is known as **pressure solution**, is one of the important processes by which sediments undergo compaction and become cemented. In Figure 11.32(B), we see the beginnings of pressure solution (the opposing arrows) where two rounded quartz grains dissolved and flattened against each other, and the dissolved quartz was deposited in the pore space to form cement. This newly precipitated quartz grew with the same crystallographic orientation as the central grain, so when observed under crossed polarized light, we see the cement and this grain are in optical continuity.

Calcareous sediment, being composed of rather soluble calcite, soon compacts into a low-porosity limestone in response to pressure solution. Even after most porosity has been eliminated, calcareous sediment may continue to compact

Figure 11.33 Horizontally bedded limestone in which vertical pressure solution generated irregular stylolites (arrows) parallel to bedding, and horizontal extension created white calcite veins along normal faults that are roughly perpendicular to bedding.

by undergoing solution along distinct planes, which are usually approximately horizontal (perpendicular to the maximum compressive stress). Fluids traveling along these planes dissolve calcite and transport it to steeply dipping fractures up which it can escape. As solution progresses, minor irregularities in the sediment cause solution to advance more rapidly in some places than in others, and the solution surface becomes extremely irregular. In addition, insoluble grains in the sediment, such as clay minerals and quartz, build up on the solution surface as insoluble residue. The resulting surface, when seen in a vertical section, looks as if it were a graph drawn with a pen or stylus, and consequently the surface is referred to as a **stylolite** (Fig. 11.33). As solutions carrying the dissolved calcite migrate up fractures, they commonly precipitate calcite to form **veins**.

11.9.4 Recrystallization, replacement, dolomitization

As temperatures increase, **recrystallization** in the solid state can occur, at least in carbonate and evaporite sediments, but quartz requires somewhat higher temperatures that are normally considered metamorphic. Recrystallization, like pressure solution, allows for the transfer of material away from points of high pressure to regions of low pressure. Carbonate grains can also change their shape by twinning. These solid-state processes further reduce porosity and help produce a hard rock from sediment.

Another important process that occurs during the diagenesis of carbonate muds is the replacement of calcite or aragonite with dolomite, which involves the substitution of Mg^{2+} in the Ca^{2+} atomic sites (Sec. 10.7). It appears that dolomite rarely forms as a primary sediment, although it may be deposited in some freshwater lakes. However, in the geologic record, rocks made of the mineral dolomite known as **dolostone**, are common. The question is, How do they form? It appears that most, if not all, are formed by a diagenetic replacement process known as **dolomitization**, in which solutions simultaneously dissolve calcium carbonate and precipitate dolomite. These solutions may come from evaporite basins in which precipitation of gypsum decreases the calcium content while increasing the magnesium content of the water. Solutions formed by the mixing of marine and fresh water may also provide the appropriate solutions to precipitate dolomite. An interesting consequence of dolomitization is that it can increase the porosity of the rock depending on the nature of the replacement reaction. This is in part because the molar volume of dolomite is 13% less than that of calcite.

Summary

The formation of sedimentary rock involves (1) the formation of sediment from the detritus of weathering, the hard parts of organisms, or chemical precipitates; (2) its transport by water, wind, or ice; (3) deposition in a basin, which is normally formed through plate tectonic processes; and (4) its compaction and cementation to form rock. The most important points covered in this chapter are the following:

- Weathering of rock is accelerated by CO_2 in the atmosphere because it combines with water to form carbonic acid. The result is that primary igneous minerals undergo weathering through a combination of oxidation, hydration, and carbonation reactions.

- Detrital grain size is an important property of sediment because it determines how easily sediment is transported.

- If grains are all of approximately the same size, sediment is said to be well sorted; if their size varies considerably, the sediment is poorly sorted.

- During transport, detrital grains are abraded and rounded. This is particularly true of wind-blown grains. Water provides a cushion that slows abrasion. Glacial sediments show the least degree of rounding. The longer sediment is transported, the better rounded and sorted it becomes.

- Many organism build hard parts of $CaCO_3$ and some of SiO_2, which can form calcareous and siliceous sediment, respectively. Many beaches in warm climates are composed of calcareous sands formed from these hard parts. The ocean floor is covered with fine sediment formed from pelagic organisms that sink to the bottom on dying. In the deepest ocean, only siliceous sediment is found, because calcareous organisms dissolve at elevated pressures.

- Accumulation of calcareous sediment to form limestone during the Phanerozoic has steadily decreased the CO_2 content of the atmosphere. Weathering of rocks through carbonation reactions also provides a sink for this greenhouse gas.

- Approximately 20% of the weathering products of igneous rocks are removed in solution. These eventually find their way to the ocean or to lakes in rift basins. The concentration of these dissolved ions is increased by evaporation in restricted marine basins or lakes, with eventual precipitation of evaporite minerals.

- Detrital sediment is transported by fluid (water, air, ice). Water and air both normally move turbulently, whereas the flow of ice is laminar. Turbulence helps suspend particles in the fluid. The rate at which particles settle through a fluid is given by Stokes' law, which shows that particle size is far more important than density in determining settling rates. As a result, sediment is sorted during transport primarily on the basis of grain size.

- Particles that are too large (or dense) to be carried in suspension are transported as bed load, which moves by a combination of rolling and jumping along the bed. Commonly, various bed forms are produced, such as ripples and dunes, depending on the flow velocity.

- Turbidity currents, which are dense suspensions of sedimentary particles in water, are important in transporting sediment to deep marine basins. They produce sedimentary beds that grade from coarse grained at the base to fine grained at the top.

- Sedimentary beds can accumulate over periods as short as hours or as long as thousands of years. Some beds can be deposited during a single storm; others may reflect seasonal variations in precipitation and may represent annual deposits (varves). Still others may reflect climatic changes over longer time periods, such as those related to the Earth's 21 000-year precession cycle (Milankovitch cycles).

- Major changes in the type of sedimentation on a still longer time scale may result from sea-level changes brought about by variations in seafloor spreading rates. When these increase, midocean ridges become larger and water is displaced onto continents, producing marine transgressions. When spreading rates slow, ridges shrink and the water drains back into the oceans producing a marine regression.

- Much of the sediment eroded from mountain ranges at convergent plate boundaries is deposited in forearc basins by turbidity currents.

- Passive margins created at divergent plate boundaries continuously subside as the crust ages and cools, making room for continuous deposition of sediment eroded from the continent.

- Crustal extension and strike-slip movement produce rift and pull-apart basins, respectively, which are sites for sedimentation on continents.

- Sediment is converted into solid rock by a combination of compaction and cementation. During compaction, grains dissolve where they impinge on one another because of high pressures at points of contact. Material removed by pressure solution precipitates in the pores where the pressure is less to form cement.

- As sediment becomes compacted and cemented, its porosity decreases. Escaping pore fluids can sometimes dissolve cement and increase porosity. Circulating groundwater can cause calcite to be replaced by dolomite. This usually involves a decrease in volume and hence an increase in porosity.

- Porosity is important in determining whether a sedimentary rock will make a good aquifer or reservoir rock for oil and natural gas.

Review questions

1. Explain how CO_2 in the atmosphere plays a role in weathering igneous rocks.

2. What are the typical weathering products of granite?

3. If sediment is well sorted, what does this tell you about its grain-size distribution?

4. What features characterize mature sediment?

5. What polymorph of $CaCO_3$ is most commonly precipitated by organisms to build their hard parts and does this change after the sediment is changed to limestone?

6. Although calcareous plankton can be found throughout the oceans, why is no calcareous ooze found on the deepest ocean floor?

7. The faunal explosion at the beginning of the Cambrian produced many organisms that built hard parts of $CaCO_3$, and limestone becomes common in the geologic record, whereas before that it is relatively rare. What effect would this faunal explosion have had on the CO_2 content of the atmosphere?

8. What is meant by the oil window?

9. Why, if NaCl is more abundant than $CaSO_4$ in seawater, is halite not more abundant than anhydrite (or gypsum) in evaporite deposits?

10. Why is glacial till so poorly sorted?

11. How can you tell from Stokes' law (Box 8.6) that grain size is more important than density in determining settling rates?

12. How might you use mud cracks in a vertically dipping folded sedimentary rock to determine the original top?

13. How might you use cross-bedding in a vertically dipping folded sedimentary rock to determine the original top?

14. How might you use a turbidity current in a vertically dipping folded sedimentary rock to determine the original top?

15. What are Milankovitch cycles, and how might they be recognized in a sequence of sedimentary rocks?

16. What would a fining upward sequence of sediments tell you about changes in sea level?

17. Why can passive continental margins have thick sequences of sediments?

18. Why do sediments in rift and pull-apart basins lack maturity?

19. Explain the role of pressure solution in the compaction and cementing of sediment.

ONLINE RESOURCES

The Society for Sedimentary Geology's Web site (http://sepmstrata.org) provides useful information about depositional environments and sedimentary petrology. It includes PowerPoint lectures on many topics related to sedimentary rocks.

Animations of the formation of bedforms and cross-bedding are given at the U.S. Geological Survey's Web site (http://walrus.wr.usgs.gov/seds/bedforms/index.html).

Videos of sedimentary processes are given at Paul Heller's (University of Wyoming) Web site (http://faculty.gg.uwyo.edu/heller/sed_video_downloads.htm).

FURTHER READING

Gould, S. J. (1989). *Wonderful Life: The Burgess Shale and the Nature of History*, W. W. Norton and Company, New York. This popular book gives a full account of the explosion of life forms that occurred in the Cambrian Period, which lead to a change in the nature of carbonate sedimentary rocks in the Phanerozoic.

Blatt, H., Tracy, R. J., and Owens, B. E. (2006). *Petrology*, W. H. Freeman and Company, San Francisco. This book provides an excellent general coverage of sedimentary processes and rocks.

Boggs, S., Jr. (2001) *Principles of Sedimentology and Stratigraphy*, Prentice Hall, Upper Saddle River, NJ. This book provides a good introduction to the physical characteristics of sedimentary deposits, their environments of deposition, and their stratigraphical relationships with each other.

Boggs, S., Jr. (2009). *Petrology of Sedimentary Rocks*, 2nd ed., Cambridge University Press, Cambridge. This book provides an in-depth coverage of sedimentary rocks.

Klein, C. (2005). Some Precambrian banded iron-formations (BIFs) from around the world: Their age, geologic setting, mineralogy, metamorphism, geochemistry, and origin. *American Mineralogist*, 90, 1473–1499. This article provides a useful survey of Precambrian banded iron-formations.

12 Sedimentary rock classification, occurrence, and plate tectonic significance

In the previous chapter, we looked at how sediment is formed, transported, deposited, and lithified to form rock. In this chapter, we discuss the classification of sedimentary rocks, which fall into three main groups: (1) the siliciclastic sedimentary rocks, which include mudrocks (shale), sandstones, and conglomerates; (2) the biogenic sedimentary rocks, which include limestones, dolostones, cherts, and coals; and (3) the chemical sedimentary rocks, which include evaporites, phosphorites, and iron-formations. We learn how each type can be identified and examine the textures of the most important types in photomicrographs of petrographic thin sections. Correct identification of sedimentary rocks is the first step to working out the geologic history of an area, because these rocks preserve a record of the source of sediment, its mode of transport, and its site of deposition, which in many cases is determined by plate tectonic processes. We examine the geologic record preserved in some specific sedimentary rock sequences. Sedimentary rocks are the source of all fossil fuels, most iron ore, material used for fertilizers, much of the world's gold ore, and many other placer mineral deposits. Most of the world's construction material comes from sediments and sedimentary rocks in the form of building stones and raw material (lime and aggregate) for making cement. Porous sedimentary rocks form important aquifers both for agricultural use and for sustaining urban areas. In this chapter, we indicate with which sedimentary rocks each of these economically important resources is associated. We also discuss the plate tectonic setting in which each sedimentary rock is formed.

Outcrop of the upper part of the late Triassic Chinle Group, consisting of alternating beds of fluvial sandstone and siltstone, capped by the lighter-colored massive bed of Middle Jurassic Entrada Sandstone, which is an eolian deposit. Light-colored layers in the Chinle sandstone are carbonate-rich caliche deposits formed by groundwater evaporation in fossil soil horizons. These rocks, which crop out near Ghost Ranch, west of Abiquiu, New Mexico, record a change in climate from semiarid to arid.

Sedimentary rocks are formed from the lithification of sediment (Fig. 11.1). They are classified into three main groups based on the type of sediment from which they are formed. Those formed from the detritus of weathered rocks are called **siliciclastic** because most of their minerals are silicates. Those formed from sediment derived from organisms are described as **biogenic**, and those formed by chemical precipitation are referred to as **chemical sedimentary rocks**.

Each of these categories is subdivided on the basis of grain size, composition, or mode of deposition. Siliciclastic sedimentary rocks are divided into **mudrocks**, **sandstones**, and **conglomerates** on the basis of grain size. The biogenic rocks are divided into **carbonates**, **cherts**, and **coals** on the basis of composition. And chemical sedimentary rocks are subdivided into **evaporites**, **carbonates**, **phosphorites**, and **banded iron-formation** on the basis of composition and mode of deposition. Each of these divisions is further subdivided into the main sedimentary rocks. Some rocks may belong to more than one of the main groups. For example, some limestones are formed from both biogenic sediment and chemical precipitates, and some organic-rich rocks such as oil shales contain considerable amounts of siliciclastic sediment. Despite these overlaps, the tripartite division provides a convenient framework in which to discuss the main types of sedimentary rock. In this chapter, we discuss the main types of sedimentary rocks, their depositional sites, and plate tectonic significance.

12.1 Siliciclastic sedimentary rocks

Siliciclastic sedimentary rocks are the most abundant of the sedimentary rocks. They are formed from the detritus left over from the weathering of igneous, metamorphic, and older sedimentary rocks. They may even include pyroclastic particles, if active volcanoes are present in their source region. The composition of the rock depends on the types of rock in the source region, the type of weathering (e.g., tropical, high latitude), and the separation of grains and minerals during transport. Most siliciclastic sediment is derived from continents and is, therefore, sometimes referred to as *terrestrial siliciclastic sediment*.

During transport, especially by water, the mud-size particles go into suspension and tend to be separated from the sand-size particles, which travel in the bedload or as desert sand, and still coarser particles may be left behind or moved only in high-energy environments. This separation leads to the division of siliciclastic sedimentary rocks into three main types: *mudrocks*, *sandstones*, and *conglomerates*. We discuss each of these rock types separately, even though there are times when they may grade from one into another.

12.1.1 Mudrocks (includes shales)

Mudrocks are formed from siliciclastic sediment with a grain size less than 0.0625 mm. If most grains are coarser than

0.004 mm, the rock is called a **siltstone**, and if most are finer, it is called a **claystone**. Mudrocks with a high clay mineral content tend to split into thin sheets parallel to the bedding. This **fissility** characterizes mudrocks known as **shales** (Fig. 12.1(A)), which are the most abundant of the mudrocks. Mudrocks are also referred to as **mudstones**, but this term is also used for fine-grained limestone (Sec. 12.2.1), so it is preferable to use mudrock for siliciclastic sedimentary rocks to avoid confusion.

Mudrocks constitute about two-thirds of all sedimentary rocks and are, therefore, by far the most abundant. They are of great economic importance as the source of organic material from which petroleum and natural gas are formed, and their low permeability makes them excellent cap rocks for oil reservoirs and for confining aquifers.

Most mudrocks are composed of variable amounts of detrital quartz and clay minerals formed from the weathering of rock. They may also contain mica grains, if they are present in the source region (e.g., low-grade metamorphic terranes). Siltstones tend to contain higher percentages of quartz and claystones higher percentages of clay minerals. Because of the fine grain size of mudrocks, individual minerals are difficult to distinguish in the hand specimen, but a simple field test that allows you to estimate the amount of detrital quartz is to bite a small piece between your teeth. If the rock contains more than

(A)

(B)   1 mm

Figure 12.1 (**A**) Fissile black shale in the Jurassic East Berlin Formation of the Hartford Basin, Connecticut, formed from mud deposited in a deep lake where conditions were anoxic (see Fig. 11.26(A) for outcrop photograph). (**B**) Photomicrograph of a thin section of shale shown in (A). Small clear grains are quartz; dark-brown laminations are organic-rich clay; light-brown laminations contain carbonate.

one-third quartz, it will feel gritty; otherwise, it feels smooth. This test works surprisingly well.

The sediment from which mudrocks are formed, especially that in the clay-size range, remains suspended in water, even when it is flowing extremely slowly (Sec. 11.6.2). As a result, clay-size particles settle only in water that has essentially stopped moving. In the ocean, this occurs below **wave base**, the effective depth to which wave action agitates water (wave base ≈ ½ wave length). On continents, stationary water is found below wave base in lakes and in temporarily ponded water on river floodplains. In these environments, the mud is commonly deposited with prominent near-horizontal bedding.

The beds in mudrocks can vary from massive to finely layered or **laminated**. When small clay platelets settle, they naturally fall with their (001) face parallel to the horizontal plane (try dropping a sheet of paper and see how it lands on the floor). During compaction, this alignment becomes even more pronounced and gives the rock a prominent horizontal foliation or fissility. The fissility shown by the shale in Figure 12.1(A) can be seen in thin section under the microscope (Fig. 12.1(B)) to result from clay-rich **laminations** alternating with laminations richer in detrital quartz grains and carbonate, which may be a chemical precipitate. Laminations can result from fluctuations in the type of sediment entering a depositional basin or to climatic changes. The rhythmic laminations in Figure 12.1(B) most likely represent annual deposits and can, therefore, be referred to as varves.

Not all mudrocks are fissile, and many show no laminations at all. Burrowing organisms in mud can rapidly destroy laminations, which is one form of the process known as **bioturbation**. The layering shown in Figure 12.1(B) was preserved because the mud accumulated in a deep lake where there was insufficient oxygen for organisms to live (anoxic). In contrast, the mudrock shown in Figure 12.2 shows no fine-scale layering. This rock was formed from mud that accumulated at the base of the Devonian Catskill Delta, as it built westward with sediment derived from the rapidly rising Acadian orogenic belt (Fig. 12.3). This mud was probably bioturbated at first but then slumped into the deeper water of the Appalachian Basin, where anoxic conditions preserved its organic content, making it a potential source of natural gas today (Sec. 16.8.1).

Sediments containing a large percentage of clay minerals can have extremely high water contents (80 volume %; Sec. 11.9.1), which makes them extremely soft and susceptible to slumping. It is not uncommon in evenly bedded mudrocks to find a single layer in which the bedding has been contorted as a result of slumping.

The mineral grains constituting mudrocks can be seen only under the microscope, and even then only the coarsest particles are distinguishable. Quartz grains tend to be angular (Fig. 12.2) because they are carried in suspension and do not experience the abrasion felt by sand grains that move in the bedload or in wind. The clay minerals have grain sizes less than 0.004 mm and can, therefore, be identified only by scanning electron

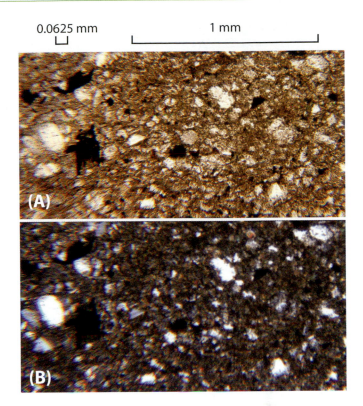

0.0625 mm   1 mm

(A)

(B)

Figure 12.2 Massive silty mudrock of the Marcellus Formation, New York State, as seen in a thin section under (**A**) plane and (**B**) crossed polarized light. The clear grains, some of which are slightly coarser than the 0.0625 mm limit for silt-size particles are quartz. The regional distribution of the Marcellus shale is shown in Figure 12.3.

microscopy or by X-ray diffraction. Under the microscope, some mudrocks can be seen to contain opaque or dark brown material, which is commonly streaked out parallel to the bedding (Fig. 12.1(B)). This is fossilized organic material referred to as **kerogen**. It does not have a fixed composition or crystal structure and, therefore, does not fit the definition of a mineral, even though it is naturally occurring (Sec. 2.1). It is commonly referred to as a mineraloid. Petroleum and natural gas are formed from kerogen by heating into the oil and natural gas windows (Sec. 11.3.2), but if its temperature goes above this, it is converted to graphite, which is also opaque. Decaying organic matter gives off hydrogen sulfide gas (the source of the rotten egg smell), which reacts with iron to form pyrite, another common opaque constituent of mudrocks. When pyrite-bearing mudrocks are exposed to weathering, the pyrite breaks down (oxidizes) to form iron oxides, limonite, or hematite, and sulfuric acid (Eq. 10.4), with limonite coating the bedding planes and fracture surfaces where water has circulated.

Mudrocks range in color from black and gray through green to brown and red. Black and gray colors indicate the presence of organic matter, which keeps the concentration of oxygen low enough that iron remains in its ferrous state. When oxygen concentrations are higher, limonite can make mudrock brown, and if hematite is present, the rock will be red. If iron oxides are not present in significant amounts, mudrocks tend to be

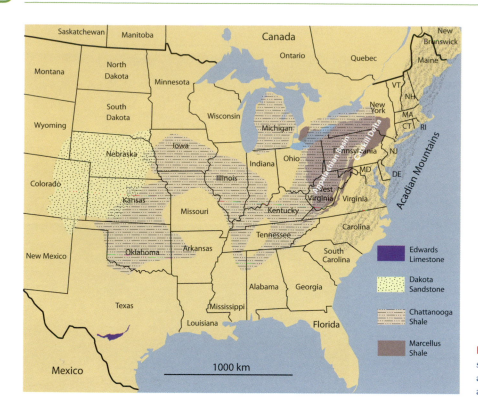

Figure 12.3 Distribution of two important black shale formations, the Marcellus and the Chattanooga, and two important aquifers, the Dakota Sandstone and the Edwards Limestone.

gray green. The variation in color from black to red indicates the increasing availability of oxygen, which typically reflects shallower water depths (Fig. 11.26(A)), but some deep water can have significant amounts of oxygen if water circulation is strong. Mudrocks deposited in shallow water are typically red and commonly contain mud cracks, which indicate that the water in which they were deposited evaporated.

The coarser mudrocks, which we refer to as siltstones, are commonly associated, and interlayered, with fluvial sandstones. Figure 12.12 shows a vertical section through a typical example of such a series of beds in the Lower Triassic Moenkopi Formation at Capitol Reef National Park, Utah. These beds consist of light-colored medium- to coarse-grained sandstones, medium-colored siltstones, and dark recessively weathered shale. The sandstones show small-scale cross-beds, channel structures, and ripple marks, and the mudrocks have mud cracks that are filled with sand from the overlying bed. The red color of the rocks is produced by hematite, which indicates oxidizing conditions during deposition. All these features are characteristic of fluvial deposition in shallow water. The Moenkopi Formation was deposited from rivers that meandered across a flat coastal plane on the western coast of North America during the early Triassic (Box 12.1).

Tectonic setting for deposition of mudrocks

Mudrocks have numerous sites of deposition. The thickest accumulations occur in forearc and foreland basins, where they are associated with turbidity current deposits (**turbidites**). When the small continental plate of **Avalonia** collided with eastern North America ~400 million years ago, sediments eroded from the newly formed **Acadian Mountains** spread westward into the **Appalachian foreland basin** forming the **Catskill Delta** (Fig. 12.3). The finest sediment was transported to the deep water of the basin, where anoxic conditions preserved organic-rich shales. The middle Devonian Marcellus black shale shown in Figure 12.2 formed at the base of this succession.

Muds also accumulate in deeper water off the coast of passive margins. Some of these muds can be transported to still greater depth on the continental shelf or slope, or abyssal plain, by turbidity currents.

Not all marine mud is deposited in deep water. Some has been deposited in shallow epeiric seas. The late Devonian Chattanooga Shale, named from its occurrence in Tennessee, is one of the most laterally extensive sedimentary units in the world, despite being no more than 10 m thick. It, and its correlatives, can be traced over thousands of square kilometers in the central part of the North American continent (Fig. 12.3), where it was deposited from a shallow sea that is believed to have been no more than 30 m deep. The Chattanooga Shale, while containing about 25% detrital quartz, does contain large amounts of organic material, so conditions must have been relatively anoxic on the floor of the shallow sea.

Another important depositional site for mud is in the lakes that form in rift valleys and pull-apart basins. Excellent modern examples of such lakes are found along Africa's Eastern Rift Valley (see Box 8.4) and in Death Valley in the United States. If the drainage is entirely internal, as in Death Valley, all sediment entering the basin must eventually accumulate there. Lacustrine mudrocks may be interlayered with evaporite deposits if the climate is arid.

Figure 12.4 Gently dipping beds of ~130 000-year-old tuff interlayered with fluvial mudrocks in the Kenya Rift Valley at Silali. (Photograph by Christian Tryon.)

Because rift valleys are also plate tectonic settings where igneous activity can be expected, lacustrine mudrocks may have associated igneous rocks. Along the East African Rift Valley system, frequent explosive volcanic activity has resulted in the deposition of layers of ash along with fluvial and lacustrine mudrocks (Fig. 12.4).

12.1.2 Sandstones

Sandstones are siliciclastic sedimentary rocks formed from sand-size detrital grains (0.0625–2.0 mm) that are predominantly quartz but may include feldspars and small fragments (clasts) of other rocks. They constitute ~25% of all sedimentary rocks and are, therefore, second to mudrocks in abundance, but because of their resistance to weathering, they are commonly the most visible sedimentary rock (Fig. 12.5). Sandstones form under all climatic conditions and, therefore, have a worldwide distribution.

Figure 12.5 Cliffs of wind-blown Permian de Chelly Sandstone in the Canyon de Chelly, Arizona. Large cross-beds were formed from migrating dunes. The dark vertical lines are formed by precipitation of iron and manganese oxides from water dripping down the cliff. The Anasazi (ancient people) cliff dwelling known as the White House was built in a recess weathered out along the base of one of the migrating dunes.

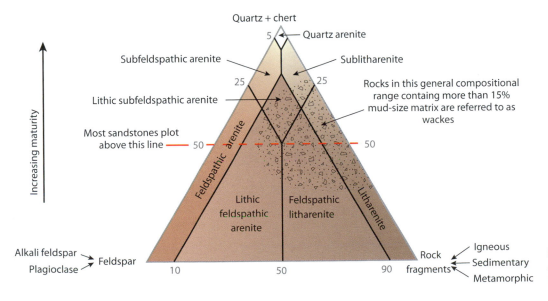

Figure 12.6 Classification of sandstones (after E. F. McBride 1963).

BOX 12.1 | THE GEOLOGIC RECORD PRESERVED IN A SUITE OF SILICICLASTIC SEDIMENTARY ROCKS

Siliciclastic sedimentary rocks are formed worldwide under all climatic conditions in rivers, estuaries, tidal flats, deltas, beaches, continental shelves, continental slopes, deep ocean basins, deserts, and glaciated regions. They have the widest distribution of all sedimentary rocks and carry a wealth of information about the source region of the sediment (**provenance**), the climate, and site of deposition. The relief necessary to cause erosion and transport of sediment and create sites of deposition is intimately related to plate tectonic processes. The composition, textures, and structures of siliciclastic sedimentary rocks can, therefore, provide a glimpse into these past tectonic processes. We present here a brief summary of the record preserved in a thick sequence of Mesozoic siliciclastic sedimentary rocks exposed in Capitol Reef National Park, Utah.

During the late Triassic, the western coast of North America was the site of a volcanic arc located approximately where California is today (Fig. A). To the east of this lay a back-arc basin, which received sediment from the arc and from uplifted areas to the east. Drainage in the basin was to the north, in general, where the basin was open to the sea. This basin was the depositional site of the Upper Triassic Chinle Formation, one of the formations exposed in the 160 km-long monocline fold in Capitol Reef National Park (Fig. B). These rocks preserve a record of plate convergence, orogenesis, volcanism, back-arc subsidence, and eventually climatic changes caused by global repositioning of tectonic plates.

The sedimentary rocks exposed in this cliff are, from the base up, the Lower Triassic Moenkopi Formation (Mk); the late Triassic Chinle Formation (C), which consists of Shinarump Conglomerate (S), the green Monitor Butte Member (M), the Petrified Forest Member (P), and the Owl Rock Member (O); and the early Jurassic Wingate Sandstone (W), capping the cliff.

The Chinle Formation unconformably overlies the Lower Triassic Moenkopi Formation (Fig. 12.14), and its base is marked by the quartz-rich Shinarump Conglomerate. This conglomerate fills valleys eroded into the underlying rocks and is, therefore, of variable thickness (0–27 m). Deposition of the conglomerate occurred when valley downcutting was stopped by a rise in the regional drainage base level. With the continued rise in the base level, rivers flowed more slowly and lakes and swamps developed; in them was deposited the grayish-green Monitor Butte claystone. This unit contains an abundance of bentonite clay, which is formed from the weathering of volcanic ash. The Monitor Butte Member, therefore, records the birth of the volcanic arc to the west (Fig. A). The color of this unit is due to its content of organic material, which keeps iron in the reduced state. This claystone is overlain by the lavender-colored Petrified Forest member, which consists of fluvial sandstones and overbank mudrocks containing an abundance of bentonite clay. This is the same unit in which the petrified trees are preserved in the Painted Desert National Park, Arizona (Fig. 11.31). The fossils in this unit indicate a wet tropical climate. The Petrified Forest member grades upward into the Owl Rock member, which consists of interlayered fluvial sandstones, mudrocks, and lacustrine limestones, the oscillations reflecting climatic fluctuations. The Chinle Formation is overlain by the eolian Lower Jurassic Wingate Sandstone, whose large cross-beds, formed by migrating dunes, record a significant change in climate. Paleomagnetic measurements indicate that this region was located between 15° and 18° north latitude during deposition of the Chinle Formation. This would have given it a wet tropical climate, which would account for its fluvial deposits and fossil plants. By the Jurassic, however, motion of the North American plate had taken this region northward into latitudes where deserts are common (20° to 30°), accounting for the eolian deposits of the Wingate Formation.

Sandstones are classified on the basis of their detrital mineral grains, which are divided into three categories: quartz (including chert), feldspars (alkali and plagioclase), and rock or lithic fragments (sedimentary, metamorphic, igneous). The percentages of these components and the names given to specific sandstone types are shown in the triangular plot of Figure 12.6 (for discussion of such plots, see Sec. 7.3). A fine-grained siliciclastic matrix may also be present but normally constitutes less than 15% (Fig. 12.7). Some sandstones, however, contain more than 15% matrix; we think of these as "dirty" sandstones. The type of cement that binds detrital grains in sandstone together is not used in their classification but may appear as a qualifier in the rock name.

The abundance of the three main types of detrital grains in sandstone depends on the composition of rocks in the source region of the sediment and the distance and time the sediment was transported prior to deposition. Quartz and chert are the most durable minerals and hence tend to become more abundant with sediment transport. The other two components are less stable and tend to be removed from the sediment with transport. Thus, in going from the bottom to the top of Figure 12.6, sandstones show what is referred to as increasing **compositional maturity**. At the same time, they are likely to show increasing textural maturity (Sec. 11.2.4).

If we think of the primary igneous rocks making up continents as having a granodiorite composition, the detritus formed from their weathering would initially contain about 30% quartz. Following transport, the quartz content of the sediment increases because of continued weathering of feldspar and abrasion of soft minerals. As a consequence, most sandstones contain more than 50% quartz, and the more mature they become, the higher is this percentage. Most sandstones, therefore, fall in the upper half of the triangular plot of Figure 12.6.

Sandstones are subdivided into a number of different types based on their position in Figure 12.6. Some of the rock names make use of the word **arenite**, which comes from the Latin for

1 mm

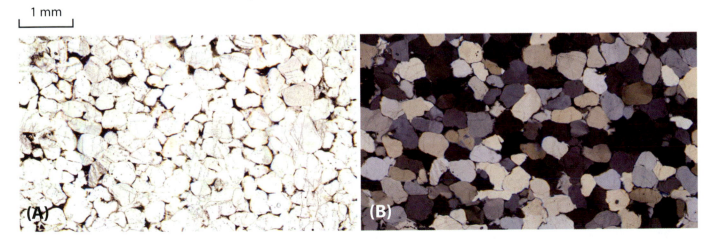

Figure 12.7 Photomicrographs of a thin section of the Potsdam Sandstone from Ausable Chasm, New York, as seen under (A) plane and (B) crossed polarized light. A small amount of dusty matrix composed of clay and iron oxides outlines the original detrital quartz grains. An outcrop photograph of this sandstone is shown in Figure 2.21(G).

sand. Thus **quartz arenite** is sandstone composed largely of quartz, whereas **litharenite** (or **lithic arenite**) is sandstone containing more than 25% lithic (rock) fragments. Sandstones containing abundant feldspar (usually the more weathering-resistant alkali feldspar) are known as **feldspathic arenites** (or more informally as **arkose** or arkosic arenite). Sandstones that contain more than 15% mud are called **wackes** (or **graywackes**).

Each of these types of sandstone has specific sites of deposition, which commonly has plate tectonic significance. Identification of the type of sandstone can, therefore, be of importance in working out the geological history of a region. We now discuss each of these sandstone types.

Quartz arenites

Sandstones that contain less than 5% feldspar or lithic fragments are called quartz arenites (Fig. 12.6). These are **mature** sandstones that have had almost all minerals other than durable quartz and chert removed by long periods of abrasion, which in most cases indicates they have been exposed to wave action along beaches or wind in deserts for extended periods of time. Transport in rivers is usually not sufficient to eliminate all of the softer minerals. Sandstones may contain small amounts of resistant dense minerals, such as magnetite, zircon, and tourmaline. Grains in these mature sandstones are usually subrounded but can be well rounded in eolian sandstones.

Quartz arenites show varying degrees of **cementation** from highly porous sandstones, with as much as 20% pore space, to ones with zero porosity. Both quartz and calcite are common cements, but limonite, hematite, gypsum, and clay minerals also occur, and two or more types of cement may be present in a given rock. As discussed in Section 11.9.3, quartz dissolves at points of grain contact where the pressure is high, and it is redeposited in the pores where the pressure is less. It normally is deposited on the surface of existing detrital grains with which it grows in crystallographic and optical continuity. Many detrital quartz grains, especially those derived from igneous rocks, appear dusty because of the presence of minute fluid inclusions. The quartz cement that grows onto these grains is relatively free of inclusions and, therefore, appears clear (Fig. 12.8). Carbonate cements are also common, which are usually composed of calcite, but dolomite or ankerite can also occur. Circulating groundwater can dissolve carbonate cements in sandstones (**decementation**). Figure 12.9 shows the Roubidoux quartz arenite of Missouri, which is cemented by calcite. In parts of this formation, solution of the cement has increased its porosity and made it an important aquifer.

Quartz arenites are mature sandstones that are commonly formed along beaches surrounding continents that have been tectonically stable for lengthy periods – so-called cratons. Most of the Canadian Shield, for example, remained tectonically stable from the end of the Grenville orogeny (1 billion years ago) until the beginning of the Paleozoic Era (545 million years ago). During this lengthy period, mature sands were formed along its beaches. When the craton began rifting apart early in the Paleozoic, this sand was deposited on the subsiding **passive margins**, to form quartz arenites, such as the Cambrian Potsdam Sandstone of northern New York State and southern Quebec (Figs. 2.21(G) and 12.7). In a similar way, but at a slightly later time, the Ordovician Table Mountain Sandstone was deposited unconformably on the Precambrian basement rocks near Cape Town, South Africa (Fig. 12.10). Quartz arenites are also formed from desert sands such as the Permian de Chelly Sandstone (Fig. 12.5) and Jurassic Navajo Sandstone (Fig. 11.17(B)) in Arizona and Utah.

1 mm

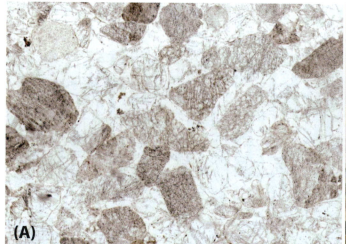

(A)   **(B)**

Figure 12.8 Photomicrographs of a thin section of Pennsylvanian Tuscarora Sandstone from Bald Eagle Mountain, Pennsylvania under **(A)** plane and **(B)** crossed polarized light. The detrital quartz grains in this quartz arenite are dusted with inclusions, whereas the quartz cement, which grows onto the grains in optical continuity, is free of inclusions.

1 mm

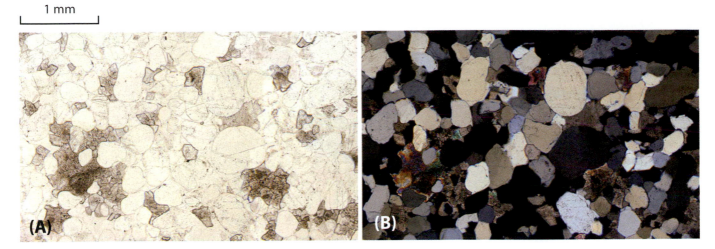

(A) (B)

Figure 12.9 Photomicrographs of a thin section of Roubidoux Sandstone from Cole County, Missouri, as seen under (**A**) plane and (**B**) crossed polarized light. This quartz arenite is cemented by calcite, which in plane light appears dusty and under crossed polars has high (pastel) interference colors.

Figure 12.10 Ordovician sandstones of the Table Mountain Group (450 million years old) unconformably overlying Precambrian Cape Granite (630 million years old), at Chapman's Peak, South Africa. These sandstones were formed on tidal flats and marginal marine deltas. Charles Darwin visited this site in 1844 while on the voyage of the *Beagle*. (Photograph courtesy of Grant Cawthorn.)

Feldspathic arenites (arkose)

Sandstones containing more than 25% feldspar are known as feldspathic arenites (informally called arkose), and those with between 25% and 5% feldspar are referred to as subfeldspathic arenites. Their grains are usually quite angular, indicating relatively short periods of sediment transport before deposition (Fig. 12.11). Although plagioclase may be present, the main feldspar is potassium feldspar, usually of the variety known as microcline, which shows characteristic grid twinning under crossed polars (Fig. 12.11(B)). Because microcline is commonly pink, most feldspathic arenite has a pinkish hue, which is made still redder when grain surfaces are coated with hematite (Fig. 11.19). Micas may also be present, and although muscovite may remain stable, biotite is usually oxidized. The mineralogical makeup of feldspathic arenite is very close to that of granite or granodiorite, which must, therefore, be present in its source region.

The composition and common lack of rounding of grains indicates that feldspathic arenites are immature sedimentary rocks. Their bedding features commonly indicate deposition in fluvial settings or on alluvial fans. As rivers meander across flood planes, sediment is eroded from the outside and redeposited on the inside of meanders as point bars. These deposits commonly show cross-bedding, a feature present in many feldspathic arenites. A meandering river may be kept in its channel by natural levees, which are banks of sand deposited along the side of a river during floods. Rivers may breach these levees during floods, and the bedload escaping through the breach fans out and is rapidly deposited as a crevasse splay, which is characterized by prominent foreset cross-beds, another common feature in feldspathic arenites (Fig. 11.19). Water that escapes from a river channel during floods spreads out across the floodplain, where it forms temporary ponds in which the suspended load is deposited as layers of mud. Many feldspathic arenites, such as those in the Moenkopi Formation in Utah (Fig. 12.12) and the Chinle Group in New Mexico (see chapter-opening photograph),

1 mm

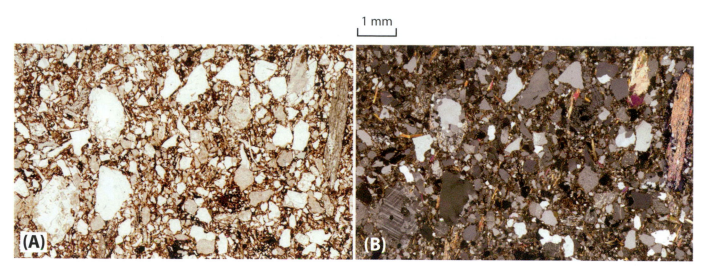

Figure 12.11 Photomicrographs of a thin section of feldspathic arenite of the Portland Formation from the Hartford Basin, Connecticut, as seen under (**A**) plane and (**B**) crossed polarized light. It contains angular grains of quartz, microcline feldspar (grid twinning), highly birefringent flakes of muscovite, and fragments of muscovite schist (upper right). Outcrop photograph shown in Figure 11.19.

Figure 12.12 Reddish-brown Lower Triassic Moenkopi feldspathic arenites and mudrocks unconformably overlain by the white Late Triassic Shinarump Conglomerate at Capitol Reef National Park, Utah.

are interlayered with mudrocks that are formed from these overbank deposits.

Lithic arenites

Sandstones containing more than 25% lithic clasts and less than 15% mud matrix are named **lithic arenites**. In contrast to feldspathic arenites, lithic arenites contain a greater abundance of sedimentary, igneous (usually volcanic), and low-grade metamorphic rock fragments than feldspar, and most do not contain potassium feldspar. Quartz grains are subangular to rounded and are usually cemented with quartz or calcite. Organic material is common, which keeps iron in a reduced state and gives most lithic arenites a gray color. Their bedding structures often indicate deposition in fluvial environments.

The immature character of lithic arenites indicates rapid erosion and deposition, which requires high relief, but the presence of clasts of volcanic and low-grade metamorphic rocks and the lack of potassium feldspar, which is unstable in low-grade metamorphic rocks, indicate source regions that have not been deeply eroded. Lithic arenites are, therefore, believed to form from the erosion of young mountain belts. As the mountains become more deeply eroded, plutonic igneous and higher-grade metamorphic rocks become available in the source region, and lithic arenites pass into feldspathic arenite.

Wackes (graywacke)

Sandstones that contain more than 15% mud matrix fit into a loosely defined group of rocks known as wackes (in the past often referred to as graywackes). These are dark-colored sandstones that are easily mistaken in hand specimens for basalt. They are always of marine origin, and most show graded bedding (Fig. 11.23), which is characteristic of deposition from turbidity currents. Many are interlayered with deepwater black shales (Fig. 14.6(A)). At the base of many wacke beds, smooth elongate depressions known as flutes, sole markings, or load casts are formed where the weight of the wacke deformed the underlying soft mud, their elongation coinciding with the flow direction of the turbidity current.

The detrital grains consist of quartz, plagioclase, and lithic fragments of fine-grained mudrocks, but fine-grained volcanic and metamorphic rocks may also be present. The mudrock particles are often difficult to distinguish from the mud matrix because they wrap around framework grains (Fig. 12.13). Cement is usually absent, and the rock is held together by the fine-grained matrix. The plagioclase in wackes is usually sodic, which is similar to the composition of plagioclase in the commonly associated hydrothermally altered pillow basalts known as spilites (Sec. 9.4.1), where the conversion of more calcic plagioclase to albite most likely is a result of reaction with sodium in seawater. Most wackes occur in regions where they have undergone some deformation and are, in fact, low-grade metamorphic rocks (metawackes). The associated mudrocks, for example, typically show the incipient development of slaty cleavage (Figs. 11.23 and 14.6(A)). The clay minerals in the mud matrix, when examined under high magnification or when studied by X-ray powder diffraction, are seen to have been converted to chlorite and an extremely fine-grained muscovite known as sericite, which gives the rock a dark green or gray color.

The composition and textures of wackes lead to the conclusion that these sandstones are formed at convergent plate boundaries, where young orogenic belts provide the necessary relief for the erosion of sedimentary, volcanic, and low-grade metamorphic rocks with rapid transport of sediment to deep sedimentary basins. As they accumulate in these basins along with black shales, they become deformed and slightly metamorphosed by continued plate convergence.

1 mm

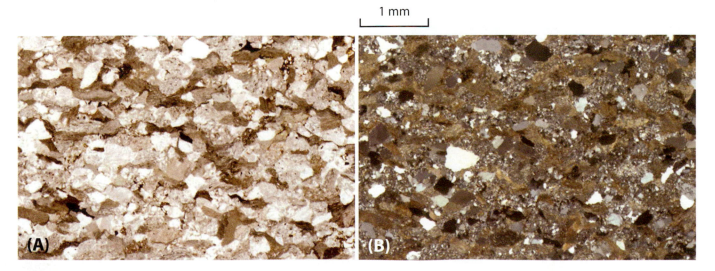

Figure 12.13 Photomicrographs of a thin section of Normanskill wacke from the Hudson Valley, New York, as seen under (**A**) plane and (**B**) crossed polarized light. This Late Ordovician sandstone contains angular quartz grains and many small flat lithic fragments of extremely fine-grained siltstone and brownish claystone that wrap around quartz grains. The rock contains abundant fine-grained matrix.

12.1.3 Conglomerates and breccias

Sedimentary rocks composed predominantly of gravel-size particles are known as conglomerates (Fig. 12.14). If the particles are particularly angular, the rock is called a **breccia**, or sedimentary breccia, to distinguish it from igneous or fault breccias. Most conglomerates are siliciclastic rocks, but some are composed of limestone fragments.

Conglomerates are formed in a number of high-energy environments. Some are formed as gravel beds in rapidly flowing rivers or on alluvial fans (Fig. 11.32). Water must flow rapidly to move gravel-size particles, but some conglomerates are formed from gravel that is the residue left over after slower-moving water has washed away the finer particles. These are known as lag deposits. Conglomerates are also formed from sediment generated along marine beaches where there is a nearby source of rock fragments, such as a cliff. These gravels grade rapidly into finer-grained sand across the width of a beach, but during transgressions and regressions of the sea (Fig. 11.27(B)), laterally extensive beds of conglomerate can result. Conglomerates are also formed from the sediment on submarine fans and in marine debris flows and mudslides. Lithified glacial till also produces conglomerate.

Conglomerates can be divided into two general types based on the relation of their coarser particles (clasts) to the finer-grained material between the clasts, which is referred to as matrix. In **clast-supported** conglomerates, the clasts rest against each other (Fig. 12.15), whereas in **matrix-supported** conglomerates, the clasts do not touch each other but instead are surrounded by matrix. On a beach, for example, wave action separates sand from pebbles, so a conglomerate formed in this environment would have pebbles resting against each other and hence would be clast supported. Some sand may wash in among the pebbles (Fig. 12.15), but the pebbles would

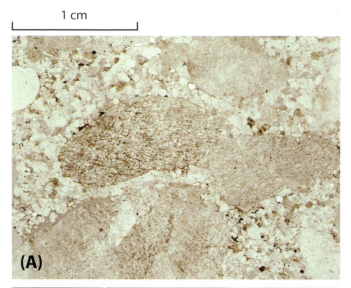

Figure 12.15 Photomicrographs of a thin section of the Olean oligomict clast-supported quartz conglomerate from Olean, New York, as seen under (**A**) plane and (**B**) crossed polarized light.

still be seen to rest against each other. In contrast, the space between the pebbles may be left empty to be later filled by cement (Fig. 12.16). Contrast this with conglomerate formed from glacial till, mudslides, or lahars, where the clasts would have been carried along by the finer sediment, so the clasts need not touch one another and the conglomerate would be matrix-supported.

If a conglomerate consists of clasts of just one rock type (Fig. 12.15), it is described as **oligomict**, from the Greek words *oligos*, meaning "little" or "scant," and *miktos*, meaning "mixed." If it contains two or more different rock types (Fig. 12.16), it is described as **polymict**, from the Greek word *polis*, meaning "many." Polymictic conglomerates are less mature than oligomictic conglomerates, and their clasts tend to be less rounded (compare the shapes of clasts in Figs. 12.15 and 12.16). Quartz-rich conglomerates commonly form on unconformities where their distribution and thickness can be highly

Figure 12.14 Interlayered fluvial conglomerate and gray sandstone of the Cantwell Formation, Denali, Alaska. The pebbles are clast supported, but finer sand filled the voids between them.

1 mm

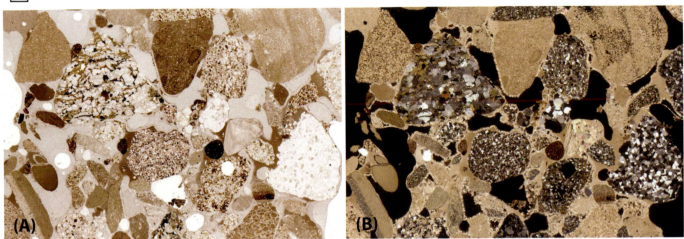

Figure 12.16 Photomicrographs of a thin section of polymict clast-supported conglomerate from Monroe County, New York, as seen under **(A)** plane and **(B)** crossed polarized light. Under crossed polars, clasts are seen to consist of extremely fine-grained limestone (pinkish yellow), fine-grained sandstone and siltstone (gray), and one clast of medium-grained metamorphic rock composed of quartz and biotite (upper left). Large voids (black under crossed polars) are partly filled with carbonate cement.

variable, reflecting the topographic relief on the unconformity. The Lower Pennsylvanian Olean quartz-rich conglomerate of New York State (Fig. 12.15) and the similar-age Pottsville conglomerate that underlies the anthracite coals of Pennsylvania, unconformably overlie older Mississippian-age rocks. The Late Triassic Shinarump Conglomerate is also quartz rich and unconformably overlies the Lower Triassic Moenkopi Formation (Fig. 12.12 and Box 12.1(B)).

Conglomerates often contain high concentrations of dense abrasion-resistant minerals, which may form economically important ore bodies (placer deposits). The Witwatersrand Conglomerate of South Africa (Fig. 12.17), for example, hosts one of the world's richest gold deposits. It has produced more than 51 000 tonnes of gold, accounting for about 40% of all the gold that has ever been mined on Earth. The gold occurs mainly near the base of this unit, where it was concentrated by sedimentary processes, but some of it was subsequently redistributed by hydrothermal solutions.

The high porosity of some conglomerates has allowed hydrothermal solutions to pass through them and deposit economically important minerals of, for example, uranium and copper. The Shinarump Conglomerate at the base of the Chinle Formation has been mined for its uranium (Fig. 12.18), which occurs along with copper minerals in pores and fractures in ancient channel fills where porosity would have been high. The uranium may initially have come from the Chinle mudrocks, which are enriched in this element, and then been redistributed by hydrothermal solution. Organic material appears to have played a role in precipitating the uranium minerals.

Figure 12.17 The 2.9 billion-year-old Witwatersrand oligomict clast-supported quartz conglomerate of South Africa hosts one of the world's richest gold deposits. The gold, occurs with pyrite between the quartz pebbles. (Photograph courtesy of Grant Cawthorn.)

Figure 12.18 Old uranium mine adits in the Shinarump Conglomerate at the base of the Chinle Formation, where it unconformably overlies the deep red Moenkopi Formation. The overlying eolian Wingate Sandstone forms the cliff at the top. Capitol Reef National Park, Utah.

12.2 Carbonate sedimentary rocks

Limestones and dolostones are estimated to constitute only 15% of all sedimentary rocks and, therefore, are the least abundant of the major sedimentary rock types. Most are of biogenic origin, but some are of chemical or mixed biogenic and chemical origin. Unlike sandstones, which form worldwide, carbonate rocks tend to form in warmer climates, but they can form as far north as 40°. However, as a result of plate motion, they are found worldwide. Despite their lesser abundance, limestones are of great economic importance. They are one of the most widely used building stones. Lime is extracted from them for the manufacture of cement that binds together concrete, today's most widely used construction material. When partially dissolved, limestones form important aquifers and host some of the world's largest oil and natural gas reservoirs and provide caves for tourism. They also host most of the marine fossil record.

Limestones and **dolostones** are rocks in which carbonate minerals predominate. The presence of calcite in a rock can be tested in the field by applying a drop of dilute hydrochloric (HCl) acid, which causes vigorous fizzing as carbon dioxide is released (Sec. 3.7). Dolomite fizzes only slowly, and usually only

after it has been ground to a fine powder (a scratched surface). Most limestones and dolostones are relatively pure carbonate rocks, so instead of using their mineralogy to classify them, we use their textures. Carbonate rocks consist of (1) primary carbonate grains referred to as **allochems** (from the Greek word *allos*, meaning "other"), which form the framework of the rock; (2) fine-grained carbonate mud matrix; and (3) pore spaces, which may be filled with coarser-grained carbonate cement, known as **spar**. Small amounts of detrital siliciclastic grains, such as quartz silt and sand, are often present in limestones. The way in which these components are assembled defines the texture of the rock. The textures convey important information about the original sediment, its site of deposition, and how it was cemented together to form rock.

The most widely used classifications for carbonate rocks are those of Dunham and Folk. Both schemes require distinguishing primary carbonate grains (allochems) from muddy matrix and cement. The Dunham classification can be used in the field on hand specimens with the aid of a hand lens, whereas the Folk classification is easier to apply if thin sections are available for study under the microscope. In this book, we use the Dunham classification. Both are described in detail and illustrated with photomicrographs of thin sections on the Web site of the Society for Sedimentary Geology (see the section "Online Resources" at the end of this chapter).

Dunham divides limestones into two major groups (Fig. 12.19), those in which the original grains (allochems as well as siliciclastic silt and quartz) were not bound together at time of deposition and those that were bound together, the latter group forming **boundstones**, as in a coral reef. The other group is subdivided into those containing mud and those that lack mud, with grains in the latter group eventually being cemented together with carbonate minerals (spar) to form **grainstone**. The carbonate rocks containing mud are further subdivided into those in which the original grains touch one another to form a framework- or grain-supported rock, which is called **packstone**, and those in which the grains are supported by the mud, which are called mudstones if the allochems constitute less than 10% and wackestones if more than 10%. The Dunham rock names are preceded by descriptive adjectives that indicate the predominant grain types, as in bioclastic wackestone and oolitic grainstone.

We now examine some specific examples of limestones and dolostones.

12.2.1 Limestones

Limestones generally form well-bedded sedimentary rocks (Fig. 12.20). The bedding may be produced by textural variations or by interlayering with other rocks. Limestones can be interlayered with dolostones, in which limestone has been replaced preferentially along certain beds by dolomite after deposition (Fig. 12.21(A)). Some fine-grained limestone mudstones are interlayered with siliciclastic mudrock, and in these the

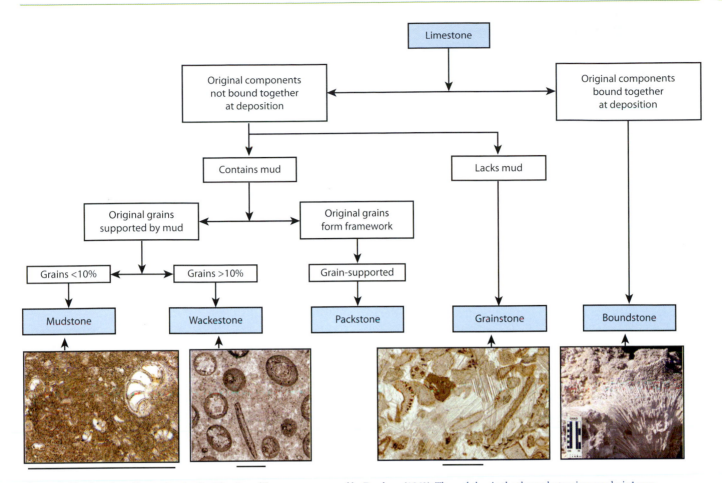

Figure 12.19 Flow chart illustrating the classification of limestones proposed by Dunham (1962). The scale bar in the three photomicrographs is 1 mm.

limestone commonly forms nodules that are surrounded by the clay fraction (Fig. 12.21(B)). Coarser allochemical limestones show many of the same bedding features shown by sandstones and conglomerates, with cross-bedding and channel structures indicating flow direction of currents. Figure 12.22 shows an example of an allochemical limestone containing intraclasts of mudstone that are analogous to the rip-up clasts in shaly rocks.

Figure 12.20 Road cut through beds of Upper Jurassic limestone that form the flat-topped mountains (i.e., causses) near Lodève, in the south of France. These limestone beds were deposited on a carbonate platform that formed on the northern side of the Tethys Ocean during the Jurassic high stand of sea level.

This limestone overlies a dolostone, which is a finely laminated mudstone that must have been deposited in quiet water, originally as a calcitic sediment, where there was no bioturbation.

Grainstone

Grainstones are formed in high-energy environments where the fine mud-size particles are winnowed out. Figure 12.23, for example, shows a grainstone formed on an ancient beach at Half Moon Cay in the Bahamas. This rock is made of broken shell fragments that were constantly agitated by the swash and backwash of waves on the beach during the high sea-level stand in the last interglacial period (120 000 years ago). The gently dipping cross-beds probably formed by erosion and deposition during storms. In a thin section, shell fragments can be seen cemented together by clear, coarse-grained sparry calcite that only partially fills the pores (Fig. 12.24). Such porous grainstones underlie much of Florida, where they form important aquifers.

Grainstones can show various degrees of sorting depending on how long the sediment was exposed to wave action or currents prior to lithification. Figure 12.25(A)–(B) shows the Mississippian Redwall Limestone from the Grand Canyon, Arizona (see Fig. 11.24(A)), which is a poorly sorted grainstone. The allochems in this rock range in size from small ooids to large circular plates consisting of single calcite crystals, which are segments of the stalklike part of crinoids, an animal whose

Figure 12.21 Ordovician limestones and dolostones from the Great Valley, near Shippensburg, Pennsylvania. (**A**) The limestone is a gray mudstone, which is interlayered with buff-weathering dolostone. (**B**) Nodular gray mudstone interlayered with dark siliciclastic mudrock.

shape resembles that of a tulip but is a relative of the modern starfish (both are echinoderms). Figure 12.25(C)–(D) shows the Ordovician Holston Limestone of Knoxville, Tennessee, in which the allochems, which consist of mollusk shell fragments and echinoderm plates, are all about the same size; hence, it is sorted. Close inspection of this image (which the reader can do by downloading the image from the textbook's Web site) reveals that many of the shell fragments are truncated against the irregular dark line that crosses the field. The fragments are truncated because the limestone has been removed by solution, leaving behind an insoluble residue of clay and iron oxide forming a stylolite. By analyzing the concentration of insoluble material in the limestone away from the stylolite and measuring the amount in the stylolite, the fraction of rock removed by solution can be estimated. This fraction can be as high as 50% in some limestones.

Mudstones to packstones

Where wave action or currents are less vigorous, the fine mud fraction formed by the abrasion of allochems, the boring of organisms, or mainly the disintegration of calcareous green algae may remain between the allochems to form a matrix. Upon lithification, this produces mudstones to packstones. Figure 12.26(A)–(B) shows, for comparison, two different oolitic limestones. The Cambrian Warrior Limestone of Pennsylvania is a grainstone in which ooids are cemented by clear calcite. This rock contains no fine-grained matrix and must, therefore, have been deposited in a high-energy environment. In contrast, the Middle Ordovician Laysburg Limestone, also from Pennsylvania, contains ooids that are embedded in a fine-grained carbonate mud matrix, so deposition must have occurred under quieter conditions. In most of the field of view shown in Figure 12.26(B), ooids rest against each other (grain supported), which makes this a packstone. On the right of the

Figure 12.22 Gray Ordovician limestone on top of buff-weathering dolostone from the Great Valley, near Shippensburg, Pennsylvania. The limestone has flat intraclasts of mudstone whose imbrication indicates flow from right to left. The dolostone is also a mudstone with fine laminations and stylolites marked by dark insoluble residue.

Figure 12.23 Grainstone formed from the lithification of carbonate beach sands on Half Moon Cay in the Bahamas. The flat cross-bedding was formed by wave action on the beach that existed during the high stand of sea level 120 000 years ago.

1 mm

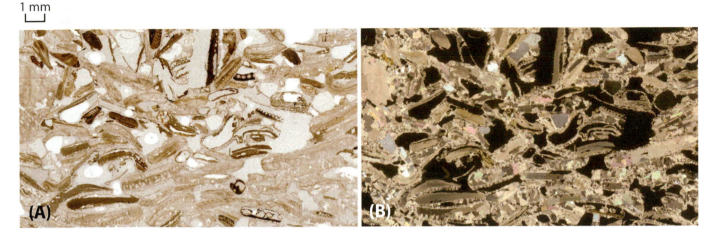

Figure 12.24 Photomicrograph of a thin section of coarse-grained grainstone under (**A**) plane and (**B**) crossed polarized light. Under crossed polars, the pores between shell fragments are seen to be partially filled with coarse birefringent crystals of calcite cement. The open pore space is black under crossed polars.

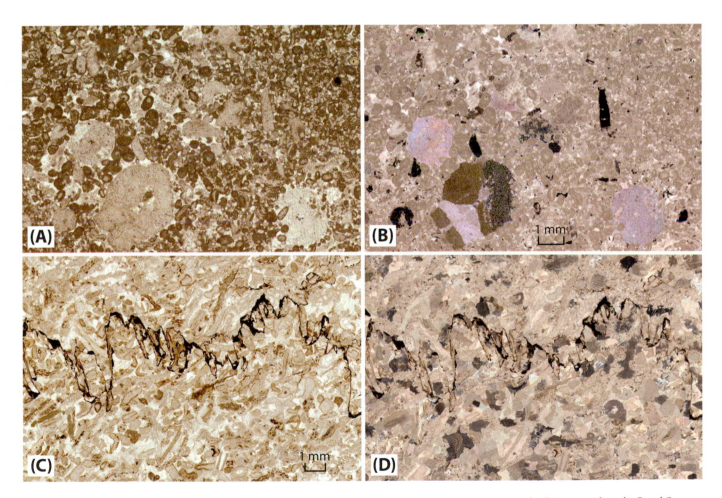

Figure 12.25 Photomicrographs of thin sections under (**A**) plane and (**B**) crossed polarized light of the Mississippian Redwall Limestone from the Grand Canyon. The large circular areas in this poorly sorted grainstone are echinoderm plates consisting of single crystals of calcite (see crossed polars in (B)), whereas the small circular allochems are ooids. (**C**) Photomicrograph of a thin section under plane polarized light of the Ordovician Holston Limestone from Knoxville, Tennessee. This well-sorted grainstone contains shell fragments and echinoderm plates. A stylolite, marked by a dark line, contains insoluble clay and iron oxide. (**D**) Same as (C) under crossed polarized light.

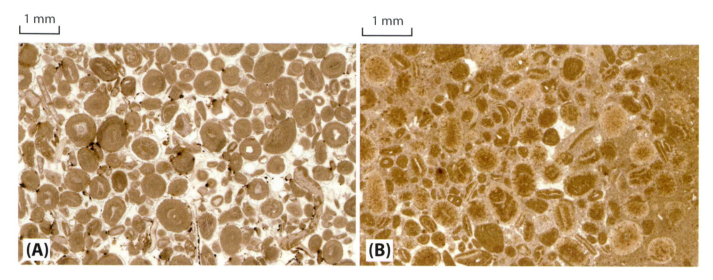

Figure 12.26 Photomicrographs of thin sections of oolitic limestones under plane polarized light. (**A**) The Cambrian Warrior Limestone of Tyrone, Pennsylvania is a grainstone cemented by clear calcite crystals. (**B**) The Middle Ordovician Laysburg Formation of central Pennsylvania is a packstone in which ooids sit in a mud matrix (dusty brown).

photomicrograph, however, the ooids are supported by the mud, so the rock grades into a wackestone (Fig. 12.19).

Gently dipping beds of limestone are shown underlying the pyramid of Khafre (Chephren) and the Sphinx at Giza, Egypt (Fig. 12.27). The limestone at the base of the Sphinx is a boundstone, consisting of hummocky reef material. This formed in agitated water no deeper than 20 m (see Sec. 11.3.1). Above this, the beds making up the body and head of the Sphinx were deposited in deeper, quieter water, so that carbonate mud was retained between the allochems, which are mainly the tests (bodies) of the foraminifera, *nummulites*. This limestone is a bioclastic wackestone to packstone.

During the Cretaceous, increased rates of seafloor spreading caused ocean ridges to grow and sea level to rise and flood low-lying parts of continents with shallow epeiric seas. In North America, a shallow sea extended from the Arctic Ocean down across the Great Plains and from what are now the Rocky Mountains to the Gulf of Mexico. At the same time, large parts of northwestern Europe and the African continent were flooded. With the rise in sea level, shorelines retreated and as a result siliciclastic sediment was deposited before traveling far into the epeiric seas. This produced relatively clear (low-turbidity), warm shallow water in which calcareous pelagic organisms thrived. The skeletons of these organisms, which in deep oceans dissolve as they sink, rained down and accumulated on the shallow sea floors and were lithified to form a fine-grained carbonate mudstone that we refer to as **chalk**. Throughout the flooded areas, thick layers of chalk were deposited, such as that in the White Cliffs of Dover on the southeastern coast of England. The name given to this geologic period is derived from the Latin word *creta*, meaning "chalk."

Figure 12.28(A) shows a photomicrograph of a thin section of the Upper Cretaceous chalk from Noxubee, Mississippi. This is a typical chalk containing many allochems (mainly foraminifera), in an extremely fine-grained carbonate mud matrix. Under high magnification, the matrix is seen to contain many small spherical coccoliths formed from the hard part of the microscopic eukaryote organism.

Chalk commonly contains nodular bodies of chert (Sec. 10.15), which is also referred to as **flint** when it occurs in chalk, especially when used in an archeological context. This cryptocrystalline variety of silica breaks with a conchoidal fracture, and because of its grain size and hardness, it can be knapped (flaked) to make sharp stone tools (Fig. 16.1(A)). Chert nodules may be distributed throughout the chalk or concentrated in beds (Fig. 12.28(B)). Nodules may show internal concentric banding (Fig. 12.28(C)), which has been interpreted to result from precipitation of amorphous silica gel

Figure 12.27 The pyramid of Khafre (Chephren) at Giza, Egypt, is built from the local Middle Eocene Mokattam Formation, a nummulitic (foram) packstone, which forms the gently dipping layers in the foreground. The Sphinx is carved from this same limestone. (Photograph courtesy of Vivian Rigg.)

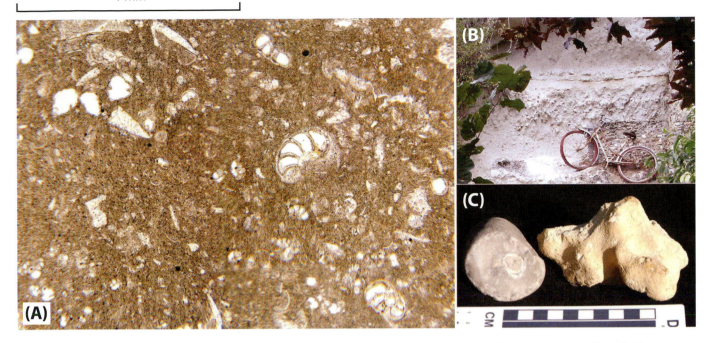

1 mm

Figure 12.28 (**A**) Photomicrograph of a thin section of Upper Cretaceous chalk from Noxubee, Missouri. This mudstone contains several larger fossil foraminifera in an extremely fine-grained calcareous matrix containing coccoliths. (**B**) Upper Cretaceous chalk from Troo in Le Loir, France, containing chert nodules that stand out in relief. (**C**) Two chert nodules from the chalk in (B).

about some nucleus. Other nodules may contain patches of chalk that are interpreted to result from incomplete replacement of chalk by silica. In a thin section, chert commonly is seen to contain siliceous fossils, such as sponge spicules, radiolaria, and diatoms, which may be the source of the silica. These organisms secrete an opaline variety of silica (amorphous), which dissolves in the pore fluids when the organisms die. This can raise the concentration of silica in the pore fluid to a level at which it may deposit chert around some object or replace the chalk.

Boundstones

Boundstones are limestones that are formed from organisms that grow in place. **Coral reefs** are an example of such limestone (Fig. 12.29). Unlike other forms of sedimentary rocks that require a depression for deposition, boundstones can form topographic highs. As such, they become a source of sediment. For example, fringing reefs are bounded on their seaward side by coarse bioclastic breccia derived from the reef and on the mainland side by finer sediment deposited in quieter lagoonal waters. The boundstone itself is composed of a

Figure 12.29 (**A**) Coral reef formed around the Caribbean island of Aruba during the last high stand of sea level in the last interglacial period (120 000 years ago). This limestone formed from organisms growing in place, such as (**B**) brain coral.

complex intergrowth of organisms between which may be a small amount of mud. Some are highly porous, which makes them excellent reservoir rocks for oil. Modern coral reefs form within ~30° of the equator, so their presence in older rocks may indicate paleolatitudes.

Another important form of boundstone is produced by blue-green algae (cyanobacteria), which form filamentous mats that trap carbonate mud as it is washed back and forth by waves in tidal flats. As the algae continue to grow, successive layers of carbonate mud are trapped to produce a finely laminated rock that commonly forms decimeter- to meter-size domes. The resulting hummocky structures are known as **stromatolites**, from the Greek word for "stony carpet" (Fig. 12.30). They can also be preserved in cherts. Stromatolites are found in rocks of all ages, with those in the Precambrian providing the earliest known fossils (3.5 billion years old). They are of great importance because during photosynthesis, cyanobacteria take in carbon dioxide and release oxygen. The growth of these organisms during the Precambrian slowly changed the Earth's atmosphere, which was initially anoxic, but by 2.2 billion years ago had been changed to one similar to today's atmosphere. In recognition of the importance of stromatolites to the evolution of life on our planet, Canada Post issued a special postal stamp that shows 2 billion-year-old stromatolites from the coast of Hudson Bay near the Belcher Islands (inset in Fig. 12.30).

Lacustrine limestones

Limestone can form in freshwater lakes, where it is commonly interlayered with siliciclastic sedimentary rocks. Freshwater limestone typically contains more clay than does marine limestone and produces a rock known as **marl**. Aragonite and calcite can precipitate directly from the water in these lakes or be secreted by organisms, or it can be derived from the breakage of freshwater shells such as those of ostrocods. Some freshwater lakes, in which carbonates precipitate, are far from "fresh" and have high concentrations of dissolved salts, as in Great Salt Lake, Utah, and the Dead Sea. In these lakes, limestone may be interlayered with evaporite beds.

High sea level during the Cretaceous Period, which resulted in the widespread deposition of chalk, began to drop by the end of the Cretaceous, and by the beginning of the Eocene had retreated from the Great Plains of North America, leaving a flat plain with meandering rivers, lakes and swamps. In many of these lakes, buff- and pink-colored marly limestones were deposited. Their color indicates deposition in an oxygen-rich environment in which iron occurs in the trivalent state. Spectacular outcrops of these limestones and interbedded sandstones and conglomerates are exposed in the Claron Formation of Bryce Canyon National Park, Utah (Fig. 12.31). The remarkable columns of rock known as hoodoos are formed by resistant siliciclastic sedimentary rocks forming protective caps over more soluble marl below.

Tufa, travertine, and caliche

Small bodies of limestone form by direct precipitation of calcite from solution, familiar examples of which are the stalactites and stalagmites that form in caves (Fig. 2.15(B)). Supersaturated **spring water** issuing onto the Earth's surface at ambient temperatures precipitates a limestone known as **tufa** (not to be confused with the volcanic rock **tuff**). The calcite often encases

Figure 12.30 Domical stromatolites in the Proterozoic Transvaal dolostone (2.3 billion years old) north of Boetsap, northern Cape Province, South Africa. Geologic hammer for scale in center of photograph. Inset photo shows a Canadian postage stamp with stromatolites from the shore of Hudson Bay taken from a photograph by Hans Hofmann, an expert on the planet's earliest fossil record.

Figure 12.31 Differential weathering of ancient lake deposits of resistant sandstone and soluble limestone of the Claron Formation in Bryce Canyon National Park, Utah.

surrounding vegetation and algae that grow in pools, which after dying and decaying leave cavities that give tufa its spongy porous appearance. As these deposits build up around a spring, the water has to rise and spill over the previously deposited layers, which creates a steep outer face (Fig. 12.32(A)). Deposits with considerable topographic relief can form in this way, as in the town of Cotignac in the Var district of southern France, where tufa completely filled a river valley and created an 80 m-high cliff. The town is nestled against this cliff, which in medieval times played important roles. The flat top of the tufa deposit is the site of two twelfth-century defensive towers, and the

soft tufa in the cliff below was tunneled into to make troglodyte dwellings and places to store provisions (Fig. 12.32(B)).

Limestone is also deposited around hot springs, and because fewer organisms live in the hotter water, a more massive but still porous variety of tufa forms, known as **travertine**. Water at depth in the Earth is under higher pressure than near the surface, and this allows more CO_2 to be held in solution, which decreases the pH (makes it more acidic) and allows more calcite to be dissolved. When heated by nearby bodies of magma or cooling igneous rock, groundwater circulation cells are set up, and as the water rises toward the surface, the decreasing pressure causes exsolution of CO_2 (some hot springs effervesce just like carbonated beverages), the pH rises, and calcite precipitates to form travertine, as in Yellowstone National Park, Wyoming (Fig. 12.33(A)). Agitation of water can also help release CO_2, just as it does when you shake a carbonated beverage. Travertine deposits are, therefore, common at waterfalls (Fig. 12.33(B)).

Travertine has been a popular building material since Roman times. The Coliseum, for example, is built largely of porous travertine, and many statues were carved from it (Fig. 2.16(A)). Today it is usually cut into thin slabs that are used as a facing stone for walls or flooring tiles.

Another calcite-rich deposit formed by direct precipitation from groundwater in arid and semiarid climates is **caliche**. In temperate or tropical climates, soluble ions produced from the weathering of rock are carried away in the groundwater. In arid and semiarid climates, groundwater can flow toward the surface as a result of evaporation, and calcium ions carried in solution combine with CO_2 from the air to precipitate as calcite. Plant

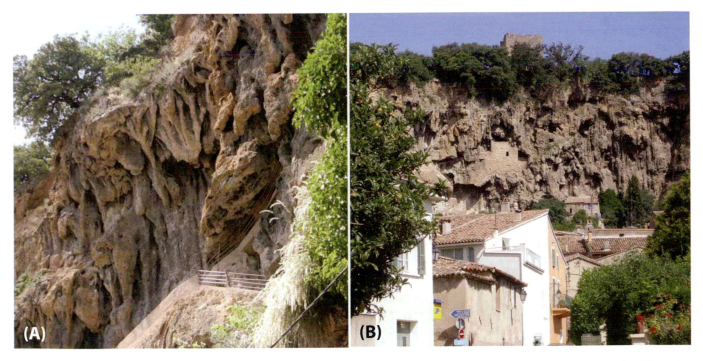

Figure 12.32 **(A)** Fluted and stalactitelike deposit of limestone forming an 80 m-high cliff of tufa in the town of Cotignac, France. **(B)** During medieval times, the tufa was tunneled into to make cliff dwellings to which people could retreat at times of danger. A twelfth-century tower can be seen on the top of the tufa deposit.

Figure 12.33 (A) Travertine deposits forming descending pools from the hot springs at Yellowstone National Park, Wyoming. (B) A hot-spring travertine deposit at a waterfall on the Jemez Creek, near Los Alamos, New Mexico.

roots can also promote precipitation of calcite because they remove water from the soil. In soils, layers of calcite precipitation are referred to as **hardpan**. In rocks, caliche layers mark ancient water tables and indicate that the climate was semiarid or arid. Caliche layers can be seen as light-colored zones in the red sandstone of the Chinle Formation shown in the chapter-opening photograph. Caliche deposits commonly have the right mixture of calcite and clay minerals to be ideal raw material for the manufacture of Portland cement, and it is used as such worldwide.

12.2.2 Dolostones

Dolostones are present in essentially all the same environments as limestones. They may be interlayered with limestones or completely replace them. Almost all dolostones were originally limestone in which the calcite was subsequently replaced by dolomite, sometimes shortly after deposition of the sediment but usually much later during deep burial (Sec. 11.9.4). They are classified by the same textural terms as limestones (Fig. 12.19). Carbonate rocks are generally either largely calcite or dolomite; intermediate members are rare because the dolomitization of limestones is a chemical reaction that usually goes to completion. However, dolomitic limestones and calcitic dolostones do occur. Dolostones are most easily distinguished from limestones in the field by their buff weathering color (Figs. 12.21(A) and 12.22), which is due to small amounts of iron housed in the dolomite structure, and by their slow reaction to dilute hydrochloric acid. In thin sections, although calcite and dolomite have the same high birefringence, dolomite typically occurs as zoned rhombohedral crystals, whereas calcite does not. If crystals forming the cement are large enough, the curved saddle shape of dolomite crystals may be visible.

Photomicrographs of a well-known dolostone are shown in Figure 12.34. This is the Lockport Dolostone, which is the resistant layer forming Niagara Falls. This Middle Silurian dolomitic packstone contains many fossil fragments, especially those of echinoderms, which can be seen under crossed polarized light to consist of single carbonate crystals. The fossil fragments touch one another and are, therefore, grain supported. The pores between the fragments are filled with a muddy matrix, which has been converted to dolomite, and a small amount of dolomite cement.

12.2.3 Tectonic settings of carbonate rocks

Unlike siliciclastic sedimentary rocks, most carbonate rocks do not have specific tectonic settings in which they are formed. Limestones tend to form at low latitudes because most originated from calcareous organisms that prefer warm, clear waters where organisms can get light. Although pelagic organisms can live in the open ocean, many of these do not form limestone because they dissolve on sinking below the carbonate compensation depth (Sec. 11.3.1). As a result, most carbonate rocks are formed on shallow platforms at low latitudes. Both passive continental margins and epeiric seas are the most common depositional sites of carbonate rocks, where they are commonly associated with mature quartz arenites.

12.3 Coals

Coal is a combustible sedimentary rock formed from fossilized plant material. Normally when plants die, they react with oxygen in the atmosphere to form carbon dioxide and water, which return to the atmosphere. If they are buried before oxidation occurs, the organic matter can be fossilized and turned to coal. The largest formation of coal occurred during the Pennsylvanian Period (~300 million years ago), where coastal plains with lush vegetation growing at low latitudes were buried beneath

1 mm

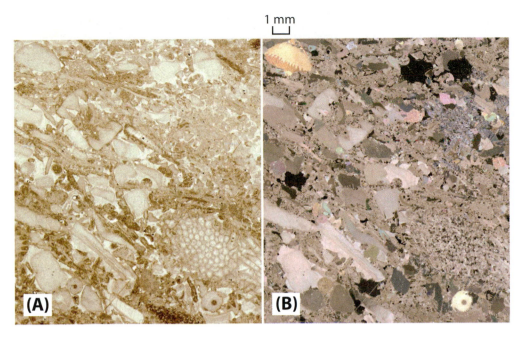

(A)   (B)

Figure 12.34 Photomicrographs of a thin section under (**A**) plane and (**B**) crossed polarized light of the Lockport dolomitic packstone. It contains abundant echinoderm fragments, which under crossed polars are seen to consist of single crystals. The small donut-shaped grain in the lower right is a fragment of an echinoderm stalk.

deltaic sands as shorelines transgressed and regressed, probably as a result of the growth and retreat of continental ice sheets in high latitudes. The fluctuating sea level produced cyclical deposits of marine and nonmarine sediments known as **cyclothems**, in which the coal is underlain by nonmarine siliciclastic mudrocks and overlain by shallow marine sandstones and conglomerates (Fig. 2.18).

When organic matter is buried in sediment, it is converted to an opaque noncrystalline solid whose composition becomes increasingly richer in carbon with rising temperature and pressure. Being noncrystalline and of variable composition, it is not considered a mineral, even though it is a naturally occurring solid. The changes that occur during burial, diagenesis, and eventually low-grade metamorphism are used to classify coal into **rank**, a name that reflects the fact that historically people were interested in how much heat it could generated when burnt. Combustion of low-rank coals generates about 2.85×10^7 J/kg, whereas high-rank coals generate as much as 3.5×10^7 J/kg.

Low-rank coal is commonly referred to as **lignite** or **brown coal**. In outcrop, it is friable and weathers easily. Fossilized plant remains are common and readily visible. Intermediate rank coal is referred to as **bituminous**. It is blacker than lignite and prominently layered on the millimeter scale with material of different reflectance and fracture. The highest-rank coal is **anthracite**, which occurs only in folded rocks and is considered a low-grade metamorphic rock. It appears almost metallic and breaks with a conchoidal fracture. It is massive and lacks the fine laminations of bituminous coal. **Cannel coal** is another variety that is fine-grained, massive, and has conchoidal fracture, but unlike anthracite, it has a greasy luster. Microscopic study under reflected light indicates that it is composed of very small particles of woody material and plant spores. These particles appear to

have been transported to a depositional site, unlike other coals, most of which appear to have accumulated in place.

Land plants evolved only in the Silurian, so coal does not occur in older rocks. In fact, it is quite rare in rocks older than the Carboniferous. It forms beds that are usually less than a meter thick but can be laterally extensive. Coal is our largest reserve of fossil fuel and will be used more in the future as reserves of oil decrease.

12.4 Oil and natural gas

This book deals with the solids that make up the Earth, so oil and natural gas do not fall under this rubric, but their origin, where they are found, and how we search for them are so intimately related to rocks that we must at least discuss them briefly.

In Section 11.3.2, we saw that kerogen in organic-rich sediment, on being buried and heated, is converted to oil and natural gas. Black shales are the most common source rock for these hydrocarbons. Although natural gas can be extracted from source rocks using special drilling techniques (Sec. 16.8.1), shales are relatively impervious, so it is difficult to extract hydrocarbons from them. Instead, a nearby **reservoir rock** is sought into which the hydrocarbons can migrate slowly from the source rock. Reservoir rocks must have high **porosity** and the pores should have a high degree of connectedness, that is, high **permeability**. Poorly cemented sandstones and carbonate rocks are the most common reservoir rocks. When we hear of the huge number of barrels of oil that are pumped from oil reservoirs or are spilled accidentally from leaking oil wells, it is easy to believe that the oil (and natural gas) sits in gigantic pools at depth just waiting to be tapped. But this is a completely wrong impression of how oil occurs in rocks. Figure 12.35 shows a coarse-grained feldspathic sandstone from the Mesozoic

1 cm

Figure 12.35 Cross-bedded feldspathic sandstone from the Jurassic Portland Formation in the Hartford Basin, Connecticut, in which the dark pores are filled with hydrocarbon. It is from such porous sandstones that much of the world's oil is produced.

Hartford Basin in Connecticut in which the pores are filled with a dark hydrocarbon. Although this rock was heated above the oil window and the oil converted to a solid hydrocarbon, it gives a good idea of how oil occurs in sandstone and why it can be difficult to extract. Imagine the rate at which oil would flow through the pores of this rock versus the speed at which you pump fuel into your car. Pumping of oil from sandstone such as this can be a slow process, with only small amounts being extracted on a daily basis. Often, detergents are pumped into the ground to loosen the oil from grain surfaces. This is a particularly com-

mon practice when oil fields are pumped for the second or third time trying to extract residual amounts of oil.

Oil and gas are both less dense than water and, therefore, tend to rise toward the surface. For them to be kept in a reservoir rock, they must be confined by an impermeable **cap rock** whose configuration creates a natural **trap** for the buoyant hydrocarbons. There are many such traps, a few of which are shown in Figure 12.36.

12.5 Evaporites

Although seawater contains many different ions in solution and many different minerals will precipitate if seawater is totally evaporated, most evaporite deposits have rocks with relatively simple mineralogy, forming beds of gypsum, anhydrite, and salt (halite). The reason for this is that rarely do large bodies of seawater completely evaporate, and many bodies probably reach some steady state between loss of water to evaporation and influx of new seawater to the basin. Once a steady state is established, the concentration of ions in the basin determines which minerals precipitate. For example, if the ion concentration in the restricted basin is increased by a factor of two over what is normal in seawater, calcite precipitates, but because it is a relatively minor constituent, no large deposits of limestone are formed. If the concentration is increased by a factor of five, gypsum and anhydrite precipitate, and because of their abundance in seawater, they produce thicker deposits. Sodium and chlorine are the most abundant ions in seawater, but halite does not precipitate until the ion concentration has increased by a factor of ten (90% of the seawater evaporated). Potassium chloride (sylvite) requires even higher levels of concentration before it precipitates. Thus, in many restricted basins, salinity may reach a concentration where only gypsum or anhydrite is deposited. Only in extremely restricted basins does halite form.

Many evaporite deposits begin with precipitation of carbonates followed by gypsum and then anhydrite. In others, anhydrite may precede gypsum. This may depend on the temperature of the water, with gypsum forming at lower temperature and anhydrite forming in water above 34°C. Some gypsum is formed by hydration of

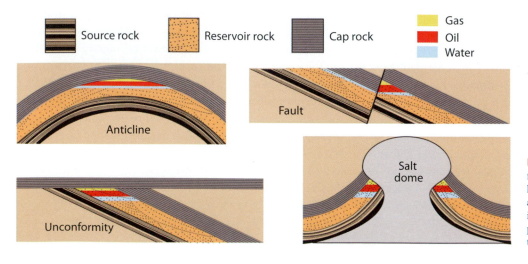

Figure 12.36 Oil and gas reservoirs are formed where a source rock supplies hydrocarbons that are trapped in a porous reservoir rock by a cap of impermeable rock whose structure prevents the hydrocarbons from rising to the surface.

anhydrite, especially at shallow burial depths. With deep burial, gypsum becomes unstable and dehydrates to anhydrite.

Evaporite beds are often deformed by flow. Halite has an extremely low density ($G = 2.16$) and recrystallizes easily. A halite bed in a sequence of sedimentary rocks invariably finds itself overlain by denser rocks, which sets up a gravitational instability with formation of **salt domes** or **diapirs**. This is an example of a **Rayleigh-Taylor instability** and commonly results in the formation of regularly spaced diapirs analogous to those formed in the oil-and-honey model shown in Figure 8.23(B).

One of the most spectacular examples of evaporite deposition occurred in the Mediterranean during the Late Miocene (10 million years ago). The Mediterranean was formed from the closing of the Tethys Ocean by the northward motion of the African plate. Today it is connected to oceans only through the Strait of Gibraltar, where Atlantic water is constantly entering to compensate for evaporation from the Mediterranean. During the Late Miocene, plate motion constricted this flow, and the Mediterranean almost completely evaporated, with deposition of up to 3 km of evaporite beds. Numerous residual bodies of water remained on the Mediterranean seafloor where the final evaporation occurred. Each of these basins contains carbonates, overlain by gypsum and topped by halite and, in a few places, even sylvite, which agrees with the expected sequence of deposition of evaporite minerals. At the end of the Miocene (5 million years ago), Gibraltar opened again and the 4000 m-deep Mediterranean basin was refilled to its present level and the evaporite deposits were covered with pelagic muds.

12.6 Phosphorites

Phosphorites are phosphorus-rich sedimentary rocks that contain >18 weight % P_2O_5 (Sec. 10.16). They are composed largely of the mineral apatite, which is often cryptocrystalline and referred to as collophane. Some clay minerals and carbonate shell fragments may also be present, so they straddle the division of sedimentary rocks between siliciclastic mudrocks, bioclastic limestones, and chemical precipitates.

Most phosphorites are black, fine grained, and well bedded on the millimeter to meter scale, and they are commonly interlayered with marine limestones, siliciclastic mudrocks, and green sandstones that contain glauconite, a type of clay mineral that grows in the sediment (**authigenic**) rather than being of detrital origin. The phosphate minerals form small pellets (>0.05 mm) to larger nodules (several centimeters), and some form ooids, but they do not form by the same process as carbonate ooids (Fig. 10.31). These are cemented together by apatite, chert, carbonate minerals, or even a clay matrix.

Phosphorus is not an abundant element in the crust, so phosphorite beds are a remarkable concentration of this element. The phosphorus is probably derived from animal remains that accumulate mainly in shallow marine environments and occasionally in lakes (recall that bones and teeth are made of apatite; see Sec. 7.33), but it moves around in pore fluids, because we find fossils that we know must have initially been composed of aragonite or calcite and are now replaced by apatite.

Phosphorites are the main source of phosphorus, which is one of the important ingredients of fertilizer (Sec. 17.2.3). They are mined in many places in Europe (France, Belgium, and Spain), North Africa (Egypt, Tunisia, Algeria, and Morocco), and in North America (Tennessee, Wyoming, Idaho, and coastal South Carolina, Georgia, and Florida).

12.7 Iron-formations

Iron-formation is defined as a chemical sedimentary rock, typically thin banded or laminated, containing 15 weight %, or more, total iron of sedimentary origin, commonly but not necessarily containing layers of chert. The average composition of a large number of these formations ranges from about 45 to 55 weight % SiO_2, a total Fe content of about 30 weight %, of which ~20% is Fe_2O_3 and 25% is FeO. The only other major chemical components are CaO and MgO, both ranging from about 3–8 weight %.

Although iron-formations are generally considered the result of inorganic precipitation, much ongoing research focuses on the possibility of their precipitation having been mediated by biogenic activity. Iron-formations have a high relative iron content, but they are not, in themselves, iron ores. Most iron ores (see Sec. 16.2) are the result of supergene enrichment of the original iron-formation together with leaching of silica, producing hematite-goethite-rich ores in iron-formation sequences. Other iron ores are obtained through beneficiation, which involves the improvement of the ore grade by milling, flotation, magnetic separation, and gravity concentration of iron-formations that have undergone metamorphism.

Summary

This chapter discussed the classification of sedimentary rocks into three main groups, the siliciclastic sedimentary rocks, the biogenic sedimentary rocks, and the chemical sedimentary rocks. The sites of deposition and plate tectonic settings of each of the main types of sedimentary rock were discussed. The value of sedimentary rocks as sources of fossil fuel, construction materials, and various ores of elements such as iron, gold, uranium, copper, phosphorus, potassium, and sodium was also discussed. The following are the main points covered in this chapter:

- Sedimentary rocks fall into three main categories based on the type of sediment from which they are formed: siliciclastic, biogenic, and chemical. These include the most abundant rock types: mudrocks, sandstones, conglomerates, limestones, dolostones, and the less abundant but important coals, evaporites, phosphorites, and iron-formations.

- Siliciclastic sedimentary rocks are the most abundant, with mudrocks forming about two-thirds of all sedimentary rocks. Mudrocks with prominent fissility are called shales.

- Mud-size particles settle only in still water, which is normally found below wave base in lakes and seas or on river floodplains.

- Fine-scale laminations in mudrocks are normally preserved only under anoxic conditions where burrowing organisms do not live.

- The color of mudrocks ranges from red through green to black. The red is due to oxidized iron, whereas the black is due to organic matter that is preserved under reducing conditions.

- The thickest mudrocks are formed at convergent plate boundaries in forearc and foreland basins.

- Mudrocks are also formed on the outer part of continental shelves along passive margins and on continents in epeiric seas.

- Mudrocks are of importance because they are the source of hydrocarbons.

- Sandstones, which constitute ~25% of all sedimentary rocks, are classified using their relative contents of detrital quartz, feldspar, and lithic fragments. The most common types are quartz arenites, feldspathic arenites, and lithic arenites. Most sandstones contain very little matrix, but if the matrix exceeds 15%, it is called a wacke or graywacke. The type of cement in sandstone is not used in classifying them, but it is added to the name as a qualifier.

- Mature sandstones are quartz-rich because unstable feldspar and lithic fragments have been removed by abrasion and weathering. In supermature sandstone, quartz grains become rounded, especially in eolian deposits.

- Mature quartz arenites characterize passive continental margins; immature wackes characterize forearc basins at convergent plate boundaries; and feldspathic arenites are typical of continental rift valleys.

- Wackes typically form graded beds, which are interpreted to indicate deposition from turbidity currents.

- Conglomerates, which consist of gravel-size particles, are formed in high-energy environments, such as river channels, alluvial fans, beaches, submarine fans, debris flows and glacial deposits. They can be clast supported or matrix supported, which indicates very different depositional environments.

- Placer deposits of gold and hydrothermal deposits of uranium and copper are associated with conglomerates.

- Carbonate sedimentary rocks are commonly formed in shallow, warm clear water where calcareous organisms flourish. They tend not to form where the input of detrital sediment is great.

- Carbonate rocks are classified on the basis of the proportions of primary carbonate grains (allochems) and the relation of these grains to the matrix or cement that binds them together.

- Most limestones are of marine origin, but they can also form in lakes where they tend to be mixed with clay to form a rock known as marl.

- Limestone also forms around both cold and hot springs to form tufa and travertine.

- Most dolostones form as a replacement of limestone, but some may form as a primary precipitate.

- Coal is classified by rank, which reflects the amount of heat released during combustion. With increasing rank, coal changes from brown lignite, to black bituminous coal, to anthracite with almost a metallic luster.

- Oil is generated in an organic-rich source rock, such as black shale. It accumulates in a reservoir rock, usually a porous sandstone, limestone or dolostones. And it is capped by an impermeable rock, such as shale.

- Evaporites commonly contain a basal layer of limestone, overlain by gypsum or anhydrite, and capped by salt, but many evaporite deposits never reach the salt-precipitating stage.

- Phosphorites are mainly marine sediments composed largely of apatite.

- Iron-formation is an iron-rich, well-banded chemical sedimentary rock that was formed only in the Precambrian when the atmosphere-ocean system was anoxic, allowing for the solution, transport, and chemical precipitation (in such oceans) of Fe^{2+} (and mixed Fe^{2+} and Fe^{3+}) mineral phases.

Review questions

1. Sedimentary rocks can be divided into three main groups. What are these groups, and what are the main rock types in each?

2. What are siliciclastic sedimentary rocks, and from what type of sediment are they formed?

3. What simple test allows you to determine whether mudrock contains silt- or clay-size particles?

4. What makes shale fissile?

5. What destroys fine-scale laminations in mudrocks?

6. What determines the color of mudrocks?

7. How do rip-up clasts form?

8. Where do the thickest mudrocks occur, and what plate tectonic significance do such deposits have?

9. The object of this question is to determine the percentage of clay matrix in the photomicrograph of Potsdam Sandstone shown in Figure 12.7(A) using the NIH ImageJ program. First, download the photomicrograph from the textbook's Web site (www.cambridge.org/earthmaterials) and import the image into the ImageJ program. Follow the instructions given in Box 6.2 to determine the percentage of clay matrix. (Answer ~9.2%)

10. If your answer to question 9 had indicated more than 15% matrix, what would you have called this sandstone?

11. What features indicate that a sandstone is chemically and texturally mature?

12. What are the differences between quartz arenite, feldspathic arenite, and litharenite?

13. What bedding differences distinguish feldspathic arenite from wacke, and what does this tell you about the site of deposition?

14. What is the difference between a clast-supported and a matrix-supported conglomerate, and what do these rocks tell you about the environment of deposition?

15. Why are mudrocks, sandstones, and conglomerates commonly interlayered in fluvial deposits?

16. What conditions favor the growth of calcareous organisms that form limestone, and what restrictions does this place on their sites of deposition?

17. How do we classify carbonate sedimentary rocks?

18. How would you distinguish carbonate cement from an allochem or calcareous mud?

19. What is the connection between the enormous amount of chalk formed during the Cretaceous Period and plate tectonics?

20. What are the three main types of coal, and which releases the most heat upon combustion?

21. What three types of rock are needed to form an oil field?

22. What is the difference between porosity and permeability?

23. What are some typical traps in which oil reservoirs may form?

24. What is the sequence of precipitation of evaporite minerals from seawater as it evaporates?

25. What is the probable source of phosphorus in phosphorites?

26. What is banded iron-formation and what must happen to it to make it iron ore?

ONLINE RESOURCES

The Society for Sedimentary Geology's Web site (http://sepmstrata.org) provides useful information about sedimentary petrology.

The National Institutes of Health's ImageJ software program can be downloaded for free from http://rsb.info.nih.gov/ij.

FURTHER READING

Blatt, H., Tracy, R. J., and Owens, B. E. (2006). *Petrology*, W. H. Freeman, San Francisco. This book provides an excellent general coverage of sedimentary processes and rocks.

Boggs, S., Jr. (2009). *Petrology of Sedimentary Rocks*, 2nd ed., Cambridge University Press, Cambridge. This book provides an in-depth coverage of sedimentary rocks.

McBride, E. F. (1963). A classification of common sandstones. *Journal of Sedimentary Research*, 33, 664–669.

Pufahl, P. K. (2010). Bioelemental sediments. In *Facies Models*, 4th ed., ed. N. P. James and R. W. Dalrymple, Geological Association of Canada, St. Johns.

Stow, D. A. V. (2006). *Sedimentary Rocks in the Field – A Color Guide*, Elsevier Academic Press, Burlington, VT.

Scoffin, T. P. (1987). *An Introduction to Carbonate Sediments and Rocks*, Chapman & Hall, New York.

Tucker, M. E. (1991). *Sedimentary Rocks: An Introduction to the Origin of Sedimentary Rocks*, 2nd ed., Blackwell Scientific Publications, Oxford, U.K.

13 Metamorphic rock-forming minerals

This chapter, like Chapters 7 and 10, provides systematic descriptions of minerals but only for those that occur most commonly in metamorphic rocks. As was noted already, the best way for you to acquire basic skills in mineral identification is to read each mineral description while you have one or several examples of this same mineral on a tabletop where you are working. Clearly, this is best accomplished in a place where your instructor has mineral collections available for study. In most courses, this is in the laboratory that accompanies the course. Observe, study, and handle as many examples of the same mineral species as are available. If possible, also look at coarse-grained examples of metamorphic rocks that may contain some representative metamorphic mineral associations.

Large euhedral kyanite and staurolite crystals in a paragonite schist (paragonite is isostructural with muscovite but with Na^+ instead of K^+ in the interlayer position). From Alpa Sponda, Switzerland. The size of the field of view is 12×19 cm. (David Nufer Photography, Albuquerque, New Mexico.)

This chapter presents the systematic mineral descriptions of 26 of the most common metamorphic minerals. All of these are silicates except for one, corundum, an oxide. Such minerals are the result of chemical reactions involving preexisting minerals in sedimentary, igneous, or metamorphic rocks. These reactions may result from changes in temperature, pressure, fluids, and shearing stress at considerable depth in the Earth (see Chapter 14). Mineral reactions that are the result of an increase in temperature are referred to as **prograde**, whereas those that result from falling temperatures are **retrograde**. Retrograde metamorphic reactions commonly involve the addition of fluids.

13.1 Systematic mineralogical descriptions of common metamorphic minerals

The order in which metamorphic minerals are discussed in this chapter reflects their relative abundance. We begin with minerals that are common in some of the most abundant metamorphic rock types. An example of such an abundant rock type would be what is known as a **pelitic schist**. A pelitic schist is a rock derived by metamorphism of an argillaceous (meaning composed of clay-sized particles or clay minerals) or fine-grained aluminous sediment. Prograde metamorphic reactions in such a rock may produce sequentially metamorphic assemblages rich in chlorite and muscovite; subsequently garnet-staurolite-biotite-muscovite assemblages; and at the highest temperature, sillimanite-garnet-cordierite-feldspar assemblages. Members of the mica group (muscovite and biotite) and the feldspars are major constituents of these metamorphic rocks but their systematic descriptions are given in Chapter 7 because they are also major constituents of igneous rocks. Similarly quartz, abundant in igneous, sedimentary, and metamorphic rocks, was discussed in detail in Chapter 7. This means that because the systematic mineralogy in this text is arranged by common occurrence in igneous, sedimentary, and metamorphic rock types, those minerals that are common to two or all three of these lithologic divisions are described only once. Therefore, although micas, quartz, and feldspar may be common constituents in metamorphic rocks, you must refer to these minerals in Chapter 7, where they were first described. Here we concentrate on those minerals that are newly formed as a result of metamorphic reactions.

13.2 Garnet: $(Mg^{2+}, Fe^{2+}, Mn^{2+})_3 Al_2Si_3O_{12}$ and $Ca_3(Fe^{3+}, Al^{3+}, Cr^{3+})_2Si_3O_{12}$

Occurrence: Garnet is one of the most common metamorphic minerals. Those rich in Fe^{2+}, known as almandine, are common in metamorphosed pelitic and mafic igneous rocks. Ca-rich garnets such as grossular occur in the metamorphosed equivalents of carbonate-rich rocks. Some aluminous granites may contain almandine.

Chemical composition and crystal structure: The highly variable chemical composition of members of the garnet group is best expressed by a generalized chemical formula, $A_3B_2(SiO_4)_3$, where the A site houses Ca^{2+}, Mg^{2+}, Fe^{2+}, or Mn^{2+} and the B site is the location for Al^{3+}, Fe^{3+}, or Cr^{3+}. In terms of end member compositions, this leads to six formulas with specific species names:

$$
\left.
\begin{array}{l}
Mg_3Al_2Si_3O_{12} - \textbf{pyrope} \\
Fe_3Al_2Si_3O_{12} - \textbf{almandine} \\
Mn_3Al_2Si_3O_{12} - \textbf{spessartine}
\end{array}
\right\} \textit{pyralspite group}
$$

$$
\left.
\begin{array}{l}
Ca_3Al_2Si_3O_{12} - \textbf{grossular} \\
Ca_3Fe_2Si_3O_{12} - \textbf{andradite} \\
Ca_3Cr_2Si_3O_{12} - \textbf{uvarovite}
\end{array}
\right\} \textit{ugrandite group}
$$

The group names are derived from parts of the three mineral names in each group: *Pyr* (from *pyrope*), *al* (from *almandine*), and *sp* (from *spessartine*). Similarly, *u* (from *uvarovite*), *gr* (from *grossular*), and *and* (from *andradite*). These group names are used because solid solution among the three member of the pyralspite group is extensive, as among the three of the ugrandite group, but little solid solution occurs between these two major groups (Fig. 13.1).

The crystal structure of one of the garnet species, grossular, $Ca_3Al_2Si_3O_{12}$, is illustrated in Figure 13.2. It consists of alternating (SiO_4) tetrahedra and (AlO_6) octahedra that share corners to form a continuous three-dimensional network in a dense structural pattern. The A site in this garnet is occupied by Ca^{2+} in an irregular 8-fold coordination polyhedron. Garnets, as well as olivine, and the three polymorphs of Al_2SiO_5 (kyanite, sillimanite, and andalusite) are part of the structural group of silicates known as **nesosilicates**. In these, the (SiO_4) tetrahedra are independent, or islands, which means that they are not linked to any other (SiO_4) tetrahedra in the structure.

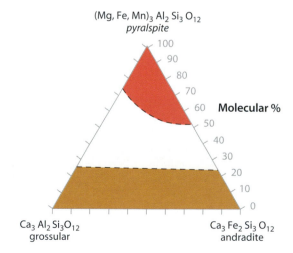

Figure 13.1 The limited extent of solid solution between the two compositionally distinct garnet groups. Pyralspite includes pyrope, almandine, and spessartine. Grossular and andradite, together with uvarovite, are part of the ugrandite group.

Grossular: Ca$_3$Al$_2$Si$_3$O$_{12}$
$I4_1/a\bar{3}2/d$

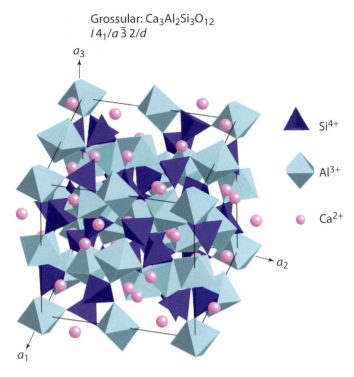

Figure 13.2 Perspective view of the crystal structure of grossular. All other garnet species have similar overall structures but different site occupancies. A cubic unit cell is outlined.

Figure 13.3 Large euhedral, trapezohedral crystals of almandine garnet in mica schist.

Crystal form: Isometric; $4/m\,\bar{3}\,2/m$. Common crystal forms are dodecahedra (Fig. 2.5(B)) and trapezohedra (Fig. 13.3). Commonly in distinct crystals but also occur as rounded grains (see photograph on front cover of this text). May be massive, granular, coarse to fine.

Physical properties: H = 6½–7½; G = 3.5–4.3. These physical parameters vary as a function of composition. Dodecahedra and trapezohedra are common crystal forms. Conchoidal fracture with many angular corners and edges. Color is highly variable but most common garnets range from red to reddish brown to brown, and black.

Distinguishing features: Good isometric crystal forms, relatively high hardness, and general red to brown to black color.

Uses: The angular fracture pattern of garnet and its relatively high hardness make it desirable for a range of abrasive purposes, including garnet paper. Because of the wide range of colors shown by garnets (most commonly red – see Fig. 2.5(A) – but also pink, orange, yellow, green and brown) due to their highly variable chemical composition, clear and transparent material of good color is extensively used in the jewelry industry. Figure 13.4 shows a very attractive orange-red crystal of grossular. All garnet species of gem quality are used except uvarovite, a brilliant, dark-green variety that occurs in small crystals generally too

Figure 13.4 An orange-red, distorted dodecahedral crystal of grossular. Although this crystal has a most attractive color, it contains many imperfections (inclusions and microfractures), which makes it unusable as a gemstone.

Figure 13.5 Crystals of the chiastolite variety of andalusite as seen in a polished piece of beige phyllite (a metamorphic rock intermediate between slate and mica schist). The black cruciform pattern results from crystallographically arranged graphite inclusions.

small for cutting. Demantoid, a dark-green to yellow-green variety of andradite (with the green color caused by small amounts of chromium) is a highly sought-after gemstone.

13.3 Andalusite: Al_2SiO_5

Occurrence: Commonly found in low- to medium-grade metamorphosed aluminous rocks referred to as pelitic schists. Also in the contact aureoles of igneous intrusions into aluminous rocks. It is one of three polymorphs of Al_2SiO_5 with a temperature-pressure stability field (Fig. 5.33(B)) in the low-temperature and low-pressure region. The higher-temperature sillimanite and higher-pressure kyanite polymorphs are discussed in Sections 13.4 and 13.5, respectively.

Chemical composition and crystal structure: Essentially of constant composition, Al_2SiO_5. The structure (shown in Fig. 5.35(A)) consists of chains of edge-sharing (AlO_6) octahedra parallel to the c axis, that are cross-linked by independent (SiO_4) tetrahedra. A second set of Al atoms occurs in 5-fold coordination between the octahedral chains.

Crystal form: Orthorhombic; $2/m\ 2/m\ 2/m$. Commonly in square prisms.

Physical properties: H = 7½; G ~3.2. Transparent to translucent. Color white when pure but commonly grayish to flesh colored to reddish-brown because of inclusions. The variety chiastolite (Fig. 13.5) contains graphite inclusions arranged in a regular pattern with a cruciform outline.

Distinguishing features: Square prisms and hardness. Chiastolite is easily recognized by the symmetrical pattern of inclusions.

Uses: All three polymorphous of Al_2SiO_5 (andalusite, sillimanite, and kyanite) are mined to produce mullite, a high-temperature reaction product that is a major component of high-temperature refractories used in the smelting and processing of ferrous metals. This reaction is as follows:

$$3Al_2SiO_5 + \text{heat} \rightarrow 3Al_2O_3 \cdot 2SiO_2 + SiO_2$$
$$\text{andalusite} \qquad\qquad \text{mullite} \qquad \text{amorphous silica}$$

Mullite is also used in glass making and in the production of ceramics such as found in spark plugs.

13.4 Sillimanite: Al_2SiO_5

Occurrence: Present in aluminous high-temperature metamorphic rocks. At lower temperatures andalusite is stable, and at lower temperatures and higher pressures kyanite occurs (Fig. 5.33(B)).

Chemical composition and crystal structure: The chemical composition is essentially Al_2SiO_5. A polyhedral representation of the structure of sillimanite is given in Figure 5.35(B). It consists of chains of edge-linked (AlO_6) octahedra parallel to the c axis cross-linked by (SiO_4) and (AlO_4) tetrahedra.

Figure 13.6 Slender, parallel crystals of beige to brown sillimanite.

Crystal form: Orthorhombic; $2/m\,2/m\,2/m$. Most commonly in long, slender crystals (Fig. 13.6). Also fibrous.

Physical properties: $H = 6.7$; $G = 3.23$. Perfect $\{010\}$ cleavage. In slender crystals; commonly fibrous. Color ranges from beige to brown, pale green, and white.

Distinguishing features: Characterized by its fibrous nature or slender crystals and one cleavage direction.

Uses: In the production of high-temperature refractory materials as discussed under uses for andalusite.

13.5 Kyanite: Al_2SiO_5

Occurrence: A mineral produced during high-pressure metamorphism of aluminous (pelitic) rocks. Commonly associated with staurolite (see chapter-opening photograph and Fig. 13.7)

Figure 13.7 Euhedral blue crystals of kyanite together with dark reddish-brown euhedral crystals of staurolite in fine-grained mica schist.

and garnet. Andalusite is the lower pressure polymorph of Al_2SiO_5 and sillimanite the high-temperature polymorph (Fig. 5.33(B)).

Chemical composition and crystal structure: The composition of kyanite is Al_2SiO_5. A polyhedral representation of its structure is shown in Figure 5.35(C). It consists of chains of (AlO_6) octahedra parallel to the c axis that are cross-linked by additional (AlO_6) octahedra and (SiO_4) tetrahedra. In this Al_2SiO_5 polymorph, all the Al is in octahedral coordination.

Crystal form: Triclinic; $\bar{1}$. Usually in long, tabular crystals (Figs. 13.7 and 3.1(E)). Also in bladed aggregates.

Physical properties: $H = 5$ (parallel to the length of crystals); $H = 7$ (at right angles to that direction); $G = 3.6$. Cleavage $\{100\}$ perfect. The color is usually light blue with a slightly darker blue color in the center of crystals. Also white and gray, the gray being due to minute graphite inclusions.

Distinguishing features: Characterized by good planar cleavage, patchy blue color and different hardnesses in two perpendicular crystallographic directions.

Uses: In the production of high-temperature refractory materials as discussed under uses for andalusite.

13.6 Staurolite: $Fe_{3-4}Al_{18}Si_8O_{48}H_{2-4}$

Occurrence: A strictly metamorphic mineral in aluminous rocks formed in prograde reactions under medium-temperature conditions. Commonly associated with kyanite (see chapter-opening photograph) and garnet.

Chemical composition and crystal structure: The formula given shows variable amounts of Fe^{2+} as a result of vacancies that replace Fe^{2+} up to several percent. Small amounts of Zn, Co, Mg, and Al may replace Fe^{2+} as well. A perspective view of a polyhedral representation of the crystal structure of staurolite is given in Figure 13.8. It consists of (SiO_4) and (FeO_4) tetrahedra connected to (AlO_6) octahedra. The (AlO_6) octahedra form edge-sharing chains parallel to the c axis but not all these Al sites are occupied. Octahedral sites that may not be fully occupied, or may contain atoms other than Fe (e.g., Zn, Ca) are shown in yellow in Figure 13.8. Silicon occurs in isolated (SiO_4) tetrahedra.

Crystal form: Monoclinic; $2/m$. Crystals appear pseudo-orthorhombic because the β angle is essentially $90°$. Prismatic crystals are common (Fig. 13.7). Also occurs as two types of cruciform twins: (1) with a twin angle of nearly $90°$ making square crosses and (2) with a twin angle of about $60°$ resulting in inclined crosses (see Fig. 5.20(B)).

Physical properties: $H = 7–7\frac{1}{2}$; $G \sim 3.7$. Commonly in prismatic crystals. Also as cruciform twins. Color reddish brown to brownish black. With a resinous to vitreous luster on a fresh broken surface.

Distinguishing features: Prismatic red-brown crystals and cruciform twins are highly diagnostic. Relatively high hardness.

Staurolite: $Fe_{3-4}Al_{18}Si_8O_{48}H_{2-4}$
$C\,2/m$

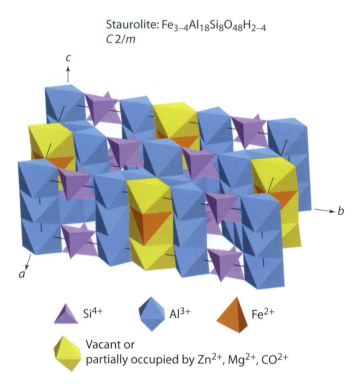

- ▲ Si^{4+}
- ◆ Al^{3+}
- ▲ Fe^{2+}
- ◆ Vacant or partially occupied by Zn^{2+}, Mg^{2+}, CO^{2+}

Figure 13.8 Perspective polyhedral representation of the structure of staurolite. A monoclinic unit cell is outlined.

Uses: It is used as an abrasive, as an alternative to garnet, such as in sand blasting and abrasive papers. Well-formed right-angle twins are sold as fairy-stone pendants and earrings; however, most of the crosses are imitations carved from fine-grained rock (i.e., subsequently dyed) or molded plastic.

In Sections 13.2–13.6, we have systematically treated five silicates (among them a group, the garnets) that are common in metamorphic alumina-rich rocks. In subsequent sections, we describe several other silicates that can be common in metamorphic assemblages as well. These are diopside, a member of the pyroxenes, and several members of the amphibole group, such as anthophyllite, cummingtonite-grunerite, tremolite-actinolite, and glaucophane. Other members of both of these chain silicate groups have already been described in Chapter 7, including enstatite, pigeonite, augite, and aegirine (members of the pyroxene group) and hornblende (a common member of the amphibole group, both in igneous and in metamorphic rocks). Related chain silicates, wollastonite and rhodonite, both members of the silicate group of pyroxenoids, follow glaucophane.

13.7 Diopside: CaMgSi₂O₆

Occurrence: Diopside is a common constituent of marble as well as other metamorphosed carbonate-rich rocks, where it occurs in assemblages such as diopside-tremolite-grossular-epidote-dolomite-calcite and diopside-forsterite-wollastonite-dolomite-calcite.

Figure 13.9 Three prismatic, dark-green diopside crystals. Two of the three (left and right front) show well-developed parting parallel to {001}.

Chemical composition and crystal structure: A complete solid solution series exists between $CaMgSi_2O_6$, diopside, and $CaFeSi_2O_6$, hedenbergite (see Fig. 7.18). The crystal structure of diopside is shown in Figure 7.23(A) and consists of infinitely extending single chains of tetrahedra parallel to the c axis that are cross-linked to octahedral chains containing Mg^{2+} and/or Fe^{2+}. The Ca^{2+} ions are located in an irregular 8-coordinated site.

Crystal form: Monoclinic; $2/m$. Commonly in prismatic crystals with square or eight-sided cross-sections (Fig. 13.9) with well-developed parting.

Physical properties: $H = 5–6$; $G \sim 3.25$. Crystals show blocky, prismatic habit. Also granular and/or massive. Generally good cleavages at 87° and 93°. Color ranges from white to light green, becoming darker green with increasing Fe^{2+} content. Hedenbergite is dark green to black.

Distinguishing features: Prismatic crystals with square or eight-sided outline. Good prismatic cleavage at 87° and 93°. Color generally white to light or dark green.

Uses: No commercial uses.

13.8 Anthophyllite: Mg₇Si₈O₂₂(OH)₂

Occurrence: A common mineral in metamorphosed Mg-rich, ultramafic rocks of medium to high grade. Unknown in igneous rocks. It also occurs in cordierite-bearing schists and gneisses.

Chemical composition and crystal structure: Anthophyllite is an end member in a partial solid solution series that extends from $Mg_7Si_8O_{22}(OH)_2$ to about $Fe_2Mg_5Si_8O_{22}(OH)_2$ (Fig. 13.10). More iron-rich compositions, from $Fe_2Mg_5Si_8O_{22}(OH)_2$ to $Fe_7Si_8O_{22}(OH)_2$, are part of a separate amphibole solid solution series known as cummingtonite-grunerite. An aluminum-rich variety of anthophyllite is known as gedrite. Members of the anthophyllite series are orthorhombic, whereas cummingtonite-grunerite compositions are monoclinic. The crystal structures

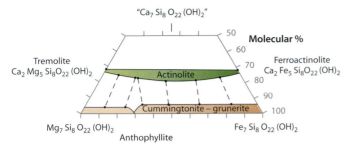

Figure 13.10 Extent of several amphibole solid-solution series in the chemical system Mg₇Si₈O₂₂(OH)₂-Fe₇Si₈O₂₂(OH)₂-"Ca₇Si₈O₂₂(OH)₂." The Ca-rich composition is given in quotation marks because it does not occur in nature. This diagram should be compared with a similar illustration for common pyroxene compositions in Figures 7.18 and 7.24. Tie lines connect coexisting amphibole compositions across a miscibility gap. (Adapted from Klein, C. (2005) Some Precambrian iron-formations (BIFs) from around the world: Their age, geologic setting, mineralogy, metamorphism, geochemistry, and origin. *American Mineralogist*, 90, 1473–1499).

of orthorhombic and monoclinic amphiboles are similar, and for this reason we refer back to an illustration of a common monoclinic amphibole, **hornblende**, in Figure 7.27(A). In the orthorhombic structure of anthophyllite, Mg^{2+} is housed in the blue sphere site that houses Ca^{2+} in hornblende. In the orthorhombic structure, this site is slightly smaller than in hornblende, so as to accommodate Mg^{2+} in a tighter, smaller 6-coordinated space.

Crystal form: Orthorhombic; $2/m\ 2/m\ 2/m$. Rarely in distinct crystals. Generally fibrous (Fig. 13.11).

Physical properties: $H = 5\frac{1}{2}–6$; $G = 2.85–3.2$. Rarely in good crystals. Commonly fibrous as well as lamellar. Color ranges from beige to brown to greenish brown.

Distinguishing features: Fibrous habit and beige to brown color are diagnostic. In hand specimens, it cannot be distinguished from members of the cummingtonite-grunerite series.

Uses: Small amounts of fibrous anthophyllite were mined in the past for the asbestos industry. Now it is of no economic importance.

13.9 Cummingtonite-grunerite: Fe₂Mg₅Si₈O₂₂(OH)₂-Fe₇Si₈O₂₂(OH)₂

Occurrence: Members of this amphibole solid solution series are of metamorphic origin. In amphibolites, cummingtonite commonly coexists with hornblende or members of the tremolite-actinolite series (see tie lines in Fig. 13.10). The more iron-rich compositions (toward the grunerite end-member) are common in metamorphosed Precambrian iron-formations.

Chemical composition and crystal structure: A complete solid solution series extends from $Fe_2Mg_5Si_8O_{22}$ to $Fe_7Si_8O_{22}(OH)_2$, (see Fig. 13.10). The crystal structure is monoclinic (that of a clinoamphibole) and is very similar to that of hornblende shown in Fig. 7.27(A). In the composition $Fe_2Mg_5Si_8O_{22}(OH)_2$ most of the Fe^{2+} is preferentially housed in the large Ca position in hornblende. In grunerite, $Fe_7Si_8O_{22}(OH)_2$, the Fe is housed across all available octahedral sites.

Crystal form: Monoclinic; $2/m$. Rarely in distinct crystals. Commonly lamellar or fibrous (Fig. 13.12).

Physical properties: $H = 5\frac{1}{2}–6$; $G = 3.1–3.6$. Most commonly fibrous but when in crystals shows perfect {110} cleavage at 56° and 124°. Color ranges from light to darker brown.

Distinguishing features: Light brown to brown color and fibrous habit are diagnostic. In hand specimens, it cannot be uniquely distinguished from anthophyllite.

Uses: A rare and highly fibrous (asbestiform) variety of cummington known as amosite was mined in the Penge region of South Africa as part of the asbestos industry. This mining district was closed in 1993 on account of the serious health hazard posed by amphibole asbestos.

13.10 Tremolite-ferroactinolite: Ca₂Mg₅Si₈O₂₂(OH)₂-Ca₂Fe₅Si₈O₂₂(OH)₂

Occurrence: Tremolite and actinolite (both compositional members of the tremolite-ferroactinolite series; see Fig. 13.10) are found in low-grade metamorphic rocks and in metamorphosed limestones and metamorphic ultramafic rocks.

Figure 13.11 Fibrous beige anthophyllite.

Figure 13.12 Fibrous light-brown cummingtonite.

Figure 13.13 Green prismatic crystals of actinolite in muscovite schist.

Hornblende, an aluminum-containing member of the series is common in igneous rocks (see Sec. 7.15) but is also a major constituent of metamorphic amphibolites.

Chemical composition and crystal structure: The magnesian end-member, $Ca_2Mg_5Si_8O_{22}(OH)_2$, is known as tremolite. An intermediate region of this solid-solution series is referred to as actinolite (see Fig. 13.10). All members of the tremolite-ferroaclinolite series are Al-poor clinoamphiboles. These compositions extend into the range of Al-rich clinoamphiboles known as hornblende. A crystal structure illustration of hornblende is shown in Figure 7.27(A).

Crystal form: Monoclinic; $2/m$. Commonly in prismatic crystals (Fig. 13.13).

Physical properties: $H = 5–6$; $G = 3.0–3.2$. Commonly in prismatic crystals. Also fibrous. Perfect {110} cleavages at 56° and 124°. Color white in tremolite but green in actinolite. Color deepens with increasing Fe^{2+} content. Hornblende is normally dark green to black (Fig. 7.28).

Distinguishing features: Good amphibole cleavage and slender prismatic crystals. Green actinolite cannot be distinguished from green hornblende in hand specimens (compare Figs. 13.13 and 13.14).

Uses: Except for earlier use of small amounts of fibrous actinolite in the asbestos industry, tremolite, actinolite, and hornblende have no industrial uses.

13.11 Glaucophane: $Na_2Mg_3Al_2Si_8O_{22}(OH)_2$

Occurrence: Glaucophane is a member of the **clinoamphibole** group and occurs in metamorphic rocks formed at low temperatures and high pressures. It is commonly found in association with lawsonite and jadeite, $NaAlSi_2O_6$ (a clinopyroxene). This assemblage characterizes the **blueschist facies**. Hand specimens rich in glaucophane are commonly referred to as **blueschists** because glaucophane is blue.

Chemical composition and crystal structure: Most glaucophane contains some Fe^{2+} substituting for Mg^{2+}, and Fe^{3+} in place of Al^{3+}. Therefore, the chemical formula given here is that of a pure end member. The structure of glaucophane is similar to that of other clinoamphiboles such as hornblende (Fig. 7.27(A)), with Na^+ in the Ca^{2+} position in hornblende and (Mg_3Al_2) in the five octahedral positions between the double tetrahedral chains.

Crystal form: Monoclinic; $2/m$. In slender, acicular crystals.

Physical properties: $H = 6$; $G = 3.1–3.4$. Color blue to blue black to black, becoming darker with increasing iron content.

Distinguishing features: In hand specimens, it is generally fibrous or acicular and shows a blue color. Distinctly blue pleochroism (Fig. 13.15) distinguishes glaucophane from most other minerals.

Uses: There is no commercial use for glaucophane. However, a similar clinoamphibole, rich in Fe^{2+} and Fe^{3+}, known as riebeckite occurs in a highly fibrous (asbestiform) habit and is known as crocidolite (Fig. 13.16). This mineral, also known as blue asbestos, represented about 2–3% of the total global production of asbestos; about 95% was chrysotile, also known as white asbestos. Crocidolite is a much greater health hazard than chrysotile (see Sec. 17.3.2). Brown crocidolite (colored brown

Figure 13.14 Dark-green acicular crystals of hornblende in a feldspar matrix.

Figure 13.15 Glaucophane in a thin section as seen in plane polarized light. It is highly pleochroic and shows the typical lavender-blue color in one crystallographic direction. Width of field of view is 3 mm.

Figure 13.16 Two broken parts of a hand specimen of crocidolite, a highly fibrous (asbestiform) variety of riebeckite known in the trade as blue asbestos.

Wollastonite: CaSiO₃
P $\bar{1}$

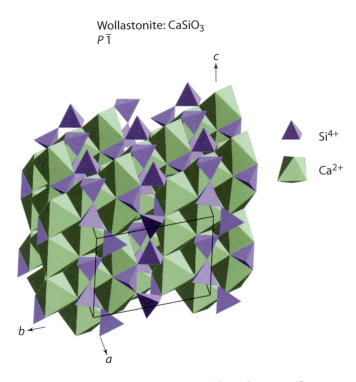

Figure 13.17 Perspective view of the polyhedral crystal structure of wollastonite. A triclinic unit cell is outlined.

due to oxidation) and replaced by quartz, known as tiger's eye, shows attractive chatoyancy and is used in jewelry.

13.12 Wollastonite: CaSiO₃

Occurrence: A high-grade metamorphic mineral commonly in contact metamorphic zones in impure limestone. It is the result of the following reaction:

$$CaCO_3 + SiO_2 \rightarrow CaSiO_3 + CO_2$$
calcite   quartz   wollastonite

Chemical composition and crystal structure: Most chemical analyses are close to the end member composition $CaSiO_3$. Its crystal structure is similar but not identical to that of the pyroxenes, with single tetrahedral chains. However, its structure with regular kinks in the tetrahedral chains, is known as a **pyroxenoid**. An illustration of the crystal structure of wollastonite is given in Figure 13.17. It consists of infinitely extending tetrahedral chains parallel to the c axis, in which the smallest repeat unit in the chains consists of three adjoining tetrahedra. In other words, there is a kink in the chain after every third tetrahedron. The chains are cross-linked by distorted CaO octahedra.

Crystal form: Triclinic; $\bar{1}$. Rarely in tabular crystals. Commonly massive.

Physical properties: H = 5–5½; G = 2.9. Perfect {100} and {001} cleavages producing splintery fragments. Commonly in cleavable masses (Fig. 13.18). Color ranges from colorless to white, or gray.

Distinguishing features: Characterized by two perfect cleavages at ~84°. May resemble white to light-colored tremolite but is distinguished by different cleavage angles.

Uses: Wollastonite with a fibrous habit, but with fibers that are not flexible as in chrysotile (i.e., white asbestos) is used commercially as a substitute for asbestos. Wollastonite is not considered a health hazard. Wollastonite is used in fire-resistant products such as floor and roofing tiles and insulation. It also has a major use in the ceramics industry and in glazes. It is a filler in paint and paper production.

Figure 13.18 Massive white wollastonite exhibiting well-developed cleavage surfaces parallel to {100} and {001}.

13.13 Rhodonite: MnSiO₃

Occurrence: Rhodonite is strictly a metamorphic mineral and is restricted to assemblages in metamorphosed manganese deposits and manganese-rich iron-formations. It is commonly the result of the following reaction:

$$\underset{\text{rhodochrosite}}{\text{MnCO}} + \underset{\text{quartz}}{\text{SiO}_2} \rightarrow \underset{\text{rhodonite}}{\text{MnSiO}_3} + \text{CO}_2$$

Chemical composition and crystal structure: Rhodonite is seldom pure MnSiO₃ and up to 20 molecular of the Mn may be replaced by Ca, Mg, or Fe. The perspective polyhedral structure illustration in Figure 13.19 is that of a calcium-containing rhodonite. It consists of tetrahedral SiO₃ chains parallel to the *c* axis with a unit repeat of five twisted tetrahedra (wollastonite has a repeat of three twisted tetrahedra; see Fig. 13.17). Both rhodonite and wollastonite are part of the silicate group known as **pyroxenoids**. The Mn is distributed over the octahedral sites, with Ca in an irregular site with 7-fold coordination.

Crystal form: Triclinic; $\bar{1}$. Crystals commonly tabular parallel to {001}.

Physical properties: $H = 5\frac{1}{2}-6$; $G = 3.4-3.7$. Two perfect cleavages, parallel to {110} and {1$\bar{1}$0} (Fig. 13.20). Color ranges from rose red to pink to pinkish brown.

Distinguishing features: Recognized by its pink color and two cleavage directions at about 90°. Distinguished from rhodochrosite by greater hardness.

Uses: Its pink color and relatively high hardness make it a desirable material for small carvings. Massive material is used to fashion beads and cabochons.

Figure 13.20 Massive rhodonite showing well-developed cleavage surfaces parallel to {110} and {1$\bar{1}$0}.

13.14 Talc: Mg₃Si₄O₁₀(OH)₂

Occurrence: Talc is common as a retrograde mineral in metamorphosed ultramafic rocks formed by the alteration of magnesium silicates, such as olivine, pyroxenes, and amphiboles. Characteristic of low-grade metamorphic rocks, where, in massive form, **soapstone**, it may compose the entire rock mass. Also in schists, such as talc schist, and in shear zones in metamorphic rocks.

Chemical composition and crystal structure: There is little variation in the chemical composition of talc as given in Mg₃Si₄O₁₀(OH)₂. The crystal structure of talc consists of infinitely extending trioctahedral sheets sandwiched between two opposing infinitely extending tetrahedral sheets resulting in *t-o-t* layers (Fig. 13.21). In a trioctahedral sheet, all

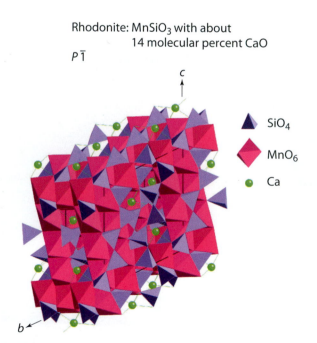

Rhodonite: MnSiO₃ with about 14 molecular percent CaO

$P\bar{1}$

SiO₄
MnO₆
Ca

Figure 13.19 Perspective view of the polyhedral crystal structure of a calcium-containing rhodonite. The *c* and *b* axes of a triclinic unit cell are shown.

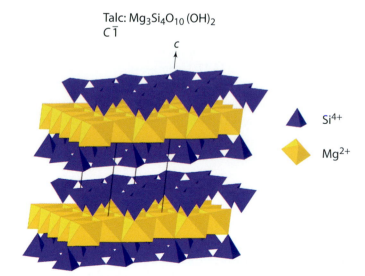

Talc: Mg₃Si₄O₁₀(OH)₂

$C\bar{1}$

Si⁴⁺
Mg²⁺

Figure 13.21 Perspective view of the polyhedral crystal structure of talc, which consists of infinitely extending *t-o-t* layers. Only the *c* axis direction of the triclinic unit cell is visible.

the octahedral positions are fully occupied by Mg^{2+} without any intervening vacancies (see Sec. 7.17 for further discussion). Talc is a member of the group of **layer silicates**, or **phyllosilicates**.

Crystal form: Triclinic; $\bar{1}$. Crystals rare. Usually in foliated masses (Fig. 13.22).

Physical properties: $H = 1$ (marks cloth); $G \sim 2.75$. Generally found in foliated masses, but also massive. Color ranges from apple green to gray and white. Commonly with a pearly luster.

Distinguishing features: It is identified by its foliated (micaceous) habit, softness, excellent cleavage, and greasy feel.

Uses: Talc is an extremely versatile mineral. In the paint industry, it is used as an extender and filler that improves the integrity of the paint. It serves as a filler in the paper industry to improve ink receptivity. It is a common constituent of ceramic tiles in the United States, where wall tiles are made of a mixture of 50–70% talc, 25–35% ball clay, with the remainder wollastonite. It is a filler in the production of plastics. In the cosmetics industry, talc is used in baby and body powders, facial products and creams and lotions. Soapstone, a metamorphic rock composed essentially of fine-grained talc, is used extensively for carving of ornamental objects such as Inuit carvings.

13.15 Chlorite: $(Mg, Al, Fe)_6(Si, Al)_4O_{10}\cdot(OH)_8$

Occurrence: A common constituent of low-grade metamorphic rocks that are part of the **greenschist facies**. The term *greenschist* describes a fine-grained, greenish rock that derives

Figure 13.22 Massive foliated talc.

its color from abundant green chlorite. Chlorite is present in many types of metamorphic rocks. It also occurs in igneous rocks as a retrograde alteration product of ferromagnesian minerals as a result of the cooling process. It is present in hydrothermal veins and ore deposits.

Chemical composition and crystal structure: A generalized formula for chlorite, in which there is considerable variation among Mg, Fe, and Al, is given here. The structure of a magnesian chlorite is shown in Figure 13.23 and consists of infinitely extending *t-o-t* layers of composition $[Mg_3(AlSi_3O_{10})(OH)_2]^-$ alternating with brucitelike sheets of composition $[Mg_2Al(OH)_6]^+$. The composition of brucite is $Mg(OH)_2$, but because chlorite has Al^{3+} as well in this sheet structure, it is referred to as brucitelike.

Crystal form: Triclinic or monoclinic depending on the stacking pattern of the layers along the *c* axis. Such stacking polymorphs are referred to as **polytypes** (see Sec. 5.8); 1, $\bar{1}$, *m*, or 2/*m*. Crystals have a habit similar to that shown by members of the mica group (Fig. 13.24). Usually in foliated masses.

Physical properties: H = 2–2½; G = 2.6–3.3. Cleavage {001} perfect. Folia flexible but not elastic. Color is commonly in various shades of green.

Distinguishing features: Characterized by its micaceous habit and green color, as well as good cleavage, and low hardness.

Uses: Chlorite is used as a catalyst or as a raw material in the commercial production of zeolites.

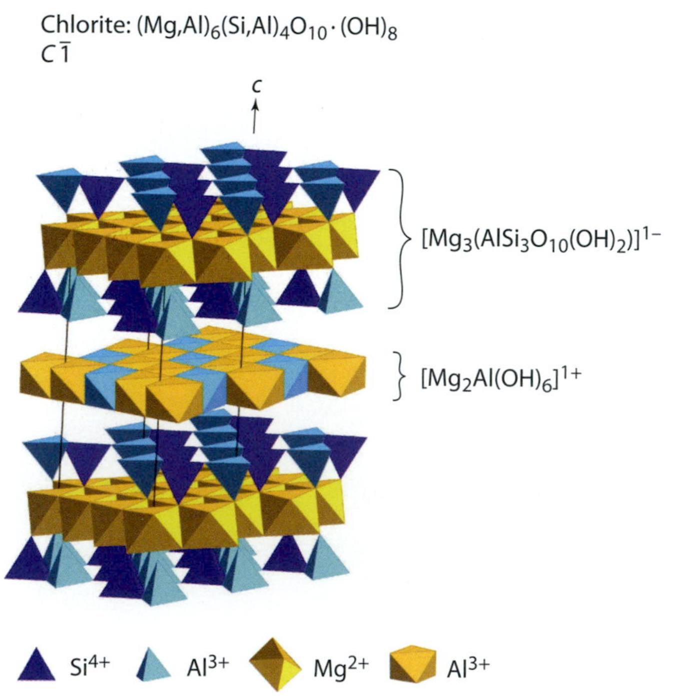

Chlorite: $(Mg,Al)_6(Si,Al)_4O_{10}\cdot(OH)_8$
$C\bar{1}$

Figure 13.24 A random intergrowth of dark green, tabular (pseudohexagonal) chlorite crystals with prominent {001} basal pinacoids. This crystal habit is similar to that of the mica group.

13.16 Antigorite: $Mg_3Si_2O_5(OH)_4$

Occurrence: Antigorite, lizardite, and chrysotile are three members of a silicate group known as serpentine minerals. Together, in various mixtures, they compose a rock known as serpentine, or **serpentinite**. This is a rock type that results from the hydrothermal alteration (and retrograde metamorphism) of olivine- and pyroxene-rich rocks such as in peridotites and pyroxenites. Serpentine may occur as pseudomorphs of these high-temperature magnesian minerals.

Chemical composition and crystal structure: Antigorite, lizardite, and chrysotile are three polymorphs of the essentially constant composition, $Mg_3Si_2O_5(OH)_4$. All three are different structural arrangements of infinitely extending tetrahedral-octahedral (*t-o*) layers (Fig. 2.4) without any interlayer cations (as in the micas, Figs. 7.29 and 7.31) or an interlayer sheet (as in chlorite, Fig. 13.23). In antigorite, the (*t-o*) layers are not straight but corrugated as in the diagrammatic sketch in Figure 13.25(A), which is based on the HRTEM image in Figure 13.25(B).

Crystal form: Monoclinic; 2/*m* or *m*. Never in well-formed crystals. Commonly massive and fine grained (Fig. 13.26).

Physical properties: H ~4; G ~2.55. Massive and fine grained. Color ranges from light to darker shades of green, commonly with a waxlike luster.

Distinguishing features: Relatively low hardness and greenish color. Occurs commonly in association with fibrous chrysotile (Fig. 13.28).

Uses: The massive variety of serpentine (a very fine-grained mixture of antigorite and lizardite) is commonly used in carvings. Verd antique is an ornamental stone consisting of serpentine mottled or veined with white marble.

c

$[Mg_3(AlSi_3O_{10}(OH)_2)]^{1-}$

$[Mg_2Al(OH)_6]^{1+}$

▲ Si^{4+} ▲ Al^{3+} ◆ Mg^{2+} ⬣ Al^{3+}

Figure 13.23 Perspective view of the polyhedral structure of chlorite of composition $(Mg, Al)_6 (Si, Al)_4O_{10}\cdot(OH)_8$. This structure consists of infinitely extending *t-o-t* layers with brucitelike interlayers of composition $(Mg_2Al(OH)_6)^+$. Only the vertical *c* axis of the triclinic unit cell is visible.

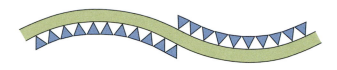

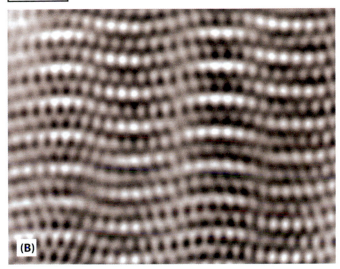

(A)

Brucite layer, Mg (OH)₂

Tetrahedral layer, Si₂O₅

1.4 nm

(B)

Figure 13.25 (A) Diagrammatic sketch of the curved and corrugated (*t-o*) layers in antigorite, as deduced from the high-resolution transmission electron microscope (HRTEM) image given in B. (B) HRTEM image as seen along the *b* axis of the antigorite structure. The vague black dots are images of the locations of Mg and Si. (Courtesy of Alain Baronnet, CINaM-CNRS, Centre Interdisciplinaire de Nanoscience de Marseilles, France.)

Figure 13.26 Hand specimen of massive antigorite with a polished and smoothly striated surface that is the result of friction along a fault plane. This surface is commonly referred to as slickensided.

13.17 Chrysotile: Mg₃Si₂O₅(OH)₄

Occurrence: Chrysotile is commonly closely associated with antigorite and lizardite in serpentine and occurs in the same geological associations as described for antigorite (Sec. 13.16).

Chemical composition and crystal structure: Chrysotile is the polymorph of Mg₃Si₂O₅(OH)₄ in which the (*t-o*) layers are tightly curled in the shape of drinking straws (Fig. 13.27). Additional illustrations of this unique crystal structure are given in Figure 2.4(B)–(D).

Crystal form: Orthorhombic or monoclinic depending on the stacking of the (*t-o*) layers. Always as fine fibers or matted intergrowths.

Physical properties: H = 2½; G ~2.5. Fibrous habit with fibers flexible but with great tensile strength. White to light gray to light green in color (Fig. 13.28).

Distinguishing features: Recognized by its fibrous and flexible habit, light greenish color and low hardness.

Uses: Chrysotile (i.e., white asbestos) constituted about 90 to 95% of all asbestos production before 1980, but with the public's concern about the risk of environmental exposure to "asbestos" in the early 1970s, consumption of chrysotile declined rapidly. At present, strict regulations by the Environmental Protection

20 nm

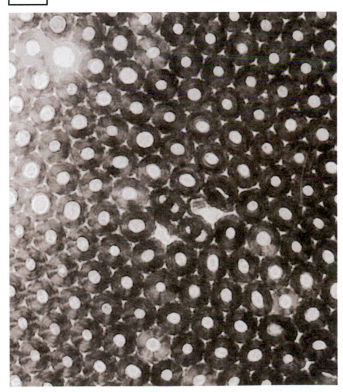

Figure 13.27 Transmission electron microscope (TEM) photograph of the cross-sections of tightly packed chrysotile fibers. (Courtesy of Alain Baronnet, CINaM-CNRS, Centre Interdisciplinaire de Nanoscience de Marseilles, France.)

Figure 13.28 Thin veins of fibrous chrysotile that cut across a specimen of massive serpentine. The nonfibrous part of the specimen consists of a mixture of antigorite and lizardite.

Epidote: $Ca_2Fe^{3+}Al_2O(SiO_4)(Si_2O_7)(OH)$
$P\,2_1/m$

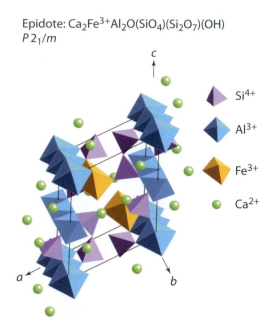

Figure 13.29 Perspective view of the polyhedral crystal structure of epidote. A monoclinic unit cell is outlined.

Agency (EPA) has stopped all mining of chrysotile in the United States. Canada, however, still has extensive chrysotile asbestos deposits in the Eastern Townships of the province of Quebec that are actively worked as open-pit mines (see Sec. 17.3.2).

13.18 Epidote: $Ca_2Fe^{3+}Al_2O(SiO_4)(Si_2O_7)(OH)$
Clinozoisite: $Ca_2Al_3O(SiO_4)(Si_2O_7)(OH)$

Occurrence: Epidote is a common mineral in medium grades of regional metamorphism, especially in mafic igneous rocks metamorphosed to amphibolites. It also occurs as the filling of vesicles and fractures in basalts. Clinozoisite shows similar occurrences and is also found as a retrograde alteration product of calcic feldspar.

Chemical composition and crystal structure: There is a complete solid-solution series from the epidote formula given here to that of clinozoisite, $Ca_2Al_3O(SiO_4)(Si_2O_7)(OH)$, through the substitution of Fe^{3+}, in epidote, by Al^{3+} in clinozoisite. The structure of epidote is illustrated in Figure 13.29. It consists of edge-sharing $(AlO)_6$ octahedra parallel to the b axis that are linked by (SiO_4) tetrahedra and (Si_2O_7) double tetrahedral groups. Ca is in 8-coordination and Fe^{3+} is in octahedral coordination. Clinozoisite is isostructural with epidote.

Crystal form: Monoclinic; $2/m$. Crystals are commonly prismatic parallel to the b axis.

Physical properties: $H = 6–7$; $G = 3.2–3.5$. {001} cleavage perfect, and {100} imperfect. Epidote ranges in color from pistachio green (Fig. 13.30) to yellowish green to black. Clinozoisite occurs in pale green to gray colors.

Distinguishing features: Epidote is characterized by its distinctive green color and one direction of perfect cleavage. In hand specimens, it is impossible to distinguish clinozoisite from epidote, unless the epidote is distinctly pistachio green in color. In thin sections, clinozoisite exhibits distinctive blue interference colors.

Uses: Some epidote, of gem quality, is cut as faceted stones or as cabochons and beads. Clinozoisite has no commercial use. Tanzanite, a purple-blue colored vanadium-bearing variety of zoisite (with orthorhombic symmetry) is a popular gemstone.

Figure 13.30 Pistachio-green epidote crystals lining a vug with white albite in its center. White calcite lines the lower left side.

13.19 Cordierite: $(Mg, Fe)_2Al_4Si_5O_{18} \cdot nH_2O$

Occurrence: A common metamorphic mineral in pelitic rocks, resulting from the metamorphism of fine-grained aluminous sediments. Its occurrence is favored by low pressures or high temperatures. It is, therefore, common in contact metamorphosed rocks and in low-pressure regional metamorphic assemblages.

Chemical composition and crystal structure: Although there is some substitution of Mg^{2+} by Fe^{2+}, most cordierites are Mg-rich. Most analyses show appreciable but variable H_2O. The structure of the common, low-temperature form of cordierite, given in Figure 13.31, is orthorhombic (pseudohexagonal) with 6-fold tetrahedral rings. Thus, cordierite is a member of the silicate group known as **ring silicates**, or **cyclosilicates**. Mg and small amounts of Fe occur in the octahedral sites, with Al and Si distributed among the tetrahedral sites. In the 6-fold rings, two of the tetrahedra are occupied by Al and the other four by Si.

Crystal form: Orthorhombic; $2/m\ 2/m\ 2/m$. Crystals are usually short prismatic.

Physical properties: $H = 7{-}7\frac{1}{2}$; $G \sim 2.65$. Cleavage {010} poor. Most commonly massive or as embedded grains. Color is generally some shade of bluish to bluish gray (Fig. 13.32).

Distinguishing features: Cordierite may resemble feldspar as well as quartz, and it is distinguished with difficulty from these two minerals in hand specimens. In thin sections, the different optical properties clearly distinguish the three.

Uses: Cordierite, because of its extremely low coefficient of thermal expansion, is used in the manufacturing of kiln furniture. It is also used for honeycomblike catalyst supports in auto-emission reduction devices.

Figure 13.32 Massive cordierite with bluish-gray color.

13.20 Vesuvianite: $Ca_{19}(Al, Mg, Fe)_{13}(Si_2O_7)_4$ $(SiO_4)_{10}(O, OH, F)_{10}$

Occurrence: Vesuvianite is a common constituent of contact metamorphosed limestones. It also occurs in regionally metamorphosed limestones and serpentinites.

Chemical composition and crystal structure: In the formula here, there is some substitution of Na for Ca and Mn^{2+} for Mg. Figure 13.33 illustrates the structure that contains independent (SiO_4) tetrahedra as well as double (Si_2O_7) tetrahedral groups. Al^{3+} is in octahedral coordination, and so is (Mg^{2+}, Fe^{2+}) in a separate octahedral site. The sites occupied by Ca^{2+} are irregular 7- and 8-fold polyhedra.

Vesuvianite: $Ca_{19}(Al,Mg,Fe)_{13}(Si_2O_7)_4(SiO_4)_{10}(O,OH,F)_{10}$
$P\,4/n\,2/n\,2/c$

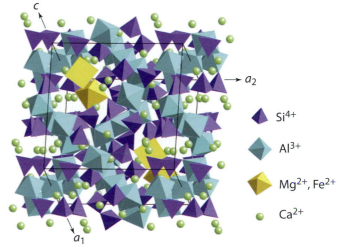

Figure 13.33 Perspective view, along the *c* axis, of the polyhedral structure of vesuvianite. A tetragonal unit cell is outlined.

Cordierite: $(Mg,Fe)_2Al_4Si_5O_{18} \cdot n\,H_2O$
$C\,2/c\,2/c\,2/m$

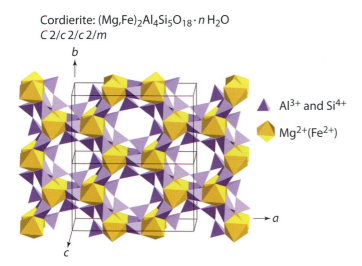

▲ Al^{3+} and Si^{4+}
◆ $Mg^{2+}(Fe^{2+})$

Figure 13.31 Polyhedral structure of cordierite in a perspective view along the *c* axis. Two adjoining orthorhombic unit cells are shown.

Figure 13.34 Prismatic, vertically striated, brown crystal of vesuvianite in a white marble matrix full of fine-grained green diopside.

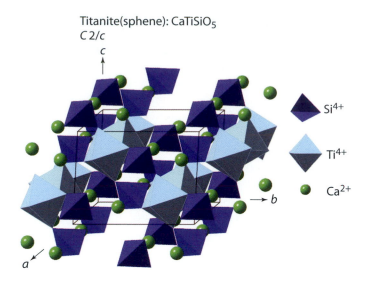

Titanite(sphene): $CaTiSiO_5$
$C\,2/c$

Si^{4+}

Ti^{4+}

Ca^{2+}

Figure 13.35 Perspective view of the polyhedral structure of titanite. A monoclinic unit cell is outlined.

Crystal form: Tetragonal; $4/m\,2/m\,2/m$. Crystals are commonly prismatic with vertical striations (Fig. 13.34).

Physical properties: $H = 6½$; $G \sim 3.4$. Poor {010} cleavage. Common as striated columnar aggregates. Color is usually brown or green. More rarely yellow, blue, or red.

Distinguishing features: Brown tetragonal prisms and striated columnar masses are characteristic.

Uses: No commercial applications.

13.21 Titanite (sphene): $CaTiSiO_5$

Occurrence: A common accessory mineral in metamorphic and igneous rocks, and as detrital grains in sedimentary rocks. It can be a major constituent in metamorphic rocks rich in Ca and Ti, such as in metabasalts and pyroxenites. Titanite is the accepted mineral name, but the name sphene is used in the jewelry industry.

Chemical composition and crystal structure: The composition is essentially that of the formula given here, with small amounts of rare earth elements and of Fe, Al, Mn, Mg, and Zr in substitution. The structure of titanite, given in Figure 13.35, consists of corner-sharing TiO_6 octahedra that form kinked chains parallel to the a axis. These chains are cross-linked by independent (SiO_4) tetrahedra. The Ca is housed in 7-coordinated sites.

Crystal form: Monoclinic; $2/m$. Wedge-shaped crystals are common (Fig. 13.36).

Physical properties: $H = 5–5½$; $G \sim 3.45$. Distinct {110} cleavage. Color ranges from gray, to brown, yellow or black. Luster is resinous to adamantine.

Distinguishing features: Characterized by wedge-shaped crystals and high luster. Hardness is greater than that of sphalerite, ZnS, and less than that of staurolite.

Uses: Sphene is a rare gem with exceptional brilliance and fire. Because of its relatively low hardness, it is not suitable for many jewelry applications but is much prized by gem collectors.

13.22 Scapolite: $3NaAlSi_3O_8 \cdot NaCl$– $3CaAl_2Si_2O_8 \cdot CaSO_4$ (or $CaCO_3$)

Occurrence: Common in marbles and some amphibolites, as well as in contact metamorphic assemblages. It is a metamorphic mineral that forms instead of feldspar in some metamorphic rocks that have a high concentration of sulfate or carbonate.

Chemical composition and crystal structure: Scapolite is a metamorphic mineral with compositions suggestive of that of the feldspars. The two end-member compositions given here have specific species names; the sodium end member

Figure 13.36 Light greenish-yellow transparent wedge-shaped crystals of titanite (sphene).

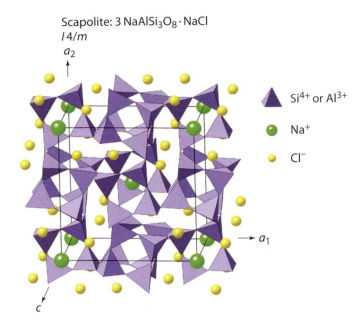

Scapolite: 3 NaAlSi$_3$O$_8$ · NaCl

I4/m

Si^{4+} or Al^{3+}

Na$^+$

Cl$^-$

Figure 13.37 A perspective view, along the *c* axis, of the crystal structure of the Na-rich end member of the scapolite group of minerals. A tetragonal unit cell is outlined.

composition is known as marialite and the calcic composition is meionite. The name *scapolite* refers to the whole series. The formulas are written such that they are immediately reminiscent of the two plagioclase feldspar end members: albite, NaAlSi$_3$O$_8$, and anorthite, CaAl$_2$Si$_2$O$_6$, with one formula weight of NaCl, CaSO$_4$, or CaCO$_3$. The crystal structure of the Na end member is given in Figure 13.37. It is a relatively open **framework (tectosilicate)** structure with large cavities that house the Na$^+$ and Cl$^-$ ions. Al and Si are housed in the tetrahedra. It is a member of the group known as **feldspathoids**.

Crystal form: Tetragonal; 4/*m*. Crystals are usually prismatic (Fig. 13.38).

Physical properties: H = 5–6; G ~2.65. {100} and {110} cleavages distinct but imperfect. Crystals exhibit a prismatic habit with square cross-section. Color ranges from white to pale green and rarely yellow or bluish.

Distinguishing features: When massive, it resembles feldspar but has a fibrous appearance on cleavage surfaces. Square crystal habit and four cleavage directions at 45° are characteristic.

Uses: Scapolite has no industrial uses, but transparent crystals have been used as gemstones.

13.23 Lawsonite: CaAl$_2$(Si$_2$O$_7$)(OH)$_2$·H$_2$O

Occurrence: Lawsonite is a common mineral in high-pressure, low-temperature metamorphosed basaltic rocks. It is typical of the blueschist facies.

Chemical composition and crystal structure: The composition of lawsonite is essentially that of anorthite, CaAl$_2$Si$_2$O$_8$ + H$_2$O. Upon heating, lawsonite converts to anorthite. The crystal

Figure 13.38 White prismatic crystals of scapolite. They are a combination of {010} and {110} prisms capped by {011}, a tetragonal dipyramid.

structure (Fig. 13.39) consists of chains of edge-sharing (AlO$_6$) octahedra parallel to the *b* axis that are linked by double (Si$_2$O$_7$) tetrahedral groups. Large cavities house the Ca^{2+} ions in 8-fold coordination with oxygen.

Crystal form: Orthorhombic; 222. Usually in tabular (Fig. 13.40) or prismatic crystals.

Physical properties: H = 6; G ~3.1. Perfect {010} and {010} cleavages that intersect at right angles. Color ranges from colorless to white, pale blue to grayish blue. Translucent.

Distinguishing features: Most easily recognized in thin sections. Association with other minerals that are part of

Lawsonite: $CaAl_2(Si_2O_7)(OH)_2 \cdot H_2O$
$C222_1$

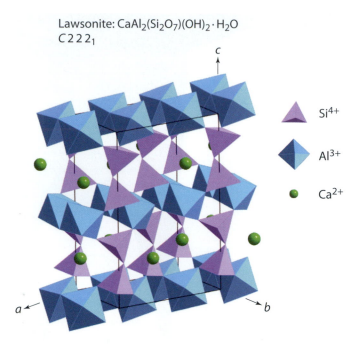

$\blacktriangle$ Si^{4+}

$\blacklozenge$ Al^{3+}

$\bullet$ Ca^{2+}

Figure 13.39 Perspective view of the crystal structure of lawsonite. An orthorhombic unit cell is outlined.

Pumpellyite: $Ca_2MgAl_2(SiO_4)(Si_2O_7)(OH)_2 \cdot H_2O$
$A 2/m$

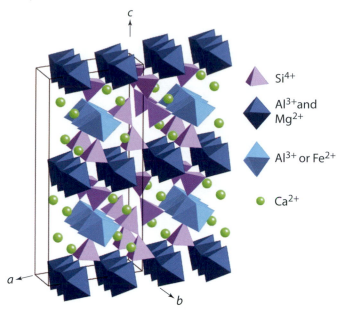

$\blacktriangle$ Si^{4+}

$\blacklozenge$ Al^{3+} and Mg^{2+}

$\blacklozenge$ Al^{3+} or Fe^{2+}

$\bullet$ Ca^{2+}

Figure 13.41 Perspective view of the polyhedral crystal structure of pumpellyite. A monoclinic unit cell is outlined.

glaucophane schists, such as glaucophane, jadeite, garnet, and pumpellyite, is most diagnostic.

Uses: None.

13.24 Pumpellyite: $Ca_2MgAl_2(SiO_4)(Si_2O_7)(OH)_2 \cdot H_2O$

Occurrence: Pumpellyite is a common constituent of glaucophane schists in association with glaucophane, lawsonite, and epidote. It also occurs as a low-grade metamorphic product in amygdaloidal basalts.

Chemical composition and crystal structure: The composition of pumpellyite is essentially that of the formula given above with some substitution of Fe^{3+} for Al, and Fe^{2+} and Fe^{3+} for Mg. The crystal structure (Fig. 13.41) consists of edge-sharing octahedra, parallel to the b axis, occupied by Mg and Al. Ca is located in an irregular 7-coordinated site. Si occurs in (SiO_4) tetrahedra and double tetrahedral groups (Si_2O_7).

Crystal form: Monoclinic; $2/m$. Crystals are bladed, commonly elongated along the b axis.

Physical properties: H = 5.5; G = 3.2. Commonly in stellate clusters (Fig. 13.42), fibrous, or in dense mats. Color ranges from green to bluish green, greenish black to brown.

Distinguishing features: Difficult to distinguish in a hand specimen from clinozoisite or epidote. Better characterized in thin sections, although X-ray diffraction may be needed for positive identification.

Uses: None.

13.25 Topaz: $Al_2SiO_4(F, OH)_2$

Occurrence: As a metamorphic mineral, it is the result of the metamorphism of bauxite. It also occurs in association with kyanite and sillimanite. It is found in both volcanic and intrusive

Figure 13.40 White to light-gray crystals of lawsonite.

Figure 13.42 Stellate, green crystals of pumpellyite filling vesicles in basalt.

Topaz: $Al_2SiO_4(F,OH)_2$
$P2_1/n\,2_1/m\,2_1/a$

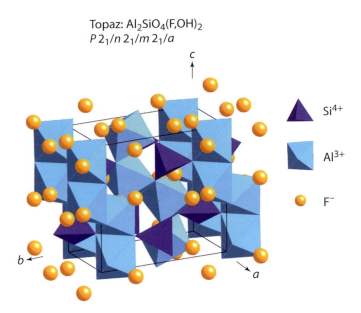

Figure 13.43 Perspective view of the crystal structure of topaz. An orthorhombic unit cell is outlined.

felsic rocks. It is also present in hydrothermal ore deposits that carry tungsten, tin, and molybdenum mineralization.

Chemical composition and crystal structure: The composition of topaz is essentially as stated in the formula here, with most of the $(OH)^-$ generally replaced by F^-. The structure (Fig. 13.43) consists of chains parallel to the c axis of (AlO_4F_2) octahedra cross-linked by independent (SiO_4) tetrahedra. F is coordinated only to Al.

Crystal form: Orthorhombic; $2/m\,2/m\,2/m$. Commonly in prismatic crystals terminated by dipyramids (Fig. 13.44).

Physical properties: $H = 8$; $G \sim 3.5$. Perfect $\{001\}$ cleavage. Transparent to translucent. Color ranges from colorless, to yellow, pink, bluish and greenish.

Distinguishing features: Recognized by its crystal form, high hardness (8), basal cleavage, and high specific gravity.

Uses: Topaz of various colors is used as a gem (see Section 15.18).

13.26 Corundum: Al_2O_3

Occurrence: Formed by extremely high-grade contact metamorphism of aluminous (pelitic) rocks. Can be associated with spinel to form **emery**. Also found in some syenites and nepheline syenites.

Chemical composition and crystal structure: The composition is essentially pure Al_2O_3 but trace amounts of Cr (in ppm) are responsible for the red gem variety known as ruby, and trace amounts of Fe and Ti are the cause for the blue color of the blue gem variety known as sapphire (see Fig. 3.7). The structure of corundum is based on oxygen in approximately hexagonal closest packing with Al in octahedral coordination (Fig. 13.45). Only two-thirds of the octahedra in each layer are occupied by Al and one-third are vacant (known as **dioctahedral**; see Sec. 7.16).

Crystal form: Hexagonal; $\bar{3}\,2/m$. Crystals are commonly hexagonal prisms that taper toward the ends. Also as tapering hexagonal dipyramids (Fig. 13.46), rounded into barrel shapes with horizontal striations.

Physical properties: $H = 9$; $G \sim 4$. Parting on $\{0001\}$ and $\{10\bar{1}1\}$. Color ranges from colorless to some shade of brown, pink, or blue.

Corundum: Al_2O_3
$R\bar{3}\,2/c$

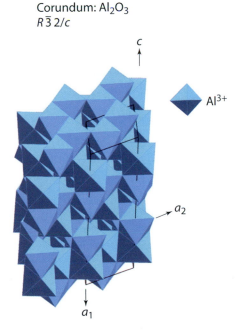

Figure 13.45 Perspective view of the polyhedral structure of corundum. A primitive hexagonal unit cell is outlined for this rhombohedral structure. See caption for Figure 5.25 for further explanation.

Figure 13.44 Prismatic crystal of translucent topaz terminated by a dipyramid, surrounded by light-gray mottled quartz.

Figure 13.46 Bluish gray barrel-shaped dipyramidal crystals of corundum with horizontal striations. The white matrix mineral is feldspar.

Distinguishing features: Characterized by its barrel-shaped crystal form, high hardness, specific gravity, and parting.

Uses: Emery, a granular intergrowth of corundum, magnetite, and spinel has been used extensively as an abrasive. The hardness of emery ranges from 7 to 9 depending on the mineral composition. Because of the much better purity and uniformity of synthetic abrasives such as silicon carbide, the commercial market for emery has declined. Ruby and sapphire are two of the most important gemstones (Section 15.18).

13.27 Chabazite: $Ca_2Al_2Si_4O_{12} \cdot 6H_2O$

Occurrence: Chabazite, in association with other members of the zeolite group of silicates, occurs in cavities (vesicles) of basalt flows. Zeolites are the result of low-temperature hydrothermal alteration and/or metamorphism in the **zeolites facies**.

Chemical composition and crystal structure: The ideal composition is as given here, but there is considerable replacement of Ca by Na and K. Zeolite compositions can be thought of as approximating hydrated feldspars and feldspathoids. All members of the zeolite group of silicates have framework structures with large openings that house the Ca(Na, and/or K) as well as variable percentages of H_2O, ranging from 10–20 weight %. The structure of chabazite is shown in Figure 13.47(A). The H_2O molecules that reside in the large openings are not shown. It shows the presence of channels with diameters of ~3.9 Å. The

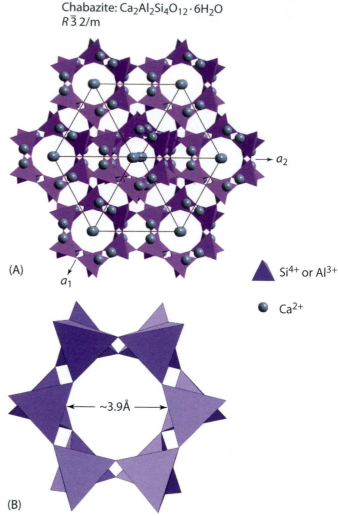

Chabazite: $Ca_2Al_2Si_4O_{12} \cdot 6H_2O$
$R\bar{3}2/m$

(A)

$\blacktriangle$ Si^{4+} or Al^{3+}

$\bullet$ Ca^{2+}

(B)

Figure 13.47 (**A**) Perspective view, along the c axis, of the polyhedral structure of chabazite. A rhombohedral unit cell is outlined. (**B**) Enlarged view of the 6-fold rings of tetrahedra in this structure.

presence of these atom-sized openings allows zeolites to be used as molecular sieves, such that molecules too large to pass through are excluded, but molecules of a size smaller than the openings can flow through (see Sec. 16.7). Figure 13.47(B) gives an enlargement of one structural ring with its internal diameter.

Crystal form: Hexagonal; $\bar{3}2/m$. Usually in rhombohedral $\{10\bar{1}1\}$ crystals (Fig. 13.48).

Physical properties: $H = 4–5$; G ~2.10. Color ranges from white to yellow, pink, and red. Transparent to translucent.

Distinguishing features: Usually recognized by its rhombohedral crystal habit. Distinguished from calcite by lack of effervescence in HCl and much poorer cleavage.

Uses: There are ~60 naturally occurring members of the zeolite group. In addition, other zeolites structure types are manufactured synthetically. The industrial applications of zeolites derive from their spacious channels and the water molecules and variable amounts of Ca^{2+}, Na^+, and K^+ that

Figure 13.48 Pink rhombohedral crystals of chabazite in a small vug that is lined with white calcite crystals.

are housed in the channels. The water molecules are so weakly bonded that when a zeolite is heated, the water in the channels is driven off easily and continuously as the temperature rises, leaving the framework structure intact. Such a dehydrated zeolite structure can be completely rehydrated when immersed in water. Zeolites in their dehydrated state can absorb molecules, other than water, as long as their overall size is compatible with the channel dimensions in the specific structure. Zeolites, as mentioned earlier, are used as molecular sieves as well, so as to separate hydrocarbon molecules that are smaller or larger than the sieve size of the channels (see Sec. 16.7).

Summary

This chapter systematically introduced 26 minerals found in metamorphic rocks. The first 18 are common constituents of metamorphic assemblages, and the last 8 represent accessory minerals. The majority are of prograde metamorphic origin, with a small number the result of retrograde reactions.

   Members of the mica group and quartz and feldspar are common constituents of metamorphic rocks as well, but because they are also abundant in igneous rocks, they are discussed in Chapter 7.

- We began with garnet, which has a wide range of chemical composition and is found in many metamorphic mineral assemblages.

- The composition Al_2SiO_5 occurs as three different polymorphs, each of which has a very different, well-determined pressure-temperature stability field. It is important that you are familiar with the phase diagram for these polymorphs, as you will use it in Chapter 14.

- Staurolite is common in association with kyanite and is best recognized by its dark-brown color, prismatic crystal habit, and occurrence in crosslike twins.

- Diopside, a member of the pyroxene group and, therefore, a chain silicate, is common in metamorphosed carbonate-rich rocks such as marble.

- Several amphiboles, also chain silicates but with double chains (instead of single chains as in pyroxenes), are common metamorphic minerals. They occur over a wide range of metamorphic conditions with the metamorphic rock known as **amphibolite** consisting mainly of hornblende and plagioclase with little or no quartz.

- Tremolite and actinolite are typical low-temperature metamorphic amphiboles. Their composition grades into that of hornblende by extensive substitution of Al^{3+} for Si^{4+}, as well as Al^{3+} for Mg^{2+} and/or Fe^{2+}. Hornblende, which is also common in igneous rocks, was systematically discussed in Chapter 7.

- Glaucophane, yet another amphibole member, is unique to metamorphic rocks. It is sodium rich and occurs in glaucophane schists, which tend to be blue in color because of the abundant glaucophane. They are also known as blueschists.

- Riebeckite, an amphibole similar in composition to glaucophane but with much Fe^{2+} and Fe^{3+} replacing Al^{3+} in the glaucophane structure, occurs in a highly fibrous (asbestiform) habit and is known as crocidolite. This is one of several asbestos minerals. Crocidolite, with an amphibole structure, is completely different in structure and composition from the much more common asbestos known as chrysotile.

- We introduced two members of the pyroxenoid group, wollastonite and rhodonite. Pyroxenoids are chain silicates that have chains with kinks at different repeat units of the vertical chains. Wollastonite is common in high-temperature metamorphic assemblages formed from calcium carbonate-rich rocks. Rhodonite is a constituent of metamorphosed manganese formations and manganese-containing iron-formation.

- Four members of the layer silicate group are talc, chlorite, antigorite, and chrysotile. Chlorite is a major component of metamorphic rocks belonging to the greenschist facies, which are green because of the chlorite.

- Antigorite and chrysotile are two polymorphs of the mineral serpentine. Antigorite has a corrugated layer structure and chrysotile consists of rolled up layers forming drinking straws. This is the most common asbestos, known in the trade as white asbestos. Crocidolite, an amphibole, which is blue, is known as blue asbestos.

- Epidote-clinozoite forms a metamorphic mineral series. Epidote is recognizable in hand specimens by its pistachio-green color.

- Cordierite is common in metamorphosed pelitic rocks but is difficult to distinguish from feldspar.

Review questions

1. How would you define the term *metamorphic mineral*?

2. List six very common metamorphic minerals and/or mineral groups.

3. What metamorphic minerals are common in Al-rich bulk compositions?

4. What metamorphic minerals develop in calcium carbonate-rich bulk compositions?

5. What type of amphibole (name and composition) might you expect in a metamorphosed iron-formation?

6. What are the acronyms for the two major compositional series of the garnets?

7. How were the acronyms for the two major groups of garnet arrived at?

8. What are some of the commercial uses of garnet?

9. What are the structural differences among the three polymorphs of Al_2SiO_5?

10. Which of the three polymorphs of Al_2SiO_5 is indicative of high-pressure metamorphism at low to moderate temperature?

11. What is chiastolite?

12. Staurolite is monoclinic in symmetry but is commonly referred to as pseudo-orthorhombic. Why?

13. What is the compositional difference between diopside and augite?

14. Give some examples of light beige to brown amphiboles.

15. Dark-green actinolite closely resembles dark-green hornblende. What are the main compositional differences between the two?

16. What is the most common amphibole asbestos?

17. Rhodonite is commonly recognized by its pink color. What is another Mn-containing pink mineral?

18. The crystal structure of talc, a layer silicate, is said to be trioctahedral. What is meant by that?

19. What are some of the commercial uses of talc?

20. What type of illustrations in this chapter show conclusively the very different crystal structures of the two polymorphs of serpentine?

21. What is chrysotile?

22. Cordierite is classified as a ring silicate. What is meant by that?

23. Among all the minerals listed as metamorphic only one is an oxide. What is its name and composition?

24. The oxide referred to in question 23 forms very high-end gemstones. Which are they?

25. One of the gemstones in question 24 is blue. What is that color due to? Refer to Figure 3.7 to obtain your answer.

FURTHER READING

Please refer to the listing at the end of Chapter 7.

14 Metamorphic rocks

Metamorphism includes all changes that affect rocks as a result of changes in pressure, temperature, or composition of fluids in the environment. These changes occur in sedimentary, igneous, and even former metamorphic rocks. The Earth's tectonic plates are constantly on the move, and consequently most rocks experience change at some time in their history. Metamorphic rocks are abundant, constituting ~60% of the continental crust. Even igneous rocks of the ocean floor are metamorphosed by circulating ocean water that cools them near midocean ridges. Metamorphic rocks preserve an important record of past conditions in the lithosphere, so it is important that we learn how to read that record. In doing this, we address the following important questions: Why do metamorphic rocks change when they find themselves in a new environment? Do these changes go to completion; that is, do they reach equilibrium under the new conditions? Do minerals in the rock tell us what those conditions were? And what are the chances that new minerals formed at depth in the Earth survive the trip to the surface during exhumation? We will learn that metamorphic rocks closely approach thermodynamic equilibrium and, as a consequence, contain only a small number of minerals. Mineral assemblages can be used to determine pressures, temperatures, and fluid compositions at the time of peak metamorphism. These assemblages are normally preserved during exhumation. Textures of metamorphic rocks provide important information about stresses in the crust during metamorphism.

Photomicrograph of a thin section of muscovite (bright colors)–chlorite (dark colors)–quartz (white-gray) schist under crossed polarized light. The prominent crystal alignment developed when this rock, which was originally a shale, was metamorphosed during the Acadian orogeny in southern New England. Width of field 5 mm.

Metamorphism is the sum of all changes that take place in a rock when it experiences changes in temperature, pressure (both lithostatic and directed), or composition of fluids in the environment. The important word in this definition is *change*. The changes may be physical, chemical, isotopic, or any combination of these. The original rock, known as the **protolith**, can be igneous, sedimentary, or a previous metamorphic rock. Most metamorphic reactions are very slow, so time is important in determining how complete a change may be. Some rocks are more reactive than others, and higher temperatures and the presence of fluids also speed up reactions. Changes that take place while rocks are heating are referred to as **prograde** and those occurring during cooling are referred to as **retrograde**. Changes that take place at low temperatures during the compaction and lithification of sediments could, in the broadest sense, be considered metamorphic, but normally only reactions taking place above about 150°C are dealt with in metamorphic studies, but the division is arbitrary. Reactions have a greater chance of going to completion at the higher temperatures experienced by metamorphic rocks than they do in sedimentary rocks. At the highest temperatures, metamorphic rocks begin to melt, producing mixtures of metamorphic and igneous rocks known as **migmatites**.

14.1 What changes occur during metamorphism?

Most metamorphic rocks undergo both physical and chemical changes during metamorphism. Physical changes may involve simple recrystallization of existing minerals, which may occur under lithostatic load or under directed pressure. These physical changes play an important role in determining the texture of the resulting rock. Chemical changes occur when new minerals form, because atoms must be exchanged between reacting minerals. In contrast, fluids passing through a rock may bring in new elements that replace former elements and change the bulk composition of the rock, a process known as **metasomatism**.

Changes in temperature, pressure, and fluid composition have numerous causes. Intrusion of magma causes heating, and changes that occur in thermal aureoles around igneous bodies are referred to as **contact metamorphism** (Fig. 2.12). Recrystallization during contact metamorphism normally occurs under lithostatic pressure, and rocks do not develop a prominent foliation. Fluids escaping from cooling magma may cause metasomatism in contact aureoles. At convergent plate boundaries, folding and thrusting thicken the crust. Rocks that were once close to the surface are buried to greater depth, where they are heated by the geothermal gradient. Crustal rocks, especially continental ones, are richer in radioactive elements than are mantle rocks. Heat generated by radioactive decay from this thickened radioactive blanket causes temperatures to rise on

a regional scale. In addition, the ascent of magmas generated by melting in the mantle wedge above subduction zones raises temperatures in orogenic belts still more. This leads to metamorphism, but because of plate convergence, new minerals grow in a directed stress field and consequently develop prominent foliations. Because this type of metamorphism occurs on a regional scale, it is called **regional metamorphism**.

The textures and mineral compositions of regional metamorphic rocks preserve a record of the processes associated with thickening of the crust in zones of plate convergence. Rocks buried by thrusting and folding experience ever-increasing pressures and temperatures, which lead to prograde metamorphic reactions. Eventually, as a result of tectonism, erosion, and isostasy, rocks find their way back to the Earth's surface, where we can examine them. On returning toward the Earth's surface, rocks experience decreasing pressures and temperatures, and we might expect reactions that occurred during burial to reverse themselves. However, most metamorphic rocks preserve the mineral assemblages they produce at maximum temperature, and retrograde reactions are often only local and commonly associated with fault zones or other prominent fractures. If it were not for this stranding of metamorphic rocks in their high-temperature form, metamorphic petrologists would have little to study, and much of the Earth's geologic history would be lost.

With the growth of new metamorphic minerals, the nature of the protolith may be lost. In some cases, the protolith is readily apparent because of the rock's bulk composition. For example, metamorphic rocks rich in carbonate minerals (**marble**) are almost certainly metamorphosed limestones, although there is a very small chance they may have been the igneous rock carbonatite. The aluminous composition of mudrocks makes them readily identifiable, even when intensely metamorphosed, because of the production of large numbers of alumina-rich minerals. These alumina-rich rocks, which are referred to as **pelites**, contain some of the most diagnostic metamorphic minerals. Metamorphosed granite and arkose, however, are difficult to distinguish. In some cases, primary igneous and sedimentary textures may be preserved. Bedding is one of the most characteristic features of sedimentary rocks. Unfortunately, the foliation that develops during regional metamorphism is commonly accompanied by a compositional layering that is difficult to distinguish from sedimentary bedding.

An example of metamorphic change

As emphasized already, the key word in the definition of metamorphism is *change*. We illustrate these changes by examining the transformation of black shale into a regional metamorphic rock known as schist. Figure 14.1(A)–(B) shows, respectively, a typical black shale and its metamorphic equivalent, a muscovite schist. This particular shale was formed from organic-rich mud that accumulated under anoxic conditions in a

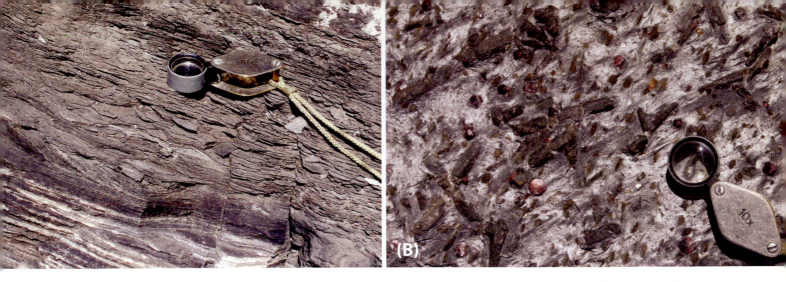

Figure 14.1 (**A**) Black shale exhibiting bedding-parallel fissility. (**B**) Garnet (red)-staurolite (dark brown)-muscovite schist produced from metamorphism of shale. The parallel alignment of muscovite grains in the plane of the photograph defines the plane of schistosity.

deep lake. It is composed largely of clay minerals but also contains fine detrital grains of quartz, alkali feldspar, muscovite, and biotite (see the photomicrograph in Fig. 12.2(A)). During compaction of the sediment, clay minerals and micas rotated into the horizontal plane, giving the shale a prominent fissility parallel to the bedding. This, then, is the protolith.

The schist was formed from black shale by regional metamorphism that accompanied the Acadian orogeny in the northern Appalachian Mountains of New England. One of the first changes to occur to shale during regional metamorphism is the reaction of clay minerals to form muscovite. Muscovite grows as platelike grains, and those oriented perpendicular to the maximum compressive stress grow more easily than do those oriented parallel to it. As a result, new muscovite crystals grow with a preferred orientation that creates a **foliation** known as **slaty cleavage** (see Sec. 14.4.2), which is usually at a high angle to the original sedimentary bedding. Thus, one of the first signs of metamorphism in shale is a **physical change**, with bedding fissility being replaced by a metamorphic foliation (see, e.g., Fig. 14.7(A)). As the grade of metamorphism increases, muscovite grains continue to grow, and when they become coarse enough to be visible to the unaided eye, the foliation they create is known as **schistosity** and the rock is called a **schist**. In Figure 14.1(B), the schistosity is the reflective surface parallel to the plane of the photograph. The chapter-opening photomicrograph is of a thin section of this same rock under crossed polarized light in which the parallel alignment of the muscovite crystals is obvious.

As black shale is heated above the oil and gas windows (Sec. 11.3.2), its organic constituents are converted to graphite. Like muscovite, graphite forms small platelike crystals that grow preferentially perpendicular to the maximum compressive stress. The gray sheen seen in the schist (Fig. 14.1(B)) is due, in part, to the presence of minute graphite crystals. If this schist is dragged across a piece of paper, its graphite leaves a gray streak.

As metamorphic temperatures and pressures increase, the growth of muscovite grains is accompanied by the appearance of other minerals that form through various chemical reactions. At the highest temperatures reached by the metapelite in Figure 14.1(B), garnet and staurolite grew to form large crystals known as **porphyroblasts**. Note that staurolite, which forms long prismatic crystals (Sec. 13.6), preferentially grows with its long axis in the plane of the micaceous foliation, perpendicular to the major compressive stress (Fig. 14.1(B)). The spacing between the porphyroblasts (~1 cm) indicates that elements needed for their growth were able to diffuse over distances of ~5 mm. Diffusion through rocks is extremely slow, so the presence of large, widely spaced porphyroblasts indicates that this schist remained at elevated temperatures for an extended period of time and only a few staurolite and garnet crystals nucleated. The resulting garnet-staurolite-muscovite schist records this maximum thermal event.

Eventually, the mountains formed during the Acadian orogeny began to erode, and isostatic adjustments brought the garnet-staurolite-muscovite schist back toward the surface. During this ascent, mineral assemblages formed at high temperatures and pressures tried to change back into lower-temperature and pressure mineral assemblages, but important ingredients needed for such reactions, namely H_2O and CO_2, had been lost from the rock during peak metamorphism and, therefore, were not available for retrograde reactions.

Although not evident from the photographs shown in Figure 14.1, the bulk composition of the shale and the schist differ in one important way: The schist contains much less water. Typical shales contain ~5 weight % H_2O. Most prograde metamorphic reactions liberate water (or carbon dioxide). The H_2O released by metamorphic reactions is a supercritical fluid (i.e., liquid and gas are indistinguishable) whose density is nearly the same as room-temperature water (~1000 kg/m^3). This fluid's density is less than that of the surrounding rock, so it rises toward the surface during metamorphism. The amount of volatiles released during metamorphism is staggering. For example, the conversion of shale into high-temperature metamorphic rock involves the loss of a volume of H_2O that is approximately equal to the volume of the rock itself. Other metamorphic reactions involving carbonate-rich sedimentary rocks liberate large volumes of CO_2.

Loss of volatiles is one of the major consequences of prograde metamorphism. As these volatiles are no longer present in the metamorphic rock, their importance is easily overlooked. When volatiles escape from rocks undergoing metamorphism, they take with them heat and dissolved constituents. The heat helps promote metamorphism in higher levels of the crust. As the fluids cool on rising through the crust, they precipitate their dissolved constituents. Fluid escaping from the garnet-staurolite-muscovite schist would have carried silica in solution, so as it traveled toward the surface, it would have precipitated quartz. Because these escaping fluids become channelized into fractures, the silica is deposited to form quartz veins. The ubiquitous occurrence of quartz veins in low-temperature metamorphic terranes is testament to the large flux of fluids generated from higher-temperature metamorphic zones beneath (Fig. 14.2).

In summary, metamorphism includes all those changes that occur to rocks in response to changes in the environmental factors of temperature, pressure, and fluid composition. Although reactions can occur during both rising and falling temperatures (prograde and retrograde, respectively) most metamorphic rocks consist of minerals formed near the maximum temperature to which they were exposed. Most prograde metamorphic reactions liberate volatiles, which migrate toward the surface, carrying heat and dissolved constituents into the overlying rocks.

Figure 14.2 Highly deformed quartz veins in low-grade metamorphic rock from Vermont. The veins are deposited by fluids escaping from higher-temperature metamorphic zones below.

14.2 Why do rocks change?

We have seen that the most important word in the definition of *metamorphism* is *change*. It is reasonable to ask why do rocks change? In fact, we can ask the more general question, why does any reaction take place? The answer is given by thermodynamics. Although a full thermodynamic explanation is beyond the scope of this book, it is possible to give a brief answer to the question in terms that are readily appreciated from everyday experience. The answer to this question is important because it determines how we study and describe metamorphic rocks.

14.2.1 Thermodynamics and the reason for change

Thermodynamics deals with the relations between energy and work. It developed during the Industrial Revolution when people were interested in the conversion of thermal energy to work. Much of its terminology relates to heat and work of expansion as might occur when coal is burned to heat water to drive steam engines. Thermodynamics was also interested in the efficiency of these processes. During the last half of the nineteenth century, Josiah Willard Gibbs of Yale University showed that thermodynamic laws developed to deal with the conversion of thermal energy to mechanical work applied equally to the energy associated with chemical reactions. Almost single handedly, he developed the field of chemical thermodynamics. In recognition of his contributions, the energy that determines the direction in which a chemical reaction proceeds is known as **Gibbs free energy**. Box 14.1 outlines how this energy determines the direction of a reaction and how it defines equilibrium, both of which are important when looking at the changes that occur in metamorphic rocks.

With the explanation of Gibbs free energy given in Box 14.1, we are in a position to answer our original question of why metamorphic changes take place. *Any reaction that can take place naturally must lower the Gibbs free energy*, and the reaction does not reach equilibrium until the free energy reaches a minimum. In our boulder analogy in Box 14.1, we see that equilibrium is achieved when the boulder reaches the valley bottom and cannot get any lower in the gravitational field. Similarly, equilibrium in a chemical reaction is achieved when the Gibbs free energy reaches a minimum.

Minimizing the free energy brings about other conditions that must be met at **equilibrium**. First, the temperature must be the same everywhere. There can be no temperature gradients; otherwise heat would flow, which lowers the free energy still more. Nor can there be any pressure gradients; otherwise material could expand or contract to eliminate the gradients, which, in turn, lowers the free energy. Finally, there can be no chemical potential gradients, for they cause material to diffuse and this also lowers free energy. This final point can be appreciated by returning to a gravitational analogy. Figure (C) in Box 14.1 shows a U-tube filled with water. If the water on either

BOX 14.1 | THERMODYNAMIC EXPLANATION FOR CHANGE

Thermodynamics is based on observations of the way nature operates. We see, for example, that heat always flows from regions of high temperature to regions of low temperature, never the reverse, and when we stir cream into coffee the two mix, and reversing the stirring direction does not cause them to unmix. Nature behaves in predictable ways. Thermodynamics quantifies these observations and, after much mathematical juggling, arrives at fundamental laws that we believe are applicable throughout the universe.

Let us illustrate these laws by examining a familiar situation in which we know the direction in which natural reactions take place. Figure (A) shows a cross-section of a valley with a large boulder resting on a ledge high on the valley side. If the stick figure is successful in pushing the boulder toward the edge of the ledge, it will roll to the valley bottom. We expect this to happen from everyday experience, so this is a *natural reaction*. We do not expect the boulder to roll uphill to the ledge; this would be an *unnatural reaction*. We accept the falling of the boulder to the valley floor as natural because the Earth's gravitational field pulls the boulder down. The boulder on the ledge has a certain **potential energy** because of its height in the Earth's gravitational field, and by falling to the valley floor the potential energy is lowered. The natural direction of this process is, therefore, to decrease the potential energy.

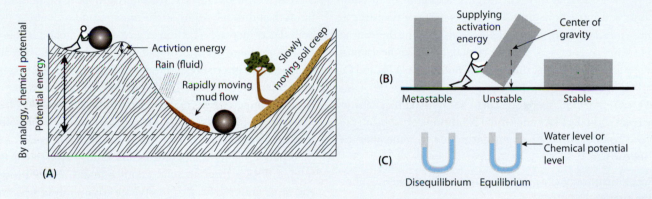

Gravitational potential energy is just one form of energy in nature. Another is *chemical potential energy*. By analogy, we can think of the boulder rolling to the valley bottom as a chemical reaction that lowers the chemical potential energy. This is the so-called *Gibbs free energy*. Just as the gravitational potential energy decreases when the boulder rolls to the valley floor, the Gibbs free energy must decrease in a natural reaction. If it were to increase, the reaction would be unnatural and, therefore, would not occur.

If the stick figure did not push the boulder over the lip of the ledge, the boulder could remain on the ledge indefinitely. The stick figure must expend a certain amount of energy to push the boulder over the lip, but as soon as the boulder crosses the threshold and begins falling, this same amount of energy is released as the boulder falls back to the elevation it had on the ledge. The amount of energy needed to cross the threshold is known as the **activation energy**. In Figure (A), the activation energy is determined by the height of the lip on the ledge. By analogy, chemical reactions also have activation energies. For example, a match must be rubbed on a rough surface to generate a small amount of heat before it ignites. The activation energy required to start metamorphic reactions is commonly provided by slight overheating or mechanical deformation.

If the activation energy is large, a reaction may never take place. Diamond, for example, should react with oxygen in air to form carbon dioxide, but diamonds are found in sediments that have been exposed to the atmosphere for extended periods of time with no signs of reaction. Material that is prevented from reacting by a large activation energy are said to be metastable. The simple mechanical model shown in Figure (B) illustrates the relation among **metastable**, **unstable**, and **stable**. If the stick figure pushes hard enough on the vertical block to make its center of gravity go beyond the lower corner, the metastable block becomes unstable and falls to the stable position, where the center of gravity is at the lowest possible position in the gravitational field.

We have argued that when a chemical reaction takes place, the chemical potential energy is lowered in an analogous manner to the decrease in the gravitational potential energy when a boulder rolls to a valley floor. Chemical potential energy is very real, as can be testified to by anyone who has accidentally touched the terminals of a car battery. The electric shock you receive is a direct result of the chemical potential energy of the reaction in the battery.

In reactions involving complex minerals such as those in metamorphic rocks, each of the individual components making up the minerals contribute to the overall chemical potential change during reactions. Gibbs recognized that each component had a certain **chemical potential**, which he defined by the Greek letter μ_i (mu), with the subscript i indicating the particular component. For example, μ_{Na} and μ_{Ca} could indicate the chemical potentials of Na and Ca in plagioclase. Chemical potential has units of joules per mole, so when multiplied by the number of moles of the particular component (n_i) we obtain the contribution that component makes to the total free energy, which can be expressed as follows:

$$\mu_a n_a + \mu_b n_b + \mu_c n_c + \ldots \ldots + \mu_i n_i = \text{Gibbs free energy}, \tag{14.1}$$

where each of the subscripts ($a, b, c \ldots i$) indicates each of the components that makes up the whole. Chemical potential is as important as temperature and pressure in driving metamorphic reactions. Diffusion, for example, is driven by chemical potential gradients; that is, atoms diffuse from regions of high chemical potential to regions of low chemical potential. Chemical potentials also determine the composition of metamorphic minerals.

side of the tube is at different heights, it cannot be at equilibrium and must flow until the heights become the same at equilibrium. Similarly, chemical potentials must be the same at equilibrium. A zoned plagioclase crystal, for example, cannot be at equilibrium because the chemical potentials of Na and Ca vary through the crystal.

We can conclude, therefore, that thermodynamics tells us why reactions take place and which conditions must be met to achieve equilibrium. The important points to remember are the following:

- Reactions always take place in a direction that lowers Gibbs free energy.
- Equilibrium is achieved only when Gibbs free energy reaches a minimum.
- At equilibrium, temperature, pressure, and chemical potentials of all components must be the same throughout.

14.2.2 Rates of metamorphic reactions

Thermodynamics explains why reactions should take place and what is required to achieve equilibrium, but it does not tell us how rapidly the reaction takes place; this is a question of **kinetics**. Let us return to the example of the boulder on the valley wall given in Figure (A) in Box 14.1. If we assume that the activation energy needed to start the boulder rolling is available, then we would expect the boulder to roll rapidly to the valley floor. Soil on the valley wall would also try to move to the valley bottom, but soil creep is a slow process. If the valley received heavy rainfall and the soil became water saturated, a mud slide could result, which would move material to the valley bottom more rapidly. In each of these cases (falling boulder, soil creep, and mud slide), material strives to lower its potential energy, but each does so at a different rate.

Metamorphic reactions similarly progress toward thermodynamic equilibrium at different rates. In general, reaction rates increase dramatically with increasing temperature, because of increased diffusion rates. Diffusion is also more rapid through fluids than through solids, so the presence of fluids along grain boundaries increases reaction rates. Prograde metamorphic reactions are, therefore, more rapid than retrograde ones because temperatures are rising and fluids are being released by reactions. Retrograde reactions, in contrast, are very much slower because temperatures are falling and fluids are generally lacking. Indeed, most retrograde reactions occur only where faults or prominent fractures allow water to infiltrate the rock.

Although higher temperatures and fluids increase reaction rates, diffusion through solid grains remains a slow process, and large grains (porphyroblasts) are commonly zoned, with cores preserving compositions generated during earlier stages of metamorphism. These zoned minerals prove useful in determining the metamorphic history of a rock.

Most environmental changes that affect rocks are sufficiently slow that metamorphic reactions have time to keep pace with them, at least during prograde metamorphism. The rate of contact metamorphism is determined by the rate at which heat is liberated from an intrusion, and as we saw in Section 8.7.1, this is a very slow process. The rate of regional metamorphism is dependent on the rates of tectonic plate convergence, folding, thrusting and erosion, which again are all slow. The slowness of these processes, consequently, allows metamorphic rocks to approach closely thermodynamic equilibrium, which greatly simplifies their study.

14.2.3 Gibbs phase rule and the number of minerals a metamorphic rock can contain

The close approach of metamorphic rocks to thermodynamic equilibrium results in their containing a relatively small number of minerals, which greatly simplifies their classification. The number of minerals that can possibly coexist at thermodynamic equilibrium is given by the **Gibbs phase rule**, which is explained in Box 14.2.

BOX 14.2 | THE GIBBS PHASE RULE

The Gibbs phase rule explains the quantitative relations between the numbers of phases, components, and degrees of freedom or variance that exist at equilibrium. We start by giving precise definitions of these three terms:

- **Phases** are individual minerals, fluids, or even magma. You can think of them as the different objects that, given a pair of tweezers or other device, you could separate one from another. We designate the number of phases in a rock by the Greek letter ϕ (phi).
- **Components** are the building units of phases, but instead of listing every element present, components are the minimum number of constituents needed to describe all phases present. This allows elements to be grouped together if they appear in the same ratios in each phase. For example, if all phases contain silicon and oxygen in the proportions of 1:2, then Si and O would be grouped together to form a single component, SiO_2. The important point is that components must be the minimum number of building units needed to account for the composition of all of the phases present. You must know what the phases and their chemical compositions are before you can choose components. We designate the number of components by the letter c.
- **Degrees of freedom** (variance) indicate how many variables can be changed without changing the phases present. For example, if we can change both the temperature and pressure without affecting the mineral assemblage, there would be two degrees of freedom. We designate the number of degrees of freedom by the letter f.
- The Gibbs phase rule, which applies to assemblages of phases at thermodynamic equilibrium, is written as follows:

$$f = c + 2 - \phi \tag{14.2}$$

The maximum number of phases a rock can contain is $(c + 2)$ when f is 0, which means that this assemblage of minerals could exist at just one pressure and temperature; that is, the mineral assemblage is invariant ($f = 0$). If the number of phases decreases to $(c + 1)$, then $f = 1$. This means that one variable could be changed, but all others would be fixed. For example, you might vary the temperature, but the pressure could not be independently changed. Such a relation is said to be univariant ($f = 1$).

We can illustrate the Gibbs phase rule and the relation between the number of components (c), phases (ϕ), and degrees of freedom (f) by examining the stability relations between an important group of metamorphic polymorphs, all of which have the formula Al_2SiO_5; these are andalusite, kyanite, and sillimanite (Secs. 13.3–13.5). If these are the only phases present, we need only one component to describe all three phases; that is, Al_2SiO_5 ($c = 1$). According to the Gibbs phase rule, the maximum number of phases that can be present is three ($c + 2$), because $c = 1$. Thus, when andalusite, kyanite, and sillimanite are all present, the assemblage is invariant; that is, this mineral assemblage can exist at only one temperature and pressure. Figure 14.3 shows this invariant triple point is at 500°C and 0.38 GPa, which corresponds to a depth of ~13 km. If only two polymorphs are present, $f = 1$, and the mineral assemblage must exist along one of the univariant lines extending from the triple point, along which either kyanite and sillimanite, andalusite and sillimanite, or kyanite and andalusite coexist. If only one polymorph is present, $f = 2$, and the polymorph can exist anywhere in the divariant fields that are bounded by the univariant lines. It is worth familiarizing yourself with this diagram and the invariant point at 500°C and a depth of 13 km, because it

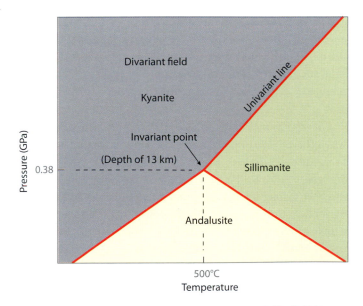

Figure 14.3 Pressure-temperature plot showing stability fields of Al_2SiO_5 polymorphs. All three polymorphs coexist at only one temperature and pressure. Surrounding this invariant point, three univariant lines indicate the conditions along which two polymorphs can coexist.

is a useful reference point when trying to determine the conditions under which a metamorphic rock formed.

In conclusion, the close approach of metamorphic rocks to thermodynamic equilibrium causes them to contain a small number of minerals, with the maximum number being two more than the number of components. Rocks containing the maximum number of minerals form under very restricted conditions of temperature, pressure, and other environmental variables such as fluid compositions.

14.3 Metamorphic grade and facies

When discussing metamorphic rocks we commonly refer to their **metamorphic grade**, which loosely expresses the intensity of metamorphism, with increasing grade being thought to relate mainly to increasing temperatures. One of the first attempts to map the grade of metamorphism was the study by Barrow (1893) in the southeastern Scottish Highlands. He noted that mudrocks (pelites) in this region become progressively more metamorphosed toward the north. He mapped the increasing metamorphic grade by recording the first appearance of particular **metamorphic index minerals**, which included chlorite, biotite, garnet, staurolite, kyanite, and sillimanite. This sequence of minerals, though extremely common and found in metamorphic terranes worldwide, is not the only sequence that can occur. The fact that other sequences are found indicates that metamorphism is not a simple phenomenon and must involve numerous factors that can differ from region to region.

Later, the boundaries between Barrow's index mineral zones were termed **isograds**, which implies that points along these lines underwent equal intensities of metamorphism. When examined in detail, however, such lines cannot truly represent equal intensities of metamorphism, because the first appearance of an index mineral is strongly dependent on a rock's bulk composition or fluid compositions. Although isograds are not rigorous lines of equal metamorphic intensity, they are still mapped and provide a useful general measure of metamorphic grade.

The big advance in measuring metamorphic intensity came when thermodynamics was first applied to the study of metamorphic rocks. In 1911, Goldschmidt, noted that contact metamorphic rocks around igneous intrusions near Oslo, Norway, rarely contained more than four or five minerals and that the mineral assemblages were consistent with the Gibbs phase rule, which implied that the assemblages approached thermodynamic equilibrium. He emphasized that it was the mineral assemblage rather than the first appearance of a single mineral that was a measure of metamorphic grade.

In 1920, Eskola, carrying out similar studies of contact metamorphic rocks around bodies of granite at Orijärvi in Finland, noted that rocks contain small numbers of minerals that are consistent with the Gibbs phase rule but that the mineral assemblages are different from those around the Oslo intrusions. He concluded that metamorphic rocks in both areas had approached thermodynamic equilibrium but under different conditions. This led him to propose the **metamorphic facies** concept. He defined a metamorphic mineral facies as comprising all rocks that have originated under temperature and pressure conditions so similar that a definite bulk rock chemical composition results in the same set of minerals.

The metamorphic facies concept has two very important consequences. If two rocks of identical bulk composition have different mineral assemblages, they must have been metamorphosed under different conditions. Conversely, if two rocks that have been exposed to the same metamorphism (e.g., found side by side in the field) contain different minerals, they must have different bulk compositions. Eskola's facies approach used the entire mineral assemblage and avoided the problems associated with different bulk compositions, which plagued the application of Barrow's index minerals.

Eskola originally proposed five different metamorphic facies to which a few extra were added subsequently. He showed the relative positions of these facies on a pressure-temperature diagram. Modern experimental studies allow us to give approximate pressure and temperature ranges for these facies (Fig. 14.4). The boundaries between facies are broad zones because of the complex nature of the reactions separating them and the fact that many of the minerals belong to solid solution series. In addition, because most prograde metamorphic rocks release water, the activity of water (relative humidity; see Sec. 8.4.2) plays an important role in determining the conditions under which the reactions occur. In Figure 14.4, the activity of water is taken to be 1; that is, the minerals coexist with a pure water fluid. If the activity were less than 1, the boundaries would occur at lower temperatures. The names of the facies are based on the typical hand specimen appearance of metamorphosed basalt in each of these facies. Thus, a metabasalt at low grade would contain abundant chlorite and so is called a **greenschist**, whereas at higher grade it would contain abundant amphibole and is called an **amphibolite**, and at the highest grade it would contain granular pyroxene and plagioclase and is called a **granulite**.

Eskola recognized that the changes in temperature and pressure associated with progressive metamorphism led to three common sequences of metamorphic facies or **metamorphic facies series**. A low-pressure, high-temperature series, including the albite-epidote hornfels, the hornblende hornfels, the pyroxene hornfels, and the sanidinite hornfels facies (Fig. 14.4), is typically associated with contact metamorphic aureoles but is also found in regional metamorphic terranes. A common metamorphic facies series found in many regionally metamorphosed orogenic belts includes the zeolite, greenschist, amphibolite, and granulite facies. This series shows a progressive increase toward both *high temperature and pressure*, with both

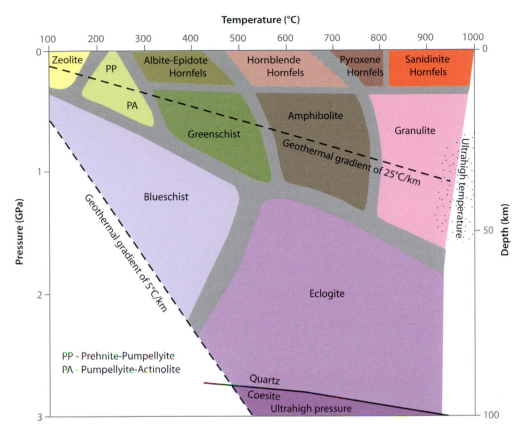

Figure 14.4 Approximate pressure and temperature ranges of metamorphic facies when the activity of water is 1 (modified from Philpotts and Ague, 2009).

rising approximately along a geothermal gradient of ~25°C/km (Fig. 14.4). A third series, marked by *high pressures at low temperatures* includes the prehnite-pumpellyite, blueschist, and eclogite facies. Although these are the most common series, intermediate series also occur. Recently, *ultrahigh-pressure* (UHP) metamorphic rocks have been found in subduction zones in which the high-pressure polymorph coesite is the stable form of silica rather than the low-pressure polymorph quartz (Sec. 7.7). These rocks form at pressures of ~3 GPa, which correspond to depths of ~100 km. Some UHP rocks have even been found to contain diamonds, which indicates depths of 200 km. How rocks from such great depth return to the surface remains a puzzle and is the subject of current research. In addition, *ultrahigh-temperature* (UHT) rocks have been found that equilibrated near 1000°C. These temperatures are attained in rocks that lack a low-melting fraction or have had all of a low-melting fraction removed at an earlier stage of metamorphism. The large latent heat of fusion of rocks normally puts a cap on metamorphic temperatures.

Assigning a rock to a particular metamorphic facies requires careful study of the mineral assemblage, usually through the use of thin sections. Some of the most common minerals found in the different metamorphic facies are given in Table 14.1. These may help you determine an approximate metamorphic grade, but it is important to emphasize that assignation

to a metamorphic facies is dependent on an entire mineral assemblage and not the presence of a single mineral. This table is given only for reference purposes; it is not something you need memorize.

The following general tips may help you focus your search in assigning a rock to a metamorphic facies. Metamorphosed basaltic rocks in the greenschist facies are characterized by an abundance of green chlorite. They may also contain the alumina-poor amphibole actinolite, but this is difficult to distinguish from the aluminous variety, hornblende, which characterizes the amphibolite facies. Plagioclase in the greenschist facies is always of albitic composition, whereas in the amphibolite and granulite facies, it can have any composition. Determining plagioclase compositions, however, requires thin sections, and even then it is not easy, because most metamorphic plagioclase does not exhibit twinning from which to measure extinction angles. The reason plagioclase is restricted to albitic compositions in the greenschist facies is that calcic plagioclase is converted to epidote at this grade, and epidote is readily identified in hand specimens by its pistachio green color (Sec. 13.18).

The metamorphic facies approach to estimating metamorphic grade was a marked improvement over earlier methods, but the complexity of metamorphic minerals and the possible variable activity of water still prevented precise metamorphic temperatures and pressures being determined. This became

Table 14.1 Common minerals in metamorphic facies.

Facies	Mafic protolith	Pelitic protolith	Carbonate protolith
Zeolite	Chlorite, serpentine, clay minerals, zeolites, analcite, albite, quartz, prehnite, pumpellyite, calcite, dolomite	Chlorite, illite, clay minerals, quartz, albite, calcite, dolomite	Calcite, dolomite, quartz, chlorite, illite, clay minerals, albite
Prehnite-pumpellyite	Chlorite, serpentine, prehnite, pumpellyite, quartz, albite, calcite, dolomite	Chlorite, muscovite, clay minerals, quartz, albite, calcite dolomite	Calcite, dolomite, quartz, clay minerals, albite
Blueschist	Glaucophane, lawsonite (or epidote), quartz, garnet	Glaucophane, Si-rich muscovite, lawsonite (or epidote), quartz, garnet	Aragonite, dolomite, glaucophane, epidote, albite
Greenschist	Chlorite, actinolite, epidote, albite, quartz	**Chlorite zone:** chlorite, muscovite, quartz, albite **Biotite zone:** chlorite, muscovite, biotite, quartz, albite **Garnet zone:** muscovite, biotite, garnet, quartz, albite	Calcite, dolomite, muscovite, quartz, albite
Albite-Epidote hornfels	Albite, pyrophyllite, epidote, actinolite, chlorite, quartz	Muscovite, chlorite, biotite, albite, quartz, pyrophyllite	Calcite, epidote, actinolite, quartz
Amphibolite	Hornblende, plagioclase, quartz, garnet	**Staurolite zone:** muscovite, biotite, quartz, garnet, staurolite, plagioclase **Kyanite zone:** muscovite, biotite, quartz, garnet, kyanite, staurolite, plagioclase **Sillimanite zone:** muscovite, biotite, quartz, garnet, sillimanite, plagioclase	Calcite, dolomite, quartz, biotite, amphibole, diopside, K-feldspar, wollastonite
Hornblende hornfels	Cordierite, plagioclase, anthophyllite, hornblende, diopside, garnet, quartz	Andalusite, muscovite, cordierite, quartz, biotite	Calcite, diopside, grossular, biotite, quartz
Pyroxene hornfels	Diopside, orthopyroxene, plagioclase, biotite, quartz	Andalusite, cordierite, orthoclase, biotite, quartz	Wollastonite, grossularite, diopside, biotite, quartz
Granulite	Clinopyroxene, orthopyroxene, plagioclase, garnet, quartz	Quartz, K-feldspar, plagioclase, sillimanite, garnet, biotite, orthopyroxene, cordierite	Calcite, dolomite, quartz, diopside, wollastonite, K-feldspar, forsterite
Sanidinite	Sanidine, clinopyroxene, orthopyroxene, plagioclase, quartz	Sillimanite or mullite, spinel, sanidine, quartz	Monticellite, melilite, diopside, calcite
Eclogite	Pyrope-rich garnet, jadeite-rich pyroxene, quartz, kyanite, rutile	Si-rich muscovite, quartz, jadeite-rich pyroxene, pyrope-rich garnet, kyanite, rutile	Aragonite, dolomite, jadeite-rich pyroxene, epidote, quartz, Si-rich muscovite, pyrope-rich garnet

possible only after sufficient thermodynamic data became available to calculate the precise conditions under which each metamorphic reaction can take place. The resulting reaction curves, when plotted on a pressure-temperature diagram, create a grid that allows a particular metamorphic mineral assemblage to be assigned a precise pressure and temperature range at any given activity of water. The resulting plot, which is referred to as a **petrogenetic grid**, is introduced in Section 14.6.

Finally, some metamorphic reactions involving minerals belonging to solid-solution series are particularly sensitive to changes in temperature, and the composition of these minerals can be used to determine metamorphic temperatures, in some cases to within a few degrees. These reactions are known as **geothermometers**. A much smaller number of reactions are sensitive to pressure, and are known as **geobarometers**. Geothermometers and geobarometers are discussed in Section 14.9.

14.4 Textures of metamorphic rocks

Metamorphism brings about the growth of new mineral grains that generate characteristic textures that can be used to distinguish metamorphic rocks from igneous and sedimentary rocks. New grains may appear as a result of recrystallization of former grains in the protolith or as products of metamorphic reaction. In either case, grain growth can occur under lithostatic pressure (pressure equal in all directions), directed pressure, or even a shear stress. Regardless of the conditions, grain growth is controlled by the general principles that we outline here.

Grain growth is a natural process and, therefore, must involve a decrease in the Gibbs free energy; that is, once a grain has grown, the free energy of the rock must be lower than it was before. Grains have all sorts of energy associated with them other than the simple chemical energy. For example, a detrital mica grain in a sedimentary rock that was bent during compaction of the sediment would have stored elastic energy. Twins in crystals are another source of excess energy. By **recrystallizing**, these excess energies are released.

All grains have boundaries, which are sources of excess energy. At grain boundaries, the crystal structure of a grain must terminate and come in contact with the crystal structure of adjoining grains. The mismatch of structures across boundaries results in **surface free energies**. The surface area of a large grain is less than the combined surface area of two smaller grains with the same total volume. Thus, by recrystallizing to a coarser-grain size, rocks reduce their surface free energies. This explains why metamorphic rocks typically become coarsergrained during metamorphism (see exception in the following paragraph).

Although recrystallization decreases the free energy of a rock by eliminating energies associated with strain, twinning, and grain surfaces, the process typically has an activation energy (see Box 14.1) that must be overcome before new strain-free grains can begin to form. The energy needed to overcome this threshold may come from rocks being overheated. In regional metamorphic rocks, deformation associated with folding and thrusting may supply the activation energy. Deformation of mineral grains can create small regions of intense strain that have a much greater tendency to recrystallize than do less deformed regions. These high-strain regions are the sites of nucleation of new grains. Once a new strain-free grain has nucleated, material diffuses to it from nearby deformed grains. We mentioned in the previous paragraph that recrystallization generally coarsens metamorphic rocks, but in zones of intense shear, so many new nucleation sites may form that the eventual

Table 14.2 Crystalloblastic series.

Magnetite, rutile, titanite, pyrite
Sillimanite, kyanite, garnet, staurolite, tourmaline
Andalusite, epidote, zoisite, forsterite, lawsonite
Amphibole, pyroxene, wollastonite
Mica, chlorite, talc, prehnite, stilpnomelane
Calcite, dolomite, vesuvianite
Cordierite, feldspar, scapolite
Quartz

strain-free rock contains so many grains that it becomes finer grained.

Finally, the shape of metamorphic mineral grains is also determined by surface free energies. Some crystal faces have lower free energies than others, and as a result, during metamorphism, these faces grow larger at the expense of the highenergy faces, thus giving mineral grains characteristic shapes. Of course, not all grains can develop crystal faces if they are to fit together with no pore space. Minerals with the highest surface energies bring about the largest decrease in surface free energy by coarsening and generating low-energy faces, so those are the ones that produce euhedral crystals. Metamorphic minerals can be arranged in what is known as the **crystalloblastic series** (Table 14.2), where a mineral in the series can form euhedral crystal faces against any mineral below it in the series. The series reflects decreasing surface free energies from top to bottom.

14.4.1 Textures of contact metamorphic rocks

Heat liberated from igneous intrusions provides the energy to drive recrystallization and create new metamorphic minerals. Activation energies are easily overcome, because of the proximity of the heat source, and the rocks recrystallize. Because the pressure is usually close to lithostatic, there is little driving force for minerals to grow with preferred orientations. Even elongate minerals formed by metamorphic reactions grow with random orientations. As a result, contact metamorphic rocks typically lack foliation, which makes them hard to break (commonly they do so with conchoidal fracture). They are referred to as **hornfels**, from the German for horn (or hard) stone (Fig. 14.5).

Many hornfels consist of polygonal-shaped grains that are bounded by faces intersecting at ~120° (Fig. 14.5(B)). This is a direct consequence of the minimization of surface free energies during recrystallization. We can think of surface energies as surface tensions pulling along grain boundaries. You actually experience such surface tension when you inflate a balloon. If three grains come together at a triple junction, the surface tensions can be represented by arrows pulling away from the triple junction, with the length of the arrows indicating the magnitudes

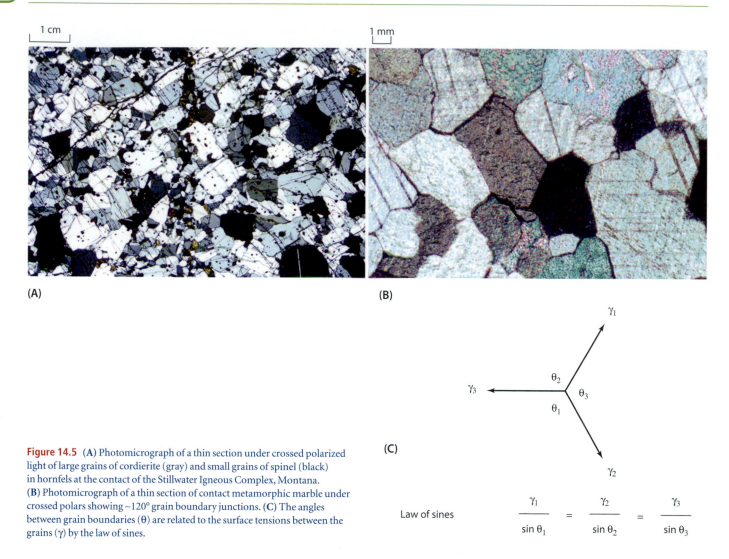

(A)

(B)

(C)

Law of sines $\qquad \dfrac{\gamma_1}{\sin \theta_1} = \dfrac{\gamma_2}{\sin \theta_2} = \dfrac{\gamma_3}{\sin \theta_3}$

Figure 14.5 **(A)** Photomicrograph of a thin section under crossed polarized light of large grains of cordierite (gray) and small grains of spinel (black) in hornfels at the contact of the Stillwater Igneous Complex, Montana. **(B)** Photomicrograph of a thin section of contact metamorphic marble under crossed polars showing ~120° grain boundary junctions. **(C)** The angles between grain boundaries (θ) are related to the surface tensions between the grains (γ) by the law of sines.

of the surface tensions (γ_1, γ_2, and γ_3 in Fig. 14.5(C)). When the rock undergoes recrystallization, the triple junction is able to move by adjusting the angles subtended by the three grains (θ_1, θ_2, θ_3) to bring surface tensions into balance. According to the *law of sines* (Fig. 14.5(C)), if the three surface tensions are of equal magnitude, the angles subtended by the grains must also be equal and hence each must be 120°.

14.4.2 Deformation and textures of regional metamorphic rocks

The characteristic feature of almost all regionally metamorphosed rocks is a prominent foliation formed by the growth of platy and needlelike minerals with a preferred orientation. Foliation first appears in the lowest grades of metamorphism and becomes more pronounced as metamorphism continues. Only at the highest grades of metamorphism, where micas and amphiboles become unstable, does foliation become less distinct.

Slate, phyllite, schist, and gneiss

Muscovite and chlorite are two of the first minerals to appear in metamorphosed pelitic rocks, and because both form platy

grains, their growth with a preferred orientation imparts to the rock a foliation known as **slaty cleavage**. Figure 14.6(A) shows an outcrop of steeply dipping Martinsburg shale near Shippensburg, Pennsylvania, that shows the first signs of regional metamorphism, as evidenced by the weak development of slaty cleavage. New muscovite and chlorite grains did not grow large enough to be visible at low magnification under the microscope, but their general alignment can be seen to have produced a foliation (Fig. 14.6(B)). This foliation, or slaty cleavage, developed perpendicular to the maximum compressive stress (Fig. 14.6(C)), which also deformed the rocks into folds with axial planes parallel to the slaty cleavage (Fig. 14.6(D)).

Although the slaty cleavage in Figure 14.6(A) is clearly evident, the foliation planes are not developed well enough to cause the rock to break into large flat slabs. With continued growth and coarsening of micaceous minerals, the characteristic slaty cleavage develops where slabs large enough to make roofing tiles or even billiard tables can be quarried (Fig. 14.7(A)). As micaceous minerals coarsen still more, the foliation planes develop a reflective sheen and the rock is referred to as **phyllite** (Fig. 14.7(B)). Individual mica grains are still not large enough

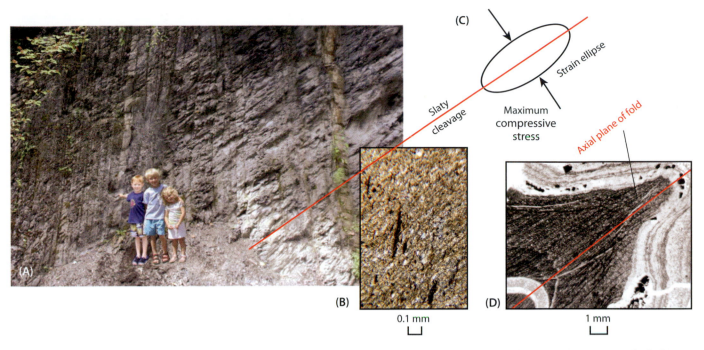

Figure 14.6 (A) Steeply dipping Martinsburg shale near Shippensburg, Pennsylvania, showing slaty cleavage dipping to the left. (B) Photomicrograph of a thin section under crossed polars of the shale in (A) where bedding is marked by opaque streaks of organic material and the slaty cleavage by pinkish birefringent muscovite. (C) A strain ellipse showing how mica grains preferentially grow in a plane perpendicular to the maximum compressive stress. (D) Slaty cleavage parallels the axial plane of folds.

to be visible to the unaided eye in phyllite, but once they are, the rock is called **schist** (Figs. 14.1(B) and 14.8(A)). Foliated metamorphic rocks that contain few micaceous or needlelike minerals are referred to as **gneiss** (Fig. 14.8(B)). The foliation in gneisses is usually defined by layers of slightly different composition. Micaceous minerals and amphiboles become unstable and react to form granular minerals at higher grades of metamorphism, so gneisses become more common in higher-grade

terranes. For example, the gneiss shown in Figure 14.8(B) is the high-grade equivalent of the schist shown in Figure 14.8(A). The distinction between schist and gneiss is arbitrary, but as a general rule, if you can pull the rock apart with your fingernails, it is a schist, but if you can't, then call it a gneiss.

Schists are usually formed from platy minerals such as micas and chlorite, but the preferred alignment of elongate minerals such as amphiboles can also create schistosity. Blueschists, for

Figure 14.7 (A) The large perfectly planar surface in this road cut is produced by breakage along slaty cleavage planes. (B) As micaceous minerals grow larger, the foliation plane develops a sheen, which is characteristic of the rock known as phyllite.

1 mm

1 cm

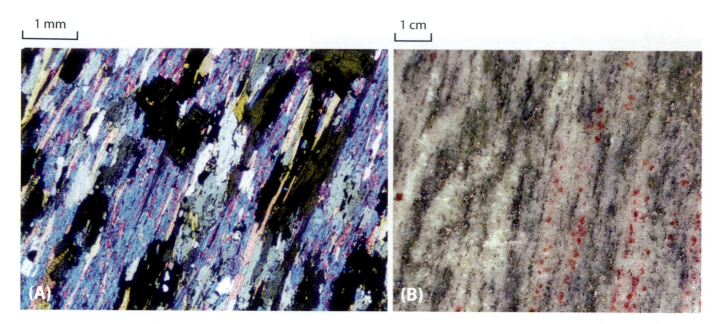

(A)

(B)

Figure 14.8 **(A)** Photomicrograph of a thin section of the schist shown in Figure 14.1(B) and chapter-opening photograph under crossed polarized light. The alignment of highly birefringent muscovite grains defines the schistose foliation. Low-birefringent grains are chlorite and gray grains are quartz. **(B)** Hand specimen of gneiss belonging to the granulite facies composed of light-colored quartz and feldspar, red garnet, and dark pyroxene. This rock is essentially the anhydrous equivalent of the schist in (A).

example, have a schistosity caused by the alignment of sodic amphibole grains (Fig. 14.9). Figure 14.10 shows a schist from western Massachusetts in which sheaflike clusters of amphibole crystals lie in the plane of the foliation. Such rocks are commonly referred to as **garbenschiefer**, from the German words *Garben* and *Schiefer*, meaning "sheaves" and "schist," respectively.

Metamorphic foliation in slates and schists can be shown to develop perpendicular to the maximum compressive stress where deformed fossils or other objects are present whose

original shape is known. For example, reduction spots in mudstones (Sec. 12.1) and oolites in limestone (Sec. 11.3.1) are initially very nearly spherical, but after deformation they become ellipsoidal. From measuring the shape of such objects, we can determine the directions of maximum and minimum compressive stresses and how much deformation the rock has undergone. Such measurements show that slaty cleavage is oriented perpendicular to the maximum compressive stress.

1 cm

1 mm

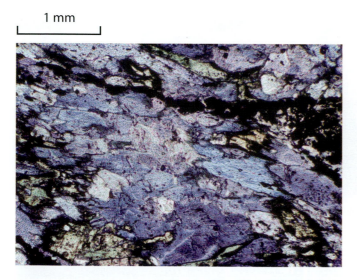

Figure 14.9 Photomicrograph of a thin section under plane polarized light of glaucophane (blue) schist from the Franciscan Complex, California. The green to beige pleochroic grains are jadeite-rich clinopyroxene ($NaAlSi_2O_6$).

Figure 14.10 Mica-garnet-hornblende schist of the Ordovician Hawley Formation of western Massachusetts with sheavelike clusters of hornblende (garbenschiefer). This rock is the same as the one shown on the cover of the book.

When a variety of different rocks are deformed together, it is quite likely that some deform more easily than others. Some rocks are strong and resist deformation, whereas others are weak and deform easily. Figure 14.11 shows a greenish-colored layer in a gray gneiss made of biotite, quartz, and feldspar. The green layer was originally an impure limestone that was metamorphosed to green calcium aluminum silicates; the rock is, therefore, referred to as a **calcsilicate**. During deformation and metamorphism, the calcsilicate resisted deformation more than the surrounding gneiss. Consequently, as the host rock flattened and developed a foliation, the resistant calcsilicate layer was stretched and began to form lenses by necking down at points where, eventually, it ruptured. As segments of the calcsilicate layer were pulled apart, the surrounding gneiss tried to flow in to fill the gaps. However, this was not fast enough to keep pace with deformation, so feldspar from the surrounding rock diffused in to fill the gaps between segments. The resulting structure resembles a string of sausages, so it is referred to as **boudinage**, as a *boudin* is a type of French sausage.

Migration of material during the development of foliation

The migration of feldspar from the gneiss into the gaps between boudins in Figure 14.11 is evidence of the mobility of some minerals during the formation of metamorphic foliation. In some rocks, the generation of foliation involves large-scale loss of material from the rock. Limestone is particularly susceptible to such losses due to pressure solution. We have already seen that during compaction, limestone dissolves along

Figure 14.11 Boudins of calcsilicate rock in gray biotite-feldspar-quartz gneiss. Feldspar (white) migrated from the gneiss to fill gaps between the boudins.

1 cm

10 cm

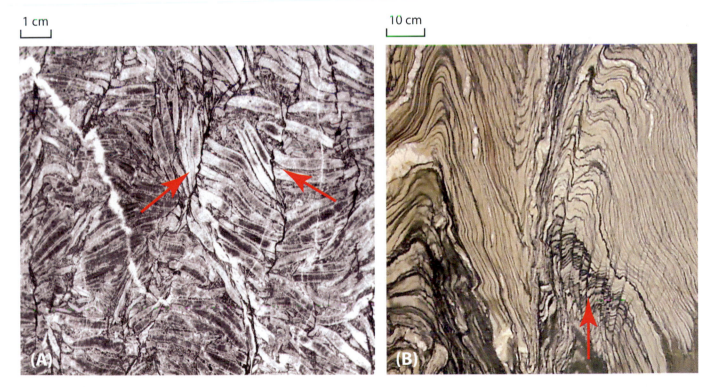

Figure 14.12 (A) Limestone containing fossil foraminifera that are truncated (arrows) by solution cleavage planes, which are marked by dark vertical lines of insoluble residue. (B) Folded marble with dark axial plane solution cleavage (arrow) developing in the marble to the right as it approaches the shear zone in the center of the photograph.

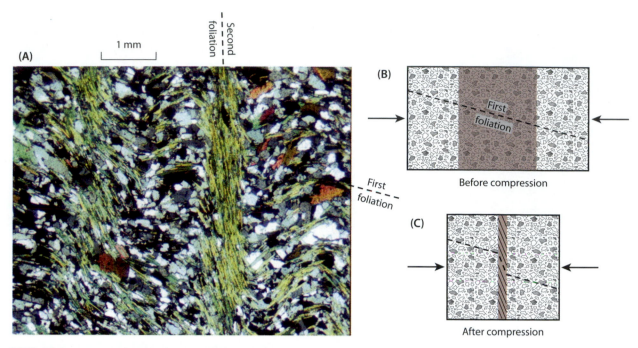

Figure 14.13 (A) Quartz-muscovite schist showing an early gently dipping foliation cut by a later steeply dipping foliation in which all quartz has disappeared. (B) Drawing showing the early foliation prior to development of the second foliation. Quartz is lost from the shaded area during formation of the second foliation. (C) Compression results in the loss of quartz and development of the second foliation.

roughly horizontal planes on which insoluble residue accumulates to form **stylolites** (Sec. 11.9.3 and Fig. 11.38). When limestone is folded and metamorphosed, solution continues but along surfaces that parallel the axial plane of folds. Figure 14.12(A) shows a foraminiferal limestone in which a steeply dipping widely spaced (~1 cm) solution cleavage has developed. The ends of foraminifera can be seen to have been removed by solution along these cleavage planes. The insoluble residue, consisting largely of clay minerals, accumulates on these planes and defines a dark axial plane foliation, as can be seen in the folded marble in Figure 14.12(B). Measurements of the amount of insoluble residue in folded limestones indicate that large volumes of rock are removed by solution during deformation, which again attests to the importance of fluids during metamorphism.

Pelitic schists can also lose large quantities of quartz during development of foliation. Figure 14.13(A) shows a quartz-muscovite schist with two ages of foliation. The earlier foliation dips gently, whereas the second dips steeply. The second foliation is more widely spaced than the first, and because it folds the first foliation, it is often referred to as **crenulation schistosity**. The second foliation is marked by zones consisting almost exclusively of muscovite; all the quartz grains have vanished from these zones. From the geometry of the two foliations, it is easy to see what volume of rock was removed during the development of the second foliation (Fig. 14.13(B)–(C)). The quartz removed from such foliation planes accounts for part of the quartz deposited in quartz veins in overlying rocks (Fig. 14.2).

Growth of metamorphic minerals during shear

Metamorphic minerals commonly are sheared and rotated during growth. Some porphyroblasts preserve a record of this rotation if they contain inclusions of minerals that were parallel to the original foliation prior to the shear. Figure 14.14(A)–(B) shows two garnet porphyroblasts that have undergone different degrees of shear during growth, as indicated by trails of quartz and opaque inclusions that were parallel to the original foliation. In Figure 14.14(A), the garnet porphyroblast grew for an extended period overgrowing layers of quartz and opaque grains that were initially parallel to the foliation. Toward the end of its growth, the garnet crystal experienced a counterclockwise shear that rotated the included quartz layers through 20°. The porphyroblast in Figure 14.14(B) also experienced counterclockwise shear, but in this case, the garnet continued growing as it rotated, with the result that it contains quartz inclusions that show folds indicating 120° of rotation. Porphyroblasts with folded trails of inclusions are said to have a **helicitic texture**, from the Greek word *helix*. Some porphyroblasts with a helicitic texture may not have been formed by rotation but may have simply grown over a preexisting crenulation. However, if the inclusions are twisted through more than 180°, the helicitic texture must have been caused by rotation.

Shear can also reduce the grain size of crystals during metamorphism. Figure 14.15 shows a metamorphosed granite composed of perthitic alkali feldspar and quartz that underwent considerable deformation, as evidenced by the large bent grain of feldspar. The deformation caused new un-deformed grains

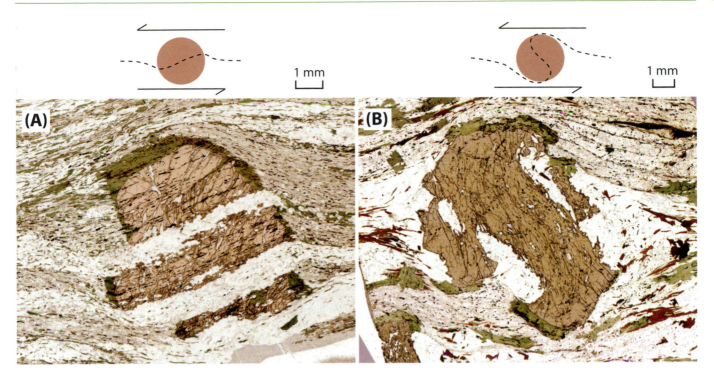

Figure 14.14 Garnet porphyroblasts with trails of quartz and opaque inclusions indicating counterclockwise shear through 20° in (A) and 120° in (B). The porphyroblasts have retrograde green chlorite developed on their rims.

to nucleate and grow. All the quartz completely recrystallized into thin lenses that define the foliation in the rock. Some of the original feldspar grains still remain, but finer grains of recrystallized feldspar surrounded them. The relict grains are shaped like eyes, so this rock is referred to as **augen gneiss** after the German word for "eye." Rocks that contain relict large grains are said to have a **porphyroclastic** texture to distinguish them from the **porphyroblastic** texture, in which grains grew to a large size during metamorphism.

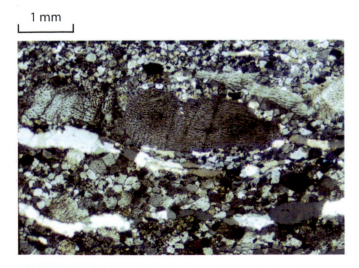

Figure 14.15 Augen gneiss, consisting of large deformed feldspar grains surrounded by finer recrystallized feldspar and lenses of recrystallized quartz that define the foliation.

When shear becomes intense near fault zones, large amounts of strain can cause so many new grains to nucleate that the rock becomes extremely fine grained. The grain size may be so fine that individual grains are not visible to the unaided eye. Such rocks are known as **mylonites**, from the Greek word *mylos*, meaning "mill," because their fine grain size was thought originally to result from grinding and breaking of grains. Fault zones near the Earth's surface may contain fine ground up rock known as **gouge**, but the fine-grained rock developed along faults in the ductile part of the crust are now known to form by recrystallization. As large grains in the protolith recrystallize to finer-grain size, they produce thin monomineralic layers that may contain relics of the original mineral (Fig. 14.16). These are streaked out along the shear plane and cause most mylonites to exhibit fine-scale layering. As a result, mylonites are easily mistaken for bedded chert.

In extreme cases, rapid shear can generate sufficient frictional heat on fault planes to cause melting. The liquid is injected into fractures in the surrounding rock where it freezes rapidly to a glass or extremely fine-grained rock (Fig. 14.17(A)–(B)). This rock is known as **pseudotachylite** because it resembles basaltic glass (tachylite) but has the composition of whatever rock was frictionally fused.

In summary, regional metamorphic rocks are characterized by foliations that can be due to the parallel alignment of platelike or needlelike minerals. Foliations typically develop perpendicular to the maximum compressive stress and, therefore, parallel to the axial plane of folds. Foliation can also be

1 mm

Figure 14.16 Plagioclase-hornblende mylonite from a thrust fault in southeastern Connecticut. Small eye-shaped grains of plagioclase are surrounded by extremely fine grains of plagioclase, which are streaked out by shear to produce the foliation.

(A)

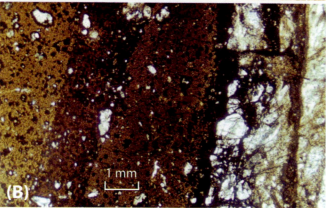

1 mm

(B)

Figure 14.17 **(A)** Black veins of pseudotachylite formed by frictional fusion along a fault 100 km northeast of Montreal, Quebec. **(B)** Photomicrograph of a thin section under plane light of pseudotachylite from (A). Unaltered yellow glass (far left) contains rounded grains of quartz. The red glass (center) at the margin of the vein has oxidized. The granite on the right is reduced to a mylonite at the contact with the pseudotachylite.

generated by shear, the sense of which may be preserved by trails of inclusions in porphyroblasts that grow during deformation. Although grain size typically increases during regional metamorphism, intense deformation can cause recrystallization to finer-grained rocks.

14.5 Simple descriptive classification of metamorphic rocks

The description of metamorphic rocks is straightforward and involves the listing of the most abundant minerals attached to a textural term, such as hornfels, slate, schist, or gneiss. For example, the rock in Figure 14.1(B) is a staurolite-garnet-muscovite schist, whereas the rock in Figure 14.5(A) is a cordierite-spinel hornfels. In addition, a number of special terms are used for rocks that have a specific protolith. Marble, for example, is used for metamorphic rock formed from relatively pure limestone. If the limestone is a marl or is interbedded with shale, it typically is converted to an assemblage of calcium aluminum silicates (anorthite, grossular, actinolite, diopside) and hence is referred to as calcsilicate. Calcsilicate-rich rocks are commonly developed at the contact of igneous bodies, where their composition may be affected by metasomatism, especially due to the introduction of iron. These rocks, which can form iron-ore deposits, are known as **skarns**. Metamorphism of quartz-rich sandstone produces a metamorphic rock known as **quartzite**. Finally, metamorphism of ultramafic igneous rocks produces a mixture of serpentine and talc, which is known as **soapstone**.

The classification of metamorphic rocks according to metamorphic facies can be made only after mineral assemblages have been carefully identified. This is normally done under the microscope, because even the presence of a small amount of a mineral is as important to defining an assemblage as a large amount. This will become apparent after we discuss the graphical representation of mineral assemblages in the following section.

14.6 Metamorphism of mudrock

Although metamorphic facies were originally named on the basis of mineral assemblages in a basaltic protolith, sedimentary rocks containing an abundance of clay minerals (pelites) are far more useful for recording metamorphism, because many more reactions take place and produce many more minerals with which to subdivide P-T space. Moreover, mudrock is the most abundant type of sedimentary rock.

Muscovite is one of the earliest minerals to form in metapelites, and it continues to exist up to the top of the amphibolite facies, where it reacts with quartz to form K-feldspar and an Al_2SiO_5 polymorph. We can write this reaction as follows:

$$KAl_2AlSi_3O_{10}(OH)_2 \ + \ SiO_2 \ = \ KAlSi_3O_8$$
$$\underset{\text{muscovite}}{} \quad \underset{\text{quartz}}{} \quad \underset{\text{K-feldspar}}{}$$
$$+ \ Al_2SiO_5 \ + \ H_2O$$
$$\underset{\text{andalusite}}{} \quad \underset{\text{fluid}}{} \qquad \qquad (14.3)$$
$$\underset{\text{kyanite}}{}$$
$$\underset{\text{sillimanite}}{}$$

As do many prograde metamorphic reactions, this one liberates a fluid. Because the fluid is pure H_2O, the activity of water in the environment affects the temperature at which the reaction occurs. For our discussion, we assume the activity to be 1, but if the metapelite were interbedded with limestone that liberated CO_2 during metamorphism, the fluid in the environment would not be pure water and, therefore, the activity would be less than 1. An Al_2SiO_5 polymorph is produced by the reaction, the specific one being determined by the pressure and temperature at which the reaction occurs relative to the stability fields of the three polymorphs (Sec. 13.3). We next develop a simple graphical way to analyze this reaction and show the relations between the various minerals in P-T space.

14.6.1 Graphical representation of a simple metamorphic reaction

Let us apply the Gibbs phase rule (Box 14.2) to a metapelite that could possibly contain all of the minerals given in Reaction (14.3); that is, quartz, muscovite, andalusite, kyanite, sillimanite, and K-feldspar. We start by writing the formula of each phase:

Quartz	SiO_2
Muscovite	$KAl_2AlSi_3O_{10}(OH)_2$
Andalusite, kyanite, sillimanite	Al_2SiO_5
Potassium feldspar	$KAlSi_3O_8$
Water	H_2O

The first question we must ask is, What is the minimum number of components needed to account for the composition of all of the phases? Inspection of the formulas reveals that we would require four components to create all of the phases. We can write these in terms of oxides containing one cation, which simplifies plotting minerals in diagrams. These components are the following:

$$HO_{1/2}, \ SiO_2, \ AlO_{3/2}, \ \text{and} \ KO_{1/2}$$

To show the composition of the phases in terms of these four components, we require a tetrahedral plot, which is given in Figure 14.18(A). The phases are shown by the black dots. Note that although the Al_2SiO_5 polymorphs, quartz, and K-feldspar plot on the front-right face of the tetrahedron, muscovite plots inside the tetrahedron because it is the only mineral that contains water. None of the minerals contains large amounts of $KO_{1/2}$, and, as a result, all plot near the front edge of the tetrahedron. We also note that the ratio of $KO_{1/2}$ to SiO_2 in K-feldspar and muscovite is the same (1:3). This allows us to make K-feldspar the potassium-bearing component rather than $KO_{1/2}$, which then expands the front part of the tetrahedron, making a more convenient plot.

Remember, there is nothing sacred about the way components are defined, as long as they are the minimum number needed to account for all the phases present. We, therefore, choose the more convenient set of components:

$$HO_{1/2}, \ SiO_2, \ AlO_{3/2}, \ \text{and} \ KAlSi_3O_8$$

We still have four components, which require a tetrahedral plot, as shown in Figure 14.18(B), but the region of interest has been expanded. Again, the Al_2SiO_5 polymorphs, quartz, and

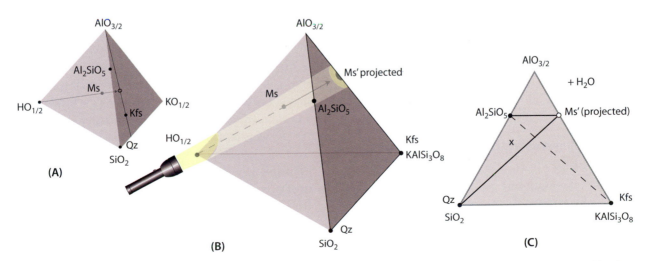

Figure 14.18 (A) Quartz (Qz), K-feldspar (Kfs), Al_2SiO_5 polymorphs, muscovite (Ms), and water ($HO_{1/2}$) plotted in the four-component compositional tetrahedron $HO_{1/2}$-SiO_2-$AlO_{3/2}$-$KO_{1/2}$. (B) Plotting of the same phases in the compositional tetrahedron $HO_{1/2}$-SiO_2-$AlO_{3/2}$-$KAlSi_3O_8$. Compositions in this tetrahedron can be projected from the $HO_{1/2}$ corner onto the $AlO_{3/2}$-SiO_2-$KAlSi_3O_8$ face of the tetrahedron as shown for the case of muscovite (Ms). (C) Minerals in (A) plotted in terms of $AlO_{3/2}$-SiO_2-$KAlSi_3O_8$ with muscovite (Ms') being projected as shown in (B).

K-feldspar plot on the front-right face of the tetrahedron, but muscovite now plots on the back face of the tetrahedron.

Three-dimensional plots such as the tetrahedron in Figure 14.18(B) are not easy to read because the positions of compositions within the tetrahedron are difficult to show. For example, if you had not been told that the dot representing muscovite plotted on the back face of the tetrahedron, it could have plotted anywhere in the tetrahedron along a line of sight toward your eye. For this reason, we search for ways to simplify three-dimensional plots by projecting them into two dimensions. We can do this if we restrict our discussion to rocks that coexist with pure water. This is equivalent to saying that the activity of water is 1. During metamorphism, Reaction (14.3) releases water. It is true that most of this escapes from the rock, but as long as small amounts of pure water remain along grain boundaries, it is valid to assume that its activity was 1. By making this assumption, we are saying that water was present as a phase, even though we may not list it in our mineral assemblage.

If all mineral assemblages coexist with water, it is unnecessary to show water in our plot. We can eliminate it by projecting mineral compositions within the tetrahedron from the water apex onto the $AlO_{3/2}$-SiO_2-$KAlSi_3O_8$ triangular face. This is shown symbolically by the flashlight shining from the $HO_{1/2}$ apex and casting a shadow of the muscovite composition on the triangular face. This shadow is then the projected composition of muscovite. Figure 14.18(C) shows the resulting triangular plot, with the minerals that actually plot on the face being shown as black dots and the projected composition of muscovite as the open circle (Ms′). We have written $+H_2O$ beside the triangle to remind us that these minerals must coexist with pure water.

We can use the triangular plot to illustrate the mineral assemblages associated with Reaction (14.3). For example, a rock composed of muscovite, andalusite, and quartz (+ water) would plot in the triangle bounded by these minerals (e.g., x in Fig. 14.18(C)). When Reaction (14.3) takes place, the tie line between muscovite and quartz is broken and switches to a tie line between K-feldspar and andalusite (dashed line). Our previous rock would, therefore, plot in the triangle quartz-andalusite-K-feldspar.

14.6.2 A simple pressure-temperature petrogenetic grid

The pressure and temperature conditions for Reaction (14.3) under an activity of water of 1 are shown in Figure 14.19 (blue line), with quartz and muscovite being stable on the low-temperature side of the reaction curve. The plot also shows the univariant lines separating the stability fields of the three Al_2SiO_5 polymorphs (red lines), which are independent of water activity. These various reactions divide pressure-temperature space into regions where particular mineral assemblages are

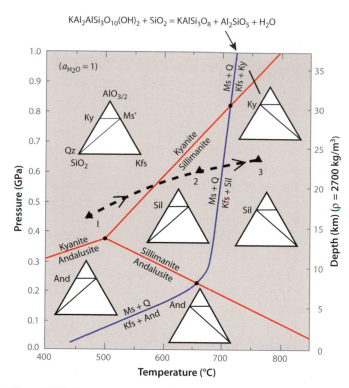

$KAl_2AlSi_3O_{10}(OH)_2 + SiO_2 = KAlSi_3O_8 + Al_2SiO_5 + H_2O$

Figure 14.19 Pressure-temperature plot of the reaction muscovite + quartz = K-feldspar + Al_2SiO_5 + H_2O (blue line) and the Al_2SiO_5 polymorphs (red lines). Triangular plots show mineral assemblages in the divariant fields between the reaction lines (mineral abbreviations are the same as in Fig. 14.18). The heavy dashed line shows a sequence of mineral assemblages that might occur across a progressively metamorphosed terrane.

stable. The intersecting reaction lines are referred to as a petrogenetic grid, because they set limits on the pressure and temperature ranges over which a given metamorphic mineral assemblage could have formed.

Reaction (14.3) involves three minerals and water, but one of the minerals (Al_2SiO_5) has three polymorphs, so the total number of phases is seven. However, according to the Gibbs phase rule, the maximum number of phases that can coexist is ($c + 2$) which in this case is six. When six phases are present, the system is invariant; that is, these six phases can exist at only one temperature and pressure. Three such invariant points are shown in Figure 14.19 by black dots. For example, where Reaction (14.3) intersects the line separating the field of andalusite and sillimanite, the six coexisting phases are quartz, muscovite, andalusite, sillimanite, K-feldspar, and water. Where this reaction intersects the line separating the field of kyanite and sillimanite, the six phases are quartz, muscovite, kyanite, sillimanite, K-feldspar, and water. And at the Al_2SiO_5 polymorph triple point, the six phases are quartz, muscovite, andalusite, kyanite, sillimanite, and water.

When a phase is not plotted in a diagram, such as water in Figures 14.18(C) and 14.19, the phase rule is commonly simplified by reducing the number of components and phases by one. If water is not plotted, one less component is required to describe the remaining phases. This reduces the number of components from four to three, and the maximum number of phases that coexist at an invariant point is five. This is just a simplification, and it is understood that water is still present as a phase and a component, as indicate by writing in Figure 14.19 that $a_{H_2O} = 1$.

Surrounding each of the invariant points in Figure 14.19 are univariant lines along which four minerals (+ water) coexist. These univariant lines separate divariant fields in which three minerals (+ water) coexist. By saying the fields are divariant, we are indicating that the pressure and temperature can be varied over some range without changing the mineral assemblage. In each of these fields, we can place a triangular plot such as that shown in Figure 14.18(C) in which we can indicate the possible mineral assemblages.

Let us start by determining the possible mineral assemblages that occur in the divariant field in the top left of Figure 14.19 (high pressure, low temperature). Because this region is on the low-temperature side of Reaction (14.3) (blue curve), quartz and muscovite are stable together. In the phase triangle, we join the minerals with a tie line to show that they stably coexist. Under these pressure and temperature conditions, kyanite is the stable polymorph of Al_2SiO_5, so we indicate this on the phase triangle by joining it with tie lines to muscovite and quartz. This creates a tie triangle; in other words, a divariant assemblage of three minerals, which could be either quartz-muscovite-kyanite or quartz-muscovite-K-feldspar depending on the bulk composition of the rock. A line does not join kyanite and K-feldspar because that tie line does not form until we cross to the high temperature side of Reaction (14.3).

If we move to the divariant field in the center of Figure 14.19, we see that the only change that occurs is that sillimanite is the stable polymorph of Al_2SiO_5, instead of kyanite. All the other minerals remain the same. This illustrates an important point in these types of diagrams; that is, only one change can occur in crossing from one divariant field to another.

Moving to higher temperature, we cross the line for Reaction (14.3), where muscovite and quartz react to form K-feldspar and sillimanite, which involves the tie line between quartz and muscovite switching to one between sillimanite and K-feldspar. On moving around the diagram, we cross other univariant lines at which a single change occurs in the mineral assemblages. It is left to the reader to work out what these changes are as indicated by the phase diagrams in each field.

14.6.3 Metamorphic field gradients

One of the uses of a petrogenetic grid, such as that shown in Figure 14.19, is that it allows us to set pressure and temperature limits for the formation of mineral assemblages found in metamorphic rocks. For example, a rock consisting of quartz, sillimanite, and muscovite and coexisting with a pure water fluid phase is stable only under the pressure and temperature conditions of the central divariant field in Figure 14.19; that

is, between 500°C and 700°C and 0.22–0.82 GPa. A rock containing quartz, kyanite and muscovite would be stable only at pressures above 0.82 GPa and temperatures above 700°C, and a rock containing andalusite and K-feldspar would be stable only below 0.22 GPa.

The petrogenetic grid in Figure 14.19 is very simple because it involves only one reaction and the three Al_2SiO_5 polymorphs. If more reactions are considered, pressure-temperature space can be subdivided into more and smaller regions, which narrows down the conditions under which any particular metamorphic rock was formed. With a more complex grid, for example, it might be possible to determine that the temperatures and pressures recorded across a progressively metamorphosed terrane, such as that mapped by Barrow in the southeastern Scottish Highlands, fell along a line such as the heavy dashed line in Figure 14.19. The triangles numbered 1, 2, and 3 would represent tightly constrained mineral assemblages. Because these assemblages indicate increasing temperature with increasing pressure and depth, they were initially interpreted as representing the geotherm during metamorphism. However, this would be true only if the assemblages in the three different areas attained equilibrium at the same time. We recognize that rocks at different depth reach their maximum temperatures at different times. The gradient shown by the dashed line in Figure 14.19 is, therefore, time transgressive and is not a fossil geotherm, so instead it is referred to as a **metamorphic field gradient**. Field gradients vary considerably between different metamorphic belts and record important information about how metamorphism and plate convergence are related (see Sec. 14.10).

14.6.4 Graphical representation of mineral assemblages in metapelites

We have seen in Figure 14.19 how a simple petrogenetic grid can be constructed to analyze the mineral assemblages in pelitic rocks. This grid was of limited use, however, because it did not include minerals such as chlorite, biotite, garnet, staurolite, cordierite, and chloritoid, all of which are common constituents of metapelites. These minerals all contain significant amounts of iron and magnesium and possibly small amounts of manganese. The question we must address is, How can we analyze rocks containing a large number of components, and can we possibly represent the mineral assemblages on simple diagrams?

Let us start by writing down all of the oxides needed to form these minerals. We know that some of these may be grouped together to define components according to the Gibbs phase rule, but the list gives us something to start with. The main oxides are the following:

$$SiO_2, TiO_2, Al_2O_3, Fe_2O_3, FeO, MnO, MgO, K_2O, \text{ and } H_2O$$

We saw in Section 14.6.1 that if pure water existed as a phase in the rock, we need not plot it or count it as a component. We

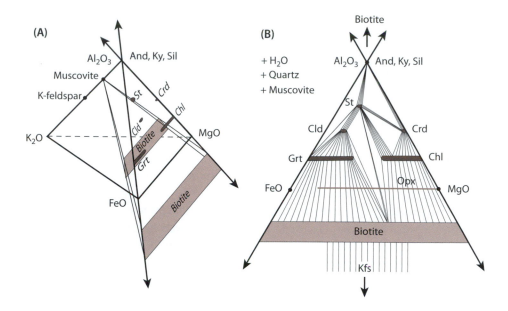

Figure 14.20 (A) Common minerals in metapelite plotted in terms of Al₂O₃, MgO, FeO, and K₂O. Minerals include muscovite, K-feldspar, andalusite (And), kyanite (Ky), sillimanite (Sil), staurolite (St), chloritoid (Cld), cordierite (Crd), biotite, garnet (Grt), and chlorite (Chl). (B) Minerals from the tetrahedral plot in (A) projected onto the Al₂O₃-FeO-MgO face of the tetrahedron. The shaded line marked Opx shows the compositional range of orthopyroxene in the granulite facies. Tie lines between coexisting phases represent the arrangement at only one grade of metamorphism (after Thompson, 1957, *American Mineralogist*).

apply the same rule to this long list of components. Most pelitic rocks also contain quartz, so we need not plot it in our assemblage and we can eliminate SiO_2 as a component. Similarly, if the rock contains rutile, as many pelites do, we need not count rutile as a phase nor TiO_2 as a component. If the rock contains hematite, then Fe_2O_3 need not be counted as well. When eliminating these components from our list, we should not forget that our arguments concerning possible mineral assemblages apply only to those rocks that contain these phases.

Many ferromagnesian minerals form solid solutions between iron and magnesium end members. Unfortunately, many of the metamorphic ferromagnesian minerals show only limited solid solution, and some such as staurolite are limited to relatively iron-rich compositions, and others such as cordierite are restricted to magnesium-rich compositions. Iron and magnesium can, therefore, not be treated as a single component. Manganese, however, does substitute readily for iron, and so the manganese can be added to the iron and treated as a single component.

Our list of components is reduced to four; that is, Al_2O_3, (FeO + MnO), MgO, and K_2O. These are shown in the tetrahedral plot in Figure 14.20(A). All the minerals except muscovite, biotite, and K-feldspar plot on the front right face of this tetrahedron, which is commonly referred to as the AFM face. As stated previously, three-dimensional tetrahedral plots are not easy to read composition from, so it is useful to project compositions from the tetrahedron onto one of the triangular faces. J. B. Thompson Jr. of Harvard University proposed that because most pelitic rocks, at least at low to medium grades of metamorphism, contain muscovite, compositions should be projected from muscovite. At higher temperatures above Reaction (14.3),

where quartz and muscovite react to form K-feldspar and an Al_2SiO_5 polymorph, projection can be made from K-feldspar. Although no projection is required for most of the minerals, as they plot on the AFM face, biotite plots within the tetrahedron and when projected from muscovite falls below the FeO-MgO base of the triangle. The resulting triangular plot is shown in Figure 14.20(B) with tie lines joining coexisting phases for just one possible assemblage in the amphibolite facies. This type of projection is the most widely used one for illustrating mineral assemblages in metapelites, and it is named the **Thompson projection** in recognition of its originator (Fig. 14.21).

Figure 14.21 J. B. Thompson Jr., who proposed the commonly used projection for pelitic rocks (see Fig. 14.20), holding a large cleavage book of muscovite.

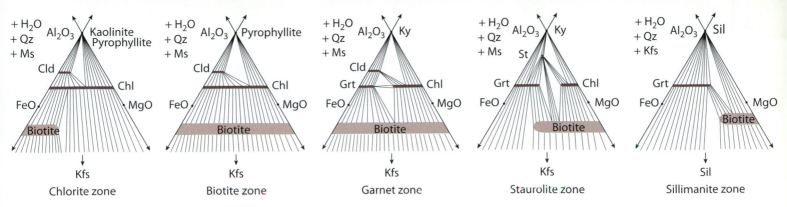

Figure 14.22 Sets of possible mineral assemblages in metapelites in Barrow's chlorite, biotite, garnet, staurolite, and sillimanite zones of regional metamorphism plotted in the Thompson projection. Minerals are projected from muscovite, except in the sillimanite zone where they are projected from K-feldspar.

14.6.5 Mineral assemblages in Barrow's metamorphic zones and part of the petrogenetic grid for metapelites

Now that we have a convenient way of plotting metamorphic mineral assemblages, we can show some of the common assemblages that form during regional metamorphism. We have chosen five assemblages from Barrow's chlorite, biotite, garnet, staurolite, and sillimanite zones (Fig. 14.22). Remember that in addition to the minerals shown in the Thompson projection, quartz and muscovite (and water) are also present, except in the sillimanite zone where quartz and K-feldspar are present and projection is from feldspar rather than from muscovite.

Three different types of compositional field are shown in these triangular diagrams. First, there are fields representing the composition of a single mineral (shaded gray). Many of the minerals belong to solid-solution series that can have a wide compositional range. Second, there are compositional fields in which two minerals coexist (+ quartz and muscovite) and these are ruled by **tie lines** joining the coexisting minerals. Third, there are compositional fields in which three minerals coexist, with their compositions joined by **tie triangles**. When three minerals coexist, their compositions are fixed (corners of the triangles), but when only two minerals coexist, the decreased number of phases increases the degrees of freedom by one, which allows the composition of one of the minerals to vary. Once the composition of that mineral is fixed, however, the composition of the other mineral is determined by the tie line (i.e., its composition is not independently variable).

Minerals that belong to Fe-Mg solid-solution series, such as chlorite, biotite, and garnet, show a progressive enrichment in magnesium with increasing metamorphic grade. For example, at low metamorphic grade, chlorite can have any composition between the iron and magnesium end members, but as the grade increases, iron-rich chlorites become unstable and react to form garnet. As the grade increases, the remaining chlorite becomes progressively richer in magnesium until eventually it totally disappears. When garnet first appears, it is iron rich (also manganese rich) and with increasing grade it can reach

progressively more magnesium-rich compositions. Biotite is also iron rich when it first appears in the chlorite zone but becomes progressively more magnesium rich with increasing grade. In the sillimanite zone, it tends to approach phlogopite in composition.

These changes have a simple explanation that we can illustrate by referring to the melting loop for olivine solid solutions (Fig. 8.9). When the temperature rises to the solidus, a reaction starts that converts solid olivine to liquid olivine, but the liquid is more iron rich than the solid and, thus, the remaining olivine becomes progressively more magnesium rich. Metamorphic reactions involving solid solutions behave in exactly the same way, except that the reaction product is an assemblage of minerals rather than a liquid. So as the temperature rises, the composition of the low-temperature mineral is forced to move to more magnesium-rich compositions. For example, in the Thompson projection shown for the sillimanite zone (Fig. 14.22), a tie triangle shows the assemblage sillimanite + biotite + garnet (+ quartz and K-feldspar), with the compositions of these minerals indicated by the corners of the triangle. We could have shown a diagram for a slightly higher temperature that would have looked similar except that the tie triangle would have shifted to the right with both garnet and biotite having more magnesium-rich compositions. As we will see in Section 14.9, this gradual shift in composition with temperature is the basis for some geothermometers.

The assemblages shown in Figure 14.22 for the various metamorphic zones are only one possible set of assemblages. In each zone, many other assemblages can occur, and many reactions separate the assemblages shown for adjacent zones, with each reaction accompanied by a new assemblage. To unravel all the possible assemblages that can occur would involve the creation of a complete petrogenetic reaction grid, which is beyond the scope of this book (for a more complete development, see chapter 18 in Philpotts and Ague, 2009). We can, however, illustrate the use of such a grid in determining metamorphic conditions by considering a very small part of the grid that involves the mineral staurolite.

Figure 14.23 shows a pressure-temperature petrogenetic grid for pelitic rocks covering a range of temperatures from 400–700°C and pressures up to 0.6 GPa (depth of ~23 km). The plot shows the stability limits of the three Al_2SiO_5 polymorphs in red lines and all of the reactions that can occur between the minerals shown in the inset Thompson projection. These reactions are shown in green except for one that is colored blue, which is the reaction of quartz with muscovite to create K-feldspar and an Al_2SiO_5 polymorph. On the high-temperature side of this reaction, compositions are projected from K-feldspar rather than from muscovite in the Thompson projections.

The minerals involved in these reactions are shown along the reaction lines. Discovering what reactions can occur is actually quite simple. Any two minerals joined by a line can react together to give two new minerals as long as the new tie line intersects the original line. For example, the line between staurolite and biotite cuts the line between chlorite and garnet, and, therefore, we can write the following reaction:

$$\text{chlorite + garnet = staurolite + biotite} \qquad (14.4)$$

If we were to write the complete chemical formulas for these minerals, we would find that it is necessary to add muscovite to the left-hand side to provide the potassium needed for biotite on the right-hand side. In the Thompson projection, quartz and muscovite are always present (or K-feldspar at high temperature) and so are available to balance reactions. Reaction (14.4) is known as a **tie-line switching reaction** for obvious reasons. Can you locate it in Figure 14.23?

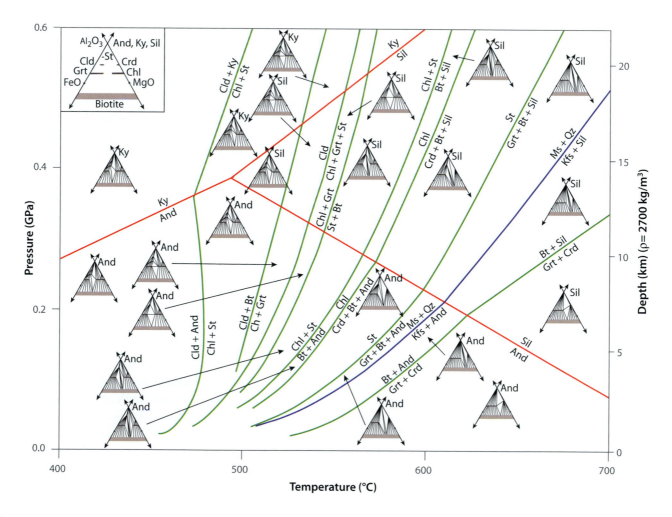

Figure 14.23 Petrogenetic grid showing reactions in metapelites in the greenschist and amphibolite facies. Mineral compositions are plotted in the Thompson projection from muscovite except above the reaction curve of muscovite + quartz = K feldspar + Al_2SiO_5 (blue line) where the projection is from feldspar. The inset at the top left shows the plotted positions of andalusite (And), kyanite (Ky), sillimanite (Sil), garnet (Grt), chloritoid (Cld), staurolite (St), cordierite (Crd), chlorite (Chl), and biotite (Bt). Reactions are written on the reaction lines, but quartz and muscovite may be needed to balance the reactions (modified from Spear and Cheney, 1989).

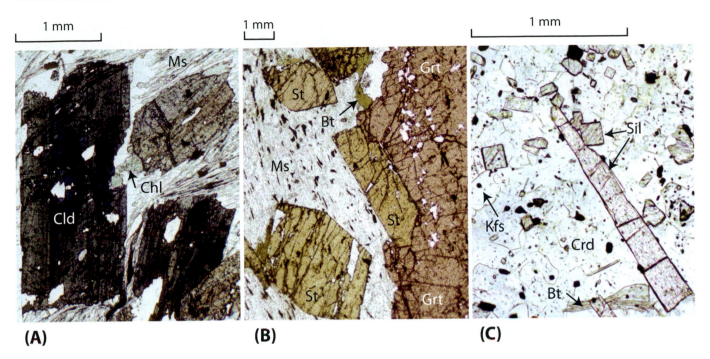

Figure 14.24 Photomicrographs of thin sections under plane light of metapelite from three different grades of metamorphism. (**A**) Chloritoid-muscovite-chlorite-quartz schist. (**B**) Staurolite-garnet-muscovite-biotite-quartz schist. (**C**) Sillimanite-cordierite-K-feldspar-biotite-quartz hornfels.

Another type of reaction can be found by locating a mineral that plots in a tie triangle between three other minerals. In this case, the mineral in the triangle can break down to form the three minerals in the surrounding triangle. For example, staurolite plots in the triangle formed by garnet, biotite, and an Al_2SiO_5 polymorph. Consequently, we can write the following reaction:

$$\text{staurolite} = \text{garnet} + \text{biotite} + Al_2SiO_5 \qquad (14.5)$$

Biotite contains potassium, so to balance this reaction, we would need to add muscovite on the left-hand side, but again this is not a problem because muscovite and quartz are always present. Reaction (14.5) is an example of what is known as a **terminal reaction**, because it terminates the existence of a mineral. This particular reaction sets the absolute upper stability limit of staurolite. You should locate it in Figure 14.23.

We do not write out all the reactions, but they can be derived in exactly the same way we derived Reactions (14.4) and (14.5). These reactions subdivide Figure 14.23 into relatively small temperature regions, but they do not limit the pressures to the same degree. A specific set of minerals coexists in each field, as indicated by tie lines and tie triangles. When you move from one field to the next, only one change can occur in the diagrams. If you wish to check this, start with the Thompson projection in the upper-left (high pressure and low temperature) field of the diagram and work your way clockwise around the diagram. Every time you cross a reaction line, one of the lines in

the Thompson projection changes, as indicated by the reaction you have just crossed. If you continue around the diagram, you eventually return to the phase triangle in the upper left corner.

A petrogenetic grid such as that in Figure 14.23 allows you to analyze mineral assemblages found in rocks and to assign pressure-temperature ranges to the assemblages. Some mineral assemblages can have restricted ranges, whereas others are stable over wide ranges. Figure 14.24 shows three photomicrographs of thin sections of pelitic rocks formed at very different temperatures. It is left to the reader to use the mineral assemblages in each of these rocks to determine their possible temperature and pressure ranges as indicated by the petrogenetic grid.

14.7 Metamorphism of impure dolomitic limestone

Impure carbonate rocks, like pelites, are useful in monitoring metamorphism, because of the large number of minerals and mineral assemblages that can form with increasing grade (Fig. 14.25). In addition, many of these reactions involve both CO_2 and H_2O; thus, they provide an ideal opportunity to illustrate the role of fluid composition in metamorphism. We discuss these reactions by using a quartz-bearing dolomitic limestone as the protolith.

Metamorphism of carbonate rocks at the margin of igneous intrusions had long been recognized to produce a variety of metamorphic minerals, but in 1940, N. L. Bowen proposed that these minerals appeared in a definite order that resulted

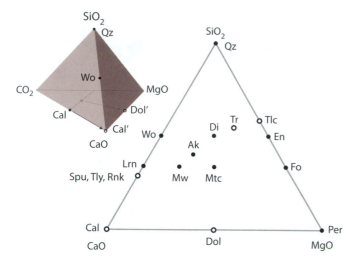

Figure 14.25 Projection of minerals in the SiO$_2$-CaO-MgO-CO$_2$ (H$_2$O) tetrahedron onto the SiO$_2$-CaO-MgO face. The minerals that are found in metamorphosed quartz-bearing dolomitic limestone, include quartz (Qz), calcite (Cal), dolomite (Dol), wollastonite (Wo), diopside (Di), tremolite (Tr), talc (Tlc), enstatite (En), forsterite (Fo), monticellite (Mtc, CaMgSiO$_4$), åkermanite (Ak, Ca$_2$MgSi$_2$O$_7$), merwinite (Mw, Ca$_3$MgSi$_2$O$_8$), larnite (Lrn, Ca$_2$SiO$_4$), spurrite (Spu, Ca$_4$Si$_2$O$_8$·CaCO$_3$), tilleyite (Tly, Ca$_3$Si$_2$O$_7$·2CaCO$_3$), and rankinite (Rnk, Ca$_3$Si$_2$O$_7$).

forsterite, diopside, periclase (MgO), wollastonite, monticellite (CaMgSiO$_4$), spurrite (Ca$_4$Si$_2$O$_8$·CaCO$_3$), merwinite (Ca$_3$MgSi$_2$O$_8$), and larnite (Ca$_2$SiO$_4$). To help remember this sequence, he suggested using the following mnemonic:

> Tremble, for dire peril walks,
> Monstrous acrimony's spurning mercy's laws.

When Professor Tilley of Cambridge University, who had studied the famous contact metamorphic rocks at Scawt Hill in Northern Ireland, pointed out to Bowen that talc actually appeared before tremolite, Bowen suggested altering the mnemonic by stuttering at the beginning; that is, "Ta … tremble." We now know that this sequence of minerals is only one of a number that can occur and that it is far more common for diopside to appear before forsterite. The question is, Why should different sequences of minerals appear at different localities? The answer to this lies in the composition of the fluids that coexist with the minerals.

Let us start to answer this question by examining one of the simplest of reactions to occur in impure limestones; that is, the reaction of calcite with quartz to form wollastonite. This reaction can be written as follows:

$$\underset{\text{calcite}}{CaCO_3} + \underset{\text{quartz}}{SiO_2} = \underset{\text{wollastonite}}{CaSiO_3} + \underset{\text{fluid}}{CO_2} \qquad (14.6)$$

This reaction releases CO$_2$ fluid that was previously in the solid state in the crystal structure of calcite (see Sec. 10.5 and Fig. 10.10). The reaction is, therefore, accompanied by a large

from a sequence of reactions that took place at progressively higher temperatures (Fig. 14.26). His pressure-temperature plot of these reactions was important because it was one of the first examples of the use of a petrogenetic grid. Bowen believed that the reactions caused the following sequence of minerals to appear with increasing temperature: tremolite,

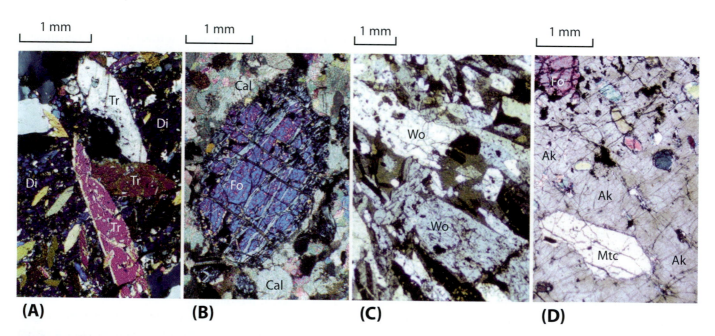

Figure 14.26 Photomicrographs of thin sections under crossed polarized light of metamorphic rocks formed from impure dolomitic limestone, with increasing grade from (A) to (D). (A) Tremolite (Tr) and diopside (Di). (B) Partially serpentinized forsterite (Fo) and calcite (Cal). (C) Wollastonite (Wo). (D) Akermanite (Ak), monticellite (Mtc), and forsterite (Fo).

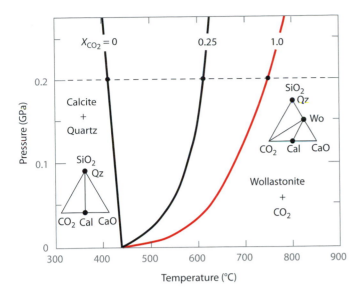

Figure 14.27 Pressure-temperature conditions for the reaction calcite + quartz = wollastonite + CO_2. The red reaction curve applies when the fluid in the environment is pure CO_2. The other two curves apply when the mole fraction of CO_2 in the fluid is 0.25 and 0. The horizontal line and black dots indicate the temperatures of the reaction at 0.2 GPa (see same dots in Fig. 14.28).

volume increase, which means that as the pressure rises, higher temperatures are needed to make the reaction proceed to the right. Figure 14.27 shows the shape of the reaction curve in a pressure-temperature plot (red line). It is typical of any metamorphic reaction releasing a fluid phase. The temperature of the reaction increases rapidly as pressure first begins to rise, because the fluid is very compressible at low pressures but becomes less compressible at higher pressures. The phase triangles (SiO_2-CaO-CO_2) show the switch in tie lines from calcite + quartz to wollastonite + CO_2 in crossing the reaction curve.

The fluid produced in this reaction is pure CO_2. If the fluid in the environment contained some water (mole fraction of CO_2 is less than 1; $X_{CO_2} < 1$), the reaction would take place at lower temperatures. For example, if the total pressure is 0.2 GPa (horizontal dashed line in Fig. 14.27), calcite and quartz react to form wollastonite at 750°C, if the fluid in the environment is pure CO_2. However, if there is enough water in the environment to reduce the mole fraction of CO_2 to 0.25, the reaction occurs at 612°C, and if the fluid in the environment is pure water ($X_{CO_2} = 0$), the reaction occurs at 412°C. Clearly, the mole fraction of CO_2 in the fluid phase makes a huge difference to the temperature at which wollastonite will form from quartz and calcite. This in part explains why Bowen's sequence of metamorphic minerals is only one of several that can occur.

The effect of the fluid phase on the temperature of a reaction at a given pressure is conveniently shown in a plot of temperature versus mole fraction of CO_2 in the fluid. Such a plot applies to only one pressure, so it is said to be isobaric. The isobaric plot in Figure 14.28 is for a pressure of 0.2 GPa, the same as the horizontal line in Figure 14.27. The other major component of

the fluid in metamorphic rocks is H_2O, so we can think of decreasing mole fraction of CO_2 as increasing mole fraction of H_2O (i.e., $X_{CO_2} = 1 - X_{H_2O}$). Figure 14.28 shows how the temperature at which wollastonite forms from calcite plus quartz decreases as the fluid phase becomes more water rich.

Let us consider another important reaction in the metamorphism of impure carbonate rocks by which tremolite and calcite react to form diopside and dolomite as follows:

$$
\begin{aligned}
&\underset{\text{tremolite}}{Ca_2Mg_5Si_8O_{22}(OH)_2} + \underset{\text{calcite}}{3CaCO_3} \\
&= \underset{\text{diopside}}{4CaMgSi_2O_6} + \underset{\text{dolomite}}{CaMg(CO_3)_2} + \underset{\text{fluid}}{H_2O + CO_2}
\end{aligned} \quad (14.7)
$$

In this reaction, CO_2 and H_2O are produced in the ratio of 1:1, so if the ratio in the fluid in the environment differs from this value, the reaction occurs at lower temperatures. The resulting reaction curve has a temperature maximum when the fluid in the environment has a CO_2:H_2O ratio of 1:1 ($X_{CO_2} = 0.5$) and falls to lower temperatures on either side of this (Fig. 14.28).

Diopside can also form from tremolite reacting with calcite and quartz according to the following reaction:

$$
\begin{aligned}
&\underset{\text{tremolite}}{Ca_2Mg_5Si_8O_{22}(OH)_2} + \underset{\text{calcite}}{3CaCO_3} + \underset{\text{quartz}}{2SiO_2} \\
&= \underset{\text{diopside}}{5CaMgSi_2O_6} + \underset{\text{fluid}}{H_2O + 3CO_2}
\end{aligned} \quad (14.8)
$$

In this reaction, CO_2 and H_2O are produced in the ratio of 3:1, so this reaction shows its maximum stability when the fluid in the environment has a CO_2:H_2O ratio of 3:1; that is, $X_{CO_2} = 0.75$. The resulting reaction curve is shown in Figure 14.28, with its maximum at a mole fraction of CO_2 of 0.75.

Let us consider one more important reaction that forms tremolite from dolomite and quartz according to the following reaction:

$$
\begin{aligned}
&\underset{\text{dolomite}}{5CaMg(CO_3)_2} + \underset{\text{quartz}}{8SiO_2} + \underset{\text{fluid}}{H_2O} \\
&= \underset{\text{tremolite}}{Ca_2Mg_5Si_8O_{22}(OH)_2} + \underset{\text{calcite}}{3CaCO_3} + \underset{\text{fluid}}{7CO_2}
\end{aligned} \quad (14.9)
$$

This reaction differs from the previous ones in that, although CO_2 is a reaction product, water appears on the left-hand side as a reactant. In Figure 14.28, this reaction produces a sigmoid-shaped curve going to progressively higher temperatures as the mole fraction of CO_2 increases.

Reactions (14.7)–(14.9) all intersect at an invariant point on the extreme right side of Figure 14.28. In each of the divariant fields between these univariant reaction curves, a triangular diagram shows the possible coexisting phases. If we start with the lowest-phase triangle and go clockwise around the invariant point, we see that when we get to the phase triangle in the upper right, it does not match the one we started with. There must still be one more reaction curve that needs to be crossed before we return to the initial phase triangle. We can determine what this

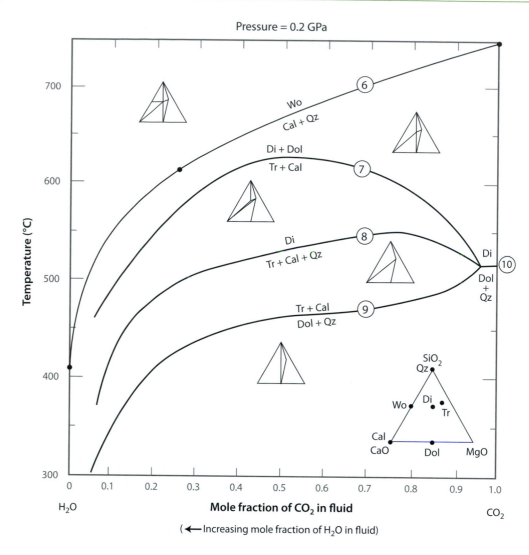

Pressure = 0.2 GPa

Figure 14.28 Temperature versus fluid composition plot for the wollastonite-forming reaction of Figure 14.27 and four other reactions in metamorphosed quartz-bearing dolomite at 0.2 GPa. Reactions are numbered according to their number in the text (black dots on reaction 6 are from Fig. 14.27).

reaction is by inspecting the two phase triangles. We can also determine it by doubling the amounts in Reaction (14.9) and then adding to it Reactions (14.7) and (14.8). This then simplifies to the following reaction:

$$CaMg(CO_3)_2 + 2SiO_2 = CaMgSi_2O_6 + 2CO_2 \quad (14.10)$$
$$\text{dolomite} \qquad \text{quartz} \quad \text{diopside} \quad \text{fluid}$$

This reaction releases only CO_2, so, like the wollastonite-forming reaction (14.6), it reaches a maximum when the mole fraction of CO_2 is one, and it passes through the invariant point.

Many more reactions involving all the minerals shown in Figure 14.25 could be added to Figure 14.28, thus producing a much denser petrogenetic grid, but this is beyond the scope of this book. The few reactions shown in this figure suffice to illustrate how the variation in the composition of fluid in the environment dramatically affects the temperature at which

metamorphic minerals form but also the sequence in which they form. For example, if the protolith consisted of calcite + dolomite + quartz and the fluid in the environment was pure CO_2, then diopside would form as soon as the temperature rose above Reaction (14.8); that is, at ~500°C at 0.2 GPa. However, if the fluid in the environment had a mole fraction of CO_2 of 0.75, dolomite and quartz would react to form tremolite and calcite at 475°C and diopside would not form until the temperature reached 550°C when Reaction (14.8) was crossed.

In summary, most prograde metamorphic reactions liberate a fluid phase, and if that fluid differs in composition from the fluid in the environment, the reaction temperature is lowered. Water and CO_2 are the main fluids released during metamorphism. In carbonate rocks, we can expect metamorphic reactions to release large amounts of CO_2, whereas pelitic rocks release primarily H_2O. The fluids are less dense than rocks and, therefore, rise toward the surface, thereby producing a regional fluid flux of highly variable composition. The composition of this environmental fluid plays an important role in determining the temperature of reactions and the sequence in which they occur during prograde metamorphism.

14.8 Metamorphism and partial melting: migmatites

Regional metamorphism is brought about by elevated temperatures and pressures that result from thickening of the radiogenic continental crust, the intrusion of mantle-derived magmas, and the flux of fluids near convergent plate boundaries. As temperatures rise, prograde reactions take place and eventually the temperature may rise high enough to cause partial melting. Only a small amount of melt normally forms because the latent heat of fusion of rocks is so large that it consumes a large quantity of heat. The rock formed from the mixture of melt and metamorphic rock is known as **migmatite**, from the Greek word *migma*, meaning "mixture."

Only the lowest possible melting fraction forms during metamorphism, and this invariably is of granitic composition. We saw in Section 8.3.3 and Figure 8.7 that partial melting occurs only at grain boundaries between quartz and alkali feldspar. This melt must soon segregate into layers, because the granitic fraction in migmatites tends to be concentrated in layers that are interspersed with layers of the refractory residue (Fig. 14.29). The granite layers are light colored and are referred to as the **leucosome**, and the refractory residue, which is usually rich in plagioclase, hornblende, and biotite, is darker colored and is referred to as the **melanosome**.

Deformation plays an important role in segregating the melt generated along grain boundaries. Experiments show

Figure 14.29 Migmatite consisting of pink granite, formed by melting of the protolith, interlayered with the refractory residue consisting of gray plagioclase-biotite-hornblende gneiss. The granite migrated into the nose of the folds.

Figure 14.30 Migmatite in which the molten granitic fraction was moved by deformation into veins and ductile fault zones.

that when partially melted crystal mush is sheared, the liquid tends to segregate into lenses. Most migmatites show evidence of considerable deformation. The migmatite in Figure 14.29 is folded, and the liquid can be seen to have moved into the nose of the folds. Ductile shear zones are also extremely common in migmatites, and the liquid can often be seen to have migrated into these zones (Fig. 14.30). As the viscous granite melt segregates into larger sheets, it is able to travel more easily, and eventually it collects to form granite batholiths that rise through the crust.

Metamorphic temperatures are normally prevented from rising higher than the onset of migmatite formation because of the large latent heat of fusion of rock. The temperature at which the first possible melt can form is at the beginning of melting (solidus) of water-saturated granite, which occurs in the middle of the amphibolite facies (Fig. 14.31). However, as soon as this melt begins to form, it absorbs all the water from the environment so that none is available to continue forming water-saturated melt. As a result, very little melt is formed at this temperature, and most probably solidifies to form pegmatites as soon as it starts to rise (Fig. 14.32).

Once all the water in the environment has been consumed by the water-saturated melting of granite, only

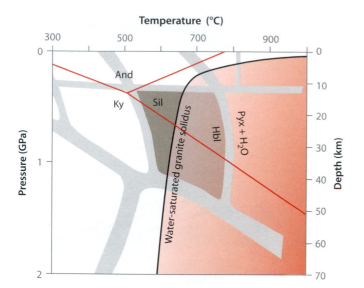

Figure 14.31 Water-saturated beginning of melting of granite relative to metamorphic facies. A broad zone of hbl ⇒ pyx + H₂O separates the amphibolite and granulite facies and provides water for melting (see Fig. 14.4).

water-undersaturated melt can form, which requires higher temperatures (see, e.g., Fig. 8.12). As a result, metamorphic temperatures continue to rise through the amphibolite facies until the boundary of the granulite facies is reached. At this boundary, hornblende reacts to form pyroxene and water (Fig. 14.31). The large amount of water released by this reaction allows for large-scale melting to occur, and this is where most migmatites form. For temperatures to rise into the granulite facies, rocks must either be extremely dry or have lost all low-melting fraction.

14.9 Geothermometers and geobarometers

One of the main goals in studying metamorphic rocks is to determine the temperatures and pressures under which they formed so that we can reconstruct past conditions in the Earth.

Figure 14.32 Deformed granite pegmatites formed by partial melting in the amphibolite facies south of Middletown, Connecticut.

By assigning rocks to metamorphic facies, we obtain approximate temperature and pressure ranges, and these can be refined by identifying where specific mineral assemblages plot in petrogenetic grids. However, we have also seen that the composition of the fluid in the environment can have a dramatic effect on metamorphic reactions, and without knowledge of the fluid composition, accurate temperatures and pressures cannot be obtained from these grids. The boundaries between the Al_2SiO_5 polymorphs are an exception to this because their positions are a function only of pressure and temperature. The question is, Can other metamorphic reactions provide independent measures of metamorphic temperatures and pressures? The answer is yes, and we refer to these reactions as geothermometers and geobarometers, respectively.

We have seen that many metamorphic minerals belong to solid-solution series and that their compositions change with metamorphic grade. Coexisting garnet and biotite, for example, become progressively more magnesium rich with increasing metamorphic grade. With geothermometers and geobarometers, we quantify this chemical change and, through the use of thermodynamic data, determine temperatures and pressure.

Metamorphic rocks containing coexisting biotite and garnet are common from the garnet zone of the greenschist facies to the granulite facies, and both minerals belong to solid-solution series, with biotite having compositions between annite ($KFe_3AlSi_3O_{10}(OH)_2$) and phlogopite ($KMg_3AlSi_3O_{10}(OH)_2$) and garnet with compositions between almandine ($Fe_3Al_2Si_3O_{12}$) and pyrope ($Mg_3Al_2Si_3O_{12}$). These compositions are shown in the Thompson projection in Figure 14.33. If biotite and garnet are both present in a rock, their compositions in the Thompson plot can be joined by a tie line. Figure 14.33 shows three such tie lines. One shows almandine coexisting with phlogopite and another shows pyrope coexisting with annite. These represent extreme compositions, and it is much more likely that the biotite and garnet would have intermediate Fe-Mg compositions such as those represented by the red tie line. Analyses of biotite and garnet in a rock using an electron microprobe (Sec. 3.8.2) would allow us to determine the precise compositions of these minerals and be able to draw the tie line. Assuming we can do this, what significance can be attached to the tie line?

To answer this, we write what is known as an Fe-Mg exchange reaction, which can be easily pictured by looking at the two extreme sets of tie lines in Figure 14.33. We have seen previously that intersecting tie lines in a diagram equate to a reaction. We can write this reaction as follows:

$$Fe_3Al_2Si_3O_{12} + KMg_3AlSi_3O_{10}(OH)_2 \Leftrightarrow$$
almandine   phlogopite
$$Mg_3Al_2Si_3O_{12} + KFe_3AlSi_3O_{10}(OH)_2)$$
pyrope   annite

(14.11)

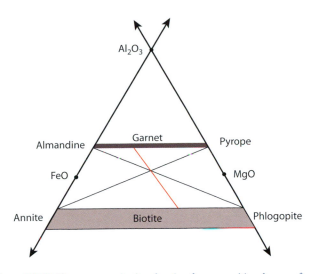

Figure 14.33 Thompson projection showing the compositional range of garnet and biotite. The compositions of coexisting garnet and biotite, shown by the tie lines, are a function primarily of temperature and, therefore, create a geothermometer.

This reaction would not go all the way to the right or all the way to the left, but it would equilibrate at some intermediate composition, with the garnet and biotite having compositions such as those represented by the red tie line in Figure 14.33. In a reaction where the reactants and products form solid solutions, we can write what is known as an *equilibrium constant (K)* for the reaction in terms of the product of the concentrations of the reaction products divided by the product of the concentration of the reactants. If we express the concentrations in terms of mole fractions (X), we can write the equilibrium constant as follows:

$$K = \frac{X_{Alm} \cdot X_{Phlog}}{X_{pyr} \cdot X_{Ann}} \qquad (14.12)$$

On rearranging, this gives the following:

$$K = \frac{X_{Alm} / X_{Pyr}}{X_{Ann} / X_{Phlog}} \qquad (14.13)$$

In Section 14.2.1, we saw that for a reaction to occur naturally, it must involve a decrease in Gibbs free energy. Reaction (14.11) would certainly occur naturally, as the Fe and Mg distributed themselves between the coexisting garnet and biotite, and this would be associated with a definite decrease in free energy. If we designate the free-energy change of this reaction by ΔG, then it can be shown that the natural logarithm of the equilibrium constant K is equal to $-\Delta G$, divided by the gas constant R and the absolute temperature T; that is:

$$\ln K = \ln \frac{X_{Alm} / X_{Pyr}}{X_{Ann} / X_{Phlog}} = -\frac{\Delta G}{RT} \qquad (14.14)$$

The volume change accompanying this reaction is very small, so pressure has little effect on the value of ΔG, which is determined mainly by temperature. The right-hand side of this equation is, therefore, predominantly a function of temperature and serves as a geothermometer known as **GARB** for coexisting garnet and biotite. Thus, from an analysis of the Fe/Mg ratios in coexisting garnet and biotite we can calculate the temperature of equilibration from the value of ΔG, which can be obtained from thermodynamic tables. Even when the total pressure is unknown, this geothermometer gives temperatures accurate to ±50°C and this can be improved considerably if estimates of the pressure can be made.

To determine pressures, we need reactions that have a large change in volume. Most metamorphic reactions that liberate a fluid phase have large volume changes, but the reactions are sensitive to the composition of the fluid in the environment, and if this is unknown, these reactions cannot be used. We, therefore, need a reaction that does not involve a fluid and that has a large volume change. One such reaction involves garnet, rutile, Al$_2$SiO$_5$, ilmenite, and quartz, and is known as **GRAIL**. The following reaction relates these minerals:

$$\begin{array}{cccc} Fe_3Al_2Si_3O_{12} & + & 3TiO_2 & \Leftrightarrow & 3FeTiO_3 \\ \text{almandine} & & \text{rutile} & & \text{ilmenite} \\ + Al_2SiO_5 & + & 2SiO_2 & & \\ \text{kyanite} & & \text{quartz} & & \end{array} \qquad (14.15)$$

Because rutile, Al$_2$SiO$_5$ polymorphs, and quartz are essentially pure minerals ($X \approx 1$), the equilibrium constant for this reaction, can be written in terms of the concentrations of almandine in garnet and ilmenite in a rhombohedral oxide (solid solution with hematite). This reaction is almost independent of temperature and, therefore, from analyses of coexisting garnet and rhombohedral oxide (hematite-ilmenite solid solution) that occur with quartz and Al$_2$SiO$_5$ polymorph pressures can be measured with an accuracy of ~0.05 GPa.

We have given only two examples of a geothermometer and a geobarometer, but they illustrate how metamorphic temperatures and pressures can be obtained from electron microprobe analyses of minerals belonging to solid-solution series.

14.10 Plate tectonic significance of metamorphism

Metamorphism is the sum of all those changes that occur to a rock in response to changes in its environment, many of which are a direct consequence of plate tectonic processes. Metamorphic rocks potentially preserve an important record of past tectonic events.

One of the largest metamorphic rock factories is at convergent plate boundaries where orogenic belts develop. During plate convergence, sedimentary rocks formed at the Earth's surface may be buried to considerable depths as a result of thrusting and folding in orogenic belts where they undergo metamorphism. Eventually, as a result of erosion and isostatic rebound, exhumation returns them to the surface of the Earth. During this history, they are exposed to changing pressures and temperatures as a function of time. With information about pressures and temperatures extracted from metamorphic rocks, we can construct what are known as **pressure-temperature-time paths** (*P-T-t*). These paths provide important information about processes at convergent plate boundaries.

14.10.1 Pressure-temperature-time (*P-T-t*) paths

Figure 14.34 shows a hypothetical *P-T-t* path that a sedimentary rock might follow during an orogenic event and the parts of the path that might be recorded in metamorphic rocks. Metamorphism that occurs during burial would be obliterated by metamorphism at higher temperatures. As the rock approaches the highest temperatures along the *P-T-t* path, the cores of large porphyroblasts, such as garnet and staurolite, might preserve some early record of the penultimate prograde metamorphism. With erosion of the orogenic belt, isostatic adjustment causes the rock to begin returning toward the surface, and this is when the rock reaches its highest temperatures and develops the mineral assemblage that we see when the rock returns to the surface. As the rock's ascent continues, retrograde reactions may occur, especially where faults and fractures allow water access. Where the rock is cracked, fluid inclusions may become trapped in mineral grains, preserving another part of the record. Eventually, the rock returns to the surface.

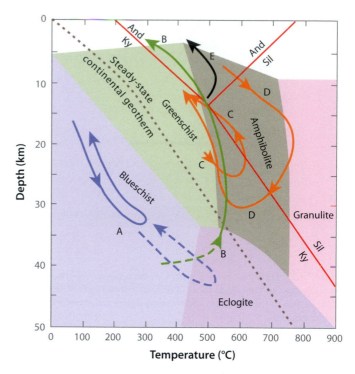

Figure 14.35 Pressure-Temperature-time paths determined from metamorphic rocks in various orogenic settings. (**A**) Low-temperature, high-pressure rocks of the Franciscan Complex, California (Ernst, 1988). (**B**) High-temperature, high-pressure rocks of the Tauern Window, Eastern Alps (Selverstone and Spear, 1985). (**C**) and (**D**) High-temperature, high-pressure rocks of central Massachusetts (Tracy and Robinson, 1980). (**E**) High-temperature, low-pressure rocks of northern New England (Lux et al., 1986) and the Pyrenees (Wickham and Oxburgh, 1985).

Through careful documentation of mineral assemblages and the use of geothermometers and geobarometers, *P-T-t* paths have been determined in a number of orogenic belts. Figure 14.35 shows a sample of the range of possible paths that have been found (A–E) relative to the Al_2SiO_5 polymorph stability fields and a steady-state continental geotherm.

P-T-t path (A) was obtained from rocks belonging to the Franciscan Complex of California. Here, graywacke type sediments were subducted by a cold oceanic plate, which also cooled the rocks above the subduction zone. Metamorphic reactions created low-temperature, high-pressure minerals of the blueschist facies (Fig. 14.9). Rocks that were subducted to the greatest depth entered the eclogite field (blue dashed line). Subduction took these rocks into the mantle, but because they were less dense than the mantle rocks, they buoyantly rose back to the surface, following almost the reverse of the pressure-temperature path that they followed during subduction. This type of *P-T-t* path is characterized by large changes in pressure with very little heating (HP/LT).

When continental crust is involved at convergent plate boundaries, the thickened crust acquires higher temperatures due to the thicker blanket of radioactive rocks, and the intrusion of magmas and flux of fluids from the lower crust. *P-T-t*

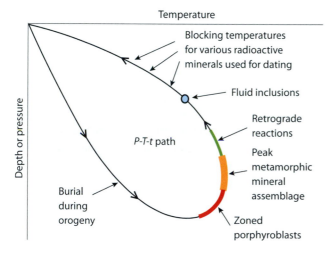

Figure 14.34 Hypothetical pressure-temperature-time path followed by a sedimentary rock during metamorphism. Only the highest temperature parts of this path are normally preserved by metamorphic rocks (adapted from Philpotts and Ague, 2009).

BOX 14.3 | HOW IS TIME OF METAMORPHISM DETERMINED

Assigning times to the various steps around a *P-T-t* path is difficult. Fossils in a sedimentary rock may tell us when the sediment was deposited, but other times along the path must be determined by absolute dating methods using radioactive elements. Most absolute dating methods rely on the decay of a radioactive parent isotope to a stable daughter element. Some of the most common pairs are: $^{238}_{92}U \Rightarrow {}^{206}_{82}Pb$, $^{235}_{92}U \Rightarrow {}^{207}_{82}Pb$, $^{87}_{37}Rb \Rightarrow {}^{87}_{38}Sr$, and $^{40}_{19}K \Rightarrow {}^{40}_{18}Ar$, where the superscript indicates the atomic mass of the isotope and the subscript, the atomic number. To determine the age of a metamorphic mineral assemblage, a mineral must form that contains one of the radioactive isotopes. Zircon and monazite (rare-earth phosphate) commonly contain small amounts of uranium, and rubidium substitutes for potassium in many potassium-bearing silicates. If the minerals containing these radioactive isotopes are formed at the same time as the metamorphic mineral assemblage, the amount of daughter product formed will be a measure of the time lapsed since metamorphism. However, metamorphic rocks typically form at temperatures where elements are able to diffuse, especially over short distances. The absolute age may, therefore, not be the age of metamorphism but the age since the rock cooled below a temperature at which diffusion became insignificant. This temperature, known as the **blocking temperature**, varies from isotope to isotope and mineral to mineral. For example, argon formed from the decay of ^{40}K is one of the least tightly held daughter products, and its blocking temperature is about 350°C but varies depending on the host mineral. The blocking temperature for rubidium-strontium dating is usually a couple of hundred degrees higher than that for K-Ar. Some minerals in metamorphic rocks, such as zircon may have been detrital grains in the original sediment and their isotopic signature may provide information about when the grain was originally formed and the age at which metamorphism occurred. Another absolute dating technique involves measuring the number of fission tracks found in minerals formed by the spontaneous fission of ^{238}U. The massive fission products damage the structure of the surrounding mineral, and when etched with appropriate chemicals, produce small tracks that can be counted. The older a mineral is, the more fission tracks it contains. However, the damage caused by fission is soon annealed at higher temperatures. This dating method, therefore, gives the age since a rock dropped below the annealing temperature. Fission tracks in apatite, for example, are annealed out above ~120°C, but they remain in zircon up to ~350°C. Fission tracks are, therefore, useful only for dating the final stages of exhumation along a *P-T-t* path. Assigning times to a *P-T-t* path is difficult, but it can be done with sufficient accuracy to provide a reliable history of the path rocks follow in orogenic belts.

paths in these regions not only pass through zones of high pressure because of the thickened crust but through zones of higher temperature (HP/HT). *P-T-t* path (B) is from the Alps where thrust faults and huge nappes (large recumbent folds) greatly thickened the continental crust. These structures would have raised pressure rapidly with little heating and the rocks would have passed through the blueschist facies, but radiogenic heating would soon have raised temperatures, obliterating most evidence of this early stage of metamorphism. Rocks that were buried deepest passed into the amphibolite facies (*P-T-t* path (B)), whereas shallower ones passed through the greenschist facies.

Continental crust was also involved in the plate convergence that formed the Devonian Acadian orogeny in New England. *P-T-t* paths from central Massachusetts (C and D) show how deformation plays an important role in controlling the paths. As in the Alps, thrust faults and large nappes formed early in the

collision and buried cold near-surface rocks to sufficient depth to place them in the kyanite stability field (beginning of *P-T-t* path (C)). These same structures brought hot rocks from depth and piled them on the cold shallow rocks, giving rise to high temperatures at shallow depths where andalusite was stable (beginning of *P-T-t* path (D)). With radiogenic heating, temperatures began to rise. In central Massachusetts, the Bronson Hill anticlinorium (large regional anticline) brought the deeply buried rocks back toward the surface, producing a typical counterclockwise *P-T-t* path (C). However, immediately to the east of this anticlinorium the high-temperature shallow rocks were lowered into the Merrimack synclinorium, which caused pressures to rise, and produced a clockwise *P-T-t* path (D).

In regions where metamorphism involves rising temperatures with little increase in pressure (LP/HT), *P-T-t* paths remain in the stability field of andalusite (E in Fig. 14.35). In

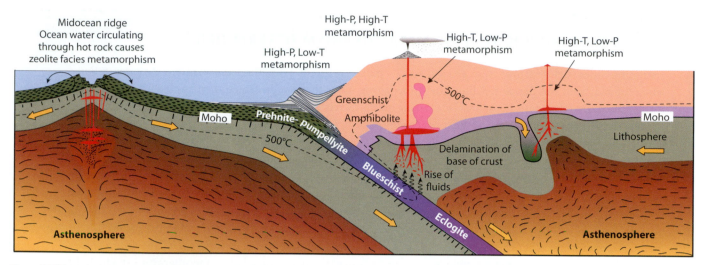

Figure 14.36 Typical plate tectonic settings for the development of metamorphic rocks shown in a cross-section of the lithosphere extending from a midocean ridge, through a convergent plate boundary, to a stable continental interior. The depth of the 500°C isotherm shows how plate motion results in very different *P-T* conditions in different parts of the plates.

many regions, such as northern New England, the high temperatures at shallow depths can be attributed to the intrusion of large quantities of magma. In the Pyrenees, however, LP/HT metamorphism is not associated with abundant igneous activity and has, instead, been linked to thinning of the crust and rise of the hot asthenosphere. Delamination and sinking of the base of the crust would also cause the asthenosphere to rise and might cause LP/HT metamorphism.

14.10.2 Plate tectonic setting of metamorphic facies

Unlike most planets, the Earth is dynamic, with a lithosphere that is in constant motion, as tectonic plates diverge, converge, and glide past one another. The motion of tectonic plates, which is driven by the cooling of the planet, results in pressure and temperature changes in rocks and also causes large fluxes of fluids to pass through the lithosphere, all of which result in metamorphism. Figure 14.36 shows a cross-section through the lithosphere passing from a midocean ridge, through a convergent plate boundary into a continental plate. The cross-section also shows an isotherm (500°C) to illustrate how plate motion causes temperatures to vary with depth. For example, 500°C is found at shallow depths beneath midocean ridges but at great depth in subduction zones and then again at shallow depths where magmas rising beneath volcanic arcs bring heat near to the surface. We now examine some of the major plate tectonic sites in which metamorphic rocks are formed.

Midocean ridges

The largest volumes of new rock are formed at midocean ridges, where basalts (MORB), dike rocks, and intrusive bod-

ies create new oceanic crust (Sec. 9.4.1). These hot rocks are cooled by ocean water circulating through them. It is estimated that the entire volume of water in the Earth's oceans circulates through this new crust every 8 million years. This huge flux of water through hot rocks brings about what we refer to as **hydrothermal alteration** but is, in fact, low-pressure, low-temperature metamorphism (Fig. 9.34). Primary igneous minerals are altered to hydrous minerals, such as chlorite, serpentine, and zeolites, all belonging to the zeolite facies of metamorphism. This metamorphism results in anhydrous igneous rocks being converted to hydrous metamorphic rocks that contain about 10 weight % H_2O. Although most of these rocks are eventually subducted at convergent plate boundaries and are not preserved in the geologic record, the water they release during subduction plays an important role in driving magmatism and metamorphism at convergent plate boundaries.

Convergent plate boundaries

At convergent plate boundaries where the ocean floor begins to subduct, zeolite facies minerals react to form assemblages in the prehnite-pumpellyite facies. Cold oceanic crust subducts faster than heat can diffuse into the slab, so pressures rise at almost constant temperature. As subduction continues, rocks develop mineral assemblages of the blueschist facies and eventually the eclogite facies (Fig. 14.36). Most of these rocks are subducted back into the mantle, but some return to the surface and do so remarkably quickly. Blueschist and eclogite facies rocks form under *P-T* conditions where aragonite is the stable polymorph of $CaCO_3$ (Sec. 10.6 and Fig. 10.11). This mineral is commonly found in these rocks, yet experimental studies show that aragonite changes to the low-pressure polymorph, calcite, if held at a couple of hundred degrees at low pressure for hundreds of years. Bodies of blueschist and eclogite facies rocks are always bounded by fault surfaces, and it appears that they are emplaced tectonically, being squeezed to the surface by the convergent plate stresses. The speed of their emplacement is commonly likened to the speed at which a grapefruit seed will shoot sideways when squeezed slowly between your fingers. Rocks containing the ultrahigh-pressure metamorphic minerals coesite and diamond have to rise still farther, and it remains a puzzle how they do so rapidly enough to preserve the high-pressure polymorphs.

Metamorphic reactions taking place in the subducting oceanic slab release water that was added to the rock while cooling at the midocean ridge, with essentially all of it being released by the time the rocks become eclogites. This water rises into the overlying mantle wedge and causes melting. These melts and fluids rise into the overlying crust, adding heat to that which already exists due to the thickened more radioactive continental crust. This brings about the typical Barrovian sequence of greenschist – amphibolite metamorphism. Where large quantities of magma rise high into the crust, low-pressure, high-temperature metamorphism results.

Rifting and delamination of continental crust

Continental rifting involves thinning of the lithosphere, and as hot asthenosphere rises to fill the gap, geothermal gradients rise and cause metamorphism at low pressure. This normally brings about metamorphism in the zeolite facies, but it has also been invoked to explain the low pressure, high-temperature metamorphism in the Pyrenees (E in Fig. 14.35). Delamination of the base of the crust caused by dense material sinking into the asthenosphere (Fig. 14.36) would also result in hot material rising, and this would raise the geotherm and cause low-pressure, high-temperature metamorphism.

Overview of metamorphism and plate tectonics

In 1961, before plate tectonics came into vogue (post 1963), Miyashiro pointed out that metamorphic rocks in Japan occurred in **paired metamorphic belts**, with a blueschist to eclogite facies series occurring on the side nearest the Pacific and a low-pressure, high-temperature series to the west of this. Granite batholiths were found in the low pressure, high-temperature belt but never in the high-pressure, low-temperature belt. He further showed that paired metamorphic belts characterized all younger orogenic belts around the Pacific. We can see how plate tectonics provides a simple explanation for these belts, with the high-pressure, low-temperature facies series occurring where ocean-floor rocks and graywacke sediments are transported into the subduction zone, and the low-pressure, high-temperature facies series occurring where magma, generated by water released from the subducting slab, rises high into the continental crust causing contact metamorphism on a regional scale (Fig. 14.36).

Miyashiro also pointed out that blueschist facies rocks become less common with increasing age and there may be none older than the late Proterozoic, although one 2.8 billion-year-old eclogite has recently been reported from the Fennoscandian shield. Ultrahigh-pressure metamorphic rocks also become scarcer with age. In contrast, rocks belonging to the low-pressure, high-temperature facies series are particularly common in the Precambrian, as are the ultrahigh-temperature metamorphic rocks. The reasons for these difference in the pressures and temperatures of metamorphic rocks with age is not yet clear but may be related to cooling of the planet and thickening of the lithosphere.

Summary

Metamorphism is the sum of all changes that occur in a rock in response to changes in pressure, temperature, and composition of fluids in the environment. In this chapter, we looked at what these changes are, why they take place, whether reactions reach equilibrium, and how we can study metamorphic rocks to determine what the pressures and temperatures were at the time of metamorphism. Most metamorphic reactions are caused by plate tectonic processes, so unraveling the metamorphic history of a rock can provide information about the Earth's dynamic past.

The following are the main points of this chapter:

- Changes in a metamorphic rock can be physical, chemical, or combinations of these. Changes that occur while a rock is heating are described as prograde and those while cooling as retrograde.

- Any rock type can be metamorphosed. The original rock type is known as the protolith.

- Changes in pressure result from plate tectonic process. In zones of plate convergence, thrusting and folding thicken the crust and increase pressure. In zones of crustal extension, rocks are brought closer to the surface and pressures decrease.

- Changes in temperature result from burial, heating from radioactive elements in a thickened crust, heat from fluids rising from below, the intrusion of magma, and exhumation.

- Contact metamorphism occurs near igneous intrusions and produces rocks that typically lack foliation and are called hornfels.

- Regional metamorphism typically occurs near convergent plate boundaries due to regional changes in pressure and temperature; the rocks typically have a prominent foliation.

- Large volumes of fluids are released during prograde metamorphism that rise toward the surface taking with them heat and dissolved constituents, which may alter overlying rocks or be deposited as veins.

- Changes in metamorphic rocks take place in order to lower the Gibbs free energy. A natural reaction must decrease this energy and the reaction reaches equilibrium when the Gibbs free energy is a minimum.

- The contribution each component in a complex mineral makes to the Gibbs free energy of that mineral is known as the chemical potential of that component.

- At equilibrium, temperature, pressure, and chemical potentials of all components must be the same throughout the rock.

- Reaction rates are more rapid at high temperature and in the presence of fluids. As a result, prograde reactions are very much faster than retrograde reactions.

- Gibbs phase rule relates the number of components needed to describe the composition of the phases to the number of phases and the number of environmental variables that can be changed at equilibrium.

- Metamorphic grade loosely refers to the intensity of metamorphism. The first appearance of certain easily recognized metamorphic index minerals was used by Barrow to map metamorphic grade. These were later equated to equal grades of metamorphism or isograds.

- A metamorphic mineral facies comprises all rocks that have originated under temperature and pressure conditions so similar that a definite bulk rock chemical composition results in the same set of minerals.

- A metamorphic facies series is a sequence of metamorphic facies encountered across a progressively metamorphosed terrane. Each series reflects different changing pressures and temperatures. Eskola recognized three main series, a low-pressure, high-temperature series; a high-pressure, high-temperature series; and a high-pressure, low-temperature series.

- Metamorphism usually results in coarsening of grain size, because this lowers the rock's overall surface free energy. Exceptions occur where rocks are strongly deformed and many new centers of recrystallization are formed that lead to a fine-grained rock.

- Minerals can be arranged in the crystalloblastic series, where any mineral in the series forms euhedral faces against minerals lower than it in the series. Garnet and staurolite, for example, are high in the series and commonly form euhedral crystals.

- Recrystallization commonly leads to grains coming together at triple junctions with angles of ~120°. This minimizes surface free energies.

- Regionally metamorphosed pelitic rocks develop a foliation, which, with increasing metamorphic grade, passes from slate, through phyllite, and schist, to gneiss. Foliation can also result from alignment of needlelike minerals, such as amphibole. These foliations develop perpendicular to the maximum compressive stress.

- During the development of foliation, metamorphic rocks undergo considerable flattening due to the change in the shape of grains and solution and removal of material from the rock.

- Graphical representations of metamorphic mineral assemblages on either side of a metamorphic reaction commonly require projecting compositions from three-dimensional tetrahedral plots into more convenient two-dimensional triangular plots. This is done by projecting from a mineral that is common to all assemblages.

- Reactions between minerals in triangular plots are indicated by the tie line between the reactants switching to form a tie line between the products.

- A petrogenetic grid is a pressure-temperature plot of all possible reactions that can take place between a given set of minerals. The univariant reaction curves intersect at invariant points to produce a grid that divides *P-T* space into divariant fields in which unique mineral assemblages are stable.

- Metamorphic minerals in pelitic rocks are conveniently projected from quartz and muscovite into a triangular diagram of Al_2O_3-FeO-MgO. This AFM projection is known as the Thompson projection.

- Metamorphic reactions that release a fluid phase are strongly dependent on the composition of the fluid phase. Only when the composition of the fluid in the environment matches the composition of the fluid produced by the reaction does the reaction take place at its highest temperature. Fluids composed of variable amounts of H_2O and CO_2 can cause metamorphic reactions in impure carbonate rocks to occur at very different temperatures and to produce very different prograde sequences of minerals.

- The normal upper limit of metamorphism occurs when melting first occurs to form migmatites. The latent heat of fusion is so high that it uses up all available heat, and temperatures do not normally rise higher.

- Geothermometers and geobarometers are assemblages of minerals that belong to solid solution series whose compositions are sensitive to temperature and pressure, respectively.

- From zoned porphyroblasts and mineral assemblages in metamorphic rocks it is possible to construct partial pressure-temperature-time (*P-T-t*) paths that rocks follow during metamorphism. The *P-T-t* paths of regionally metamorphosed rocks are determined largely by plate tectonic processes.

- At midocean ridges, ocean water circulates through, and cools, newly formed crust and in so doing adds up to 10 weight % water, as previously anhydrous igneous rocks are converted to zeolite facies metamorphic rocks. This water is later released when oceanic rocks are subducted and converted to blueschist and eclogite facies rocks. As the water rises back toward the surface, it causes melting in the mantle wedge and metamorphism in the overlying orogenic belt.

- At convergent plate boundaries, metamorphic rocks commonly form two belts, one close to the subduction zone where high-pressure, low-temperature metamorphic rocks are formed (blueschists), and another a little farther into the overriding plate where low-pressure, high-temperature metamorphic rocks are commonly associated with abundant igneous intrusions.

Review questions

1. What is metamorphism, and what are the main factors that cause it?

2. What is the difference between regional metamorphic rocks and contact metamorphic rocks?

3. What major constituents are lost during prograde regional metamorphism and what important roles do they play?

4. What is the Gibbs free energy and what happens to it during a metamorphic reaction?

5. How can we define equilibrium in the case of a metamorphic reaction?

6. What determines the rate of a metamorphic reaction?

7. What does Gibbs phase rule tell you about the number of minerals you might expect to find in a metamorphic rock?

8. What are metamorphic index minerals and how do they relate to isograds?

9. What is a metamorphic mineral facies, and how is it used to determine metamorphic grade?

10. What is a metamorphic facies series?

11. What causes most rocks to become coarser-grained during metamorphism and are there exceptions?

12. Why do contact metamorphic rocks known as hornfels commonly consist of polygonal grains that are bounded by faces that intersect at ~120°?

13. What textural feature characterizes most regionally metamorphosed rocks?

14. What evidence indicates rocks can lose material during the development of metamorphic foliation?

15. Although the rock name *mylonite* comes from the Greek word for a "mill," the fine grain size of this rock is not due to breaking and grinding of grains. What causes mylonite to be extremely fine-grained?

16. What is a petrogenetic grid, and how can it be used to determine the conditions under which a metamorphic rock formed?

17. Starting with the triangular diagram indicating the stable mineral assemblages in any divariant field in Figure 14.19 work your way clockwise around the diagram until you come back to the starting diagram and check that only one change occurs in the mineral assemblages in crossing from one divariant field to the next.

18. The Thompson projection is a convenient way of showing pelitic mineral assemblage in terms of Al_2O_3-FeO-MgO. From what mineral is the projection made, and what other minerals must be present in the rock in addition to those shown in the projection?

19. From the Thompson projections for any two of Barrow's adjoining zones in Figure 14.22, write a reaction that must take place in going from one grade of metamorphism to another. Recall that reactions can be spotted from a switch in tie lines or if a mineral plots within a phase triangle.

20. Using the petrogenetic grid in Figure 14.23, determine the possible pressure and temperature ranges over which the mineral assemblages in the three different pelites in Figure 14.24 are stable.

21. Write a balanced reaction for tremolite reacting with calcite to produce diopside and forsterite. From the proportions of CO_2 and H_2O produced in the reaction, calculate the mole fraction of CO_2 in the fluid at the reaction's maximum stability, and what would be the shape of the reaction in Figure 14.28? To balance the reaction, let the amounts of tremolite, calcite, diopside and forsterite be a, b, c, and d respectively. Then write equations equating the amounts of Ca, Mg, and Si on both sides of the reaction, and solve for the coefficients.

22. Why does migmatite formation normally put a cap on metamorphic temperatures?

23. What are Fe-Mg exchange reactions, and how are they used as geothermometers?

24. Why do metamorphic rocks in some young mountain belts occur in paired metamorphic belts, with a high-pressure, low-temperature belt nearest the subduction zone and a low-pressure, high-temperature belt farther into the overriding plate?

FURTHER READING

The following three articles are of historical interest to the development of ideas about metamorphism:

Barrow, G. (1983). On an intrusion of muscovite-biotite gneiss in the southeast Highlands of Scotland, and its accompanying metamorphism. *Quarterly Journal of the Geological Society of London*, 49, 330–358.

Bowen, N. L. (1940). Progressive metamorphism of siliceous limestone and dolomite. *Journal of Geology*, 48, 225–274.

Eskola, P. (1920). The mineral facies of rocks. *Norsk Geologisk Tidsskrift*, 6,143–194.

Philpotts, A. R., and Ague, J. J. (2009). *Principles of Igneous and Metamorphic Petrology*, 2nd ed., Cambridge University Press, Cambridge. This book presents a detailed and quantitative coverage of metamorphic petrology.

Spear, F. S., and Cheney, J. T. (1989). A petrogenetic grid for pelitic schists in the system SiO_2-Al_2O_3-FeO-MgO-K_2O-H_2O. *Contributions to Mineralogy and Petrology*, 101, 149–164. This article presents a full development of a petrogenetic grid for pelitic rocks.

Thompson, J. B. Jr. (1957). The graphical analysis of mineral assemblages in pelitic schists. *American Mineralogist*, 42, 842–858. This article describes the projection scheme used to analyze mineral assemblages in metamorphic rocks.

Vernon, R. H. and Clarke, G. L. (2008). *Principles of Metamorphic Petrology*, Cambridge University Press, Cambridge. This book provides an in-depth treatment of metamorphic petrology.

Yardley, B. W. D., MacKenzie, W. S., and Guilford, C. (1990). *Atlas of Metamorphic Rocks and Their Textures*, Longman, Harlow, U.K. This book contains beautiful colored illustrations of metamorphic rocks and their textures as seen under the microscope.

15 Some economic minerals, mainly from veins and pegmatites

In Chapters 7, 10, and 13, we provided systematic treatments of minerals according to their most common origin, such as igneous, sedimentary, and metamorphic. This classification omits a number of minerals that are of major economic importance, such as some native elements, sulfides, and a few silicates, as well as barite, fluorite, and some gem minerals.

The minerals in the prior three mineralogical chapters include a large number of silicates, with fewer carbonates, oxides, and sulfides. Many of these are of economic importance and are referred to as **industrial minerals**. An industrial mineral includes any rock, mineral, or other naturally occurring substance of economic value, exclusive of metallic ores, mineral fuels, and gemstones. In this chapter we include a variety of metallic **ore minerals** consisting of native elements and sulfides. The term *ore mineral* refers usually to a metallic mineral that is part of an **ore** and is economically desirable, in contrast to those minerals that are part of the **gangue**. Gangue is the valueless rock or mineral aggregate in an ore deposit that is not economically desirable but cannot be avoided in mining.

We also discuss several nonmetallic minerals that are of economic importance but could not be included in the mineral classification used prior. We close the chapter with examples of several important gem minerals.

Coarse-grained vein mineralization consisting of white quartz, brown sphalerite and metallic gray galena. From Lautenthal, Hartz Mountains, Germany. The size of the field of view is 8 cm by 13 cm. (David Nufer Photography, Albuquerque, New Mexico.)

The minerals discussed in this chapter are commonly found in either **hydrothermal veins** or **pegmatites**. Hydrothermal deposits result from precipitation of ore and gangue minerals in fractures, faults, breccia openings, and other spaces, by replacement or open-space filling, from water-rich fluids ranging in temperature from 50° to 700°C but generally below 400°C and ranging in pressure from 0.1 to 0.3 GPa; the fluids are of diverse origin. Pegmatites are exceptionally coarse-grained igneous rocks (Fig. 8.24), usually found as irregular dikes (Fig. 9.6), lenses or veins, commonly at the edges of batholiths. Their most common overall composition is that of granite, with grains 1 cm or more in size. Pegmatites represent the last and most hydrous portion of a magma to crystallize and, therefore, may contain high concentrations of minerals that occur only in trace amounts in granites and are concentrated in the hydrous-rich residue.

The first five minerals of this chapter are native elements, Au, Ag, Cu, C (diamond), and S. Gold in its native state is the most important ore mineral in gold occurrences. However, Ag as a native element, is less important as an ore mineral than acanthite, Ag_2S; proustite, Ag_3AsS_3; and pyrargyrite, Ag_3SbS_3. Similarly, native copper is a minor ore mineral when compared with several copper sulfides described in this chapter. Carbon, as diamond, is the only source of diamond and bort. Native sulfur is the major source for the production of sulfuric acid, H_2SO_4, and hydrogen sulfide, H_2S, but large amounts are produced as the byproduct from the processing of numerous sulfide ore minerals.

After these native elements come seven sulfides (one sulf-arsenide among them), six of which are the major ore minerals for Pb, Zn, Cu, Mo, and As. One of these sulfides, marcasite (one of two polymorphs of FeS_2; the much more common being pyrite), is present in many hydrothermal sulfide vein associations.

This is followed by descriptions of bauxite (a mineral mixture that is the main source of Al), fluorite (the main source of fluorine for the production of hydrofluoric acid, HF), and barite (a major component of drilling muds). After that come two lithium-containing minerals, spodumene, and lepidolite.

We end the chapter with a selection of some important gem minerals: ruby and sapphire, gem varieties of corundum; topaz; tourmaline; emerald, heliodore, and aquamarine (gem varieties of beryl); precious opal; jade; and turquoise.

15.1 Gold: Au

Occurrence: Most gold occurs in the native state and is of hydrothermal origin in gold-quartz veins, together with pyrite and other sulfides. Such gold is the result of deposition from ascending mineral solutions. When such primary gold occurrences are weathered the liberated gold remains as an alluvial deposit as part of the soil or is transported to form a **placer deposit**, which may be the source of **nuggets** (a nugget is a large

lump of placer gold). A placer deposit is defined as a surficial mineral deposit formed by the mechanical concentration of mineral particles from weathered debris. Much of the world's gold production is from deposits where the gold is classified as invisible gold because it is so fine grained that it cannot be seen by the naked eye.

Chemical composition and crystal structure: There is a complete solid solution series between Au and Ag, and most native gold contains some Ag, Cu, or other metals. **Electrum** is the name given to a solid solution series between Au and Ag. Small amounts of Cu and Fe may also be present. The proportion of pure gold in an alloy is expressed in **karat** (k). Pure gold is 24 karat, whereas 18 k gold consists 18 parts gold and 6 parts of other metals. The crystal structure of gold is given in Figure 15.1. It is isometric with an all-face-centered lattice (*F*) and consists of Au atoms in cubic closest packing (see Box 4.2). Gold, silver, and copper are isostructural.

Crystal form: Isometric; $4/m\,\bar{3}\,2/m$. Gold is seldom in good crystals such as well-developed octahedra. Commonly in irregular plates, scales, or masses. Very attractive, museum-quality specimens consist of arborescent crystal groups and irregular dendritic shapes, as shown in Figure 15.2.

Physical properties: H = 2½–3; G = 19.3 (when pure) with the presence of other metals in solid solution lowering the **G** to as low as 15. Malleable and ductile. Fracture is hackly. Metallic luster and opaque. Color ranges from various shades of yellow to very pale yellow as a function of increasing Ag content.

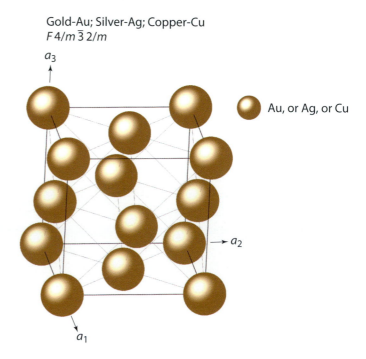

Gold-Au; Silver-Ag; Copper-Cu
$F\,4/m\,\bar{3}\,2/m$

Au, or Ag, or Cu

Figure 15.1 Perspective view of the crystal structure of gold as seen in a ball-and-stick representation. A cubic unit cell is outlined. Each gold atom is surrounded by 12 closest neighbors in cubic closest packing (see Box 4.2). Silver and copper have identical structures and are, therefore, referred to as **isostructural**.

Figure 15.3 Native silver in a complex intergrowth of arborescent and wiry forms.

Figure 15.2 A highly valuable, museum-quality exhibit specimen of native gold. It consists of a branching network with some sporadic, poorly developed octahedral crystals. The transparent, colorless-to-white crystals are quartz.

Distinguishing features: Gold is distinguished from yellow sulfides such as pyrite and chalcopyrite by its high specific gravity and sectility. Pyrite is sometimes referred to as fool's gold.

Uses: Gold has drawn the attention of many cultures since the earliest of time because of its color, indestructibility, and ease of fabrication into beautiful objects. About 51% of all gold exists as jewelry and decorative artifacts, about 34% is retained by investment houses and/or individuals as bullion and coins, and about 12% is used in industry. About half of this amount is used in electrical components because of its resistance to corrosion and high electrical and heat conductivity. In dentistry, alloyed with silver and other metals, it is used in dental crowns.

15.2 Silver: Ag

Occurrence: Native silver deposits are the result of deposition from hydrothermal solutions. These may consist of native silver in association with other silver minerals, sulfides, zeolites, calcite, barite, fluorite, and quartz. Also with silver arsenides and sulfides of nickel and cobalt. Although native silver is

mined as an ore, most of the world's supply of silver is extracted from minerals such as acanthite, Ag_2S, and proustite, Ag_3AsS_3.

Chemical composition and crystal structure: Native silver is commonly alloyed with some Au, Cu and Hg. Amalgam is a solid solution between Ag and Hg. The crystal structure of Ag is the same as that of gold (Fig. 15.1) with Ag atoms in cubic closest packing.

Crystal form: Isometric; $4/m\ \bar{3}\ 2/m$. Crystals commonly malformed and generally in branching, reticulated, or arborescent groups (Fig. 15.3). Also in irregular masses and in wirelike forms.

Physical properties: $H = 2\frac{1}{2}–3$; $G = 10.5$ (when pure). Malleable and ductile. Fracture hackly. Color and streak silver white, commonly tarnished to brown or gray black. When untarnished, it has a metallic luster.

Distinguishing features: Native silver is characterized by the silver-white color on a fresh surface, high specific gravity, and malleability.

Uses: Silver is extensively used in decorative items such as jewelry and silverware. Its use in coinage has been much reduced because of its price and been replaced by metals such as copper and nickel. In dentistry, it is used in the alloy amalgam for dental fillings. It is applied as a coating to mirrors that require exceptional reflectivity of visible light such as in solar reflectors. Some mineral instruments such as flutes are made of silver or are silver plated. It is also used as a catalyst in the chemical industry.

15.3 Copper: Cu

Occurrence: Native copper is only a minor ore of copper worldwide. Most copper ores are of hydrothermal origin and consist of various mixtures of copper sulfides such as chalcopyrite, $CuFeS_2$; bornite, Cu_5FeS_4; and chalcocite, Cu_2S. Most native copper is associated with basaltic lavas where

Figure 15.4 Native copper in an arborescent intergrowth.

copper was deposited because of the interaction of hydrothermal fluids with iron oxide minerals. The famous native copper deposits in the Lake Superior district of the United States have copper in interstices of sandstones and conglomerates and filling the cavities of amygdaloidal lavas.

Chemical composition and crystal structure: Native copper may contain small amounts of Ag, As, and Fe. The crystal structure of native copper is the same as that of gold and silver (Fig. 15.1) and is based on cubic closest packing of Cu atoms.

Crystal form: Isometric, $4/m\,\bar{3}\,2/m$. Crystals are distorted cubes, octahedrons, and dodecahedrons. Commonly in branching and arborescent groups (Fig. 15.4). Usually in irregular masses and scales.

Physical properties: H = 3½–3; G = 8.9. Highly malleable and ductile. Hackly fracture. Color is copper red on a fresh surface but is usually dark as a result of tarnish. It has a metallic luster and is opaque.

Distinguishing features: Native copper is recognized by its high specific gravity, malleability, and red color on a fresh surface.

Uses: Copper finds many applications because it is malleable and ductile, is a good conductor of heat and electricity, and is resistant to corrosion. It is widely used in the piping of water supplies and in refrigeration and air conditioning. It is extensively used in electrical applications and in electronics, as electrical wire, and in integrated circuits and printed circuit boards. In architecture, it has been used as roofing material since ancient times. It is a component in ceramic glazes, gives color to glass, and is a constituent of many copper-based alloys such as bronze and brass.

15.4 Diamond: C

Occurrence: Diamonds originate below 200 km in the Earth's upper mantle. They are transported to the Earth's surface by unique volcanic eruptions that drill narrow explosive vents or pipes through the Earth's crust. The rock types that fill such vents are known as kimberlites and lamproites (Sec. 8.2.1),

which consist of peridotitic or ultramafic rock fragments set in a fine-grained groundmass (Fig. 15.5). These fragmental rock types may contain olivine, phlogopite, spinel, and diopside (Fig. 9.49).

Large quantities of synthetic diamonds, for use in drilling and cutting tools, are produced commercially.

Chemical composition and crystal structure: The composition of diamond is pure carbon. The crystal structure, illustrated in Box 5.7, shows that each carbon atom is surrounded by four neighboring carbons in tetrahedral coordination with covalent bonding. Carbon is common in two different polymorphs, diamond and graphite, with diamond the high-pressure and graphite the low-pressure polymorph. The structure of graphite, shown in Figure 4.20, is completely different from that of diamond.

Crystal form: Isometric; $4/m\,\bar{3}\,2/m$. Crystals are commonly octahedral with curved crystal faces (Fig. 15.5). Also found in cubes and dodecahedrons. Gem-quality diamonds are classified according to the four Cs: color, clarity, cut, and carat weight (1 carat = 0.2 gram). **Bort**, a gray to brown variety of diamond, is poorly crystallized and has rounded forms and a rough exterior (see opening photograph for Chapter 5).

Physical properties: H = 10 (the hardest known mineral); G = 3.52. Perfect octahedral {111} cleavage. Adamantine luster. The very high refractive index, 2.42, and strong dispersion of light are the reason for the brilliancy and **fire** of cut gem-quality diamonds. Color ranges most commonly from pale yellow to colorless, but may also be gray to black. The term **fancy color** is used for strongly colored (pink, orange, green, and blue) gem-quality diamonds.

Distinguishing features: Diamond has great hardness, commonly octahedral crystal habit, excellent octahedral cleavage, and adamantine luster.

Uses: Apart from the use of diamond as a gem, there are many industrial applications in which natural and synthetic diamonds are both used. Diamond crystals, fragments, and powders are used in drill bits for oil and mineral exploration as well as dental work. It is impregnated into wheels for cutting dimension stone for building purposes. Diamond grit and powder are extensively used as abrasives. Very fine diamond powder is part of lapping compounds and fine finishing tools. The high thermal conductivity of diamond finds its application as a heat-sink in electronics.

Synthesis and imitation: There is no law against the imitation of gemstones and synthetic gemstones are an important part of the gemstone trade. Synthetic diamonds are extensively used in industry. Synthetic diamonds of considerable size and exhibiting all the required properties of a gemstone are also being produced. In addition, there is an excellent diamond imitation known as cubic zirconia, synthetic ZrO_2, also known as CZ. Its various physical properties make it visually essentially indistinguishable from natural gem diamond.

Figure 15.5 Clear, transparent, distorted octahedron of diamond (top center) in a matrix of kimberlite, which consists of angular fragments or **xenoliths** of mantle-derived rocks.

15.5 Sulfur: S

Occurrence: Sulfur, of volcanic origin, is found in almost all volcanic regions of the world (Fig. 8.15(A)). In such settings it occurs as replacement deposits in lens-shaped ore bodies in tuffs and volcanic breccias, and as sublimation deposits resulting from fumarolic activity (Fig. 15.6) near volcanic craters. Such volcanic deposits are generally small compared to those that are part of evaporite basins.

It is commonly present in the cap rock of salt domes where it is thought to be the result of the action of sulfate-reducing bacteria in the presence of a sulfate source, consisting of gypsum and anhydrite, and hydrocarbons. It is of the same origin in bedded (stratiform) anhydrite-gypsum deposits.

Chemical composition and crystal structure: Native sulfur may contain small amounts of selenium in solid solution. The crystal structure of the most common orthorhombic polymorph is shown in Figure 15.7 and consists of covalently bonded, puckered S_8 rings. The unit cell contains 128 sulfur atoms distributed over 16 such rings.

Figure 15.6 An irregular intergrowth of euhedral sulfur crystals inside the neck of a small fumarole. The beige rock on the outside is what the fumarolic vent intruded into.

Sulfur - S
F 2/d 2/d 2/d

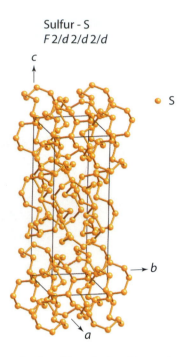

Figure 15.7 Perspective view of the orthorhombic crystal structure of sulfur. An orthorhombic unit cell is outlined which contains 128 sulfur atoms in 16 puckered springs.

Crystal form: Orthorhombic; $2/m\,2/m\,2/m$. Occurs in crystals with orthorhombic dipyramids. Most commonly in irregular masses and as encrustations.

Physical properties: $H = 1\frac{1}{2}–2\frac{1}{2}$; $G = 2.07$. Fracture conchoidal to uneven. Color sulfur-yellow, ranging from yellow to green, and red shades on account of impurities. It has a resinous luster and ranges from transparent to translucent.

Distinguishing features: Sulfur is recognized by its yellow color, low hardness and brittleness, and lack of good cleavage.

Uses: Most sulfur is used in the production of sulfuric acid, H_2SO_4, which is a primary component in the production of sulfur-bearing phosphatic fertilizers. It is also used in the production of inorganic chemicals, in synthetic rubber and plastic materials, and in petroleum and coal refining.

15.6 Galena: PbS

Occurrence: Galena is a common sulfide in hydrothermal veins in association with sphalerite (see opening photograph to this chapter), pyrite, chalcopyrite, marcasite, calcite, quartz, barite, and fluorite. It also occurs in veins associated with silver minerals, in which case it may contain silver as an admixture, and is then mined as a source of silver as well as lead. It also occurs in low-temperature hydrothermal replacement deposits, in association with sphalerite, in limestone.

Chemical composition and crystal structure: Most commonly, the composition is close to PbS but may contain small amounts of Zn, Cd, Sb, and As in inclusions. The crystal structure of galena, which is the same as that of halite, NaCl, with Pb

Galena - Pb S
F 4/m $\overline{3}$ 2/m

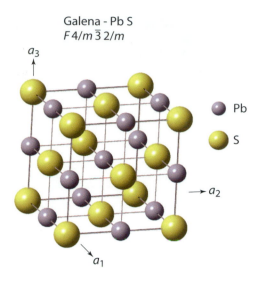

Figure 15.8 Perspective view of the crystal structure of galena in a ball-and-stick model. An isometric unit cell is outlined.

in place of Na and S in place of Cl, is shown in Figure 15.8. Both Pb and S are in 6-fold (octahedral) coordination with their closest neighbors.

Crystal form: Isometric; $4/m\,\overline{3}\,2/m$. The most common form is that of the cube, frequently truncated at the corners by the octahedron, as shown in Figure 15.9.

Physical properties: $H = 2\frac{1}{2}$; $G = 7.4–7.6$. Perfect cubic {100} cleavage. Opaque. Color and streak lead-gray. Luster bright metallic if not tarnished.

Distinguishing features: Galena is characterized by good cubic crystal habit, excellent cubic cleavage, low hardness, high specific gravity, and lead-gray color and streak.

Uses: The most important commercial use of lead is in the manufacture of electric storage batteries for fossil-fuel powered cars and trucks. It is part of alloys such as solder and type metal, and it is a principal component in ammunition. It is also used as a sheathing of cables and is employed as protective shielding against X-rays and radiation from nuclear reactors. The "lead" of a pencil does not contain lead but is a mixture of graphite and clay.

Because lead can be a highly toxic element to human health, lead-based paints and lead as an antiknock additive in gasoline have been outlawed. Lead is also no longer used in plumbing.

15.7 Sphalerite: ZnS

Occurrence: Sphalerite and galena are most commonly found together in hydrothermal ore deposits, and, therefore, the occurrence and mode of origin of sphalerite are essentially the same as those given for galena (Sec. 15.6). It is found in assemblages consisting of galena, sphalerite, marcasite, chalcopyrite, pyrrhotite, pyrite, calcite, dolomite, and quartz.

Chemical composition and crystal structure: The chemical composition of sphalerite commonly reflects the presence of

Figure 15.9 An intergrowth of euhedral crystals of galena showing well-developed, large cubes truncated by octahedral faces. The purple color of several cubes on the right is a natural tarnish.

some iron in substitution for zinc, as shown by a range of colors from colorless for pure ZnS to dark brown or black with considerable Fe content. The crystal structure of isometric sphalerite is shown in Figure 15.10. It has a high-temperature hexagonal polymorph known as wurtzite, which occurs in various polytypes. The isometric structure of sphalerite is similar to that of diamond, in which one-half the carbon is replaced by Zn and the other half by S. Zn is in tetrahedral coordination with S.

Crystal form: Isometric; $4/m\,\bar{3}\,2/m$. Tetrahedrons, dodecahedrons, and cubes are common forms. Some crystals may show highly complex morphology.

Physical properties: H = 3½–4; G = 3.9–4.1. Perfect dodecahedral {011} cleavage. Luster nonmetallic and resinous. Color ranges from colorless when pure ZnS to various shades of yellow, yellow-brown, to brown, and black as a function of

increasing iron content (Fig. 15.11). Streak ranges from white, yellow, to pale brown. Transparent to translucent.

Distinguishing features: Sphalerite has perfect dodecahedral {011} cleavage and resinous luster. Very dark varieties can be recognized by a reddish-brown streak.

Uses: About half of all zinc produced is used to make galvanized steel, in which thin coatings of zinc prevent corrosion. Galvanized steel is a major component of power transmission towers, tanks, culverts, and nails, and it is extensively used in automobiles. A large percentage of zinc production goes into brass (copper alloyed with between 3% and 45% Zn) and other zinc alloys.

It is also used in the pharmaceutical and cosmetic industries. Zinc is an essential element of biologic and public health importance. Zinc deficiency affects many people in the developing world and is associated with many diseases.

Sphalerite - Zn S
$F\bar{4}3m$

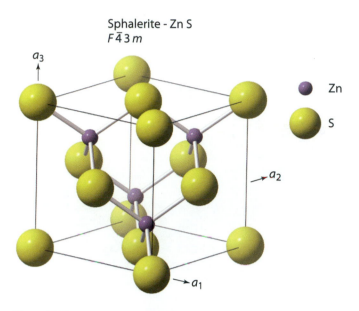

Figure 15.10 Perspective view of the crystal structure of sphalerite. A cubic unit cell is outlined.

15.8 Bornite: Cu₅FeS₄

Occurrence: Bornite is a widely distributed copper ore mineral commonly associated with other sulfides such as chalcopyrite, chalcocite, pyrrhotite, and pyrite. Two major types of copper deposits are **porphyry copper** and **volcanogenic massive sulfide (VMS) deposits**. Porphyry copper deposits formed around intrusions that fed volcanoes as a result of magmatic waters expelled from the intrusions, whereas VMS deposits consist of massive lenses of Cu, Zn, and Pb sulfides deposited as sediments from hydrothermal systems (as hot springs) that vented onto the ocean floor (Fig. 9.39 and Sec. 16.4).

Chemical composition and crystal structure: Bornite shows considerable variability in its contents of Cu and Fe, and is, therefore, somewhat variable in composition. It occurs in two polymorphic forms; a tetragonal polymorph at low temperature

Bornite - Cu₅FeS₄
$P\bar{4}2_1c$

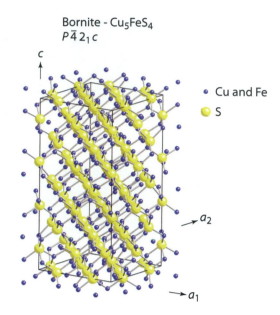

Figure 15.12 A perspective view of the tetragonal crystal structure of the low-temperature form (polymorph) of bornite. A tetragonal unit cell is outlined. All the cation sites are colored blue, which means that Cu and Fe are randomly distributed (disordered) over these sites. They are in tetrahedral and trigonal (planar) coordination with S. The coordination number of S with adjoining Cu and Fe ranges from 4 to 6.

(below 228°C) and an isometric form above that. Figure 15.12 is a perspective view of the tetragonal, low-temperature polymorph of bornite, with Cu and Fe distributed randomly (disordered) over the tetrahedrally coordinated sites.

Crystal form: Tetragonal; $\bar{4}2m$. Good crystals are uncommon. Usually massive (Fig. 15.13).

Physical properties: H = 3; **G** ~5.1. Usually massive with brownish-bronze color on the fresh surface which is quickly tarnished to a mixture of purple and blue colors (**peacock ore**; see Fig. 15.13). May become almost black on further exposure. It has a metallic luster and gray-black streak. Bornite is commonly in association with other Cu-Fe sulfides (Fig. 15.14).

Figure 15.11 An intergrowth of coarsely crystalline sphalerite showing a range of color from yellow brown on the inside to dark brown (to black) on the outer surface.

Figure 15.13 Massive bornite with a fresh surface of brownish-bronze color. The tarnished portion shows variegated purples and blues, also known as peacock ore.

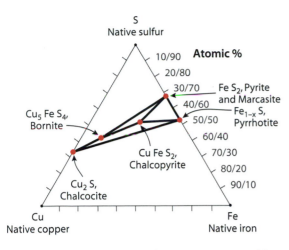

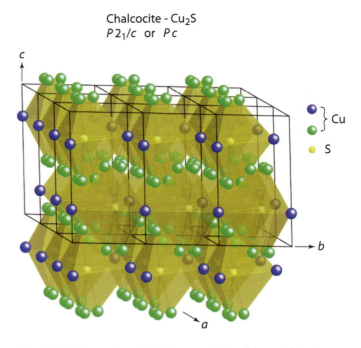

Chalcocite - Cu$_2$S
$P2_1/c$ or Pc

Figure 15.14 A triangular composition diagram using Cu, Fe, and S as end members. This allows for the graphical representation of the compositions of several Cu-Fe sulfides. Tie lines (bold lines) connect coexisting minerals.

Distinguishing features: Most easily recognized by the unique variegated colors on a tarnished surface with the brownish-bronze color of a fresh surface (see Fig. 15.13).

Uses: Bornite being a major copper ore mineral has the same uses as those given for native copper (Sec. 15.3).

15.9 Chalcocite: Cu$_2$S

Occurrence: Chalcocite is an important copper ore mineral that is of **supergene** origin. This means that it is the result of oxidation, at the surface, of primary copper sulfides, and the subsequent downward movement of soluble sulfates reacting with primary copper sulfides to form chalcocite. This produces a zone of supergene enrichment where the water table is the lower limit of such a zone. This zone is commonly referred to as **chalcocite blanket**.

Chemical composition and crystal structure: The composition of chalcocite is generally close to Cu$_2$S, with only minor amounts of Ag and Fe in substitution. A perspective view of the polyhedral crystal structure is given in Figure 15.15, with Cu atoms distributed over two different sites (colored blue and green). Sulfur atoms are in octahedral coordination with neighboring Cu atoms. A group of nine monoclinic unit cells is shown.

Crystal form: Monoclinic; $2/m$ or 2. Crystals are uncommon; generally massive and fine grained.

Physical properties: H = 2½–3 (imperfectly sectile); G ~5.7. Poor {110} cleavage. Conchoidal fracture. Color shining lead gray on fresh surfaces, but tarnishes to dull black on exposure (Fig. 15.16). Metallic luster.

Distinguishing features: Chalcocite is recognized by its lead-gray color on a fresh surface and its low hardness (somewhat sectile).

Uses: Chalcocite being a copper ore mineral has the same uses as those given for native copper (Sec. 15.3).

Figure 15.15 Perspective polyhedral representation of the crystal structure of chalcocite. Nine adjoining monoclinic unit cells are outlined. Copper atoms are distributed over the two colored (green and blue) sites.

15.10 Marcasite: FeS$_2$

Occurrence: Marcasite and pyrite are two polymorphs of FeS$_2$ (pyrite is discussed in Sec. 7.30). Marcasite is less stable and more easily decomposed than pyrite and, therefore, less common. It is part of primary lead-zinc deposits and is also found in supergene ore deposits. It occurs as flattened spheres in black shales, referred to as **pyrite suns**. It is also found as nodules or concretions in clays and shales with a radiating fibrous structure.

Figure 15.16 Massive black chalcocite with fractures filled by dark-green brochanthite, Cu$_4$SO$_4$(OH)$_6$, an alteration product.

Marcasite - FeS$_2$
$P\,2_1/n\,2_1/n\,2_1/m$

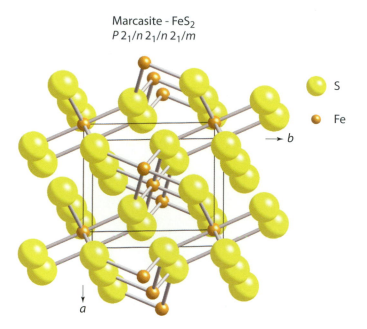

S

Fe

Figure 15.17 Perspective view, essentially along the *c* axis, of the crystal structure of marcasite. Fe atoms are in 6-fold coordination with S. An orthorhombic unit cell is outlined.

Figure 15.18 Platy, spear-shaped twin crystals of marcasite in clay.

Chemical composition and crystal structure: Of constant composition as one of two polymorphs of FeS$_2$, the other being pyrite. The stability relations between marcasite and pyrite are not well established, but geological occurrences suggest that marcasite is the polymorph stable at low temperature. The crystal structure of marcasite is shown in Figure 15.17. Iron atoms are in octahedral coordination with sulfur, the same as in pyrite. In marcasite, the octahedra are linked in chains parallel to the *c* axis, joined along their edges.

Crystal form: Orthorhombic; $2/m\,2/m\,2/m$. Crystals commonly tabular; also twinned giving cockscomb and spear-shaped groups (Fig. 15.18).

Physical properties: **H** = 6–6½; **G** = 4.9. Metallic luster. Opaque. Color pale bronze yellow to almost white on a fresh surface; yellow to brown when tarnished. Streak grayish black.

Distinguishing features: Distinguished from pyrite by its paler yellow color, its very different crystal form (pyrite is very common in cubic crystals), and a fibrous or radiating habit.

Uses: May be mined locally as a source of sulfur. The uses of sulfur are given in Section 15.5.

15.11 Molybdenite: MoS$_2$

Occurrence: Molybdenite is the most common ore mineral of molybdenum, which is a very scarce element in the Earth's crust, with an average abundance of 1 ppm. Molybdenite is present in some porphyry copper deposits associated with felsic intrusive rocks. These have a porphyritic texture consisting of larger crystals in a fine-grained matrix. Molybdenite may also be an accessory mineral in granites and pegmatites, and occurs in high-temperature vein deposits.

Chemical composition and crystal structure: It is essentially MoS$_2$ in composition. A ball-and-stick view of the crystal structure of molybdenite is given in Figure 15.19 with sheets of Mo atoms sandwiched between two sheets of S atoms. The bond strength in these S-Mo-S sheets is strong but adjacent sheets are weakly bonded by van der Waals bonds, which gives rise to excellent cleavage along the basal {0001} plane.

Crystal form: Hexagonal; $6/m\,2/m\,2/m$. Crystals occur as hexagonal plates grouped in foliated masses (Fig. 15.20).

Molybdenite - MoS$_2$
$P\,6_3/m\,2/m\,2/c$

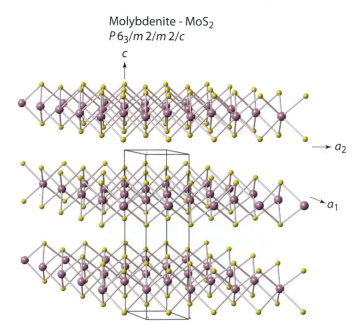

Figure 15.19 Perspective view of a ball-and-stick representation of the crystal structure of molybdenite. The large spacing, along the *c* axis direction, reflects the weak van der Waals bonding between successive layers of S-Mo-S. A primitive hexagonal unit cell is outlined.

Figure 15.20 Crystals of molybdenite with hexagonal crystal outlines as part of a larger composite crystal that shows slightly rounded hexagonal edges.

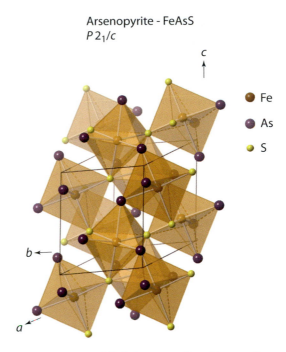

Arsenopyrite - FeAsS
$P\,2_1/c$

- Fe
- As
- S

Figure 15.21 Perspective polyhedral representation of the crystal structure of arsenopyrite. A monoclinic unit cell is outlined. Iron atoms are at the center of octahedral coordination polyhedra with three arsenic and three sulfur atoms at the corners.

Physical properties: H = 1–1½; **G** ~4.7. Perfect {0001} cleavage. Sectile with a greasy feel. Laminae are flexible but not elastic. Color lead gray with a bright metallic luster. Streak grayish black. Opaque.

Distinguishing features: Resembles graphite, C, but is distinguished from it by its higher specific gravity (graphite: **G** = 2.2). The molybdenite color has a blue tone, whereas graphite has a brown tinge. On a porcelain streak plate molybdenite leaves a greenish streak and graphite a black streak.

Uses: Molybdenum is used in steel alloys for its high corrosion resistance and weldability. High strength steel alloys contain between 0.25% and 8% molybdenum. Examples of such are stainless steel, tool steel, cast iron, and high-temperature superalloys. The sulfide itself, MoS_2, is used as a solid lubricant and a high-pressure, high-temperature, antiwear agent, forming strong films on metallic surfaces. Small amounts are used in fertilizers.

15.12 Arsenopyrite: FeAsS

Occurrence: Arsenopyrite is the most common As-containing mineral. It is found in high-temperature deposits that contain tungsten and tin minerals, and in copper and silver ores in association with sulfides such as chalcopyrite, pyrite, galena, and sphalerite. Because of its relatively widespread occurrence in association with sulfides in various ore types, arsenic is commonly recovered as a by-product in the smelting of such ores.

Chemical composition and crystal structure: Close to FeAsS in most occurrences but may show some variability in both As and S contents. A perspective polyhedral view of the crystal structure of arsenopyrite is given in Figure 15.21.

Crystal form: Monoclinic; 2/*m*. Commonly in prismatic crystals (Fig. 15.22).

Physical properties: H = 5½–6; **G** = 6.1. Poor cleavage. Silver-white color with metallic luster. Opaque. Streak black.

Distinguishing features: Distinguished from marcasite by its silver-white color. Unique identification may require a chemical test for arsenic.

Uses: Arsenic and its compounds are mainly used as an alloying agent for lead batteries, but historically it was part of pesticides, insecticides, herbicides, and used in the preservation of wood. All these applications have declined because of toxicity. Another use of arsenic is as a doping agent in solid state devices such as transistors. Doping is the process by which an impurity such as arsenic is added to a semiconductor to increase its conductivity. Arsenic is also used in some medical treatments. The arsenic content of public water supplies where arsenic is derived from surrounding rocks and soils is carefully monitored because of potential health effects (see Section 17.3.4).

Figure 15.22 Short prismatic crystals of arsenopyrite randomly distributed throughout a schist.

15.13 Bauxite: a mixture of diaspore, gibbsite, and boehmite

Occurrence: Bauxite is a mixture of hydrous aluminum oxides in varying proportions, and as such, it is not a mineral but a rock name. The three minerals are diaspore, α AlO(OH); gibbsite, Al(OH)$_3$; and boehmite, γ AlO(OH). Diaspore and boehmite are two polymorphs of AlO(OH). Bauxite is the result of supergene enrichment under tropical to subtropical climatic conditions, by prolonged weathering and leaching of silica of aluminum-bearing rocks such as granites. The original feldspar and other minerals initially break down to form clay, commonly kaolinite, which, after silica removal, leaves behind the three insoluble Al hydroxides.

Chemical composition and crystal structure: Because bauxite consists of a mixture of minerals it does not have a specific, or predictable, chemical composition. Most bauxites range between 40% and 60% weight percentage Al$_2$O$_3$, with the rest made up of clay minerals, iron hydroxides, silt, and some silica.

State of aggregation: Commonly pisolitic in round concretionary grains (Fig. 15.23) but also massive, claylike.

Physical properties: H = 1–3 (highly variable as a function of mineral content); G ~2–2.5. Dull to earthy luster. Color white, gray, yellow brown, red.

Distinguishing features: Best recognized by its pisolitic character.

Uses: Bauxite is the ore from which aluminum metal is produced. Aluminum metal is remarkable because of its low density and resistance to corrosion. Structural components made from aluminum and aluminum alloys are extensively used in the aerospace industry, in transportation, and building. Examples of use in transportation are aircraft, automobiles, railway cars, and bicycles. It is extensively used in packaging, such as in cans and aluminum foil. In construction, it appears as frames for windows and sliding doors and as aluminum siding for covering outer walls. It is also found in a wide range of cooking utensils. In the chemical industry, it is used to produce a large variety of Al-containing chemical compounds.

15.14 Fluorite: CaF$_2$

Occurrence: Fluorite is common in hydrothermal mineral deposits in association with sulfides such as galena, sphalerite, and/or pyrite. Other associations are with calcite, dolomite, gypsum, barite, and quartz. It also occurs in granites and pegmatites, where it is a late crystallization product. It may be present in geodes.

Chemical composition and crystal structure: The composition is generally close to CaF$_2$. A perspective view of a ball-and-stick illustration of the crystal structure is shown in Figure 15.24. The Ca^{2+} ions are arranged at the corners and face centers of a cubic unit cell and the F$^-$ ions are at the centers of the eight small cubes into which the unit cell may be divided. Each Ca^{2+} is coordinated to eight F$^-$ ions at corners of a cube; the Ca^{2+} ions surround F$^-$ at the corners of a tetrahedron.

Crystal form: Isometric; $4/m\ \bar{3}\ 2/m$. Commonly in good cubic crystals (Fig. 15.25).

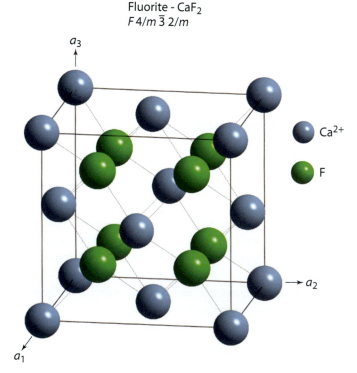

Fluorite - CaF$_2$
$F\ 4/m\ \bar{3}\ 2/m$

Figure 15.24 Perspective view of the crystal structure of fluorite. A cubic unit cell is outlined with Ca$^+$ at the corners of the cube and at the centers of each of the cube faces. This results in an all-face centered unit cell, F.

Figure 15.23 A sawed block of pisolitic bauxite.

Figure 15.25 Euhedral, light-purplish cubes of fluorite with fine-grained encrustations (known as a drusy overgrowth) of milky quartz crystals.

Physical properties: H = 4; **G** = 3.2. Perfect octahedral {111} cleavage. Color varies widely from colorless to wine yellow, green, blue, and violet blue, but also white or gray. Vitreous luster.

Distinguishing features: Distinguished by cubic crystals, perfect octahedral cleavage, vitreous luster, and commonly fine coloring. Its hardness of 4 allows it to be scratched by a knife.

Uses: The commercial name for fluorite is **fluorspar**. It is the main source for the production of hydrofluoric acid, HF. It is used as a flux in the steel and ceramic industries, in iron foundries, and in the making of ferroalloys. Fluorides are also used in insecticides, preservatives, and antiseptics, and in fluoridating public water supplies. Fluorine is part of many compounds produced by the chemical industry.

15.15 Barite: BaSO$_4$

Occurrence: Barite is a common mineral associated with many types of ore deposits, especially in hydrothermal veins that are mined principally for other minerals, such sphalerite, galena, gold, silver, fluorite, and rare earth elements. The commercially most valuable deposits occur in sequences of sedimentary rocks in association with what are known as massive sulfide deposits.

Chemical composition and crystal structure: Most barite is close to the composition of BaSO$_4$, but a complete solid solution series extends to SrSO$_4$, celestite. The crystal structure is given in Figure 15.26. The (SO$_4$)$^{2-}$ groups are represented by the tetrahedra. The large Ba^{2+} ions, with a radius of 1.75 Å, are in 12-fold coordination with surrounding oxygens.

Crystal form: Orthorhombic; $2/m\,2/m\,2/m$. Crystals are commonly tabular on {001} and may be complex. May occur in divergent groups of tabular crystal forming **crested barite** or barite roses (Fig. 15.27).

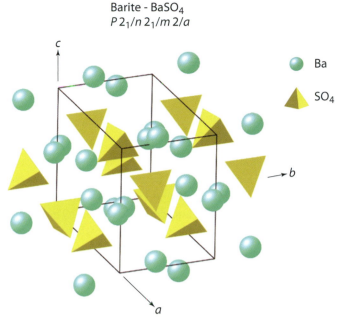

Barite - BaSO$_4$
$P\,2_1/n\,2_1/m\,2/a$

Figure 15.26 Perspective view of the crystal structure of barite. The yellow tetrahedra represent (SO$_4$)$^{2-}$ groups. An orthorhombic unit cell is outlined.

Physical properties: H = 3–3½; **G** = 4.5 (heavy for a nonmetallic mineral). Perfect {001} cleavage. Vitreous luster. Color ranges from colorless to white to light shades of blue, yellow, or red.

Distinguishing features: Recognized by its high specific gravity, tabular crystal habit, and excellent {001} cleavage.

Uses: About 90% of the world's production of barite is used in the petroleum industry as a major ingredient of the heavy fluid called **drilling mud**. This is a carefully formulated heavy suspension, commonly in water but may be in oil, used in rotary drilling of oil and gas wells. It commonly consists of clays, chemical additives, and barite. This mud is used to lubricate and cool the drill bit; to carry the rock cuttings up from the bottom; to prevent blowouts and cave-ins of friable or porous

Figure 15.27 An intergrowth of tabular barite crystals known as crested barite.

rock formations; and for maintaining a hydrostatic pressure in the borehole. The remaining 10% of barite production is used in glass, and as a pigment, and in the manufacture of barium chemicals. Some of these chemicals are used in medicine, as in barium milk shakes or barium enemas for obtaining X-ray images of the intestinal tract.

15.16 Spodumene: LiAlSi$_2$O$_6$

Occurrence: Spodumene is a relatively rare mineral found mainly in lithium-bearing pegmatites, in which it may occur with quartz, albite, lepidolite, muscovite, and locally tourmaline.

Chemical composition and crystal structure: The chemical composition of spodumene is generally close to that of the formula LiAlSi$_2$O$_6$, with some substitution of Na for Li. Spodumene is a monoclinic member of the common silicate group known as pyroxenes. As such, it is a clinopyroxene of which the structure is shown in Figure 7.23 with the typical single tetrahedral silicate chains parallel to the c axis and both Li and Al in 6-fold (octahedral) coordination with oxygen.

Crystal form: Monoclinic; 2/m. Commonly in prismatic crystals with deep vertical striations.

Physical properties: H = 6½–7; G ~3.2. Perfect prismatic {110} cleavage with typical pyroxene cleavage angles close to 90° (actually, 87° and 93°). Color ranges from colorless to grayish white to pale pink, green, and yellow. A complex crystal of a lilac-colored, gem-quality spodumene is given in Figure 15.28.

Distinguishing features: It has relatively high hardness compared with that of another pyroxene, diopside (with H = 5.6), and excellent {110} cleavage.

Figure 15.28 A complex prismatic crystal of the lilac-colored gem variety of spodumene known as kunzite.

Uses: Lithium is used commercially in three different forms: as a lithium ore or concentrate, in the making of glass, ceramics, and porcelain enamel; as lithium metal in lithium-aluminum and lithium-magnesium alloys, in which it imparts high-temperature strength, and improves elasticity; and in lithium battery applications (see Section 16.5). It is used in the production of lithium-based grease for the lubrication of automotive and other industrial and military equipment. It is also used in various lithium compounds made by the chemical industry, in sanitation and bleaching products, and in pharmaceuticals.

The pink, transparent variety of spodumene, named kunzite, is used as a gemstone (Fig. 15.28).

15.17 Lepidolite: K(Li, Al)$_{2-3}$(AlSi$_3$O$_{10}$)(OH, F)$_2$

Occurrence: Lepidolite is a relatively rare mineral found in pegmatites, together with another Li-bearing mineral, spodumene, and quartz, albite, microcline, and muscovite.

Chemical composition and crystal structure: Chemical analyses show that there is a range of Al and Li contents in the octahedrally coordinated site. Lepidolite is a member of the layer silicate group known as micas. Its structure is similar to that of phlogopite (Fig. 7.31), with an octahedral sheet sandwiched between two opposing tetrahedral sheets forming a t-o-t layer. This layer is trioctahedral because all adjacent octahedral sites are occupied by Al^{3+} (see Sec. 7.17).

Crystal form: May be monoclinic or hexagonal depending on the polytype. For the monoclinic form: 2/m. Crystals are platelike or in books with hexagonal outlines.

Physical properties: H = 2½–4; G = 2.8. Perfect {001} cleavage. Color ranges from grayish white to pink and lilac (Fig. 15.29). Pearly luster.

Distinguishing features: Characterized by the lilac to pink color and excellent micaceous cleavage.

Uses: Same as those given for spodumene (Sec. 15.16).

15.18 Several gem minerals

In prior sections (Secs. 15.1–15.17), we presented systematic treatments of several important ore and industrial minerals. Two of these, diamond and spodumene, are used as gems in addition to their industrial applications. Diamond is generally considered the most highly prized of gemstones, but pink spodumene, as kunzite, is a rarely used gem. Here, we provide brief discussions of some other selected minerals that are retrieved from the Earth mainly because of their aesthetic quality.

A gem mineral becomes a gem only after careful cutting and polishing. The most valuable and desirable gems are rare and durable, as well as beautiful. Factors that contribute to beauty are color; luster; transparency; fire and brilliance. Most gems have several of these qualities, but in some, as in turquoise, color alone is the basis of its attractiveness.

Figure 15.29 A random intergrowth of books and plates of light pink to lilac lepidolite.

Because most gems are used as objects for personal adornment, they must resist abrasion and scratching so as not to lose the reflectivily and luster of their polished surfaces. This means that they must be hard, and the most desirable gems have a hardness of that of quartz ($H = 7$) or more. Other factors that affect the desirability of a gem are rarity and fashion. Some of the most prized gems such as emerald, ruby, and sapphire are also rare. Turquoise, however, is not so rare but depends much on the whims of fashion, as seen in the demand for silver-turquoise jewelry that compliments the U.S. Southwestern cowboy look. Here follow brief descriptions of seven selected gem minerals.

Ruby and sapphire (gem varieties of corundum, Al_sO_3)

Corundum is described in Section 13.26 because of its occurrence mainly in high-grade contact metamorphosed Al-rich rocks. Its most important industrial use, on account of its high hardness ($H = 9$), is as an abrasive, or as emery. Both red rubies and sapphires, of various colors, but most commonly blue (see Figs. 15.30(A) and 3.7 for an optical spectrum of blue sapphire), are recovered from sands and gravels, as well as soils where they have been preserved because of their hardness and chemical inertness. Famous sources of ruby are Burma (officially known as Myanmar), Thailand, and Sri Lanka. The most desirable blue sapphires come from Kashmir in the northwestern region of the Indian subcontinent.

Ruby and sapphire are July and September birthstones, respectively.

Topaz

Topaz is described in Section 13.25 because it is a mineral that results from the metamorphism of bauxite, but it is also found in volcanic and intrusive felsic rocks. Its composition is $Al_2SiO_4(F, OH)_2$. The colors of topaz range from light blue to light pink, to yellowish orange brown (Fig. 15.30(B)). Trace amounts of iron and chromium are the coloring agents.

Topaz is recovered from pegmatites and placer deposits. The state of Minas Gerais, Brazil, is the most important source. A highly prized yellow-orange to orange-brown variety known as imperial topaz is found near Ouro Prêto, Minas Gerais, Brazil. Topaz is the birthstone for November.

Tourmaline

Tourmaline was introduced in Section 7.21 because of its common occurrence in granite pegmatites. It has a complex chemical composition that allows for many elemental substitutions causing a wide range of colors. This has resulted in a number of varietal names used in the jewelry industry such as rubellite (for orange to pink and red), verdelite (for green to yellow green), and indicolite (for violet blue to greenish blue). The name watermelon is a term applied to tourmaline that is pink to red in the center and green on the edge (Fig. 15.30(C)). Tourmaline and opal are both birthstones for October.

Emerald, heliodor, and aquamarine (gem varieties of beryl, $Be_3Al_2Si_6O_{18}$)

Beryl can be an accessory mineral in granitic rocks and it may be a major constituent of Be-rich pegmatites. The crystal structure of beryl is given in Figure 15.31 (see also Fig. 4.28). It is a ring (or cyclo) silicate with vertically stacked Si_6O_{18} rings inside of which small amounts of H_2O, $(OH)^-$, F^-, Cs^+, and Na^+ may be housed. It occurs frequently in hexagonal ($6/m\ 2/m\ 2/m$) prismatic crystals. $H = 7\frac{1}{2}-8$; $G = 2.65-2.8$. Common colors range from light bluish green to white or colorless.

Emerald, the dark green and most highly prized gem variety of beryl (Fig. 15.30(D)), owes its color to trace amounts of Cr^{3+} in the crystal structure. The yellow color of golden beryl, the darker yellow of heliodor, and the blue of the aquamarine variety of beryl are due to trace amounts of Fe.

The most famous emerald mines are Chivor and Muzo, Colombia. The main sources of aquamarine and other gem-colored varieties of beryl are from pegmatites in Brazil and Madagascar. Emerald and aquamarine are May and March birthstones, respectively.

Beryl is the main source of the element Be, which is used extensively for its light weight, high strength, and high thermal conductivity, in a large number of industrial applications such as metal alloys and oxides.

Figure 15.32(A) is an example of the use of a variety of colored gemstones that are available at very reasonable prices in a

Figure 15.30 (**A**) Ruby crystal in a white calcite matrix and a sapphire crystal showing tapering hexagonal pyramids. (**B**) Three crystals of topaz. A multifaceted crystal of blue topaz; a prismatic crystal of pink topaz; and a yellow-brown prismatic crystal. (**C**) Prismatic tourmaline crystals commonly have trigonal (3-fold) cross-sections, as seen in the slice cut across the length of a prismatic crystal. This slice is an example of watermelon tourmaline with a pink core and green margin. The central crystal is solid green, and the crystal on the right has a pink zone at the base. (**D**) A hexagonal prismatic crystal of golden beryl, and a dark-green prismatic crystal with a basal pinacoid of emerald in calcite.

modern design pendant. In addition to several of the gems discussed above, it contains two gem varieties of quartz (amethyst and citrine) as well as the gem variety of olivine known as peridot.

Opal

The play of color that precious opal is so famous for was discussed in Section 3.3.1 together with the arrangement of large (~3000 Å in diameter) amorphous spheres of composition $SiO_2 \cdot nH_2O$ responsible for this unique phenomenon (Figs. 3.10 and 3.11). **Precious opal** includes material that displays play of color as well as transparent to translucent opal of a single color but without play of color (Fig. 15.32(B)). Common opal is generally milky white without internal reflections.

Opal occurs in near-surface deposits resulting from precipitation from groundwater or low-temperature hydrothermal solutions. It occurs as fillings and linings of cavities in the host rocks. The largest source of precious opal is Australia, but some important locations occur in Mexico as well. Opal is the October birthstone.

Jade

The gem name *jade* includes two distinct and unrelated species, jadeite and nephrite. Jadeite is discussed here and is a member of the pyroxene group with the chemical formula $NaAlSi_2O_6$. Nephite is a member of the amphibole group. The crystal structure of jadeite is very similar to that of the clinopyroxene diopside, $CaMgSi_2O_6$, shown in Figure 7.23 and discussed in Section 13.7. The Na^+ in jadeite is housed in the same atomic location as Ca^{2+} in diopside, and Al^{3+} is housed in the same site as Mg^{2+}. Jadeite is a metamorphic mineral formed in terrains that have undergone

Beryl - $Be_3Al_2Si_6O_{18}$
$P\,6/m\,2/c\,2/c$

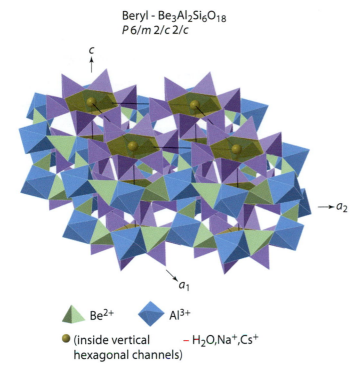

c

a_2

a_1

▲ Be^{2+} ◆ Al^{3+}

● (inside vertical — H_2O, Na^+, Cs^+
hexagonal channels)

Figure 15.31 Perspective view of a polyhedral image of the crystal structure of beryl. A primitive hexagonal unit cell is outlined.

(A)

(B)

(C)

(D)

high pressures but relatively low temperatures (Fig. 14.9). It has a **H** = 6½–7 and is extremely tough and difficult to break. Its primary occurrence is in nodular and rounded masses in serpentine. It is mined from alluvial deposits, where it occurs as boulders in rivers (Fig. 15.32(C)). The most prized jadeite comes from Burma (Myanmar). A vivid green, transparent to translucent variety is most valued. It is used both in carvings and in jewelry.

Turquoise

The chemical formula of turquoise is $CuAl_6(PO_4)(OH)_8{\cdot}5H_2O$ in which considerable Fe^{2+} may substitute for Cu. Turquoise with a strong sky-blue (Fig. 15.32(D)), bluish-green, to apple-green color is easily recognized by its color and is most commonly used as a gem.

It is a secondary mineral, generally formed in arid regions by the interaction of surface waters with high-alumina igneous or sedimentary rocks. It occurs most commonly as small strings or veins traversing more or less decomposed volcanic rocks (Fig. 15.31(D)). Since antiquity, fine-quality material with a vivid, evenly colored medium blue color, known as Persian turquoise, has originated from near Nishapur in the province of Khorasan, Iran (Persia). In the United States, localities in Arizona and Nevada produce high-quality material. Although turquoise can be used in any type of jewelry, it is probably best known in the United States because of the distinctive silver-turquoise designs produced by Native Americans.

Turquoise, together with zircon and tanzanite (a gem variety of clinozoisite), are December birthstones.

Figure 15.32 (A) A contemporary pendant designed by Horst Lang, Idar-Oberstein, Germany. It consists of two pink tourmalines, two blue topaz, one golden beryl, one amethyst (purple variety of quartz; Sec. 7.7), one citrine (yellow-brown variety of quartz), and two green peridots (the gem variety of olivine; Sec. 7.19). (B) Two types of precious opal. On the left, a fire opal with an orange body color; on the right, a black opal with play-of-color and a dark body color. (C) An irregular boulder of jade with a grayish weathering surface. The sawed and polished cut shows the dark-green color of the gem material. (D) Turquoise in small veins traversing rock matrix.

Summary

This chapter introduced 17 ore and industrial minerals that did not fit our earlier classification of minerals being of igneous, sedimentary, or metamorphic origin. These minerals are prevalent in either hydrothermal vein deposits or pegmatites.

- Native gold is the prime ore mineral in the production of gold, although some gold is recovered from gold tellurides, such as $AuTe_2$, calaverite.

- Native silver is an ore of silver although most of the world's supply is derived from Ag sulfides and arsenides.

- Native copper is only a minor ore of copper, with Cu sulfides being the principal Cu ore.

- Diamond is used as an industrial mineral and is the most highly prized gemstone.

- Native sulfur is mined mainly for use in the chemical industry for the manufacture of sulfuric acid, H_2SO_4.

After these five native elements, we discussed six sulfides that are the main ore minerals of lead, zinc, copper, molybdenum, and arsenic:

- Lead and zinc mineralization occurs commonly together because of the association of galena and sphalerite (see chapter-opening photograph).

- Major copper ore minerals are bornite; chalcopyrite (introduced in Sec. 7.32); and chalcocite, in common association with pyrite, marcasite, and pyrrhotite in hydrothermal vein deposits (see Fig. 15.14).

- Pyrite was introduced in Section 7.30, pyrrhotite in Section 7.31, and marcasite was discussed here. These three iron sulfides are commonly closely associated with the copper sulfides but are not mined for their iron content.

- The last mineral in this group is arsenopyrite, an ore of arsenic.

This was followed by the discussion of five important industrial minerals: bauxite, fluorite, barite, spodumene, and lepidolite:

- Bauxite, a mixture of three aluminum hydroxides, is the major Al ore worldwide.

- Fluorite, commonly found in vein deposits associated with sulfides, is the major source of fluorine used in the chemical industry to produce hydrofluoric acid, HF.

- Barite is used mainly as an additive to make heavy mud for oil- and gas-well drilling, to support drill rods and to prevent the blowing out of gas.

- Spodumene and lepidolite are both industrial minerals that are mined for their Li content. Lithium is a component of lithium batteries and is used in a range of electronic devices. With the development of electric cars that run on lithium batteries, Li mining can be expected to increase (see Sec.16.5).

The last seven minerals are gem minerals, which in most instances are the gem-quality equivalents of rock-forming minerals, most of which were discussed in prior chapters:

- Ruby and sapphire are the gem varieties of corundum (Sec. 13.26).

- Topaz is a well-known gem but also occurs as a metamorphic rock-forming mineral (Sec. 13.25).

- Tourmaline is an accessory mineral in granite, granodiorite, and other felsic rocks but occurs in various gem varieties mainly in pegmatites (Sec. 7.21).

- Beryl was not described in prior chapters but is fairly common as an accessory in granitic rocks and Be-rich pegmatites. Gem varieties of this ring silicate include emerald, aquamarine, golden beryl, and heliodor.

- Opal is classified as either **common opal** or precious opal. Common opal is generally milky white in color without internal reflections. Precious opal has a brilliant internal play of color or an intense orange to red body color. The origin of play of color in opal was discussed in Section 3.3.1.

- Jade, a tough and hard-to-break gem variety of jadeite, a member of the pyroxene group of silicates, is used extensively in carved oriental ornaments.

- Turquoise, a hydrous copper-aluminum phosphate is a blue to blue-green gem that originates from Iran (Persia) as well as Arizona and Nevada in the U.S. Southwest, where it is most commonly seen in Native American jewelry designs set in sterling silver. In Europe, Persian turquoise is commonly set in gold.

Review questions

1. Which five chemical elements are found in the native state, namely as native elements?

2. Three of the native elements are referred to as native metals. These have unusual physical properties. What are these, and what type of chemical bond do these properties reflect?

3. Diamond is one of the above five native elements. What is its hardness and what is the bonding type that causes the hardness?

4. The space group of diamond is $F4_1/d\,\bar{3}\,2/m$. What is the equivalent point group (crystal class)?

5. What is the meaning of F in this notation? Explain the arrangement of the carbon atoms.

6. What do 4_1 and d, respectively, represent?

7. What are the two main uses of diamond?

8. What are the two main ore minerals for lead and zinc? Give their names and formulas.

9. What are the two copper sulfides described in this chapter? Give their names and formulas.

10. What is an additional copper sulfide that was described earlier in Chapter 7 among igneous minerals? Give its name and formula.

11. The foregoing three Cu-Fe sulfides commonly occur together in hydrothermal vein deposits. What is meant by the phrase "of hydrothermal origin"?

12. We described two additional metallic, gray minerals. One is a sulfide and the other a sulf-arsenide. Give their names and formulas.

13. What is the name of the Earth material that is the main source of aluminum? Explain its makeup with names and formulas.

14. Give the chemical formula for fluorite and describe its most common crystal habit.

15. Which mineral is the main source of barium? Give its name and formula.

16. What is barite most commonly used for?

17. We described two lithium silicates. Give their names.

18. For what application will lithium be in great demand in the near future?

19. Give the names of three gem minerals that are commonly green?

20. Ruby and sapphire are gem varieties of which relatively common oxide? Give its name and formula.

21. Emerald is a highly prized variety of which rock-forming mineral? Give its name and formula.

22. Which rock-forming mineral, in its gem varieties, displays a very wide range of colors?

23. Precious opal is commonly referred to as amorphous in the literature. This is not really correct. Explain why with reference to its internal architecture.

FURTHER READING

Please refer to the works listed at the end of Chapter 7 as well as the following:

Gems (2009). *Elements* 5, 3, 147–180.

Gold (2009). *Elements* 5, 5, 277–313.

Diamonds (2005). *Elements* 1, 2, 73–108.

London, D. (2008). Pegmatites, Special Pub. 10, Mineralogical Society of Canada, Quebec City, Canada

Schumann, W. (2009). *Gemstones of the World*, 4th ed., Sterling Publishing, New York.

16 Some selected Earth materials resources

In four prior chapters that deal with systematic aspects of minerals (Chapters 7, 10, 13, and 15) each discussion begins with occurrence and ends with uses. The paragraphs on occurrence dealt mainly with the types of rocks in which a specific mineral occurs most commonly, and under uses we listed some of the most important applications of that same mineral. Here, we give a few brief descriptions of some mineral and ore deposits that are the source of rocks, minerals, and chemical elements commonly used in everyday life.

A **mineral deposit** is a concentration of minerals that was formed by geologic processes. An **ore deposit** is a geologic occurrence of minerals (and chemical elements) from which minerals (or elements) of economic value can be extracted at a reasonable profit. Our most basic essential resources are water and soil.

It is instructive to assess what materials you are surrounded by in your own house that are the result of mining of Earth materials and their subsequent transformation into metals (or alloys and other materials) used in the house. The house's foundation, or basement, is probably poured concrete, which consists of an aggregate of sand and gravel in Portland cement. Steel beams that support the structure are derived from iron ore, which consists mainly of hematite, magnetite, and/or goethite. The brick, drain pipes, and roof tiles are made from common kaolinitic clays and shales. The ceramic products such as bathroom toilets, sinks, and wall tiles are made from essentially pure white kaolin clays. So is the chinaware in the kitchen cabinets. The stainless steel cooking pots and utensils, as well as kitchen and bathroom fixtures, are made of various metal alloys, and much of the plumbing and all electrical wiring is made of copper. There may be aluminum frames around windows, the glass of which is the product of the high-temperature melting in furnaces of pure quartz sands. The kitchen and bathroom counters may be fitted with dimension stone, such as granite or marble. Even the gems and gold used for adornment are extracted from the Earth. The fertilizer used in the yard, consisting of various ratios of P_2O_5, K_2O,

Examples of earthenware in which members of the clay mineral group (also known as kaolin, and including the mineral kaolinite) are major components. From top left in a clockwise direction: a pot by Wendell Kowemy, Laguna Pueblo, New Mexico; a dinner plate from Fez, Morocco; a black pottery jar by Maria Martinez (1887–1980), San Ildefonso Pueblo, New Mexico; and a small terrine in a classic Wedgwood design, United Kingdom. For further discussion, see Box 16.1. (David Nufer Photography, Albuquerque, New Mexico.)

and S, is extracted from apatite, silvite, and native sulfur, respectively. The house may be heated with a furnace that runs on natural gas, and the electricity is probably supplied by a coal-fired or nuclear power plant. The car in the garage probably runs on gasoline, which is refined from crude oil. A barrel of crude oil (42 gallons) makes about 19.5 gallons of gasoline, in addition to numerous other products such as heating oil, plastics, and fertilizer.

Most of these examples illustrate the intensive use of what are called **nonrenewable resources**. These are natural resources that cannot be produced, grown, generated, or used on a scale that sustains their consumption rate. Timber, when harvested sustainably, or metals, when recycled, are considered **renewable resources**.

In this chapter, we provide short discussions of some selected major Earth material resources that are used in our everyday existence.

The construction industry is the biggest user by far of sand and gravel, together with cement, dimension stone, and common clay. Sand and gravel is mined in every country in the world.

Steel is a major component in high-rise construction, ships, cars, bridges, appliances, and armaments. Energy supplies cannot be effective without pipelines, furnaces, and engines, which are usually made of steel. The original source of the iron used in steel making is found in iron-ore deposits, which are the iron-enriched products of sedimentary iron-formations. Iron deposits occur in most countries but large to very large iron deposits are restricted in number.

Clay minerals are hydrous aluminum silicates with layered structures that occur with a grain size of <2 μm in the largest dimension. Kaolinite (Sec. 10.4), $Al_2Si_2O_5(OH)_4$, is a primary constituent of kaolin. Another clay mineral, montmorillonite, with the approximate formula of $(Na, Ca)_{0.3}(Al, Mg)_2 Si_4O_{10}(OH)_2 \cdot nH_2O$, is the dominant clay mineral of bentonite, a major constituent of volcanic ash. This material has many technical applications. Vermiculite with an idealized formula of $Mg_3(Si,Al)_4O_{10}(OH)_2 \cdot 4.5H_2O[Mg]_{0.35}$ is another clay used in various commercial applications.

Copper is extensively used in electrical wiring, generators, and motors. It was smelted from its ore in the Middle East by about 4000 BCE (Fig. 9.32(D)). Most of the present-day copper is extracted from deposits that consist of several copper sulfides such as chalcopyrite, bornite, and chalcocite (see Fig. 15.14).

Lithium is used as a glass additive, in lubricants, and in storage batteries. With the active development of electrical cars large lithium storage batteries are needed, and the biggest lithium mineral resources are in Chile and Bolivia.

Rare earth elements (REEs) are those with atomic numbers 57–71, known as the lanthanide series (see the periodic table on the inside front cover), together with yttrium (atomic number 39) and scandium (atomic number 21). The REEs are critical to a range of high-technology products.

Zeolites are a large group of hydrous silicates with important technical applications that result from the spacious open channels and the water molecules and Na^+, Ca^{2+}, and K^+ that can be housed in these channels. We described one zeolite, chabazite, in Section 13.27.

The modern world has enormous energy demands, which currently are met primarily through the use of fossil fuels (coal, oil, and gas) and to a lesser extent hydroelectric and nuclear energy. Oil has been the most convenient of these energy sources to use, but its reserves are rapidly dwindling, thus making it necessary to exploit other energy sources.

16.1 Construction materials

From the Stone Age to the space age, Earth materials have played a defining role in human cultural evolution (Fig. 16.1). Our history is divided into ages based on the dominant Earth material used during that period; that is, *Stone Age* (both Paleolithic and Neolithic), *Chalcolithic* (copper and stone), *Bronze Age* (copper and tin), *Iron Age*, and today the *Nuclear Age* (some might say the *Plastic Age*). Precisely where and when each new material first came into use is uncertain, but each new technology spread rapidly across most of the world. In this section, we deal with Earth materials that have been used in the construction of buildings and other edifices.

16.1.1 Building stones

The earliest period in human history, the Stone Age, is divided into the Paleolithic and Neolithic. Paleolithic people were nomadic hunter-gatherers, and most of their record is preserved in the stone tools they used. The two most widely used stones were chert (Secs. 10.15, 12.3.1, and Fig. 12.35(C)) and

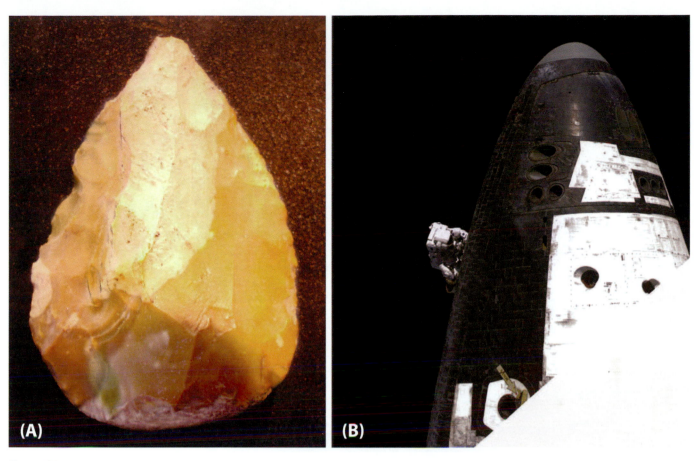

Figure 16.1 **(A)** Neanderthal projectile point made of flint (chert). (Photo courtesy of Museum of Natural History, Amherst College, Amherst, Massachusetts.) **(B)** Heat-resistant tiles on NASA's space shuttle *Discovery* are made with fused silica fibers, and their temperatures are monitored by computers using silicon chips. (Photo courtesy of NASA.)

obsidian (Sec. 9.2.3 and Fig. 9.43), but other fine-grained hard rocks, such as basalt and graywacke, and even quartz (because of its conchoidal fracture), were used for making tools on which a sharp edge was created by knapping (process of chipping conchoidal flakes from the edge of a stone to create a sharp edge). The Neolithic Age (~8000 BCE) saw the beginning of agriculture and the development of permanent settlements where houses were built with stone (Fig. 16.2). These stones were initially rough hewn blocks from nearby outcrops, but by the time the pyramid of Khafre at Giza, Egypt (Fig. 12.34), was built around 2550 BCE, limestone was being quarried and cut into large blocks. Limestone has remained one of the most common building stones throughout history, because it is easily worked and has considerable strength. For example, the tallest structure ever built by the Romans, the Pont du Gard in the south of France, was built of limestone (Fig. 16.3(A)–(B)). This aqueduct, which was built around 14 CE, brought spring water to the city of Nimes. Where the aqueduct crosses the Gard River, it is 55 m high. The limestone used in its construction is a coarse-grained Miocene biosparite cut from a nearby quarry. The fact that the structure is still standing 2000 years after its construction

is testament to the skill of the Roman engineers and to the durability of limestone.

Harder rocks than limestone have been quarried throughout history, even as early as Egyptian times when huge granite

Figure 16.2 Stone house built during the seventh millennium BCE at Khirokitia, Cyprus, a UNESCO World Heritage Site. The carved stone vessel (arrow) in the left foreground is typical of the early Neolithic; ceramic vessels become important later in the Neolithic.

Figure 16.3 (**A**) The Pont du Gard, a Roman aqueduct, brought water to nearby Nimes in southern France. (**B**) Limestone used in the construction of the Pont du Gard. (**C**) Bridge at Millau, France, is made of concrete and steel. Inset shows the heights of the Statue of Liberty and the Empire State Building, New York City, relative to the Millau Bridge.

statues and pillars were carved and polished. One can only marvel at the labor that must have gone into creating these stone objects. Today, the quarrying of granite is far easier with modern cutting tools. Granite tends to break along three mutually perpendicular directions, known as the rift, grain, and hardway, which quarriers take advantage of (Fig. 16.4). In addition, **sheet joints** roughly parallel the topographic surface (Fig 9.10(A)). Although historically granite has been used as a building stone, most granite is now used as a decorative facing stone on buildings or for making countertops.

Many other hard igneous rocks have been used for building purposes throughout history. For example, the street in Pompeii shown in Figure 16.5 is paved with basalt. Here, natural columnar jointing (Sec. 9.2.3) was taken advantage of to create the paving stones. The stepping-stones, which are also made of basalt, were necessary to keep pedestrians out of the water and manure that would have washed along the street, and to allow chariots to pass through. Despite the hardness of basalt, the chariot wheels can be seen to have worn groves into the basalt. Passage along such streets required standardizing axel lengths, a dimension that has carried on to the present day in standard gauge railways (1.44 m).

16.1.2 Bricks, cement, and concrete

Another building material evident along the Pompeii street in Figure 16.5 is **brick**. Romans used brick extensively for all types of structures. Bricks had been used since the Neolithic, but they were made of sun-dried clay and sand that was not very durable. The Romans mass produced bricks in large furnaces at temperatures above 1000°C. Modern bricks, though thicker than Roman bricks, are made by exactly the same process. At these high temperatures, clay along with an admixture of quartz-bearing sand, is converted into a material resembling a high-temperature contact metamorphic rock. A variety of minerals are formed, the most important of which is mullite, $3Al_2O_3 \cdot 2SiO_2$, a mineral first described from the Isle of Mull, Scotland, where a cone sheet had metamorphosed shale (Fig. 9.7(B)). It is a needlelike mineral, similar to sillimanite. In bricks it grows in random orientations and gives bricks considerable strength.

Bricks require mortar to hold them together, and here, again, Romans made a significant contribution to building technology. During the Neolithic, stones were either uncemented or mortared with a mixture of mud and straw. By 800 BCE, the

Figure 16.4 One of the world's largest granite quarries at Barre Vermont. Note man (arrow) for scale.

Figure 16.5 A street in Pompeii, Italy, that was buried during the 79 CE eruption of Vesuvius (in background) is paved with columnar joints of basalt. The bricks in walls are bound together with cement made from a mixture of lime and volcanic ash, which is similar to modern Portland cement. (Photograph courtesy of Liane Philpotts.)

Greeks and others were using lime mortar. The lime was prepared by heating limestone, $CaCO_3$, which drove off carbon dioxide, CO_2. When mixed with water, the lime reacted to form calcium hydroxide, $Ca(OH)_2$, which, over time, reacted with CO_2 in the air to revert back to calcium carbonate. Lime cement became the most common cement for the following two millennia. The Romans discovered that if volcanic ash was mixed with lime, a much harder and more rapidly setting cement was formed, which would harden even under water. The lime reacted with the clay minerals in the ash to form hydrous

calcium-aluminum-silicate minerals. This cement was far superior to lime mortar, but unfortunately, following the decline of the Roman Empire, its recipe was lost and rediscovered only in the 1790s, when patents were issued for what we now refer to as **Portland cement**.

Today, Portland cement is mixed with sand and gravel in roughly 30:70 proportions to form **concrete**, which is the most widely used construction material in the world, especially for large structures, where it is reinforced with steel rods (rebar) or sometimes with organic or inorganic fibers. For example, the tallest bridge in the world, at Millau, France (not far from the Roman Pont du Gard), is held aloft by 343 m-high piers of reinforced concrete (Fig 16.3(C)). The piers have a horizontal cross-section at their base equivalent to the area of a tennis court, and the tallest ones are the same height as the Empire State Building. Huge amounts of concrete are also used in constructing hydroelectric dams. The worldwide production of concrete is in excess of 8 billion tonnes per year; that is, more than 1 tonne of concrete per year for every person on the planet. As discussed in Section 12.3.4, the manufacture of cement requires the firing of limestone to drive off CO_2 and then mixing the lime with an approximately equal proportion of shale. Its manufacture, therefore, releases CO_2, which is of concern because of its greenhouse effect on the climate (Sec. 17.3.6). Cement production accounts for as much as 7% of anthropogenic CO_2 in the atmosphere.

16.1.3 Crushed stone, sand, and gravel

We have seen that concrete is composed of approximately two-thirds sand and rock fragments (**aggregate**) and one-third cement. Similarly, the asphalt used to surface roads consists of two-thirds aggregate and one-third asphalt. Crushed stone and gravel are also used by themselves for the foundations of many structures such as highways and railway beds. The construction industry, therefore, uses an enormous amount of aggregate each year.

Aggregate can be either crushed rock or naturally occurring sands and gravels. The type of aggregate used depends on the properties desired in the final product and on local availability. For example, fine-grained basalt is the most durable of all the common rocks, and consequently, it is the preferred aggregate in asphalt on roads. However, the same basalt aggregate would produce a much denser concrete than one in which limestone or granite aggregate was used, and the increased weight would make this concrete undesirable for structures such as bridges and dams. The higher density of basalt, however, would be a desirable property where crushed stone was placed along the base of an unstable slope to prevent slumping.

Because of the strength and durability of basalt, it accounts for the largest amount of crushed stone produced. The preferred type of basalt forms thick flood-basalt flows, where slow cooling rids the lava of vesicles, and yet the rock is still fine grained. This type of basalt is known as **trap rock** (Sec. 2.5). These flows are commonly thick and laterally extensive, so quarry operations can be large. For example, the trap-rock quarry shown in Figure 2.9 produces 20 000 tonnes of crushed stone per day.

In many areas, **sand** and **gravel** provide the main source of aggregate, where it can be mined from stream, beach, and glacial deposits. It is common along rivers and streams, where it collects along point bars on the inside of river bends. During the waning stages of continental glaciation, meltwaters deposit large amounts of sand and gravel in **kame** terraces, **eskers**, and **outwash plains**, and where meltwater enters glacial lakes, **deltas** can form. In glaciated regions, these types of deposits form valuable sources of sand and gravel. For example, during the retreat of the Wisconsin ice sheet from New England, a large glacial lake (Hitchcock) occupied what is now the Connecticut River valley in Connecticut and Massachusetts, and wherever rivers entered into this lake deltas were created. These now form important sand and gravel deposits (Fig. 16.6). Almost all sand and gravel produced come from open-pit mines. Such mines, as well as processing facilities, are usually located as close as possible to the market in which the sand or gravel is to be used, because transportation is a significant fraction of the cost of this bulky material.

The total value of the annual production of sand and gravel, crushed stone, gypsum, dimension stone, and common clay (all of which are used in the making of cement) in the United States is the highest of the industrial mineral group. The average U.S. home contains about 60 tonnes of concrete; 25 tonnes of sand, gravel, and stone; 7 tonnes of gypsum products (in plaster and wallboard), and 0.1 tonne of glass.

Figure 16.6   Foreset beds of sand in a delta that built out into the former Glacial Lake Hitchcock at South Hadley, Massachusetts, forms one of many valuable sand and gravel deposits in the northeastern United States. (Photograph courtesy of Janet Radway Stone, U.S. Geological Survey.)

16.2 Iron ore

The amount of steel used worldwide in high-rise construction, and in the making of ships, bridges, cars, and armaments is enormous. The iron ore needed to feed the steel mills worldwide is mined from Precambrian terrains of several continents. Some of the largest producers are the Hamersley Range of Western Australia (Fig. 16.7), the Quadrilátero Ferrífero of the state of Minas Gerais in Brazil, and the Labrador region of eastern Canada. In the United States, iron is mined in the Lake Superior region of Minnesota and Michigan. The Precambrian ages of these very large iron-formation sequences from which the iron ores are mined range in age from 2.7 to about 1.8 billion years ago. Because of the enormous recent boom in construction and industrialization of China, the demand for steel has increased. All the above iron regions are expanding their mining operations and exploring for new deposits.

The largest and most productive iron ore deposits such as those in Western Australia and in the Carajás area of Brazil, are the product of supergene alteration of the original banded iron-formation that was rich in oxides such as magnetite and hematite, and quartz. This chemical alteration process involves the leaching of silica (the original chert and quartz), thus resulting in the residual enrichment of iron oxides. Oxidation and hydration occur as well, forming high-grade iron ores consisting of hematite (the hematite in the original banded iron-formation as well as newly formed hematite from magnetite) and goethite. The original banded iron-formations that consisted mainly of carbonates (siderite and ankerite) and iron silicates with only minor iron oxides are not the source for the present iron ores.

Iron-formations that have not benefited from supergene enrichment are mined as well. The banded iron-formations in the Lake Superior region of the United States, and the Labrador region of eastern Canada are examples of such. The grinding and extensive beneficiation of such iron-formations produces a clean

Figure 16.7 Large open pits for the mining of iron ore in the Mount Newman area in the Hamersley Range of Western Australia. The overall red color is a result of the presence of much hematite.

concentrate of iron oxide minerals, magnetite and hematite. If such iron-formations have been metamorphosed, as those of the Labrador region, they consist of relatively coarse-grained hematite, magnetite, and quartz, which are easier to concentrate than finer-grained iron minerals. If the average Fe content of banded iron-formations is about 20 weight %, iron concentrate shipped to steel mills has a total Fe content of at least 66%.

The four largest iron ore producers (in 2007), in decreasing order of production, are China, Australia, Brazil, and India. The iron ores are shipped to steel mills to fabricate steel. Steel is an alloy that consists mostly of iron with a carbon content ranging from 0.2 to 2.1 weight %. Other alloying elements used in steel production are manganese, chromium, vanadium, and tungsten. These alloying elements control the hardness, ductility, corrosion resistance, and strength of the final steel product.

16.3 Clay minerals

Of the various clay minerals known, we will discuss here only some aspects of kaolinite; montmorillonite, the dominant clay mineral in bentonite, formed from alteration of volcanic ash; and vermiculite.

Kaolinite, $Al_2Si_2O_5(OH)_4$ (Sec. 10.4), is the alteration product, in the weathering zone and in hydrothermal occurrences,

of feldspar-rich rocks and clay-rich soils, especially in humid, subtropical, and tropical climates. Kaolinite-rich deposits (kaolin) were formed in large quantities during Cretaceous to Eocene time (144 million to 44 million years ago). The kaolin deposits of Georgia were formed during this time, producing lenses of pure kaolinite (Fig. 16.8) that were deposited in rivers

Figure 16.8 Large white heaps of freshly mined kaolin clay in Washington County, Georgia. (Photograph courtesy of Thiele Kaolin Company, Sandersville, Georgia, United States.)

BOX 16.1 | SOME PRACTICAL ASPECTS OF POTTERY MAKING

Before introducing the next clay mineral, montmorillonite, let's consider a few technical and aesthetic aspects of the production of earthenware pottery, as shown in the opening photograph of this chapter. Two contemporary Native American pots (1 and 3) were fired once and are unglazed; that is, they lack the transparent glossy surface produced by melted compositions of silica, borates, and other fluxes common to ceramic dinnerware. Fabricated from coils of clay, the surfaces of these pots are sanded, rubbed with mineral oil, and then burnished (polished) with a smooth stone. The artist paints geometric and stylized designs onto the polished unfired surface with paintbrushes comprised of about five to seven human hairs using iron oxide, hematite, for reds and browns, and manganese oxide for black.

Pot 3 was probably originally light-colored clay (white or gray) but became completely blackened by large amounts of free carbon in the kiln environment during firing. Carbon in the kiln atmosphere, produced by the combustion of hydrocarbons (e.g., wood, oil, gas) is absorbed into the hot, porous clay body and becomes trapped as the pot cools. After firing, the nonpainted areas are shiny (the feather motifs on pot 3), and the areas that were painted with black manganese oxide paint appear dull.

The other two pieces (2 and 4) were fired twice. The first firing drives the structural water (OH) from the kaolinite and other clay minerals, and provides a hard, resistant surface on which painted designs and glaze are applied. These two pieces were fired a second time to make permanent the painted design on object 3, and to melt the glaze, producing smooth glossy surfaces on both 2 and 4.

The complex aspects of the compositions of ceramic clay bodies and glazes are well described in a book by Mimi Obstler, listed in the "Further Reading" section.

---

that drained the Appalachian Mountains. The kaolin mining activities are restricted to near-surface deposits that are less than 150 m down from the Earth's surface and between 3 m and 15 m thick. After extraction of the ore, mine reclamation restores the landscape to its original state.

The main use of kaolin is in the paper-coating industry, as filler, and as a coating pigment. Paper is essentially a thin sheet of intertwined cellulose fibers with voids and surface irregularities. These surfaces are coated with a slurry of kaolin and some adhesive such as starch. This improves the smoothness, brightness, opacity, and ink absorption of high-quality papers. Kaolin is also a filler in plastics and rubber compounds and an additive in paints. In ceramics, it is the main component of bathroom fixtures (toilets and sinks), tile, porcelain, dinnerware, and enamels. In all of these applications, kaolin clay is highly desirable because of its whiteness, particle size and shape, and lack of contaminants such as iron.

Montmorillonite is a clay mineral, and the term *bentonite* refers to a rock made up mainly of montmorillonite. Bentonites are the result of in situ alteration of volcanic ash or tuff. This alteration occurs when volcanic ash falls into shallow oceans or lakes, or by hydrothermal action on a felsic host rock such as rhyolite. It involves the devitrification (the conversion of glass to crystalline material) and accompanying chemical alteration of glassy igneous material, thus producing a soft, plastic light-colored rock, named bentonite. The idealized formula for montmorillonite is $(Na, Ca)_{0.3}(Al, Mg)_2Si_4O_{10}(OH)_2 \cdot nH_2O$. This layer structure can swell as a result of incorporating water molecules between the tetrahedral-octahedral-tetrahedral (*t-o-t*) sheets in association with the Na^+ and Ca^{2+} interlayer cations. The structure of this clay mineral is overall somewhat electrically unbalanced, with a slight negative charge, which allows for the absorption of exchangeable cations around the edges of the fine clay particles. Bentonite, therefore, has many commercial applications, in drilling muds, as a bonding agent in the production of hematite-rich iron ore pellets, and as an absorbent and desiccant. It is also used in foundry molds for the metal casting of engine blocks, brake drums, and manhole covers and as an environmental barrier material in industrial spills and a base liner in municipal solid-waste landfills.

Vermiculite is another clay mineral with many commercial applications. Its idealized formula is $Mg_3(Si, Al)_4O_{10} \cdot 4.5H_2O[Mg]_{0.35}$, in which [Mg] represents exchangeable ions in the structure. Vermiculite is the supergene alteration product of phlogopite and/or biotite. The structure of vermiculite can expand as a result of the absorption of H_2O. When it is rapidly heated, it produces a lightweight expanded product that is widely used in thermal insulation and potting soil. This exfoliated product is used in agriculture for soil conditioning as a carrier in fertilizers, herbicides, and insecticides. It is also applied as an absorbent for environmentally hazardous liquids.

Figure 16.9 The Kennecott Utah Copper's Bingham Canyon Mine, now under ownership of Rio Tinto, is the world's largest manmade excavation. It is located 28 miles southwest of Salt Lake City, Utah. (Photograph courtesy of Rio Tinto, Kennecott Utah Copper.)

16.4 Copper ore

Copper, as well as gold and silver, occur in nature in their native metallic states. Copper's ready-made presence in some rock types such as mafic extrusive rocks, its softness, and malleability have made it a choice metal for the production of objects by humans as far back as 6000 BCE. Although some native copper is still being mined, the bulk of all copper used globally is extracted from various copper- and copper-iron sulfides. The top five copper-producing countries in 2007, in decreasing order of production, are Chile, Peru and the United States, China, and Australia.

The most important copper deposits are of hydrothermal origin and are known as **porphyry copper deposits**. These are closely associated with intrusions of monzonite or diorite and consist of closely spaced veinlets of quartz, pyrite, chalcopyrite, and bornite with lesser sphalerite and molybdenite. In the weathering zone, these sulfides are largely removed leaving behind voids filled with limonite. Below this zone, supergene enrichment produces copper sulfides such as chalcocite (see Fig. 15.14) and covellite, CuS.

Other major copper deposits, produced by seawater hydrothermal systems, are known as **volcanogenic massive sulfides** (**VMS**). These consist of massive lenses of sulfide minerals that were deposited as sediments as a result of hydrothermal activity that vented onto the seafloor (Fig. 9.34).

Very large porphyry copper deposits occur in the United States, Chile, and Peru. One of the largest of these deposits is known as the Chuquicamata mine, 215 km northeast of Antogafasta, Chile. This is the second deepest open-pit mine in the world (after the Bingham Canyon Mine in Utah, United States). The primary ore minerals, sulfides, are associated with granodioritic and dioritic intrusions. These have been oxidized in the upper regions to unusual copper sulfates.

An enormous open-pit copper mine is the Bingham Canyon Mine, also known as the Kennecott Copper Mine, located southwest of Salt Lake City, Utah, now owned by the Rio Tinto Group. The ore minerals are part of a quartz monzonite porphyry that intruded into sedimentary rocks. Underground mining began in the 1860s, and the open-pit operation began in 1906; it is now represented by an open pit more than 1.2 km deep and 4 km wide (Fig. 16.9). It has produced about 16.9 million tonnes of copper ore.

The sulfide minerals in this ore deposit are present in very low concentrations and are disseminated throughout the host monzonite as tiny grains, seams, and fracture coatings and occur in a concentric zoning pattern in and around the quartz monzonite porphyry body. These zones, going from the interior outward, are molybdenite, bornite-chalcopyrite, chalcopyrite-pyrite, pyrite, and galena-sphalerite.

The extracted sulfide-containing ore is run through a concentrator, where huge grinding mills reduce it to the consistency of very fine sand with 70% of the particle size less than 150 μm. This is followed by a floatation process that separates the sulfide particles from the gangue minerals. This leads to the recovery of a copper concentrate that also contains small amounts of silver, gold, lead, molybdenum, platinum, and palladium.

The very long mining history at Bingham Canyon, first as underground mines and later as open pits, has produced immense amounts of mine waste which can be divided into (1) waste rock, which consists of rocks that were extracted but are not economical to process, and (2) tailings, which are left over from grinding and processing of the ore-containing rock (Fig. 16.10). Much of the waste with contaminants such as arsenic, lead, and copper was placed close to Bingham Creek and its banks, and eventually contaminated soil and sediment were deposited close to some residential areas.

With the mine waste and tailings exposed to natural precipitation and oxygen from the atmosphere slow dissolution of the sulfides occurs, which results in the production of sulfate- and metal-rich waters, also known as **acid mine drainage** (**AMD**). A key reaction in this process is the following:

$$FeS_2 + 7/2\ O_2 + H_2O \rightarrow Fe^{2+} + 2(SO_4)^{2-} + 2H^+$$
pyrite   aqueous   aqueous   aqueous   aqueous

$$(16.1)$$

The increasing acidity of the water results in a decrease in pH. These metal-rich solutions may enter nearby streams as well as the groundwater system. All this occurred at the Bingham Canyon district over the many years of mining where contaminated surface water from the mining operations migrated into the underlying groundwater, which is a principal aquifer in the region. The waste-rock dumps were also irrigated for many years to artificially leach copper; some of these highly acidic waters percolated into the underlying groundwater as well. As a result, the then owner of the mine, Rio Tinto, which became the owner of Kennecott Utah Copper in 1989, proposed on enormous and very costly program of remediation to the Environmental Protection Agency (EPA) in 1991. This groundbreaking proposal was to avert the EPA's intent to list the area as a Superfund site. The cleanup, which cost more than $500 million, removed historical mine waste and facilities and remediated the groundwater supply in the area. The source of the groundwater contamination became controlled through the continuous underground pumping and reverse osmosis treatment of the toxic acidic and sulfate-rich contaminant plume associated with the waste-rock dumps. Mine waste rock and contaminated soils were cleaned up as well. This enormous and sustained remediation program prevented this very large mining site from being listed in the Superfund National Priorities List. As part of all this, Rio Tinto continues to develop the master-planned community of Daybreak, which incorporates a renewable environment and a vibrant local economy built on remediated historic evaporation ponds. A Web site given in the "Online Resources" section describes Rio Tinto's efforts in cleaning up this historic mining waste.

All of the above is in stark contrast to the large open-pit copper mine, the Berkeley Pit, in Butte, Montana, which is filled with highly toxic wastewater. This is one of the largest bodies of seriously contaminated water in the United States and is listed as a Superfund site.

16.5 Lithium ore

In Sections 15.16 and 15.17, we described two lithium minerals, spodumene and lepidolite. Originally, these were two of the main ore minerals of lithium that occur mainly in granitic pegmatites. But lithium is also found in brine deposits, which are the present and future major sources of lithium globally. These brines are volcanic in origin and occur in desert areas, in playas, and saline lakes.

Lithium has a wide range of industrial applications. Spodumene and lepidolite concentrates are used along with silica sands for the manufacture of lithium glass. Lithium-containing glasses have low melting points, low annealing temperatures, and reduced coefficients of expansion. In the chemical industry, lithium is extracted from the host Li minerals, as well as brines, for the manufacture of lithium carbonate, which is the starting material for the production of various chemicals that are used in lubricating greases, ceramics,

Figure 16.10 An active tailings pond in the vicinity of the Bingham Canyon Mine. (Photograph courtesy of Rio Tinto, Kennecott Utah Copper.)

air-conditioning and refrigeration, bleaching, chlorination for water in swimming pools, and in lithium-ion batteries used in cell phones and laptop computers. The lithium-ion batteries that are presently under development for use in electric vehicles are designed to be considerably lighter than lead-acid or nickel-based batteries. They also feature high-energy density and high capacity. Lithium metal is used to a small extent in alloys of chromium, aluminum, copper, and zinc, where it imparts tensile strength and toughness. Lithium is also used in medicine in various Li salts for the treatment of bipolar disorder and in various antidepressant medicines.

With the present accelerating development of electric vehicles, the demand for lithium in the manufacture of rechargeable storage batteries will see dramatic increases.

As noted already, brines are, and will be, the main source for the production of Li chemicals. Lithium concentrations in various brine locations are as follows:

Searles Lake, California: 70 ppm
Clayton Valley, Nevada: 100–300 ppm
Salar de Hombre Muerto, Argentina: 100–700 ppm
Salar de Uyuni, Bolivia: 100–500 ppm
Salar de Atacama, Chile: 1000–5000 ppm.

At present, the three largest producers of lithium are, in decreasing order, Chile, Australia, and Argentina. The largest reserves, however, are found in Argentina, Chile, and Bolivia. Exact reserve estimates are not available for any of these countries, but it may be that the vast salt flats in southeastern Bolivia, known as the Salar de Uyuni, represent the world's largest reservoir of lithium, with significant amounts of boron and potassium as well (Wright, 2010). Bolivians refer to it as "the Saudi Arabia of lithium." The Salar looks like a frozen sea of salt covering 4000 square miles. In the brief rainy season, a sheet of water covers the surface. Underneath the salt crust is a layer of brine that carries lithium in solution. Pilot studies for extracting a lithium concentrate from the brine involve cutting shallow pools into the salt in which the brine is allowed to become more concentrated through evaporation driven by sun and wind. Although Bolivia contains this enormous lithium resource, it must first build an infrastructure that supports mining of the brines and the transportation of the mined products to the rest of the world where electric cars are being manufactured. This will involve not only very large financial investments but also corporate demands and governmental agreements and oversight.

16.6 Rare earth elements (REEs)

The rare earth elements (REEs) are a group of 17 closely related elements from lanthanum, La (atomic number 57), to lutetium, Lu (atomic number 71), referred to as the lanthanide series (see the periodic table on the inside cover of this text), as well as yttrium, Y (atomic number 39) and scandium, Sc (atomic

number 21). These two elements are grouped with the REEs because of their chemical similarities and tendency to occur in the same mineral deposits.

Rare earths are on average relatively abundant in the Earth's crust (cerium is present at 60 ppm, compared with Cu, at 55 ppm) but are generally not concentrated enough to make them easily exploitable commercially.

Four of the minerals in which REEs are housed are

Bastnasite, $(Ce, La) CO_3F$
Monazite, $(La, Ce, Nd) PO_4$
Xenotime, YPO_4
Allanite, $(Ce, La, Ca)_2(Al, Fe^{2+}, Fe^{3+})_3(SiO_4)(Si_2O_7)(OH)$
  (Sec. 7.22)

The world demand for REEs has increased rapidly since about 1965, with total global production of the elements now nine times greater than in 1965. They are critical in a wide range of high-technology products and manufacturing processes, including superconductors, catalytic converters, color televisions and flat-panel displays, batteries for hybrid and electric vehicles, petroleum refining, permanent magnets, and medical devices. Several defense-related products include REEs as enhancers in night-vision goggles, range finders, and some radar systems.

The present demand for REEs has strained the world's supply. The United States was self-reliant until about 1995 in domestically produced REEs but since then has become totally dependent on imports, mainly from China. About 95% of the world's production of REEs is controlled by China. Not only does China have extensive REE resources, but other countries, especially in the West, have avoided or abandoned production because the mining and processing of REE resources can be environmentally risky, creating toxic and even radioactive wastes. Much of the radioactivity associated with rare earths comes from the element thorium, which is not a rare earth but is typically found in the same ore. Virtually all commercial grade rare-earth deposits contain thorium.

Most rare-earth mining is by open-pit or dredge operations. Because of the world's rising demand for REEs, China has increased production for domestic use while limiting exports. In most of China's REE mining regions, the tailings that are left after the REEs have been extracted consist of a slightly radioactive sludge laced with toxic chemicals held in artificial reservoirs or lakes. The radioactivity and toxicity of the waste products of mining of REEs is a global environmental challenge.

16.7 Zeolites

The zeolite group of hydrous tectosilicates (framework structures) includes about 60 naturally occurring minerals and more than 100 that have been produced synthetically. Their crystal structures consist of a framework of SiO_4 and AlO_4

tetrahedra that is very open with channels and large interconnecting spaces.

Zeolites occur typically in fissures and amygdules in mafic volcanic rocks as well as in diagenetic and low-grade metamorphic settings. Figure 14.4 shows the zeolites facies of metamorphism in the left, lower corner, at the lowest temperature-pressure conditions.

Four of the most common naturally occurring zeolites are

Chabazite: $Ca_2Al_2Si_4O_{12} \cdot 6H_2O$
Natrolite: $Na_2Al_2Si_3O_{10} \cdot 2H_2O$
Heulandite: $(Na, Ca)_{2-3}Al_3(Al, Si)_2Si_{13}O_{36} \cdot 12H_2O$
Stilbite: $NaCa_2Al_5Si_{13}O_{36} \cdot 14H_2O$

Chabazite is described in Section 13.27, and its open crystal structure with large channels is shown in Figure 13.47.

The substitution of Al^{3+} for Si^{4+} in the tetrahedral sites causes a charge imbalance to the overall crystal structure that is satisfied by other cations such as Ca^{2+}, Na^+, and K^+ that are loosely bonded to the tetrahedral framework and can be removed, or exchanged, with a strong solution of another ion. This is known as the **cation exchange** property that allows for the absorption of ions and gases. This is employed in water softeners, where Na^+ in the zeolite structure exchanges places with Ca^{2+} in the hard water. The zeolites that fill the water tank in a water-softening system may well be synthetic $Na_2Al_2Si_3O_{10} \cdot 2H_2O$. The natural hard water with Ca^{2+} in solution is passed through this tank, and in the process the Na^+ in the zeolite structure is replaced by (exchanged with) the Ca^{2+} from the water, producing $CaAl_2Si_3O_{10} \cdot 2H_2O$. After time, when the zeolite in the tank is saturated with Ca^{2+}, a strong brine of NaCl is passed through the tank, forcing the reverse reaction, producing the original $Na_2Al_2Si_3O_{10} \cdot 2H_2O$.

The ability to exchange cations is one of the most important properties of natural and synthetic zeolites. Principal uses are in the removal of ammonium in sewage treatment and pet litter; in the sequestration of heavy metal ions in nuclear, mine, and industrial waste; and in odor control.

The water molecules in zeolite structures are weakly bonded to the framework atoms. This allows the structure to slowly lose its water upon heating, thus leaving the overall structure intact. A dehydrated structure can be rehydrated by immersion in water. This property of rehydration allows zeolites to be used as desiccants, as in the removal of water from petroleum and other hydrocarbons. Dehydrated zeolites are also used to absorb other molecules as long as the overall size of channels in the structure is large enough to accommodate such molecules. This leads to a major use of zeolites as **molecular sieves**, in which the size of the holes in a specific structure can be used to separate gases and ions on the basis of their molecular or ionic size. For example, a specific zeolite with a channel diameter of about 4.5 Å can filter hydrocarbons with cross-sectional diameters that are less, but it will exclude other hydrocarbons that have larger effective diameters. This is called molecular sieving.

Synthetic zeolites can be made with a greater range of crystal structures than are found in natural zeolites. They may have wider channels that are applicable to the separation of larger organic molecules. These properties are extensively used in oil refining and in the manufacture of petrochemicals. Synthetic zeolites are also produced to supplement shortfalls of naturally occurring zeolites that are used in various industrial applications.

16.8 Energy resources

Energy makes the modern world function. It is used for transportation, manufacturing, heating, lighting, agriculture, communications, and numerous other less energy consuming functions. As the world's population grows, and as many less-developed nations modernize, the demand for energy grows rapidly. Its main source has been fossil fuels, but their rapidly dwindling reserves will necessitate use of other energy sources as we progress through the century. Renewable sources, such as solar, wind, and hydroelectric will undoubtedly play a role in filling this demand, but they are not likely to be sufficient, except in certain favorable localities. Nuclear energy is likely to be called on to fill the gap, despite the problems with disposal of nuclear waste and major nuclear accidents as happened in Japan in March 2011. Geothermal energy may also make a larger contribution than at present. Our knowledge of Earth materials and how they behave will be used extensively in trying to solve this pending energy crisis.

16.8.1 Oil, natural gas, and coal reserves

In Section 11.3.2, we discussed the origin of the hydrocarbons that constitute fossil fuels, and in Section 12.4, we saw how they occur in sedimentary sequences. Their mode of occurrence is well understood, so estimates of their reserves can be made with considerable confidence. Only rising prices are likely to affect reserves significantly; that is, as prices rise, it may become economical to mine lower-grade coal or search for smaller, more elusive oil and gas fields.

The current demand for fossil fuels is staggering (Fig. 16.11). According to the U.S. Department of Energy, 85 million barrels of oil are consumed worldwide every day (1 barrel = 42 U.S. gallons or 159 liters). To give some idea of the magnitude of this consumption, we can compare it to the volume of water flowing in the Mississippi past New Orleans every minute or the volume flowing in the Thames past London every hour and 20 minutes. The United States is the largest consumer, at 20 million barrels per day, almost half of which is used to produce gasoline. China is the second largest consumer, at 8 million barrels per day. Next comes Japan at 4.8 million, followed by India, Russia, and Germany, each at just under 3 million, and

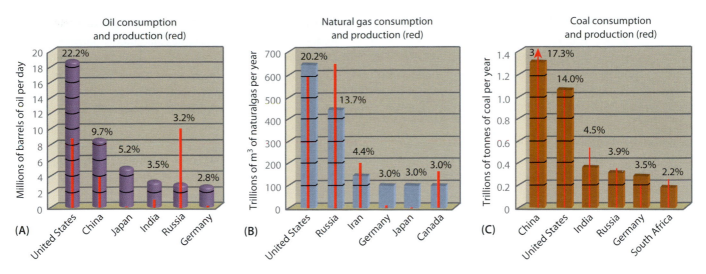

Figure 16.11 The world's major consumers of **(A)** oil, **(B)** natural gas, and **(C)** coal. At the top of the column for each country, the percentage indicates that country's share of the world's total consumption of that fossil fuel. The red lines indicate the amount of fossil fuel each of these countries produces. Data are from the U.S. Central Intelligence Agency for years 2008–2010.

then a large number of developed countries each using between 1 million and 2 million barrels per day. The worldwide annual natural-gas consumption is 3.2 trillion m³, with the United States being the largest consumer, at 657 billion m³, followed by the European Union and Russia at ~490 billion m³ each. The worldwide annual consumption of coal is 4.2 billion tonnes (1 tonne = 1000 kg), which would create a 3 km³ hole if it were to be mined all in one place. China is the largest consumer at 1.2 billion tonnes per year, with the United States a close second at 1 billion tonnes per year. The other major consumers are India, Russia, and Germany, each using about 0.3 billion tonnes per year.

These huge demands for fossil fuels require enormous production. The leading oil producer is Saudi Arabia, producing ~10.8 billion barrels of oil per day, followed by Russia, at 9.9 billion; the United States, at 6.7 billion; Iran, at 4.3 billion; and then China, Canada, Mexico, United Arab Emirates, Kuwait, and Venezuela, each at slightly less than 3 billion barrels of oil per day. China is by far the largest producer of coal, followed by the United States, Australia, and India. Russia and the United States are by far the largest natural gas producers.

More important than current production rates, however, are the proven reserves of fossil fuels and their geographic distribution. As of 2010, worldwide oil reserves are estimated to be 1.39 trillion barrels, which is sufficient to last about 45 years at current usage rates. Saudi Arabia has 19% of these reserves, followed by Canada (12.6%), Iran (9.9%), Iraq (8.3%), Kuwait (7.5%), United Arab Emirates (7.0%), Venezuela (7.0%), Russia (5.3%), and Libya (3.4%). Natural gas reserves are estimated at 187 trillion m³, which is sufficient to last about 60 years. Russia has 22% of these reserves, with Iran (17.6%), Qatar (13.5%), Turkmenistan (4.2%), Saudi Arabia (3.9%), United

States (3.6%), and United Arab Emirates (3.2%) also having substantial reserves. The world's coal reserves are estimated at 847 billion tonnes, which is enough to last 200 years at current usage rates, but this rate is likely to increase as other fossil fuels become scarcer, and 150 years is a more likely estimate. The United States and Canada have 29% of these reserves, followed by Russia (23%), China (14%), Australia (9%), and southern Asia (8%).

As reserves of these fossil fuels dwindle, less conventional types of fossil fuel will be used. Already, large bitumen-rich deposits of what are known as **tar sands** in the Athabaska region of northern Alberta have come into production. These sands have been known of for many years, with the native people having used the tarlike material to caulk their canoes. However, extraction of petroleum from this heavy crude oil is more expensive than from traditional light crude oil, and it was not until the rise in the price of crude oil that it become economically feasible to extract this oil. Currently 1.4 million barrels of oil are being extracted per day. The reserves are estimated at 170 billion barrels, which places them second largest in the world after Saudi Arabia.

Some shales known as **oil shales** are so rich in organic matter that the rock is combustible and can be used directly as fuel. In some countries, such as Estonia, China, Brazil, Germany, Israel, and Russia, oil shales are burned as fuel in electric-power-generating plants in the same way that coal is, but the large quantities of waste product make coal a far preferable fuel, if available. At present, Estonia accounts for 80% of the world's use of oil shale for this purpose.

A petroleumlike fuel can be extracted from oil shales by heating. This process has been known for several hundred years but was developed on an industrial scale only in the mid-nineteenth century. The process, however, is expensive and has serious environmental side effects. First, the oil shale has to be mined in large open pits, and after the oil is extracted, the large quantities of solid waste product are used to reconstruct the former topography. This, however, causes problems because oil shales have a high sulfur content, which generates **acid mine drainage**.

Although organic-rich mudrocks have not commonly been explored directly for oil or natural gas, rising prices of fossil fuel and recent technological advances involving **hydraulic fracturing** have made it possible to extract gas from them economically. A vertical hole is drilled down into the mudrock bed and then deflected to parallel the bed. Next, a slurry of water and sand is injected under high pressure into the hole, which fractures the surrounding rock. After the pressure is released, the fractures are held open by the sand that was injected into them, making a network of permeable fractures through which the gas can flow into the hole. The Marcellus Shale of New York and Pennsylvania (Figs. 12.2 and 12.3) is just one of several organic-rich shales that is currently producing gas using this hydraulic fracturing technique. Recent estimates have set the amount of recoverable gas from this one shale at 50 trillion cubic feet. The Barnett Shale in Texas is another black shale that is producing gas by this technique. The hydraulic fracturing that is required to produce gas from such shales, however, has associated environmental problems, with groundwater possibly becoming contaminated with the fluids used to cause fracturing or the brines released from these deep formations.

16.8.2 Nuclear energy

Dwindling fossil fuel reserves may force the world to depend more heavily on nuclear energy in the future. Unfortunately, many people are apprehensive about nuclear power generation because of the danger of exposure to radiation (see Sec. 17.3.5). Despite accidents at three nuclear power plants (Three Mile Island near Harrisburg, Pennsylvania; Chernobyl in the Ukraine; and Fukushima in Japan) nuclear power generation has had a remarkably safe record compared with, for example, coal mining. However, all nuclear power plants generate highly radioactive waste, which to date has been stored mainly on site. Research is currently under way to determine the safest way of providing long-term storage for this material (see Sec. 17.3.5). Convincing the public that nuclear power can be generated safely will be a major hurdle to increasing its use. At present, France, which generates 79% of its electrical power from nuclear power plants, appears to have won the public over to this form of energy.

The use of nuclear power is not limited by reserves, which are large. Uranium ore is mined in many countries, but Canada is the largest producer, being responsible for almost one-third of the world's production. Uranium is associated with granitic rocks and their hydrothermal veins and with clastic sediments formed from these rocks (conglomerate; e.g., Fig. 12.18). It also occurs in black shales, such as the Chattanooga Shale (Fig. 12.3), which typically contains more than 60 ppm uranium and in places can exceed 1000 ppm. In southern Sweden, black shale containing up to 7000 ppm uranium has been mined as uranium ore. It has also been recovered as a by-product from the oil-shale mining in Estonia. It is mined underground and in open pits, as well as by in situ leaching, all of which produce potential hazards in the form of acid mine drainage from tailings and groundwater contamination from solution mining. Keeping tailings in lined storage cells minimizes the hazards.

16.8.3 Geothermal energy

Energy extracted from heat within the Earth is known as **geothermal energy**. In contrast to energy generated by the burning of fossil fuels, it causes little pollution, although some CO_2 may be released from hydrothermal waters. In addition, it generates none of the hazardous materials produced by nuclear power plants. Geothermal energy is, therefore, a relatively "green" source of energy.

Most geothermal power is currently produced near tectonic plate boundaries where near-surface magma bodies heat water that circulates through prominent fracture zones. Shallow drill holes tap into these hydrothermal systems, and the super heated steam is fed into power plants where turbines generate electrical power (Fig. 16.12). Currently the world produces 8000 MW (megawatts) of geothermal energy. These geothermal fields may be near convergent plate boundaries, as in the Philippines and New Zealand, or at divergent plate boundaries as in Iceland. These countries generate, respectively, 27%, 10%, and 17% of their electrical power in this way, and Iceland provides 90% of the central heating for homes from geothermal sources. The Geysers Geothermal Field, north of San Francisco, is the world's largest geothermal energy producing field, generating 2555 MW of electricity, or 4.8% of California's electricity.

Geothermal power generation is not necessarily restricted to hydrothermal systems in active volcanic areas near plate boundaries. Temperature increases with depth in the Earth, and hence hot rock can be reached anywhere if holes are drilled deep enough. Extraction of heat from hot rocks, however, normally requires fracturing the rock by pumping water down under high pressure (hydraulic fracturing or hydrofracking) and then removing heat from water that is circulated through injection and recovery wells. In some hot-rock geothermal pilot projects, such as at Basel, Switzerland, the hydrofracturing caused small earthquakes, which resulted in suspension of the project. Whether energy can be generated economically in this way depends on how close the hot rock is to the surface (geothermal gradient). One case where a pilot plant indicates this can be done economically is at Innamincka in South Australia. Here, Carboniferous (320 million years) granite containing high concentrations of radioactive elements generates heat (7–10 μwatts/m^3) whose escape

Figure 16.12 Geothermal well head (red geodesic dome) with steam rising from the adjoining relief valve and pipes feeding the superheated steam into the power-generating plant at the Krafla Volcano, Iceland.

to the surface has been retarded by overlying Cretaceous sedimentary rocks that provide an insulating blanket. The amount of heat generated by radioactive decay from this granite has been sufficient to metamorphose the unconformably overlying sedimentary rocks. The steep geothermal gradient (60°C/km) at this locality has resulted in temperatures of 220°C occurring at a depth of only 4.2 km, which has allowed for economical generation of power. With increasing fossil fuel prices, we can expect more hot-rock geothermal plants to be built. For a review of the potential for such enhanced geothermal power generation in the United States, see the article by Blackwell and others in the "Further Reading" section.

Summary

This chapter discussed some important Earth materials that we make use of in everyday life. No attempt was made to cover all useful Earth materials but simply to give a sample of some of the most important ones that are used in construction, in making tools, and in making possible modern technologies. The main points discussed in this chapter are the following:

- Human cultural evolution is divided into ages based on the dominant Earth material used for making tools – Stone Age, Chalcolithic, Bronze Age, Iron Age, Nuclear Age.

- Stone has been one of the most widely used building materials since the Neolithic Age. Limestone has been favored because of its durability and the ease with which it can be worked.

- Bricks, formed from the high-temperature firing of clay-rich mud, have been a common building material since the Roman times.

- Building stones and bricks are commonly held together by cement, which is made by heating mixtures of limestone and shale, which drives off CO_2 and H_2O. When water is added to this powdered mixture, it hardens through formation of hydrous calcium aluminum silicates.

- Today, concrete, which is approximately a 30:70 mixture of cement to sand and gravel, is the most commonly used construction material for large buildings, especially when reinforced with steel rods (rebar).

- Enormous amounts of sand, gravel, and crushed stone (especially basalt or trap rock) are used for concrete, roadbeds, and asphalt for highway construction.

- In addition to construction materials, there are some ore types that provide us with metals and silicates (clays) that are commonly used in construction, appliances, earthenware, and ceramics.

- Iron ore deposits are the source of the materials that are shipped to steel mills for the production of various types of steel.

- Clay (especially kaolin) deposits provide the raw materials for the production of ceramic products, but a large part of kaolin production is used in the coating of paper to improve its smoothness, opacity, and ink absorption. Other clays, such as montmorillonite and vermiculite, are mined for specialized applications. Montmorillonite is used in drilling muds, in the production of hematite-rich iron ore pellets (for the steel mills), and as an absorbent and desiccant. Vermiculite is used in thermal insulation, in potting soil, and as an absorbent of environmentally hazardous liquids.

- Copper ores are mined because of the extensive use of copper piping for water supplies, in refrigeration and air-conditioning, and in electrical applications.

- Lithium and REE (rare earth element) deposits are important because these chemical elements are basic to present-day industrial and high-technology applications.

- Lithium ores will be much in demand because of the need for lithium-ion batteries that are being developed for use in electric vehicles.

- REEs are critical elements in high-technology products such as superconductors, catalytic converters, color televisions and flat-panel displays, and in batteries for hybrid and electric vehicles.

- Zeolites are a unique group of framework Al-Si silicates with a wide range of technological and engineering applications. Their relatively open structures have channels filled with water and exchangeable ions. Upon heating, these structures lose all their water by about 250°C, which can be reversibly reabsorbed at room temperature. This allows them to be used as desiccants. The cation exchange capacity allows zeolites to be used in the removal of toxic ions in a range of environmental applications. And their various channel sizes are used as molecular sieves that allow for the separation of gases and ions on the basis of their molecular or ionic size.

- Earth materials also supply most of the energy needed to make the modern world function. Fossil fuels provide most of this energy.

- The world's reserves of fossil fuels are rapidly dwindling, with oil becoming essentially depleted in 45 years, natural gas in 60 years, and coal in 150 years.

- The largest reserves of oil are in the Middle East and Canada. The Middle East and Russia have the largest reserves of natural gas, and the United States, Canada, and Russia have the largest coal reserves.

- Dwindling supplies of traditional fossil fuels will increase the production of fuels from more difficult sources, such as oil shales, tar sands, and gas from hydraulic fracturing of black shales, all of which have associated environmental problems.

- Nuclear energy will be used more in the future despite concerns about nuclear accidents. The most serious problem with its use is finding safe ways of storing the highly radioactive spent nuclear fuel.

- The use of geothermal power can be expected to increase as fossil fuel reserves decrease. In addition to power generated from volcanic fields, we can expect power to be generated from hot rocks that are heated by the decay of radioactive elements. These may have a far wider distribution than hydrothermal fields associated with volcanic centers.

Review questions

1. What are the main ages of human cultural evolution, and what Earth material characterizes each?

2. What has been the most commonly used building stone throughout history, and why?

3. What is the most commonly used construction material in large buildings today, and what Earth materials are used in its preparation?

4. Why are sand, gravel, and crushed stone such valuable natural resources?

5. Name three countries that are very large producers of iron ore.

6. Iron ore is the product of leaching and supergene enrichment of iron-formation assemblages. What is the geologic age of such iron-formations?

7. What three iron minerals make up the bulk of iron ore?

8. The clay mineral kaolinite is an alteration product of what primary rock-forming mineral?

9. Montmorillonite is also a clay mineral. This is the alteration product of what volcanic material?

10. Porphyry copper deposits are mined because of the presence of which two primary copper minerals?

11. These deposits may also contain one or two supergene copper minerals. Which are these?

12. Acid mine drainage (AMD) can be a serious environmental problem as a result of large piles of waste rock and tailing ponds. What is meant by AMD, and which mineral group is mainly responsible for this problem? Try to give a chemical equation that defines this process.

13. Lithium-containing silicates may be present in pegmatites. Give the names of two of these that were described in Chapter 15.

14. Which three countries have the largest-known reserves of lithium brines?

15. Which elements are part of what is referred to as the rare earth elements (REEs)?

16. What radioactive element is commonly present in REE ores that makes their mining and mining waste disposal environmentally dangerous?

17. Give the name and chemical formula of a zeolite that was described in Chapter 13.

18. What is meant by cation exchange property as applied to zeolites? Give an example of a common commercial application.

19. A zeolite structure can lose all of its water but can regain it all as well when immersed in water after complete dehydration. How is this property used in commercial applications?

20. What is meant by the term *molecular sieving*?

21. What is meant by *fossil fuel*? Give the main types of fossil fuels.

22. Give the number of years the present reserves of each of the main types of fossil fuels are likely to last at current usage rates.

23. Why will environmental concerns become important as reserves of conventional fossil fuels dwindle?

24. Although nuclear energy is considered green, what serious environmental problems are associated with its use?

25. From where does geothermal power derive its energy?

26. How is geothermal power extracted from hot dry rocks?

27. How might you explore for hot rocks in nonvolcanic regions? What geologic conditions would be needed to create a suitable geothermal source in such a region?

ONLINE RESOURCES

Rio Tinto's Web site at http://www.kennecott.com/sustainable-development/environmental-stewardship/legacy-clean-up-issues describes the efforts made in cleaning up 150 years of historic mining waste.

The Web site of Geodynamics Ltd. at http://www.geodynamics.com.au provides a useful survey of the geology and engineering associated with extracting geothermal energy from hot fractured rock.

FURTHER READING

Rapp, G. R. (2002). *Archaeomineralogy*, Springer-Verlag, Berlin.

Bentonites, versatile clays. (2009). *Elements* 5, 2, 83–116.

Bish, D. L., and Ming, D. W. (2001). Natural Zeolites: Occurrences, Properties, Applications. *Reviews in Mineralogy and Geochemistry*, vol. 45, Mineralogical Society of America, Chantilly, VA.

Blackwell, D. D., Negraru, P. T., and Richards, M. C. (2006). Assessment of the enhanced geothermal system resource base of the United States. *Natural Resources Research* 15, 283–308. htpp://www.springerlink.com/content/b0772778x9106566.

Industrial Minerals and Rocks: Commodities, Markets, and Uses, 7th ed., Kogel, J. E., Trivedi, N. C., Baker, J. M., Krukowsk, S. T., eds. (2006). Society for Mining, Metallurgy, and Exploration, Littleton, CO.

Industrial Minerals and Rocks, 6th ed., Carr, D. M., ed. (1994). Society for Mining, Metallurgy, and Exploration, Littleton, CO.

Kesler, S. E. (1994). *Mineral Resources, Economics, and the Environment*, Macmillan College Publishing, New York.

Obstler, M. (2000). *Out of the Earth and into the Fire: A Course in Ceramic Materials for the Studio Potter*, 2nd ed., American Ceramic Society, Westerville, OH.

Mine Wastes. (2011). *Elements* 7, 6.

Wright, L. (2010). Lithium Dreams: Can Bolivia Become the Saudi Arabia of the Electric Car Era? *New Yorker*, March 22, 48–59.

17 Earth materials and human health

In this chapter, we briefly examine how Earth materials affect human health. A moment's reflection will convince us that our bodies and their well-being are intimately related to Earth materials. Most of the elements in our bodies are derived from soil, which in turn is derived from the weathering of rock. Humans have evolved to be in harmony with what the Earth's surface provides. Our food ultimately comes from vegetation, which depends on the soil and climate for growth. On occasions, and in certain regions, conditions can change from the norm as a result of natural processes or stresses placed on the planet by human activity, and these changes can adversely affect our health. When Earth materials are mentioned in connection with human health, we immediately think of major environmental problems, because these attract so much attention in the press, but Earth materials affect our health daily and, for the most part, beneficially. In some areas, however, certain Earth materials pose a serious health hazard, and on rare occasions, such as during the eruption of a volcano, earth materials can pose serious and catastrophic hazards to humans.

The effect of Earth materials on human health is a huge topic, and we can address only a few important highlights in this chapter. We start by looking at how Earth materials are beneficial to our health, and then we examine some serious problems that they can pose. We finish with catastrophic hazards resulting from volcanic eruptions and the extremely rare occurrence of meteorite impacts.

Two very different minerals classified as asbestos. The light-colored specimen (in the foreground) with long, wool-like fibers is chrysotile, a layer silicate structure. The blue specimen (in the background) with a more splintery nature is crocidolite, a fibrous variety of the amphibole known as riebeckite, a tetrahedral chain structure. The chrysotile specimen is from Thetford Mines, Quebec, Canada; the crocidolite is from Kuruman, Northern Cape Province, South Africa. (David Nufer Photography, Albuquerque, New Mexico.)

In this book, we have discussed the important minerals and rocks that constitute our planet, how they form, and where they occur. We have also discussed their use as ore minerals, building materials, and sources of energy. However, Earth materials affect us in a more direct and personal way through our health. Our bodies, like those of all other living organisms, have evolved to use Earth materials for their construction and function. In this chapter, we briefly examine which Earth materials are important for good health and which can cause chemical and physical threats to our well-being. We have room to touch on only a few highlights of this extremely important field. To underline its importance, we can point out that organizations, such as the U.S. Geological Survey, the British Geological Survey, and the French Geological Survey have divisions devoted to health and the environment and medical geology, and all have Web sites geared to providing important related information to the public (see "Online Resources" at the end of the chapter).

17.1 The human body's need for Earth materials

The human body relies on Earth materials for its growth and well-being. We are composed of elements that were previously part of the soil, air, or water, and those elements are continuously cycled through our body during our life as old material is removed and new material grows. Even the bones in our skeleton are replaced every five to ten years. We obtain the elements with which our body is made from food, which also supplies the energy to make our bodies function. We briefly consider the most important of these elements before discussing how they affect our health.

The most abundant of these elements is phosphorus, which constitutes ~1.2 weight % of the human body, 80% of it being in bones. Bones are a composite material composed of cells that control the growth of proteins (collagen) that, in turn, create a fibrous substrate on which inorganic **hydroxylapatite** is deposited. Apatite in rocks has a Ca/P ratio of 5/3 or 1.67 (Sec. 7.33), but apatite in bones has a Ca/P ration closer to 1.5 due to substitution of many ions, in particular carbonate ions, and the presence of vacancies in the structure. Its formula can be expressed as $(Ca,Na,Mg,K,Sr,Pb)_5(PO_4,CO_3,SO_4)_3(OH,F,Cl,CO_3)$. The apatite in tooth enamel comes closest to geological apatite, especially when exposed to fluoridated drinking water, which, through ionic substitution, can form fluorapatite, a major component of apatite in rocks.

The complex composition of bone apatite is important because bones provide a source of chemicals needed for the function of cells. Bones are very much alive, with their apatite crystals giving up these elements when needed and then repositing them at other times. When this process of give-and-take gets out of balance, diseases can result, such as *osteoporosis*, in which loss of bone mass results in porous weak bones. The

Table 17.1 U.S. Department of Agriculture recommended daily intake of nutrients.

Nutrient	Amount (mg)
Potassium	4044
Sodium	1779
Phosphorus	1740
Calcium	1316
Magnesium	380
Iron	18
Zinc	14
Copper	2

Source: http://www.cnpp.usda.gov/Publications/DietaryGuidelines/2005/2005DGPolicyDocument.pdf

extremely small size of apatite crystals in bone (tens of nanometers; 10^{-9} m) results in a large fraction of the atoms being on, or near, the surface of grains, where they are readily exchanged with the surroundings.

Many chemicals are required for a healthy body, and bones help maintain an appropriate balance of these elements. Phosphorous is needed for the function of cells and for building *deoxyribonucleic acid* or *DNA*, which contains the code by which all living organisms replicate themselves. Magnesium is needed for the transmission of nerve signals and muscle movement. Potassium is also important for cell functions and muscular activity. The U.S. Department of Agriculture's recommended daily dietary amounts of these important elements are given in Table 17.1.

17.2 Soils and human health

Life on the planet is a process that requires energy, most of which comes from the Sun. Humans derive this energy from food, which ultimately comes from vegetation that stores the solar energy through photosynthesis, the process of converting CO_2 in the atmosphere to organic compounds, such as sugars. The vegetation, in turn, grows in soil, where certain elements provide essential nutrients for cell growth (Table 17.1). These elements eventually help build our bodies and sustain our health. Soils, therefore, play an all-important role in the growth of a healthy human.

17.2.1 What constitutes a fertile soil?

The fertility of soil depends on many factors, including the type of bedrock from which it is formed, the climate, and the age of the soil. Not all soils are equally fertile, and the fertility can change as soil ages. Soil is a complex mixture of minerals and organic material (Sec. 10.17 and Fig. 10.32) that is continuously changing and can be dramatically affected by human activity. The minerals are formed by the weathering of rock, a process whose rate can vary considerably with climate, increasing with temperature and rainfall. Unfortunately, heavy rainfall also washes away elements needed for crops. As soils age, many of the elements that are important for plant growth and food production are lost as they are removed in the groundwater. This is why old tropical soils are not suitable for crop production.

In some parts of the world, soils are young and have high concentrations of nutritional elements, and these form the most productive agricultural lands. Fertile soils are normally formed by an influx of small relatively unweathered rock particles that are able to provide quickly their nutritional elements. Common sources of such material are volcanoes, youthful mountain ranges, continentally glaciated terranes, and some deserts.

Volcanic soils

A large fraction of volcanic ash consists of small glassy fragments with a large surface area (e.g., pumice) that weathers rapidly and releases nutritional elements to the soil. Explosive volcanism is commonly associated with silica-rich magma (Sec. 9.2.3), which typically has a high potassium content. In Central America, for example, potassium-rich volcanic soils produce bumper crops of bananas, which are shipped all over the world. Another feature of such volcanoes is that they remain active for long periods and erupt with frequencies ranging from tens to hundreds of years, renewing the soil with each eruption.

Soil from young tectonic regions

Erosion from young mountain chains produces large quantities of fresh sediment. Fertile agricultural land is created if this sediment is deposited in deltas. The Nile Delta, for example, owes its fertility to the annual floods that bring sediment from the highest part of the African continent, where it is being rifted apart by a possible mantle plume beneath the Afar region (Box 8.4). The world's largest mountain chains are created at convergent plate boundaries, and rivers draining these areas can form large agriculturally rich deltas. Sediment derived from the Himalayas, for example, has created the fertile Ganges and Mekong deltas.

Glacial soil

Much of the agriculturally rich soil in the Northern Hemisphere owes its productivity to the input of Pleistocene glacial deposits. Continental glaciation removes old soils and produces a large quantity of fresh rock particles that range in size from boulders to clay (Fig. 11.13(A)). Rivers and wind carry large quantities of the finer particles far beyond the glaciated regions. Much of the agricultural land of North America, for example, owes its richness to particles that were initially ground up by glacial action but later transported farther south by rivers.

Wind-blown soil (loess)

Wind can transport fine silt-size particles that accumulate to form the sedimentary deposit known as loess (Sec. 11.6.2).

The breakage caused by impacting grains creates new surfaces from which nutritionally beneficial elements can be released to the soil. Loess deposits are some of the world's most fertile soils. In North America and Europe, most of the loess deposits were formed from wind blowing across glacial meltwater deposits, as the continental ice sheets retreated and before vegetation bound soils in place. The world's largest loess deposits are in central China, where archeologists have found some of the earliest records of agriculture. These deposits were blown from deserts in northern China and Mongolia, but much of the sediment may have originally been derived from mountain glaciers.

17.2.2 Increasing crop production from agricultural land and soil depletion

Changes in agricultural practices throughout human history have lead to ever-increasing crop production from arable land, and for food to continue containing adequate levels of nutritional elements fertilizers must be added to soil.

Agriculture began at the transition from the Paleolithic Age to the Neolithic Age about 9500 BCE. By 3000 BCE, agricultural practices had become quite advanced, as indicated in some of the earliest written human records found at Ur in Mesopotamia, where calendars indicated when to plant and harvest crops and when to leave land fallow. These practices remained essentially unchanged until the eighth century CE, when an agricultural revolution took place that dramatically increased crop production, which in turn meant that more nutrients were extracted from the soil. This revolution involved three important advances. First, instead of leaving land fallow half of the time, land was divided into thirds (the *three-field system*), with one field being planted with a crop; another field being left fallow; and another being planted with peas, beans, or vetches. This immediately decreased the percentage of land left fallow, and the roots of the legumes introduced symbiotic bacteria that form nitrogen compounds that are essential to plant growth. The second advance involved the invention of a new harness for horses and oxen where the yoke distributed the load to the animal's shoulders rather than choking it when it pulled a heavy load. This resulted in a huge increase in the use of animals for farming, which in turn created large amounts of manure, which became fertilizer. Third, the new harness was accompanied by another important invention, the *moldboard plow*. Previously, the *scratch plow* had simply broken up a shallow furrow in the soil, but the moldboard penetrated to a considerable depth and completely overturned the soil, leaving a deep furrow and bringing fresh nutrients to the surface.

This agricultural revolution increased crop production so much that the population of Europe is estimated to have doubled between 1000 and 1300 CE. The world's population

Table 17.2 Element concentrations in a McDonald's Quarter Pounder with Cheese and in a similar volume (145 g) of IUGS average granite (see Fig. 17.1).

Element	Cheeseburger (mg)	Granite (mg)	Granite/Cheeseburger
K	455	4959	10.9
Na	1250	3969	3.2
Ca	360	1886	5.2
Mg	52	621	11.9
P	360	76	0.2

continues to grow at a staggering rate, and by the middle of the twentieth century, many countries were facing serious food shortages. Disaster was averted, however, by the introduction of specially developed highly productive cereal grains, synthetic fertilizers, and pesticides. This change in agricultural practice is commonly referred to as the **Green Revolution**. It is estimated that half of the world's population is fed with food that has benefited from these new agricultural techniques.

These changes in agricultural practices have resulted in ever-increasing crop production, which runs the danger of depleting soils in elements that are essential to our health (Table 17.1). If plants grow and die in situ, the elements they contain are returned to the soil, but crops that are harvested remove these elements from the soil. To illustrate this point, Table 17.2 shows the quantities of potassium, sodium, calcium, magnesium, and phosphorus in a McDonald's Quarter Pounder with Cheese. For comparison, we include an analysis of a hamburger-size patty (similar volume) of granite (145 g). Granite is often used as an approximation to the composition of the upper continental crust and can, therefore, be taken as the raw material from which soil must form. We have used the International Union of the Geological Sciences average granite composition (average

of 2485 samples; see Sec. 9.3.1) for comparison. Table 17.2 also shows the ratio of the concentrations of elements in the granite to those in the cheeseburger.

The average of the element ratios in Table 17.2 show that for every eight cheeseburgers eaten, one hamburger-size piece of granite would have to weather to soil if the concentration of nutritional elements were to be maintained (Fig. 17.1). We have used a hamburger simply as a convenient quantity of food, but loaves of bread or bowls of rice could have been used with similar results. Considering the rate at which cheeseburgers or other foods are consumed, granite could never form new soil fast enough to maintain the abundance of these nutritional elements. Phosphorus is far more abundant in most plants and other foods than it is in granite, so a granite-derived soil could never provide sufficient phosphorus, and it must be added from other sources, such as phosphorite beds (Secs. 10.16 and 12.6). This simple equating of elemental abundances in food and granite makes it clear why fertilizers need to be applied if agricultural land is to continue producing nutritional crops.

17.2.3 The need for fertilizers

Since the Green Revolution, agricultural productivity has increased dramatically, in part, because of the use of synthetic fertilizers. Modern fertilizers are a mixture of chemicals that are specially blended for the particular crop, the type of soil, and the climate. Most of these chemicals are created from material extracted from the Earth and, therefore, have a finite supply. As the world's population continues to grow, shortages of these resources are bound to develop.

The 14 elements that have been identified as necessary for healthy plant growth fall into three groups based on the amounts needed. The *primary macronutrients* include nitrogen, phosphorus, and potassium. The **NPK numbers** on a bag of fertilizer (e.g., 20–27–5 in Fig. 17.2) indicate its weight

145 g of Granite   1/4 lb Cheeseburger

1   ≈   8

Figure 17.1 Equating the trace element content of a hamburger-size piece of granite to eight cheeseburgers.

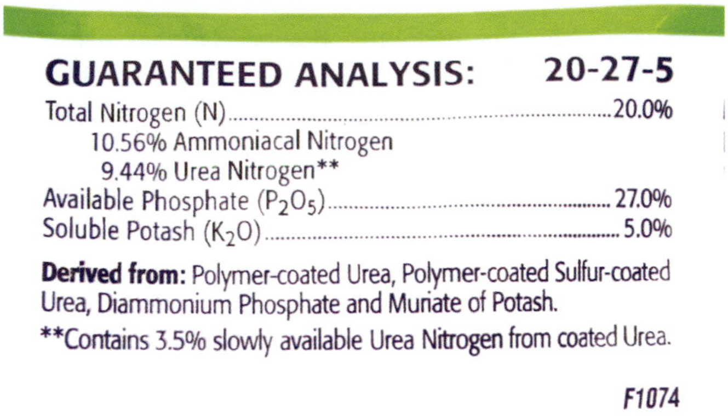

GUARANTEED ANALYSIS: **20-27-5**

Total Nitrogen (N)...20.0%
    10.56% Ammoniacal Nitrogen
    9.44% Urea Nitrogen**
Available Phosphate (P_2O_5)........................ 27.0%
Soluble Potash (K_2O)................................... 5.0%

Derived from: Polymer-coated Urea, Polymer-coated Sulfur-coated Urea, Diammonium Phosphate and Muriate of Potash.

**Contains 3.5% slowly available Urea Nitrogen from coated Urea.

F1074

Figure 17.2 Label on a bag of fertilizer indicating its content of nitrogen, phosphorus, and potassium (NPK).

percentages of N, P_2O_5, and K_2O, respectively. Nitrogen is part of all living cells. It is also part of chlorophyll, which is responsible for photosynthesis. Phosphorus is needed in every cell and participates in photosynthesis. Potassium does not become part of a plant but plays important roles in regulating water balance, protecting against disease, and driving nitrogen fixation. The *secondary macronutrients* include calcium, which is an essential component for the strength of plant cell walls; sulfur, which is essential for production of protein; and magnesium, which is part of chlorophyll. The *micronutrients* include boron, chlorine, iron, manganese, zinc, copper, molybdenum, and selenium.

The need for fertilizer in modern agriculture puts an enormous demand on the supply of its three major constituents; that is, N, P, and K. Most nitrogen is obtained from ammonia (NH_3), and nitrate (NO_3^-) salts. Ammonia is produced by the fractional distillation of nitrogen from liquid air, which is combined with hydrogen from natural gas. Russia, China, the United States, and India are the world's major producers. Phosphorus is mined from sedimentary phosphorite deposits (Secs. 10.16 and 12.6 and Fig. 10.31). The United States accounts for 30% of the world's production of phosphorus, but Morocco has large reserves. Potassium is obtained from evaporite deposits (Sec. 12.5) that contain the mineral sylvite, KCl. The province of Saskatchewan, Canada, has the world's largest deposits, which account for 35% of the world's production.

Nitrogen, phosphorous, and potassium are the main ingredients in fertilizers because these are the elements that promote growth. But with increased growth rates and crop production, soils may become depleted in the trace elements that are considered important to our health (Table 17.1). Although these elements can be added to fertilizers, in many cases they are not. When soils are deficient in trace elements, serious health problems can arise. Zinc deficiency in soils has been identified as the cause of one of the world's most serious dietary problems, with at least 25% of the world's population being at risk. Both plants and humans need zinc for healthy growth. Zinc deficiency in soils stunts crop growth, and in our diets, it can lead to hair loss, skin lesions, and diminished appetite, which in turn lead to weight loss and anorexia in extreme cases. Increasing the zinc content of soils, especially in developing countries, is one of agriculture's greatest global challenges.

Since the Green Revolution, application of fertilizers has become an important part of agriculture (Fig. 17.3). Application of fertilizer and planting of seeds on modern farms is a highly technical process commonly referred to as **precision agriculture**. First, information concerning the composition of the soils, maps of previous crop yields, pest infestations, and so on, are stored in a computer using *Geographic Information System* (*GIS*) software. This information is then fed to a computer in

Figure 17.3 (**A**) A modern farm vehicle carrying three separate sources of nitrogen, phosphorus, and potassium fertilizer that are blended and applied at a specified rate determined by the vehicle's position as determined by a global position system (GPS). (**B**) Interior of the cab of a modern farm vehicle showing the GPS unit, which actually governs the navigation of the vehicle on the left and the computer controlling the mixing and application of fertilizer and seeds on the right. (Photographs courtesy of Merlin Anderson.)

the vehicle applying the seed or fertilizer. A *global positioning system* (*GPS*) controls the location of the vehicle and the composition of the fertilizer and how much to apply (Fig. 17.3(A)) or how much seed to sow. The farmer no longer needs to drive a tractor but watches a computer screen instead (Fig. 17.3(B)). Precision agriculture optimizes the use of fertilizer and maximizes crop production, the results of which are entered into the GIS data bank for the following planting season.

As the world's population continues to grow, the demand for fertilizer will increase. The phosphorous and potassium used in its production are mined from nonrenewable resources whose estimated reserves at current usage rates are 100 and 600 years, respectively. We will not run out of these materials, but as their reserves dwindle, poorer-grade deposits will have to be mined, which will increase the price of fertilizer and the price of food.

17.3 Carcinogenic and chemical hazards posed by Earth materials

Some Earth materials are hazardous if inhaled as dust particles or consumed in our food or drinking water. Medical evidence indicates that exposure to dust from only a very few minerals is clearly linked to health problems. Three of these minerals, erionite (a zeolite), asbestos, and silica are discussed here. We then consider the example of one element, arsenic, that can pose a chemical threat to our health, as well as radon gas, which may be responsible for some radioactivity in the home environment.

17.3.1 Erionite

Erionite is a fibrous zeolite with composition $(K_2, Na_2, Ca)_2Al_4Si_{14}O_{36} \cdot 15H_2O$. Its prismatic crystals may be as long as 15 mm, but typically it is finely fibrous and wool-like. The crystal structure consists of an open framework with columns of cages, parallel to the *c* axis, that are linked by hexagonal tetrahedral rings. Such open framework structures are typical of zeolites, as discussed in Sections 13.27 and 16.7.

Erionite is a major constituent of soils and rocks in the central mountainous Anatolian Plateau, called Cappadocia, in Turkey. This region is underlain by recent volcanic ash deposits (tuff) in which the original volcanic glass, feldspar, and ash were altered by the local groundwater and in alkali lakes to a mixture of montmorillonite and erionite. This material is relatively easy to cut and fashion into bricks and dimension stone. Inhabitants of the village of Karain in that region have used these materials as building stones, and occasionally have ground the tuffs into a fine powder for use in stucco paste and whitewash for the interiors and exteriors of buildings. Some have carved their houses and other structures out of the soft outcroppings. In medical surveys of the population of the village of Karain taken between 1970 and 1977, it was found that the town had a large excess mortality from malignant pleural mesothelioma, a highly

malignant cancer that occurs on the mesothelical surface of the chest cavity (the mesothelium is a membrane that forms the lining of several body cavities). The cause of this incurable and quickly advancing cancer is attributed to background and occupational exposure to erionite-rich mineral dust. The toxicity of erionite is attributed to its needlelike morphology and to the chemical interaction of the particle surface with biological matter. This clear-cut connection between cancer and long-term exposure to this fibrous mineral dust has led to the classification of erionite as a carcinogen by the *World Health Organization* (WHO) and the U.S. *Environmental Protection Agency* (EPA). However, erionite is not currently regulated by the EPA, unlike six asbestos minerals discussed in the following section.

17.3.2 Asbestos minerals

The chapter-opening photograph illustrates the two most important asbestos minerals. The light-colored specimen (in the front of the figure) with very long flexible fibers is chrysotile, $Mg_3Si_2O_5(OH)_4$ (Figs. 2.4, 13.27, and 13.28 and Sec. 13.17), one of three polymorphs (antigorite, lizardite, and chrysotile) of serpentine. It is a member of the group of layer silicates (phyllosilicates). The blue specimen, with shorter, straighter fibers, is referred to as crocidolite, the asbestiform variety of the amphibole riebeckite, $Na_2Fe_3^{2+}Fe_2^{3+}Si_8O_{22}(OH)_2$ (Sec. 13.11). Its structure consists of infinitely extending double tetrahedral chains that are unique to the group of silicates known as chain silicates (or inosilicates), which includes all amphiboles, pyroxenes, and pyroxenoids. Here, both specimens are described as **asbestiform**, a term in common use by Earth scientists. The adjective *asbestiform* refers to highly fibrous minerals in which the fibers readily separate into thin, strong fibers that are flexible. This general usage has been superseded by a definition established in the early 1970s by the National Institute for Occupational Safety and Health and enforced by the EPA. This definition states that the term **asbestos** includes six naturally occurring fibrous minerals, namely chrysotile, crocidolite, amosite, anthophyllite, tremolite, and actinolite. Of these, the last five are all amphiboles. The definition of asbestos also includes a statement that a **fiber** is a particle with a length-to-width ratio (known as the **aspect ratio**) of at least 3:1 and a length of at least 5 μm. Most asbestos fibers have an aspect ratio of >20:1, with chrysotile fibers reaching 10 000:1.

Amosite is the asbestiform variety of cummingtonite-grunerite, $(Fe, Mg)_7Si_8O_{22}(OH)_2$ (Sec. 13.9). Anthophyllite, $Mg_7Si_8O_{22}(OH)_2$ (Sec. 13.8), and tremolite-actinolite, $Ca_2(Mg, Fe)_5Si_8O_{22}(OH)_2$ (Sec. 13.10) are major rock-forming silicates with a slender prismatic habit or, more rarely, a fibrous habit. Amosite (also known as brown asbestos in the trade) was commercially produced from 1914 until 1992 in the region of Penge in Limpopo, South Africa; anthophyllite and tremolite-actinolite, as well as chrysotile asbestos, made only small contributions to the South African asbestos industry. The major mining regions for crocidolite (also known as blue asbestos) were in the Wittenoom area of the Hamersley Range of Western Australia and the Kuruman region of the Northern Cape Province of South Africa. Both of these mining districts are associated with Precambrian banded iron-formation (Secs. 11.4 and 12.7). Amosite occurs in amphibolite facies rocks, and crocidolite is of low-grade metamorphic origin. Mining of crocidolite was stopped in 1966 in Western Australia and in 1995 in South Africa. The main mining regions for chrysotile are in the Thetford Mines area of the eastern townships of the province of Quebec, Canada, and in the central and southern Urals of Russia. Chrysotile occurs as hydrothermal alteration of ultramafic rocks, such as dunite, in veins and mats of fibers replacing serpentine host rock (Fig. 13.28).

Approximately 95% of asbestos that was produced commercially is chrysotile, also known in the trade as white asbestos. About 5% or less was made up of crocidolite (blue asbestos).

In the legal and regulatory definition of asbestos given already, a layer silicate (chrysotile) and five chain silicates (amphibole species) are grouped together. This grouping causes a serious fundamental problem. Chrysotile is a hydrous magnesium silicate with the roll structure of a drinking straw (Figs. 2.4(B)–(D) and 13.27), whereas the five amphiboles have very different and more variable chemical compositions and a totally different crystal structure (Fig. 7.27). These basic chemical and structural differences result in very different toxicologies. A pertinent example of the different behaviors of chrysotile and crocidolite is provided by experimental studies of **biopersistence**, which defines the time of residence of a particulate material in the body. Chrysotile fibers that are introduced into solutions that replicate those of the biological environment of the lung, dissolve completely in about 9 ($\pm$4.5) months, whereas crocidolite fibers may persist for more than several years, releasing Fe to the lung during this time. This biopersistence is much longer than that estimated for chrysotile fibers and is consistent with the lifetimes of these two very different minerals observed in asbestos lung-burden studies.

Why did asbestos become such a contentious and overly expensive problem? Let us begin with some examples of the earlier wide range of applications of asbestos commercially. Long fibers of chrysotile asbestos can be woven into cloth that resists heat and corrosion as in fireproof suits, asbestos gloves used in high-temperature experiments (Fig. 2.4(E)), and theater curtains. Short fibers are used to increase tensile strength in cement pipe for transporting water in municipal waterworks, asphalt shingles, automotive brake linings, electrical insulation materials, in wall and ceiling coatings, as filler in vinyl and asphalt floor tile, in joint cements, roof coatings, plastics, and caulking compounds. Both chrysotile and crocidolite were extensively used as insulating materials for the steel structures inside high-rise buildings and inside navy vessels, as a protection against the heat generated by fires. This insulation consisted of a mix of asbestos fiber and a cementlike binder that was sprayed onto the steel structural elements. Amosite was used in loosely compacted mats as a covering for steam locomotives, jet engines, and marine turbines. Long-fiber crocidolite was used mainly in asbestos cement pipe, in acid-resistant packings and gaskets, and to cover steam boilers and pipes. During World War II, it was used extensively in navy shipyards in Britain to fireproof the inside of naval vessels.

Although federal policy in the United States does not differentiate the various asbestos types (as noted earlier), medical studies on the pathogenicity of the different types of asbestos show that crocidolite (blue asbestos) poses much greater health hazards in occupational settings (e.g., asbestos mining, the manufacture of asbestos-containing products) than chrysotile. Nonoccupational exposure to chrysotile asbestos has not been shown by epidemiological studies to be a significant health hazard. Occupational exposure to asbestos can cause the following types of disorders:

> *asbestosis* a respiratory disease brought on by the inhaling of asbestos fibers causing scar tissue inside the lung,
> *lung cancer*,
> *malignant mesothelioma*, an uncommon cancerous tumor of the lining of the lung, chest cavity, or lining of the abdomen that results typically from long-term as well as short-term asbestos exposure,
> *benign changes in the pleura*,

Crocidolite fibers, as well as other fibrous amphiboles, appear to be the most pathogenic, especially with respect to mesothelioma. Smoking is a strong contributor to the incidence of the above diseases, especially lung cancer.

Although asbestos has caused disease in occupational settings (in the workplace), one must consider the following three questions:

1. What was (or still is) the main type of asbestos in U.S. buildings?
2. Does airborne asbestos dust present in schools and other buildings present a risk to the occupants?
3. What does the natural ambient air (from outside the building) contribute to the fiber count inside a building?

The answer to question 1 is simple. The asbestos fiber found in buildings is about 95% chrysotile.

The answer to question 2 is also simple. Available data do not support the concept that low-level (nonoccupational) exposure to chrysotile asbestos is a health hazard in buildings and schools.

The answer to question 3 is that natural erosional processes contribute background concentrations of fibers to the global air mass. Abelson (1990) stated: "We live on a planet on which there is an abundance of serpentine – and amphibole-containing rocks. Natural processes have been releasing fibers (mainly chrysotile) throughout Earth history. We breathe about 1 million fibers per year" (see also Klein 1993).

The U.S. government decision to ban asbestos is not simple, but misunderstanding how very different chrysotile is from amphibole asbestos and the public's fear of asbestos ("one fiber kills") generated an explosive growth in asbestos identification and removal companies over the past 30 years. During that same period, there has been a total reduction in asbestos mining (at least in the United States) and in the production of asbestos-containing products. Chrysotile mining, from very large open pits, continues in the eastern townships of Quebec, Canada, in the vicinity of two mining towns, named Thetford Mines and Asbestos. The chrysotile mine in Thetford Mines is almost exhausted, and the Jeffrey open-pit mine in Asbestos has been operating only sporadically over the past few years.

Asbestos litigation in the United States has been enormously expensive, with about 35% of the total cost having gone to plaintiffs and the rest going to legal costs. Such litigation is ongoing, and the future of asbestos-related products is extremely uncertain.

17.3.3 Silica minerals

In Section 7.7, we discussed quartz, SiO_2, and four polymorphs, tridymite, cristobalite, coesite, and stishovite. Of these, quartz is the second most common mineral after feldspar in the Earth's crust (Fig. 7.1). Quartz is a major constituent of many rock types and may constitute 100% of many beach sands.

In some occupational settings, such as the mining of granite, and in the use of quartz powder in sandblasting, workers are exposed to large quantities of respirable quartz dust. The main disease related to long-term (occupational) exposure to quartz dust is **silicosis**, a progressive lung disease characterized by the development of fibrotic (scar) tissue. Some lung cancer cases also appear to have been associated with exposure to quartz dust.

In 1987, the International Agency for Research on Cancer (IARC) declared quartz a carcinogen on the basis of some animal experiments in which rats, mice, and hamsters were exposed to intense airborne quartz dust. The results of these experiments were highly inconsistent, and there is always the fundamental question of how well such animal studies relate to human beings. As a result of the IARC decision, the U.S. Occupational Safety and Health Administration (OSHA) requires that any U.S. product with more than 0.1% "free silica" display a warning label. For example, bags of sand bought at a construction supply or hardware store for use in a children's sandbox or in cement mix may have the following label:

"This product may contain silica. Silica dust if inhaled may cause respiratory or other health problems.

Prolonged inhalation may cause delayed lung injury, including silicosis and possibly cancer."

Ross et al. (1993) review the epidemiology of workers employed in the mining industry since 1950 and in sandblasting. These high-dust occupational settings can clearly lead to silicosis. However, they found no evidence for malignant chronic disease, such as silicosis, and no justification for the prediction of cancer risk due to nonoccupational (ambient) silica dust exposure.

There is no evidence of excess mortality among humans who live long-term near quartz-rich beaches or in the desert regions of Arizona, Nevada, California, New Mexico, and Mexico, which commonly have fierce dust storms. The declaration of quartz as a carcinogen is confusing and alarming to the public at large.

17.3.4 Arsenic, an example of a chemically hazardous Earth material

A number of elements, such as arsenic, selenium, mercury, lead, manganese, chromium, and copper, can occur in soils or groundwater at concentrations that are high enough to cause health problems. Hazardous concentrations of these elements often result from human activities, such as mining and various industrial processes, but in other cases they can be natural in origin. As an example of a natural occurring problem, we discuss arsenic poisoning in Southeast Asia, which the World Health Organization has declared "the greatest mass poisoning in human history." Much of the material in this discussion is from the 2006 issue of *Elements* (vol. 2, no. 2) referred to in "Further Reading" at the end of this chapter, and interested readers should refer to that material for further details.

Arsenic in the general environment

Arsenic is one of the most poisonous elements known, and ingestion of no more than 100 mg of colorless and odorless As_2O_3 is lethal. At concentrations well below the lethal dosage, arsenic still poses serious health threats. Long-term exposure to ingestion or inhalation of low doses can result in lesions on the hands and feet (black-foot disease – gangrene); cirrhosis of the liver; heart muscle degeneration leading to cardiac arrest; and cancer of the bladder, kidney, lung, and skin.

Arsenic can occur naturally as a native element, but this is rare. Instead, most occurs as sulfides, sulfarsenides, and arsenides, combined with other elements such as iron, as in arsenopyrite (Sec. 15.12). These minerals occur along with other sulfides in many hydrothermal vein deposits (Sec. 16.4). Because of its position in the periodic table (see inside front cover of the book), arsenic has five different valence states, which allows it to have a wide range of properties in the environment. In groundwater, arsenic can be present under reducing conditions as arsenide, $H_3As^{3+}O_3$, and under oxidizing conditions as arsenate, $H_3As^{5+}O_4$. Under oxidizing conditions, the soluble arsenate bonds strongly with $Fe^{3+}(OH)_3$, and as a result, ferric hydroxide is commonly used to remove arsenic in drinking water remediation schemes.

Given its toxicity, arsenic has been used widely in agriculture as an insecticide, pesticide, and weed killer. Large quantities have been used in preserving pressure-treated lumber. This treatment makes use of chromated copper arsenate (*CCA*), which gives the preserved wood a greenish color. It is also used in chemotherapy for treatment of some types of cancer. In addition to arsenic's toxic uses, it has application in small quantities as an additive to change the physical properties of some materials. Since antiquity, it has been used to harden bronze, an otherwise soft metal. For the same reason, arsenic is used to strengthen the lead in car batteries. It is used as a doping agent in many solid-state electronic devices and is used to clarify glass.

As knowledge of the toxicity of low dosages of arsenic has grown, health organizations have warned of impending health crises, and governments have passed legislation controlling its use. For example, CCA pressure-treated lumber can no longer be used for residential construction in North America and Europe, but many other countries still use it. Even where it is banned, old CCA-treated wood can cause problems if, for example, it is burned. The amount of arsenic used in agriculture has also been dramatically reduced. In the United States, in 2002, the EPA reduced the permissible level of arsenic in public drinking water from 0.05 to 0.01 mg/l (milligrams per liter) or 10 parts per billion (ppb), which is three orders of magnitude less than is common in most rocks.

Arsenic in groundwater in Southeast Asia

In some parts of the world, concentrations of arsenic in groundwater are so high as to cause major public health problems, and nowhere is this problem more serious than in Southeast Asia. In Bangladesh alone, 60 million people are exposed to drinking water with high arsenic contents. The most seriously affected areas are the deltas of the Ganges, Mekong, and Red rivers, where there has been a rapid influx of new sediment from the young Himalayan Mountains. The rocks in the source region are not unusually rich in arsenic, and much of the arsenic formed from weathering of primary minerals is adsorbed in ferric hydroxide coatings on sedimentary grains. However, in the deltas, rapid sedimentation traps large amounts of organic material along with the sand grains (see discussion of the origin of coal in Sec. 11.3.2), and this produces anoxic conditions. When the ferric hydroxide coating on sand grains is reduced, it goes into solution and releases adsorbed arsenic.

Although the arsenic in these deltas is not a result of industrial pollution, the hazardous conditions have, strangely enough, developed as a result of endeavors to improve people's health. Much of the drinking water in these regions previously came from hand-dug wells and ponds that were shared with cattle and were full of pathogens, which caused an enormous number of deaths from cholera and diarrhea. During the 1960s and 1970s, governments and international agencies funded the drilling of shallow wells from which clean, bacteria-free groundwater could be extracted. These wells, unfortunately, tapped more reducing groundwater at depth where arsenic is more soluble. Precisely how the arsenic gets into the groundwater is the subject of much current research.

Arsenic can be removed from water supplies, but at a cost. The EPA Web site given at the end of this chapter provides information about the processes and costs. Commonly, the arsenic is removed by being adsorbed onto activated alumina, iron or manganese hydroxides, or anion exchange resins. Glauconitic sands (Sec. 12.6) can also be used, and recent research indicates that bottom ash from coal-fired power plants is effective and available at a fraction of the cost of the other arsenic scavengers. Water softening to remove calcium and magnesium also removes arsenic. These methods remove As^{5+} more effectively than As^{3+}, so water needs to be oxidized before treatment. Although cities can remediate their water supplies, small villages and farms may find the cost prohibitive, especially because some of the countries affected by arsenic poisoning are among the poorest.

17.3.5 Health hazards due to radioactivity

Radon gas

Radioactive elements are normally very minor constituents of rocks and consequently do not pose hazards to most people. For example, the average continental crust contains only 1.55 ppm uranium. Granite, however, contains on average 4.7 ppm uranium, and some black shales and coal deposits can contain as much as 100 ppm uranium. Uranium miners, of course, are exposed to greater risks, especially because of the accumulation in the mine of dense radon gas, one of the daughter products of uranium decay. Inhalation of this gas gave these miners an exceptionally high incidence of lung cancer until regulations forced mines to install adequate ventilation, and then the incidence of lung cancer attributed to mining dropped to the same level as in the general population.

Radon (^{222}Rn) is considered the second leading cause of lung cancer after smoking, and it is the number one natural cause of lung cancer. It is an extremely dense inert gas that can accumulate in the basement of poorly ventilated houses. Based on the early studies of the exposure of uranium miners to radon gas, the U.S. EPA considers levels of radon gas in a house above 4.0 picocuries per liter to be hazardous; this limit is based on a 70-year residence time with 75% of the time spent indoors. Tests for radon can be performed inexpensively, and problems can be solved with appropriate ventilation. The amount of radon escaping from the ground depends on the uranium content of underlying rocks. Some granites, for example, have high uranium contents, and among sedimentary rocks black shale and coal are high radon producers. Many hot springs have high concentrations of radon gas. It is ironic that many hot springs in Europe, around which huge health spas have developed, contain extremely high levels of radon. Maps showing the potential hazard from radon are commonly available. In the United States, for example, the EPA has a map showing the radon risk across the country, and individual states have detailed radon maps. The risk from radon can be extremely local, depending on fractures in the bedrock and even within a single dwelling can vary from one room to another depending on air circulation. Inexpensive testing is really the only way to be certain that radon does not pose a hazard.

Radioactive waste disposal from nuclear power plants

Nuclear power plants, while providing electrical power, pose numerous health hazards because of the highly radioactive nature of the fuel they use. In 1979, a major accident occurred in the reactor at *Three Mile Island* near Harrisburg, Pennsylvania. Fortunately, the problem was confined to the reactor, and no one was injured. However, in 1986, two explosions at the *Chernobyl* nuclear power plant in the Ukraine sent radioactive material over large parts of Eastern Europe and Russia and were blamed for many deaths due to cancer. In 2011, a tsunami (see Sec. 17.3.6) generated by an offshore magnitude-9.0 earthquake destroyed the nuclear reactor at *Fukushima*, Japan, causing considerable release of radioactive material. These have been the only serious accidents, but all nuclear power plants generate **high-level radioactive waste**, which, to date, no one has successfully devised a method of getting rid of. At present, most nuclear waste is stored at the power plant where it is generated, where it awaits a solution to the problem of its long-term storage.

After 5% of a nuclear fuel rod has reacted, it can no longer be used because of the buildup of fission products. It, therefore, becomes high-level radioactive waste, which remains dangerously radioactive for thousands of years. In France, Britain, and Russia, **spent nuclear fuel** (**SNF**) is reprocessed to make new fuel rods, which reduces the amount of radioactive waste, but the amount of high-level nuclear waste still remains large. The United States is opposed to this reprocessing, because it believes that the waste could be used to create nuclear weapons. Whether the SNF is recycled or disposed of directly, the waste must eventually be buried in a permanent repository. In what form should the waste be sequestered and where it should be buried are two of the most important questions being investigated at present.

Long-term storage of SNF is an engineering problem that requires geological input and knowledge of Earth materials. How minerals behave in the presence of radioactive material and how tight a seal entombing rocks can provide against circulating groundwater are important factors in designing a successful long-term repository. In addition, the use of nuclear power requires the mining of uranium, which presents its own special environmental problems, which again requires knowledge of Earth materials.

The storage of SNF presents a far more serious problem than the hazards associated with mining, because of the much higher level of radiation from the extremely radioactive fission products in the SNF. Several hundred thousand years are required for the radiation from SNF to decay to the level in the uranium ore used to make the fuel. Predicting what will happen to SNF over thousands of years is a complex problem because of the ever-changing nature of the material as a result of radioactive decay, changes in its chemical stability with respect to groundwater, and changes in its mobility both through diffusion and transport in groundwater.

The main component of SNF is UO_2, which is unstable under oxidizing conditions, reacting to form U^{6+} minerals and releasing fission products into the environment. Reducing conditions are, therefore, preferred in long-term storage sites. High-level radioactive liquid waste from reprocessing of SNF has been immobilized by dissolving it in molten borosilicate liquids that solidify to form glass, which is essentially insoluble in groundwater. The glass does crack during cooling, which increases the surface area from which radioactive elements could be leached. Research is also being done to immobilize these elements in ceramics that contain such minerals as perovskite, $CaTiO_3$, which is known to remain stable and be insoluble in groundwater for geologically long periods of time despite high contents of radioactive elements. These ceramics may prove more stable than glasses for long-term storage of waste products.

High-level radioactive waste eventually needs a repository for long-term storage. Several of these sites have been proposed and are undergoing testing. In Finland, a repository is being built 500 m beneath the granite island of *Olkiluoto* in the Gulf of Bothnia. In the United States, considerable research was done in evaluating *Yucca Mountain*, Nevada, as a possible nuclear waste repository, but funding for the project was terminated in 2010. Currently, the United States has no long-term storage site for SNF, although several localities are being considered. Predicting how material will behave and how geological conditions may change at any given site over the next 10 000 years, which is the regulatory compliance period, is presenting Earth scientists with a major challenge. Until satisfactory answers are given to these questions, radioactive waste continues to accumulate on the Earth's surface.

17.3.6 Carbon sequestration to mitigate climate change

Although CO_2 in the atmosphere is not directly hazardous to human health, the burning of fossil fuels is increasing its atmospheric concentration, which is generally accepted as

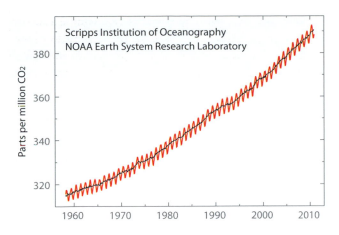

Figure 17.4 Monthly mean atmospheric CO_2 content at the Mauna Loa Observatory, Hawaii. The black curve is the corrected seasonal average. (Graph courtesy of NOAA.)

contributing to global warming. Measurements over the past half century at the National Oceanic and Atmospheric Administration's (NOAA) observatory on Mauna Loa, Hawaii, indicate a steady increase in the atmosphere's CO_2 content (Fig. 17.4). Although CO_2 has natural sinks, these are currently unable to cope with the quantity of CO_2 being released. As a result, major efforts are being made to use more efficient vehicles and to have cleaner-burning fossil fuel power plants. In addition, considerable research is being done to find ways in which CO_2 could be removed from the atmosphere and stored in the Earth, a process known as **carbon sequestration**. These sequestration schemes require an extensive knowledge of the behavior of Earth materials and of Earth systems.

Geological carbon sequestration involves taking CO_2 released from fossil fuel-burning power plants or captured from the air, and pumping it into the ground, where it can be trapped either in the same way that natural gas is trapped beneath impervious layers of shale (Sec. 12.4 and Fig. 12.36), dissolved in formations containing dense saline water, or reacted with primary igneous minerals to create carbonates. This latter process mimics the weathering of igneous rocks (Sec. 11.2.1), with magnesium in ferromagnesian minerals reacting with CO_2 to produce magnesite, $MgCO_3$ (Eq. 11.2), and calcium in plagioclase reacting to form calcite. Another proposal is to pump CO_2 deep into the ocean, where it would dissolve in the ocean water. At depths below 3000 m in the ocean, CO_2 forms a liquid that is denser than seawater, so it would sink to the bottom where it could react with water to form solid CO_2 hydrates. The study of these various techniques is in its infancy, but with the continued burning of fossil fuels, we can expect interest in them to grow. The Web site provided at the end of this chapter will lead interested readers to efforts of the U.S. Geological Survey in this field.

17.4 Hazards from volcanic eruptions

Nine percent of the world's population lives within 100 km of an active volcano and is at risk of being affected directly by an eruption. The deleterious effects of volcanoes, however, can extend far beyond the local environment and can even be global, depending on the magnitude and style of the eruption. When this particular paragraph was being written in April 2010, the eruption of the Eyjafjallajökull volcano in Iceland had emitted so much ash that airports across most of Europe were forced to close, resulting in 17 000 flight cancellations in a single day. Although this was a relatively minor eruption, it illustrates how widespread the effects of an eruption can be.

Volcanoes are not equally hazardous, because of differences in eruptive style, which depends largely on magma viscosity (Sec. 9.2.3). The more viscous a magma the more hazardous is the volcano, because of the increased probability of explosive eruptions. Habitations near volcanoes can be slowly engulfed by lava (Fig. 17.5(A)) or buried rapidly by ash falls or ash flows. Lava flows slowly enough that people can escape from it, but this is not possible with ash flows, which can have velocities as high as 100 m/s. Inhalation of sharp particles of pumice causes death through laceration of the throat and lungs (Fig. 17.5(B)). Volcanic gases are also noxious. For example, in 1986, CO_2 that had slowly accumulated in the bottom of crater Lake Nyos in Cameroon suddenly escaped as a result of a density inversion in the lake and killed 1700 people and 8000 animals in the immediate surroundings. Volcanic gases also kill vegetation and crops, which can lead to starvation. Sulfur dioxide ejected into the stratosphere (>10 km) combines with water to form frozen aerosol droplets (0.1–1 μm) of sulfuric acid that backscatter and absorb incoming solar radiation, which results in global cooling. Finally, lahars (mud slides; Sec. 9.2.3) descending the slopes of volcanoes are one of the greatest causes of volcano-related deaths. The effects of all these hazards are magnified by the fact that soils surrounding volcanoes are fertile and form prime agricultural land, which explains why such a large fraction of the world's population lives near volcanoes.

We have clear historic evidence that volcanic eruptions can have global effects. El Chichón volcano in southern Mexico is unusual in that it erupts extremely sulfur-rich magma, which is able to crystallize anhydrite, $CaSO_4$, a mineral that normally forms in evaporite deposits (Sec. 11.4). The SO_2 released during the eruptions produces sulfuric acid droplets that cause global cooling and produce acid rain. The volcano last erupted in 1982, killing 2000 local inhabitants, but its effects were detected over the entire Northern Hemisphere. The NOAA's observatory at the summit of Mauna Loa, Hawaii, in addition to monitoring the composition of the Earth's atmosphere monitors the influx of solar radiation. The 1982 eruption of El Chichón caused the solar radiation received at this observatory to decrease by 16%. In addition, ice cores collected in Greenland record the El

Figure 17.5 **(A)** The roof of a car that was slowly engulfed by the basaltic lava that descended the slope in the background from Pu'u 'O'o, Hawaii. **(B)** A fossil cast of a resident of Pompeii who died apparently while putting hands to mouth to prevent inhalation of the volcanic ash and noxious gases.

Chichón eruptions with a sudden spike of high acidity caused by the fallout of sulfuric acid aerosols. Deeper samples from the ice core record earlier eruptions of this volcano extending back through the Holocene. This volcano clearly affected the entire Northern Hemisphere.

The damage done by a volcanic eruption depends on the scale of the eruption, and this can vary considerably (see VEI in Sec. 9.2.3). We have historical records of some large eruptions, but these are dwarfed by the magnitude of eruptions in the geologic record. These huge eruptions come from what have come to be known as *supervolcanoes*, which spew out huge volumes of volcanic ash and leave behind large calderas where the crust has sunk into the underlying magma chamber (e.g., Yellowstone; see Sec. 9.2.3 and Fig. 9.24). Although no such volcano has erupted in historic times, their occurrence in the geologic record indicates an eruption frequency of from 10 000 to 100 000 years.

The largest eruption of lava in historic times was the 1783 basaltic fissure eruption at Lakagigar in Iceland (Sec. 9.2.1 and Fig. 9.15(B)). Here, 14 km³ of basalt erupted over several months, releasing as much as 120 megatons of SO_2. Although few people were killed directly by the eruption, the gas released stunted vegetation, which caused the starvation of 24% of Iceland's population. In 1784, Benjamin Franklin attributed the "fog" that hung over Europe to this eruption in Iceland. This "fog" resulted in the lowest mean temperatures since records

have been kept, and ice cores from Greenland indicate that the eruption created the highest acidity in the last 1000 years.

In 1815, Tambora, in Sumbawa, Indonesia, erupted with a VEI of 7 to create the most deadly eruption in recorded history. This composite volcano killed 12 000 people directly with ash falls and ash flows, but the resulting devastation to agricultural land caused starvation, which raised the death toll to more than 70 000. The gas and ash sent into the stratosphere took years to settle. The resulting cooling in the Northern Hemisphere made 1816 the "year without a summer." Crop failures and animal deaths resulted in the worst famine of the nineteenth century. That year, Mary Shelley was vacationing in Switzerland with her husband, the poet and philosopher Percy Shelley, and Lord Byron, and the weather was so bad that they had to stay inside, and to keep themselves entertained they told "ghost stories." Later, Mary's story was published as the now-famous *Frankenstein* (one of the benefits of a volcanic eruption).

The eruption of the Toba volcano, in nearby Sumatra, ~75 000 years ago was very much larger than Tambora and is estimated to have had a VEI of 8, making it a supervolcano. It erupted 2800 km³ of magma and, hence, was two orders of magnitude greater than the Lakagigar eruption. Ash and aerosols erupted into the atmosphere from this eruption are estimated to have caused significant global cooling, creating winterlike conditions for at least ten years. It has been proposed that the climatic changes brought about by this eruption may have drastically reduced the human population.

Even the amount of magma erupted from a supervolcano pales by comparison with the volumes of basaltic magma erupted as flood-basalt flows in large igneous provinces (Sec. 9.4.3). Admittedly, these lavas are erupted over a considerable time span (<1 million years), but their cumulative effect on the atmosphere and climate could have been significant. These eruptions have been proposed as the cause for mass extinctions at the end of the Permian, the end of the Triassic, and the end of the Cretaceous periods, because of the close correlation in time to the eruptions of the Siberian traps, the flood basalts associated with the opening of the Atlantic, and the Deccan traps of India, respectively. By comparing the volume of the flows in these provinces with the volume of the Lakagigar eruption, we conclude that enormous amounts of SO_2 and CO_2 would have been released. These two gases have opposite effects on climate, the SO_2 forming sulfuric acid aerosols that reduce the amount of sunlight received and causing global cooling, and the CO_2 acting as a greenhouse gas that causes global warming. The CO_2 has a relatively short life in the atmosphere, so SO_2 may have been more important. Whether large igneous provinces play a role in mass extinctions is still a hotly debated topic, and as we will see in Section 17.6, catastrophic meteorite impacts provide a compelling alternative explanation.

17.4.1 Monitoring active volcanoes

We cannot stop volcanoes from erupting, but given their hazardous nature, it is important that we monitor them so that eruptions can be predicted. Such monitoring has already paid untold dividends. Studies carried out by the U.S. Geological Survey at Mount St. Helens in Washington State (Fig. 9.12(C)) both before and after the 1980 catastrophic eruption led to the development of techniques that were instrumental in predicting the 1991 eruption of Mount Pinatubo in the Philippines. This eruption was the world's second largest of the twentieth century, with its SO_2 cloud shrouding almost the entire Earth within three months of the eruption and cooling the Earth for at least three years. Fortunately, the eruption was predicted in time to evacuate tens of thousands of people and save equipment from nearby Clark Air Base.

Some volcanoes have been monitored for many years with instruments such as tiltmeters, seismographs, and gas analyzing equipment. The U.S. Geological Survey's Hawaiian Volcano Observatory (HVO), which is located on the rim of the Kilauea caldera (Fig. 9.16(B)), has monitored activity of this volcano and adjoining Mauna Loa since 1912. These are shield volcanoes, so their activity is relatively peaceful as far as volcanoes go and, therefore, make excellent subjects for study. However, these volcanoes do not present a major hazard to the public, even though their flows have slowly destroyed homes and villages along the south shore of the Big Island (Fig. 17.5(A)).

In contrast, volcanoes such as those around Naples, Italy, present a far greater threat because of their explosive character and the huge population in the area (1 million in Naples alone). Vesuvius buried Pompeii (Fig. 9.24) and Herculaneum in 79 CE, and as recently as 1944, the observatory that was built on the volcano's flank to monitor activity was almost destroyed as lava flowed around both sides of the building. Just to the west of Naples, a large number of volcanoes form the Phlegraean Fields (from the Greek for *flaming*), the most recent eruption being at Monte Nuovo in 1538. This eruption was accompanied by dramatic elevation changes, which were particularly noticeable in the harbor of nearby Pozzuoli. Historically, this harbor has had its ups and downs. Around 550 CE, the harbor subsided by tens of meters, and the Roman Temple of Serapis was flooded by the sea, as evidenced by the barnacles that bored into its pillars (Fig. 17.6). Around 1500 CE, the coastline rose by 5 m, and the Monte Nuovo eruption ensued. These fluctuations in elevation have been modeled as resulting from changes in pressure in a magma chamber located 4 km beneath the surface. This region is again beginning to inflate, and in 2009 the harbor was rising at a rate of 1.6 mm/month. With such intense

Figure 17.6 The pitted surface of pillars (white arrow) at the Temple of Serapis in Pozzuoli, Italy, is the result of burrowing by barnacles when the temple sank below the sea in the sixth century CE, probably as a result of a decrease in pressure in the underlying magma chamber.

magmatic-tectonic activity and the potential for a catastrophic eruption in this densely populated area, the entire region is carefully monitored with seismographs, tiltmeters, strainmeters, and GPS stations.

Installing and maintaining monitoring equipment is expensive and obviously cannot be done on all active volcanoes. In recent years, satellite imagery has allowed monitoring of essentially all volcanoes in the world. This involves a technique known as **interferometric synthetic aperture radar** (**InSAR**), in which differences in elevation of the land surface in satellite radar images collected at different times are shown as interference colors on an *interferogram*. Just as light reflected from the top and bottom of a thin film of oil spilled on a wet road creates interference colors that are indicative of the film's thickness (Fig. 17.7(A)), so the interference colors in the radar interferogram indicate the change in elevation of the land between the time the two satellite images were obtained. Figure 17.7(B) shows an interferogram of the caldera at the top of Sierra Negra volcano on Isabela Island in the Galápagos created from images, collected in September 1998 and March 1999. Each order of color in the interferogram corresponds to a difference in elevation of 5 cm. During the six-month period between these two images, the center of the caldera domed upward 0.3 m. This deformation was successfully modeled as resulting from infla-

tion of a magma lens beneath the caldera. The inflation was then used to predict an impending eruption, which occurred in October 2005 from the north side of the caldera.

Frequent passes by orbiting satellites provide radar imagery of most of the Earth's surface, allowing changes in elevation on active volcanoes to be detected. InSAR imagery will be important in the future in predicting volcanic eruptions. Indeed, it may even be able to predict the next eruption of a supervolcano such as Yellowstone, where InSAR studies are constantly keeping track of changes in elevation within the very active caldera.

17.4.2 Lahars

One of the biggest hazards posed by composite volcanoes are mudslides known as lahars (Sec. 9.2.3 and Fig. 9.18(C)–(D)). These can occur at any time but are commonly brought on when pyroclastic material on the flanks of volcanoes becomes water-saturated following heavy rains or when triggered by an earthquake. For example, in 1985, seismic tremors associated with the eruption of the Nevado del Ruiz volcano in Colombia triggered lahars that buried the town of Armero, killing more than 20 000 inhabitants. Lahars can travel so rapidly that people have little chance of escape. Because we lack the ability to predict lahars, our only defense is to map ancient lahars and avoid building in areas where they are common.

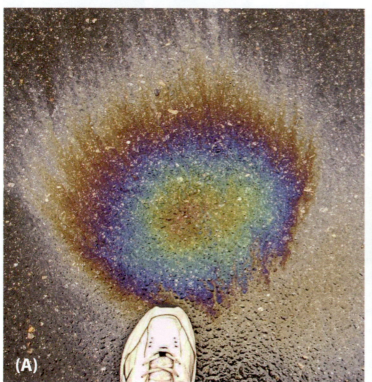
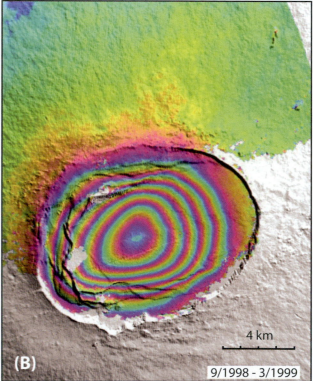

Figure 17.7 (A) Interference colors created by a thin film of oil spilled on a wet road (shoe for scale). (B) InSAR interferogram of the caldera at the summit of Sierra Negra volcano on Isabela Island, Galápagos. Each order of color indicates a change in elevation of 5 cm between September 1998 and March 1999. (InSAR image courtesy of Sigurjón Jónsson; see Jónsson, Zebker, and Amelung, 2005.)

17.5 Tsunamis

Tsunamis, the giant ocean waves caused by earthquakes, pose the greatest hazard of all natural disasters. The magnitude-9.2 earthquake that occurred on December 26, 2004, off the western coast of northern Sumatra Island in Indonesia triggered 30 m-high tsunamis around the Indian Ocean that killed more than 226 000 people. Although seismologists around the world were aware that the earthquake had occurred, no tsunami warning system existed in the Indian Ocean, as it does in the Pacific, so people were caught by surprise. An adequate warning system could have greatly diminished the death toll from this tsunami. Even when tsunami-warning systems are in place, there may be insufficient time to evacuate people from low-lying coastal areas. This was the case on March 11, 2011, when a magnitude-9.0 earthquake off the east coast of Honshu, Japan, created a 30 m-high tsunami that reached the town of Sendai only a few minutes after the quake occurred, killing more than 20 000 people and destroying a nuclear power plant in nearby Fukushima, from which large amounts of radioactive material were released into the environment.

Although the 2004 Indian Ocean tsunami was one of the deadliest natural disasters on record, the geologic record has evidence of far larger tsunamis. The rock shown in Figure 17.8, consisting of broken shell, coral, and basalt fragments, is a 100 000-year-old tsunami deposit found 40 m above present sea level on the island of Molokai, Hawaii. This island has been sinking since it moved off the Hawaiian hot spot, so when the tsunami deposit was formed, it would have been still higher above sea level. What could have caused such a gigantic tsunami, and could a similar magnitude event occur again?

The answer to this question may have been found on the floor of the ocean surrounding the Hawaiian Islands. As each Hawaiian volcano has built up from the ocean floor, its weight has depressed the surrounding lithosphere, so that each volcano is surrounded by a topographic low on the ocean floor. As the volcanoes grow higher, they become unstable and shed material into the surrounding moat as either slowly moving *slumps* or rapidly moving *debris avalanches*. The slumps are the larger of the two types of deposit, being up to 120 km long. Their movement is either by creep or sudden small movements associated with earthquakes. Debris avalanches are smaller than slumps, being up to 2 km thick, but they travel rapidly as far as 230 km. Evidence for their speed comes from the momentum they must have had not only to descend into the moat surrounding the volcanoes but also to climb the slope on the far side. Such rapid movement would have been accompanied by tsunamis. Similar submarine landslides have now been identified around major volcanoes in all of the world's oceans. Collapse of any of these volcanoes could result in devastating tsunamis that would affect the shoreline communities around entire oceans.

17.6 Ejecta from meteorite impacts

In Section 9.5.3, we saw that the Earth has been impacted numerous times by large meteorites, whose explosion ejecta have orbited the Earth and probably cooled the planet. Although the frequency of such impacts has decreased with time, the record indicates that they have occurred throughout geologic time (Fig. 17.9) and will probably occur again in the future. In fact, while the authors were preparing this book, a large explosion occurred on the surface of Jupiter on July 20, 2009, producing a scar with a diameter almost equal to the diameter of the Earth. This probably resulted from a large meteorite impact. Given

Figure 17.8 A tsunami deposit consisting of fragments of shells, coral, and basalt from Kaunakakai, Molokai, Hawaii.

50 km

Figure 17.9 The 214 million-year-old Manicouagan meteorite impact structure, Quebec, Canada. (Image from NASA Earth Observatory, http://www.visibleearth.nasa.gov/view_rec.php?id=2261.)

the rarity of such occurrences, it is surprising that exactly 15 years earlier, astronomers had watched the Shoemaker-Levy 9 comet collide with the surface of Jupiter. We must be thankful that these impacts did not involve the Earth, because they were large enough to have had devastating effects.

The discovery of the iridium anomaly on the Cretaceous-Tertiary (K-T) boundary led Luis and Walter Alvarez to hypothesize that the mass extinction that occurred at this boundary was related to a meteorite impact (Sec. 9.5.3). This iridium anomaly has been found worldwide wherever the K-T boundary is exposed. In addition, it was found to be accompanied by small glassy spheres and particles of soot, the glass being formed by the flash fusion of rocks and the soot presumably from burning vegetation. Subsequently, this ejecta material was found to have come from the Chicxulub impact structure on the northern tip of Mexico's Yucatán Peninsula.

The correlation of the products of the Chicxulub impact structure with the K-T boundary is convincing, but how the impact would have led to the mass extinction is not fully understood. The most likely explanation is that it resulted from a dramatic climate change. The common occurrence of soot on the boundary suggests that the heat from the explosion may have burned an enormous amount of vegetation. The rocks impacted at Chicxulub include gypsum, which would have generated an enormous amount of SO_2, which would subsequently have created a layer of sulfuric acid aerosols around the Earth, which, along with other ejecta and soot from the explosion, would have dramatically reduced the solar radiation reaching the Earth's surface. Vegetation that escaped the fires may have subsequently perished in the cold. The oceans may even have frozen over completely. Eventually, as the sulfuric acid aerosols settled out, enormous amounts of acid rain could have further devastated surviving vegetation.

Meteorite impacts have been proposed as the cause of the mass extinctions at the Permian-Triassic boundary and the Triassic-Jurassic boundary. Supporting evidence is the presence of iridium anomalies and mineral grains exhibiting shock features (Sec. 9.5.3). At these same boundaries, however, huge eruptions of flood basalts (Sec. 9.2.3) may have had equally devastating effects on the climate. Debate will undoubtedly continue before we fully understand the cause of these extinctions.

Summary

This chapter examined how Earth materials affect human health. For the most part, their effects are beneficial, as they supply us with essential nutrients for the growth of a healthy body. Some materials, however, are hazardous when inhaled or ingested and must be monitored carefully in the atmosphere and food supply. Still other hazardous Earth materials are generated during catastrophic events, such as volcanic eruptions, earthquakes, and meteorite impacts.

- Phosphorus is the most abundant of the elements derived from Earth materials in our bodies, being used to make hydroxylapatite in bones and important parts of cells, including DNA. Other important elements are K, Na, Ca, Mg, Fe, Zn, and Cu.

- These elements are derived from the soil through the food we eat. Whether the food contains sufficient amounts of these nutrient elements depends on their concentration in the soil.

- Soils are not all equally fertile. Young soils produced by the input of fresh rock particles from volcanoes, newly eroded mountainous regions, glacial outwash, and deserts have the highest concentrations of nutritional elements.

- Continuous agricultural production can deplete soil of important elements that must be replaced with fertilizer.

- The main ingredients of fertilizers are N, P, and K, but it may also contain Ca, S, and Mg and still other less abundant elements, including Zn, Cu, Mo, and Se.

- Modern agriculture makes use of computers with GIS software to blend and spread the fertilizer to maximize crop production.

- A number of Earth materials are hazardous to our health if inhaled. We discuss two of the most important of these, erionite, a member of the zeolite group of silicates, and the asbestos group of minerals.

- In Turkey, a direct correlation was found between the occupational and background exposure to erionite dust and the occurrence of malignant mesothelioma, a virulent type of cancer.

- The name asbestos, as mandated by the National Institute for Occupational Safety and Health and enforced by the U.S. EPA, applies to six minerals. These are chrysotile (a layer silicate), with the roll structure of a drinking straw, and five members of the amphibole group of silicates (chain silicates) with more complex chemical compositions and very different internal structures from that of chrysotile. These basic chemical and structural differences result in different toxicologies.

- Health experts have shown that of the six asbestos minerals, the five amphibole types are the most carcinogenic, with crocidolite the most common in some asbestos applications. Exposure to crocidolite dust can lead to lung cancer as well as mesothelioma. However, crocidolite was used in less than 5% of commercial products. Chrysotile, which appears to be much less toxic, was used in about 95% of all asbestos products and applications.

- Quartz, the second most common mineral after feldspar in the Earth's crust, can be the cause of silicosis, and in some cases lung cancer, when humans are exposed to large quantities of respirable quartz dust in occupational settings. There is no evidence of the development of such malignant chronic disease as a result of nonoccupational (ambient) silica dust exposure.

- The 1987 declaration by the International Agency for Research on Cancer that quartz is a carcinogen is, therefore, highly confusing and alarming to the public.

- Ingestion of even very small amounts of arsenic can pose a serious health hazard. As a result, its use as an agricultural insecticide and pesticide and as a preservative in pressure treated lumber has been drastically reduced. The U.S. EPA has set the permissible level of arsenic in drinking water at 10 ppb.

- Widespread arsenic poisoning is occurring in parts of Southeast Asia, in particular in the deltas of the Ganges, Mekong, and Red rivers. The arsenic occurs in the groundwater, where rapid deposition of considerable organic material in the sediments of the deltas creates reducing conditions in which arsenic becomes soluble.

- Radon is a dense radioactive gas produced during the decay of uranium. It can accumulate in the basement of poorly ventilated houses and is known to cause lung cancer. It is easily detected and can be removed by adequate ventilation.

- The highly radioactive waste products from nuclear power plants are currently stored on the surface of the Earth at the plants. They will remain highly radioactive for 10 000 years, so their long-term storage presents a major problem that must be solved if nuclear power is to see greater use.

- Carbon dioxide is produced in large quantities by the burning of fossil fuels. This greenhouse gas is steadily increasing its concentration in the atmosphere, which contributes to global warming. Under consideration are various ways of lowering this concentration by removing the CO_2 from the atmosphere and sequestering it in geological formations or in the deep ocean.

- Volcanic eruptions, especially of volcanic ash, pose catastrophic hazards to people living near composite volcanoes (high silica). Lahars (mud slides) off these volcanoes pose an even bigger threat. Gases from volcanoes can kill vegetation but can also cause global cooling if erupted into the stratosphere. Large volcanic eruptions can have global effects and may even have caused mass extinctions in the geologic past.

- Forecasting eruptions can now be done from satellite imagery using interferometric synthetic aperture radar (InSAR).

- Tsunamis are the most lethal natural phenomenon. They cannot be predicted, but warnings of their approach can be given. Mapping of the telltale deposits of tsunamis would allow hazard-prone areas to be recognized. Catastrophic collapse of oceanic volcanic islands has caused tsunamis in the geologic past and will undoubtedly occur in the future.

- Large meteorite impacts have resulted in catastrophic climate changes in the geologic past that have probably affected the evolution of life on the planet. Although rare, such impacts will likely occur at some time in the future.

Review questions

1. What is the most abundant element obtained from Earth materials in our body, and what function does it play there? What are some other important elements that our bodies derive from the Earth?

2. What makes a soil fertile?

3. Why is it necessary to add fertilizer to land that is heavily cultivated? What elements must be added?

4. To which mineral group does erionite belong?

5. Where in the world is there a clear connection between abundant exposure to erionite dust and mesothelioma?

6. This erionite was formed as an alteration product of what volcanic material?

7. What is meant by the adjective *asbestiform*?

8. Which six minerals are part of the federally mandated definition of asbestos?

9. What is meant by aspect ratio?

10. What are the structural differences between chrysotile and members of the amphibole group?

11. How do the chemical compositions of chrysotile and crocidolite differ?

12. Which of the six minerals in the definition of asbestos is the most carcinogenic?

13. What asbestos mineral made up about 95% of all asbestos insulation inside buildings?

14. In what country is chrysotile asbestos still being mined?

15. Long-term occupational exposure to quartz dust leads to what pulmonary disease?

16. Arsenic, even in very small concentrations in the soil, can cause serious health problems. Why is arsenic causing widespread poisoning in Southeast Asia?

17. What steps could you take to avoid the hazards posed by radon gas in your house?

18. What is the most serious problem associated with the use of nuclear power, and how might it be solved?

19. The burning of fossil fuels increases the CO_2 content of the atmosphere. Describe ways to remove this greenhouse gas from the atmosphere.

20. What types of volcanoes pose the greatest hazard? Which are more dangerous, ash falls or ash flows?

21. How can you recognize an ancient lahar?

22. How is interferometric synthetic aperture radar (InSAR) used to monitor volcanoes from satellites?

23. How might you recognize an ancient tsunami deposit?

24. What evidence found on the Cretaceous-Tertiary boundary indicates that the mass extinction at this boundary may be related to meteorite impact?

ONLINE RESOURCES

USGS health and environment

The U.S. Geological Survey has an excellent Web site devoted to human health and environment, in which topics such as acid mine drainage, mercury and selenium in soils, groundwater chemistry, and CO_2 sequestration are discussed. This Web site is found at http://energy.usgs.gov/HealthEnvironment.aspx.

Arsenic

The following U.S. EPA Web site provides information about arsenic in the environment and how it can be reduced to safe levels in drinking water: http://water.epa.gov/lawsregs/rulesregs/sdwa/arsenic/index.cfm.

This EPA Web site provides specific information concerning the removal of arsenic from drinking water:

http://www.epa.gov/safewater/arsenic/pdfs/treatments_and_costs.pdf.

Acid mine drainage

This U.S. Geological Survey Web site deals specifically with acid mine drainage: http://energy.usgs.gov/HealthEnvironment/EcosystemsHumanHealth/AcidMineDrainage.aspx/

Carbon dioxide

Concentration of CO_2 in the atmosphere: http://www.esrl.noaa.gov/gmd/ccgg/trends/

Carbon dioxide sequestration: http://energy.er.usgs.gov/health_environment/co2_sequestration/

FURTHER READING

The following references are articles that deal with health issues resulting from exposure to asbestos and other Earth material dusts discussed in Sections 17.3.1–17.3.3:

Abelson, P. H. (1990). The asbestos removal fiasco (editorial). *Science*, 247, 1017.

Alleman, J. E., and Mossman, B. T. (1977). Asbestos revised. *Scientific American*, July, 70–75.

Bish, D. L., and Ming, D. W. (2001). Natural zeolites: occurrence, properties, applications. *Reviews of Mineralogy and Geochemistry*, 45. Mineralogical Society of America, Chantilly, VA.

Fubini, B., and Fenoglio, I. (2007). Toxic potential of mineral dusts. *Elements*, 3, 407–414.

Guthrie, G. D., and Mossman, B. T., eds. (1993). Health effects of mineral dusts. *Reviews in Mineralogy*, 28. Mineralogical Society of America, Chantilly, VA.

Klein, C. (1993). Rocks, minerals, and a dusty world. *Reviews in Mineralogy*, 28, 7–59.

Mossman, B. T., and Gee, J. B. L. (1989). Asbestos-related diseases. *New England Journal of Medicine*, 320, 1721–1829.

Mossman, B. T., Bignon, J., Corn, M., Seaton, A., and Gee, J. B. L. (1990). Asbestos: scientific developments and implications for public policy. *Science*, 247, 294–301.

Nolan, R. P., Langer, A. M., Ross, M., Wicks, F. J., and Martin, R. F. (2001). The health effects of chrysotile asbestos. *Canadian Mineralogist*, Special Publication, 5.

Ross, M. (1984). A survey of asbestos-related disease in trades and mining occupations and in factory and mining communities as a means of predicting health risks of non-occupational exposure of fibrous minerals. *American Society for Testing and Materials*, Special Technical Publication, 834, 51–104.

Ross, M., Nolan, R. P., Langer, A. M., and Cooper, W. C. (1993). Health effects of mineral dusts other than asbestos. *Reviews in Mineralogy*, 28, 361–407.

Skinner, H. C. W. (2005). Biominerals. *Mineralogical Magazine*, 69, 621–641. This article gives a general discussion of minerals that form in a biological setting.

The following issues of *Elements* provide excellent surveys of topics addressed in this chapter:

Large Igneous Provinces: Origin and Environmental Consequences. *Elements* 1, no. 5 (2005).

Arsenic. *Elements* 2, no. 2 (2006).

The Nuclear Fuel Cycle: Environmental Aspects. *Elements* 2, no. 6 (2006).

Medical Mineralogy and Geochemistry. *Elements* 3, no. 6 (2007).

Supervolcanoes. *Elements* 4, no. 1 (2008).

Carbon Dioxide Sequestration. *Elements* 4, no. 5 (2008).

Mine Wastes. (2011). *Elements* 7, no. 6.

de Boer, J. Z., and Sanders, D. T. (2002). *Volcanoes in Human History*, Princeton University Press, Princeton, NJ. This book gives an excellent account of the effects nine major volcanic eruptions had on human history.

The following two articles provide an excellent example of the use of interferometric synthetic aperture radar (InSAR) to predict the eruption of a volcano.

Jónsson, S., Zebker, H., and Amelung, F. (2005). On trapdoor faulting at Sierra Negra Volcano, Galápagos. *Journal of Volcanology and Geothermal Research*, 144, 59–71.

Chadwick, W. W., Jr., Geist, D. J., Jónsson, S., et al. (2006). A volcano bursting at the seams: inflation, faulting, and eruption at Sierra Negra Volcano, Galápagos. *Geology*, 34, 1025–1028.

Glossary

Aa: A basaltic lava flow with an extremely rough surface that is broken into clinkerlike pieces of rubble.

Ablation till: Till formed from sediment distributed throughout glacial ice and released as the ice melts and ablates. Unlike lodgement till formed beneath the ice, it is not as compact.

Absolute plate velocity: The velocity of a plate relative to hot spots or mantle plumes, which provide a fixed reference frame with which to measure plate velocities relative to the deep mantle.

Absorption of light: When light passes through a mineral, certain wavelengths of light may be absorbed, which affects its color in transmitted light. This absorption may vary with crystallographic direction in anisotropic minerals, thereby resulting in pleochroism.

Absorption spectrum: This is the experimental result produced by absorption spectroscopy, which measures the absorption of radiation as a function of frequency or wavelength, as a result of its interaction with a sample. An example would be a visible and infrared absorption spectrum of a specific mineral.

Accessory mineral: A mineral present in only small amounts in a rock.

Accessory plate: A small plate inserted into the optical path of a petrographic microscope. It creates optical effects that help identify fast and slow vibration directions and the order of interference colors in anisotropic minerals.

Accretionary wedge: A wedge of highly deformed sedimentary rocks from the ocean floor, the trench slope, and older forearc basins that are accreted onto a continent at a convergent plate boundary.

Acicular: A mineral with a needlelike shape.

Acid igneous rocks: A general term indicating igneous rocks that are silica rich.

Acid mine drainage (AMD): The slow dissolution of sulfides in waste dumps (generated by mining) and the subsequent production of acidic sulfate- and metal-rich waters.

Activation energy: The energy needed to initiate a reaction but which is returned to the system once the reaction begins.

Activity: The activity of a chemical component in a solution is its effective mole fraction. If the solution is ideal, as in the case of ideal gases, the activity is the same as the mole fraction. For nonideal solutions, it can differ significantly.

Adamantine luster: A bright, sparkly luster as seen in diamonds.

Adiabatic: A process in which no heat is lost or gained.

Agate: A type of chalcedony that exhibits alternating layers or bands of different thickness or color that tend to parallel the walls of the rock cavity in which it was deposited.

Aggregate: In the construction industry, it is a loose mixture of sand, gravel, pebbles, or crushed stone. It can be used by itself as road and railway ballast, or to stabilize slopes. It is also mixed with approximately one-third Portland cement to form concrete.

A horizon: The uppermost soil horizon.

Albite-epidote hornfels facies: A set of metamorphic mineral assemblages characterized by the coexistence of albite, epidote, and actinolite, formed at low temperatures and pressures in the outer part of contact metamorphic aureoles.

Albite twin: A twin type that is common in and highly diagnostic of triclinic feldspars. It is a polysynthetic twin with the twin plane parallel to {010}.

Alkali olivine basalt (AOB) series: A subdivision of the alkaline igneous rock series.

Alkaline igneous rocks: One of the major divisions of igneous rocks. They tend to be rich in alkalis and most do not have sufficient silica to crystallize quartz but instead crystallize nepheline or other feldspathoids.

Allochem: Primary carbonate sedimentary particles that form the framework of a limestone. They may include, for example, shell fragments, ooids, and pellets.

Alluvial deposit: Sediment that is deposited by water in a nonmarine environment.

Alluvial fan: A flat cone-shaped deposit of sediment formed where a stream issues forth from mountains into a valley. They are common in arid environments. The grain size of the sediment can vary from extremely coarse near the apex to fine near the distal ends of the fan.

Amorphous: A solid that lacks an ordered crystalline structure.

Amphibole: A major rock-forming silicate group with infinitely extending double tetrahedral chains. Hydrous with Ca, Mg, Fe, Na, Al, and Si as major constituents.

Amphibolite: A metamorphic rock that is composed mainly of amphibole and plagioclase.

Amphibolite facies: A set of metamorphic mineral assemblages that indicate regional metamorphic conditions of moderate to high pressures with temperature in a range of 450–750°C. Metabasalt in this grade contains abundant amphibole, hence the name of the facies.

Amygdale (diminutive is **amygdule**): A gas bubble or vesicle in a volcanic rock that is later filled with minerals deposited from hydrothermal solutions. Quartz, calcite, and zeolites are common fillings.

Andesite: A volcanic rock that characterizes composite volcanoes formed along island arcs above subduction zones. It is composed of approximately equal proportions of pyroxene and plagioclase, whose average composition is less than An_{50}, although more calcic compositions may occur in the cores of phenocrysts.

Anhedral: A mineral grain that lacks well-formed crystal faces as a result of having grown against adjacent minerals.

Anion: An ion with a negative charge.

Anisodesmic: Describes a crystal structure in which the bond strengths are unequal.

Anisotropic: In optics, it indicates that the refractive index of a mineral varies with crystallographic direction.

Anorthosite: An igneous rock composed of more than 90% plagioclase.

Anthracite: The highest-rank coal, which is slightly metamorphosed.

Apex: In geometry, it is the highest or most distant point in a triangle, a tetrahedron, a pyramid, and so on.

Arborescent: Said of an aggregate of the same mineral grains in a treelike pattern (also known as dendritic).

Arenite: A sedimentary rock composed of sand-size grains (0.0625–2.0 mm).

Arkose: See feldspathic arenite.

Asbestiform: Said of a highly fibrous mineral in which the fibers readily separate into thin, strong fibers that are flexible.

Asbestos: Includes minerals that are highly fibrous in which the fibers readily separate into thin, strong fibers that are flexible. Also, as defined by National Institute for Occupational Safety and Health (NIOSH), includes six naturally occurring fibrous minerals: chrysotile, crocidolite, amosite, anthophyllite, tremolite, and actinolite.

Ash: See volcanic ash.

Ash fall: A deposit formed from volcanic ash that is ejected high into the air and is commonly carried by prevailing winds some distance from its source. The deposits tend to blanket the countryside.

Ash flow (nuée ardente): A suspension of hot particles of volcanic ash that flows as a dense highly fluid mass to topographic lows, where it ponds and cools slowly, often welding itself together to form a welded ash-flow tuff or ignimbrite.

Aspect ratio: The length to width ratio of a mineral grain or fiber.

Assimilation: The process by which magma is contaminated by including fragments of foreign rock (xenoliths) and partially melting or reacting with them.

Association (mineral): A group of minerals found together in a rock.

Asterism: An optical phenomenon that results from light reflected by minute inclusions, inside a mineral, in a starlike (six-rayed) pattern. Best seen in star rubies and sapphires.

Asthenosphere: A layer in the upper mantle extending from depths between 20 km and 50 km beneath oceans and 70 km and 220 km beneath continents that is marked by low seismic velocities. Compared with the lithosphere above, this layer is weak and provides the zone on which lithospheric plates can move.

Atom: A basic unit of matter that consists of a dense, central nucleus surrounded by a cloud of negatively charged electrons.

Atomic radius: The radius of an atom, expressed in Angstroms or nanometers, where 1 nanometer = 10 Å.

Augen gneiss: A gneiss containing eye-shaped feldspar crystals that have been deformed by shear and surrounded by finer-grained recrystallized feldspar.

Authigenic: Formed or generated in place. Said of a mineral that formed in the spot where it is now found and that came into existence at the same time as, or subsequent to, the formation of the rock of which it is a constituent.

Axis (crystallographic): One of three imaginary lines in a crystal (four in the hexagonal system) that pass through its center.

Azimuthal quantum number: This is symbolized by l and represents a quantum number for an atomic orbital and determines its angular momentum and describes its shape.

Back-arc basin: Often, but not always, coupling between subducting and overriding tectonic plates at a convergent plate boundary creates extensional stresses in the upper plate that results in formation of a rift valley, or even opening of new ocean floor.

Banded: Applies to a mineral or rock specimen that shows bands of different color, texture, or mineral makeup.

Banded iron-formation (BIF): A prominently layered sedimentary rock of Precambrian age consisting of alternating silica- and iron-rich beds. The silica-rich beds consist of chert, quartz, or jasper, and the iron-rich beds contain magnetite, hematite, siderite, ankerite, or hydrous iron silicates.

Bar diagram: A graphical illustration of the extent of solid solution between two end-member compositions.

Basal pinacoid: An open crystal form consisting of two parallel faces that, in this instance, cut only the vertical crystallographic axis.

Basalt: A fine-grained igneous rock composed predominantly of equal proportions of plagioclase feldspar and pyroxene (or olivine). It is the most common igneous rock and forms most of the ocean floor.

Basic igneous rocks: A general term covering rocks of roughly basaltic composition.

Batholith: An extremely large intrusive igneous body with an outcrop area exceeding 100 km². Usually composed of granite and granodiorite.

Becke line: A line of bright light near the boundary between materials of different refractive index caused by the refracting of transmitted light toward the mineral of higher refractive index.

Bed (sedimentary): The main sedimentary unit that produces layering at the outcrop scale.

Bedforms: Constructional features formed by the bedload in a river channel, such as ripples and dunes.

Bedload: The sediment that is carried at the bottom of a river and is moved by rolling or saltation.

Benioff seismic zone: The zone of earthquakes that are generated in a subducting plate.

B horizon: A soil horizon below the A horizon.

Big Bang: The term used in cosmology to describe the beginning of the universe ~14 billion years ago, when it suddenly started expanding from an extremely dense state. That expansion continues today.

Binary: A chemical system (or solid solution series) that can be described by two end members.

Biogenic sediment: Sediment derived from the remains of organisms.

Biopersistence: The time of residence of a particulate material in the body, such as that of an asbestos fiber in the lung environment.

Bioturbation: Any process whereby organisms disturb sediment after it has been deposited.

Birefringence: The difference between the maximum and minimum refractive indexes in an anisotropic mineral.

Bituminous coal: An intermediate-rank coal.

Black shale: An organic-rich shale deposited under anoxic conditions such as occur in deep lakes and marine basins.

Black smoker: A hydrothermal vent on the ocean floor around which a chimney is built by the precipitation of sulfides, which are leached from the cooling igneous rocks by circulating ocean water.

Bladed: Crystals with the shape of blades, as in a knife blade.

Blocking temperature: The temperature at which radioactive parent and daughter isotopes are no longer able to diffuse through a rock. An absolute age determination is the age when a rock passed below the blocking temperature.

Blocky: A term used to describe the rubbly surface of many andesite lava flows. The blocks are coarser (20–40 cm) and smoother than the clinkerlike particles in aa flows.

Blue asbestos: A trade term used for crocidolite, the asbestiform variety of riebeckite.

Blueschist: A schistose metamorphic rock with a bluish color owing to the presence of glaucophane, an Na-rich amphibole.

Blueschist facies: A set of metamorphic mineral assemblages characterized by the mineral glaucophane, formed at low temperatures but high pressures.

Body centered: Describes a unit-cell choice in which there are eight corner nodes as well an additional node in the center. The node count of such a cell is 2. Abbreviated by *I*.

Body diagonal: An imaginary line passing between the opposite corners of a cube.

Bohr model: A model of the atom, developed by Niels Bohr, in which a positively charged nucleus is surrounded by electrons that travel in circular orbits about the nucleus.

Bomb (volcanic): Particle ejected from a volcano with a diameter greater than 64 mm.

Bort: A granular to very finely crystalline aggregate of imperfectly crystallized diamonds used as an abrasive and in other industrial applications.

Boudinage: A deformation structure in which competent layers of rock neck down and pull apart to form lenses in less competent (more plastic) surrounding rock.

Boundstone: A limestone in which the original carbonate grains were bound together at the time of their formation, as in a coral reef.

Botryoidal: Describes a mineral specimen that has the outward appearance of a bunch of grapes.

Bragg equation: A mathematical equation that expresses the relationship between the angle of diffraction (θ), X-ray wavelength (λ), and atomic spacing (d), as in $n\lambda = 2d \sin \theta$.

Bravais lattices: The 14 different ways in which nodes (or atoms) can be arranged periodically in three dimensions.

Breccia: A rock composed of angular rock fragments. It can be of sedimentary, igneous, tectonic, or meteorite impact origin.

Brick: A block of construction material that is formed from a mixture of clay and sand, which normally has been fired at a high temperature to produce a strong material consisting of a number of aluminum silicates.

Bridging oxygen: An oxygen ion or atom that links two tetrahedral structural units.

Brittle: A term that applies to minerals that break or shatter easily when struck with an object.

Brown asbestos: A trade term used for amosite, the asbestiform variety of grunerite.

Buoyancy: The force that causes material to rise in a gravitational field as a result of its density contrast with the surroundings.

Cabochon: Refers to gems fashioned with curved surfaces.

Calcalkaline igneous rocks: Subalkaline igneous rocks that tend to contain less iron and more alumina than tholeiitic rocks and form a series of volcanic rocks ranging from high-alumina basalt to andesite, dacite, and rhyolite and their plutonic equivalents. They are characteristic of igneous rocks associated with convergent plate boundaries.

Calcsilicate: A metamorphic rock composed of calcium aluminum silicates formed from the metamorphism of impure limestone.

Caldera: A large, roughly circular part of the Earth's surface that collapses into a near-surface magma chamber.

Caliche (hardpan): A deposit in soil formed by the precipitation of calcite from groundwater as it evaporates in arid and semiarid climates.

Calorie: The amount of heat required to raise 1 gram of water 1°C. It is equivalent to 4.184 joules in SI units.

Cannel coal: A massive coal with conchoidal fracture and greasy luster.

Capillary: Said of crystals with a hairlike or threadlike appearance.

Cap rock: An impervious rock, such as shale, that prevents oil or natural gas from escaping from a reservoir rock beneath.

Carat: A unit of mass equal to 200 mg that is used to measure gemstones and pearls. For example, a 10 carat (ct) diamond weighs 2 grams.

Carbon sequestration: The process by which the greenhouse gas CO_2 is removed from the atmosphere and stored in the ground, as in the reservoirs from which oil and gas have previously been extracted.

Carbonate: A mineral in which the anionic group is $(CO_3)^{2-}$ as in calcite, $CaCO_3$.

Carbonate compensation depth: The depth in the ocean below which carbonate sediment becomes soluble and therefore cannot accumulate.

Carbonation reaction: A reaction with CO_2 to produce a carbonate mineral. High levels of CO_2 in the atmosphere promote such reactions.

Carbonatite: An igneous rock composed primarily of calcite.

Carlsbad twin: A twin law in feldspar, especially orthoclase, resulting in a penetration twin, on account of rotation being the twin operator.

Cation: An ion with a positive charge.

Cation exchange: The exchange of cations between a solution and a solid (mineral).

Cement: In a sedimentary rock, any mineral that cements together the original sedimentary grains.

Cement (Portland): See *Portland cement*.

Cementation: The process by which sedimentary grains are cemented together to form a solid rock.

Center of symmetry (also known as inversion, *i*): This exists if every face on a crystal has an equivalent identical face on the opposite side.

C-face centered: Describes a unit-cell choice in which the top and bottom faces of the unit cell carry nodes, in addition to the eight corner nodes. The node content of such a cell is 2. Abbreviated by *C*.

Chain silicate: A silicate structure type with infinitely extending tetrahedral chains. Also known as inosilicate.

Chalcocite blanket: A zone of supergene enrichment in a mineral or ore deposit in which chalcocite is a major constituent.

Chalcophile: An adjective describing those elements, such as copper, that prefer to bond with sulfur rather than silica.

Chalk: A carbonate mudstone (usually white) that under high magnification can be seen to contain small spherical particles known as coccoliths formed from the hard parts of eukaryote organisms.

Chatoyancy: An optical phenomenon, seen in some minerals in reflected light, in which a movable wavy or silky sheen is concentrated in a narrow band of light that changes its position as the mineral is turned.

Chemical potential: The partial molar free energy of a chemical component in a system. When multiplied by its number of moles, it gives that component's contribution to the total Gibbs free energy of the system. At equilibrium, the chemical potential of a component must be the same in all phases.

Chemical precipitate: A mineral that is the result of chemical precipitation from an aqueous solution.

Chemical sediment: Sediment that is formed by the chemical precipitation of minerals from water.

Chert (flint): A sedimentary rock composed primarily of extremely fine-grained quartz. *Flint* is synonymous with *chert* but is used more commonly in archeological literature.

Chilled margin: The margin of an igneous body that cools rapidly and consequently is extremely fine grained.

China clay: A commercial term for kaolin that is obtained from china clay rock after washing and is suitable for the use in the manufacture of chinaware.

Chondrite: A type of stony meteorite (in contrast to an iron meteorite) that contains small spherical bodies known as chondrules.

Chondrule: A small millimeter-size sphere that characterizes chondritic meteorites. The spheres are composed of extremely fine-grained grains of magnesium-iron silicates (olivine and pyroxene) and glass. They were formed by flash heating and rapid cooling of dust particles in the early solar nebula.

C horizon: A mineral horizon of soil, beneath the A and/or B horizons, consisting mainly of unconsolidated rock material.

Chromated copper arsenate (CCA): A chemical compound used in preserving pressure treated lumber. Its use in North America and Europe for residential construction is banned because of the toxic effects of arsenic.

Chromitite: A plutonic igneous rock composed primarily of chromite.

Chromophore elements: Chemical elements that are responsible for the color of many rock-forming minerals.

CIPW norm: See *norm*.

Clastic: Refers to a rock or sediment composed principally of broken fragments that are derived from preexisting rocks or minerals.

Clast supported: An adjective indicating that sedimentary detrital particles, such as pebbles in a conglomerate, touch one another and form an interconnected framework. Contrast this with *matrix supported*.

Clay: A sedimentary particle with a grain size less than 0.0039 mm.

Clay mineral: A term used for a small number of minerals that are hydrous aluminum silicates with layered structures.

Claystone: A siliciclastic sedimentary rock composed of particles with a grain size of less than 0.0039 mm.

Cleavage: The breaking of a mineral along crystallographic planes.

Clinoamphibole: A subgroup of the amphiboles with monoclinic symmetry.

Clinographic projection: An oblique projection used for the representation of crystals so as to arrive at a portraitlike image.

Clinopyroxene: A subgroup of the pyroxenes with monoclinic symmetry.

Closed form: A crystal form that encloses space, such as a cube in the isometric system.

Closest packing: A type of crystal structure that provides the tightest possible packing, which results in a central atom having 12 closest neighbors.

Coal: A sedimentary rock composed predominantly of fossilized plant remains.

Cobble: A sedimentary particle that has a grain size between 64 mm and 256 mm.

Colonnade: See *columnar joints*.

Color: The response of the eye to electromagnetic radiation. The human eye responds to a limited range of wavelengths, from about 3,500 to 7,500 Å.

Color center: A defect in a crystal structure that is the cause of color.

Color index: The modal abundance of dark minerals in a rock.

Columnar: A mineral occurrence in which the crystals are rounded columns, commonly in parallel arrangement.

Columnar joints: Regular five- to six-sided joint-bounded columns formed by the shrinkage of lava as it cools. These joints propagate up from the bottom and down from the top of a lava flow. The columns in the lower part of the flow are extremely regular and are known by the Greek architectural term *colonnade*. Those that propagate down from the top are smaller and less regular and form what is called the *entablature*.

Common opal: A milky-white variety of opal without internal reflections (no play of color).

Compact: Describes a mineral specimen that is so fine-grained that individual particles cannot be recognized by the unaided eye.

Compaction: The process by which solids are gravitationally compacted together while interstitial liquid is expelled upward. Compaction takes place when sediment is buried and when crystal mush is solidifying in a magma chamber.

Complete order: This is said of a crystal structure when the location of specific atoms (or ions) is in specific atomic sites of the structure.

Component: In thermodynamics, this refers to the smallest number of chemical components needed to make all the phases present in a system.

Composite volcano: A symmetrical conical volcano whose slopes become steeper toward the summit crater. It is composed of alternating layers of lava and ash, which is usually andesitic. Also known as a strato-volcano, because of the stratified alternation of lava and ash.

Conchoidal fracture: The fracturing of a mineral along curved surfaces, as seen in the breakage of glass.

Concordant: A term indicating that an intrusive igneous body parallels the layering in the rocks it intrudes.

Concrete: A mixture of approximately two-thirds crushed stone or gravel and one-third Portland cement, which, when mixed with water, hardens to form a strong building material especially when reinforced with steel bars (rebar).

Conduction: See *thermal conduction*.

Cone sheet: A thin conical sheetlike discordant intrusive igneous body.

Conglomerate: A sedimentary rock composed of gravel-size particles.

Congruent melting: Melting of a solid to form a liquid of its own composition.

Contact metamorphism: Metamorphism caused by heat from a cooling body of magma.

Contact twin: A twin with a regular composition surface (the twin plane) separating two identical crystals.

Continental flood basalt: A type of basalt, usually of tholeiitic composition, that erupts onto continents in large volumes from long fissures that are usually related to plate divergence or doming over a mantle plume. Repeated eruptions of these basalts build up thick sequences of flows that fill in and flood the previous topography.

Continuous reactions: Reactions that take place over a range of temperatures between minerals or liquids that form continuous solutions.

Convection: See *thermal convection*.

Convergent plate boundary: Where two tectonic plates converge, with one subducting beneath the other.

Coordination number (C.N.): The number of closest neighbors that surround a central atom or ion.

Coordination polyhedron: The geometric shape outlined by anions surrounding a central cation.

Coordination principle: One of Linus Pauling's five rules. It states that a cation's coordination number is determined by the ratio of the radius of the cation to that of the anion.

Coquina: A loosely cemented fragmental shelly limestone.

Coral reef: A complex carbonate structure built from corals and many other organisms in, and just beneath, the surf zone near beaches in warm, clear marine waters.

Core formation: The process by which the Earth's iron-rich core was formed in the first 30 million years after accretion of the planet.

Coupled substitution: The simultaneous substitution of two or more different ions in a crystal structure so as to maintain electrical neutrality, also known as charge balance. This occurs in albite, $NaAlSi_3O_8$, when $Ca^{2+}Al^{3+}$ substitute for Na^+Si^{4+}.

Covalent bond: A chemical bond in which electrons are shared between adjoining atoms, or ions.

Crater: A depression on the Earth's surface formed by the explosive venting from a volcano or the impact and explosion of a meteorite.

Crenulation schistosity: A second and younger plane of schistosity that cuts and crenulates the earlier schistosity.

Crested barite: Barite in divergent groups of tabular crystals

Cross-bedding: Bedding that is at a significant angle to the main bedding in a sedimentary rock. It can be formed, for example, by erosion and deposition in stream channels, deltas, and sand dunes, and by the swash and backwash of waves on a beach.

Crushed stone: A type of construction aggregate produced by crushing certain rock types into fragmental pieces.

Crust: The outer part of the Earth above the Mohorovičić discontinuity. It is typically about 30 km thick on continents but can be up to 70 km thick beneath mountain chains. Beneath oceans it is 7–10 km thick.

Cryoscopic equation: An equation that relates the lowering of the melting point of a substance to its dilution in a solution.

Crystal: A mineral, or other crystalline chemical compound, with an external shape bounded by smooth plane surfaces (crystal faces).

Crystal chemistry: The science that relates the chemical composition, the internal structure, and the physical properties of crystalline materials.

Crystal class (same as point group): One of 32 symmetry groups to which a crystal is assigned on the basis of its translation-free symmetry content.

Crystal field transitions: Interactions between the energy of white light and the *d* orbitals of certain elements known as chromophores.

Crystalline: Having an ordered internal crystal structure. Applies to well-developed (euhedral) crystals as well as poorly formed (anhedral) mineral grains.

Crystalloblastic series: A series indicating the ease with which metamorphic minerals develop crystal faces. A mineral in the series tends to develop euhedral crystal faces against any mineral below it in the series.

Crystal plane: An external face of a crystal and/or a planar direction inside a crystal structure.

Crystal structure: The ordered arrangement of chemical elements into a specific atomic architecture. A crystal structure provides information on the location of all the atoms, bond positions and bond types, space group symmetry, and the chemical content and size of the unit cell.

Crystal system: One of six different geometric coordinate systems to which all crystals, and their atomic structures, are assigned: isometric, hexagonal, tetragonal, orthorhombic, monoclinic, and triclinic with a rhombohedral subdivision in the hexagonal system.

Crystal zone: This is a collection of crystal faces with parallel edges. The direction of such a zone is expressed by a Miller index inside square brackets, such as [010].

Crystallographic axes: A set of axes that is compatible with each of the six crystal systems, or point groups.

Crystallography: The science that deals mainly with the arrangement of atoms inside crystal structures. Before the development of X-ray diffraction techniques, crystallography dealt mainly with the external geometry of crystals.

Cube: A form in the isometric system with six crystal faces.

Cubic cleavage: Breaking of a mineral along three different planar directions at 90° to each other.

Cubic coordination: Describes a cation with eight closest anion neighbors in the geometric shape of a cube.

Cubic closest packing (CCP): The packing of atoms of the same size (as spheres) in a three-dimensional array with cubic symmetry. Results in an ABCABC sequence.

Cumulate: An igneous rock formed by the accumulation of early crystallizing minerals from a magma.

Cyclic twin: A repeated twin of three or more individual crystals in which the twin axes or twin planes are not parallel.

Cyclosilicate: A silicate crystal structure in which there are 4- or 6-fold tetrahedral rings. Also known as ring silicate.

Cyclothem: A sequence of sedimentary rocks in which coal is underlain by nonmarine siliciclastic mudrocks and overlain by shallow marine sandstones and conglomerates. The sequence commonly repeats, which indicates that sea level was in a state of flux.

D″ (D double prime) layer: A 100–300 km-thick layer immediately above the core-mantle boundary. It may be composed of subducted tectonic plates.

Dacite: A volcanic rock (commonly partially glassy) that has the same composition as the plutonic rock granodiorite.

Darcy's law: The rate at which a fluid flows through a porous material is proportional to the pressure gradient.

Debris flow: A sedimentary deposit formed by the flow of a mixture of particles ranging in size from mud to boulders, which may move as slowly as 1 m/year or as fast as 1 km/hour, if it becomes water saturated.

Defect: An atomic flaw in an otherwise well-ordered crystal structure.

Deformation twinning: Twinning as a result of deformation.

Degrees of freedom (variance): In thermodynamics, it indicates the number of variables that can be independently changed without changing the phases that are present.

Delamination: A process whereby the base of the crust sinks into the mantle possibly as a result of the development of dense metamorphic minerals.

Delta: A deposit of sediment formed where rivers enter standing bodies of water. They fan out from the river, with a gently sloping upper surface and steep outer face.

Dendritic: A term describing the occurrence of a mineral (or crystal) aggregate in a treelike branching pattern.

Density: The density of a material is defined as its mass per unit volume. The most common units used are kilogram per cubic meter (kg/m^3) or gram per cubic centimeter (g/cm^3).

Desilication reaction: A reaction that removes silica from magma, as happens, for example, when limestone xenoliths react with magma.

Detrital: A term that applies to any particle resulting from the disintegration (weathering) of a preexisting rock.

Detritus: Sediment formed from detrital grains.

Diabase: A dike rock of basaltic composition. Referred to as dolerite in Britain.

Diagenesis: Chemical, physical, and biological changes that affect sediment or sedimentary rock after initial deposition but excluding weathering or metamorphism.

Diagenetic: Pertaining to or caused by diagenesis.

Diagnostic properties: A combination of physical properties that may lead to the identification of a mineral or rock.

Diapir: A large dome-shaped body that buoyantly rises toward the Earth's surface. It can be composed of low-density sedimentary rock, such as salt, hot mantle, or magma in the crust.

Diatreme: A carrot- or pipelike body of broken rock fragments formed by the escape of volcanic gases. It usually has no igneous matrix. Fragments may come from as deep as the upper mantle and may contain diamonds.

Differentiation: See *magmatic differentiation.*

Diffraction: This refers to various phenomena that occur when a wave encounters an obstacle. Diffraction occurs with all waves, such as sound waves, water waves, and electromagnetic waves such as visible light, X-rays, and radio waves.

Dihexagonal dipyramid: A closed crystal form consisting of 12 faces at the top and 12 identical faces at the bottom, related by a horizontal mirror. The general form in 6/*m* 2/*m* 2/*m*.

Dihexagonal prism: A prism that consists of 12 vertical faces instead of 6.

Dike (dyke): An igneous sheetlike body that cuts across the layering in the surrounding rocks. Most dikes have steep dips.

Dike swarm: A large number of subparallel or radial dikes. Swarms are common at divergent plate boundaries or over mantle plumes.

Dioctahedral: Refers to a layer silicate (phyllosilicate) structure in which only two of three available octahedral sites are occupied.

Diorite: A plutonic igneous rock composed predominantly of pyroxene and plagioclase whose average composition is less calcic than An_{50}. Andesite is its volcanic equivalent.

Dipyramid: A closed crystal form consisting of 6, 8, 12, 16, or 24 faces. It is made up of two pyramids related by a horizontal mirror plane.

Discontinuous reactions: Reactions that take place at a given temperature, as, for example, at a peritectic.

Discordant: A term indicating that an intrusive igneous body cuts across the layering in the rocks it intrudes.

Disequilibrium: In thermodynamics, a term indicating that a system is not at equilibrium (see *equilibrium*).

Disilicate: A silicate crystal structure in which there are double tetrahedral groups of $(Si_2O_7)^{4-}$ composition. Also known as sorosilicate.

Displacive polymorphism: In this type of polymorphism, the overall structure is left completely intact and no bonds between atoms are broken. There is only a sight displacement of the atoms (or ions) and some readjustment of bond angles between the atoms.

Ditetragonal dipyramid: A closed form consisting of 16 triangular faces, of which 8 are at the top and 8 at the bottom. This is the general form in 4/*m* 2/*m* 2/*m*.

Ditetragonal prism: A form consisting of eight rectangular vertical faces, each of which intersects the two horizontal axes unequally.

Divergent plate boundary: A boundary at which tectonic plates move apart and new crust is created between them, as occurs at a midocean ridge.

Dodecahedron: A closed form in the isometric system with 12 crystal faces.

Dolerite: See *diabase.*

Dolomitization: The replacement process whereby half of the Ca^{2+} in the calcite structure is replaced by Mg^{2+}, thus converting calcite into dolomite.

Dolostone: A sedimentary rock composed of the mineral dolomite. Most dolostones are formed by replacement of limestone.

Dome (volcanic): A dome-shaped body of silica-rich volcanic rock that is formed by eruption of lava that is so viscous that it is unable to flow far from the vent.

Drilling mud: A carefully formulated heavy suspension, usually in water but sometimes in oil, used in rotary drilling. It commonly consists of bentonitic clays, chemical additives, and weighting materials such as barite.

Ductile: Said of a mineral or rock that can sustain deformation before fracturing or faulting.

Dune: A moundlike sedimentary structure formed by flowing water or wind, in which sand particles are eroded from the upstream side and carried up and over the structure and deposited on the downstream side.

Dunite: A plutonic igneous rock composed primarily of olivine.

Eclogite: A high-pressure metamorphic rock containing jadeite and pyrope-rich garnet.

Eclogite facies: A set of metamorphic mineral assemblages characterized by the presence of jadeite and pyrope-rich garnet, formed at pressures in excess of 1 GPa.

Edge diagonal: An imaginary line passing between the centers of opposing edges of a cube.

Effervescence: The reaction that releases CO_2 gas bubbles when calcite reacts with hydrochloric acid.

Electromagnetic spectrum: This is the range of all possible frequencies of electromagnetic radiation. It ranges from gamma rays through X-rays, ultraviolet, visible, infrared and microwaves, to radio waves. This sequence represents a range of wavelengths from 0.1 Å to 1000 m.

Electron: A subatomic particle with a negative electric charge. It has a mass that is 1/1836 that of a proton.

Electron microprobe analysis (EMPA): A chemical analytical instrument that deploys a finely focused electron beam that generates characteristic X-ray emissions from the chemical elements housed in the material under the beam.

Electrostatic valency principle: One of Linus Pauling's five rules. It states that the strength of an ionic bond is equal to the ionic charge divided by the coordination number.

Electrum: A naturally occurring alloy of gold and silver, containing more than 20% silver.

Element: A chemical substance consisting of atoms of the same atomic number that cannot be decomposed into simpler materials by chemical reactions.

Emery: A gray to black, granular rock comprised of corundum and magnetite with some additional trace impurities, used as an abrasive.

Enantiomorphous: Said of a pair of screw axes in which the screw operations are in opposite directions, such as in 3_1 and 3_2, or 4_1 and 4_3.

End member: The chemical formula of the mineral that is at the end of a solid solution series. For example, in the olivine series, $(Mg, Fe)_2SiO_4$, Mg_2SiO_4 is the forsterite end member on the Mg side of the series.

Entablature: See *columnar joints*.

Epeiric sea: A shallow sea formed by the flooding of a continent during periods of high sea-level stand.

Equilibrium: In thermodynamics, indicates that no further change or reaction is needed because the temperature, pressure, and chemical potential are everywhere the same throughout the system, and the Gibbs free energy is at a minimum.

Esker: A stratified meltwater deposit formed in a river channel flowing in a tunnel in glacial ice.

Euhedral: A crystal bounded by well-developed crystal faces.

Eutectic: The lowest possible melting mixture of substances that do not form a complete solid solution.

Eutectic intergrowth: The intergrowth of mineral grains that crystallize together at a eutectic.

Evaporite: A nonclastic sedimentary rock composed primarily of minerals that resulted from the extensive or total evaporation of a saline liquid.

Exsolution: The process by which an originally homogeneous solid solution separates into two (or sometimes more) distinct crystalline materials. These are commonly oriented along specific crystallographic directions in the host mineral.

Exsolution lamellae: A microstructure inside minerals that consists of a regular pattern of closely spaced, parallel planar lamellae, with the lamellae having a chemical composition different from that of the host mineral.

Extinction: In optics, it occurs when the principle vibration directions in an anisotropic mineral parallel the polarization planes in the microscope. A mineral in extinction appears totally black under crossed polarized light.

Extinction angle: The angle measured between an extinction position and a prominent crystallographic direction, such as the length of a crystal or a cleavage direction.

Extrusive: Indicates that an igneous body or igneous rock is formed by eruption of magma onto the Earth's surface.

Face centered: Describes a unit-cell choice in which there are nodes at all eight corners of the unit cell as well as in the center of each of the six faces. The node content of such a cell is 4. Abbreviated by *F*.

Face pole: A direction perpendicular to a crystal face.

Facet: A polished face on a gemstone.

Facies: See *metamorphic facies*.

Failed rift: One of the divergent plate boundaries at a triple junction that widens very slowly or ceases to widen at all.

Fancy color (fancy diamond): A diamond with a strong natural body color so as to be attractive rather than off color.

Feldspar: A major rock-forming silicate mineral group with compositions rich in K, Na, Ca, Al, and Si.

Feldspathic arenite (arkose): Sandstone containing more than 25% feldspar (usually alkali feldspar).

Feldspathoid: A comparatively rare rock-forming silicate mineral group rich in K, Na, Ca, Al, and Si but with less silica than feldspar.

Fiamme: A term describing the appearance of flattened pumice fragments in welded ash-flow tuffs where the elongate bubbles give the appearance of flames (from the Italian word *fiamme* meaning "flames").

Fiber: A particle with a length-to-width ratio of at least 3:1.

Fibrous: Crystals that form a fibrous mass, as in asbestos.

Fire: Flashes of various spectral colors seen in diamond and other gemstones.

First-order red: An interference color produced by two light waves that are out of phase by 550 nm.

Fissility: The property of shale to split into thin sheets.

Flint: See *chert*.

Flood basalt: A type of basalt that is erupted in large volumes from long fissures and tends to form large flat lava flows. They are commonly associated with large igneous provinces.

Fluorescence: The emission of visible light, by some minerals, in response to exposure to ultraviolet (UV) light.

Fluorspar: Commercial name for fluorite.

Foliated: A mineral (or crystal) that consists of a stack of thin leaves or plates that can be separated from each other.

Foliation: The property of a rock to break into thin sheets that are bounded by planes along which typically platy minerals are aligned. This alignment may result from sedimentation of clay minerals or from the growth of micas in a preferred orientation in metamorphic rocks.

Forearc basin: A basin formed between an oceanic trench and a volcanic arc at a convergent plate boundary.

Form (in crystals): A form consists of a group of like crystal faces, all of which have the same relation to the inherent symmetry elements.

Fossil fuel: Any fuel derived from rocks that are formed from fossil organisms (e.g., coal, oil, natural gas).

Fourier's law: The rate at which heat is transferred by conduction through a material is proportional to the temperature gradient.

Framework: A term applied to silicates with crystal structures that are infinitely extending tetrahedral frameworks. Also known as tectosilicate.

Gabbro: A plutonic igneous rock composed of approximately equal amounts of pyroxene (and olivine) and plagioclase whose average composition is more calcic than An_{50}. Its volcanic equivalent is basalt.

Gangue: The valueless rock or mineral aggregates in an ore that are not economically desirable but cannot be avoided in mining.

GARB: A geothermometer based on the compositions of coexisting garnet and biotite.

Garbenschiefer: A metamorphic term indicating sheaflike clusters of elongate mineral grains such as amphibole (see cover of book).

Garnet: A major rock-forming silicate mineral group with extensive solid solution among individual species of widely varying compositions.

Gas giant planets: The large outer planets in the solar system composed predominantly of condensed gases. They include Jupiter, Saturn, Uranus, and Neptune.

Gemstone: A term applied to those minerals that usually, after they are cut and polished, are sufficiently attractive and durable to be used for personal adornment. Also referred to as a gem.

General form: A form of a crystal in which the faces intersect all of the crystallographic axes at different lengths. This is expressed in Miller index notation as $(h\,k\,l)$ or $(h\,k\,\bar{i}\,l)$.

General Miller index notation: When a crystal face appears to be in a pretty general position with respect to the appropriate crystal axes, it is given a nonspecific Miller index such as $(h\,k\,l)$ or $(h\,k\,\bar{i}\,l)$.

Geobarometer: A mineral or mineral assemblage, where the compositions of the minerals are sensitive to changing pressure and can be used to determine the pressure at which a rock formed.

Geode: A hollow or partially filled rock cavity commonly lined with crystals.

Geotherm: The measured or calculated temperature as a function of depth in the Earth.

Geothermal energy: Heat extracted from inside the Earth. The heat can be extracted from hydrothermal systems usually located in active volcanic regions. It can also be extracted from hot, dry rock that has been hydraulically fractured by circulating water through injection and recovery wells.

Geothermal gradient: The rate of temperature rise with depth in the Earth; typically about 25°C/km.

Geothermometer: A mineral or mineral assemblage, where the compositions of the minerals are sensitive to changing temperature and can be used to determine the temperature at which a rock formed.

Gibbs free energy (G): The energy that determines the direction of a chemical reaction. Reactions proceed in a direction that lowers the Gibbs free energy, and equilibrium is achieved when it is minimized.

Gibbs phase rule: A rule that relates the number of phases (ϕ) to the number of components (c) and number of variables that can be changed (f) at equilibrium. The relation is $f = c + 2 - \phi$. This rule, for example, sets a limit on the number of minerals that could form in a metamorphic rock.

Glaucophane schist facies: A set of metamorphic mineral assemblages in which basic rocks contain glaucophane, reflecting high-pressure and low-temperature conditions.

Glide plane: A mirror plane combined with translation.

Gneiss: A metamorphic rock in which prominent layers are produced by variations in the abundance of minerals. Often platy minerals, such as micas, alternate with more granular minerals, such as quartz and feldspar.

Gouge: A fine-grained material in fault zones formed by the grinding of rocks together along the fault.

Gossan: An iron-bearing weathered product overlying a sulfide deposit. It is formed by the oxidation of sulfides and the leaching of sulfur and most metals.

Graded layer: A layer in which mineral grains gradually change size and or density from bottom to top of the layer. In sedimentary rocks, graded beds result from settling of grains in turbidity currents.

GRAIL: A geobarometers based on the compositions of coexisting garnet, rutile, Al_2SiO_5 polymorph, and ilmenite.

Grain size: The average diameter of a grain, usually measured in millimeters.

Grainstone: A limestone in which the original carbonate grains (allochems) were not bound together at the time of deposition but were later cemented together by carbonate cement.

Granite: A medium- to coarse-grained igneous rock composed of approximately one-third quartz and two-thirds alkali feldspars.

Granodiorite: A plutonic igneous rock containing approximately one-third quartz and two-thirds feldspar, of which plagioclase is more abundant than alkali feldspar. It contains a higher percentage of mafic minerals than does granite and hence is slightly darker than granite.

Granophyre: A fine-grained igneous rock of granitic composition that is characterized by an intergrowth of quartz and feldspar that resembles graphic granite but is finer grained and slightly less regular.

Granophyric intergrowth: The texture that characterizes granophyre.

Granular: Composed of mineral grains of approximately the same size.

Granulite facies: A set of metamorphic mineral assemblages that indicate regional metamorphic conditions of high pressures and temperatures (>800°C). Metabasalt in this grade consists of granular pyroxene and plagioclase, hence the name of the facies.

Graphic granite: An intergrowth of small quartz crystals in large crystals of alkali feldspar, named for its resemblance to cuneiform writing. It may be a eutectic intergrowth of these minerals. It is common in pegmatites.

Gravel: Sedimentary particles that have a grain size greater than 2 mm.

Graywacke: See *wacke*.

Greenschist facies: A set of metamorphic mineral assemblages containing chlorite, muscovite, epidote, and/or actinolite, reflecting metamorphic conditions of about 300–500°C and 2–8 Kb.

Greenhouse gas: Any gas in the atmosphere that can absorb infrared radiation and by so doing causes global warming (e.g., carbon dioxide).

Green Revolution: The twentieth-century agricultural revolution that dramatically increased world crop production through the introduction of new types of cereal grains and the use of fertilizers and pesticides.

Groundmass: The fine-grained crystals that surround phenocrysts in a porphyritic rock.

Growth twin: A twin that is the result of nucleation errors, or accidents, during the free growth of the crystals involved.

Habit: The characteristic appearance of a mineral reflecting crystal form, crystal intergrowth, or other physical aspects of its occurrence.

Halide: A mineral with halogen anions such as F^-, Cl^-, or I^-.

Hand specimen: A piece of mineral or rock that can be held in the hand for evaluation of its macroscopic properties.

Hard mineral: Any mineral with a hardness (**H**) greater than 7.

Hardness: The resistance of a mineral to scratching by a sharp point, edge, or other mineral. Designated by **H**.

Harzburgite: A variety of peridotite composed of olivine and orthopyroxene.

Hawaiian eruption: A style of volcanic eruption involving mainly lava with very little explosive activity.

Heat conduction: See *thermal conduction*.

Heft: The estimation of the specific gravity (**G**) of a mineral specimen by holding it in the hand.

Helicitic texture: Curving inclusion trails in metamorphic porphyroblasts such as garnet and staurolite. These trails may indicate that the porphyroblast rotated during growth.

Hermann-Mauguin notation: A symbolic representation for the symmetry content of a point group, or crystal class, as well as space group. Also known as the international notation.

Hexagonal closest packing (HCP): The packing of atoms of the same size (as spheres) in a three-dimensional array with hexagonal symmetry. Results in an ABAB sequence.

Hexagonal dipyramid: A closed form consisting of 12 faces at the top and 12 identical faces at the bottom, related by a horizontal mirror plane.

Hexagonal prism: In the hexagonal system, there are two prisms of six vertical faces that intersect the horizontal axes at different locations. The Miller-Bravais indices for these two prisms are $\{10\bar{1}0\}$ and $\{11\bar{2}0\}$.

Hexagonal scalenohedron: The general form in point group $\bar{3}\,2/m$. It consists of 12 scalene triangular faces with a zigzag appearance of the middle edges.

Hexagonal system: The crystal system in which a crystal is classified if it contains one unique 6-fold rotation axis, 6-fold rotoinversion axis, or one unique 3-fold rotation axis, or 3-fold rotoinversion axis. Those crystals with unique 3-fold symmetry are commonly referred to as rhombohedral (a subdivision of the hexagonal system). A hexagonal point group example is $6/m\,2/m\,2/m$.

Hexoctahedral class: The highest symmetry class (point group) in the isometric system, $4/m\,\overline{3}\,2/m$, in which the hexoctahedron is the general form.

Hexoctahedron: A form in the isometric systems with 48 crystal faces.

High-alumina basalt: Basalt containing more than 16 weight % Al_2O_3. It is typical of the calcalkaline series of rocks that characterize convergent plate boundaries.

High-level radioactive waste: After 5% of a nuclear fuel rod has reacted, the buildup of highly radioactive fission products makes the fuel no longer usable. This spent nuclear fuel is referred to as high-level radioactive waste, to contrast it with low-level radioactive waste, which could be produced, for example, from medical procedures.

High-resolution transmission electron microscopy (HRTEM): This is an imaging mode of the transmission electron microscope (TEM) that allows the imaging of a crystal structure of a material at the atomic scale.

Hornblende hornfels facies: A set of metamorphic mineral assemblages characterized by aluminous hornblende, formed at moderate temperature and low pressure in contact metamorphic aureoles.

Hornfels: A contact metamorphic rock lacking foliation.

Hot spot: A region on the Earth's surface marked by a lengthy period of igneous activity. They have been interpreted to be located above plumes that have sources deep in the mantle.

Hot-spot track: A chain of volcanoes or igneous centers created by the motion of a tectonic plate across a fixed hot spot (e.g., Hawaiian-Emperor seamount chain).

Hubble space telescope: NASA's space telescope orbits 575 km above the Earth's surface, where it is free of the distorting effects of the atmosphere.

Hydraulic fracturing (hydrofracking or fracking): The fracturing of rock caused by pumping water into a well under high pressure. It is used to increase extraction rates of oil, natural gas, and water, and to increase the volume of dry rock from which geothermal energy can be extracted.

Hydraulic radius: A term used in fluid mechanics in calculating the resistance to flow caused by friction of a fluid against the banks and bed of a channel. The hydraulic radius is equal to the cross-sectional area of the channel divided by its wetted perimeter, which is normally taken to be twice the depth of the channel plus its width.

Hydrolysis reaction: Reactions that involve the breaking of the water molecule into H^+ and OH^- ions.

Hydrostatic pressure: The pressure at depth in a fluid (such as water) that has no shear strength. The pressure is due entirely to the weight of the overlying fluid. At depth in the Earth, rocks have little shear strength, so pressures can be calculated assuming rocks behave like a fluid (see *lithostatic pressure*).

Hydrothermal: Refers to hot, water-rich fluids that are commonly the source of mineral and/or ore deposits.

Hydrothermal alteration: The process by which minerals in a rock are altered, usually to hydrous minerals, by reaction with hot water.

Hydroxide: A mineral in which the anionic group is (OH) as in brucite, $Mg(OH)_2$.

Hydroxylapatite: The type of apatite found in bones, which differs from that in rocks by having considerable substitution of hydroxyl ions for chlorine and fluorine ions.

Hypabyssal: Is an adjective that indicates that an intrusive body or rock is formed at shallow depth in the Earth. Hypabyssal rocks are fine grained and resemble volcanic rocks more than they do plutonic igneous rocks.

Hypersolvus: A term used to describe crystallization of liquids above a solvus. Hypersolvus granites, for example, crystallize above the alkali feldspar solvus and therefore contain only one type of alkali feldspar (contrast with subsolvus).

Igneous: An adjective describing rocks, minerals, gases, textures, structures, and processes that are related to magmas.

Igneous rocks: A rock formed by the solidification of molten rock (magma).

Ignimbrite: Another word for ash flow.

Immiscibility (liquid): The process whereby a homogeneous liquid splits into two liquids of different compositions.

Incongruent melting: Melting in which a solid breaks down to a liquid and another solid, both of which have different compositions from that of the initial solid.

Industrial mineral: Any mineral of economic value except mineral ores, fuels, and gemstones.

Inner core: The inner core of the Earth is solid and is composed primarily of iron, but small amounts of lighter elements must also be present. Its radius is 1220 km.

Inorganic: Not of biological origin.

Inosilicate: A silicate crystal structure that contains infinitely extending tetrahedral chains. Also known as chain silicate.

Interference colors: Colors that result from the interference of two waves of light that are slightly out of phase with each other as a result of passing through the two different vibration directions in an anisotropic mineral or from being reflected off the top and bottom of a thin film.

Interferometric synthetic aperture radar (InSAR): An image created by superimposing two satellite radar images of the Earth taken at different times, and in which differences in elevation are rendered as interference colors.

International Heat Flow Commission: A commission that has the task of compiling heat-flow measurements from the Earth.

International notation: A symbolic representation for the symmetry content of a point group, or crystal class, as well as space group. Also known as the Hermann-Mauguin notation.

Interstice: A space between close-packed atoms (ions) in a crystal structure.

Interstitial solid solution: The substitution (entry) of an element into an atomic site in a crystal structure that is normally empty.

Intrusive: An adjective indicating that an igneous body or igneous rock is formed in the Earth's interior by the intrusion of magma.

Inversion twinning: Twinning that happens when an earlier-formed crystal transforms to a lower symmetry, as a result of decreasing temperature.

Inverted pigeonite: An orthorhombic pyroxene containing exsolution lamellae of augite whose orientation indicates that the orthopyroxene initially crystallized as the higher-temperature monoclinic pyroxene pigeonite. On cooling, the pigeonite inverts to orthopyroxene and exsolves augite lamellae.

Ion: An atom that as a result of gaining or losing one or more electrons has acquired a positive or negative electric charge.

Ionic bond: A chemical bond in which electrons are exchanged between adjoining ions, thus resulting in electrostatic forces between two oppositely charged ions.

Iridescence: An optical phenomenon in which the hue of a surface changes as a function of the angle at which the surface is seen.

Isinglass: Muscovite in thin transparent sheets used in furnace and stove doors.

Island arc (volcanic arc): An arc of regularly spaced composite volcanoes that form above a subducting oceanic plate when it reaches a depth of ~100 km.

Isodesmic: Describes a crystal structure in which all the bonds are of equal strength.

Isogonal: Describes the relationship between screw axes and the equivalent rotational axes. For example, a 3-fold screw axis has the same angle of rotation (120°) as the equivalent 3-fold rotational axis.

Isograd: A line drawn on a geological map indicating the first appearance of a metamorphic index mineral that is believed to join points of equal grade of metamorphism.

Isogyre: A dark, diffuse line seen in an interference figure in the polarizing microscope when using strongly convergent light. The isogyre indicates those parts of the optical figure where light is vibrating parallel to the polarizers in the microscope.

Isometric system: The crystal system in which a crystal is classified if it contains four 3-fold rotation or four 3-fold rotoinversion axes. As in the crystal class (point group) $4/m\,\overline{3}\,2/m$.

Isostasy: The state of buoyant equilibrium between masses of rock with different densities. Lower-density continental rocks, for example, float higher than denser oceanic rocks in the asthenosphere.

Isostructural: A term used to describe that two minerals of different chemical composition have identical structural arrangements.

Isotherm: A line on a map, cross-section, or phase diagram joining points of equal temperature.

Isotopes: One or several species of the same chemical element. They have the same number of protons in the nucleus but differ from each other in having different numbers of neutrons.

Isotropic: In optics, it indicates that the refractive index of a substance remains constant regardless of the direction of the passage of light. Isometric minerals and glass are isotropic.

IUGS: International Union of the Geological Sciences.

Jolly balance: A balance used to determine the specific gravity (**G**) of a solid material.

Joule (J): The SI unit of energy equal to a newton·meter. It is equivalent to 0.239 calories in the cgs system of units.

Kame: A stratified meltwater deposit formed in a river flowing along the side of a glacier or in a crevasse.

Kaolin: A soft, earthy, usually white or nearly white clay composed principally of kaolinite.

Karat (k): The proportion of gold in an alloy. Pure gold is 24k; 18k gold is 18 parts pure gold and 6 parts of other metals. Not to be confused with carat.

Karst: A type of topography characterized by solution features that develops on carbonate rocks.

Kerogen: A complex organic substance formed from fossilized organic matter. It is from kerogen that oil and natural gas are formed.

Kimberlite: A phlogopite-rich igneous rock that commonly contains mantle xenoliths in which diamonds occur.

Kinetics: The rate of processes, such as the rate of a metamorphic reaction or rate of cooling of magma.

Komatiite: Ultramafic lava that grows long bladelike crystals of olivine on cooling. Most are of Archean age.

Labradorescence: An optical phenomenon consisting of flashes of laminated iridescence of a single bright hue that changes gradually as the mineral is moved about in reflected light. Seen commonly in labradorite, a member of the plagioclase series.

Laccolith: A near-surface, mushroom-shaped concordant intrusive igneous body.

Lahar: A mudflow that descends the flanks of a volcano.

Laminae: In sedimentary rocks, a term used to describe beds that are thinner than 1 cm.

Laminar flow: The property of a fluid to flow as if it were a series of sheets or laminae parallel to the surface over which it flows and whose relative distances from that surface do not change.

Lamination or laminae: Sedimentary beds thinner than 1 cm.

Lamprophyre: A basaltic dike rock characterized by large phenocrysts of amphibole or biotite.

Lapilli: Particles ejected from a volcano that have diameters between 2 mm and 64 mm.

Large igneous province (LIP): A large region ($>100\,000$ km^2) within a tectonic plate characterized by extensive and long lasting (50 million years) igneous activity, which is usually predominantly basaltic in composition. Believed to form over large mantle plumes and to often lead to the rifting apart of tectonic plates (e.g., the Central Atlantic Magmatic Province was the precursor to the breakup of Pangaea).

Latent heat of crystallization: The amount of heat that must be dissipated for a liquid to change into a crystalline solid at constant temperature: ~ 400 kilojoules/kg for most magmas.

Latent heat of fusion: The amount of heat needed to convert a crystalline solid into a liquid at constant temperature.

Lattice: An imaginary pattern of points (or nodes) in which every point (or node) has an environment that is identical to that of every other point (or node) in the pattern. A lattice has no specific origin as it can be shifted parallel to itself.

Lava: Magma that is erupted onto the Earth's surface and flows as a coherent mass.

Lava tube: A tunnel formed when the supply of lava to a crusted-over lava flow is cut off and the molten lava drains from the flow's interior.

Law of superposition: Sedimentary layers are deposited in a sequence with the oldest at the bottom and youngest at the top.

Layer silicate: A silicate crystal structure with infinitely extending tetrahedral sheets. Also known as a sheet silicate, or phyllosilicate.

Leachate: A solution that results from leaching. It involves the dissolving out of the soluble constituents of a rock or ore body by percolating water.

Left-handed: Describes a screw axis operation in which the motif (e.g., unit of structure, atom) moves away from the observer in a counterclockwise direction.

Lenses: In sedimentary rocks, this refers to layers that do not have great lateral extent (less than few meters).

Leucosome: In a migmatite, the light-colored igneous part is referred to as the leucosome.

Level of neutral buoyancy: The level at which the density of magma buoyantly rising through the density stratified crust matches the density of its surroundings. Magmas are thought to spread laterally at this level.

Lever rule: A rule that makes use of the principles of a lever to determine the quantities of coexisting phases in a given bulk composition in a phase diagram.

Lherzolite: A variety of peridotite containing approximately equal proportions of orthopyroxene and clinopyroxene along with olivine.

Lignite (brown coal): A low-rank coal.

Liquid immiscibility: See *immiscibility (liquid)*.

Liquidus: The temperature above which a mixture of materials is completely liquid.

Limestone: A sedimentary rock composed predominantly of calcite or more rarely aragonite.

Linear coordination: Describes a cation with only two adjoining anion neighbors.

Lithic arenite (litharenite): Sandstone containing more than 25% rock fragments.

Lithification: The process by which unconsolidated sediment is converted into sedimentary rock. It involves compaction, cementation, and recrystallization.

Lithosphere: The relatively rigid outer part of the Earth, including the crust and upper mantle down to the asthenosphere. It is broken into large plates that move relative to one another and generate plate tectonics.

Lithostatic pressure: The calculated pressure at depth in the Earth based on the assumption that rocks have almost no shear strength (i.e., they behave the same as water).

Lodgement till: Till deposited beneath glacial ice and hence extremely compact.

Loess: A sedimentary deposit of fine-grained wind-blown sand and silt formed on the lee side of a desert.

Lopolith: A large saucer-shaped igneous intrusion.

Luster: The reflection of light from the surface of a mineral, described by its quality and intensity. Examples are metallic or resinous luster.

Maar: A shallow volcanic explosion crater surrounded by fragmental material blown from the crater. Typically the explosion is not accompanied by the eruption of lava.

Magma: Molten rock. Most magma is not totally molten but contains some solids that are carried along in the liquid.

Magma chamber: A body of magma inside the Earth. Magma chambers can have many different shapes from thin sheets (e.g., dikes, sills) to large diapirs (e.g., batholiths).

Magma ocean: The name given to the huge body of molten rock that is believed to have covered the outer part of the Earth early in its history.

Magmatic differentiation: A process that causes magma to change its composition. This is most commonly the result of mineral grains separating from the magma.

Magnetic quantum number: This is the third of a set of quantum numbers symbolized by m. It defines the energy levels in an atomic subshell.

Magnetism: The physical phenomenon of attraction of a mineral to a magnet.

Major element: A chemical element that constitutes over one weight % of a mineral analysis.

Malleable: Said of a mineral that can be plastically deformed under compressive stress such as hammering, such as gold, silver, and copper.

Mammillary: Applies to a mineral specimen that has a breastlike appearance.

Mantle: That part of the Earth lying between the Mohorovičić discontinuity and the core-mantle boundary. It is divided into the upper and lower mantle by the seismic discontinuity at a depth of ~660 km.

Mantle plume: A narrow column of rock that is hypothesized to rise buoyantly from deep in the mantle (possibly as deep as the core-mantle boundary) and on reaching the lithosphere mushrooms out to form a plume head. Hot spots are believed to be located above mantle plumes (e.g., Hawaiian hot spot).

Marble: A rock formed from the metamorphism of limestone and composed predominantly of calcite, but it may also contain dolomite.

Marl: A limestone containing a significant proportion of siliciclastic mud.

Massif-type anorthosites: Large Proterozoic intrusive igneous rock bodies composed of anorthosite, a rock composed largely of intermediate composition plagioclase.

Massive: A mineral, or mineral aggregate, without obvious crystal habit, or other distinguishable physical characteristics.

Maturity: When applied to sediment, it can refer to textural or chemical maturity. As sediment is transported and abraded, less chemically resistant minerals are weathered out, leaving mainly quartz, and grains become rounder.

Matrix-supported: An adjective indicating that sedimentary detrital particles, such as pebbles in a conglomerate, do not touch one another but are supported by the matrix that surrounds them. Contrast this with *clast-supported*.

Mélange: A mixture of highly deformed marine sedimentary rocks that form accretionary wedges at convergent plate boundaries.

Melanosome: In a migmatite, the refractory part of the rock is typically dark colored and consequently is referred to as the melanosome.

Mesodesmic: Describes a crystal structure in which the bond strengths of all the bonds reaching a cation from surrounding anions are exactly one-half the charge of the anions.

Metagraywacke: A metamorphosed graywacke (wacke). Most graywackes are slightly metamorphosed.

Metallic bond: A chemical bond between atoms in which their outer valence electrons are free to drift through the metal crystal structure.

Metallic luster: A luster that results from strong reflection of white light by the bright surface of metals and metallic minerals.

Metamict: Describes a mineral that has undergone structural damage from radioactive elements housed in its structure but has retained its original external morphology.

Metamorphic: An adjective describing rocks, minerals, textures, structures, and processes that involve a change in a previously existing rock due to changes in temperature, pressure, or fluid composition in the environment.

Metamorphic facies: A metamorphic mineral facies comprises all rocks that have originated under temperature and pressure conditions so similar that a definite bulk rock chemical composition results in the same set of minerals.

Metamorphic facies series: A series of metamorphic facies that indicate specific conditions of pressure and temperature in progressively metamorphosed regional metamorphic terranes. Three main series are recognized: low pressure and high temperature, high pressure and high temperature, and high pressure and low temperature.

Metamorphic field gradient: The gradient of temperature and pressure indicated by the mineral assemblages preserved in metamorphic rocks.

Metamorphic grade: A loose term that refers to the intensity of metamorphism. Higher-grade rocks have experienced higher temperatures and pressures.

Metamorphic index minerals: A sequence of common metamorphic minerals that indicates increasing metamorphic grade. The most common sequence in pelitic rocks include from lowest to highest grade: chlorite, biotite, garnet, staurolite, kyanite, and sillimanite.

Metamorphic rock: Any rock that has been changed in texture, mineralogy, or chemical composition as a result of changes in environmental factors such as temperature, pressure, directed stress or shear, and composition of fluids.

Metamorphism: The change that takes place in rocks either in texture, mineralogy, or chemical composition as a result of changes in temperature, pressure, or composition of fluids in the environment.

Metasomatism: A metamorphic process whereby the bulk chemical composition of a rock is changed.

Metastable: A condition that is not ultimately stable but does not result in change because of a large activation energy that prevents the system from changing into the stable form.

Metapelite: One of the most common types of metamorphic rock, formed from the metamorphism of alumina-rich sedimentary rocks such as mudrock and shale.

Meteorites: Natural objects from space that impact the Earth's surface. A meteor is a similar object that burns up while passing through the atmosphere and does not reach the Earth's surface.

Microcrystalline: Said of crystals that can be seen only with the aid of a microscope.

Microlite: Extremely small crystals found in volcanic glass.

Midocean ridge: The topographic high produced by the creation of new hot oceanic crust at a divergent plate boundary. Ridges are often located near the center of an ocean, as in the case of the Mid-Atlantic Ridge, but they need not be, as in the case of the East Pacific Rise.

Midocean ridge basalt (MORB): A type of basalt erupted at midocean ridges and forming most of the ocean floor.

Migmatite: A mixed metamorphic and igneous rock formed when the rock was raised to a temperature at which partial melting occurred. The melt tends to be granitic in composition and forms light-colored layers in the darker refractory metamorphic rock.

Milankovitch cycles: Cyclical climate changes caused by variations in the Earth's orbital eccentricity about the Sun (~100 000 years), and the tilt (~40 000 years) and precession (19 000–23 000 years) of its axis.

Miller-Bravais index: This index system was developed for the hexagonal system only and is different from the Miller index system because it refers to four crystallographic axes instead of just three. An example is $(11\bar{2}1)$.

Miller index: A set of three digits (or four, as in the hexagonal system) that describe the orientation of a crystal plane with respect to a set of crystallographic axes. These digits are normally inside parentheses, for example (110).

Mineral: A naturally occurring solid with an ordered atomic arrangement and a definite (but commonly not fixed) chemical composition, and of inorganic origin.

Mineral deposit: A concentration of a mineral formed by geologic processes. If the mineral can be extracted at a profit, the deposit is called an ore deposit.

Minimum: The liquidus of some solid solutions passes through a minimum, toward which liquids tend to fractionate. Many granites have a composition close to a minimum in the alkali feldspar and quartz system.

Minor element: A chemical element that constitutes between 1.0 and 0.1 weight % of a mineral analysis.

Mirror plane: Causes the reflection of a specific crystal face, or a unit of structure, into its mirror image.

Miscibility gap: A compositional range in a temperature-composition diagram in which a single homogeneous mineral is not stable.

Modal analysis: Determination of the mode of a rock.

Modal layering: Layering in rocks resulting from variations in the modal abundance of minerals.

Mode: The volumetric abundance of minerals constituting a rock. In a microscope thin section, this is equivalent to the areal abundance of the minerals.

Mohorovičić discontinuity (also **Moho** or **M discontinuity**): The seismic discontinuity defining the base of the Earth's crust. The sharp increase in seismic velocities across this boundary is believed to result from a change from crustal rocks to mantle peridotite.

Mohs hardness scale: A relative hardness scale of 1–10, based on the hardness of ten common minerals used to evaluate a mineral's hardness.

Molecular sieve: A mineral (or material) with tiny pores, of uniform size, used as an absorbant for gases and liquids. Molecules that are small enough pass through the pores and are absorbed, whereas larger molecules are not.

Molecular orbital transition: The result of the transfer of electrons between cations, of variable charges, in a crystal structure. They result in the blue color of several rock-forming minerals.

Monoclinic system: The crystal system in which a crystal is classified if it contains a single 2-fold axis and/or a single mirror. An example is $2/m$.

Monzonite: A plutonic igneous rock composed of approximately equal proportions of alkali feldspar and plagioclase.

Moonstone: Refers to an albite-rich plagioclase that shows iridescence caused by exsolution lamellae across the peristerite miscibility gap.

Morphology: Includes an assessment of the geometric aspects of the crystal faces and their symmetry as seen on the outside of crystals. It also refers to the prediction of the external shape (morphology) of crystals based on their internal structure.

Mud: Sedimentary particles that have a grain size less than 0.0625 mm.

Mud cracks: A polygonal pattern of cracks generated in a layer of mud as it dries and shrinks.

Mudflow: A flow of debris with particles of various size carried along in a mud matrix.

Mudstone: A calcareous sedimentary rock composed of more than 90% mud-size particles (<0.0625 mm).

Mudrock: A siliciclastic sedimentary rock composed of mud-size particles (<0.0625 mm).

Multiple twin: A twin that consists of three or more individual, identical crystals that are related by composition surfaces (twin planes).

Mylonite: An extremely fine-grained metamorphic rock formed by recrystallization of rock in a fault zone.

Native element: Any element found uncombined in a nongaseous state in nature.

Nebula: A cloud of interstellar gas and dust particles that can collapse under its own gravitational attraction to form a star and planetary system.

Nepheline syenite: An igneous rock composed essentially of alkali feldspar and nepheline.

Nephelinitic-leucitic-analcitic series: A subdivision of the alkaline igneous rock series.

Nesosilicate: A silicate crystal structure in which there are independent $(SiO_4)^{4-}$ tetrahedra that do not link to other silica tetrahedra but instead link their corner oxygens to other cations. Also known as orthosilicate.

Network structure: A common silicate structure in which tetrahedra that house Si^{4+} and Al^{3+} form an infinite three-dimensional network. Also known as framework structure or tectosilicate.

Neutral: A term applied to a screw axis that relates atoms (ions, atomic units) in such a way that their location is irrespective of the direction of rotation (either right- or left-handed).

Neutron: A subatomic particle with no net electric charge and a mass slightly larger than that of a proton.

Newton (N): The SI unit of force. It is the force needed to give a mass of 1 kg an acceleration of 1 m/s^2.

Nonmetallic luster: A mineral with a luster that is nonmetallic and that is described by such terms as *vitreous*, *resinous*, or *adamantine*.

Nonrenewable resource: A resource that was formed over very long geological periods. Minerals and fossil fuels fall into this category.

Norite: A plutonic igneous rock of gabbroic composition in which the pyroxene is predominantly orthopyroxene.

Norm (CIPW): A calculated mineral composition of an igneous rock based on its chemical analysis. The most widely used norm is that created by Cross, Iddings, Pirsson, and Washington (hence, CIPW).

Normal fault: A fault in which the hanging wall (upper block) moves downward relative to the footwall. Normal faults result from crustal extension.

NPK number: A number indicating the weight percentages of N, P_2O_5, and K_2O in fertilizer.

Nuée ardente: See *ash flow*.

Nugget: A large lump of placer gold or other metal.

Obduction: The process by which a slice of subducting ocean crust is thrust up onto a continent.

Obsidian: A glassy volcanic rock, usually of granitic composition.

Occurrence (also known as paragenesis): The characteristic association or occurrence of minerals or mineral assemblages in rock types and ore deposits.

Octahedral cleavage: The breaking of a mineral along four different planar directions that form the shape of an octahedron.

Octahedral coordination: Depicts a cation with six closest anion neighbors.

Octahedron: An eight-faced closed form in the isometric system with Miller index {111}.

Oil or natural gas trap: Any structure that provides an impermeable cap over a porous and permeable reservoir rock in which oil or natural gas can be trapped.

Oil shale: An organic-rich black shale that can be burned as fuel or from which oil can be extracted by appropriate heating.

Oil window: The temperature interval (60–120°C) in which kerogen breaks down to form oil.

Oligomict: A term indicating that a conglomerate is composed of clasts of just one rock type.

Omission solid solution: The partial emptying out of an atomic site in a crystal structure. The omission, as the term implies, results in a defect structure. The best mineralogical example of this is seen in pyrrhotite, $Fe_{(1-x)}S$.

Ooid (oolith): A spherical millimeter-sized particle formed by the precipitation of calcite or aragonite in concentric layers around a nucleus grain as it is washed back and forth in shallow marine waters.

Oolitic: Describes a mineral aggregate consisting of grains in rounded masses the size of fish roe.

Oolitic limestone: A limestone containing ooids.

Ooze, calcareous and siliceous: Extremely fine sediment formed on the ocean floor by the accumulation of the tests of pelagic organisms. Only siliceous ooze forms in the deepest ocean below the carbonate compensation depth.

Opaque: Describes a material that is impervious to visible light and cannot be seen through.

Open form: A crystal form that does not enclose space, such as a four-faced prism.

Operator (symmetry): Any symmetry element that repeats a motif (atom, ion, ionic group) or crystal face in a symmetrical pattern. Examples are rotation, mirror, inversion, rotoinversion, translation, glide plane, and screw axis.

Ophiolite: A suite of rocks that is thought to represent a slice of the oceanic crust that is obducted onto the continent. It normally contains, from top to bottom, radiolarian chert, pillow basalt, sheeted dikes, layered gabbro, and serpentinized harzburgite.

Ophitic texture: An intergrowth involving numerous small crystals of plagioclase embedded in large crystals of pyroxene or olivine. It is believed to be a eutectic intergrowth.

Optical indicatrix: An ellipsoid representing the variation in refractive index with direction in an anisotropic crystal.

Optic angle (2V): The angle between the two optic axes in a biaxial mineral.

Optic axis: A direction perpendicular to a circular section through the optical indicatrix. A uniaxial indicatrix has only one such direction, but a biaxial indicatrix has two.

Order-disorder polymorphism: This type of polymorphism ranges from a state of perfect order (at 0 Kelvin), through partial disorder, to total disorder and relates to the site occupancy by a specific atom (or ion) in a specific atomic site.

Ordered: A crystal structure in which the atoms (or ions) are located in specific crystallographic sites of the structure.

Ore deposit: A concentration of ore minerals that can be extracted and sold for a profit.

Ore mineral: That part of an ore, or ore deposit, usually metallic, that is economically desirable, as contrasted with the gangue.

Orthopyroxene: A subgroup of the pyroxenes with orthorhombic symmetry.

Orthorhombic system: The crystal system in which a crystal is classified if it contains binary symmetry elements such as 2-fold axes or mirrors along three mutually perpendicular directions. An example is $2/m\ 2/m\ 2/m$.

Orthosilicate: A silicate crystal structure in which there are independent $(SiO_4)^{4-}$ tetrahedra that do not link to other silicate tetrahedra, but instead link their corner oxygens to other cations. Also known as nesosilicate.

Outer core: The outer core of the Earth extends from a depth of 2888 km to 5151 km. It is composed primarily of molten iron, whose convection creates the Earth's magnetic field.

Outwash plain: A gently sloping plain in front of a glacier where sediment is deposited by melt water.

Oxide: A mineral in which the anion is oxygen, as in hematite, Fe_2O_3.

Oxygen factor: A number that is the ratio of the number of oxygens in a chemical formula (of a mineral) to that of the oxygen summation arrived at in the recalculation of a chemical analysis (of a mineral).

Packstone: A grain-supported limestone that has a carbonate mud matrix.

Pahoehoe: A basaltic lava flow with a smooth glassy surface, which is commonly deformed into ropelike wrinkles.

Paired metamorphic belts: Near convergent plate boundaries, metamorphic rocks commonly form paired belts, with a high-pressure low-temperature series forming closest to the subducting plate and a low-pressure high-temperature series forming farther from it.

Partial melting: The process whereby mixtures of material, such as rocks, melt only partly because of insufficient heat.

Partial order: This describes the distribution of a specific atom (or ion) equally over two equivalent atomic sites in a crystal structure.

Pascal (Pa): The SI unit of pressure, which is equal to a newton/m^2 (~10^{-5} atmospheres).

Passive margin: A continental margin created when plate divergence rifts a continent apart. As the plate cools on moving away from the divergent boundary, the passive continental margin slowly subsides.

Pauling's rules: Five empirical rules formulated by Linus Pauling that relate to structural aspects, packing, and bonding of ions in crystal structures.

Peacock ore: An informal name for an iridescent, tarnished surface with variegated colors as in bornite.

Pebble: A sedimentary particle that has a grain size between 4 mm and 64 mm.

Pegmatite: An exceptionally coarse-grained igneous rock, usually of granitic composition and forming irregular dikes and lenses.

Pelagic organisms: Organisms that live in the water column (sea or lake) rather than near the bottom (benthic) or near the shore (littoral).

Peléan eruption: A style of volcanic eruption accompanied by ash flows.

Pelite: An alumina-rich sedimentary rock.

Pelitic schist: A metamorphic rock derived from a pelite, an argillaceous or fine-grained aluminous sedimentary rock.

Pellet (peloid): Fecal pellets formed of carbonate minerals that are produced by many different organisms.

Penetration twin: A twin in which two identical crystals are related by a rotational axis, thus resulting in an irregular composition surface between the two.

Peralkaline igneous rock: An igneous rock in which the molecular amounts of $Na_2O + K_2O$ exceed the molecular amount of Al_2O_3. The excess alkalis enter sodic pyroxenes and amphiboles.

Perfect order: This describes the 100% probability of finding a specific atom (or ion) in a specific atomic site in a crystal structure.

Pericline twin: A twin type that leads to a characteristic tartan pattern (as seen under the microscope) in microcline feldspar, $KAlSi_3O_8$.

Peridotite: A plutonic igneous rock composed essentially of olivine and pyroxene.

Peristerite gap: A compositional gap, also known as a miscibility gap, expressed by the presence of closely spaced exsolution lamellae in plagioclase of An_0 and An_{25} composition.

Peritectic: A reaction whereby a solid melts incongruently to produce a liquid and another solid, both of different composition from the initial solid.

Permeability: A measure of how connected the pores in a rock are. Not to be confused with porosity.

Perthite: A parallel to subparallel intergrowth of a potassium-rich feldspar (usually microcline) with a sodium-rich feldspar (usually albite), due to exsolution. The exsolved regions are visible to the naked eye.

Perthitic: Said of a texture produced by parallel to subparallel intergrowths of microcline and albite.

Petrified wood: Wood that has been replaced by silica by burial in sediment. Much of the detailed structure of the wood is often preserved by the silica that replaces it.

Petrogenetic grid: A pressure-temperature phase diagram in which metamorphic reactions have been plotted from which limits can be set on the conditions under which any given metamorphic mineral assemblage formed.

Petrographic microscope: A microscope that has a rotating stage and uses polarized light to examine thin slices of rock in transmitted light. By using a second polarizer above the section, interference colors are created that are often characteristic of the minerals.

Phase: In thermodynamics, this refers to any material that can be physically distinguished from any other material. For example, each different mineral in a rock is a phase, as are liquids and gases. Iron and magnesium atoms substituting for one another in olivine, however, are not phases.

Phase diagram: A graphical representation of the stability fields of phases as a function of temperature, pressure, and composition.

Phenocrysts: Crystals in igneous rocks that are significantly larger than most of the crystals in the rock. Igneous rocks containing phenocrysts are described as porphyritic.

Phosphate: A mineral with the anionic group of $(PO_4)^{3-}$, as in fluorapatite, $Ca_5(PO_4)_3F$.

Phosphorite: A sedimentary rock with a high-enough phosphate mineral content to be of economic interest.

Photosphere: The outer light-emitting part of the Sun.

Phyllite: A metamorphic rock with a prominent foliation. The foliation surfaces are less regular than in slate and have a sheen. Individual mineral grains on the foliation plane are not visible to the unaided eye.

Phyllosilicate: A silicate crystal structure with infinitely extending tetrahedral sheets. Also known as sheet silicate.

Physical property: A measurable property the value of which describes a certain aspect of the state of a physical system. Examples of physical properties are hardness, melting point, refractive index, and color.

Pillows: Rounded sacklike bodies of lava that are erupted beneath water. They have a glassy shell but their interiors cool slowly enough to form fine-grained rock. Much of the ocean floor is underlain by pillowed basalt.

Pinacoid: An open crystal form that consists of two parallel crystal faces.

Pisolitic: Describes a mineral aggregate consisting of grains in rounded masses the size of peas.

Placer deposit: A surficial mineral deposit formed by mechanical concentration of mineral particles from weathered debris.

Planar cleavage: The breakage of a mineral along a single planar direction, as shown by mica.

Plankton: Organisms that are free floating in oceans, seas, or lakes and live in the pelagic zone, that is, away from the bottom or shore. They are carried wherever currents take them.

Plate tectonics: The deformation of the Earth as a result of the motion of lithospheric plates.

Play of color: The interaction of white light with a mineral resulting in the separation of visible, individual colors, as seen best in precious opal.

Pleochroic halo: A thin zone of darkening, as seen under the microscope, surrounding and produced by a radioactive mineral.

Pleochroism: The variation in the color of a mineral in plane polarized transmitted light due to absorption of different wavelengths of light in different crystallographic directions in a mineral.

Plinian eruption: A style of volcanic eruption where a column of ash rises and then spreads into a mushroom-shaped cloud at its peak, which may reach as high as the stratosphere.

Plutonic: An adjective that indicates an igneous body or rock is formed at depth in the Earth. Most plutonic rocks are relatively coarse grained.

Poikilitic texture: An igneous texture in which smaller crystals are embedded in larger crystals of a different mineral.

Point group (same as crystal class): One of 32 symmetry groups to which a crystal is assigned on the basis of its translation-free symmetry content. The word *point* indicates that the symmetry elements inside the crystal all intersect at its center, a central point.

Polarized light: Light whose waves vibrate in only one plane, the plane of polarization.

Polarizing filter: A filter that allows light to pass through it vibrating in only one plane.

Polyhedron: A three-dimensional geometric solid with flat faces and straight edges. Examples of specific polyhedra are tetrahedron, octahedron, and cube.

Polymict: A term indicating that a conglomerate is composed of clasts of two or more rock types.

Polymorph: Refers to minerals with the same chemical composition but different crystal structures. Quartz, tridymite, and cristobalite are three of several polymorphs of SiO_2.

Polymorphism: The ability of a specific mineral (or chemical compound) to occur in more than one type of structure as a function of changes in temperature, or pressure, or both.

Polysynthetic twin: A repeated or multiple twin in which the composition surfaces are parallel.

Polytype: A type of polymorph in which different structural arrangements result from the different stackings of nearly identical atomic structural units, such as layers.

Polytypism: The property of a mineral to crystallize in more than one crystal structure, in which the basic structural elements (commonly atomic layers) are stacked in different sequences.

Porosity: The percentage of void space in a rock.

Porphyroblast: A crystal in a metamorphic rock that is significantly larger than the other mineral grains in the rock, and commonly euhedral and has grown during metamorphism.

Porphyroclast: A large grain in a metamorphic rock that is a relic of a previously existing large grain.

Porphyry: An igneous rock that contains crystals that are significantly larger (phenocrysts) than other crystals in the rock. Most igneous rocks are porphyritic.

Porphyry copper deposit: A large body of rock, typically a porphyry, that contains disseminated chalcopyrite and other sulfide minerals.

Portland cement: The most common type of cement found in concrete, mortar, stucco, and grout. The raw materials for Portland cement production are a mixture of lime, silica, alumina, and iron oxides.

Potential energy: The energy that matter has because of its position in a potential field, such as the Earth's gravitational field.

Powder diffraction: An X-ray analytical technique that uses a finely powered mineral sample.

Precious opal: Opal with good play of color, or transparent to translucent opal with a single body color but no play of color.

Precision agriculture: The modern method of farming that uses global position systems to maneuver farm vehicles and to mix and apply fertilizers using computer-stored geographic information about soils and previous crop production.

Prehnite-pumpellyite facies: A set of metamorphic mineral assemblages characterized by the coexistence of prehnite and pumpellyite formed at temperatures between 200°C and 250°C.

Pressure solution: The process by which grains in rocks dissolve at points of contact where the pressure is relatively high and redeposit in the pore spaces where the pressure is relatively low.

Pressure-temperature-time path (P-T-t): The pressure-temperature-time path followed by a rock during regional metamorphism as deduced from mineral assemblages, geothermometers, geobarometers, and fluid inclusions. Usually only the last part of this path is preserved.

Primitive: Describes a unit-cell choice in which there are nodes only at the eight corners. The node content of such a cell choice is only 1. Abbreviated by P.

Primitive circle: The equatorial plane onto which face poles are projected in the process of stereographic projection.

Principal quantum number: It is symbolized by n and is the first of a set of quantum numbers. It can have only positive integer values, which relate to electron shells in the Bohr atomic model.

Principle of parsimony: One of Linus Pauling's five rules. It states that crystal structures tend to have a limited set of distinctly different cation and anion sites.

Prism: An open crystal form that consists of three or more crystal faces parallel to a common direction.

Prismatic: A crystal bounded by a prism or prisms, commonly giving an elongate appearance.

Prismatic cleavage: Breakage of a mineral along two different cleavage directions whose lines of intersection commonly parallel a specific crystallographic direction.

Prograde: Refers to metamorphic reactions that take place as a rock is heated either because of burial or proximity to an igneous body.

Protolith: The original rock from which a metamorphic rock was formed.

Proton: A subatomic particle with an electric charge of +1. It occurs in the nucleus of an atom, along with neutrons.

Provenance: The source of a substance, as in the source of the sediment in a sedimentary rock.

Pseudotachylite: A black fine-grained or glassy rock formed by frictional fusion in a fault zone.

Pull-apart basin: A basin formed by strike-slip motion at an irregularity along a transform fault.

Pumice: An extremely vesicular volcanic rock, usually of rhyolite (granite) composition.

Pumpellyite-actinolite facies: A set of metamorphic mineral assemblages characterized by the coexistence of pumpellyite and actinolite formed at temperatures near 300°C.

Pyramid: An open crystal form consisting of three or more crystal faces that meet at a point.

Pyroclastic deposit: A deposit of fragmental volcanic particles, many of which have been disrupted by the expansion of gas bubbles in viscous lava.

Pyrite sun (also known as a pyrite dollar): A flat aggregate of small pyrite crystals in the shape of a round dollar coin, found between layers of coal and black shale.

Pyroxene: A major rock-forming silicate mineral group with infinitely extending chains of tetrahedral $(SiO_4)^{4-}$ groups and compositions rich in Mg, Fe, Ca, Al, and Si.

Pyroxene hornfels facies: A set of metamorphic mineral assemblages characterized by the presence of pyroxene, formed at high temperature and low pressure in contact metamorphic aureoles.

Pyroxenoid: A silicate mineral group very similar to the pyroxenes with tetrahedral $(SiO_4)^{4-}$ single chains but with repeat units in the chains of 3, 5, 7, or 9.

Quartz arenite: A mature sandstone containing more than 95% quartz.

Quartzite: A rock formed by the metamorphism of quartz-rich sandstone.

Quicklime: Calcium oxide, CaO, a widely used chemical compound that is produced by the thermal decomposition of calcium carbonate in a lime kiln.

Radiation cooling: The loss of heat by the emission of electromagnetic radiation.

Radioactivity: The spontaneous decay of atoms of some isotopes into daughter elements, accompanied by emission of high-energy particles and gamma rays.

Radiolaria: Minute pelagic organisms that build their tests of silica.

Radius ratio: The ratio of the radius of a cation to that of an anion.

Radon gas: A radioactive gas produced as a decay product of uranium. Its inhalation has been linked to lung cancer.

Rank: A term that refers to the amount of heat released by burning coal. Low-rank coal gives off ~2.85×10^7 J/kg, whereas high-rank coal can give off as much as 3.5×10^7 J/kg.

Rapakivi granite: A common decorative facing stone in which large pink alkali feldspar phenocrysts are rimmed by gray or white plagioclase.

Rare earth elements (REE): Seventeen closely related elements from lanthanum, La (atomic number 57) to lutetium (atomic number 71), referred to as the lanthanide series, as well as yttrium, Y (atomic number 39), and scandium, Sc (atomic number 21).

Rayleigh-Taylor instability: A dense fluid overlying a less dense fluid is gravitationally unstable, and the less dense fluid rises through the denser fluid in a series of regularly spaced domes.

Reaction texture: A texture in which one mineral rims another as a result of a reaction. In igneous rocks, reaction textures are formed at peritectics, and in metamorphic rocks, reaction rims are commonly the result of incomplete reaction because of slow diffusion rates.

Reconstructive polymorphism: This involves extensive structural rearrangement in going from one atomic structure to another.

Recrystallization: The solid-state process whereby new crystals grow at the expense of old ones to get rid of strain or to decrease surface energies by increasing grain size.

Refractive index (RI): The refractive index of a substance is the ratio of the velocity of light in air to the velocity of light in the substance.

Refractometer: An instrument for measuring refractive index.

Regional metamorphism: Metamorphism that takes place on a regional scale as a result of increasing temperature and pressure on a regional scale. It is usually associated with deformation related to tectonic plate convergence.

Regolith: The unconsolidated material above bedrock.

Regression of the sea: The emergence of land due to a fall in sea level or a rise of the land.

Relative plate velocities: The rates at which one plate diverges from another, as indicated by the age of magnetic anomalies on the ocean floor created between diverging plates (contrast this with absolute plate velocities).

Relief: When applied to the optical properties of minerals, it is a comparison of a mineral's refractive index to that of the mounting material in which the section is glued to the glass slide (usually 1.537). Minerals with higher refractive index than the glue have positive relief, those with lower refractive index, negative relief.

Renewable resource: One that can be replenished or reproduced easily. Examples are agricultural crops, water, and forests.

Reniform: Describes a mineral having the outward appearance in the shape of a kidney.

Reservoir rock: A sedimentary rock with high porosity and permeability in which oil and natural gas can accumulate.

Resinous luster: Describes a mineral with the luster of resin.

Retrograde: Refers to metamorphic reactions that occur while a rock is cooling down, because of either exhumation or the cooling of a body of igneous rock.

Reynolds number (Re): A numerical ratio between inertial forces and viscous forces in a fluid, which allows us to predict whether flow will be laminar or turbulent.

Rhombic dipyramid: A closed crystal form in the orthorhombic system consisting of eight inclined faces.

Rhombic prism: In the orthorhombic system, there are three different prisms, all of which have four faces that are parallel to one axis and intersect the other two. Their Miller indices are: $\{0\,k\,l\}$, $\{h\,0\,l\}$, and $\{h\,k\,0\}$.

Rhombohedral cell: Describes a unit cell choice with the geometric outline of a rhombohedron.

Rhombohedral cleavage: The breakage of a mineral along three planar directions that outline the shape of a rhombohedron.

Rhombohedron: A closed crystal form consisting of six rhomb-shaped faces, with three faces at the top and three at the bottom, with an offset between top and bottom faces of 60°.

Rhyolite: A volcanic rock of granitic composition. It is often glassy, and when completely glassy is called obsidian.

Rift valley (basin): A valley or basin formed by the down faulting of a block along normal faults as a result of crustal extension (e.g., East African Rift Valley).

Rift zone: A prominent fissure on the flanks of a volcano.

Right-handed: Describes a screw-axis operation in which the motif (e.g., unit of structure, atom) moves away from the observer in a clockwise direction.

Ring dike: An arcuate to circular steeply outward-dipping sheetlike discordant intrusive igneous body.

Ring silicate: A silicate crystal structure in which there are 4- or 6-fold tetrahedral rings. Also known as cyclosilicate.

Ripples: A wavelike sedimentary bedform, commonly formed by the movement of currents (asymmetric), but it can also form through the oscillation of waves (symmetrical) generated by wind blowing across water.

Rip-up clast: Fragments of mud, such as those formed between mud cracks, that are ripped up and carried along as sedimentary particles.

Rock: A rock is a naturally occurring consolidated mixture of minerals. It is the solid material that makes up the Earth.

Rock-forming mineral: A mineral that is a common constituent of various rock types found in the Earth's crust.

Rotation: A symmetry operation that relates objects (atoms, ions, ionic groups) or crystal faces about a rotational symmetry axis.

Rotation axis: An imaginary line about which an object, or a crystal face, or an atomic arrangement (inside a crystal structure), is rotated and repeats itself 1, 2, 3, 4, or 6 times.

Rotoinversion axis: An imaginary line about which a crystal face or an atom (or atomic cluster) in a crystal is rotated as well as inverted. The angle of rotation is represented by digits ($\bar{1}$, $\bar{2}$, $\bar{3}$, $\bar{4}$, and $\bar{6}$) used also for rotation axes, but with overbars.

Salt dome (diapir): A domelike structure formed by the buoyant rise of a bed of salt through overlying denser sedimentary rocks.

Saltation: The process by which a particle moves along a surface in a series of short jumps, as does a sand grain in the bed of a river.

Sand: Sedimentary particles that have a grain size between 0.0625 mm and 2.0 mm.

Sandstone: A sedimentary rock composed predominantly of sand-sized grains (0.0625–2.0 mm), which are usually made of quartz.

Sand waves: A wavelike sedimentary bedform similar to current ripples but larger (1–10 m), which are formed by stronger currents than those that form ripples.

Sanidinite facies: The highest-temperature contact metamorphic facies, characterized by the presence of sanidine.

Scalenohedron: The general form in the hexagonal (rhombohedral) point group $\bar{3}\,2/m$. See *hexagonal salenohedron*.

Scanning electron microscopy (SEM): An instrumental electron beam technique that allows for magnification as high as about 1 million times.

Schist: A metamorphic rock in which medium- to coarse-grained platy minerals, such as micas, or needlelike minerals, such as amphiboles, create a prominent foliation or schistosity.

Schistosity: See *schist*.

Schrödinger wave equation: A mathematical model that describes the wavelike motions of electrons and which is the basis of quantum mechanics.

Sclerometer: An instrument that microscopically measures the absolute hardness of a material.

Scoria: An extremely vesicular volcanic rock of basaltic composition.

Screw axis: A rotational axis combined with translation.

Secondary twin: Twinning that results from mechanical deformation or in the process of displacive deformation of one polymorph to another.

Sedimentary: An adjective describing rocks, minerals, textures, structures, and processes that involve the deposition of sediment.

Sedimentary bed: A sedimentary layer that is distinguishable from layers above and below it on the basis of rock type, grain size, or physical property.

Sedimentary rocks: Rocks that are formed from sediment. The aggregation of loose sedimentary particles to form a solid rock involves burial, compaction, cementing, and recrystallization. The sediment can be derived from the weathering of rock, the remains of organisms, or a chemical precipitate.

Serpentinite: A rock composed primarily of the mineral serpentine formed by the hydrothermal alteration of ultramafic rocks.

Shale: An extremely fine-grained (<0.004 mm) sedimentary rock that is characterized by a prominent foliation or fissility parallel to the bedding.

Shard (volcanic): A Y-shaped particle of glassy volcanic ash formed from the liquid between bubbles, when the lava is disrupted by bubble expansion.

Shatter cone: A conical set of fractures formed by the passage of a shock wave from the explosion accompanying the impact of a meteorite. The apex of the cones points in the direction from which the shock wave came.

Sheeted dike complex: A series of parallel dikes that intrude other dikes formed above a divergent plate boundary.

Sheet joints: Gently dipping joints formed in massive igneous rocks that roughly parallel the surface topography.

Sheet silicate: A silicate crystal structure with infinitely extending tetrahedral sheets. Also known as phyllosilicate.

Shield volcano: A broad, gently sloping volcano composed predominantly of basaltic lava flows.

SI: Système International d'Unités. The international system of units used in science.

Sieve texture: A texture found in some igneous rocks where phenocrysts have been partially melted along intersecting cleavage planes to produce a meshlike or sievelike pattern.

Silicate: A mineral with the anionic $(SiO_4)^{4-}$ tetrahedral group, as in olivine, $(Mg, Fe)_2\,SiO_4$.

Siliciclastic sediment: Sediment formed from the solid weathering products of rocks, which consist primarily of silicate minerals.

Silicosis: A progressive lung disease characterized by the development of fibrotic (scar) tissue caused by the inhalation of quartz dust.

Sill: A sheetlike intrusive igneous body that parallels the layering in the intruded rock. Most large sills have shallow dips.

Silt: Sedimentary particles that have a grain size between 0.0625 mm and 0.0039 mm.

Siltstone: A siliciclastic sedimentary rock composed of particles with a grain size between 0.0625 mm and 0.0039 mm.

Skarn deposit: A contact metamorphic rock containing calcsilicate minerals and iron-bearing minerals that may be in high-enough concentration to be mined.

Slaked lime: Calcium hydroxide, $Ca(OH)_2$. It is obtained when CaO (quicklime) is mixed, or slaked, with water.

Slate: A fine-grained metamorphic rock characterized by a prominent foliation resulting from the preferred orientation of platy minerals such as muscovite and chlorite.

Slaty cleavage: The extremely planar cleavage that is characteristic of slate and is caused by the preferred alignment of platy minerals such as muscovite and chlorite. The cleavage typically parallels the axial plane of folds.

Snell's law: The law relating the angle of incidence to the angle of refraction when light passes from one medium to another of different refractive index.

Soapstone: A rock composed of a mixture of serpentine and talc formed by the metamorphism of peridotite or other ultramafic igneous rock.

Soil: The unconsolidated materials above the bedrock but also the medium for growth of plants.

Soil horizon: A layer of soil that is distinguishable from adjacent layers by characteristic physical properties or chemical composition.

Solid solution: The extent of substitution of one ion (or atom), or ionic group, for another. Adjectives that apply are *complete*, or *partial*.

Solidus: The temperature below which a mixture of materials is completely solid.

Solvus: A line in a phase diagram indicating the temperature and composition below which a solid solution becomes unstable and splits into separate phases.

Sorosilicate: A silicate crystal structure in which there are double tetrahedral groups of $(Si_2O_7)^{6-}$ composition. Also known as disilicate.

Sorting: If sedimentary grains are all of approximately the same size, the sediment is said to be well sorted, but if they have a wide range of sizes, it is poorly sorted.

Space group: Symbolic representation of the translational and symmetry operations reflecting the various ways in which atoms (or ions, or ionic groups) are related in a homogeneous array inside a crystal structure. There are 230 such possible arrangements.

Space lattices: The 14 different ways in which nodes (atoms) can be arranged periodically in three dimensions. Also known as Bravais lattices.

Spar: A term used to describe the coarser (and usually clearer) carbonate crystals that fill pore spaces and form cement in limestones and dolostones.

Spatter cone: A steep-sided cone built around a volcanic vent by lava splatter that is ejected by the bursting of large gas bubbles.

Specific gravity: The ratio of the weight of a volume of a substance to the weight of the same volume of water at 1 atm, 4°C. Designated by **G**.

Specular: Refers to the black or gray variety of hematite with a splendent metallic luster.

Spent nuclear fuel (SNF): When nuclear fuel rods have undergone 5% reaction, the buildup of fission products makes them no longer usable, and they become spent nuclear fuel. Safely disposing of this highly radioactive material is a major problem faced by the nuclear power industry.

Spherical projection: The first step in the process of stereoprojection. This involves locating the face poles (perpendiculars to crystal faces) on an imaginary sphere that surrounds the crystal.

Spicule (sponge): Small rods composed of opaline silica (amorphous) that help support the structure of sponges. They are commonly found in chert.

Spillite: A hydrothermally altered ocean floor basalt in which the plagioclase has been altered to albite and the pyroxenes to hydrous minerals.

Spinel group: A group of oxide minerals with oxygen in cubic closest packing (CCP).

Spinel twin: A contact twin seen in spinel with the octahedral {111} plane the twin plane.

Spinifex texture: A texture resulting from the growth of long blades of olivine crystals down from the surface of an ultramafic lava flow. The blades form clusters resembling tufts of grass (upside down).

Stability series: A grouping of minerals according to their persistence in nature, that is, their resistance to alteration or destruction by weathering, and abrasion during transportation.

Stable: A condition where no change takes place because potential energy or Gibbs free energy has been minimized (see *equilibrium*).

Stacking polymorph: This is the equivalent of a polytype. The structures of stacking polymorphs differ from one another by the manner in which individual atomic layers are arranged in space.

Stalactite: A conical or cylindrical mineral deposit that hangs from the ceiling of a cave.

Stalactitic: A mineral with the appearance of a stalactite.

State of aggregation: Most minerals, unless unusually well-crystallized, occur as aggregates of smaller grains. Various adjectives, such as granular, or compact distinguish different states of aggregation.

Stereogram: A circular diagram that depicts the projected positions of perpendiculars to planes, such as crystal faces, which are known as face poles.

Stereographic net: A commercially available two-dimensional projected image of a sphere with meridians and parallels as a coordinate system.

Stereographic projection: A projection procedure in which point locations in the upper (and lower) hemisphere are projected onto the equatorial plane by viewing from the south pole (or north pole).

Stock: A discordant, intrusive igneous body that has an areal extent of less than 100 km^2.

Stokes' law: The velocity at which a spherical particle settles through a fluid is proportional to the gravitational acceleration, density contrast, and the square of the grain diameter, and inversely proportional to the fluid's viscosity.

Strata-bound sulfide deposit: A sulfide deposit that is conformable with layering in the surrounding volcanic and sedimentary rocks and is typically formed by deposition of sulfides on the ocean floor.

Stratified meltwater deposit: Stratified sediment deposited by water from melting ice.

Strato-volcano: See *composite volcano*.

Streak: The color of the finely powdered mineral on an unglazed white porcelain plate, called a streak plate.

Stromatolite: A boundstone formed by filamentous mats of blue-green algae that trap carbonate mud, producing a laminated domelike structure.

Strombolian eruption: A style of volcanic eruption involving frequent ejection of volcanic bombs from the vent.

Structure: In describing rocks, it is used for features that are on a larger scale than the relations between individual mineral grains, which is referred to as texture.

Stylolite: A dark irregular line in limestone formed where insoluble residue has accumulated on a surface where solution of the rock has taken place.

Subalkaline igneous rocks: See *tholeiitic igneous rocks*.

Subhedral: A crystal that is only partly bounded by good crystal faces. The remaining faces are less well developed because of crowding by adjacent mineral grains.

Subsolvus: A term used to describe crystallization of liquids below a solvus. Subsolvus granites, for example, crystallize below the alkali feldspar solvus and therefore contain two types of alkali feldspar (contrast with hypersolvus).

Substitution: Refers to the extent of atomic (or ionic) substitution in a specific site of a crystal structure. Analogous to the term *solid solution*.

Substitutional solid solution: The substitution of one element for another in a crystal structure.

Sulfate: A group of minerals with the anionic group $(SO_4)^{2-}$ as in gypsum, $CaSO_4 \cdot 2H_2O$.

Sulfide: A group of minerals in which the anion is sulfur, as in pyrite, FeS_2.

Sulfosalt: A type of sulfide in which both a metal and semimetal are present, forming a double sulfide, as in enargite, Cu_3AsS_4.

Surface free energy: The Gibbs free energy associated with the surface of a grain or droplet. By coarsening the grain size, this energy can be reduced.

Supergene: Describes a mineral or ore deposit that formed near the surface from descending solutions.

Superheated liquids: Liquids that are heated to temperatures above the liquidus. Most magmas are not superheated.

Supernova: An explosion of a massive star, which is accompanied by a dramatic and sudden increase in its luminosity (brightness). Elements heavier than iron are produced during these explosions.

Suspended load: The sediment carried in water or air by suspension.

Syenite: A plutonic igneous rock composed primarily of alkali feldspar.

Sylvinite: A mixture of halite and sylvite mined as a potassium ore.

Symmetry (in a crystal): The regular repeat pattern of identical crystal faces that results from the ordered internal atomic arrangement.

Symmetry element: This includes mirror planes, rotational axes, center of symmetry (or inversion, *i*), and rotoinversion axes.

Synthetic: Applies to a substance produced in a laboratory and having the same crystal structure and chemical and physical properties as the naturally occurring equivalent (e.g., synthetic quartz).

System: In thermodynamics, *system* is used to define the volume of interest being considered.

Tabular: A mineral or crystal mass with pronounced flat surfaces as in a board.

Tar sand: A bitumen-rich sand from which oil can be extracted, such as that in the Athabaska region of Alberta, Canada, which is the world's second-largest oil reserve.

Tectosilicate: A silicate crystal structure that consists of an infinitely extending tetrahedral framework. Also known as framework silicate.

Tektite: A small (~1 cm) streamlined or fluted glassy particle that falls to Earth from space. The composition of tektites indicates that they are formed from melted crustal rocks that are ejected into space from the Earth as a result of explosions accompanying the impact of large meteorites. Large clusters of tektites form what are known as *fields*, which in most cases can be correlated with known meteorite impact craters.

Temperature-composition diagram: A two-dimensional diagram of composition versus temperature. Such a diagram depicts the state of a chemical system as a function of increasing temperature.

Tephra: Lava that is broken apart into separate particles by the expansion of gas.

Tephrochronolgy: The dating and correlation of rocks over wide areas through the use of layers of volcanic ash that can be identified through composition or mineralogy.

Terminal reaction: A reaction that terminates the existence of a mineral.

Ternary diagram: Same as a triangular diagram.

Terrestrial planets: The small rocky planets closest to the Sun, including Mercury, Venus, Earth, and Mars. Though not large enough to be considered planets, the asteroids are commonly grouped with the terrestrial planets.

Tetragonal dipyramid: A closed crystal form with eight isosceles triangular faces.

Tetragonal prism: In the tetragonal system there are two prisms, both of which are parallel to the vertical c axis but that intersect the horizontal a_1 and a_2 differently. Their Miller indices are {010} and {110}.

Tetragonal system: The crystal system in which a crystal is classified if it contains a unique 4-fold rotation or 4-fold rotoinversion axis. An example is $4/m\ 2/m\ 2/m$.

Tetrahedral coordination: Describes a cation with four closest anion neighbors.

Tetrahedron: A crystal form that has four equilateral triangular faces. Also the geometric arrangement of Si^{4+} and O^{2-} into the tetrahedral $(SiO_4)^{4-}$ group.

Tetrahexahedron: A closed crystal form with 24 isosceles triangular faces in the isometric system.

Textural maturity: A term describing the degree of rounding and sorting of detrital sedimentary particles. Detrital sediment can range from having poorly sorted, angular grains when immature, through submature, mature, to supermature, where grains are well sorted and rounded.

Texture: In describing rocks, it is used to indicate how individual mineral grains are related to one another.

Thermal conduction: The vibrational transfer of thermal energy from one atom to an adjoining atom down a temperature gradient.

Thermal conductivity: The property of a material to conduct heat. Rocks and magmas have low thermal conductivity and hence cool very slowly.

Thermal convection: The movement of material due to density differences that result from thermal expansion and contraction. As material cools, it shrinks and becomes denser, setting up a gravitational instability.

Tholeiitic igneous rocks (also known as subalkaline): One of the main divisions of igneous rocks. They tend to have low alkalis and have sufficient silica to crystallize free quartz.

Thompson projection: A triangular plot of Al_2O_3, FeO, and MgO into which the mineral assemblages of quartz- and muscovite-bearing metapelites are projected.

Tie line: A line in a phase diagram joining coexisting minerals.

Tie-line switching reaction: A reaction that in a compositional phase triangle involves switching of tie lines.

Tie triangle: A triangle joining three coexisting phases in a phase diagram.

Till: A glacial sediment deposited directly from ice and consisting of poorly sorted material ranging in size from boulders to clay.

Total disorder: This describes the equal probability of finding one of two different atoms (or ions) in a specific atomic site in a crystal structure.

Trace element: A chemical element that constitutes less than 0.1 weight % of a mineral analysis. Such a small amount is normally reported as parts per million (ppm) or parts per billion (ppb).

Trachyte: A volcanic rock composed primarily of alkali feldspar, which commonly forms a felted mass of crystals known as a trachytic texture.

Trachytic texture: A parallel arrangement of feldspar crystals found in some volcanic and hypabyssal rocks.

Transformation twinning: A twin pattern that results when a crystal (or mineral grain) that was formed at high temperature is cooled and rearranges its crystal structure from a high- to a low-temperature architecture. A polymorphic transition.

Transform plate boundary: A plate boundary at which one plate grinds past the other, as, for example, along the San Andreas Fault.

Transgression of the sea: Flooding of the land due to a rise in sea level or a sinking of the land.

Transition elements: Chemical elements with atomic numbers between 21 and 30 in the periodic table of the elements: Sc, Ti, V, Cr, Mn, Fe, Co, Ni, Cu, and Zn

Transition zone: A zone in the lower part of the upper mantle between ~410 km and ~660 km at which numerous mineral changes probably occur as a result of increasing pressure.

Translation: A shift in position of an atom (or a node or a lattice point) without rotation.

Translation-free: A point group is the isogonal, translation-free residue of the equivalent space group.

Translucent: Applies to a mineral that can transmit light but is not transparent.

Transmission electron microscope (TEM): An analytical instrument in which a beam of electrons is transmitted through an ultrathin specimen, thus producing a highly magnified structural image on a fluorescent screen or camera.

Transparent: Applies to a mineral that is capable of transmitting light and through which an object can be seen.

Trapezohedron: A closed crystal form with 24 trapezium-shaped faces in the isometric system.

Trap: Any geological structure in which an impervious cap rock prevents oil or natural gas from escaping from a reservoir rock.

Trap rock: The name given to massive flat-lying basaltic lava flows that are quarried for the rock's durability, which makes it particularly desirable for highway construction. Its name comes from the steplike topographic expression of the lava flows, *trap* being the Dutch word for "stairs."

Travertine: A porous boundstone formed by precipitation of calcite from hot springs.

Triangular coordination: Describes a cation with three closest anion neighbors.

Triangular diagram: A diagram that allows for the graphical representation of mineral chemistry in terms of three end-member components.

Triclinic system: The crystal system in which a crystal is classified if it contains only a center of symmetry (equivalent to inversion, i), or no symmetry. An example is $\bar{1}$.

Trioctahedral: Refers to a layer silicate (phyllosilicate) structure in which all three available octahedral sites are occupied.

Triple junction: The point at which three divergent plate boundaries meet. Typically two of the boundaries remain active and form new ocean floor, whereas the third is less active or stops moving, in which case it becomes a failed arm.

Triple point in Al$_2$SiO$_5$ phase diagram: The temperature and pressure at which all three Al$_2$SiO$_5$ polymorphs (andalusite, kyanite, sillimanite) can coexist (500°C, 0.38 GPa or a depth of 13 km).

Trisoctahedron: A closed crystal form with 24 isosceles triangular faces in the isometric system.

Tsunami: A large destructive ocean wave caused by an earthquake.

Tungstate: A group of minerals with the anionic group $(WO_4)^{2-}$ as in scheelite, $CaWO_4$.

Turbulent flow: The flow of a fluid in which it is impossible to predict the path taken by any given particle of fluid.

Tufa: A highly porous boundstone formed by the deposition of calcite from springwater.

Tuff: A volcanic rock formed from the solidification of tephra or ash.

Turbidite: A sedimentary rock formed from sediment deposited from a turbidity current.

Turbidity current: A rapidly moving suspension of sedimentary particles. The suspension travels as a coherent body, and when it comes to rest, the larger and denser mineral grains settle first to produce a graded sedimentary bed.

Twin: A symmetrical intergrowth of two or more crystals of the same substance.

Twin axis: The rotational axis about which two identical crystals (or identical lattices) are related.

Twin element: The crystallographic element (or operator) that relates two (or more) identical crystals. There are three of these: a twin plane (or mirror), a twin axis, and a center of symmetry (inversion, i).

Twin lamellae: Thin parallel lamellae that are part of a polysynthetic twin, such as in albite twinning.

Twin law: A statement about how two or more identical crystals are related to each other by a new symmetry element. This element (or operator) can be a twin plane, a twin axis, or a center of symmetry (inversion, i).

Twinning: The development of a twin crystal during growth or as a result of inversion, or gliding.

Twin plane: The mirror plane by which two identical crystals (or two identical lattices) are related.

Two V (2V): See *optic angle*.

Udden-Wentworth scale: A scale for measuring the grain size of sedimentary particles.

Ultrabasic igneous rock: Same as ultramafic.

Ultramafic rock: An igneous rock composed of more than 90% olivine and pyroxene.

Unconformity: A boundary in a sequence of rocks formed by either a period of erosion or a lack of deposition before overlying sedimentary beds were deposited.

Underplating: The process by which the crust may have thickened by intrusion of mantle-derived magma along its base.

Unit cell: The smallest unit of structure that can be infinitely repeated to generate the entire structure.

Unit face: A crystal face (or plane) that intersects all crystallographic axes at unit distances.

Unstable: A condition in which change to a stable condition takes place spontaneously.

Valence electrons: The outermost electrons in an atom that play a major part in its bonding to other atoms.

Van der Waals bond: A weak chemical bond resulting from residual electrostatic charges on units of structure. It has a pronounced influence on the crystal structure of graphite and talc.

Variation diagram: A binary or ternary diagram that depicts the variation in some physical property (e.g., index of refraction) as a function of composition.

Varves: Sedimentary layers whose rhythmic banding results from annual fluctuations in the rate of deposition.

Vein: A fracture that is filled with a mineral, which is commonly quartz or calcite, but can include minerals of economic value.

Vesicle: A gas bubble in a volcanic or hypabyssal rock.

Viscosity: The measure of a fluid's resistance to flow when exposed to a shear stress. The SI unit of viscosity is the pascal·second (Pa·s).

Vitreous luster: A luster that is similar to that of glass.

Volcanic arc: See *island arc*.

Volcanic ash: Particles ejected from a volcano that have diameters less than 2 mm.

Volcanic explosivity index (VEI): A scale that quantifies the explosive power of volcanoes. It is logarithmic, with the most powerful eruptions in the geologic record having a value of 8.

Volcano: A body of igneous rock formed on the Earth's surface by the eruption of magma mainly from a central vent. Its shape depends on the magma viscosity and the proportion of lava to ash.

Volcanogenic massive sulfide (VMS) deposit: A type of metal sulfide ore deposit (mainly Cu-Pb-Zn) that is associated with and created by volcanic-associated hydrothermal vents in submarine environments (see *black smoker*).

Vulcanian eruption: A style of volcanic eruption in which many solid fragments are ejected from the vent.

Vug: A small cavity in a rock, or in a vein, usually lined with crystals.

Wacke: Sandstone containing more than 15% muddy matrix.

Wackestone: A limestone in which the allochems, which constitute more than 10% of the rock, are supported by a carbonate mud matrix.

Walther's law: The lateral changes in sediment (e.g., sand to mud) due to changes in depositional environment at any given time are found in vertical successions of sediment formed by transgressions and regressions of the sea.

Watt (W): An SI unit of power equal to 1 joule/second.

Wave base: The depth to which water is agitated by surface waves. It is equal to approximately half the wavelength.

Weathering: The processes by which rocks are decomposed or broken down chemically and physically by reaction with the atmosphere, water, and ice.

Welded ash-flow tuff: See *ash flow*.

White asbestos: A trade term used for chrysotile, one of three polymorphs of serpentine.

White light: Light composed of the complete spectrum of wavelengths that are in the visible range.

Wulff net: A stereonet, used in crystallography, for stereoprojection of the location of crystal faces.

Xenolith: A foreign inclusion in an igneous rock.

X-ray powder diffraction: An instrumental method in which an X-ray beam that impinges on a powdered mineral sample produces distinctive X-ray diffraction patterns that can be recorded for unique mineral identification.

X-ray powder diffraction pattern: A pattern that results when X-rays strike a crystalline powdered material. It consists of diffraction peaks, of differing intensities, as a function of the θ angle in the Bragg equation.

X-rays: A form of electromagnetic radiation. The wavelengths of X-rays range from about 10 nm to 0.10 nm (classified as soft X-rays) to about 0.10 nm to 0.01 nm (classified as hard X-rays).

X-ray spectrum: An intensity distribution of characteristic wavelengths superimposed on a continuous background (a continuum) that is the result of high velocity electrons striking the atoms of the element in the target region of an X-ray tube.

Yield strength: Some liquids, including many magmas, require a minimum shear stress before they will flow.

Zeolite facies: A set of metamorphic mineral assemblages that contain zeolites and reflect very low pressure-temperature conditions.

Zeolites: A group of hydrous framework silicates (tectosilicates) with generally large voids as part of the framework structure. These voids may house H_2O, Na, Ca, and/or K atoms.

Zone: The arrangement of a group of faces, on a well-formed crystal, with parallel intersection edges.

Zone axis: The direction along which a group of crystal faces are parallel. This is symbolically expressed by a Miller index inside square brackets such as [010].

Zoning of crystals: Variation in composition in a crystal of a mineral belonging to a solid solution series. If the composition from core to rim indicates falling temperature, the crystal is normally zoned, and if it indicates increasing temperature, it is reversely zoned. If the composition fluctuates, the zoning is said to be oscillatory.

Minerals and varieties

Primary page numbers for their descriptions

Common igneous, sedimentary, and metamorphic rocks

Primary page numbers for their descriptions

Igneous rocks

Plutonic

Anorthosite 262, 281
Diorite 260, 264, 279
Gabbro 260, 264, 271
Granite 28, 204, 210, 260, 264, 271, 279
Granodiorite 260, 264, 279
Harzburgite 262, 268
Hornblendite 262
Lherzolite 262
Monzonite 260, 264
Nepheline syenite 204, 260, 264
Norite 262
Peridotite 195, 262
Pyroxenite 262
Syenite 260
Tonalite 260, 264
Troctolite 262, 268
Websterite 262
Wehrlite 262, 268

Volcanic and hypabyssal

Andesite 260, 265
Basalt 28, 203, 260, 265
Basanite 265
Dacite 260, 265
Diabase 271
Dolerite 271
Komatiite 280
Lamprophyre 273
Latite 260
Midocean ridge basalt (MORB) 33, 266

Phonolite 260
Picrite 265, 281
Rhyolite 215, 260, 265
Tephrite 260, 265
Trachyte 260, 265

Sedimentary rocks

Arkose 332, 346
Arenite 346
Banded iron-formation (BIF) 319, 363
Boundstone 357
Breccia 350
Caliche 359
Chalk 356
Chert 268, 356, 358, 363
Coal 30, 360
Conglomerate 30, 350
Dolostone 360
Evaporite 319, 362
Feldspathic arenite 347
Grainstone 353
Graywacke 349
Litharenite 349
Limestone 29, 352
Marl 358
Mudrock 340
Mudstone 354
Packstone 354
Phosphorite 363
Quartz arenite 346
Sandstone 30, 343

Shale 29, 340
Siliciclastic sedimentary rocks 340
Siltstone 340
Tar sand 463
Tufa 358
Travertine 29, 359
Turbidite 327, 342
Wacke 349
Wackestone 352, 356

Metamorphic rocks

Amphibolite 398
Augen gneiss 407
Blueschist 399, 403
Calcsilicate 405, 408
Eclogite 399
Garbenschiefer 404
Gneiss 30, 403
Granulite 398, 399
Greenschist 398, 399
Hornfels 401
Marble 27, 31, 408, 416
Migmatite 31, 419
Mylonite 407
Phyllite 402
Pseudotachylite 407
Quartzite 408
Schist 30, 403
Skarn 408
Slate 30
Soapstone 408

Common units of measure

Measure	SI unit	CGS units	Conversions
Basic units			
Length	meter (m)	centimeter (cm)	1 cm = 0.01 m
Mass	kilogram (kg)	gram (g)	1 g = 0.001 kg
Time	second (s)	second (s)	
Temperature	kelvin (K)	kelvin (K) and degrees centigrade (°C)	K = °C + 273.15
Derived units			
Force	newton (N) kg m s^{-2}	dyne (dyn) g cm s^{-2}	1 dyn = 10^{-5} N
Pressure	pascal (Pa) N m^{-2}	bar (b) 106 dyn cm^{-2}	1 bar = 10^5 Pa
Energy	joule (J) N m	calorie (cal)	1 cal =4.184 J
Power	watt (W) J s^{-1}	erg/s dyn cm s^{-1}	1 erg/s = 10^{-7} W
Viscosity	Pa s	poise (P) g cm^{-1} s^{-1}	1 P = 0.1 Pa s
Density (mass/volume)	Kg m^{-3}	g cm^{-3}	
Other units			
Ångstrom (Å)	1 Å = 10^{-8} m		10 Å = 1 nanometer
Atmosphere (atm)	1 atm ~10^5 Pa	1 atm = 1.01325 bar	
Foot (ft)	1 ft = 0.3048 m	1 ft = 30.48 cm	
Inch (in)	1 in = 0.0254 m	1 in = 2.54 cm	
Liter (l)		1 l = 1000 cm^3	
U.S. gallon (US gal)			3.785 liter
Imperial gallon (Imp gal)			4.546 liter
Barrel of oil (bbl)			1 bbl = 42 US gal
Tonne, metric (t)	1000 kg		
Karat (purity of gold)			24 karat = pure Au
Carat (weight of gemstone)			1 carat = $\frac{1}{5}$ gram
Parts per million (ppm)	kg/10^6 kg		
Parts per billion (ppb)	kg/10^9 kg		

Note: SI = Système International d'unités. CGS = centimeter gram second.

Common prefixes that are added to units

pico (p)	10^{-12}	tera (T)	10^{12}
nano (n)	10^{-9}	giga (G)	10^9
micro (μ)	10^{-6}	mega (M)	10^6
milli (m)	10^{-3}	kilo (k)	10^3

Other quantitative terms

Refractive index (RI): Speed of light in air divided by speed of light in substance

Miller indices and diffraction notation:

Face symbol: ($h\ k\ l$); form symbol: {$h\ k\ l$}; edge or zone symbol: [$h\ k\ l$];

Diffraction symbol: $h\ k\ l$; interplanar spacing: (d); e.g. $d_{h\ k\ l}$

Index

Bold numbers indicate pages that give the primary definition of a term; italicized numbers indicate references to figures.

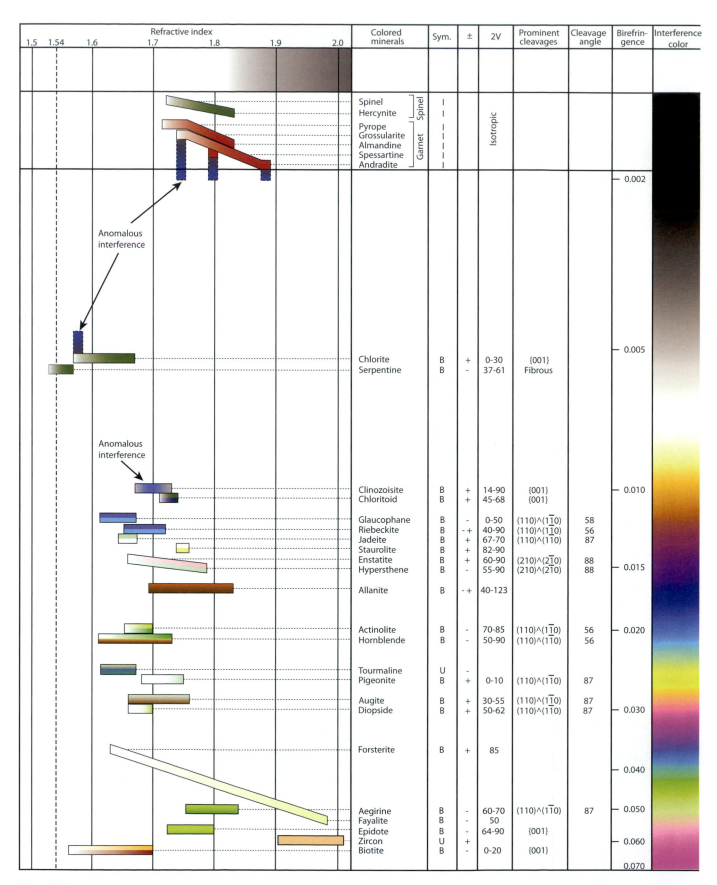

Figure 6.10 Optical properties of common minerals, arranged according to increasing refractive index left to right, and increasing birefringence top to bottom. Minerals are divided into colored to the left of the interference color chart and colorless to the right. Interference colors are for 30 μm-thick sections. The color and pleochroism of colored minerals is indicated in the bars showing their range of refractive index. The relief of minerals relative to a mounting medium of R.I. = 1.54 (dashed line) is shown across the top of the figure. Under the column for symmetry (Sym.), I, U, and B stand for isotropic, uniaxial, and biaxial, respectively. The

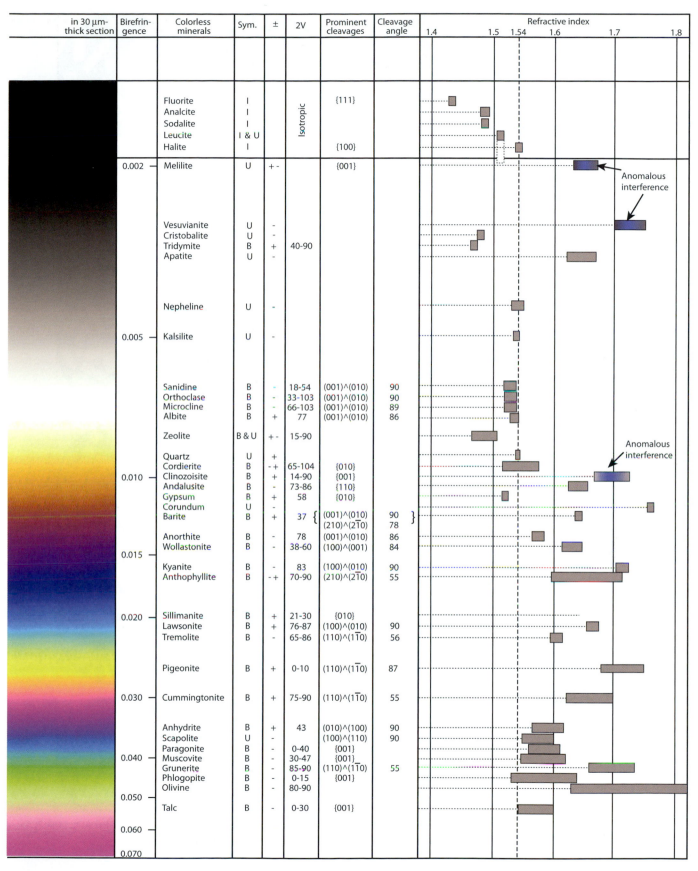

in 30 μm-thick section	Birefrin-gence	Colorless minerals	Sym.	±	2V	Prominent cleavages	Cleavage angle	Refractive index
		Fluorite	I		Isotropic	{111}		
		Analcite	I					
		Sodalite	I					
		Leucite	I & U					
		Halite	I			{100}		
	0.002	Melilite	U	+ -		{001}		Anomalous interference
		Vesuvianite	U	-				
		Cristobalite	U	-				
		Tridymite	B	+	40-90			
		Apatite	U	-				
		Nepheline	U	-				
	0.005	Kalsilite	U	-				
		Sanidine	B	-	18-54	(001)^(010)	90	
		Orthoclase	B	-	33-103	(001)^(010)	90	
		Microcline	B	-	66-103	(001)^(010)	89	
		Albite	B	+	77	(001)^(010)	86	
		Zeolite	B & U	+ -	15-90			
		Quartz	U	+				Anomalous interference
		Cordierite	B	- +	65-104	{010}		
	0.010	Clinozoisite	B	+	14-90	{001}		
		Andalusite	B	-	73-86	{110}		
		Gypsum	B	+	58	{010}		
		Corundum	U	-				
		Barite	B	+	37 {	(001)^(010)	90 }	
						(210)^(2̄10)	78	
		Anorthite	B	-	78	(001)^(010)	86	
		Wollastonite	B	-	38-60	(100)^(001)	84	
	0.015	Kyanite	B	-	83	(100)^(010)	90	
		Anthophyllite	B	- +	70-90	(210)^(2̄10)	55	
	0.020	Sillimanite	B	+	21-30	{010}		
		Lawsonite	B	+	76-87	(100)^(010)	90	
		Tremolite	B	-	65-86	(110)^(11̄0)	56	
		Pigeonite	B	+	0-10	(110)^(11̄0)	87	
	0.030	Cummingtonite	B	+	75-90	(110)^(11̄0)	55	
		Anhydrite	B	+	43	(010)^(100)	90	
		Scapolite	U	-		(100)^(110)	90	
		Paragonite	B	-	0-40	{001}		
	0.040	Muscovite	B	-	30-47	{001}		
		Grunerite	B	-	85-90	(110)^(11̄0)	55	
		Phlogopite	B	-	0-15	{001}		
		Olivine	B	-	80-90			
	0.050							
		Talc	B	-	0-30	{001}		
	0.060							
	0.070							

optic sign is indicated by + or –, and the optic angle is given under 2V. Miller indices of prominent cleavages are listed, and if more than one is present, the angle between them is given. The only common minerals not shown on the chart are the carbonates and titanite (sphene), which typically have sixth-order interference colors and appear a pinky shade of white under crossed polars. Titanite is normally brown under plane light and appears almost the same color under crossed polars. From Philpotts (2003).

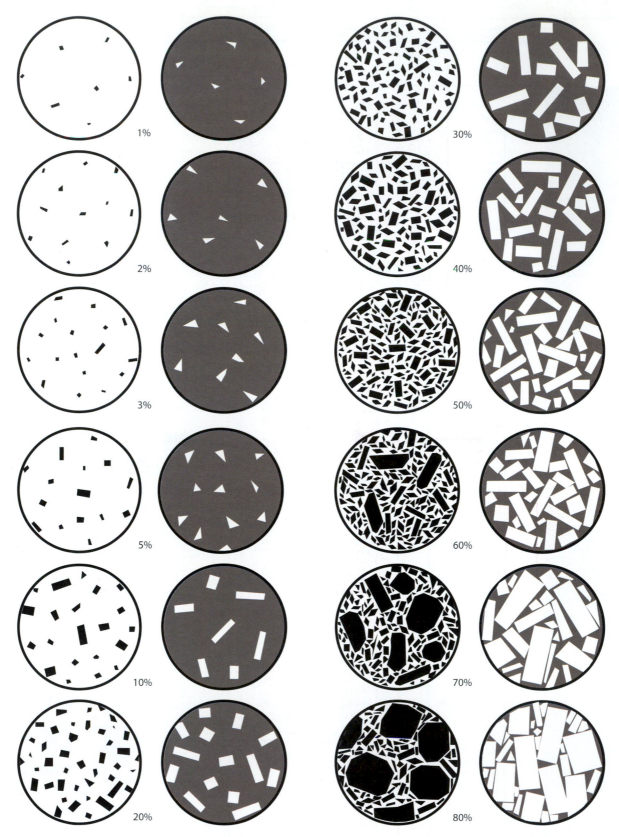

Figure 6.17 Chart for determining the approximate modal abundance of minerals in rocks (from Philpotts 2003).

Periodic Table of the Elements

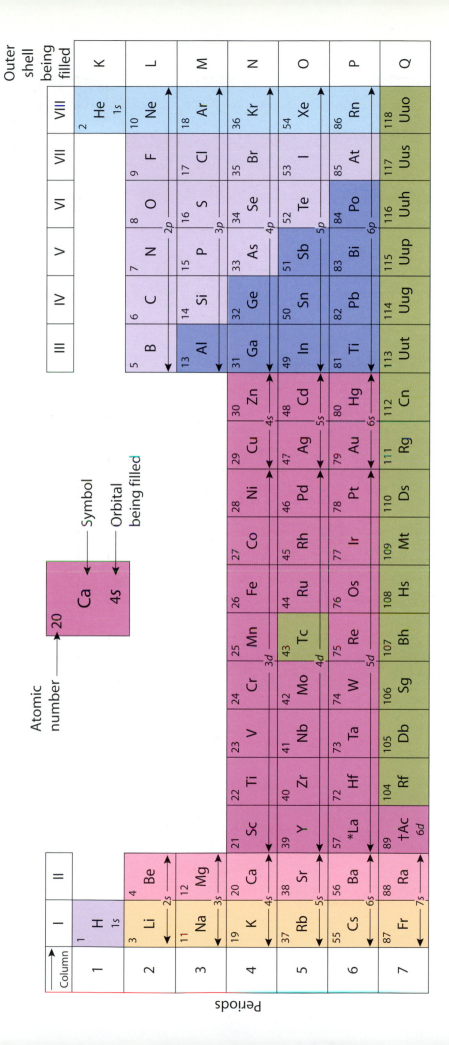